Lecture Notes in Artificial Intelligence 11431

Subseries of Lecture Notes in Computer Science

More information about this series at http://www.springer.com/series/1244

Ngoc Thanh Nguyen · Ford Lumban Gaol ·
Tzung-Pei Hong · Bogdan Trawiński (Eds.)

Intelligent Information and Database Systems

11th Asian Conference, ACIIDS 2019
Yogyakarta, Indonesia, April 8–11, 2019
Proceedings, Part I

 Springer

Editors
Ngoc Thanh Nguyen 🆔
Ton Duc Thang University
Ho Chi Minh City, Vietnam

Wrocław University of Science
and Technology
Wrocław, Poland

Tzung-Pei Hong
National University of Kaohsiung
Kaohsiung, Taiwan

Ford Lumban Gaol 🆔
Bina Nusantara University
Jakarta, Indonesia

Bogdan Trawiński 🆔
Wrocław University of Science
and Technology
Wrocław, Poland

ISSN 0302-9743   ISSN 1611-3349 (electronic)
Lecture Notes in Artificial Intelligence
ISBN 978-3-030-14798-3   ISBN 978-3-030-14799-0 (eBook)
https://doi.org/10.1007/978-3-030-14799-0

Library of Congress Control Number: 2019932916

LNCS Sublibrary: SL7 – Artificial Intelligence

This Springer imprint is published by the registered company Springer Nature Switzerland AG
The registered company address is: Gewerbestrasse 11, 6330 Cham, Switzerland

Preface

ACIIDS 2019 was the 11th event in a series of international scientific conferences on research and applications in the field of intelligent information and database systems. The aim of ACIIDS 2019 was to provide an international forum of research workers with scientific background on the technology of intelligent information and database systems and its various applications. The ACIIDS 2019 conference was co-organized by BINUS University (Indonesia) and Wrocław University of Science and Technology (Poland) in co-operation with the IEEE SMC Technical Committee on Computational Collective Intelligence, European Research Center for Information Systems (ERCIS), University of Newcastle (Australia), Yeungnam University (South Korea), Leiden University (The Netherlands), Universiti Teknologi Malaysia (Malaysia), Quang Binh University (Vietnam), Ton Duc Thang University (Vietnam), and Vietnam National University, Hanoi (Vietnam). It took place in Yogyakarta in Indonesia during April 8–11, 2019.

The ACIIDS conference series is already well established. The first two events, ACIIDS 2009 and ACIIDS 2010, took place in Dong Hoi City and Hue City in Vietnam, respectively. The third event, ACIIDS 2011, took place in Daegu (South Korea), followed by the fourth event, ACIIDS 2012, in Kaohsiung (Taiwan). The fifth event, ACIIDS 2013, was held in Kuala Lumpur in Malaysia while the sixth event, ACIIDS 2014, was held in Bangkok in Thailand. The seventh event, ACIIDS 2015, took place in Bali (Indonesia), followed by the eighth event, ACIIDS 2016, in Da Nang (Vietnam). The ninth event, ACIIDS 2017, was organized in Kanazawa (Japan). The 10th jubilee conference, ACIIDS 2018, was held in Dong Hoi City (Vietnam).

For this edition of the conference, we received more than 300 papers from 38 countries all over the world. Each paper was peer-reviewed by at least two members of the international Program Committee and the international reviewer board. Only 124 papers with the highest quality were selected for an oral presentation and publication in these two volumes of the ACIIDS 2019 proceedings.

Papers included in the proceedings cover the following topics: knowledge engineering and Semantic Web; text processing and information retrieval; machine learning and data mining; decision support and control systems; computer vision techniques; databases and intelligent information systems, collective intelligence for service innovation, technology management, e-learning and fuzzy intelligent systems, data structures modelling for knowledge representation, advanced data mining techniques and applications; intelligent information systems; intelligent methods and artificial intelligence for biomedical decision support systems, intelligent and contextual systems, intelligent systems and algorithms in information sciences, intelligent supply chains and e-commerce, sensor networks and internet of things, analysis of image, video, movements and brain intelligence in life sciences, computer vision and intelligent systems.

The accepted and presented papers focus on new trends and challenges facing the intelligent information and database systems community. The presenters showed in what way research work could stimulate novel and innovative applications. We hope you will find these results useful and inspiring for your future research work.

We would like to extend our heartfelt thanks to Jarosław Gowin, Deputy Prime Minister of the Republic of Poland and Minister of Science and Higher Education, for his support and honorary patronage of the conference.

We would like to express our sincere thanks to the CEO of Bina Nusantara Group, Ir Bernard Gunawan, honorary chairs, Prof. Harjanto Prabowo (Rector of BINUS University, Indonesia), and Prof. Cezary Madryas (Rector of Wrocław University of Science and Technology, Poland), for their support.

Our special thanks go to the program chairs, special session chairs, organizing chairs, publicity chairs, liaison chairs, and local Organizing Committee for their work for the conference. We sincerely thank all the members of the international Program Committee for their valuable efforts in the review process, which helped us to guarantee the highest quality of the selected papers for the conference. We cordially thank the organizers and chairs of special sessions, who contributed to the success of the conference.

We would like to express our thanks to the keynote speakers: Prof. Stephan Chalup from the University of Newcastle, Australia, Prof. Hamido Fujita from Iwate Prefectural University, Japan, Prof. Tokuro Matsuo from Advanced Institute of Industrial Technology, Japan, Prof. Michał Woźniak from Wrocław University of Science and Technology, Poland, and Dr. Alfred Budiman, CEO of Samsung R&D Indonesia (SRIN), for their world-class plenary speeches.

We cordially thank our main sponsors, BINUS University (Indonesia), Wrocław University of Science and Technology (Poland), IEEE SMC Technical Committee on Computational Collective Intelligence, European Research Center for Information Systems (ERCIS), University of Newcastle (Australia), Yeungnam University (South Korea), Leiden University (The Netherlands), Universiti Teknologi Malaysia (Malaysia), Quang Binh University (Vietnam), Ton Duc Thang University (Vietnam), and Vietnam National University, Hanoi (Vietnam). Our special thanks are due also to Springer for publishing the proceedings and sponsoring awards, and to all the other sponsors for their kind support.

We wish to thank the members of the Organizing Committee for their excellent work and the members of the local Organizing Committee for their considerable effort. We cordially thank all the authors, for their valuable contributions, and the other participants of this conference. The conference would not have been possible without their support. Thanks are also due to many experts who contributed to making the event a success.

April 2019

Ngoc Thanh Nguyen
Ford Lumban Gaol
Tzung-Pei Hong
Bogdan Trawiński

Organization

Honorary Chairs

Harjanto Prabowo   Rector of BINUS University, Indonesia

Cezary Madryas   Rector of Wrocław University of Science and Technology, Poland

General Chairs

Ngoc Thanh Nguyen   Wrocław University of Science and Technology, Poland

Ford Lumban Gaol   BINUS University, Indonesia

Program Chairs

Spits Warnars Harco Leslie Hendric   BINUS University, Indonesia

Tzung-Pei Hong   National University of Kaohsiung, Taiwan

Edward Szczerbicki   University of Newcastle, Australia

Bogdan Trawiński   Wrocław University of Science and Technology, Poland

Steering Committee

Ngoc Thanh Nguyen (Chair)   Wrocław University of Science and Technology, Poland

Longbing Cao   University of Science and Technology Sydney, Australia

Suphamit Chittayasothorn   King Mongkut's Institute of Technology Ladkrabang, Thailand

Ford Lumban Gaol   Bina Nusantara University, Indonesia

Tu Bao Ho   Japan Advanced Institute of Science and Technology, Japan

Tzung-Pei Hong   National University of Kaohsiung, Taiwan

Dosam Hwang   Yeungnam University, South Korea

Bela Stantic   Griffith University, Australia

Geun-Sik Jo   Inha University, South Korea

Hoai An Le-Thi   University of Lorraine, France

Zygmunt Mazur   Wrocław University of Science and Technology, Poland

Toyoaki Nishida   Kyoto University, Japan

Leszek Rutkowski Częstochowa University of Technology, Poland
Ali Selamat Universiti Teknologi Malaysia, Malyasia

Special Session Chairs

Maciej Huk Wrocław University of Science and Technology,
 Poland
Yaya Heriyadi BINUS University, Indonesia

Liaison Chairs

Suphamit Chittayasothorn King Mongkut's Institute of Technology
 Ladkrabang, Thailand
Quang-Thuy Ha VNU University of Engineering and Technology,
 Vietnam
Mong-Fong Horng National Kaohsiung University of Applied
 Sciences, Taiwan
Dosam Hwang Yeungnam University, South Korea
Le Minh Nguyen Japan Advanced Institute of Science
 and Technology, Japan
Ali Selamat Universiti Teknologi Malaysia, Malyasia

Organizing Chairs

Harisno BINUS University, Indonesia
Adrianna Kozierkiewicz Wrocław University of Science and Technology,
 Poland

Publicity Chairs

Agung Trisetyarso BINUS University, Indonesia
Marek Kopel Wrocław University of Science and Technology,
 Poland
Marek Krótkiewicz Wrocław University of Science and Technology,
 Poland

Publication Chairs

Marcin Maleszka Wrocław University of Science and Technology,
 Poland
Andrzej Siemiński Wrocław University of Science and Technology,
 Poland

Webmaster

Marek Kopel   Wrocław University of Science and Technology,
                               Poland

Keynote Speakers

Alfred Budiman   CEO of Samsung R&D Indonesia (SRIN),
                               Indonesia
Stephan Chalup   University of Newcastle, Australia
Hamido Fujita   Iwate Prefectural University, Japan
Tokuro Matsuo   Advanced Institute of Industrial Technology, Japan
Michał Woźniak   Wrocław University of Science and Technology,
                               Poland

Special Session Organizers

1. *Special Session on Intelligent Systems and Algorithms in Information Sciences (ISAIS 2019)*

Martin Kotyrba   University of Ostrava, Czech Republic
Eva Volná   University of Ostrava, Czech Republic
Ivan Zelinka   VŠB – Technical University of Ostrava,
                               Czech Republic
Pavel Petr   University of Pardubice, Czech Republic

2. *Special Session on Intelligent and Contextual Systems (ICxS 2019)*

Maciej Huk   Wrocław University of Science and Technology,
                               Poland
Keun Ho Ryu   Chungbuk National University, South Korea
Goutam Chakraborty   Iwate Prefectural University, Japan

3. *Intelligent Methods and Artificial Intelligence for Biomedical Decision Support Systems (IMAIBDSS 2019)*

Jan Kubicek   VŠB – Technical University of Ostrava,
                               Czech Republic
Marek Penhaker   VŠB – Technical University of Ostrava,
                               Czech Republic
Ondrej Krejcar   University of Hradec Kralove, Czech Republic
Kamil Kuca   University of Hradec Kralove, Czech Republic
Ali Selamat   Universiti Teknologi Malaysia, Malaysia

4. *Special Session on Computer Vision and Intelligent Systems (CVIS 2019)*

Van-Dung Hoang	Quang Binh University, Vietnam
Wahyono	Universitas Gadjah Mada, Indonesia
Kang-Hyun Jo	University of Ulsan, South Korea
Hyun-Deok Kang	Ulsan National Institute of Science and Technology, South Korea
Thi-Lan Le	Hanoi University of Science and Technology, Vietnam

5. *Special Session on Advanced Data Mining Techniques and Applications (ADMTA 2019)*

Chun-Hao Chen	Tamkang University, Taiwan
Bay Vo	Ho Chi Minh City University of Technology, Vietnam
Tzung-Pei Hong	National University of Kaohsiung, Taiwan

6. *Special Session on Intelligent Supply Chains and e-Commerce (ISCEC 2019)*

Arkadiusz Kawa	Poznań University of Economics and Business, Poland
Bartłomiej Pierański	Poznań University of Economics and Business, Poland

7. *Special Session on Intelligent Information Systems (IIS 2019)*

Urszula Boryczka	University of Silesia, Poland
Andrzej Najgebauer	Military University of Technology, Poland
Dariusz Pierzchała	Military University of Technology, Poland

8. *Special Session on Collective Intelligence for Service Innovation, Technology Management, E-learning and Fuzzy Intelligent Systems (CISTEF 2019)*

Pen-Choug Sun	Aletheia University, Taiwan
Gong-Yih Hsieh	Aletheia University, Taiwan

9. *Analysis of Image, Video, Movements and Brain Intelligence in Life Sciences (IVMBI 2019)*

Andrzej Przybyszewski	Polish-Japanese Academy of Information Technology, Poland
Jerzy Nowacki	Polish-Japanese Academy of Information Technology, Poland
Konrad Wojciechowski	Polish-Japanese Academy of Information Technology, Poland

Marek Kulbacki   Polish-Japanese Academy of Information
                                 Technology, Poland
Jakub Segen   Polish-Japanese Academy of Information
                                 Technology, Poland

10. *Data Structures Modelling for Knowledge Representation (DSMKR 2019)*

Marek Krótkiewicz   Wrocław University of Technology, Poland
Piotr Zabawa   Cracow University of Technology, Poland

International Program Committee

Ajith Abraham   Machine Intelligence Research Labs, USA
Muhammad Abulaish   South Asian University, India
Andrew Adamatzky   University of the West of England, UK
Waseem Ahmad   Waiariki Institute of Technology, New Zealand
Bashar Al-Shboul   University of Jordan, Jordan
Lionel Amodeo   University of Technology of Troyes, France
Toni Anwar   Universiti Teknologi Malaysia, Malaysia
Taha Arbaoui   University of Technology of Troyes, France
Ahmad Taher Azar   Benha University, Egypt
Thomas Bäck   Leiden University, The Netherlands
Amelia Badica   University of Craiova, Romania
Costin Badica   University of Craiova, Romania
Kambiz Badie   ICT Research Institute, Iran
Zbigniew Banaszak   Warsaw University of Technology, Poland
Dariusz Barbucha   Gdynia Maritime University, Poland
Ramazan Bayindir   Gazi University, Turkey
Juri Belikov   Tallinn University of Technology, Estonia
Maumita Bhattacharya   Charles Sturt University, Australia
Leon Bobrowski   Białystok University of Technology, Poland
Bülent Bolat   Yildiz Technical University, Turkey
Veera Boonjing   King Mongkut's Institute of Technology
                                 Ladkrabang, Thailand
Mariusz Boryczka   University of Silesia, Poland
Urszula Boryczka   University of Silesia, Poland
Zouhaier Brahmia   University of Sfax, Tunisia
Stephane Bressan   National University of Singapore, Singapore
Peter Brida   University of Zilina, Slovakia
Andrej Brodnik   University of Ljubljana, Slovenia
Grażyna Brzykcy   Poznań University of Technology, Poland
Robert Burduk   Wrocław University of Science and Technology,
                                 Poland
Aleksander Byrski   AGH University of Science and Technology,
                                 Poland
David Camacho   Universidad Autonoma de Madrid, Spain

Tru Cao   Ho Chi Minh City University of Technology,
                                    Vietnam
Frantisek Capkovic   Institute of Informatics, Slovak Academy
                                    of Sciences, Slovakia
Oscar Castillo   Tijuana Institute of Technology, Mexico
Dariusz Ceglarek   Poznań High School of Banking, Poland
Zenon Chaczko   University of Technology, Sydney, Australia
Goutam Chakraborty   Iwate Prefectural University, Japan
Somchai Chatvichienchai   University of Nagasaki, Japan
Chun-Hao Chen   Tamkang University, Taiwan
Rung-Ching Chen   Chaoyang University of Technology, Taiwan
Shyi-Ming Chen   National Taiwan University of Science
                                    and Technology, Taiwan
Suphamit Chittayasothorn   King Mongkut's Institute of Technology
                                    Ladkrabang, Thailand
Sung-Bae Cho   Yonsei University, South Korea
Kazimierz Choroś   Wrocław University of Science and Technology,
                                    Poland
Kun-Ta Chuang   National Cheng Kung University, Taiwan
Dorian Cojocaru   University of Craiova, Romania
Jose Alfredo Ferreira Costa   Federal University of Rio Grande do Norte
                                    (UFRN), Brazil
Ireneusz Czarnowski   Gdynia Maritime University, Poland
Theophile Dagba   University of Abomey-Calavi, Benin
Quang A Dang   Vietnam Academy of Science and Technology,
                                    Vietnam
Tien V. Do   Budapest University of Technology
                                    and Economics, Hungary
Grzegorz Dobrowolski   AGH University of Science and Technology,
                                    Poland
Habiba Drias   University of Science and Technology Houari
                                    Boumediene, Algeria
Maciej Drwal   Wrocław University of Science and Technology,
                                    Poland
Ewa Dudek-Dyduch   AGH University of Science and Technology,
                                    Poland
El-Sayed M. El-Alfy   King Fahd University of Petroleum and Minerals,
                                    Saudi Arabia
Keiichi Endo   Ehime University, Japan
Sebastian Ernst   AGH University of Science and Technology,
                                    Poland
Nadia Essoussi   University of Carthage, Tunisia
Rim Faiz   University of Carthage, Tunisia
Simon Fong   University of Macau, SAR China
Dariusz Frejlichowski   West Pomeranian University of Technology,
                                    Szczecin, Poland

Takuya Fujihashi	Ehime University, Japan
Hamido Fujita	Iwate Prefectural University, Japan
Mohamed Gaber	Birmingham City University, UK
Ford Lumban Gaol	BINUS University, Indonesia
Dariusz Gąsior	Wrocław University of Science and Technology, Poland
Janusz Getta	University of Wollongong, Australia
Daniela Gifu	University Alexandru Ioan Cuza of Iasi, Romania
Dejan Gjorgjevikj	Ss. Cyril and Methodius University in Skopje, Republic of Macedonia
Barbara Gładysz	Wrocław University of Science and Technology, Poland
Daniela Godoy	ISISTAN Research Institute, Argentina
Ong Sing Goh	Universiti Teknikal Malaysia Melaka, Malaysia
Antonio Gonzalez-Pardo	Universidad Autonoma de Madrid, Spain
Yuichi Goto	Saitama University, Japan
Manuel Graña	University of Basque Country, Spain
Janis Grundspenkis	Riga Technical University, Latvia
Jeonghwan Gwak	Seoul National University, South Korea
Quang-Thuy Ha	VNU University of Engineering and Technology, Vietnam
Dawit Haile	Addis Ababa University, Ethiopia
Pei-Yi Hao	National Kaohsiung University of Applied Sciences, Taiwan
Marcin Hernes	Wrocław University of Economics, Poland
Francisco Herrera	University of Granada, Spain
Koichi Hirata	Kyushu Institute of Technology, Japan
Bogumiła Hnatkowska	Wrocław University of Science and Technology, Poland
Huu Hanh Hoang	Hue University, Vietnam
Quang Hoang	Hue University of Sciences, Vietnam
Van-Dung Hoang	Quang Binh University, Vietnam
Jaakko Hollmen	Aalto University, Finland
Tzung-Pei Hong	National University of Kaohsiung, Taiwan
Mong-Fong Horng	National Kaohsiung University of Applied Sciences, Taiwan
Jen-Wei Huang	National Cheng Kung University, Taiwan
Yung-Fa Huang	Chaoyang University of Technology, Taiwan
Maciej Huk	Wrocław University of Science and Technology, Poland
Zbigniew Huzar	Wrocław University of Science and Technology, Poland
Dosam Hwang	Yeungnam University, South Korea
Roliana Ibrahim	Universiti Teknologi Malaysia, Malaysia

Dmitry Ignatov	National Research University Higher School of Economics, Russia
Lazaros Iliadis	Democritus University of Thrace, Greece
Hazra Imran	University of British Columbia, Canada
Agnieszka Indyka-Piasecka	Wrocław University of Science and Technology, Poland
Mirjana Ivanovic	University of Novi Sad, Serbia
Sanjay Jain	National University of Singapore, Singapore
Jarosław Jankowski	West Pomeranian University of Technology, Szczecin, Poland
Joanna Jędrzejowicz	University of Gdańsk, Poland
Piotr Jędrzejowicz	Gdynia Maritime University, Poland
Janusz Jeżewski	Institute of Medical Technology and Equipment ITAM, Poland
Geun Sik Jo	Inha University, South Korea
Kang-Hyun Jo	University of Ulsan, South Korea
Jason Jung	Chung-Ang University, South Korea
Przemysław Juszczuk	University of Economics in Katowice, Poland
Janusz Kacprzyk	Systems Research Institute, Polish Academy of Sciences, Poland
Tomasz Kajdanowicz	Wrocław University of Science and Technology, Poland
Nadjet Kamel	University Ferhat Abbes Setif1, Algeria
Hyun-Deok Kang	Ulsan National Institute of Science and Technology, South Korea
Mehmet Karaata	Kuwait University, Kuwait
Nikola Kasabov	Auckland University of Technology, New Zealand
Arkadiusz Kawa	Poznań University of Economics and Business, Poland
Rafal Kern	Wrocław University of Science and Technology, Poland
Zaheer Khan	University of the West of England, UK
Manish Khare	Dhirubhai Ambani Institute of Information and Communication Technology, India
Chonggun Kim	Yeungnam University, South Korea
Marek Kisiel-Dorohinicki	AGH University of Science and Technology, Poland
Attila Kiss	Eotvos Lorand University, Hungary
Jerzy Klamka	Silesian University of Technology, Poland
Goran Klepac	Raiffeisen Bank, Croatia
Shinya Kobayashi	Ehime University, Japan
Marek Kopel	Wrocław University of Science and Technology, Poland
Raymondus Kosala	BINUS University, Indonesia
Leszek Koszałka	Wrocław University of Science and Technology, Poland

Leszek Kotulski	AGH University of Science and Technology, Poland
Martin Kotyrba	University of Ostrava, Czech Republic
Jan Kozak	University of Economics in Katowice, Poland
Adrianna Kozierkiewicz	Wrocław University of Science and Technology, Poland
Bartosz Krawczyk	Virginia Commonwealth University, USA
Ondrej Krejcar	University of Hradec Kralove, Czech Republic
Dalia Kriksciuniene	Vilnius University, Lithuania
Dariusz Król	Wrocław University of Science and Technology, Poland
Marek Krótkiewicz	Wrocław University of Science and Technology, Poland
Marzena Kryszkiewicz	Warsaw University of Technology, Poland
Adam Krzyzak	Concordia University, Canada
Jan Kubicek	VSB - Technical University of Ostrava, Czech Republic
Tetsuji Kuboyama	Gakushuin University, Japan
Elżbieta Kukla	Wrocław University of Science and Technology, Poland
Julita Kulbacka	Wrocław Medical University, Poland
Marek Kulbacki	Polish-Japanese Academy of Information Technology, Poland
Kazuhiro Kuwabara	Ritsumeikan University, Japan
Halina Kwaśnicka	Wrocław University of Science and Technology, Poland
Annabel Latham	Manchester Metropolitan University, UK
Bac Le	University of Science, VNU-HCM, Vietnam
Hoai An Le Thi	University of Lorraine, France
Yue-Shi Lee	Ming Chuan University, Taiwan
Florin Leon	Gheorghe Asachi Technical University of Iasi, Romania
Horst Lichter	RWTH Aachen University, Germany
Kuo-Sui Lin	Aletheia University, Taiwan
Igor Litvinchev	Nuevo Leon State University, Mexico
Rey-Long Liu	Tzu Chi University, Taiwan
Doina Logofatu	Frankfurt University of Applied Sciences, Germany
Edwin Lughofer	Johannes Kepler University Linz, Austria
Lech Madeyski	Wrocław University of Science and Technology, Poland
Bernadetta Maleszka	Wrocław University of Science and Technology, Poland
Marcin Maleszka	Wrocław University of Science and Technology, Poland
Petra Maresova	University of Hradec Kralove, Czech Republic

Urszula Markowska-Kaczmar	Wrocław University of Science and Technology, Poland
Mustafa Mat Deris	Universiti Tun Hussein Onn Malaysia, Malaysia
Takashi Matsuhisa	Karelia Research Centre, Russian Academy of Science, Russia
Tamás Matuszka	Eotvos Lorand University, Hungary
Joao Mendes-Moreira	University of Porto, Portugal
Mercedes Merayo	Universidad Complutense de Madrid, Spain
Jacek Mercik	WSB University in Wrocław, Poland
Radosław Michalski	Wrocław University of Science and Technology, Poland
Peter Mikulecky	University of Hradec Kralove, Czech Republic
Marek Miłosz	Lublin University of Technology, Poland
Kazuo Misue	University of Tsukuba, Japan
Jolanta Mizera-Pietraszko	Opole University, Poland
Leo Mrsic	IN2data Ltd. Data Science Company, Croatia
Agnieszka Mykowiecka	Institute of Computer Science, Polish Academy of Sciences, Poland
Paweł Myszkowski	Wrocław University of Science and Technology, Poland
Saeid Nahavandi	Deakin University, Australia
Kazumi Nakamatsu	University of Hyogo, Japan
Grzegorz J. Nalepa	AGH University of Science and Technology, Poland
Mahyuddin K. M. Nasution	Universitas Sumatera Utara, Indonesia
Richi Nayak	Queensland University of Technology, Australia
Fulufhelo Nelwamondo	Council for Scientific and Industrial Research, South Africa
Huu-Tuan Nguyen	Vietnam Maritime University, Vietnam
Le Minh Nguyen	Japan Advanced Institute of Science and Technology, Japan
Loan T. T. Nguyen	Ton Duc Thang University, Vietnam
Quang-Vu Nguyen	Korea-Vietnam Friendship Information Technology College, Vietnam
Thai-Nghe Nguyen	Cantho University, Vietnam
Van Du Nguyen	Wrocław University of Science and Technology, Poland
Yusuke Nojima	Osaka Prefecture University, Japan
Jerzy Nowacki	Polish-Japanese Academy of Information Technology, Poland
Agnieszka Nowak-Brzezińska	University of Silesia, Poland
Mariusz Nowostawski	Norwegian University of Science and Technology, Norway
Alberto Núñez	Universidad Complutense de Madrid, Spain
Manuel Núñez	Universidad Complutense de Madrid, Spain
Richard Jayadi Oentaryo	Singapore Management University, Singapore

Kouzou Ohara	Aoyama Gakuin University, Japan
Tarkko Oksala	Aalto University, Finland
Shingo Otsuka	Kanagawa Institute of Technology, Japan
Marcin Paprzycki	Systems Research Institute, Polish Academy of Sciences, Poland
Rafael Parpinelli	Santa Catarina State University (UDESC), Brazil
Danilo Pelusi	University of Teramo, Italy
Marek Penhaker	VSB - Technical University of Ostrava, Czech Republic
Hoang Pham	Rutgers University, USA
Maciej Piasecki	Wrocław University of Science and Technology, Poland
Bartłomiej Pierański	Poznań University of Economics and Business, Poland
Dariusz Pierzchała	Military University of Technology, Poland
Marcin Pietranik	Wrocław University of Science and Technology, Poland
Elias Pimenidis	University of the West of England, UK
Jaroslav Pokorný	Charles University in Prague, Czech Republic
Nikolaos Polatidis	University of Brighton, UK
Elvira Popescu	University of Craiova, Romania
Piotr Porwik	University of Silesia, Poland
Radu-Emil Precup	Politehnica University of Timisoara, Romania
Małgorzata Przybyła-Kasperek	University of Silesia, Poland
Andrzej Przybyszewski	Polish-Japanese Academy of Information Technology, Poland
Paulo Quaresma	Universidade de Evora, Portugal
Ngoc Quoc Ly	Ho Chi Minh City University of Science, Vietnam
David Ramsey	Wrocław University of Science and Technology, Poland
Mohammad Rashedur Rahman	North South University, Bangladesh
Ewa Ratajczak-Ropel	Gdynia Maritime University, Poland
Leszek Rutkowski	Częstochowa University of Technology, Poland
Tomasz M. Rutkowski	University of Tokyo, Japan
Henryk Rybiński	Warsaw University of Technology, Poland
Alexander Ryjov	Lomonosov Moscow State University, Russia
Keun Ho Ryu	Chungbuk National University, South Korea
Virgilijus Sakalauskas	Vilnius University, Lithuania
Daniel Sanchez	University of Granada, Spain
Rafał Scherer	Częstochowa University of Technology, Poland
Juergen Schmidhuber	Swiss AI Lab IDSIA, Switzerland
Jakub Segen	Polish-Japanese Academy of Information Technology, Poland
Ali Selamat	Universiti Teknologi Malaysia, Malaysia
S. M. N. Arosha Senanayake	Universiti Brunei Darussalam, Brunei Darussalam

Tegjyot Singh Sethi	University of Louisville, USA
Andrzej Siemiński	Wrocław University of Science and Technology, Poland
Dragan Simic	University of Novi Sad, Serbia
Paweł Sitek	Kielce University of Technology, Poland
Adam Słowik	Koszalin University of Technology, Poland
Vladimir Sobeslav	University of Hradec Kralove, Czech Republic
Kulwadee Somboonviwat	Kasetsart University, Thailand
Kamran Soomro	University of the West of England, UK
Zenon A. Sosnowski	Białystok University of Technology, Poland
Bela Stantic	Griffith University, Australia
Stanimir Stoyanov	University of Plovdiv Paisii Hilendarski, Bulgaria
Ja-Hwung Su	Cheng Shiu University, Taiwan
Libuse Svobodova	University of Hradec Kralove, Czech Republic
Tadeusz Szuba	AGH University of Science and Technology, Poland
Julian Szymański	Gdańsk University of Technology, Poland
Krzysztof Ślot	Łódź University of Technology, Poland
Jerzy Świątek	Wrocław University of Science and Technology, Poland
Andrzej Świerniak	Silesian University of Technology, Poland
Ryszard Tadeusiewicz	AGH University of Science and Technology, Poland
Yasufumi Takama	Tokyo Metropolitan University, Japan
Maryam Tayefeh Mahmoudi	ICT Research Institute, Iran
Zbigniew Telec	Wrocław University of Science and Technology, Poland
Aleksei Tepljakov	Tallinn University of Technology, Estonia
Dilhan Thilakarathne	Vrije Universiteit Amsterdam, The Netherlands
Satoshi Tojo	Japan Advanced Institute of Science and Technology, Japan
Krzysztof Tokarz	Silesian University of Technology, Poland
Bogdan Trawiński	Wrocław University of Science and Technology, Poland
Krzysztof Trojanowski	Cardinal Stefan Wyszyński University in Warsaw, Poland
Ualsher Tukeyev	al-Farabi Kazakh National University, Kazakhstan
Aysegul Ucar	Firat University, Turkey
Olgierd Unold	Wrocław University of Science and Technology, Poland
Joost Vennekens	Katholieke Universiteit Leuven, Belgium
Jorgen Villadsen	Technical University of Denmark, Denmark
Bay Vo	Ho Chi Minh City University of Technology, Vietnam
Eva Volná	University of Ostrava, Czech Republic

Wahyono	Universitas Gadjah Mada, Indonesia
Lipo Wang	Nanyang Technological University, Singapore
Junzo Watada	Waseda University, Japan
Konrad Wojciechowski	Silesian University of Technology, Poland
Krystian Wojtkiewicz	Wrocław University of Science and Technology, Poland
Krzysztof Wróbel	University of Silesia, Poland
Marian Wysocki	Rzeszow University of Technology, Poland
Farouk Yalaoui	University of Technology of Troyes, France
Xin-She Yang	Middlesex University, UK
Tulay Yildirim	Yildiz Technical University, Turkey
Piotr Zabawa	Cracow University of Technology, Poland
Sławomir Zadrożny	Systems Research Institute, Polish Academy of Sciences, Poland
Drago Zagar	University of Osijek, Croatia
Danuta Zakrzewska	Łódź University of Technology, Poland
Constantin-Bala Zamfirescu	Lucian Blaga University of Sibiu, Romania
Katerina Zdravkova	Ss. Cyril and Methodius University in Skopje, Republic of Macedonia
Vesna Zeljkovic	Lincoln University, USA
Aleksander Zgrzywa	Wrocław University of Science and Technology, Poland
Qiang Zhang	Dalian University, China
Beata Zielosko	University of Silesia, Poland
Maciej Zięba	Wrocław University of Science and Technology, Poland
Adam Ziębiński	Silesian University of Technology, Poland
Marta Zorrilla	University of Cantabria, Spain

Program Committees of Special Sessions

Special Session on Intelligent Systems and Algorithms in Information Sciences (ISAIS 2019)

Martin Kotyrba	University of Ostrava, Czech Republic
Eva Volná	University of Ostrava, Czech Republic
Ivan Zelinka	VŠB – Technical University of Ostrava, Czech Republic
Hashim Habiballa	Institute for Research and Applications of Fuzzy Modeling, Czech Republic
Alexej Kolcun	Institute of Geonics, Czech Republic
Roman Senkerik	Tomas Bata University in Zlin, Czech Republic
Zuzana Kominkova Oplatkova	Tomas Bata University in Zlin, Czech Republic
Katerina Kostolanyova	University of Ostrava, Czech Republic
Antonin Jancarik	Charles University in Prague, Czech Republic
Petr Dolezel	University of Pardubice, Czech Republic

Igor Kostal	University of Economics in Bratislava, Slovakia
Eva Kurekova	Slovak University of Technology in Bratislava, Slovakia
Leszek Cedro	Kielce University of Technology, Poland
Dagmar Janacova	Tomas Bata University in Zlin, Czech Republic
Martin Halaj	Slovak University of Technology in Bratislava, Slovakia
Radomil Matousek	Brno University of Technology, Czech Republic
Roman Jasek	Tomas Bata University in Zlin, Czech Republic
Petr Dostal	Brno University of Technology, Czech Republic
Jiri Pospichal	University of Ss. Cyril and Methodius (UCM), Slovakia
Vladimir Bradac	University of Ostrava, Czech Republic
Roman Jasek	Tomas Bata University in Zlin, Czech Republic
Petr Pavel	University of Pardubice, Czech Republic
Jan Capek	University of Pardubice, Czech Republic

Special Session on Intelligent and Contextual Systems (ICxS 2019)

Adriana Albu	Politehnica University Timişoara, Romania
Basabi Chakraborty	Iwate Prefectural University, Japan
Dariusz Frejlichowski	West Pomeranian University of Technology, Poland
Erdenebileg Batbaatar	Chungbuk National University, South Korea
Goutam Chakraborty	Iwate Prefectural University, Japan
Ha Manh Tran	Ho Chi Minh City International University, Vietnam
Hong Vu Nguyen	Ton Duc Thang University, Vietnam
Hideyuki Takahashi	RIEC, Tohoku University, Japan
Jerzy Świątek	Wrocław University of Science and Technology, Poland
Józef Korbicz	University of Zielona Gora, Poland
Keun Ho Ryu	Chungbuk National University, South Korea
Kilho Shin	University of Hyogo, Japan
Maciej Huk	Wrocław University of Science and Technology, Poland
Marcin Fojcik	Western Norway University of Applied Sciences, Norway
Masafumi Matsuhara	Iwate Prefectural University, Japan
Michael Spratling	University of London, UK
Musa Ibrahim	Chungbuk National University, South Korea
Nguyen Khang Pham	Can Tho University, Vietnam
Plamen Angelov	Lancaster University, UK
Qiangfu Zhao	University of Aizu, Japan
Quan Thanh Tho	Ho Chi Minh City University of Technology, Vietnam

Rashmi Dutta Baruah	Lancaster University, UK
Tadahiko Murata	Kansai University, Japan
Takako Hashimoto	Chiba University of Commerce, Japan
Tetsuji Kubojama	Gakushuin University, Japan
Tetsuo Kinoshita	RIEC, Tohoku University, Japan
Thai-Nghe Nguyen	Can Tho University, Vietnam
Tsatsral Amarbayasgalan	Chungbuk National University, South Korea
Zhenni Li	University of Aizu, Japan

Special Session on Intelligent Methods and Artificial Intelligence for Biomedical Decision Support Systems (IMAIBDSS 2019)

Ani Liza Asmawi	International Islamic University, Malaysia
Martin Augustynek	VŠB – Technical University of Ostrava, Czech Republic
Martin Cerny	VŠB – Technical University of Ostrava, Czech Republic
Klara Fiedorova	VŠB – Technical University of Ostrava, Czech Republic
Habibollah Harun	Universiti Teknologi Malaysia, Malaysia
Lim Kok Cheng	Universiti Tenaga Nasional, Malaysia
Roliana Ibrahim	Universiti Teknologi Malaysia, Malaysia
Jafreezal Jaafar	Universiti Teknologi Petronas, Malaysia
Vladimir Kasik	VŠB – Technical University of Ostrava, Czech Republic
Ondrej Krejcar	University of Hradec Kralove, Czech Republic
Jan Kubicek	VŠB – Technical University of Ostrava, Czech Republic
Kamil Kuca	University of Hradec Kralove, Czech Republic
Petra Maresova	University of Hradec Kralove, Czech Republic
David Oczka	VŠB – Technical University of Ostrava, Czech Republic
Sigeru Omatu	Osaka Institute of Technology, Japan
Marek Penhaker	VŠB – Technical University of Ostrava, Czech Republic
Lukas Peter	VŠB – Technical University of Ostrava, Czech Republic
Chawalsak Phetchanchai	Suan Dusit University, Thailand
Antonino Proto	VŠB – Technical University of Ostrava, Czech Republic
Naomie Salim	Universiti Teknologi Malaysia, Malaysia
Ali Selamat	Universiti Teknologi Malaysia, Malaysia
Imam Much Subroto	Universiti Islam Sultan Agung, Indonesia
Lau Sian Lun	Sunway University, Malaysia

Takeru Yokoi	Tokyo Metropolitan International Institute of Technology, Japan
Hazli Mohamed Zabil	Universiti Tenaga Nasional, Malaysia

Special Session on Computer Vision and Intelligent Systems (CVIS 2019)

Kang-Hyun Jo	University of Ulsan, South Korea
Van-Dung Hoang	Quang Binh University, Vietnam
The-Anh Pham	Hong Duc University, Vietnam
Thi-Lan Le	Hanoi University of Science and Technology, Vietnam
Wahyono	Universitas Gadjah Mada, Indonesia
Alireza Ghasempour	University of Applied Science and Technology, Iran
Afiahayati	University Gadjah Mada, Indonesia
Byeongryong Lee	University of Ulsan, South Korea
Cheolgeun Ha	University of Ulsan, South Korea
Chi-Mai Luong	University of Science and Technology of Hanoi, Vietnam
Christos Bouras	University of Patras, Greece
Heejun Kang	University of Ulsan, South Korea
Hyun-Deok Kang	Ulsan National Institute of Science and Technology, South Korea
Moh Edi Wibowo	Universitas Gadjah Mada, Indonesia
Mu-Song Chen	Da-Yeh University, Taiwan
My-Ha Le	HCMC University of Technology and Education, Vietnam
Ngoc-Son Pham	HCMC University of Technology and Education, Vietnam
Nobutaka Shimada	Ritsumeikan University, Japan
Pavel Loskot	Swansea University, UK
Thanh-Hai Tran	Hanoi University of Science and Technology, Vietnam
Thanh-Truc Tran	Danang ICT, Vietnam
Trung-Duy Tran	Posts and Telecommunications Institute of Technology, Vietnam
Van-Huy Pham	Tong Duc Thang University, Vietnam
Van Mien	University of Exeter, UK
Vu-Viet Vu	Hanoi National University, Vietnam
Yoshinori Kuno	Saitama University, Japan
Youngsoo Suh	University of Ulsan, South Korea
Yuansong Qiao	Athlone Institute of Technology, Ireland

Special Session on Advanced Data Mining Techniques and Applications (ADMTA 2019)

Tzung-Pei Hong	National University of Kaohsiung, Taiwan
Tran Minh Quang	Ho Chi Minh City University of Technology, Vietnam
Bac Le	University of Science, VNU-HCM, Vietnam
Bay Vo	Ho Chi Minh City University of Technology, Vietnam
Chun-Hao Chen	Tamkang University, Taiwan
Chun-Wei Lin	Harbin Institute of Technology Shenzhen Graduate School, China
Wen-Yang Lin	National University of Kaohsiung, Taiwan
Yeong-Chyi Lee	Cheng Shiu University, Taiwan
Le Hoang Son	Vietnam National University, Vietnam
Thi Ngoc Chau	Ho Chi Minh City University of Technology, Vietnam
Van Vo	Ho Chi Minh University of Industry, Vietnam
Ja-Hwung Su	Cheng Shiu University, Taiwan
Ming-Tai Wu	University of Nevada, USA
Kawuu W. Lin	National Kaohsiung University of Applied Sciences, Taiwan
Tho Le	Ho Chi Minh City University of Technology, Vietnam
Dang Nguyen	Deakin University, Australia
Hau Le	Thuyloi University, Vietnam
Thien-Hoang Van	Ho Chi Minh City University of Technology, Vietnam
Tho Quan	Ho Chi Minh City University of Technology, Vietnam
Ham Nguyen	University of People's Security Hochiminh City, Vietnam
Thiet Pham	Ho Chi Minh University of Industry, Vietnam
Nguyen Thi Thuy Loan	International University, VNU-HCMC, Vietnam

Special Session on Intelligent Supply Chains and e-Commerce (ISCEC 2019)

Bartłomiej Pierański	Poznań University of Economics and Business, Poland
Justyna Światowiec-Szczepańska	Poznań University of Economics and Business, Poland
Carlos Andres Romano	Polytechnic University of Valencia, Spain
Davor Dujak	University of Osijek, Croatia
Paulina Golińska-Dawson	Poznań University of Technology, Poland
Paweł Pawlewski	Poznań University of Technology, Poland
Jakub Bercik	Slovak University of Agriculture in Nitra, Slovakia
Adam Koliński	Poznań School of Logistics, Poland

Special Session on Intelligent Information Systems (IIS 2019)

Ryszard Antkiewicz	Military University of Technology, Poland
Leon Bobrowski	Białystok University of Technology, Poland
Urszula Boryczka	University of Silesia, Poland
Mariusz Chmielewski	Military University of Technology, Poland
Jan Hodický	University of Defense in Brno, Czech Republic
Rafal Kasprzyk	Military University of Technology, Poland
Jacek Koronacki	Polish Academy of Sciences, Poland
Rafał Ładysz	George Mason University, USA
Andrzej Najgebauer	Military University of Technology, Poland
Ewa Niewiadomska-Szynkiewicz	Warsaw University of Technology, Poland
Dariusz Pierzchała	Military University of Technology, Poland
Jarosław Rulka	Military University of Technology, Poland
Khalid Saeed	Białystok University of Technology, Poland
Zenon A. Sosnowski	Białystok University of Technology, Poland
Zbigniew Tarapata	Military University of Technology, Poland

Special Session on Collective Intelligence for Service Innovation, Technology Management, E-learning and Fuzzy Intelligent Systems (CISTEF 2019)

Albim Y. Cabatingan	University of the Visayas, Philippines
Teh-Yuan Chang	Aletheia University, Taiwan
Chi-Min Chen	Aletheia University, Taiwan
Chih-Chung Chiu	Aletheia University, Taiwan
Wen-Min Chou	Aletheia University, Taiwan
Chao-Fu Hong	Aletheia University, Taiwan
Chia-Lin Hsieh	Aletheia University, Taiwan
Gong-Yih Hsieh	Aletheia University, Taiwan
Chia-Ling Hsu	Tamkang University, Taiwan
Fang-Cheng Hsu	Aletheia University, Taiwan
Chi-Cheng Huang	Aletheia University, Taiwan
Rahat Iqbal	Coventry University, UK
Huan-Ting Lin	The University of Tokyo, Japan
Kuo-Sui Lin	Aletheia University, Taiwan
Min-Huei Lin	Aletheia University, Taiwan
Yuh-Chang Lin	Aletheia University, Taiwan
Shin-Li Lu	Aletheia University, Taiwan
Janet Argot Pontevedra	University of San Carlos, Philippines
Shu-Chin Su	Aletheia University, Taiwan
Pen-Choug Sun	Aletheia University, Taiwan
Ai-Ling Wang	Tamkang University, Taiwan
Leuo-Hong Wang	Aletheia University, Taiwan
Hung-Ming Wu	Aletheia University, Taiwan
Feng-Sueng Yang	Aletheia University, Taiwan

Hsiao-Fang Yang	National Chengchi University, Taiwan
Sadayuki Yoshitomi	Toshiba Corporation, Japan

Special Session on Analysis of Image, Video, Movements and Brain Intelligence in Life Sciences (IVMBI 2019)

Andrei Barborica	Research & Compliance and Engineering, FHC, Inc., USA
Artur Bąk	Polish-Japanese Academy of Information Technology, Poland
Konrad Ciecierski	Warsaw University of Technology, Poland
Leszek Chmielewski	Warsaw University of Life Sciences, Poland
Zenon Chaczko	University Technology of Sydney, Australia
Christopher Chiu	University of Technology Sydney, Australia
Aldona Barbara Drabik	Polish-Japanese Academy of Information Technology, Poland
Marcin Fojcik	Sogn og Fjordane University College, Norway
Adam Gudyś	Silesian University of Technology, Poland
Ryszard Gubrynowicz	Polish-Japanese Academy of Information Technology, Poland
Piotr Habela	Polish-Japanese Academy of Information Technology, Poland
Celina Imielińska	Vesalius Technolodgies LLC, USA
Henryk Josiński	Silesian University of Technology, Poland
Mark Kon	Boston University, USA
Wojciech Knieć	Polish-Japanese Academy of Information Technology, Poland
Ryszard Klempous	Wrocław University of Science and Technology, Poland
Ryszard Kozera	The University of Life Sciences - SGGW, Poland
Julita Kulbacka	Wrocław Medical University, Poland
Marek Kulbacki	Polish-Japanese Academy of Information Technology, Poland
Krzysztof Marasek	Polish-Japanese Academy of Information Technology, Poland
Majaz Moonis	UMass Medical School, USA
Radoslaw Nielek	Polish-Japanese Academy of Information Technology, Poland
Peter Novak	Brigham and Women's Hospital, USA
Wieslaw Nowinski	Cardinal Stefan Wyszynski University, Poland
Jerzy Paweł Nowacki	Polish-Japanese Academy of Information Technology, Poland
Eric Petajan	LiveClips LLC, USA
Andrzej Polański	Silesian University of Technology, Poland
Andrzej Przybyszewski	Polish-Japanese Academy of Information Technology, Poland

Zbigniew Ras   University of North Carolina at Charlotte,
                                USA & PJAIT, Poland
Joanna Rossowska   Polish Academy of Sciences, Institute of
                                Immunology and Experimental Therapy, Poland
Jakub Segen   Gest3D LLC, USA
Aleksander Sieroń   Medical University of Silesia, Poland
Dominik Ślęzak   University of Warsaw, Poland
Michał Staniszewski   Polish-Japanese Academy of Information
                                Technology, Poland
Zbigniew Struzik   RIKEN Brain Science Institute, Japan
Adam Świtoński   Silesian University of Technology, Poland
Agnieszka Szczęsna   Silesian University of Technology, Poland
Kamil Wereszczyński   Polish-Japanese Academy of Information
                                Technology, Poland
Alicja Wieczorkowska   Polish-Japanese Academy of Information
                                Technology, Poland
Konrad Wojciechowski   Polish-Japanese Academy of Information
                                Technology, Poland
Sławomir Wojciechowski   Polish-Japanese Academy of Information
                                Technology, Poland

Special Session on Data Structures Modelling for Knowledge Representation (DSMKR 2019)

Marek Krótkiewicz   Wrocław University of Science and Technology,
                                Poland
Piotr Zabawa   Cracow University of Technology, Poland
Jerzy Tomasik   CNRS - LIMOS University Clermont-Auvergne,
                                France
Helena Dudycz   Wrocław University of Economics, Poland
Robert Sochacki   University of Opole, Poland
Wojciech Hunek   Opole University of Science and Technology,
                                Poland
Sophia Katrenko   University of Amsterdam, The Netherlands
Krystian Wojtkiewicz   Wrocław University of Science and Technology,
                                Poland
Marcin Jodłowiec   Wrocław University of Science and Technology,
                                Poland

Contents – Part I

Machine Learning and Data Mining

Databases and Intelligent Information Systems

Contents – Part II

Intelligent Information Systems

Intelligent and Contextual Systems

Intelligent Systems and Algorithms in Information Sciences

Computer Vision and Intelligent Systems

Knowledge Engineering and Semantic Web

Algorithms for Merging Probabilistic Knowledge Bases

Van Tham Nguyen[1,3]($\boxtimes$), Ngoc Thanh Nguyen[2],
and Trong Hieu Tran[1]

[1] University of Engineering and Technology, Vietnam National University,
Hanoi, Vietnam
{17028002,hieutt}@vnu.edu.vn
[2] Wroclaw University of Science and Technology, Wrocław, Poland
Ngoc-Thanh.Nguyen@pwr.edu.pl
[3] Nam Dinh University of Technology Education, Nam Dinh, Vietnam
thamnv.nute@gmail.com

Abstract. There is an increasing need to develop appropriate techniques for merging probabilistic knowledge bases (PKB) in knowledge-based systems. To deal with merging problems, several approaches have been put forward. However, in the proposed models, the representation of the merged probabilistic knowledge base is not similar to the representation of original knowledge bases. The drawback of the solutions is that probabilistic constraints on the set of input knowledge bases must have the same structure and there is no algorithm for implementing the merging process. In this paper, we proposed two algorithms for merging probabilistic knowledge bases represented by various structures. To this aim, the method of constraint deduction is investigated, a set of mean merging operators is proposed and several desirable logical properties are presented and discussed. These are the basis for building algorithms. The complexity of algorithms as well as related propositions are also analysised and discussed.

Keywords: Probabilistic knowledge bases · Merging operator ·
Merging algorithms

1 Introduction

The ability to merge knowledge bases is essential for establishing and upholding actions in knowledge-based systems. Knowledge merging is an important problem and applied in many fields [1]. One of the first challenges to have this ability is the handling of the inconsistency in knowledge bases coming from various systems. The method of solving the inconsistency of probabilistic knowledge bases has been discussed in more detail in [2–7]. Knowledge base revision is the process wherein knowledge base is modified to incorporate the new received knowledge base while knowledge base merging could be understood as the process wherein an intelligent agent is given to one or more knowledge bases to merge. Knowledge base merging is the process wherein a new knowledge base is resulted from a set of knowledge bases [8].

© Springer Nature Switzerland AG 2019
N. T. Nguyen et al. (Eds.): ACIIDS 2019, LNAI 11431, pp. 3–15, 2019.
https://doi.org/10.1007/978-3-030-14799-0_1

In the information combining field such as statistics, decision theory, and economic sciences, the problem of combining subjective probabilities has long been studied in [9]. An approache is performed by using Dempster's rule to combine subjective probabilistic knowledge bases [10]. In order to handle the fusion problem, the maximum entropy (ME) techniques were applied to probabilities represented as random variables [11]. With the help of the ME approach, ME-principles have been used to represent incomplete probabilistic knowledge [12]. ME-revision operators and how ME-reasoning works have been also presented and discussed. However, probability values are likewise not within the interval of two original probabilities. Another method for probabilistic belief revision and information fusion, developing and overcoming the disadvantages in [12] was introduced in [13]. The main idea of this method is that elaborating fusion operator to handle probabilistic information faithfully by making use of optimum entropy principles. This operator satisfies basic demands and calculate the information theoretical mean of probability values. Probabilistic constraint deduction rules were proposed in [12, 14]. These rules were based on the ME and Grobner basis techniques to obtain reduced constraints and compute new probability value for these constraints.

A general model of merging knowledge could be found in [8]. The merging problem has been defined and analyzed and a set of postulates of merging process as well as merging algorithms have been also proposed in that work. Another algorithm for merging knowledge bases has been presented and discussed in [15]. The main idea of this algorithm is that using distance function to determine subprofiles for attributes and each subprofile was employed to find its merging fulfilling several merging postulates. However, these algorithms are employed for a consensus-based model.

In a probabilistic environment, the merging problem has been defined and discussed in [16–20]. In knowledge-based systems, the main idea of the method for merging probabilistic knowledge bases was based on the KL-projection means [16]. A family of probabilistic merging operators could be found in [17–19]. Merging operators presented in [19] were based on convex Bregman divergences satisfied several logical properties. However, the input of the models is the set of low-dimensional distributions representing the original probabilistic knowledge bases. Another merging model has been proposed in [20]. This merging process was divided into two stages. The first stage finds the representing of original probabilistic knowledge bases including probabilistic functions. This stage related to the calculation inconsistency measures [5–7] and probabilistic vectors [6, 7, 20]. The second stage is to compute the merging result by using merging operators SQ, SE, and LE that was based divergence distance functions [18, 19]. However, the representation of merged probabilistic knowledge base differs from the representation of input ones.

When knowledge is represented as a set of facts and rules, several approaches have been developed to handle the uncertainty by means of fuzzy logic. Implication operators and a method for fixpoint queries which contains the facts and fulfills the rules which were proposed in [21, 22] to compute new value probability. This idea continues to be developed and presented in [23]. The model proposed in [23] was based on the concept of multivalued knowledge-base.

The main contribution of this paper is threefold. First, we make a deeply survey on the way to represent a probabilistic knowledge base, how to reduce constraints on it, how to obtain the mean value of probability values. Second, two merging operators are proposed to compute new probability of common constraint. They ensure that the resulting probability lies within the interval between the smallest and the largest value of probabilities. The first merging operator is based on the probability value of constraints themselves in knowledge bases while the second merging operator is built by more utilizing a coefficient. Then, we propose a probabilistic constraint deduction algorithm (PCDA) and two merging algorithms, namely, Mean Merging Algorithm (MMA) and Coefficient Mean Merging Algorithm (CMMA), respectively. MMA Algorithm is designed to merge probabilistic knowledge bases having different structures by employing the first merging operator and PCDA algorithm whereas CMMA algorithm is designed by employing the second merging operator and PCDA algorithm. The complexity of algorithms and propositions ensuring the consistency of the resulting knowledge base also are investigated and discussed.

This paper is organized as follows. In Sect. 2 we present some necessary preliminaries about the probability, probabilistic constraint, probabilistic knowledge base, probabilistic constraint deduction rules, and probabilistic knowledge base profile. Afterwards, in Sect. 3 we propose two probabilistic merging operators, the logical properties for each operator as well as several related propositions are also introduced and discussed. Two main algorithms for merging probabilistic knowledge bases are proposed in Sect. 4. This section also analyses and discusses the complexity of algorithms, and several related propositions. Section 5 highlights our main contributions and gives an outlook on further work.

2 Probabilistic Knowledge Bases

A set containing all possible outcomes of a statistical experiment is called a sample space, signified by $\mathcal{S}$. A subset of the sample space $\mathcal{S}$ is called an event, signified by E, then $\mathcal{E} = \{E_1, \ldots, E_n\}$. For $F, G \in \mathcal{E}$, the intersection of two events F and G, signified by FG, is the event, including all the components featured for both F and G; negation of F, signified by $\neg F$, is abbreviated by $\overline{F}$. Let $\Theta = \widehat{E}_1 \widehat{E}_2 \ldots \widehat{E}_n \left(\widehat{E}_i \in \{E_i, \overline{E}_i\} \right)$ be a complete conjunction of $\mathcal{E}$. A set containing all complete conjunctions of $\mathcal{E}$, signified by $\Gamma(\mathcal{E})$, therefore $\Gamma(\mathcal{E}) = \{\Theta_1, \ldots, \Theta_{2^n}\}$. A complete conjunction $\Theta \in \Gamma(\mathcal{E})$ fulfills an event F, signified by $\Theta \models F$, iff F positively appears in Θ Let $\Lambda(H) = \{\Theta \in \Gamma(\mathcal{E}) | \Theta \models H\}$ with H denote an event or a set of events. Let $\mathbb{R}_{\geq 0}$ be a set of all non-negative real values from 0 to $+\infty$. Let $\mathbb{R}_{[0,1]}$ be a set of all real values from 0 to 1.

Definition 1. *Let $m = |\Delta(\mathcal{E})|$ be the number of complete conjunctions of $\mathcal{E}$. A probability function is a map $\mathcal{P} : \Delta(\mathcal{E}) \to \mathbb{R}_{[0,1]}$ such as*

$$\sum_{i=1}^{m} \mathcal{P}(\Theta_i) = 1$$

A set of all probability functions $\mathcal{P}$ over set of events $\mathcal{E}$, signified by $\widehat{P}(\mathcal{E})$, therefore $\widehat{P}(\mathcal{E}) = \{\mathcal{P}(\Theta_1), \ldots, \mathcal{P}(\Theta_{2^n})\}$. With $F \in \mathcal{E}$, $\mathcal{P}(F) = \sum_{\Theta \in \Delta(\mathcal{E}): \Theta \models F} \mathcal{P}(\Theta)$.

Definition 2. *Let $F, G \in \mathcal{E}$ and $\rho \in \mathbb{R}_{[0,1]}$. A probabilistic constraint is represented by form $\kappa[\rho]$, where $\kappa = (F|G)$.*

Tautologies and contradictions will be denoted by $\top$ and $\bot$, respectively. If G is tautological, $G \equiv \top$, we abbreviate $(F|\top)[\rho]$ by $(F)[\rho]$. For $\kappa = (F|G)$ and $G \not\equiv \top$, we let $\text{Left}(\kappa) = F$ and $\text{Right}(\kappa) = G$.

Definition 3. *Two probabilistic constraints κ_1 and κ_2 are considered equivalent, signified by $\kappa_1 \approx \kappa_2$, iff $\text{Left}(\kappa_1) = \text{Left}(\kappa_2)$ and $\text{Right}(\kappa_1) = \text{Right}(\kappa_2)$.*

Definition 4. *A probabilistic knowledge base, signified by $\mathcal{K}$, whenever it is defined as follows:*

$$\mathcal{K} = \langle \kappa_1[\rho_1], \ldots, \kappa_n[\rho_n] \rangle$$

Let $n = |\mathcal{K}|$ be number of constraints in $\mathcal{K}$.

Definition 5. *(Probabilistic constraint deduction rule).*

Let $\mathcal{K} = \langle (F_1|G_1)[\rho_1], \ldots, (F_n|G_n)[\rho_n] \rangle$. A probabilistic knowledge base $\mathcal{K}$ is reduced to $\mathcal{K}^* = \langle (F_1^*|G_1^*)[\rho_1^*], \ldots, (F_m^*|G_m^*)[\rho_m^*] \rangle$ such that $m \leq n$, signified by

$$\frac{\mathcal{K} : (F_1|G_1)[\rho_1], \ldots, (F_n|G_n)[\rho_n]}{(F_1^*|G_1^*)[\rho_1^*], \ldots, (F_m^*|G_m^*)[\rho_m^*]}$$

Proposition 1. *Let $F, G, H \in \mathcal{E}$ and $\rho_1, \rho_2 \in \mathbb{R}_{[0,1]}$. Then*

$$\frac{\mathcal{K} : (F|G)[\rho_1], (H|F)[\rho_2]}{(H|G)\left[\frac{1}{2}(2\rho_1\rho_2 - \rho_1 + 1)\right]}$$

Definition 6. *A probabilistic knowledge base $\mathcal{K}$ is consistent, signified by $\mathcal{K} \not\models \bot$, iff $\Lambda(\mathcal{K}) \neq \emptyset$. Otherwise, $\mathcal{K}$ is inconsistent, denoted by $\mathcal{K} \models \bot$.*

Definition 7. *A probabilistic knowledge base profile, signified by $\mathcal{B} = \{\mathcal{K}_1, \ldots, \mathcal{K}_n\}$, is a finite set of probabilistic knowledge bases such that $\forall \mathcal{K}_i \in \mathcal{B} : \mathcal{K}_i \not\models \bot$.*

For $\forall \mathcal{K}_i \in \mathcal{B}$, let $\mathcal{K}_i = \left\langle \kappa_1^{(i)}\left[\rho_1^{(i)}\right], \ldots, \kappa_{n_i}^{(i)}\left[\rho_{n_i}^{(i)}\right] \right\rangle$, where $n_i = |\mathcal{K}_i|$.

3   Probabilistic Merging Operators

3.1   Mean Merging Operator (MMO) $\oplus$

Let $\rho^+ = \rho^{1-\rho}$, $\rho^- = \rho^{-\rho}$, $\widehat{\rho} = (1-\rho)^{\rho-1}$

Definition 8. *Let $\mathcal{K}_1 = \langle (F|G)[\rho_1] \rangle; \mathcal{K}_2 = \langle (F|G)[\rho_2] \rangle$. Operator $\oplus$ is called a mean merging operator of two probabilities ρ_1, ρ_2 whenever it is defined as follows:*

$$\oplus(\rho_1, \rho_2) = \frac{\rho_1^+ \widehat{\rho}_1 + \rho_2^+ \widehat{\rho}_2}{\rho_1^- \widehat{\rho}_1 + \rho_2^- \widehat{\rho}_2}$$

Proposition 2. *Mean merging operator of two probabilities* ρ_1, ρ_2 *satisfies the following properties:*

  (CMT) *Commutativity.* $\oplus(\rho_1, \rho_2) = \oplus(\rho_2, \rho_1)$
  (IDP) *Idempotence.* $\oplus(\rho_1, \rho_1) = \rho_1$
  (MVP) *Mean Value Property. If* $\rho_1 < \rho_2$ *then* $\rho_1 < \oplus(\rho_1, \rho_2) < \rho_2$
  (SFS) *Self-Symmetry.* $\oplus(\rho_1, 1 - \rho_1) = 0.5$
  (SM) *Symmetry.* $\oplus(1 - \rho_1, 1 - \rho_2) = 1 - \oplus(\rho_1, \rho_2)$

  (SIS) *Semi-Symmetry.* $\oplus(1 - \rho_1, 0.5) = \frac{1 + \rho_1^+ \widehat{\rho}_1}{2 + \rho_1^- \widehat{\rho}_1}$

Definition 9. *Let* $\mathcal{K}_1 = \langle (F|G)[\rho_1] \rangle; \ldots; \mathcal{K}_n = \langle (F|G)[\rho_n] \rangle$. *Operator* $\oplus^n$ *is called a mean merging operator of n probabilities* $\rho_1, \ldots, \rho_n$ $(n > 2)$ *whenever it is defined as follows:*

$$\oplus^n(\rho_1, \ldots, \rho_n) = \frac{\sum_{i=1}^n \rho_i^+ \widehat{\rho}_i}{\sum_{i=1}^n \rho_i^- \widehat{\rho}_i}$$

Proposition 3. *Operator* $\oplus^n$ *satisfies properties* **CMT, IDP, MVP, SFS, SM, SIS**

Proposition 4. $min_{1 \le i \le n} \rho_i \le \oplus^n(\rho_1, \ldots, \rho_n) \le max_{1 \le i \le n} \rho_i$

3.2 Coefficient Mean Merging Operator (CMMO) $\ominus$

Definition 10. *Let* $\mathcal{K}_1 = \langle (F|G)[\rho_1] \rangle; \mathcal{K}_2 = \langle (F|G)[\rho_2] \rangle$. *Operator* $\ominus$ *is called a coefficient mean merging operator with respect to coefficient c of two probabilities* ρ_1, ρ_2 *whenever it is defined as follows:*

$$\ominus(\rho_1, \rho_2) = \frac{\rho_1^c \rho_2^{1-c}}{\rho_1^c \rho_2^{1-c} + (1 - \rho_1)^c (1 - \rho_2)^{1-c}}$$

where $c \in [0; 1]$ *is a coefficient.*

Proposition 5. *Mean merging operator of two probabilities* ρ_1, ρ_2 *satisfies the following properties:*

  (IDP1) *Idempotence.* $\ominus(\rho_1, \rho_1) = \rho_1$
  (MVP1) *Mean Value Property. If* $\rho_1 < \rho_2$ *then* $\rho_1 < \ominus(\rho_1, \rho_2) < \rho_2$
  (SM1) *Symmetry.* $\ominus(1 - \rho_1, 1 - \rho_2) = 1 - \ominus(\rho_1, \rho_2)$

Definition 11. *Let* $\mathcal{K}_1 = \langle (F|G)[\rho_1] \rangle; \ldots; \mathcal{K}_n = \langle (F|G)[\rho_n] \rangle$. *Operator* $\ominus^n$ *is called a coefficient mean merging operator with respect to coefficient c of n probabilities* $\rho_1, \ldots, \rho_n$ *(n > 2) whenever it is defined as follows:*

$$\ominus^n(\rho_1, \ldots, \rho_n) = \ominus\left(\ominus^{n-1}, \rho_n\right)$$

Proposition 6. *Operator* $\ominus^n$ *satisfies properties **IDP1, MVP1, SM1***
It is straightforward to see that operator $\ominus^n$ satisfies IDP1, MVP1, SM1. So, operator $\oplus^n$ appears to be more well behaved.

Proposition 7. $min_{1 \le i \le n} \rho_i \le \ominus^n(\rho_1, \ldots, \rho_n) \le max_{1 \le i \le n} \rho_i$
Propositions 4 and 7 ensure that the new probabilities are kind of mean values, so that the resulting probabilistic knowledge base reflects average degrees of knowledge as a compromise.

4   Proposed Merging Algorithms for PKB

4.1   Algorithms

In this section, we propose two main algorithms for merging probabilistic knowledge bases. The PCDA algorithm describes a probabilistic knowledge base PCDA($\mathcal{K}$) resulting from the deduction of probabilistic constraints in $\mathcal{K}$. For each pair of two probabilistic constraints (line 1), if event at the right of one constraint equals event at the left of the other (line 2) then we employ Proposition 1 to replace such two probabilistic constraints with new one (lines 3 to 6). The algorithm ends when all pairs of constraints in $\mathcal{K}$ are considered.

Algorithm 1. The PCDA algorithm for PKB
Input: A probabilistic knowledge base $\mathcal{K}$
Output: A probabilistic knowledge base **PCDA($\mathcal{K}$)=$\mathcal{K}_0$**

1:	**For** each $(\kappa_i, \kappa_j) \in \mathcal{K}$ and $i \ne j$ **do**
2:	**If** $left(\kappa_i) = right(\kappa_j)$ **then**
3:	$\mathcal{K} \leftarrow \mathcal{K} \backslash \{(\kappa_i)[\rho_i], (\kappa_j)[\rho_j]\}$;
4:	$\kappa_h \leftarrow \left(right(\kappa_j) \big| left(\kappa_i)\right)$
5:	$\rho_h \leftarrow \frac{1}{2}\left(2\rho_i\rho_j - \rho_i + 1\right)$
6:	$\mathcal{K} \leftarrow \mathcal{K} \cup \{(\kappa_h)[\rho_h]\}$;
7:	**End if**;
8:	**End for**;
9:	**Return** $\mathcal{K}$;

The PCDA algorithm leads to the following proposition stating that if the input $\mathcal{K}$ is consistent, then The PCDA algorithm returns a consistent probabilistic knowledge base such that $|\text{PCDA}(\mathcal{K})| \le |\mathcal{K}|$.

Proposition 8. *If $\mathcal{K} \not\models \perp$ then PCDA $(\mathcal{K}) \not\models \perp$.*

The MMA algorithm describes a probabilistic knowledge base MMA($\mathcal{B}$) resulting from the merging of probabilistic knowledge bases in $\mathcal{B}$. The MMA algorithm consists of two stages, (i) finding common constraints with new probability values, and followed by (ii) PCDA algorithm to reduce probabilistic constraints. The first stage employs MMO operator to compute new probabilities value for resulting constraint. Therefore, probability of constraints in resulting knowledge base fulfills Proposition 4. The second stage uses the knowledge base obtained at the first stage to reduce probabilistic constraints, so to as decrease storage space for resulting knowledge base. The details are described below.

Lines 1 to 14 present the first stage. The algorithm first initializes the resulting probabilistic knowledge base to an empty set (line 1). Lines 2 to 14 add the common constraints to the resulting knowledge base. For each probabilistic knowledge base $\mathcal{K}_i$ in $\mathcal{B}$ (line 2), while this knowledge base is not an empty set (line 3), we employ Definition 3 to consider the equivalence between the first constraints in $\mathcal{K}_i$ and each constraint in the remaining knowledge bases (lines 4 to 6). If the equivalence is fulfilled then we use Definition 8 to compute the new probability of common constraint (line 7). Line 8 and line 11 remove the considered constraint from knowledge bases while line 12 adds a common constraint to the resulting base. The second stage is presented in line 15. The algorithm ends when all knowledge bases in $\mathcal{B}$ are empty and the resulting knowledge base is reduced by using the PCDA algorithm (line 15).

Algorithm 2. The MMA algorithm for PKB

Input: - A probabilistic knowledge base *profile* $\mathcal{B} = \{\mathcal{K}_1, \ldots, \mathcal{K}_n\}$
　　　　 - MMO Operator　　　$\oplus$

Output:　A probabilistic knowledge base MMA($\mathcal{B}$) = $\mathcal{K}_0$

1:　　　　　$\mathcal{K}_0 \leftarrow \emptyset$; $h \leftarrow 0$;
2:　　　　**For each** $\mathcal{K}_i \in \mathcal{B}$ **do**
3:　　　　　**While** $\mathcal{K}_i \neq \emptyset$ **do**
4:　　　　　　$h{+}{+}$; $\kappa_h^{(0)} \leftarrow \kappa_1^{(i)}$; $\rho_h^{(0)} \leftarrow \rho_1^{(i)}$;
5:　　　　　　**For each** $\kappa_t^{(j)} \in \mathcal{K}_j$ and $\mathcal{K}_j \neq \mathcal{K}_i$ **do**
6:　　　　　　　**If** $\kappa_t^{(j)} \approx \kappa_1^{(i)}$ **then**
7:　　　　　　　　$\rho_h^{(0)} \leftarrow \oplus \left(\rho_h^{(0)}, \rho_t^{(j)} \right)$;
8:　　　　　　　　$\mathcal{K}_j \leftarrow \mathcal{K}_j \backslash (\kappa_t^{(j)})[\rho_t^{(j)}]$;
9:　　　　　　　**End if**;
10:　　　　　**End for**;
11:　　　　　$\mathcal{K}_i \leftarrow \mathcal{K}_i \backslash (\kappa_1^{(i)})[\rho_1^{(i)}]$;
12:　　　　　$\mathcal{K}_0 \leftarrow \mathcal{K}_0 \cup (\kappa_h^{(0)})[\rho_h^{(0)}]$;
13:　　　　**End while**;
14:　　　**End for**;
15:　　　**Return PCDA($\mathcal{K}_0$)**;

The MMA algorithm leads to the following proposition stating that the MMA algorithm returns a consistent probabilistic knowledge.

Proposition 9. *MMA*($\mathcal{B}$) $\not\models \perp$.

The CMMA algorithm is implemented similar to the MMA algorithm. The main idea is that instead of using MMO operator for line 7, we compute a new probability by using CMMO operator according to Definition 10.

Algorithm 3. The CMMA algorithm for PKB

Input: - A probabilistic knowledge base *profile* $\mathcal{B} = \{\mathcal{K}_1, \dots, \mathcal{K}_n\}$
        - CMMO Operator $\ominus$
Output: A probabilistic knowledge base **CMMA**($\mathcal{B}$) = $\mathcal{K}_0$

1:	$\mathcal{K}_0 \leftarrow \emptyset$; h$\leftarrow$0;
2:	**For each** $\mathcal{K}_i \in \mathcal{B}$ **do**
3:	**While** $\mathcal{K}_i \neq \emptyset$ **do**
4:	h++; $\kappa_h^{(0)} \leftarrow \kappa_1^{(i)}$; $\rho_h^{(0)} \leftarrow \rho_1^{(i)}$;
5:	**For each** $\kappa_t^{(j)} \in \mathcal{K}_j$ and $\mathcal{K}_j \neq \mathcal{K}_i$ **do**
6:	**If** $\kappa_t^{(j)} \approx \kappa_1^{(i)}$ **then**
7:	$\rho_h^{(0)} \leftarrow \ominus (\rho_h^{(0)}, \rho_t^{(j)})$;
8:	$\mathcal{K}_j \leftarrow \mathcal{K}_j \backslash (\kappa_t^{(j)})[\rho_t^{(j)}]$;
9:	**End if;**
10:	**End for;**
11:	$\mathcal{K}_i \leftarrow \mathcal{K}_i \backslash (\kappa_1^{(i)})[\rho_1^{(i)}]$;
12:	$\mathcal{K}_0 \leftarrow \mathcal{K}_0 \cup (\kappa_h^{(0)})[\rho_h^{(0)}]$;
13:	**End while;**
14:	**End for;**
15:	**Return PCDA**($\mathcal{K}_0$);

The CMMA algorithm leads to the following proposition stating that the CMMA algorithm returns a consistent probabilistic knowledge.

Proposition 10. *CMMA*($\mathcal{B}$) $\not\models \perp$.

4.2 The Computational Complexity

In this section, we will analyze and assess the computational complexity of merging algorithms by looking at its phases separately.

Proposition 11. *The computational complexity of the PCDA algorithm is* $\mathcal{O}\left(|PCDA(\mathcal{K}_0)|^2\right)$.

Proof. The complexity of the PCDA algorithm in the worst case depends on the number of constraints in $\mathcal{K}_0$. Therefore, the cost of the second stage is $\mathcal{O}\left(|\text{PCDA}(\mathcal{K}_0)|^2\right)$

Proposition 12. *The computational complexity of the MMA algorithm is* $\mathcal{O}\Big(max\big\{|\mathcal{B}|max\{|\mathcal{K}_i|\}, |PCDA(\mathcal{K}_0)|^2\big\}\Big)$

Proof. In the first stage, the cost is estimated with respect to the number of probabilistic knowledge bases in $\mathcal{B}$ and the number of constraints of each knowledge base $\mathcal{K}_i \in \mathcal{B}$.

- Two assignments in line 1 take $\mathcal{O}(1)$. For the process of considering each knowledge base $\mathcal{B}$, the time needed takes $\mathcal{O}(|\mathcal{B}|)$ where $|\mathcal{B}|$ is the number of iterations of for loop (in line 2). Three assignments in line 4 take $\mathcal{O}(1)$.
- The process of calculating the probability of constraint in $\mathcal{K}_0$ (in lines 5 to 10) needs:
+ The number of iterations of for loop (in line 5) depend on the number of constraints in $\mathcal{K}_j$. It can be done in $\mathcal{O}(max(|\mathcal{K}_j|))$.
+ The equivalence consideration between two constraints in line 6 includes two assignments, so it takes $\mathcal{O}(1)$; calculating the new probability (in line 7) can be done in $\mathcal{O}(1)$.
+ Removing the considered constraint from knowledge base $\mathcal{K}_j$ (in line 8) can be computed in time $\mathcal{O}(1)$.

Therefore, this process runs in $\mathcal{O}(max(|\mathcal{K}_j|))$.

- Removing the considered constraint (in line 11) can be computed in time $\mathcal{O}(1)$.
- Adding common constraint to the new knowledge base (in line 12) can be computed in time $\mathcal{O}(1)$. Therefore, the cost of the first stage is $\mathcal{O}(|\mathcal{B}|max\{|\mathcal{K}_i|\})$.

By Proposition 11, the cost of the second stage is $\mathcal{O}\Big(|PCDA(\mathcal{K}_0)|^2\Big)$.

Therefore, the cost of the MMA algorithm in the worst case is:

$$\mathcal{O}\Big(max\big\{|\mathcal{B}|max\{|\mathcal{K}_i|\}, |PCDA(\mathcal{K}_0)|^2\big\}\Big).$$

Proposition 13. *The computational complexity of the CMMA algorithm is* $\mathcal{O}\Big(max\big\{|\mathcal{B}|max\{|\mathcal{K}_i|\}, |PCDA(\mathcal{K}_0)|^2\big\}\Big)$

Proof. Propositions 12 and 13 are similar in proof.

Example:

A hospital makes a survey of heart-related diseases. It assigns four experts being cardiologists to this survey.

- The first doctor provides the results in which the probability that people has heart disease (denoted by H) is $\mathcal{P}(H) = 0.7$; the probability that people have difficulty breathing (denoted by B) is $\mathcal{P}(B) = 0.5$; and the probability that a person has difficulty breathing, given that he has heart disease, is $\mathcal{P}(B|H) = 0.6$.
- The second doctor provides the results in which the probability that people have heart disease (denoted by H) is $\mathcal{P}(H) = 0.7$; the probability that people have chest

pain (denoted by C) is $\mathcal{P}(C) = 0.64$; and the probability that a person has heart disease when that person who has chest pain, is $\mathcal{P}(H|C) = 0.9$.

- The third doctor provides the results in which the probability that people have difficulty breathing is $\mathcal{P}(B) = 0.7$; the probability that people have chest pain is $\mathcal{P}(C) = 0.5$; the probability that people have heart disease is $\mathcal{P}(H) = 0.8$.

- The fourth doctor provides the results in which the probability that people have difficulty breathing (denoted by B) is $\mathcal{P}(B) = 0.8$; the probability that a person has heart disease when that person who has chest pain, is $\mathcal{P}(H|C) = 0.7$.

According to the above results, we have the probabilistic knowledge bases as follows:

$$\mathcal{K}_1 = \langle (H)[0.7], (B)[0.5], (B|H)[0.6] \rangle; \quad \mathcal{K}_2 = \langle (H)[0.7], (C)[0.64], (H|C)[0.9] \rangle;$$
$$\mathcal{K}_3 = \langle (B)[0.7], (C)[0.5], (H)[0.8] \rangle; \quad \mathcal{K}_4 = \langle (B)[0.8], (H|C)[0.7] \rangle$$

Therefore, by using Algorithm 2.

- Consider $\kappa_1^{(1)} \in \mathcal{K}_1 \colon \mathcal{K}_1 = \left\langle \kappa_2^{(1)} \left[\rho_2^{(1)} \right], \kappa_3^{(1)} \left[\rho_3^{(1)} \right] \right\rangle$

+ For $\kappa_1^{(2)} \in \mathcal{K}_2$, we have $\kappa_1^{(1)} \equiv \kappa_1^{(2)}$,

therefore $\rho_1^{(0)} = \oplus \left(\rho_1^{(1)}, \rho_1^{(2)} \right) = 0.7$, $\mathcal{K}_2 = \left\langle \kappa_2^{(2)} \left[\rho_2^{(2)} \right], \kappa_3^{(2)} \left[\rho_3^{(2)} \right] \right\rangle$

+ For $\kappa_3^{(3)} \in \mathcal{K}_3$, we have $\kappa_1^{(1)} \equiv \kappa_3^{(3)}$,

therefore $\rho_1^{(0)} = \oplus \left(\rho_1^{(0)}, \rho_3^{(3)} \right) = 0.75$, $\mathcal{K}_3 = \left\langle \kappa_1^{(3)} \left[\rho_1^{(3)} \right], \kappa_2^{(3)} \left[\rho_2^{(3)} \right] \right\rangle$

$$\mathcal{K}_0 = \left\langle \left(\kappa_1^{(1)} \right) \left[\rho_1^{(0)} \right] \right\rangle = \langle (H)[0.75] \rangle$$

- Consider $\kappa_2^{(1)} \in \mathcal{K}_1 \colon \mathcal{K}_1 = \left\langle \kappa_3^{(1)} \left[\rho_3^{(1)} \right] \right\rangle$

+ For $\kappa_1^{(3)} \in \mathcal{K}_3$, we have $\kappa_2^{(1)} \equiv \kappa_1^{(3)}$,

therefore $\rho_2^{(0)} = \oplus \left(\rho_2^{(1)}, \rho_1^{(3)} \right) = 0.6$, $\mathcal{K}_3 = \left\langle \kappa_2^{(3)} \left[\rho_2^{(3)} \right] \right\rangle$

+ For $\kappa_1^{(4)} \in \mathcal{K}_4$, we have $\kappa_2^{(1)} \equiv \kappa_1^{(4)}$,

therefore $\rho_2^{(0)} = \oplus \left(\rho_2^{(0)}, \rho_1^{(4)} \right) = 0.69$, $\mathcal{K}_4 = \left\langle \kappa_2^{(4)} \left[\rho_2^{(4)} \right] \right\rangle$

$$\mathcal{K}_0 = \left\langle \left(\kappa_1^{(0)} \right) \left[\rho_1^{(0)} \right], \left(\kappa_2^{(0)} \right) \left[\rho_2^{(0)} \right] \right\rangle = \langle (H)[0.75], (B)[0.69] \rangle$$

- Consider $\kappa_3^{(1)} \in \mathcal{K}_1 \colon \mathcal{K}_1 = \emptyset$

$$\mathcal{K}_0 = \left\langle \left(\kappa_1^{(0)} \right) \left[\rho_1^{(0)} \right], \left(\kappa_2^{(0)} \right) \left[\rho_2^{(0)} \right], \left(\kappa_3^{(0)} \right) \left[\rho_3^{(0)} \right] \right\rangle$$
$$= \langle (H)[0.75], (B)[0.69], (B|H)[0.6] \rangle$$

- Consider $\kappa_2^{(2)} \in \mathcal{K}_2 \colon \mathcal{K}_2 = \left\langle \kappa_3^{(2)} \left[\rho_3^{(2)} \right] \right\rangle$

For $\kappa_2^{(3)} \in \mathcal{K}_3$, we have $\kappa_2^{(2)} \equiv \kappa_2^{(3)}$ therefore $\rho_4^{(0)} = \oplus \left(\rho_2^{(2)}, \rho_2^{(3)} \right) = 0.57$, $\mathcal{K}_3 = \emptyset$

$$\mathcal{K}_0 = \left\langle \left(\kappa_1^{(0)} \right) \left[\rho_1^{(0)} \right], \left(\kappa_2^{(0)} \right) \left[\rho_2^{(0)} \right], \left(\kappa_3^{(0)} \right) \left[\rho_3^{(0)} \right], \left(\kappa_4^{(0)} \right) \left[\rho_4^{(0)} \right] \right\rangle$$
$$= \langle (H)[0.75], (B)[0.69], (B|H)[0.6], (C)[0.57] \rangle$$

- Consider $\kappa_3^{(2)} \in \mathcal{K}_2$: $\mathcal{K}_2 = \emptyset$

For $\kappa_2^{(4)} \in \mathcal{K}_4$, we have $\kappa_3^{(2)} \equiv \kappa_2^{(4)}$ therefore $\rho_5^{(0)} = \oplus \left(\rho_3^{(2)}, \rho_2^{(4)} \right) = 0.79$,

$$\mathcal{K}_0 = \left\langle \left(\kappa_1^{(0)} \right) \left[\rho_1^{(0)} \right], \left(\kappa_2^{(0)} \right) \left[\rho_2^{(0)} \right], \left(\kappa_3^{(0)} \right) \left[\rho_3^{(0)} \right], \left(\kappa_4^{(0)} \right) \left[\rho_4^{(0)} \right], \left(\kappa_5^{(0)} \right) \left[\rho_5^{(0)} \right] \right\rangle$$
$$= \langle (H)[0.75], (B)[0.69], (B|H)[0.6], (C)[0.57], (H|C)[0.79] \rangle$$

- $\mathcal{K}_0 = \text{PCDA}(\mathcal{K}_0) = \langle (H)[0.75], (B)[0.69], (C)[0.57], (B|C)[0.71] \rangle$.

Table 1 shows the new probability of knowledge bases $\mathcal{K}_0$ after using the MMA algorithm and the CMMA algorithm. However, the process of probabilistic constraints deduction is not implemented. It is easy to see that the smaller the value of c is, the larger the value of probability is. The probability values always fulfill Propositions 3 and 7.

Table 1. New probabilistic of $\mathcal{K}_0$

(κ)	$\mathcal{K}_1$	$\mathcal{K}_2$	$\mathcal{K}_3$	$\mathcal{K}_4$	$MMA(\mathcal{B})$ $\mathcal{K}_0$	$CMMA(\mathcal{B})$ $\mathcal{K}_0$				
						$c = 0$	$c = 0.3$	$c = 0.5$	$c = 0.7$	$c = 1$
(H)	0.70	0.70	0.80		0.75	0.80	0.77	0.75	0.73	0.70
(B)	0.50		0.70	0.80	0.69	0.80	0.75	0.71	0.67	0.60
(B, H)	0.60				0.60	0.60	0.60	0.60	0.60	0.60
(H, C)		0.90		0.70	0.79	0.82	0.82	0.82	0.82	0.82
(C)		0.64	0.50		0.57	0.57	0.57	0.57	0.57	0.57

After implementing the MMA algorithm, we obtain knowledge base as shown in Table 2.

Accordding to the Definition 6 and Proposition 5 in [20], it is easy to see that $\Lambda(\mathcal{K}_0) \neq \emptyset$. Therefore, $\mathcal{K}_0 \nvDash \bot$. Accordding to Sect. 4.2, the computational complexity of the MMA algorithm as well as CMMA algorithm in the worst case is:

$$\mathcal{O}\left(max\left\{ |\mathcal{B}| max\{|\mathcal{K}_i|\}, |PCDA(\mathcal{K}_0)|^2 \right\} \right) = \mathcal{O}(max\{4 \times 3, 4^2\}) = \mathcal{O}(16)$$

Table 2. Knowledge base after reducing constraints

(κ)	$MMA(\mathcal{B})$	$CMMA(\mathcal{B})$				
	$\mathcal{K}_0$	$\mathcal{K}_0$				
		c = 0	c = 0.3	c = 0.5	c = 0.7	c = 1
(H)	0.75	0.80	0.77	0.75	0.73	0.70
(B)	0.69	0.80	0.75	0.71	0.67	0.60
(B, C)	0.67	0.69	0.69	0.69	0.69	0.69
(C)	0.57	0.57	0.57	0.57	0.57	0.57

5   Conclusion

In this paper, we have investigated the way to reduce probabilistic constraints, the way to compute common probability of each constraint such that they are likewise within the interval of two original probabilities. To ensure this requirement, we also proposed two mean merging operators. Two main algorithms for merging probabilistic knowledge bases have been put forward by obeying constraint deduction rules and condition property of merging operators. However, algorithms are only implemented and evaluated on a small and assumed set of data. Therefore, we will go on investigating to show experimental results of proposed algorithms on real-world datasets.

Acknowledgment. This study was fully supported by Science and Technology Development Fund from Vietnam National University, Hanoi (VNU) under grant number QG.19.23 (2019-2020). The authors would like to thank Professor Quang Thuy Ha and Knowledge Technology and Data Science Lab, Faculty of Information Technology, VNU - University of Engineering and Technology for expertise support.

References

1. Bloch, I., et al.: Fusion: general concepts and characteristics. Int. J. Intell. Syst. **16**(10), 1107–1134 (2001)
2. Potyka, N., Thimm, M.: Consolidation of probabilistic knowledge bases by inconsistency minimization. In: Proceedings ECAI 2014, pp. 729–734. IOS Press (2014)
3. Potyka, N., Thimm, M.: Probabilistic reasoning with inconsistent beliefs using inconsistency measures. In: International Joint Conference on Artificial Intelligence 2015 (IJCAI 2015), pp. 3156–3163. AAAI Press (2015)
4. Potyka, N.: Solving Reasoning Problems for Probabilistic Conditional Logics with Consistent and Inconsistent Information. FernUniversitat, Hagen (2016)
5. Nguyen, V.T., Tran, T.H.: Inconsistency measures for probabilistic knowledge bases. In: Proceedings KSE 2017, pp. 148–153. IEEE Xplore (2017)
6. Nguyen, V.T., Tran, T.H.: Solving inconsistencies in probabilistic knowledge bases via inconsistency measures. In: Nguyen, N.T., Hoang, D.H., Hong, T.-P., Pham, H., Trawiński, B. (eds.) ACIIDS 2018. LNCS (LNAI), vol. 10751, pp. 3–14. Springer, Cham (2018). https://doi.org/10.1007/978-3-319-75417-8_1

7. Nguyen, V.T., Nguyen, N.T., Tran, T.H., Nguyen, D.K.L.: Method for restoring consistency in probabilistic knowledge bases. J. Cybern. Syst. 1–22 (2018)
8. Nguyen, N.T.: Advanced Methods for Inconsistent Knowledge Management, pp. 1–351. Springer, Heidelberg (2008). https://doi.org/10.1007/978-1-84628-889-0
9. Genest, C., Zidek, J.V.: Combining probability distributions: a critique and an annotated bibliography. Stat. Sci. **1**, 114–135 (1986)
10. Dempster, A.P.: Upper and lower probabilities induced by a multivalued mapping. In: Yager, R.R., Liu, L. (eds.) Classic Works of the Dempster-Shafer Theory of Belief Functions, vol. 219, pp. 52–72. Springer, Heidelberg (2008). https://doi.org/10.1007/978-3-540-44792-4_3
11. Levy, W.B., Deliç, H.: Maximum entropy aggregation of individual opinions. IEEE Trans. Syst. Man Cybern. **24**(4), 606–613 (1994)
12. Kern-Isberner, G.: Conditionals in Nonmonotonic Reasoning and Belief Revision. LNCS (LNAI), vol. 2087. Springer, Heidelberg (2001). https://doi.org/10.1007/3-540-44600-1
13. Kern-Isberner, G., Rödder, W.: Belief revision and information fusion on optimum entropy. Int. J. Intell. Syst. **19**(9), 837–857 (2004)
14. Kern-Isberner, G., Wilhelm, M., Beierle, C.: Probabilistic knowledge representation using the principle of maximum entropy and Gröbner basis theory. Ann. Math. Artif. Intell. **79**(1–3), 163–179 (2017)
15. Nguyen, N.T.: Methods for Consensus Choice and their Applications in Conflict Resolving in Distributed Systems. Wroclaw University of Technology Press (2002). (in Polish)
16. Vomlel, J.: Methods of probabilistic knowledge integration. Ph.D. thesis, Czech Technical University, Prague (1999)
17. Wilmers, G.: The social entropy process: axiomatising the aggregation of probabilistic beliefs. In: Probability, Uncertainty and Rationality (2010)
18. Adamcík, M., Wilmers, G.: The irrelevant information principle for collective probabilistic reasoning. Kybernetika **50**(2), 175–188 (2014)
19. Adamcik, M.: Collective reasoning under uncertainty and inconsistency. Ph.D. thesis, University of Manchester, UK (2014)
20. Nguyen, V.T., Nguyen, N.T., Tran, T.H.: Framework for merging probabilistic knowledge bases. In: Nguyen, N.T., Pimenidis, E., Khan, Z., Trawiński, B. (eds.) ICCCI 2018. LNCS (LNAI), vol. 11055, pp. 31–42. Springer, Cham (2018). https://doi.org/10.1007/978-3-319-98443-8_4
21. Achs, Á., Kiss, A.: Fixed point query in fuzzy datalog Programs. In: Annales Univ. Sci. Budapest, Sect., no. 15, pp. 223–231 (1995)
22. Achs, Á., Kiss, A.: Fuzzy extension of datalog. Acta Cybern. **12**(2), 153–166 (1995)
23. Achs, Á.: A multivalued knowledge-base model. CoRR, abs/1003.1658 (2010)

A Formal Framework for the Ontology Evolution

Adrianna Kozierkiewicz[ID] and Marcin Pietranik[✉][ID]

Faculty of Computer Science and Management,
Wroclaw University of Science and Technology,
Wybrzeze Wyspianskiego 27, 50-370 Wroclaw, Poland
{adrianna.kozierkiewicz,marcin.pietranik}@pwr.edu.pl

Abstract. Ontologies are a formal tool for expressing knowledge. They are based on providing a logic theory about some selected universe of discourse, which defines a decomposition of a topic of interest into its basic elements and formally describes certain restrictions and constraints that need to be met. In the modern days, users cannot expect that ontologies in their initial states will be resistant to changes and remain static throughout the lifespan of their particular application. In this paper we want to focus on expressing alterations, that can be applied to ontologies in time, due to changing business requirements, emerging new knowledge or modification of the scope that an ontology needs to cover.

1 Introduction

Ontologies are a formal tool for expressing knowledge, based on providing a logic description of a decomposition of a topic of interest into its basic elements (referred to as concepts) and formally describing certain restrictions and constraints that need to be met along with mutual interactions these concepts may involve. Such approach serves as a flexible, expressive tool that can also assert its inner semantic consistency.

However, in the modern days, users cannot expect that a domain of discourse will be fixed. Ontologies in their initial states won't remain static throughout the lifespan of their application. Therefore, a tool for managing appearing alterations is necessary. At the simplest level, it is as issue related to tracking changes in a source code of complex computer systems created by a group of developers. At this level, one can confine on a basic approach based on tracking changes of OWL files (which is only one of the format of representing ontologies) using some kind of a version control system, such as Git[1]. However, ontologies are complex knowledge structures and therefore, the eventual approach to managing their evolution, cannot only compare them on a pure syntactical level. It is expected to compare changes entailed by evolving semantics.

---

[1] https://git-scm.com.

© Springer Nature Switzerland AG 2019
N. T. Nguyen et al. (Eds.): ACIIDS 2019, LNAI 11431, pp. 16–27, 2019.
https://doi.org/10.1007/978-3-030-14799-0_2

In our previous publication [15], we have approached the topic of detecting changes introduced to maintained ontologies that are significant enough to trigger revalidation of existing mappings between them. We have build this tool on the top of our ontology alignment framework described in [14].

In this article, we would like to propose a different, more thorough, approach by enhancing our base definition of an ontology and its internal components with a time factor and a mathematical model of time. The main goal (along a detailed review of the state of the art of ontology evolution) is preparing a solid, formal and flexible foundation that can be used to express changes that appear while maintained ontologies evolve.

Informally speaking - we would like to prepare a tool that can answer a question "what changed when". Formally, our approach to the problem of maintaining evolving ontologies can be described as follows: *For a given ontology O, one should determine a set of functions that could be used to extract changes applied to O in certain, discrete moments in time on a level of concepts, a level of instances, a level of concepts' relations and a level of instances' relations.*

The further parts of the paper are organised as follows. In Sect. 2 we describe related research that can be found in the literature. Section 3 contains basic ontology definitions that are a mathematical foundation for our work. Section 4 provides a detailed overview of a set of functions that can find and describe changes applied in time to ontologies. In Sect. 5 some illustrative examples are given. A summary and a short overview of our upcoming research can be found in Sect. 6.

2   Related Works

Throughout the years, the topic of ontology evolution has gained interest. High complexity of maintained, large ontologies and their even bigger repositories [1] raises problems related to assessing a consistent ontology development [12]. Aspects that need to be covered are identified and explained, among others, in [3] and [4].

One of the approaches is based on a set of triggering rules that can detect potential changes applied to ontologies over time [16]. A solution based on a model theory can be found in [6]. Interesting and more practical ideas can be found in [8], where authors describe a change history management framework for evolving ontologies. The paper addresses several subproblems such as ontology versioning (also covered in [9]), tracking a change's provenance, a consistency assertion, a recovery procedure, a change representation and a visualisation of the ontology evolution. Experimental results show that the described approach provide accurate outcomes in terms of a semantical consistency.

In [7] authors approach the given task, by analysing available change operators and eventually, identify patterns that appear during the ontology evolution. A pattern-based layered operator framework is developed and further applied in content management systems ([13]). A more practical approach to the topic can be found in [2] where authors describe GALILEO - a system developed in order to automate the evolution of higher-order logic ontologies by incorporating users' interactions for diagnosing and repairing faults. Its formal foundations can be found in [11].

The topic of maintaining ontology changes has been also applied in the context of ontology alignment. In [10], authors analyse a potential impact that alteration applied to ontologies, can have on their alignment. Several publications address problems related to calculation complexity minimisation, while updating ontology mappings, in order to adjust them to changing ontologies [5,17].

3   Basic Notions

By "real world" we define a pair (A, V), where A is a set of attributes describing objects and V is a set of valuations of such attributes (their domains). Obviously, assuming that V_a is attribute's a domain, a following property occurs: $V = \bigcup_{a \in A} V_a$. We define the ontology as a quintuple:

$$O = (C, H, R^C, I, R^I) \tag{1}$$

where:

- C is a finite set of concepts,
- H is a concepts' hierarchy, that may be treated as a distinguished relation between concepts. Therefore, $H \subset C \times C$. However, for clarity purposes, it is excluded from the set R^C.
- R^C is a finite set of relations between concepts $R^C = \{r_1^C, r_2^C, ..., r_n^C\}$, $n \in N$, such that every $r_i^C \in R^C$ ($i \in [1, n]$) is a subset of a cartesian product, $r_i^C \subset C \times C$,
- I denotes a finite set of instances' identifiers,
- $R^I = \{r_1^I, r_2^I, ..., r_n^I\}$ symbolises a finite set of relations between concepts' instances.

A concept c taken from the set C is defined as:

$$c = (id^c, A^c, V^c, I^c) \tag{2}$$

where: id^c is an identifier of the concept c, A^c is a set of its attributes, V^c is a set of attributes domains (formally: $V^c = \bigcup_{a \in A^c} V_a$), and I^c is a set of concepts' c instances. For short, we can write $a \in c$ which denotes the fact that the attribute a belongs to the concept's c set of attributes A^c.

Referring to the real world, an ontology is called (A, V)-based if the following conditions are met: (i) $\forall_{c \in C} A^c \subseteq A$ and (ii) $\forall_{c \in C} V^c \subseteq V$. A set of all (A, V)-based ontologies is denoted as $\tilde{O}$.

To give attributes additional meaning, we assume an existence of a sublanguage of the sentence calculus (denoted as L_s^A) to define a function $S_A : A \times C \to L_s^A$ which assigns certain attributes within concepts logic sentences.

The overall meaning of a concept is given by its context, which is a conjunction of semantics of each of its attributes. Formally, for a concept c such that $A^c = \{a_1, a_2, ..., a_n\}$, its context is given as $ctx(c) = S_A(a_1, c) \wedge S_A(a_2, c) \wedge ... \wedge S_A(a_n, c)$.

We define instances of a concept c, from the set I^c as a tuple:

$$i = (id^i, v_c^i) \tag{3}$$

where id^i is an instance identifier, and v_c^i is a function with a signature v_c^i : $A^c \rightarrow V^c$. In terms of a consensus theory, the function v_c^i can be interpreted as a tuple of type A^c. A set of instances' identifiers from the Eq. 1 can be therefore formally defined as:

$$I = \bigcup_{c \in C} \{id^i | (id^i, v_c^i) \in I^c\} \tag{4}$$

For short, we can write $i \in c$ which denotes the fact that the instance i belongs to the concept c.

Every relation from the set R^C has a complementary relation from the set R^I. Obviously, $|R^C| = |R^I|$. In other words, a relation $r_j^C \in R^C$ describes potential connections that may occur between instances of concepts from the set C. On the other hand, $r_j^I \in R^I$ describes what is actually connected. For example, the set R^C may contain relations *is_husband* or *is_wife* and in R^I statements like *Dale is a husband of Laura* or *Jane is a wife of David* can be found. To denote this, we use the same index of relations taken from both sets - a relation $r_j^I \in R_I$ contains pairs of instances, which are included within concepts that are mutually connected by a relation $r_j^C \in R^C$. Additionally, relations also carry some intended semantics defined as a function S_R that similarly to the function S_A assigns logical sentences to relations. Its formal definitions and entailed properties are beyond the scope of this paper. For broad description please refer to our previous publications [14].

4 Tracking Changes During Ontology Evolution

By $\overline{TL}$ we will denote a universal timeline of every tracked evolving ontology. It can be understood as an ordered set of discrete moments in time:

$$\overline{TL} = \{t_n | n \in N\} \tag{5}$$

A pair $\langle TL, \prec \rangle$ (where $\prec$ is an ordering relation, that can be identified with a time sequence) forms a strict partial order, which is irreflexive, transitive and antisymmetric.

For a given ontology O, by $TL(O)$, we will denote a timeline of the ontology which is a set of moments in time in which the ontology O has been somehow modified. Obviously $TL(O) \subseteq \overline{TL}$. Thanks to this notion, we can define an ontology repository, as an ordered set of states of the ontology O. In other words, its subsequent versions in time. Formally, the repository is defined below:

$$Rep(O) = \left\{ O^{(m)} | \forall m \in TL(O) \right\} \tag{6}$$

In further parts of this paper, a superscript $O^{(m)} = (C^{(m)}, H^{(m)}, R^{C(m)},$ $I^{(m)}, R^{I(m)})$ will denote the ontology O (along with its internal elements) in a given moment in time t_m. We will also assume that $O^{(m-1)} \prec O^{(m)}$ denotes the fact that $O^{(m-1)}$ is an earlier version of the ontology O than the version $O^{(m)}$. This notation can be extended for particular elements of the given ontology, e.g. $c^{(m-1)} \prec c^{(m)}$ denotes that $c^{(m-1)}$ is an earlier version of c than the version $c^{(m)}$.

For convenience, to acquire some starting point of a reference, within every timeline $TL(O)$ we distinguish an initial moment, in which the ontology's O evolution began being tracked. We denote this moment as 0 and the ontology O in that state is denoted as $O^{(0)} = (C^{(0)}, H^{(0)}, R^{C(0)}, I^{(0)}, R^{I(0)})$.

In the next parts of the following section we will define a tuple of functions $\left\langle diff_C, diff_I, diff_{R^C}, diff_{R^I} \right\rangle$ that extract changes applied to O on a level of concepts, a level of instances, a level of concepts' relations and a level of instances' relations.

4.1   Level of Concepts

A function $diff_C : Rep(O) \times Rep(O) \rightarrow Rep(O) \times Rep(O) \times Rep(O)$ takes as an input two consecutive states of the ontology O (namely $O^{(m-1)}$ and $O^{(m)}$) and returns a triple of sets, that describe concepts added to the ontology, concepts removed from the ontology and concepts that have been altered. Formally, for two ontology states $O^{(m-1)}$ and $O^{(m)}$, such that $O^{(m-1)} \prec O^{(m)}$, these sets are defined as follows:

$$diff_C(O^{(m-1)}, O^{(m)}) = \left\langle new_C(C^{(m-1)}, C^{(m)}), \right.$$
$$del_C(C^{(m-1)}, C^{(m)}), \tag{7}$$
$$\left. alt_C(C^{(m-1)}, C^{(m)}) \right\rangle$$

where:

1. $new_C(C^{(m-1)}, C^{(m)}) = \left\{ c | c \in C^{(m)} \wedge c \notin C^{(m-1)} \right\}$

2. $del_C(C^{(m-1)}, C^{(m)}) = \left\{ c | c \in C^{(m-1)} \wedge c \notin C^{(m)} \right\}$

3. $alt_C(C^{(m-1)}, C^{(m)}) = \left\{ (c^{(m-1)}, c^{(m)}) | c^{(m-1)} \in C^{(m-1)} \wedge c^{(m)} \in C^{(m)} \wedge \right.$
$c^{(m-1)} \prec c^{(m)} \wedge (A^{c^{(m-1)}} \neq A^{c^{(m)}} \vee V^{c^{(m-1)}} \neq V^{c^{(m)}} \vee I^{c^{(m-1)}} \neq I^{c^{(m)}}) \vee$
$\left. ctx(c^{(m-1)}) \neq ctx(c^{(m)}) \right\}$

The first two sets' definitions are self explanatory. The last one describes alterations applied to the ontology O, by returning a set of pairs of concept's c versions, that have neither been added nor deleted, but differ in their structures (according to Eq. 2) or contexts.

4.2   Level of Instances

By analogy to the function defined on Eq. 7, a function $diff_I : Rep(O) \times Rep(O) \to Rep(O) \times Rep(O)$ takes as an input two consecutive states of the ontology O and returns a pair of sets, which describe instances that have been added to the ontology, and instances removed from it. Formally, for two ontology states $O^{(m-1)}$ and $O^{(m)}$, such that $O^{(m-1)} \prec O^{(m)}$:

$$diff_I(I^{(m-1)}, I^{(m)}) = \left\langle new_I(I^{(m-1)}, I^{(m)}), del_I(I^{(m-1)}, I^{(m)}) \right\rangle \tag{8}$$

where:

1. $new_I(I^{(m-1)}, I^{(m)}) = \left\{ i \middle| i \in I^{(m)} \wedge i \notin I^{(m-1)} \right\}$

2. $del_I(I^{(m-1)}, I^{(m)}) = \left\{ i \middle| i \in I^{(m-1)} \wedge i \notin I^{(m)} \right\}$

These definitions simply define what have been added or removed from the set I. Note that on this level, we do not consider instances' alterations - the set I contains only atomic instances' identifiers, therefore, their strict modification is not possible.

4.3   Level of Concepts' Relations

A function $diff_{R^C} : Rep(O) \times Rep(O) \to Rep(O) \times Rep(O) \times Rep(O)$ takes as an input two consecutive states of the ontology O and returns three sets describing concepts' relations added to the ontology, concepts' relations removed from the ontology and changed concepts' relations. Formally, for two ontology states $O^{(m-1)}$ and $O^{(m)}$, such that $O^{(m-1)} \prec O^{(m)}$, the function $diff_{R^C}$ is defined below:

$$diff_{R^C}(O^{(m-1)}, O^{(m)}) = \left\langle new_{R^C}(R^{C(m-1)}, R^{C(m)}), \right.$$

$$del_{R^C}(R^{C(m-1)}, R^{C(m)}), \tag{9}$$

$$\left. alt_{R^C}(R^{C(m-1)}, R^{C(m)}) \right\rangle$$

where:

1. $new_{R^C}(R^{C(m-1)}, R^{C(m)}) = \left\{ r \middle| r \in R^{C(m)} \wedge r \notin R^{C(m-1)} \right\}$

2. $del_{R^C}(R^{C(m-1)}, R^{C(m)}) = \left\{ r | r \in R^{C(m-1)} \wedge r \notin R^{C(m)} \right\}$

3. $alt_{R^C}(R^{C(m-1)}, R^{C(m)}) = \left\{ (r^{(m-1)}, r^{(m)}) | r^{(m-1)} \in R^{C(m-1)} \wedge r^{(m)} \in \right.$
$\left. R^{C(m)} \wedge (r^{(m-1)} \oplus r^{(m)} \neq \phi \vee S_R(r^{(m-1)}) \neq S_R(r^{(m)})) \right\}$

As in previous sections, addition or subtraction of elements considered on this level, are simple. The last set alt_{R^C} results in a set of pairs of relation's r versions, that have neither been added nor deleted, but connections between involved concepts have been changed. In other words, some pairs of concepts within such relations have been added or removed. To achieve this we use a symmetrical difference between sets of concepts' pairs, that obviously exclude not altered relations. The last part of alt_{R^C} considers alterations applied to semantics of relations expressed using the function S_R, defined in Sect. 3, which may give different results over time.

4.4   Level of Instances' Relations

A function $diff_{R^I} : Rep(O) \times Rep(O) \rightarrow Rep(O) \times Rep(O) \times Rep(O)$ takes as an input two consecutive states of the ontology O and returns sets that describe instances' relations added to the ontology, instances' relations removed from the ontology and altered instances' relations. Formally, for two ontology states $O^{(m-1)}$ and $O^{(m)}$ such that $O^{(m-1)} \prec O^{(m)}$ the definition is as follows:

$$diff_{R^I}(O^{(m-1)}, O^{(m)}) = \left\langle new_{R^I}(R^{I(m-1)}, R^{I(m)}), \right.$$
$$del_{R^I}(R^{I(m-1)}, R^{I(m)}), \qquad (10)$$
$$\left. alt_{R^I}(R^{I(m-1)}, R^{I(m)}) \right\rangle$$

where:

1. $new_{R^I}(R^{I(m-1)}, R^{I(m)}) = \left\{ r_j^I | r_j^C \in new_{R^C}(R^{C(m-1)}, R^{C(m)}) \right\}$

2. $del_{R^I}(R^{I(m-1)}, R^{I(m)}) = \left\{ r_j^I | r_j^C \in del_{R^C}(R^{C(m-1)}, R^{C(m)}) \right\}$

3. $alt_{R^I}(R^{I(m-1)}, R^{I(m)}) = \left\langle new_r(R^{I(m-1)}, R^{I(m)}), del_r(R^{I(m-1)}, R^{I(m)}) \right\rangle$

As previously, additions and removals of instances relations appearing in sets new_{R^I} and del_{R^I} are straightforward. However, the set alt_{R^I} needs to include two sets of instances pairs that have been added or removed from the ontology in the state m, entailing alteration of instances relations. New instances connections can be described within a set new_r that collects instances pairs added to the ontology O using a simple set difference between two states of a considered ontology. It is defined below:

$$
\begin{aligned}
new_r(R^{I(m-1)}, R^{I(m)}) = \Big\{ r_j^{I(m)} \setminus r_j^{I(m-1)} | \\
r_j^{I(m)} \notin new_{R^I}(R^{I(m-1)}, R^{I(m)}) \wedge \\
r_j^{I(m-1)} \notin del_{R^I}(R^{I(m-1)}, R^{I(m)}) \wedge \\
r_j^{I(m)} \nsubseteq r_j^{I(m-1)} \Big\}
\end{aligned}
\tag{11}
$$

By analogy we define a set containing those instances pairs that have been removed:

$$
\begin{aligned}
del_r(R^{I(m-1)}, R^{I(m)}) = \Big\{ r_j^{I(m-1)} \setminus r_j^{I(m)} | \\
r_j^{I(m)} \notin new_{R^I}(R^{I(m-1)}, R^{I(m)}) \wedge \\
r_j^{I(m-1)} \notin del_{R^I}(R^{I(m-1)}, R^{I(m)}) \wedge \\
r_j^{I(m-1)} \nsubseteq r_j^{I(m)} \Big\}
\end{aligned}
\tag{12}
$$

4.5 Ontology Change Log

Having formal definitions of $\left\langle diff_C, diff_I, diff_{R^C}, diff_{R^I} \right\rangle$ we are able to define an ontology log, which is a complete illustration of changes applied to a given ontology O. We define it as a set containing sets of results of all *diff* functions, calculated for every pair of subsequent moments taken from the ontology's timeline $TL(O)$. Formally:

$$
\begin{aligned}
Log(O) = \Big\{ \big\langle diff_C(O^{(m-1)}, O^{(m)}) \\
diff_I(O^{(m-1)}, O^{(m)}) \\
diff_{R^C}(O^{(m-1)}, O^{(m)}) \\
diff_{R^I}(O^{(m-1)}, O^{(m)}) \big\rangle | \forall m \in TL(O) \setminus \{0\} \Big\}
\end{aligned}
\tag{13}
$$

The next part of the paper will provide illustrative examples of our approach to the ontology evolution and its aspects covered in the current section.

5 Illustration of the Ontology Evolution

Let us illustrate by simple examples how ontologies may evolve. The ontology is a complicated knowledge structure, thus our use case scenarios are divided into

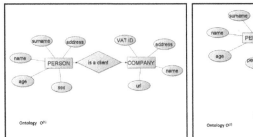 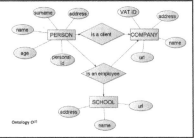

Fig. 1. A tracking changes in ontologies for the concept and the concepts' relations level;

two parts. The first (presented in Fig. 1) is devoted to concepts and relations between them. The second (Fig. 2) considers instances and their relations. As we can see, in the both figures, we have two ontology states $O^{(1)}$ and $O^{(2)}$, such that $O^{(1)} \prec O^{(2)}$. Our aim is to find the respective ontology log. Therefore, we need to determine the following sets: $diff_C(O^{(1)}, O^{(2)})$, $diff_I(O^{(1)}, O^{(2)})$, $diff_{R^C}(O^{(1)}, O^{(2)})$, $diff_{R^I}(O^{1)}, O^{(2)})$.

It is easy to show that $diff_C(O^{(1)}, O^{(2)}) = \left\langle new_C(C^{(1)}, C^{(2)}), del_C(C^{(1)}, C^{(m)}), \right.$

$\left. alt_C(C^{(1)}, C^{(2)}) \right\rangle = \left\langle \{SCHOOL\}, \emptyset, \{PERSON\} \right\rangle$. The concept "SCHOOL" appears only in the ontology $O^{(2)}$ and does not exist in the ontology $O^{(1)}$. No concept is removed from $O^{(2)}$. The concept "PERSON" appears in both ontologies, however its structure has been changed - an attribute "sex" has been substituted by attribute "personal id".

The Fig. 1 presents also concepts' relations. Thus, $diff_{R^C}(O^{(1)}, O^{(2)}) = \left\langle new_{R^C}(R^{C(1)}, R^{C(2)}), del_{R^C}(R^{C(1)}, R^{C(2)}), alt_{R^C}(R^{C(1)}, R^{C(2)}) \right\rangle = \left\langle \{\text{"is} \right.$

$an \ employee"\}, \emptyset, \emptyset\} \Big\rangle$. The relation "is an employee" is new in the ontology $O^{(2)}$, but no relation has neither been deleted nor modified.

Based on the Fig. 2 we determine $diff_I(I^{(1)}, I^{(2)})$. This part of the log determination is as follows: $diff_I(I^{(1)}, I^{(2)}) = \left\langle new_I(I^{(1)}, I^{(2)}), del_I(I^{(1)}, I^{(2)}) \right\rangle =$

$\left\langle \{Brian, Anna\}, \{Tom\} \right\rangle$. Two instances "Brian" i "Anna" have been added to the ontology $O^{(2)}$ and only one has been deleted (namely "Tom").

The last part of our ontology change log is the set $diff_{R^I}(O^{(1)}, O^{(2)} = \left\langle new_{R^I}(R^{I(1)}, R^{I(2)}), del_{R^I}(R^{I(1)}, R^{I(2)}), alt_{R^I}(R^{I(m-1)}, R^{I(m)}) \right\rangle$. In our example $diff_{R^I}(O^{(1)}, O^{(2)} = \left\langle \{Anna \ is \ an \ employee \ ABC \ Company\}, \emptyset, \right.$

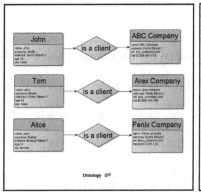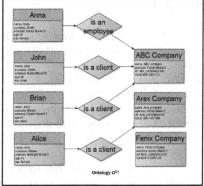

Fig. 2. Tracking changes in ontologies for the instances' and the instances' relations level

$\{Brian \quad is \quad a \quad client \quad ABC \quad Company, Alice \quad is \quad a \quad client \quad Arex \quad Company,$
$Tom \quad is \quad a \quad client \quad Arex \quad Company, Alice \quad is \quad a \quad client \quad Fenix \quad Company\}\Big\rangle.$

6 Future Works and Summary

Ontologies are complex structures, which allow storing and processing knowledge. However, they can become stale and outdated. This entails the necessity of tracking ontology changes. Due to this fact, this paper is devoted to proposing a formal framework for the ontology evolution that occur on four levels of its expressivity: concepts, relations between them, instances and instances' relations. For each level we defined a separate function, which takes as an input two consecutive states of ontology and returns sets describing spotted changes. We distinguished three type of changes: adding, subtractions and alteration of some ontology elements or their structures. Based on the formal definition $\Big\langle diff_C, diff_I, diff_{R^C}, diff_{R^I} \Big\rangle$ we defined the ontology change log as a set of results of $diff$.

In our future work we will apply the defined ontology change log in tasks related to the ontology integration. We will answer a question, how to update the output ontologies being the results of the integration of some input ontologies, if those input ontologies have changed. The ontology integration is a difficult, time- and cost-consuming process. Thus, performing the integration all over again every time ontologies alter is not effective. In our nearest research, we would like to develop an efficient ontology integration algorithm that takes into the account, that ontologies may evolve.

Acknowledgement. This research project was supported by grant No. 2017/26/D/ST6/00251 from the National Science Centre, Poland.

References

1. Allocca, C., d'Aquin, M., Motta, E.: Detecting different versions of ontologies in large ontology repositories. In: Proceedings of IWOD 2009, Washington, D.C., USA (2009)
2. Chan, M., Lehmann, J., Bundy, A.: GALILEO: a system for automating ontology evolution. In: ARCOE-11, p. 46 (2011)
3. De Leenheer, P., Mens, T.: Ontology evolution. In: Hepp, M., De Leenheer, P., De Moor, A., Sure, Y. (eds.) Ontology Management. Computing for Human Experience, vol. 7, pp. 131–176. Springer, Boston (2008). https://doi.org/10.1007/978-0-387-69900-4_5
4. Flouris, G., Manakanats, D., Kondylaiks, H., Plexousakis, D., Antoniou, G.: Ontology change: classification and survey. Knowl. Eng. Rev. **23**, 117–152 (2008)
5. Hartung, M., Terwilliger, J., Rahm, E.: Recent advances in schema and ontology evolution. In: Bellahsene, Z., Bonifati, A., Rahm, E. (eds.) Schema Matching and Mapping. DCSA, vol. 21, pp. 149–190. Springer, Heidelberg (2011). https://doi.org/10.1007/978-3-642-16518-4_6
6. Heflin, J., Pan, Z.: A model theoretic semantics for ontology versioning. In: McIlraith, S.A., Plexousakis, D., van Harmelen, F. (eds.) ISWC 2004. LNCS, vol. 3298, pp. 62–76. Springer, Heidelberg (2004). https://doi.org/10.1007/978-3-540-30475-3_6
7. Javed, M., Abgaz, Y.M., Pahl, C.: A pattern-based framework of change operators for ontology evolution. In: Meersman, R., Herrero, P., Dillon, T. (eds.) OTM 2009. LNCS, vol. 5872, pp. 544–553. Springer, Heidelberg (2009). https://doi.org/10.1007/978-3-642-05290-3_68
8. Khattak, A.M., Latif, K., Lee, S.: Change management in evolving web ontologies. Knowl.-Based Syst. **37**, 1–18 (2013)
9. Klein, M., Fensel, D., Kiryakov, A., Ognyanov, D.: Ontology versioning and change detection on the web. In: Gómez-Pérez, A., Benjamins, V.R. (eds.) EKAW 2002. LNCS (LNAI), vol. 2473, pp. 197–212. Springer, Heidelberg (2002). https://doi.org/10.1007/3-540-45810-7_20
10. Kondylakis, H., Plexousakis, D.: Ontology evolution without tears. Web Semant.: Sci. Serv. Agents World Wide Web **19**, 42–58 (2013). https://doi.org/10.1016/j.websem.2013.01.001
11. Lehmann, J., Chan, M., Bundy, A.: A higher order approach to ontology evolution in physics. J. Data Semant. **2**(4), 163–187 (2013)
12. Noy, N.F., Kunnatur, S., Klein, M., Musen, M.A.: Tracking changes during ontology evolution. In: McIlraith, S.A., Plexousakis, D., van Harmelen, F. (eds.) ISWC 2004. LNCS, vol. 3298, pp. 259–273. Springer, Heidelberg (2004). https://doi.org/10.1007/978-3-540-30475-3_19
13. Pahl, C., Javed, M., Abgaz, Y.M.: Ontology evolution for learning content management systems. In: EdMedia: World Conference on Educational Media and Technology, pp. 1431–1436. Association for the Advancement of Computing in Education (AACE) (2013)
14. Pietranik, M., Nguyen, N.T.: A multi-atrribute based framework for ontology aligning. Neurocomputing **146**, 276–290 (2014). https://doi.org/10.1016/j.neucom.2014.03.067

15. Pietranik, M., Nguyen, N.T.: Framework for ontology evolution based on a multi-attribute alignment method. In: 2015 IEEE 2nd International Conference on Cybernetics (CYBCONF), pp. 108–112. IEEE (2015)
16. Plessers, P., De Troyer, O., Casteleyn, S.: Understanding ontology evolution: a change detection approach. Web Semant.: Sci. Serv. Agents World Wide Web **5**, 39–49 (2007)
17. dos Reis, J.C., et al.: Understanding semantic mapping evolution by observing changes in biomedical ontologies. J. Biomed. Inform. **47**, 71–82 (2014)

Logical Problem Solving Framework

Kiyoshi Akama[1], Ekawit Nantajeewarawat[2(✉)], and Taketo Akama[3]

[1] Information Initiative Center, Hokkaido University, Sapporo, Japan
akama@iic.hokudai.ac.jp
[2] Computer Science Program, Sirindhorn International Institute of Technology,
Thammasat University, Pathumthani, Thailand
ekawit@siit.tu.ac.th
[3] Modeleet Labs, Sapporo, Japan
taketo.akama@gmail.com

Abstract. We propose Logical Problem Solving Framework (LPSF), an axiomatic structure for generating methods of logical problem solving. Input parameters of LPSF consist of (1) a canonical logical structure, (2) a set of equivalent transformation rules (ET rules), (3) a control, and (4) an answer mapping. Given these input parameters, LPSF provides a logical problem solver, which receives an original problem, sets an initial state in a formula on the logical structure, makes a computation path, and if the computation path reaches the domain of the answer mapping, it outputs an answer, which is guaranteed to be correct. By taking input parameters such as KR-Logic, a set of extended clauses obtained through meaning-preserving Skolemization, ET rules including unfolding in the extended clause-set space constructed on KR-Logic, we can solve, with strict guarantee of correctness, a larger class of logical problems, compared to conventional methods.

Keywords: Proof problem · Query-answering problem ·
Meaning-preserving Skolemization · Equivalent transformation rule ·
Computation control · Logical structure · KR-Logic · First-order logic

1 Introduction

Proof problems and query-answering (QA) problems are important classes of logical problems. Many solutions to their subclasses have been intensively studied [7–10]. Many of them are resolution-based and inference-based methods. By these methods, however, we still cannot solve many practical logical problems. Even logical puzzles such as Agatha proof problems and Agatha QA problems cannot be solved correctly by conventional methods. Innovation of the theory of logical problem solving is necessary. To compare and evaluate various methods of logical problem solving from theoretical and correctness-based viewpoints, we propose Logical Problem Solving Framework (LPSF) in this paper.

LPSF is a generator of logical problem solving methods. Input parameters of LPSF consist of (1) a canonical logical structure $\mathcal{L}$, (2) a set R of equivalent

© Springer Nature Switzerland AG 2019
N. T. Nguyen et al. (Eds.): ACIIDS 2019, LNAI 11431, pp. 28–40, 2019.
https://doi.org/10.1007/978-3-030-14799-0_3

transformation rules (ET rules), (3) a control *ctrl*, and (4) an answer mapping A. LPSF with the input parameters $\mathcal{L}$, R, *ctrl*, and A works as follows:

- A problem q formalized in $\mathcal{L}$ is received.
- The problem q is taken as the initial state S_0.
- S_0 is successively transformed by using ET rules in R determined by *ctrl*.
- Computation S_1, S_2, ... is a sequence of states in $\mathcal{L}$.
- If the computation S_1, S_2, ... reaches the domain of A at S_i, LPSF outputs $A(S_i)$ as an answer.

The correctness of computation is strictly guaranteed.

LPSF is useful for (1) comparison of logical problem solving methods, and (2) creation of new methods. To explain the possibility of LPSF, we compare a conventional method based on Skolemization (CSK) and resolution on the first-order logic with a new method based on meaning-preserving Skolemazation (MPS) and equivalent transformation on a new logic, called KR-Logic. Experiments on Agatha proof and QA problems support our claim that LPSF provides better methods of logical problem solving with KR-Logic and MPS than the conventional method with the first-order logic and CSK.

The rest of this paper is organized as follows: Sect. 2 proposes Logical Problem Solving Framework (LPSF). Section 3 recalls the famous Agatha puzzle and presents some experiments on this puzzle, giving successful results using KR-Logic and MPS, along with failure when using the first-order logic and CSK. Section 4 explains the canonical logical structure used for the first parameter of LPSF for solving Agatha puzzle successfully. Section 5 compares two logical problem solving methods, one on the first-order logic, and the other on KR-Logic. Section 6 concludes the paper.

The notation that follows holds thereafter. Given a set A, $pow(A)$ denotes the power set of A. Given two sets A and B, $Map(A, B)$ denotes the set of all mappings from A to B, and for any partial mapping f from A to B, $dom(f)$ denotes the domain of f, i.e., $dom(f) = \{a \mid (a \in A) \;\&\; (f(a) \text{ is defined})\}$.

2   Logical Problem Solving Framework

An axiomatic theory of logical problem solving is developed. All concepts in the theory are only required to satisfy the relations described in Sect. 2. The sets such as $\mathcal{K}$, $\mathcal{G}_u$, and W can be chosen arbitrarily.

2.1   Canonical Logical Structures and Basic Related Concepts

The notion of a logical structure is reviewed below.

Definition 1. A logical structure $\mathcal{L}$ is a triple $\langle \mathcal{K}, \mathcal{I}, \nu \rangle$, where

1. $\mathcal{K}$ and $\mathcal{I}$ are sets,
2. $\nu : \mathcal{K} \to Map(\mathcal{I}, \{true, false\})$.

An element of $\mathcal{K}$ is called a *description* and that of $\mathcal{I}$ is called an *interpretation*.
□

A typical description is a logical formula in the first-order syntax. A set of clauses on the first-order logic is also a typical example of a description.

Despite its simplicity, the notion of a logical structure provides a sufficient structure for defining the concepts of logical equivalence, models, satisfiability, and logical consequence, which are given below.

Definition 2. Let $\mathcal{L} = \langle \mathcal{K}, \mathcal{I}, \nu \rangle$ be a logical structure. Two descriptions $k_1, k_2 \in \mathcal{K}$ are *logically equivalent* iff $\nu(k_1) = \nu(k_2)$. An interpretation $I \in \mathcal{I}$ is a *model* of a description $k \in \mathcal{K}$ iff $\nu(k)(I) = true$. The set of all models of a description $k \in \mathcal{K}$ is denoted by *Models*(k). A description $k \in \mathcal{K}$ is *satisfiable* iff there exists a model of k. A description $k_1 \in \mathcal{K}$ *entails* a description $k_2 \in \mathcal{K}$ (i.e., k_2 is a *logical consequence* of k_1), denoted by $k_1 \models k_2$, iff every model of k_1 is a model of k_2.
□

Each logical structure defines one logic. Two logical structures with the same set of descriptions but different sets of interpretations are considered to be different. In this sense, the first-order logic with standard semantics and the first-order logic with Herbrand interpretations are different logical structures.

Let $\mathcal{G}_u$ be a set, which represents user-defined ground atoms. $\mathcal{L} = \langle \mathcal{K}, \mathcal{I}, \nu \rangle$ is a canonical logical structure iff $\mathcal{I} = pow(\mathcal{G}_u)$. We assume that all logical structures in this paper are canonical logical structures.

2.2   Model-Intersection Problems

A *model-intersection problem* (for short, *MI problem*) on a logical structure $\mathcal{L} = \langle \mathcal{K}, \mathcal{I}, \nu \rangle$ is a pair $\langle Cs, \varphi \rangle$, where $Cs \in \mathcal{K}$ and φ is a mapping from $pow(\mathcal{G}_u)$ to some set W. Intuitively, given $Cs \in \mathcal{K}$, we need to find some element w in the set W, where W is a given set of all possible objects to find. The mapping φ is called an *extraction mapping*. The answer to this problem, denoted by $ans_{MI}(Cs, \varphi)$, is defined by

$$ans_{MI}(Cs, \varphi) = \varphi(\bigcap Models(Cs)),$$

where $\bigcap Models(Cs)$ is the intersection of all models of Cs. Note that when $Models(Cs)$ is the empty set, $\bigcap Models(Cs) = \mathcal{G}_u$.

2.3   Answer Mappings

An answer mapping is a partial mapping that gives the answer to an MI problem whenever it is applicable to that problem. When a problem description reaches the domain of an answer mapping, we compute the answer by applying the answer mapping to the final problem description.

Definition 3. Let W be a set. A partial mapping A from

$$\mathcal{K} \times Map(pow(\mathcal{G}_u), W)$$

to W is an *answer mapping* iff for any $\langle Cs, \varphi \rangle \in dom(A)$, $A(Cs, \varphi) = ans_{MI}(Cs, \varphi)$.
□

2.4   ET Steps and ET Rules

Let STATE be the set of all MI problems on $\mathcal{L}$. Elements of STATE are called *states*.

Definition 4. Let $\langle S, S' \rangle \in$ STATE $\times$ STATE. $\langle S, S' \rangle$ is an *ET step* iff if $S = \langle Cs, \varphi \rangle$ and $S' = \langle Cs', \varphi' \rangle$, then $ans_{\mathrm{MI}}(Cs, \varphi) = ans_{\mathrm{MI}}(Cs', \varphi')$. $\square$

Definition 5. A sequence $[S_0, S_1, \ldots, S_n]$ of elements of STATE is an *ET sequence* iff for any $i \in \{0, 1, \ldots, n-1\}$, $\langle S_i, S_{i+1} \rangle$ is an ET step. $\square$

Given an answer mapping A, the role of ET computation constructing an ET sequence $[S_0, S_1, \ldots, S_n]$ is to start with S_0 and to reach $S_n \in dom(A)$. The concept of ET rule on STATE is defined by:

Definition 6. An *ET rule* r on STATE is a partial mapping from STATE to STATE such that for any $S \in dom(r)$, $\langle S, r(S) \rangle$ is an ET step. $\square$

2.5   Control

Let R be a set of ET rules. A partial mapping $ctrl \colon$ STATE $\rightarrow R$ is a control iff for any $Cs \in \mathcal{K}$, if $ctrl(S) = r$, then $S \in dom(r)$. Given a state $S_0 \in$ STATE, a control $ctrl$ determines a finite or infinite sequence of states as follows:

1. $ctrl$ determines a finite sequence $[S_0, S_1, \ldots, S_n]$ if
    (a) $ctrl$ is applicable to S_i for each $i \in \{0, 1, \ldots, n-1\}$,
    (b) $ctrl(S_i) = r$ and $r(S_i) = S_{i+1}$ for each $i \in \{0, 1, \ldots, n-1\}$, and
    (c) there is no $r \in R$ such that $S_n \in dom(r)$.
2. $ctrl$ determines an infinite sequence $[S_0, S_1, \ldots]$ if
    (a) $ctrl$ is applicable to S_i for each $i \in \{0, 1, \ldots\}$, and
    (b) $ctrl(S_i) = r$ and $r(S_i) = S_{i+1}$ for each $i \in \{0, 1, \ldots\}$.

Let A be an answer mapping. Assume that for any $i \in \{1, 2, \ldots, m\}$ and any state S, if $S \in dom(A)$, then r_i is not applicable to S. Consequently, if a control $ctrl$ produces an infinite computation $[S_0, S_1, \ldots]$, then none of $S_0, S_1, \ldots$ is in $dom(A)$. We have the following three cases:

1. If $ctrl$ determines a finite sequence $[S_0, S_1, \ldots, S_n]$, $S_n \in dom(A)$, and $S_n = \langle Cs_n, \varphi_n \rangle$, then the computed answer is $A(Cs_n, \varphi_n)$.
2. If $ctrl$ determines a finite sequence $[S_0, S_1, \ldots, S_n]$ and $S_n \notin dom(A)$, then no answer is obtained.
3. If $ctrl$ determines an infinite sequence $[S_0, S_1, \ldots]$, then no answer is obtained.

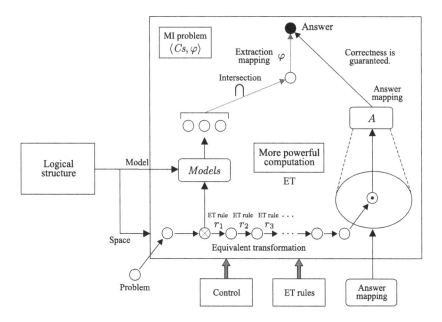

Fig. 1. Logical problem solving framework

2.6   Logical Problem Solving Framework

Logical Problem Solving Framework (LPSF) is proposed, which is illustrated in Fig. 1. Input parameters of LPSF consist of (1) a canonical logical structure $\mathcal{L} = \langle \mathcal{K}, \mathcal{I}, \nu \rangle$, (2) a set R of ET rules, (3) a control *ctrl*, and (4) an answer mapping A.

Given a set Cs of clauses and an extraction mapping φ, the answer to the MI problem $\langle Cs, \varphi \rangle$ is $\varphi(\bigcap Models(Cs))$, which is called the definition path and is represented by the left upward path (from $\otimes$ to $\bullet$) in Fig. 1. The definition path is usually not suitable for computing the answer since it may take huge cost. Instead of taking this definition path, we take a computation path consisting of (i) the lowest path (from $\otimes$ to $\odot$) for ET computation and (ii) the right path (from $\odot$ through the answer mapping A upwards to the answer $\bullet$) in Fig. 1.

Using LPSF, an MI problem $\langle Cs, \varphi \rangle$, where $Cs \in \mathcal{K}$, is solved as follows:

1. Let $S_0 = \langle Cs, \varphi \rangle$.
2. Construct an ET sequence $[S_0, \ldots, S_n]$ by (i) taking S_0 as a initial state, (ii) applying ET rules in R determined by *ctrl*, and (iii) stopping at S_n.
3. Assume that $S_n = \langle Cs_n, \varphi_n \rangle$. If the computation reaches the domain of A, i.e., $\langle Cs_n, \varphi_n \rangle \in dom(A)$, then compute the answer by using the answer mapping A, i.e., output $A(Cs_n, \varphi_n)$.

Every output computed by using any arbitrary ET sequence is correct.

Theorem 1. *Let A be an answer mapping. When an ET sequence starting from $S_0 = \langle Cs, \varphi \rangle$ reaches S_n in dom(A), the above procedure gives the correct answer to $\langle Cs, \varphi \rangle$.*   □

2.7   Solution Methods Based on LPSF

Let FOL_c be the set of all first-order formulas possibly with constraint atoms. Given a QA problem on FOL_c or a proof problem on FOL_c, we try to solve it using LPSF as follows:

1. Set a logical structure $\mathcal{L} = \langle \mathcal{K}, \mathcal{I}, \nu \rangle$.
2. Prepare a set R of ET rules.
3. Determine a control *ctrl*.
4. Determine an answer mapping A.
5. Convert the given problem into an MI problem $\langle Cs, \varphi \rangle$, where $Cs \in \mathcal{K}$.
6. Solve the MI problem $\langle Cs, \varphi \rangle$ using LPSF with the parameters $\mathcal{L}$, R, *ctrl*, and A.

The solution method above is summarized as follows:

$$\text{LPSF}(\mathcal{L}, R, ctrl, A) + (\text{Problem formalization}) = (\text{A solution method})$$

The problem formalization part supplies an MI problem to $\text{LPSF}(\mathcal{L}, R, ctrl, A)$, and typically consists of

1. representation of a problem in natural language as a QA or proof problem on FOL_c, and
2. conversion of the resulting QA or proof problem into an MI problem on $\mathcal{L}$ by using Skolemization.

3   Experiments

We solve the Agatha puzzles (proof and QA versions) based on LPSF by deciding parameters such as logical structures, rules, and controls. The logical structure used for solving the Agatha puzzles will be explained in the next section.

3.1   Agatha Puzzle

Consider the "Dreadsbury Mansion Mystery" problem, which was given by Len Schubelt and can be described as follows: Someone who lives in Dreadsbury Mansion killed Aunt Agatha. Agatha, the butler, and Charles live in Dreadsbury Mansion, and are the only people who live therein. A killer always hates his victim, and is never richer than his victim. Charles hates no one that Aunt Agatha hates. Agatha hates everyone except the butler. The butler hates everyone not richer than Aunt Agatha. The butler hates everyone Agatha hates. No one hates everyone. The Agatha QA problem is to find all persons who killed Agatha. The Agatha proof problem is to show that Agatha killed herself.

$$F_1: \quad \exists x : (live(x, D) \wedge kill(x, A))$$
$$F_2: \quad \forall x : (live(x, D) \leftrightarrow (eq(x, A) \vee eq(x, B) \vee eq(x, C)))$$
$$F_3: \quad \forall x : \forall y : (kill(x, y) \rightarrow hate(x, y))$$
$$F_4: \quad \forall x : \forall y : (kill(x, y) \rightarrow \neg richer(x, y))$$
$$F_5: \quad \neg \exists x : (hate(C, x) \wedge hate(A, x) \wedge live(x, D))$$
$$F_6: \quad \forall x : ((neq(x, B) \wedge live(x, D)) \rightarrow hate(A, x))$$
$$F_7: \quad \forall x : ((\neg richer(x, A) \wedge live(x, D)) \rightarrow hate(B, x))$$
$$F_8: \quad \forall x : ((hate(A, x) \wedge live(x, D)) \rightarrow hate(B, x))$$
$$F_9: \quad \neg \exists x : (live(x, D) \wedge \forall y : (live(y, D) \rightarrow hate(x, y)))$$

Fig. 2. Background knowledge of Agatha puzzle in first-order formulas

3.2 Formalization by First-Order Formulas

Assume that eq and neq are predefined binary constraint predicates and for any ground usual terms t_1 and t_2, (i) $eq(t_1, t_2)$ is true iff $t_1 = t_2$, and (ii) $neq(t_1, t_2)$ is true iff $t_1 \neq t_2$. The background knowledge of this mystery is formalized as the conjunction of first-order formulas in Fig. 2, where (i) the constants A, B, C, and D denote "Agatha," "the butler," "Charles," and "Dreadsbury Mansion," respectively, and (ii) for any terms t_1 and t_2, $live(t_1, t_2)$, $kill(t_1, t_2)$, $hate(t_1, t_2)$, and $richer(t_1, t_2)$ are intended to mean "t_1 lives in t_2," "t_1 killed t_2," "t_1 hates t_2," "t_1 is richer than t_2," respectively.

3.3 Meaning-Preserving Skolemization

We take KR-Logic as a logical structure and use the space of extended clauses constructed on KR-Logic as a canonical logical structure, ECLS_{FC}, which will be explained in Sect. 4.

Consider the formulas F_1–F_9 in Fig. 2. The clauses C_1–C_{14} in Fig. 3 are formulas in ECLS_{FC} such that $\text{MPS}(F_1) = \{C_1, C_2\}$, $\text{MPS}(F_2) = \{C_3, C_4, C_5, C_6\}$, $\text{MPS}(F_3) = \{C_7\}$, $\text{MPS}(F_4) = \{C_8\}$, $\text{MPS}(F_5) = \{C_9\}$, $\text{MPS}(F_6) = \{C_{10}\}$, $\text{MPS}(F_7) = \{C_{11}\}$, $\text{MPS}(F_8) = \{C_{12}\}$, and $\text{MPS}(F_9) = \{C_{13}, C_{14}\}$. Let $K = F_1 \wedge F_2 \wedge \cdots \wedge F_9$. Let $Cs = \{C_1, C_2, \ldots, C_{14}\}$. Then $\text{MPS}(K) = Cs$.

The formalization consists of the following two steps:

- From the Agatha problem described by sentences (Sect. 3.1), we obtain K in the first-order formulas.
- We apply MPS to K to obtain a set of extended clauses Cs.

3.4 Computation Control with Prioritized ET Rules

ET rules together with a binary relation $>$ are referred to as prioritized ET rules or a prioritized ET rule set. Given ET rules r, r', and r'', if $r > r' > r''$, then r has priority over r' and r' has priority over r''. For ease of explanation, a prioritized ET rule set

$$C_1: \quad live(x, D) \leftarrow func(f_0, x)$$
$$C_2: \quad kill(x, A) \leftarrow func(f_0, x)$$
$$C_3: \quad \leftarrow live(x, D), neq(x, A), neq(x, B), neq(x, C)$$
$$C_4: \quad live(A, D) \leftarrow$$
$$C_5: \quad live(B, D) \leftarrow$$
$$C_6: \quad live(C, D) \leftarrow$$
$$C_7: \quad hate(x, y) \leftarrow kill(x, y)$$
$$C_8: \quad \leftarrow kill(x, y), richer(x, y)$$
$$C_9: \quad \leftarrow hate(A, x), hate(C, x), live(x, D)$$
$$C_{10}: \quad hate(A, x) \leftarrow neq(x, B), live(x, D)$$
$$C_{11}: \quad richer(x, A), hate(B, x) \leftarrow live(x, D)$$
$$C_{12}: \quad hate(B, x) \leftarrow hate(A, x), live(x, D)$$
$$C_{13}: \quad \leftarrow hate(x, y), func(f_1, x, y), live(x, D)$$
$$C_{14}: \quad live(y, D) \leftarrow live(x, D), func(f1, x, y)$$

Fig. 3. Background knowledge of Agatha puzzle in Clauses

$$r_1 > r_2 > \ldots > r_m$$

is also represented as a sequence R_p of ET rules such that $R_p = [r_1, r_2, \ldots, r_m]$.

Prioritized ET rules are useful to specify a control. Given prioritized ET rules $R_p = [r_1, r_2, \ldots, r_m]$, we have a control $ctrl$, which is called a control determined by R_p, by defining $ctrl(\langle Cs, \varphi \rangle) = r_j$ if $Cs \in dom_j$ for all $j = 1, 2, 3, \ldots, m$, where $dom_j = dom(r_j) - dom(r_1) - dom(r_2) - \cdots - dom(r_{j-1})$.

Prioritized ET rules can be used for directly selecting an ET rule at each computation step as follows: Let S be a state. R_p is applicable to S with r_j if

1. none of $r_1, r_2, \ldots, r_{j-1}$ is applicable to S, and
2. r_j is applicable to S.

Note that j is determined uniquely by S and R_p. If R_p is applicable to S with r_j, then the result of the application of R_p to S, denoted by $R_p(S)$, is $r_j(S)$.

3.5   ET Rules and Computation Control for Agatha Problems

ET rules and computation control for Agatha proof and QA problems are specified by the following prioritized ET rules:

(eq) > (neq) > (subsumed) > (dup) > (erase) > (posfvr) > (negfvr) > (funcEv)
> (udi 1) > (udi 3) > (udi 10) > (specAtom) > (chSide).

The followings are short explanation of the above ET rules. More detailed explanation can be found in our previous papers, e.g., [2, 4–6].

(eq)	constraint solving for *eq*,
(neq)	constraint solving for *neq*,
(subsumed)	elimination of subsumed clauses,

(dup)	elimination of duplicate atoms,
(erase)	elimination of independent satisfiable atoms,
(posfvr)	positive function-variable restriction,
(negfvr)	negative function-variable restriction,
(funcEv)	*func*-atom evaluation,
(udi i)	unfolding and definite-clause removal,
(specAtom)	finite specialization of atoms,
(chSide)	side-change transformation.

3.6   Solution of Agatha QA Problem

Every QA problem on FOL_c can be converted into an MI problem on $ECLS_{FC}$ [3]. Let $F_{qa} = (\forall x : (kill(x, A) \rightarrow killer(x)))$. Let $C_{qa} = (killer(x) \leftarrow kill(x, A))$. Then $MPS(F_{qa}) = \{C_{qa}\}$, and $MPS(K \wedge F_{qa}) = Cs \cup \{C_{qa}\}$. According to Theorem 2 in [3], we have $Models(K \wedge F_{qa}) = Models(Cs \cup \{C_{qa}\})$. Given a set G of ground atoms, let φ_{qa} be a mapping to obtain the set of all *killer* atoms in G. According to Theorem 4 in [3], $ans_{QA}(K \wedge F_{qa}, kill(x, A)) = ans_{MI}(Cs \cup \{C_{qa}\}, \varphi_{qa})$.

We have a 114-step solution, in which (udi 1) is applied 25 times, (udi 3) is applied 3 times, (udi 10) is applied 8 times, (neq) 10 times, (chSide) 5 times, (specAtom) 1 time, (dup) 4 times, (subsumed) 23 times, (erase) 6 times, (funcEV) 22 times, (posfvr) 5 times, and (negfvr) 2 times. After 114 application of ET rules, we have a singleton clause set $\{(killer(A) \leftarrow)\}$, from which the answer can be readily obtained, i.e., only Agatha is the killer of Agatha. This solution is guaranteed to be correct based on Theorem 1 in Sect. 2.6 and model preservation of MPS.

3.7   Solution of Agatha Proof Problem

Every proof problem on FOL_c can be converted into an MI problem on $ECLS_{FC}$ [3]. Let $F_{pr} = kill(A, A)$. Let $C_{pr} = (\leftarrow kill(A, A))$. Then $MPS(\neg F_{pr}) = C_{pr}$, and $MPS(K \wedge \neg F_{pr}) = Cs \cup \{C_{pr}\}$. According to Theorem 2 in [3], we have: $Models(K \wedge \neg F_{pr}) = \varnothing$ iff $Models(Cs \cup \{C_{pr}\}) = \varnothing$. Given a set G of ground atoms, let φ_{pr} be a mapping to output "yes" if $G = \mathcal{G}_u$, and "no" otherwise. According to Theorem 5 in [3], $ans_{Pr}(K, F_{pr}) = ans_{MI}(Cs \cup \{C_{pr}\}, \varphi_{pr})$.

We have a 77-step solution, in which (udi 1) is applied 9 times, (udi 3) is applied 2 times, (udi 10) is applied 7 times, (neq) 6 times, (chSide) 2 times, (specAtom) 1 time, (dup) 3 times, (subsumed) 18 times, (erase) 6 times, (funcEV) 14 times, (posfvr) 6 times, and (negfvr) 3 times. After 77 application of ET rules, we have a empty clause ($\leftarrow$), thereby proving that Agatha is a killer of herself. This solution is guaranteed to be correct based on Theorem 1 in Sect. 2.6 and model preservation of MPS.

3.8   Failures of Conventional Methods

The conventional formalization consists of the following two steps:

- From the Agatha problem described by sentences (Sect. 3.1) we obtain K in the first-order formulas.
- To K we apply CSK to have a set of clauses Cs'.

The conventional Skolemization (CSK) is a major hindrance to developing a general solution for proof and QA problems on ECLS_{FC} since it does not generally preserve satisfiability nor logical meanings by Theorem 1 in [3], and thus does not generally preserve the answers to proof and QA problems.

When we transform the Agatha background knowledge K into a set Cs' of usual clauses using the conventional Skolemization, we know that $Models(K) \neq \varnothing$ and $Models(Cs') = \varnothing$, hence neither satisfiability nor logical meanings is preserved in this case, and further computation starting from Cs' to find answers is thus useless.

4 Canonical Logical Structure in KR-Logic

To solve the Agatha puzzle in the general axiomatic framework of LPSF, we propose a new logical structure, consisting of extended clauses with existential quantification of function variables, which is constructed from a new logic, called *KR-Logic*.

4.1 Alphabet of KR-Logic

An alphabet $\langle \mathbb{F}, \mathbb{V}, \mathbb{FC}, \mathbb{FV}, Pred, Pred_C \rangle$ is assumed, where $\mathbb{F}$ is a set of usual function symbols, $\mathbb{V}$ is a set of usual variables, $\mathbb{FC}$ is a set of function constants, $\mathbb{FV}$ is a set of function variables, $Pred$ is a set of usual predicate symbols, and $Pred_C$ is a set of built-in constraint predicate symbols.

4.2 Terms and Atoms

Definition 7. A *term* on $\langle \mathbb{F}, \mathbb{V}, \mathbb{FC}, \mathbb{FV} \rangle$, which is also simply called a term, is inductively defined as follows:

1. A 0-ary element in $\mathbb{F} \cup \mathbb{FC} \cup \mathbb{FV}$ is a term.
2. If $v \in \mathbb{V}$ is a variable, then v is a term.
3. If $f \in \mathbb{F} \cup \mathbb{FC} \cup \mathbb{FV}$, the arity of f is $n > 0$, and $t_1, \ldots, t_n$ are terms, then $f(t_1, \ldots, t_n)$ is a term. □

The set of all terms on $\langle \mathbb{F}, \mathbb{V}, \mathbb{FC}, \mathbb{FV} \rangle$ is denoted by $\text{T}(\mathbb{F}, \mathbb{V}, \mathbb{FC}, \mathbb{FV})$. Let $\text{T}(\mathbb{F}) = \text{T}(\mathbb{F}, \varnothing, \varnothing, \varnothing)$. A term in $\text{T}(\mathbb{F})$ is called a *ground term*.

A *user-defined atom* takes the form $p(t_1, \ldots, t_n)$, where p is a user-defined predicate and the t_i are terms. A *built-in constraint atom*, also simply called a *constraint atom* or a *built-in atom*, takes the form $c(t_1, \ldots, t_n)$, where c is a predefined constraint predicate and the t_i are terms. A *func-atom* [1] is an expression of the form $func(f, t_1, \ldots, t_n, t_{n+1})$, where f is either an n-ary function constant or an n-ary function variable, and the t_i are terms.

An atom is a *ground* atom if all arguments are either function constants or ground terms. Let $\mathcal{A}_u$ be the set of all user-defined atoms, $\mathcal{A}_c$ the set of all constraint atoms, and $\mathcal{F}$ the set of all func-atoms.

4.3 Extended Clauses

There are two types of variables: usual variables and function variables. A function variable can be instantiated into only function constants.

An *extended clause* C on $\mathcal{A}_u \cup \mathcal{A}_c \cup \mathcal{F}$ is a formula of the form

$$a_1, \ldots, a_m \leftarrow b_1, \ldots, b_n, \mathbf{f}_1, \ldots, \mathbf{f}_p,$$

where each of $a_1, \ldots, a_m, b_1, \ldots, b_n$ is a user-defined atom in $\mathcal{A}_u$ or a constraint atom in $\mathcal{A}_c$, and $\mathbf{f}_1, \ldots, \mathbf{f}_p$ are *func*-atoms. All usual variables occurring in C are implicitly universally quantified and their scope is restricted to the extended clause C itself. All function variables occurring in C are implicitly existentially quantified at the top level of Cs, and their scope covers all the extended clauses in Cs.

4.4 Constraints for Function Variables

We use constraints for function variables to increase the expressive power of clauses [4].

A constraint set S is attached to each occurrence of a function variable $f \in \mathbb{FV}$ and such an occurrence is denoted by the pair $\langle f, S \rangle$. We assume that if there are two occurrences of $\langle f, S_1 \rangle$ and $\langle f, S_2 \rangle$, then $S_1 = S_2$.

When $\langle f, S \rangle$ exists, instantiation of f is restricted to the extent specified by S.

When the attached constraint set S is empty, f can be instantiated freely, and is simply represented by f itself.

4.5 Canonical Logical Structure

Let $\mathrm{ECLS}_{\mathrm{FC}}$ be the set of all extended clauses, possibly containing function variables with constraints. $\mathrm{ECLS}_{\mathrm{FC}}$ is an extension of $\mathrm{ECLS}_{\mathrm{F}}$, which is the space of extended clauses with only function variables without constraints for function variables.

Let $\mathcal{G}_u$ be defined as the set:

$$\{p(t_1, \ldots, t_m) \mid (m \text{ is a nonnegative integer}) \,\&$$
$$(p \text{ is an } m\text{-ary predicate in } Pred) \,\&$$
$$(t_1, \ldots, t_m \in \mathrm{T}(\mathbb{F}))\}.$$

Let $\mathcal{S}_\theta$ be the set of all substitutions for usual variables in $\mathbb{V}$. Let $\mathcal{S}_\sigma(Cs)$ be the set of all substitutions for function variables in $\mathbb{FV}$ that satisfy all constraints for function variables in Cs. Let GCL be the set of all ground atoms with only user-defined atoms. For any $Cs \subseteq \mathrm{ECLS}_{\mathrm{FC}}$, for any $G \subseteq \mathcal{G}_u$, $\nu(Cs)(G) = true$ iff there exists $\sigma \in \mathcal{S}_\sigma(Cs)$ such that for any $C \in Cs$, for any $\theta \in \mathcal{S}_\theta$, if $C\sigma\theta \in GCL$, then $C\sigma\theta$ is true with respect to G. Letting $a_1, a_2, \ldots, a_n \in \mathcal{G}_u$, and $b_1, b_2, \ldots, b_m \in \mathcal{G}_u$, $(a_1, a_2, \ldots, a_n \leftarrow b_1, b_2, \ldots, b_m)$ is true with respect to G iff $a_1 \in G$ or $a_2 \in G$ or $\ldots$ or $a_n \in G$ or $b_1 \notin G$ or $b_2 \notin G$ or $\ldots$ or $b_m \notin G$. Then we have a canonical logical structure $\langle pow(\mathrm{ECLS}_{\mathrm{FC}}), pow(\mathcal{G}_u), \nu \rangle$, which is the canonical logical structure used as the first parameter of LPSF.

5   Improvement of Logical Problem Solving

Let CLS_B be the set of all usual clauses possibly with built-in constraint atoms. Obviously, CLS_B is a proper subset of ECLS_FC. Roughly speaking, $CLS_B = ECLS_{FC} -$ (function constant and function variables) $-$ (constraints for function variables). Note that the conventional first-order logic has domain-based interpretations and has not canonical interpretations. However, from CLS_B, we can make a canonical logical structure $\langle pow(\text{CLS}_\text{B}), pow(\mathcal{G}_\text{u}), \nu_1 \rangle$, where ν_1 is a restriction of ν to CLS_B. The conventional resolution-based proof methods [7,11] are considered as instance procedures of LPSF, where input parameters and problem formalization are basically given as follows:

1. Canonical logical structure $= \langle pow(\text{CLS}_\text{B}), pow(\mathcal{G}_\text{u}), \nu_1 \rangle$, where a description is a subset of the set of all normal clauses with universal quantifications of usual variables.
2. Formalization = Representation and transformation using CSK.
3. ET rules = The resolution and factoring inference rules.

It has been turned out that such conventional methods with these parameters have fatal limitations when considering the full class of proof and QA problems on FOL_c. The conventional Skolemization, by which neither the satisfiability nor the logical meaning of a formula in FOL_c can be preserved in general, has been a major hindrance to solving problems in these two classes correctly. For example, neither Agatha proof nor QA problem can be transformed equivalently into clausal form by the conventional Skolemization, which result in wrong answers.

To overcome the difficulty, a new logic, called KR-Logic, and new Skolemization, called meaning-preserving Skolemization (MPS), were introduced [1,3]. Our new parameters and problem formalization are basically as follows:

1. Canonical logical structure $= \langle pow(\text{ECLS}_\text{FC}), pow(\mathcal{G}_\text{u}), \nu \rangle$, where a description is a set (1) consisting of universal quantified extended clauses, and (2) possibly being quantified existentially by function variables.
2. Formalization = Representation and transformation using MPS.
3. ET rules = The unfolding and definite clause removal rules and many other ET rules.

With these parameters, we are successful in giving a more general methodology for logical problem solving.

For example, Agatha proof and QA problem can be transformed equivalently into extended clausal form by MPS, which result in correct answers.

6   Conclusions

Logical Problem Solving Framework (LPSF) is a generator of logical problem solving methods. Inputs of LPSF consist of (1) a canonical logical structure, (2) ET rules, (3) a control, and (4) an answer mapping. The core computation structure of LPSF is solving MI problems by ET, and the correctness of computation is strictly guaranteed. We can compare two logical problem solving methods by formalizing them on LPSF. We can develop a new logical problem solving method by taking new parameters. Main evaluation of logical problem solving methods should be the range of guarantee of correct problem solving. We have illustrated using Agatha proof and QA problems that KR-Logic together with MPS is better than usual first-order logic with CSK.

References

1. Akama, K., Nantajeewarawat, E.: Meaning-preserving Skolemization. In: 3rd International Conference on Knowledge Engineering and Ontology Development, Paris, France, pp. 322–327 (2011)
2. Akama, K., Nantajeewarawat, E.: Equivalent transformation in an extended space for solving query-answering problems. In: Nguyen, N.T., Attachoo, B., Trawiński, B., Somboonviwat, K. (eds.) ACIIDS 2014. LNCS (LNAI), vol. 8397, pp. 232–241. Springer, Cham (2014). https://doi.org/10.1007/978-3-319-05476-6_24
3. Akama, K., Nantajeewarawat, E.: Model-intersection problems with existentially quantified function variables: formalization and a solution schema. In: 8th International Joint Conference on Knowledge Discovery, Knowledge Engineering and Knowledge Management, Porto, Portugal, vol. 2, pp. 52–63 (2016)
4. Akama, K., Nantajeewarawat, E.: Solving query-answering problems with constraints for function variables. In: Nguyen, N.T., Hoang, D.H., Hong, T.-P., Pham, H., Trawiński, B. (eds.) ACIIDS 2018. LNCS (LNAI), vol. 10751, pp. 36–47. Springer, Cham (2018). https://doi.org/10.1007/978-3-319-75417-8_4
5. Akama, K., Nantajeewarawat, E., Akama, T.: Computation control by prioritized ET rules. In: 10th International Joint Conference on Knowledge Discovery, Knowledge Engineering and Knowledge Management, KEOD, Seville, Spain, vol. 2, pp. 84–95 (2018)
6. Akama, K., Nantajeewarawat, E., Akama, T.: Side-change transformation. In: 10th International Joint Conference on Knowledge Discovery, Knowledge Engineering and Knowledge Management, KEOD, Seville, Spain, vol. 2, pp. 237–246 (2018)
7. Chang, C.-L., Lee, R.C.-T.: Symbolic Logic and Mechanical Theorem Proving. Academic Press, Cambridge (1973)
8. Donini, F.M., Lenzerini, M., Nardi, D., Schaerf, A.: $\mathcal{AL}$-log: integrating datalog and description logics. J. Intell. Coop. Inf. Syst. **10**, 227–252 (1998)
9. Lloyd, J.W.: Foundations of Logic Programming, 2nd edn. Springer, Heidelberg (1987). https://doi.org/10.1007/978-3-642-83189-8
10. Motik, B., Sattler, U., Studer, R.: Query answering for OWL-DL with rules. J. Web Semant. **3**, 41–60 (2005)
11. Robinson, J.A.: A machine-oriented logic based on the resolution principle. J. ACM **12**, 23–41 (1965)

Term Rewriting that Preserves Models in KR-Logic

Kiyoshi Akama[1], Ekawit Nantajeewarawat[2(✉)], and Taketo Akama[3]

[1] Information Initiative Center, Hokkaido University, Sapporo, Japan
akama@iic.hokudai.ac.jp
[2] Computer Science Program, Sirindhorn International Institute of Technology,
Thammasat University, Pathumthani, Thailand
ekawit@siit.tu.ac.th
[3] Modeleet Labs, Sapporo, Japan
taketo.akama@gmail.com

Abstract. In human proofs of mathematical problems, such as proofs in a group theory, term rewriting is usually used. When we consider Herbrand semantics for the first-order logic with constraints (FOL_c), correct representation of evaluable terms cannot be obtained due to lack of representation power of the logic. In place of FOL_c with Herbrand semantics, we use KRL_c (KR-Logic with built-in constraints). We propose a class of term rewriting rules, and prove that they preserve the sets of all models in KR-Logic. Representation and computation by the rewriting rules in KR-Logic is well established in the space of KRL_c. This paper opens a new method of logical problem solving, with KRL_c being the representation space and ECLS_N being the computation space. This theory integrates logical inference and functional rewriting under the broader concept of equivalent transformation.

Keywords: Proof problem · Query-answering problem ·
Function variable · Built-in equality · KR-Logic ·
Equivalent transformation · Constructor · Term rewriting rule ·
Model preservation

1 Introduction

A proof problem is a "yes/no" problem; it is concerned with checking whether or not one given logical formula entails another given logical formula. A query-answering (QA) problem is an "all-answers finding" problem, i.e., all ground instances of a given query atom satisfying given requirements are to be found. The resolution method was invented to give solutions to all proof problems on first-order formulas [8,13]. The success of the resolution method and research on automated theorem proving on first-order formulas [14] formed the satisfiability-based and resolution-centered approach, which provided a foundation for the research on solving QA problems. Much work in logic programming has been

© Springer Nature Switzerland AG 2019
N. T. Nguyen et al. (Eds.): ACIIDS 2019, LNAI 11431, pp. 41–52, 2019.
https://doi.org/10.1007/978-3-030-14799-0_4

done under the satisfiability-preserving Skolemization, usual clauses with universal quantification of usual variables, and inference rules [11].

Serious limitation of such a proof-based conventional approach has been revealed in a broader viewpoint [1,10,12]. Satisfiability cannot be preserved in general when we consider formulas that may contain built-in constraint atoms. Built-in constraint atoms play a crucial role in knowledge representation and are essential for practical applications. It is inevitable to consider logical problems that are formalized by first-order formulas with built-in constraint atoms. The set of all such formulas is denoted by FOL_c. The conventional Skolemization does not preserve satisfiability nor logical meanings in general in the space of FOL_c. Therefore, the conventional Skolemization (CSK) does not provide a transformation process towards correct solutions for proof problems and QA problems on FOL_c.

To overcome the difficulty, meaning-preserving Skolemization (MPS) was invented and used in place of CSK [1,2]. All proof problems and all QA problems on FOL_c are mapped, preserving their answers, into logical problems on an extended space, called $ECLS_F$, and are solved by repeated problem simplification using equivalent transformation (ET) rules [3–6].

In this paper, the space FOL_c is extended into KRL_c, which is the set of all formulas in KR-Logic that may contain function variables at any position (see Fig. 1). In the space of FOL_c, function variables are used only in *func*-atoms in the body of clauses, while they are freely used, i.e., any atom may include function variables, in the space of KRL_c. Together with the extension of FOL_c into KRL_c, the clause space $ECLS_F$ is extended into $ECLS_N$. These extensions increase the power of representation and computation in logical problem solving.

In the space of KRL_c, term rewriting (TR) is considered as one of the main transformations with strict correctness of computation (see Fig. 1). We propose a class of term rewriting rules on $ECLS_N$, and prove that they preserve the sets of all models of formulas in the power set of $ECLS_N$. This computation is similar to term rewriting and function rewriting [7,9] in the research domain of functional programming. This theory is expected to naturally integrate the representation and computation in functional and logic programming.

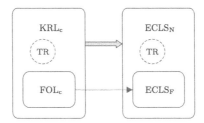

Fig. 1. Extending representation and computation spaces

The rest of the paper is organized as follows: Sect. 2 takes a proof problem of a group theory as an example, and explains its solution outline. Section 3 introduces a new logic, called KR-Logic. Section 4 constructs a canonical logical structure of clauses in KR-Logic. Section 5 proposes a class of term rewriting rules. Section 6 proves that the proposed term rewriting rules preserve models. Section 7 concludes the paper.

Given a set A, $pow(A)$ denotes the power set of A. Given two sets A and B, $Map(A, B)$ denotes the set of all mappings from A to B, and for any partial mapping f from A to B, $dom(f)$ denotes the domain of f, i.e., $dom(f) = \{a \mid (a \in A)$ & $(f(a)$ is defined$)\}$.

2 An Introductory Example

2.1 Example: Proof Problem in Group Theory

Consider the following proof problem: Let G be a set, $(\cdot)$ a mapping from $G \times G$ to G, and $(^{-1})$ a mapping from G to G that satisfy the following conditions:

A1: $\forall x, y, z \in G : x \cdot (y \cdot z) = (x \cdot y) \cdot z$
A2: $\forall x \in G : x \cdot e = e \cdot x = x$
A3: $\forall x \in G, \exists x^{-1} \in G : x \cdot x^{-1} = x^{-1} \cdot x = e$
A4: $\forall x \in G : x \cdot x = e$

The problem is to prove B: $\forall x, y \in G : x \cdot y = y \cdot x$

2.2 Example: Usual Proof by Rewriting Terms

A1 $\wedge$ A2 $\wedge$ A3 $\wedge$ A4 $\wedge$ $\neg$B gives the conjunction of the following formulas:
$f(f(x, y), z) = f(x, f(y, z))$, $f(x, e) = x$, $f(e, x) = x$, $f(x, x^{-1}) = e$, $f(x^{-1}, x) = e$, $f(x, x) = e$, $f(a, b) \neq f(b, a)$. A usual proof is shown as follows:

$f(a, b)$
$=$ {by application of $(r_1 : x \to f(e, x))$}
$=$ $f(e \cdot f(a, b))$
$=$ {by application of $(r_2 : e \to f(x, x))$}
$=$ $f(f(f(b, a), f(b, a)), f(a, b))$
$=$ {by application of $(r_3 : f(f(x, y), z) \to f(x, f(y, z)))$}
$=$ $f(f(b, a), f(f(b, a), f(a, b)))$
$=$ {by application of $(r_3 : f(f(x, y), z) \to f(x, f(y, z)))$}
$=$ $f(f(b, a), f(b, f(a, f(a, b))))$
$=$ {by application of $(r_4 : f(x, f(y, z)) \to f(f(x, y), z))$}
$=$ $f(f(b, a), f(b, f(f(a, a), b)))$
$=$ {by application of $(r_5 : f(x, x) \to e)$}
$=$ $f(f(b, a), f(b, f(e, b)))$
$=$ {by application of $(r_6 : f(e, x) \to x)$}
$=$ $f(f(b, a), f(b, b))$
$=$ {by application of $(r_5 : f(x, x) \to e)$}
$=$ $f(f(b, a), e)$
$=$ {by application of $(r_7 : f(x, e) \to x)$}
$=$ $f(b, a)$.

Hence $f(a, b) = f(b, a)$, which contradicts $f(a, b) \neq f(b, a)$.

2.3 Example: Formalization with Formulas in the First-Order Logic

We assume Herbrand semantics for first-order logic with constraints (FOL$_c$).

The alphabet of FOL$_c$ is $\langle \mathbb{F}, \mathbb{V}, Pred, Pred_C \rangle$, where $\mathbb{F}$ is a set of usual function symbols, $\mathbb{V}$ is a set of usual variables, $Pred$ is a set of usual predicate symbols, and $Pred_C$ is a set of built-in constraint predicate symbols. Each element in $\mathbb{F}$ has an arity, which is a non-negative integer.

Let $g \in Pred$ and $g(x)$ represents the information that x is in G. Let $eq \in Pred_C$ be a built-in predicate defined by: $eq(t_1, t_2)$ is true iff $t_1 = t_2$ for any ground terms t_1 and t_2. Assume that f, i, and e are binary, unary, and 0-ary function symbols, respectively, in $\mathbb{F}$, and x, y, and z are usual variables in $\mathbb{V}$. Let K denote the conjunction of the following formulas:

$$F_1 : \forall x \forall y : (g(x) \wedge g(y) \rightarrow g(f(x, y))) \quad F_2 : \forall x : (g(x) \rightarrow g(i(x))) \quad F_3 : g(e)$$
$$F_4 : \forall x \forall y \forall z : ((g(x) \wedge g(y) \wedge g(z)) \rightarrow eq(f(x, f(y, z)), f(f(x, y), z)))$$
$$F_5 : \forall x \forall y \forall z : ((g(x) \wedge g(y) \wedge g(z)) \rightarrow eq(f(f(x, y), z), f(x, f(y, z))))$$
$$F_6 : \forall x : (g(x) \rightarrow eq(f(x, e), x)) \qquad F_7 : \forall x : (g(x) \rightarrow eq(f(e, x), x))$$
$$F_8 : \forall x : (g(x) \rightarrow eq(f(x, i(x)), e)) \qquad F_9 : \forall x : (g(x) \rightarrow eq(f(i(x), x), e))$$
$$F_{10} : \forall x : (g(x) \rightarrow eq(f(x, x), e))$$

We want to prove $K \rightarrow (\forall x \forall y : (g(x) \wedge g(y)) \rightarrow eq(f(x, y), f(y, x)))$. Its negation is $K \wedge \neg(\forall x \forall y : (g(x) \wedge g(y)) \rightarrow eq(f(x, y), f(y, x)))$, which is equivalent to $K \wedge (\exists x \exists y : (g(x) \wedge g(y) \wedge \neg eq(f(x, y), f(y, x))))$.

2.4 Inappropriateness of Formalization in the First-Order Logic

The Herbrand semantics is taken for the first-order logic with equality constraints (FOL$_c$), which seems to be natural for most mathematicians. The predicate eq is defined based on the relation '=' on ground terms. Since '=' means syntactical coincidence, e.g., if $cons$ and nil are ground terms, we have $eq(nil, nil)$ and $eq(cons(nil, nil), cons(nil, nil))$.

The fact that f, i, and e are function symbols in $\mathbb{F}$, however, results in inconvenience as follows: Let's take an atom $eq(f(x, e), x)$ as an example. What is the meaning of this equality? It says the term $f(x, e)$ is equal to the another term x. Whatever value the variable x may take, however, $f(x, e)$ cannot be equal to x due to the inclusion of x inside $f(x, e)$.

The intuitive meaning of the equality $eq(f(x, e), x)$ should be as follows: f computes a value, say a term t, from x and e, and the resulting term t is equal to x. f should not be a constant in $\text{TERM}(\mathbb{F})$, but an evaluable function.

Hence intuitively we need two kinds of "functions":

1. constructors for constructing compound terms, and
2. evaluable functions for mapping ground terms.

FOL$_c$ (together with Herbrand semantics) supports only one kind of "functions". This is the reason why we cannot represent this proof problem correctly in FOL$_c$.

2.5   A Logical Problem Solving Method

We assume basic familiarity of model-based semantics of logic. A set G is a model of a formula E iff E is true with respect to G. Let E be a formula and $Models(E)$ denote the set of all models of E.

We solve the proof problem by the following steps in this paper.

1. (Section 3) We take a logical structure $\mathcal{L}$. We formalize the problem by making a formula E in $\mathcal{L}$.
2. (Section 4) We construct a logical structure $\mathcal{L}'$ of clauses from $\mathcal{L}$. We transform E in $\mathcal{L}$ into Cs in $\mathcal{L}'$.
3. (Section 5) We prepare a set R of rewriting rules on clauses from Cs.
4. (Section 6) We construct an rewriting sequence $[S_0, \ldots, S_n]$ by (i) taking $S_0 = Cs$ as a initial state, (ii) applying rewriting rules in R to each S_i ($i = 1, 2, \ldots, n - 1$), and (iii) stopping at S_n. Let $Cs' = S_n$. By showing that $Models(Cs') = \varnothing$, we prove the given problem with strict correctness theory.

3   Representation in KR-Logic

We introduce a new logic, called *KR-Logic*. It is considered as an extension of usual first order logic by introduction of existential quantification of function variables, which is essential for representation of proof problems with equality constraints.

3.1   Alphabet and Terms of KR-Logic

An alphabet $\langle \mathbb{F}, \mathbb{V}, \mathbb{FC}, \mathbb{FV}, Pred, Pred_C \rangle$ is assumed, where $\mathbb{F}$ is a set of constructors, $\mathbb{V}$ is a set of usual variables, $\mathbb{FC}$ is a set of function constants, $\mathbb{FV}$ is a set of function variables, $Pred$ is a set of user-defined predicate symbols, and $Pred_C$ is a set of built-in constraint predicate symbols. Each element in $\mathbb{F} \cup \mathbb{FC} \cup \mathbb{FV}$ is associated with a non-negative integer, called its arity.

Definition 1. A *term* on $\langle \mathbb{F}, \mathbb{V}, \mathbb{FC}, \mathbb{FV} \rangle$, which is also simply called a term, is inductively defined as follows:

1. A 0-ary element in $\mathbb{F} \cup \mathbb{FC}$ is a term.
2. If $v \in \mathbb{V} \cup \mathbb{FV}$, then v is a term.
3. If $f \in \mathbb{F} \cup \mathbb{FC} \cup \mathbb{FV}$, the arity of f is $n > 0$, and $t_1, \ldots, t_n$ are terms, then $f(t_1, \ldots, t_n)$ is a term. □

The set of all terms on $\langle \mathbb{F}, \mathbb{V}, \mathbb{FC}, \mathbb{FV} \rangle$ is denoted by $T(\mathbb{F}, \mathbb{V}, \mathbb{FC}, \mathbb{FV})$. Let $T(\mathbb{F}) = T(\mathbb{F}, \varnothing, \varnothing, \varnothing)$. Let $T(\mathbb{F}, \mathbb{FC}) = T(\mathbb{F}, \varnothing, \mathbb{FC}, \varnothing)$. A term in $T(\mathbb{F}, \mathbb{FC})$ is called a *ground term*.

3.2   Atoms and WFFs

A *user-defined atom* takes the form $p(t_1, \ldots, t_n)$, where p is a user-defined predicate and the t_i are terms. A *built-in constraint atom*, also simply called a *constraint atom* or a *built-in atom*, takes the form $c(t_1, \ldots, t_n)$, where c is a predefined constraint predicate and the t_i are terms. A *func-atom* [1] is an expression of the form $func(f, t_1, \ldots, t_n, t_{n+1})$, where f is either an n-ary function constant or an n-ary function variable, and the t_i are terms. An atom is a *ground* atom if all arguments are ground terms. Let $\mathcal{A}_u$ be the set of all user-defined atoms, $\mathcal{A}_c$ the set of all built-in constraint atoms, and $\mathcal{F}$ the set of all *func*-atoms.

   A *well-formed formula* (for short, *wff*) in KRL_c is constructed inductively by finitely many applications of the following rules:

1. *Atoms:* If α is an atom in $\mathcal{A}_u \cup \mathcal{A}_c \cup \mathcal{F}$, then α is a wff.
2. *Negation:* If α is a wff, then $\neg\alpha$ is a wff.
3. *Binary connectives:* If α and β are wffs, then $\alpha \wedge \beta$, $\alpha \vee \beta$, $\alpha \rightarrow \beta$, and $\alpha \leftrightarrow \beta$ are wffs.
4. *Quantifiers for usual variables:* If α is a wff and v is a variable in $\mathbb{V}$, then $\forall v : \alpha$ and $\exists v : \alpha$ are wffs.
5. *Quantifiers for function variables:* If α is a wff and f_v is a variable in $\mathbb{FV}$, then $\forall_F f_v : \alpha$ and $\exists_F f_v : \alpha$ are wffs.

3.3   Example: Formalization with Formulas in KR-Logic

Let $g \in Pred$, with $g(x)$ being intended to mean that x is in G. Let $eq \in Pred_C$ be a built-in predicate defined by: $eq(t_1, t_2)$ is true iff $t_1 = t_2$, for any $t_1, t_2 \in \mathrm{T}(\mathbb{F})$. Assume that x, y, and z are variables in $\mathbb{V}$. Assume also that $\$f$, $\$i$, and $\$e$ are binary, unary, and 0-ary function variables, respectively, in $\mathbb{FV}$. Let K denote the conjunction of the following formulas:

$$F_1 : \forall x \forall y : (g(x) \wedge g(y) \rightarrow g(\$f(x,y)))$$
$$F_2 : \forall x : (g(x) \rightarrow g(\$i(x)))$$
$$F_3 : g(\$e)$$
$$F_4 : \forall x \forall y \forall z : ((g(x) \wedge g(y) \wedge g(z)) \rightarrow eq(\$f(x, \$f(y,z)), \$f(\$f(x,y),z)))$$
$$F_5 : \forall x \forall y \forall z : ((g(x) \wedge g(y) \wedge g(z)) \rightarrow eq(\$f(\$f(x,y),z), \$f(x, \$f(y,z))))$$
$$F_6 : \forall x : (g(x) \rightarrow eq(\$f(x, \$e), x))$$
$$F_7 : \forall x : (g(x) \rightarrow eq(\$f(\$e, x), x))$$
$$F_8 : \forall x : (g(x) \rightarrow eq(\$f(x, \$i(x)), \$e))$$
$$F_9 : \forall x : (g(x) \rightarrow eq(\$f(\$i(x), x), \$e))$$
$$F_{10} : \forall x : (g(x) \rightarrow eq(\$f(x,x), \$e))$$

   We want to prove $K \rightarrow (\forall x \forall y : (g(x) \wedge g(y)) \rightarrow eq(\$f(x,y), \$f(y,x)))$. Its negation is $K \wedge \neg(\forall x \forall y : (g(x) \wedge g(y)) \rightarrow eq(\$f(x,y), \$f(y,x)))$, which is equivalent to $K \wedge (\exists x \exists y : (g(x) \wedge g(y) \wedge \neg eq(\$f(x,y), \$f(y,x))))$. By meaning-preserving Skolemization (MPS), this formula is transformed into

$$E : \quad \exists \$f \exists \$i \exists \$e \exists \$a \exists \$b : \quad K \wedge g(\$a) \wedge g(\$b) \wedge \neg eq(\$f(\$a, \$b), \$f(\$b, \$a)),$$

where $\$a$ and $\$b$ are new function variables.

Note that the only difference between the representation in Sect. 2.3 and that in this section is the change of function symbols such as f, i, and e in $\mathbb{F}$ into function variables in $\mathbb{FV}$. Function symbols such as *cons* and *nil* in $\mathbb{F}$ remain in $\mathbb{F}$. Syntactically, the prefix '$\$$' is added to f, i, and e to make $\$f$, $\$i$, and $\$e$, while the others are unchanged.

4   A Canonical Logical Structure in KR-Logic

We propose a new logical structure, consisting of extended clauses with existential quantification of function variables, which is constructed from KR-Logic.

4.1   Extended Clauses in KR-Logic

There are two types of variables: usual variables and function variables. An *extended clause* C on $\mathcal{A}_\mathrm{u} \cup \mathcal{A}_\mathrm{c} \cup \mathcal{F}$ is a formula of the form

$$a_1, \ldots, a_m \leftarrow b_1, \ldots, b_n, \mathbf{f}_1, \ldots, \mathbf{f}_p,$$

where each of $a_1, \ldots, a_m, b_1, \ldots, b_n$ is a user-defined atom in $\mathcal{A}_\mathrm{u}$ or a built-in constraint atom in $\mathcal{A}_\mathrm{c}$, and $\mathbf{f}_1, \ldots, \mathbf{f}_p$ are *func*-atoms.

All usual variables occurring in C are implicitly universally quantified and their scope is restricted to the extended clause C itself. Given a set Cs of extended clauses, all function variables occurring in Cs are implicitly existentially quantified at the top level of Cs, and their scope covers all the clauses in Cs. The set of all extended clauses is denoted by $\mathrm{ECLS_N}$. A function variable can be instantiated into function constants. A ground term in $\mathrm{T}(\mathbb{F}, \mathbb{FC})$ containing a function constant in $\mathbb{FC}$ and its argments in $\mathrm{T}(\mathbb{F})$ at the top level can be reduced to a ground term in $\mathrm{T}(\mathbb{F})$, the operation of which is denoted by fev. Let $bind(\mathbb{V})$ be the set of all bindings on $\mathbb{V}$. A substitution on $\mathbb{V}$ is a sequence of elements in $bind(\mathbb{V}) \cup \{fev\}$. Let $\mathcal{S}_\theta$ be the set of all substitutions on $\mathbb{V}$. Let $bind(\mathbb{FV})$ be the set of all bindings on $\mathbb{FV}$. A substitution on $\mathbb{FV}$ is a sequence of elements in $bind(\mathbb{FV})$. Let $\mathcal{S}_\sigma$ be the set of all substitutions on $\mathbb{FV}$. Let GCL_σ be the set of all clauses that contain no atom with occurrences of function variables. Let GCL be the set of all ground clauses consisting of only user-defined ground atoms with no occurrences of function constants in $\mathbb{FC}$. A clause in GCL is obtained from an extended clause in $\mathrm{ECLS_N}$ by (1) instantiation of variables in $\mathbb{FV}$, (2) instantiation of variables in $\mathbb{V}$, (3) evaluation of all evaluable subterms in $\mathrm{T}(\mathbb{F}, \mathbb{FC})$ by fev into ground terms in $\mathrm{T}(\mathbb{F})$, and (4) removal of true ground built-in atoms and true ground *func*-atoms in the clause body.

4.2   Example: Formalization in Clausal Form in KR-Logic

The formula E is represented in a clausal form by:

C_1: $g(\$f(x,y)) \leftarrow g(x), g(y)$ C_2: $g(\$i(x)) \leftarrow g(x)$ C_3: $g(\$e) \leftarrow$
C_4: $eq(\$f(\$f(x,y),z), \$f(x,\$f(y,z))) \leftarrow g(x), g(y), g(z)$
C_5: $eq(\$f(x,\$e),x) \leftarrow g(x)$ C_6: $eq(\$f(\$e,x),x) \leftarrow g(x)$
C_7: $eq(\$f(x,\$i(x)),\$e) \leftarrow g(x)$ C_8: $eq(\$f(\$i(x),x),\$e) \leftarrow g(x)$
C_9: $eq(\$f(x,x),\$e) \leftarrow g(x)$ C_{10}: $g(\$a) \leftarrow$
C_{11}: $g(\$b) \leftarrow$ C_{12}: $\leftarrow eq(\$f(\$a,\$b), \$f(\$b,\$a))$

Let $Cs = \{C_1, C_2, \ldots, C_{12}\}$. Then we solve the given proof problem in Sect. 3 by proving that $Models(Cs) = \varnothing$.

4.3   A Canonical Logical Structure of Extended Clauses

Let $\mathcal{G}_t$ be defined as $T(\mathbb{F})$. Let $\mathcal{G}_u$ be defined as the set of all ground atoms of the form $p(t_1, \ldots, t_m)$, where p is an m-ary user-defined predicate in $Pred$ and $t_1, \ldots, t_m$ are terms in $\mathcal{G}_t$.

The main role of a logical structure is to determine the truth or falsity of each formula with respect to an interpretation $G \in pow(\mathcal{G}_u)$. We define a mapping $\nu : pow(\text{ECLS}_N) \rightarrow Map(pow(\mathcal{G}_u), \{true, false\})$ as follows: For any $Cs \subseteq \text{ECLS}_N$ and any $G \subseteq \mathcal{G}_u$, $\nu(Cs)(G) = true$ iff there exists $\sigma \in \mathcal{S}_\sigma$ such that for any $C \in Cs$ and any $\theta \in \mathcal{S}_\theta$, if $C\sigma\theta \in GCL$, then $C\sigma\theta$ is true with respect to G. Hence, for completion of defining ν, it is enough to define the following case: Letting $a_1, a_2, \ldots, a_n \in \mathcal{G}_u$ and $b_1, b_2, \ldots, b_m \in \mathcal{G}_u$, the clause $(a_1, a_2, \ldots, a_n \leftarrow b_1, b_2, \ldots, b_m)$ is true with respect to G iff $a_1 \in G$ or $a_2 \in G$ or $\ldots$ or $a_n \in G$ or $b_1 \notin G$ or $b_2 \notin G$ or $\ldots$ or $b_m \notin G$.

$\mathcal{L} = \langle pow(\text{ECLS}_N), pow(\mathcal{G}_u), \nu \rangle$ is a canonical logical structure, i.e., (1) $\mathcal{L}$ is a logical structure, and (2) $\mathcal{L}$ has $pow(\mathcal{G}_u)$ as the interpretation domain.

5   Term Rewriting Rules

5.1   Term Contexts, Atom Contexts, and Clause Contexts

Assume that $\square$ is a function symbol with arity 0 that does not belong to $\mathbb{F} \cup \mathbb{V} \cup \mathbb{FC} \cup \mathbb{FV}$. A subterm is a part of a term, and a subterm is extended by a *term-context* into a term. Let t be a term and tc a term context. $t \triangleright tc$ is the term $tc\{\square/t\}$. A term is a part of an atom, and a term is extended by an *atom context* into an atom. Let t be a term and ac an atom context. $t \triangleright ac$ is the atom $ac\{\square/t\}$. An atom is a part of a clause, and an atom is extended by a *clause context* into a clause. Let a be an atom and con a clause context. $a \triangleright con$ is the clause $con\{\square/a\}$. The sets of all *term contexts*, all *atom contexts*, and all *clause contexts* are denoted, respectively, by Con_T, Con_A, and Con_C.

5.2   Conditional Term Rewriting Rules

Definition 2. (*Conditional term rewriting rule*) Let *conds* be a finite sequence of atoms. $(t_1 \rightarrow t_2 : conds)$ is a relation r on ECLS_N defined as follows:

$$r = \{(Cs_1, Cs_2) \mid (t_1, t_2 \in T(\mathbb{F}, \mathbb{V}, \mathbb{FC}, \mathbb{FV})) \ \&$$
$$(tc \in Con_T) \ \& \ (ac \in Con_A) \ \& \ (con \in Con_C) \ \&$$
$$(\theta \in S_\theta) \ \& \ (conds\theta = true) \ \&$$
$$(C_0 = (eq(t_1, t_2) \leftarrow conds)) \ \& \ (Cs \subseteq \text{ECLS}_N) \ \&$$
$$(Models(Cs) = Models(Cs \cup \{C_0\})) \ \&$$
$$(C_1 = ((((t_1\theta) \rhd tc) \rhd ac) \rhd con)) \ \&$$
$$(C_2 = ((((t_2\theta) \rhd tc) \rhd ac) \rhd con)) \ \&$$
$$(Cs_1 = \{C_1\} \cup Cs) \ \& \ (Cs_2 = \{C_2\} \cup Cs)\}. \quad \square$$

The *conditional term rewriting rule r* that is represented by $(t_1 \to t_2 : conds)$ is denoted by $r : (t_1 \to t_2 : conds)$. When Cs_1 is transformed into Cs_2 by a *conditional term rewriting rule r*, we write $(Cs_1, Cs_2) \in r$ or $Cs_1 \xrightarrow{r} Cs_2$.

Proposition 1. *Assume that C_1 and C_2 are clauses in ECLS$_N$. Assume that $\sigma_1 \in S_\sigma$ and $C_1\sigma_1 \in GCL_\sigma$. Then there is $\sigma_{12} \in S_\sigma$ such that $C_1\sigma_{12} = C_1\sigma_1$, $C_1\sigma_{12} \in GCL_\sigma$, and $C_2\sigma_{12} \in GCL_\sigma$.* $\square$

Proof. Consider $C_1\sigma \in GCL_\sigma$. Then, let σ be a restriction of σ_1 into the domain of the function variables in C_1. Then $C_1\sigma = C_1\sigma_1$, and $C_1\sigma \in GCL_\sigma$. Furthermore, let σ_{12} be the extension of σ to the domain of the function variables in C_1 and C_2. Then $C_1\sigma_{12} = C_1\sigma = C_1\sigma_1$, $C_1\sigma_{12} \in GCL_\sigma$, $C_2\sigma_{12} \in GCL_\sigma$. $\square$

Proposition 2. *Assume that (1) $\theta \in S_\theta$, (2) $conds\theta = true$, (3) $C_0 = (eq(t_1, t_2) \leftarrow conds)$, (4) $Models(Cs) = Models(Cs \cup \{C_0\})$, (5) $Cs_1 = \{C_1\} \cup Cs$, (6) $Cs_2 = \{C_2\} \cup Cs$, (7) $C_1 = ((((t_1\theta) \rhd tc) \rhd ac) \rhd con)$, and (8) $C_2 = ((((t_2\theta) \rhd tc) \rhd ac) \rhd con)\}$. Also assume that (9) $\sigma_{12} \in S_\sigma$, (10) $C_1\sigma_{12} \in GCL_\sigma$, and (11) $C_2\sigma_{12} \in GCL_\sigma$. If $m \in Models(C_1\sigma_{12})$, then $m \in Models(C_2\sigma_{12})$.* $\square$

Proof. Assume that $m \in Models(C_1\sigma_{12})$. Let θ_2 be an arbitrary substitution in S_θ that satisfies $C_2\sigma_{12}\theta_2 \in GCL$. Then $t_2\theta\sigma_{12}\theta_2 \in \mathcal{G}_t$. Hence there is $\rho \in S_\theta$ such that $t_1\theta\sigma_{12}\theta_2\rho \in \mathcal{G}_t$ and $t_2\theta\sigma_{12}\theta_2\rho \in \mathcal{G}_t$. From $conds\theta = true$, $C_0 = (eq(t_1, t_2) \leftarrow conds)$ and $Models(Cs) = Models(Cs \cup \{C_0\})$, we have $Models(Cs) = Models(Cs \cup \{(eq(t_1\theta, t_2\theta) \leftarrow)\})$. It follows that $t_1\theta\sigma_{12}\theta_2\rho = t_2\theta\sigma_{12}\theta_2\rho \in \mathcal{G}_t$. Hence $C_1\sigma_{12}\theta_2\rho = C_2\sigma_{12}\theta_2\rho \in GCL$. Since $m \in Models(C_1\sigma_{12})$, we have $C_1\sigma_{12}\theta_2\rho$ is true with respect to m. Hence $C_2\sigma_{12}\theta_2\rho$ is true with respect to m. Since $C_2\sigma_{12}\theta_2\rho = C_2\sigma_{12}\theta_2$, we have $C_2\sigma_{12}\theta_2$ is true with respect to m. Hence $m \in Models(C_2\sigma_{12})$. $\square$

5.3 Example: Term Rewriting Rules for the Sample Problem

Since each equality produces two rules, we obtain twelve rewriting rules with respect to Cs:

$r_1 : (x \to \$f(x, \$e) : \{g(x)\})$ $r_2 : (\$f(x, \$e) \to x : \{g(x)\})$
$r_3 : (x \to \$f(\$e, x) : \{g(x)\})$ $r_4 : (\$f(\$e, x) \to x : \{g(x)\})$
$r_5 : (\$e \to \$f(x, \$i(x)) : \{g(x)\})$ $r_6 : (\$f(x, \$i(x)) \to \$e : \{g(x)\})$
$r_7 : (\$e \to \$f(\$i(x), x) : \{g(x)\})$ $r_8 : (\$f(\$i(x), x) \to \$e : \{g(x)\})$

$r_9 : (\$e \to \$f(x,x) : \{g(x)\})$ $r_{10} : (\$f(x,x) \to \$e : \{g(x)\})$
$r_{11} : (\$f(\$f(x,y),z) \to \$f(x,\$f(y,z)) : \{g(x),g(y),g(z)\})$
$r_{12} : (\$f(x,\$f(y,z)) \to \$f(\$f(x,y),z) : \{g(x),g(y),g(z)\})$

6   Term Rewriting that Preserves Models

6.1   Model-Preservation Theorem

We prove that the term rewriting rules introduced in this paper preserve models when they are applied to clause sets.

Theorem 1. *Let Cs be a set of clauses. Let $r : (t_1 \to t_2 : conds)$ be a conditional term rewriting rule. If $(Cs_1, Cs_2) \in r$, then $Models(Cs_1) = Models(Cs_2)$.*   □

Proof. Since $(Cs_1, Cs_2) \in r$, we have $\{t_1, t_2\} \subseteq \mathrm{T}(\mathbb{F}, \mathbb{V}, \mathbb{FC}, \mathbb{FV})$, $tc \in Con_T$, $ac \in Con_A$, $con \in Con_C$, $\theta \in \mathcal{S}_\theta$, $conds\theta = true$, $C_0 = (eq(t_1, t_2) \leftarrow conds)$, $Models(Cs) = Models(Cs \cup \{C_0\})$, $C_1 = ((((t_1\theta) \rhd tc) \rhd ac) \rhd con)$, $C_2 = (((((t_2\theta) \rhd tc) \rhd ac) \rhd con)$, $Cs_1 = \{C_1\} \cup Cs$, and $Cs_2 = \{C_2\} \cup Cs$. Now we start with the assumption that $m \in Models(Cs_1)$. Then $m \in Models(\{C_1\} \cup Cs)$. Hence there exists $\sigma_1 \in \mathcal{S}_\sigma$ such that (1) $\{C_1\sigma_1\} \cup Cs\sigma_1 \subseteq GCL_\sigma$ and (2) $m \in Models(\{C_1\sigma_1\} \cup Cs\sigma_1)$. It follows that there exists $\sigma_1 \in \mathcal{S}_\sigma$ such that (1) $C_1\sigma_1 \in GCL_\sigma$, (2) $Cs\sigma_1 \subseteq GCL_\sigma$, (3) $m \in Models(Cs\sigma_1)$, and (4) $m \in Models(C_1\sigma_1)$. By Proposition 1, we take a substitution $\sigma_{12} \in \mathcal{S}_\sigma$ such that (1) $C_1\sigma_{12} = C_1\sigma_1$, (2) $C_1\sigma_{12} \in GCL_\sigma$, and (3) $C_2\sigma_{12} \in GCL_\sigma$. Hence there exists $\sigma_{12} \in \mathcal{S}_\sigma$ such that (1) $C_1\sigma_{12} \in GCL_\sigma$, (2) $C_2\sigma_{12} \in GCL_\sigma$, (3) $Cs\sigma_{12} \subseteq GCL_\sigma$, (4) $m \in Models(Cs\sigma_{12})$, and (5) $m \in Models(C_1\sigma_{12})$. By Proposition 2, we have $m \in Models(C_2\sigma_{12})$. It follows that there exists $\sigma_{12} \in \mathcal{S}_\sigma$ such that (1) $C_1\sigma_{12} \in GCL_\sigma$, (2) $C_2\sigma_{12} \in GCL_\sigma$, (3) $Cs\sigma_{12} \subseteq GCL_\sigma$, (4) $m \in Models(Cs\sigma_{12})$, (5) $m \in Models(C_1\sigma_{12})$, and (6) $m \in Models(C_2\sigma_{12})$. Then we have that $m \in Models(\{C_2\} \cup Cs)$, which is equivalent to $m \in Models(Cs_2)$.

Similarly, we next start with $m \in Models(Cs_2)$. By Proposition 1, we take a substitution $\sigma_{21} \in \mathcal{S}_\sigma$ such that (1) $C_2\sigma_{21} = C_2\sigma_2$, (2) $C_2\sigma_{21} \in GCL_\sigma$, and (3) $C_1\sigma_{21} \in GCL_\sigma$.

By Proposition 2, we have $m \in Models(C_1\sigma_{21})$. Finally, we reach $m \in Models(Cs_1)$.

From these, it follows that $m \in Models(Cs_1)$ iff $m \in Models(Cs_2)$.   □

6.2   Example: Model-Preserving Rewriting for the Sample Problem

Using the rewriting rules in Sect. 5.3 and a constraint solving rule for equality, the sample problem is solved. The computation starts with $S_0 = \{C_0\} \cup Cs$, where

$$C_0 = (\leftarrow eq(\$f(\$a, \$b), \$f(\$b, \$a))).$$

After 3 steps, we reach $S_3 = \{C_3\} \cup Cs$, where

$$C_3 = (\leftarrow eq(\$f(\$f(\$b, \$a), \$f(\$f(\$b, \$a), \$f(\$a, \$b))), \$f(\$b, \$a))).$$

Consider the rule r_{11}. Since $g(\$b)$, $g(\$a)$, and $g(\$f(\$a,\$b))$ are true, r_{11} can be applied to S_3 with $\theta = \{x/\$b, y/\$a, z/\$f(\$a,\$b)\}$. The next state is $S_4 = \{C_4\} \cup Cs$, where

$$C_4 = (\leftarrow eq(\$f(\$f(\$b,\$a), \$f(\$b, \$f(\$a, \$f(\$a, \$b))))), \$f(\$b, \$a))).$$

All steps of computation are shown below.

$Models(\{(\leftarrow eq(\$f(\$a, \$b), \$f(\$b, \$a)))\} \cup Cs)$
$= \ (r_3$ is applied with $x = \$f(\$a, \$b)$ (since $g(\$f(\$a, \$b))$ is true))
$= \ Models(\{(\leftarrow eq(\$f(\$e, \$f(\$a, \$b)), \$f(\$b, \$a)))\} \cup Cs)$
$= \ (r_9$ is applied with $x = \$f(\$a, \$b)$ (since $g(\$f(\$a, \$b))$ is true))
$= \ Models(\{(\leftarrow eq(\$f(\$f(\$f(\$b, \$a), \$f(\$b, \$a)), \$f(\$a, \$b)), \$f(\$b, \$a)))\} \cup Cs)$
$= \ (r_{11}$ is applied with $x = y = z = \$f(\$a, \$b)$ (since $g(\$f(\$a, \$b))$ is true))
$= \ Models(\{(\leftarrow eq(\$f(\$f(\$b, \$a), \$f(\$f(\$b, \$a), \$f(\$a, \$b))), \$f(\$b, \$a)))\} \cup Cs)$
$= \ (r_{11}$ is applied with $x = \$b$, $y = \$a$, and $z = \$f(\$a, \$b)$ (see the above explanation)
$= \ Models(\{(\leftarrow eq(\$f(\$f(\$b, \$a), \$f(\$b, \$f(\$a, \$f(\$a, \$b)))), \$f(\$b, \$a)))\} \cup Cs)$
$= \ (r_{12}$ is applied with $x = y = \$a$ and $z = \$b$ (since $g(\$a)$ and $g(\$b)$ are true))
$= \ Models(\{(\leftarrow eq(\$f(\$f(\$b, \$a), \$f(\$b, \$f(\$f(\$a, \$a), \$b))), \$f(\$b, \$a)))\} \cup Cs)$
$= \ (r_{10}$ is applied with $x = \$a$ (since $g(\$a)$ is true))
$= \ Models(\{(\leftarrow eq(\$f(\$f(\$b, \$a), \$f(\$b, \$f(\$e, \$b))), \$f(\$b, \$a)))\} \cup Cs)$
$= \ (r_4$ is applied with $x = \$b$ (since $g(\$b)$ is true))
$= \ Models(\{(\leftarrow eq(\$f(\$f(\$b, \$a), \$f(\$b, \$b)), \$f(\$b, \$a)))\} \cup Cs)$
$= \ (r_{10}$ is applied with $x = \$b$ (since $g(\$b)$ is true))
$= \ Models(\{(\leftarrow eq(\$f(\$f(\$b, \$a), \$e), \$f(\$b, \$a)))\} \cup Cs)$
$= \ (r_2$ is applied with $x = \$f(\$b, \$a)$ (since $g(\$f(\$b, \$a))$ is true))
$= \ Models(\{(\leftarrow eq(\$f(\$b, \$a), \$f(\$b, \$a)))\} \cup Cs)$
$= \ $(the $eq(x, x)$ atom is removed since it is true.)
$= \ Models(\{(\leftarrow)\} \cup Cs)$
$= \ \varnothing.$

In the last step of transformation, a constraint solving rule for equality is used. This rule is not a term rewriting rule. However, it obviously preserves models. In our computation model, logical inference and term rewriting are used together under the principle of equivalent transformation.

7   Conclusion

The space FOL_c is the set of all first-order formulas that may contain built-in constraint atoms. Previously, logical problems on FOL_c were target problems, and their solution paths were constructed in the space of $ECLS_F$.

In this theory, we have considered logical problems on KRL_c, and they are converted into problems on $ECLS_N$, i.e., each computation is a sequence of clause sets in the $ECLS_N$ space. The representation space FOL_c has been extended into KRL_c, and the computation space $ECLS_F$ has been extended into $ECLS_N$.

We have proposed a class of term rewriting rules in the space of $ECLS_N$, and have strictly proved that they preserve the sets of all models of formulas. This computation is similar to term rewriting/function rewriting in the research domain of functional programming. In this theory, logical inference and functional rewriting co-exist, both of them being instances of a broader concept of equivalent transformation.

References

1. Akama, K., Nantajeewarawat, E.: Meaning-preserving Skolemization. In: 3rd International Conference on Knowledge Engineering and Ontology Development, Paris, France, pp. 322–327 (2011)
2. Akama, K., Nantajeewarawat, E.: Equivalent transformation in an extended space for solving query-answering problems. In: Nguyen, N.T., Attachoo, B., Trawiński, B., Somboonviwat, K. (eds.) ACIIDS 2014. LNCS (LNAI), vol. 8397, pp. 232–241. Springer, Cham (2014). https://doi.org/10.1007/978-3-319-05476-6_24
3. Akama, K., Nantajeewarawat, E.: Model-intersection problems with existentially quantified function variables: formalization and a solution schema. In: 8th International Joint Conference on Knowledge Discovery, Knowledge Engineering and Knowledge Management, Porto, Portugal, vol. 2, pp. 52–63 (2016)
4. Akama, K., Nantajeewarawat, E.: Solving query-answering problems with constraints for function variables. In: Nguyen, Ngoc Thanh, Hoang, Duong Hung, Hong, Tzung-Pei, Pham, Hoang, Trawiński, Bogdan (eds.) ACIIDS 2018. LNCS (LNAI), vol. 10751, pp. 36–47. Springer, Cham (2018). https://doi.org/10.1007/978-3-319-75417-8_4
5. Akama, K., Nantajeewarawat, E., Akama, T.: Computation control by prioritized ET rules. In: 10th International Joint Conference on Knowledge Discovery, Knowledge Engineering and Knowledge Management, KEOD, Seville, Spain, vol. 2, pp. 84–95 (2018)
6. Akama, K., Nantajeewarawat, E., Akama, T.: Side-change transformation. In: 10th International Joint Conference on Knowledge Discovery, Knowledge Engineering and Knowledge Management, KEOD, Seville, Spain, vol. 2, pp. 237–246 (2018)
7. Bird, R., Wadler, P.: Introduction to Functional Programming. Prentice-Hall Inc, New Jersey (1988)
8. Chang, C.-L., Lee, R.C.-T.: Symbolic Logic and Mechanical Theorem Proving. Academic Press, Cambridge (1973)
9. Dershowitz, N., Jouannaud, J.-P.: Rewrite Systems, Handbook of Theoretical Computer Science, Volume B: Formal Models and Semantics, Chap. 6, pp. 243–320. MIT Press, Cambridge (1990)
10. Donini, F.M., Lenzerini, M., Nardi, D., Schaerf, A.: $\mathcal{AL}$-log: integrating datalog and description logics. J. Intell. Coop. Inf. Syst. **10**, 227–252 (1998)
11. Lloyd, J.W.: Foundations of Logic Programming, 2nd edn. Springer, Heidelberg (1987). https://doi.org/10.1007/978-3-642-83189-8
12. Motik, B., Sattler, U., Studer, R.: Query answering for OWL-DL with rules. J. Web Semant. **3**, 41–60 (2005)
13. Robinson, J.A.: A machine-oriented logic based on the resolution principle. J. ACM **12**, 23–41 (1965)
14. Fitting, M.: First-Order Logic and Automated Theorem Proving. Springer, Heidelberg (1996). https://doi.org/10.1007/978-1-4612-2360-3

Towards Knowledge Formalization
and Sharing in a Cognitive Vision Platform
for Hazard Control (CVP-HC)

Caterine Silva de Oliveira[1(✉)], Cesar Sanin[1(✉)],
and Edward Szczerbicki[2(✉)]

[1] The University of Newcastle, Newcastle, NSW, Australia
caterine.silvadeoliveira@uon.edu.au,
cesar.maldonadosanin@newcastle.edu.au
[2] Gdansk University of Technology, Gdansk, Poland
edward.szczerbicki@zie.pg.gda.pl

Abstract. Hazards can be found in all work environments and may cause injuries, illnesses, or fatalities. In this context, controlling of risks and safety management has become indispensable to guarantee the laborers wellbeing in worksites. Aiming to achieve a systematic, explicit and comprehensive system for managing safety risks, a Cognitive Vision Platform for Hazard Control (CVP-HC) has been proposed. This platform is designed to automatically detect unsafe activities and improve the decision making process when they occur in different workplace scenarios, while attending specific safety requirements of organizations by adapting its behavior accordingly. To meet generality, the CVP-HC utilizes the Set of Experience Knowledge Structure (SOEKS) and Decisional DNA (DDNA) to administrate knowledge. To ensure scalability and adaptability, a loosely coupled communication model, the publishing/subscribe interaction scheme is used over the Robot Operating System (ROS) framework.

Keywords: Cognitive vision · SOEKS · DDNA · ROS framework ·
Hazard control

1 Introduction

Hazards can be found in all work environments and may cause injuries, illnesses, or fatalities [1]. In this context, controlling of risks and safety management has become indispensable to guarantee the laborers wellbeing in worksites [2]. An analysis of the efficacy of a variety of injury prevention interventions has found that the two most effective methodologies are the culture-change and behavior-based safety approaches [3]. Systems and tools to support the practical implementation of either rubrics have emerged as a need; however, the current technologies available perform far from satisfactory [4–7].

Aiming to achieve a systematic, explicit and comprehensive system for managing safety risks, a Cognitive Vision Platform for Hazard Control (CVP-HC) has been proposed. The CVP-HC is designed to automatically detect unsafe activities and improve the decision making process when they occur in different workplace scenarios,

© Springer Nature Switzerland AG 2019
N. T. Nguyen et al. (Eds.): ACIIDS 2019, LNAI 11431, pp. 53–61, 2019.
https://doi.org/10.1007/978-3-030-14799-0_5

while meeting specific safety requirements of companies by adapting its behavior accordingly [8]. To meet generality, the platform is based on the Set of Experience Knowledge Structure (SOEKS) and Decisional DNA (DDNA), which were first introduced by Sanin and Szczerbicki and later expanded for a number of dedicated domains and applications [9, 10].

To ensure scalability and adaptability the publishing/subscribe interaction scheme is used over the Robot Operating System (ROS) framework [11]. This loosely coupled communication model is able to provide availability of information, offering better generality than traditional client-server architectures [12]. In addition, it enables heterogeneous interacting applications to properly retrieve and comprehend exchanged knowledge in the platform.

This paper is structured as follow: In Sect. 2 the idea of Cognitive Systems and its functionalities is introduced. In Sect. 3 the fundamentals of Knowledge Engineering is explained as well as the chosen knowledge representation. In Sect. 4 the Communication System chosen to pass, store and share knowledge is described, aiming to achieve a flexible, general-purpose, and adaptable system. In Sect. 5 the Cognitive Vision Platform for Hazard Control (CVP-HC) and details of the SOEKS and DDNA implementation over ROS is given. Lastly, in Sect. 6, conclusions obtained from the configuration model design is presented together with next steps for future work.

2 Cognitive Systems

The design of a general-purpose vision system with the robustness and resilience of the human vision is still a challenge [13]. One of the latest trends in computer vision research to mimic the human-like capabilities is the joining of cognition and computer vision into cognitive computer vision. For instance, a flexible and automatic semantic annotation system has been proposed by Zambrano et al. [6]. This service is able to perform in a variety of video analysis with little or no modification to the code by combining detection algorithms and an experience-based approximations, and it is considered a pathway towards cognitive vision. Results have demonstrated scalability enhancements; however, the tool has been designed to execute when offline, not being applicable to real time video analysis.

In cognitive vision systems knowledge and learning are central elements for reasoning about events and for the decision making process. To be readily articulated, codified, accessed and shared, knowledge must be represented in an explicit and structured way [14]. This knowledge, stored as experience, can be used for learning purposes, improving the decision making at some defined situation. A knowledge representation that carries the decision making in its structure can be helpful in this case and contribute for the development of a better cognitive vision system.

3 Knowledge Engineering

There are two main phases that can be associated with knowledge engineering: acquisition and representation within a formalism [15].

3.1 Knowledge Acquisition

The data sources that will be integrated into the CVP-HC may differ from one configuration to another. The more sources incorporated into the systems the more complex scenarios and events may be managed by the central reasoning (depending also on the computer processing power available). However, to ensure minimal operation, it must be composed by, at least, two main data sources modules: RGB Camera and a computer system.

The RGB camera gathers the visual knowledge for creation of visual variables in the system (which are detected by a Convolutional Neural Network – for tests, the Tensorfow has been used [16]). The computer system is accountable for the production of a minimum context needed to improve the interpretation of detected events (context variables). These two components compose the first layer of integration.

On the second layer of integration a Lidar sensor is included [17]. The Lidar sensor feeds the system with depth data (M8 Quanergy has be utilized), which supports tracking of workers, creation of more complex relationship among visual variables and identifications of activities.

Lastly, the third layer is composed by any extra sensor data available (such as temperature, illumination, oxygen level, etc.), input signals (to indicate if machines are in operation), GPS, etc. If the integration reaches the third layer, unique context may be produced. In this case, set of specific and detailed rules might be created, as well as reasoning about complex events will become possible.

3.2 Knowledge Formalism

As a computational process, reasoning requires systemic techniques and data structures to be feasible. A number of techniques have been designed trying acquire, structure and formalize knowledge for reasoning purposes [18–20]. These technologies attempt to automatically collect and administrate knowledge, but lack of keeping experiences of the formal decision events they participate on and rarely reuse them [21]. These decisional experiences are frequently overlooked and not shared [22–25]. A small part of these acquired experiences survive over time, and in most cases, become inaccessible due to poor knowledge engineering or changes in software, hardware or storage media technologies [26].

To administrate knowledge and manage decision events the platform is based on The Set of Experience Knowledge Structure (SOEKS) and Decisional DNA (DDNA). SOEKS is a knowledge representation structure capable of storing formal decision events in an explicit way. It composed by four elements: variables, functions, constraints, and rules. Variables are used to represent knowledge as an attribute-value. Functions is used to create relationships among these variables and to build multi-objective goals. Constraints are also a type of function, but acts to limit the set of possible solutions and control the performance of the system in relation to its goals. Rules gives support for flexibility of the system by creating relationships among variables as a condition-consequence connection ("if-then-else") and can be easily created and modified by the user to attend the conditions under which the platform should operate [9]. The DDNA is a unique and single structure that captures decisional fingerprints of the company and has the SOEKS as its basis. Multiple Sets of Experience can be created, classified, and according to their characteristics grouped into decisional chromosomes. They store decisional strategies for a specific area of the organization. The set of chromosomes constitutes what is called the Decisional DNA (DDNA) and may differ from one company to another, creating uniqueness decisional events that characterize the safety culture of that company [10].

4   Communication System for Knowledge Sharing

Design of a communication model is one of the first needs when developing an application which objective is to use, share, store, and/or retrieve knowledge [27]. In this work, a messaging system implemented in ROS (a framework composed by group of tools, libraries, and standardized procedures that simplifies the creation of complex robotic systems behavior [11]) is used to manage the details of communication between distributed nodes in the publish/subscribe scheme. In this framework, clear interfaces between the nodes is implemented to improve encapsulation and promote reuse of code [12].

4.1   Real Time Publisher/Subscriber Communication Model

ROS uses the subject-based model, one of the earliest and matured among the publisher/subscriber configurations. In this composition, each unit of knowledge belongs to one of a set of topics [12]. Publishers label each of these topics with a subject, characterizing them. Consumers, then can subscribe to topics within a particular subject of interest. For instance, a general subject-based publisher/subscriber system for human activity recognition may define different groups according to the nature of data and their source; publishers in a node might publish information to the appropriate groups (topics), and subscribers from another node may subscribe to the topics relevant for the recognition process of a specific activity. Nodes can also exchange a request and response message as part of a ROS service call. This general concept is represented in Fig. 1. In the past ten years, systems supporting this paradigm have developed significantly, resulting in numerous academic and powerful industrial solutions [28–32].

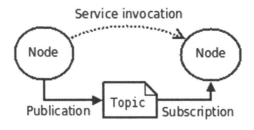

Fig. 1. ROS basic concepts [33].

5   CVP-HC

Knowledge representation has been brought to focus of attention in artificial intelligence and associated areas. Designers and programmers face the representation issue [34] even before acquiring knowledge. Current computers systems are still not able to shape representations of the world as efficient as human beings, or even simpler, representations of ordinary formal decision events.

Choosing the most appropriate and effective formalization of knowledge for a given system is not trivial, and the consideration of using SOEKS in the proposed platform is founded in the fact that experience must considered if the objective is to mimic the human intelligence capabilities. Therefore, SOEKS is used as carrier for decision making. This explicitly represented knowledge is used for two main purposes in our platform

- For recommendation purposes. On configuration process, the existing variables, functions, constraints and rules are available for selection, promoting code and knowledge reuse.
- To retrain the platform from time to time. This will ensure that the system learns as it runs improving the DDNA of each application profile without affecting the general-purpose characteristic of the platform.

SOEKS and DDNA is implemented over ROS framework, by defining general and multi-purpose message data structures represented as variables, functions, constraints and rules. These messages are used by nodes to communicate with each other by publishing/subscribing to topics of interest. Depending on the learning rate chosen by the user, the messages' content are stored as experiences to be used for the two main purposes described previously.

5.1   Message Configuration

A message is basically a data structure, containing different type fields. A group of primitive types, such as boolean, integer, float, etc., are supported to comprise data structures, as well as arrays of these primitive elements. Messages can contain single or multiple elements and arrays.

In the given platform, a generic mechanism for expressing machine-readable semantics of data has been designed using SOEKS. This configuration offers the possibility of representing simple and complex knowledge structure in a standardized way. SOEKS is implemented over ROS to facilitate knowledge sharing among different technologies and applications and the reuse of this knowledge for the given purposes previously mentioned. Figure 2 shows the components of SOEKS in the *msg* file specification.

In Fig. 2 the main elements are variables, functions, constraints and rules. They may contain primitive types (integer, float, boolean, etc.) or structured elements (which can also contain other elements and so on) which are defined as Factors, Simfactors, Assofactors, Terms, Joints, Conditions, Values and Consequences.

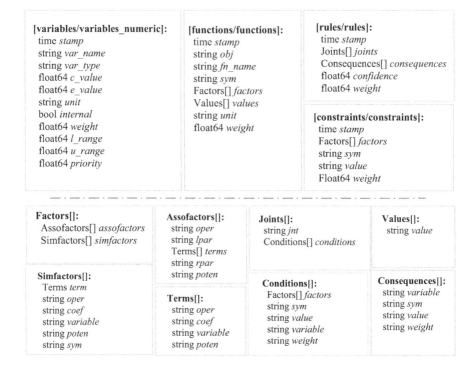

Fig. 2. Example of SOEKS message types.

5.2   Communication Model Structure

In the proposed communication model messages are sent and received via a transport system with publish/subscribe scheme using ROS framework [12]. A node sends messages by publishing them to topics. Another node receives these messaged by subscribing to that topic. Topic is a designation that characterize the content of messages. There may be a single or multiple publishers and subscribers for a single topic. In the same way, a node may publish or subscribe to single or multiple topics at the same time. In general, as the publishers and subscribers are loosely coupled, they are not aware of each other's existence. The idea is to decouple the production of knowledge from the consumption of knowledge to ensure the desired flexibility. A representation of the communication model of the CVP-HC, including nodes, topics and the publishing/subscribing is shown on Fig. 3.

In Fig. 3, *nodes* are represented by round shapes and *topics* by square shapes. When the platform has been configured and is in operation, the *rgb_cam node* publishes to the *rgb_cam topic*, which *variables node* or any another *node* can subscribe to. After subscribing to *rgb_cam topic*, variables created or reused by past configurations are published to the *soeks topic* (in a hierarchical structure under/*soeks/variables*). *Constraints node* and *functions node* may be launched to subscribe *soeks topic* and create (or selected already crated) constraints and functions to publish over/

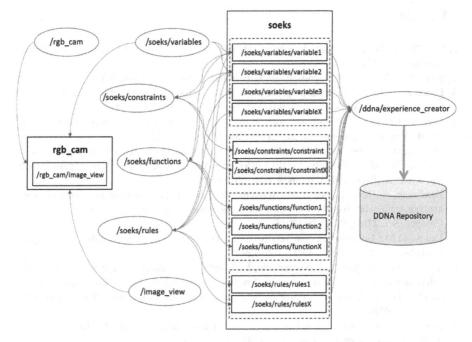

Fig. 3. Nodes, topics and publishing/subscribing representation of the communication model of CVP-HC.

soeks/constraints and/*soeks/functions topics*). Finally *rules node* may be launched to create (or reuse) the rules of operation of the system. As the system runs, the *experience_creator node* may be launched and according to the learning rate given, subscribe to the *soeks topic* to create experiences and save them in the DDNA repository. In the given example, the platform is in the first layer of integration, i.e. only the visual and context information is being manipulated by the system.

6   Conclusions

In this paper, the idea of Cognitive Systems and its functionalities have been presented. The fundamentals of Knowledge Engineering (required to structure and represent the knowledge shared in the proposed platform) has been explained and the chosen knowledge representation used is shown. In addition, the Communication System chosen to pass, store and share knowledge is presented, aiming to achieve a flexible, adaptable and general purpose system; and some details of the SOEKS and DDNA implementation over ROS framework is provided. At this point the implementation of the communication model has been tested over the creation of a few variables, constraints, functions, and the platform is able to recognize an unsafe situation for simple given rules on the first layer of integration. Following steps, second or third layer of integration will be examined where more complex scenarios can be analyzed, and results will be presented. Suitability of reusing experiences will also be explored.

References

1. Safe Work Australia: Australian Work Health and Safety Strategy 2012–2022. Creative Commons (2012)
2. Safetycare Australia: Recognition, evaluation and control of hazards, VIC (2015)
3. Lund, J., Aarø, L.E.: Accident prevention. Presentation of a model placing emphasis on human, structural and cultural factors. Saf. Sci. **42**(4), 271–324 (2004)
4. Han, S., Lee, S.: A vision-based motion capture and recognition framework for behavior-based safety management. Autom. Constr. **35**, 131–141 (2013)
5. Mosberger, R., Andreasson, H., Lilienthal, A.J.: Multi-human tracking using high-visibility clothing for industrial safety. In: IEEE/RSJ International Conference on Intelligent Robots and Systems (IROS 2013), pp. 638–644 (2013)
6. Zambrano, A., Toro, C., Nieto, M., Sotaquirá, R., Sanín, C., Szczerbicki, E.: Video semantic analysis framework based on run-time production rules – towards cognitive vision. J. Univ. Comput. Sci. **21**(6), 856–870 (2015)
7. DeJoy, D.M.: Behavior change versus culture change: divergent approaches to managing workplace safety. Saf. Sci. **43**(2), 105–129 (2005)
8. de Oliveira, C.S., Sanin, C., Szczerbicki, E.: Cognition and decisional experience to support safety management in workplaces. In: Świątek, J., Borzemski, L., Wilimowska, Z. (eds.) ISAT 2018. AISC, vol. 853, pp. 266–275. Springer, Cham (2019). https://doi.org/10.1007/978-3-319-99996-8_24
9. Sanin, C., Szczerbicki, E.: Set of experience: a knowledge structure for formal decision events. Found. Control Manage. Sci. **3**, 95–113 (2005)
10. Sanin, C., Szczerbicki, E.: Decisional DNA and the smart knowledge management system: a process of transforming information into knowledge. In: Gunasekaran, A. (ed.) Techniques and Tool for the Design and Implementation of Enterprise Information Systems, pp. 149–175. IGI Global, New York (2008)
11. Quigley, M., et al.: ROS: an open-source Robot Operating System. In: ICRA Workshop on Open Source Software, vol. 3, no. 3.2, p. 5, May 2009
12. Banavar, G., Chandra, T., Mukherjee, B., Nagarajarao, J., Strom, R. E., Sturman, D.C.: An efficient multicast protocol for content-based publish-subscribe systems. In: Proceedings of the 19th IEEE International Conference on Distributed Computing Systems, pp. 262–272. IEEE (1999)
13. Vernon, D.: The space of cognitive vision. In: Christensen, H.I., Nagel, H.-H. (eds.) Cognitive Vision Systems. LNCS, vol. 3948, pp. 7–24. Springer, Heidelberg (2006). https://doi.org/10.1007/11414353_2
14. Brézillon, P., Pomerol, J.C.: Contextual knowledge and proceduralized context. In: Proceedings of the AAAI-99 Workshop on Modeling Context in AI Applications, Orlando, Florida, USA, July. AAAI Technical Report (1999)
15. Berthe, D., Dowson, D., Godet, M., Taylor, C.M. (eds.): Tribological Design of Machine Elements, vol. 14. Elsevier, Amsterdam (1989)
16. Abadi, M., et al.: Tensorflow: a system for large-scale machine learning. In: OSDI, vol. 16, pp. 265–283, November 2016
17. M8: A Proven LiDAR Powerhouse. 2016 Quanergy Systems, Inc. (2016). https://quanergy.com/m8/
18. Moore, R.C.: A formal theory of knowledge and action. In: Allen, J.F., Hendler, J., Tate, A. (eds.) Readings in Planning, pp. 480–519. Morgan Kaufmann Publishers, San Mateo (1990)

19. Morgenstern, L.: A first-order theory of planning, knowledge, and action. In: Halpern, J.Y. (ed.) Proceedings of the 1986 Conference on Theoretical Aspects of Reasoning About Knowledge, pp. 99–114. Morgan Kaufmann Publishers, San Mateo (1986)

20. Morgenstern, L.: Knowledge preconditions for actions and plans. In: Proceedings of the Tenth International Joint Conference on Artificial Intelligence (IJCAI 1987), Milan, Italy, pp. 867–874 (1987)

21. Sanin, C., Szczerbicki, E.: Experience-based knowledge representation: SOEKS. Cybern. Syst. Int. J. **40**(2), 99–122 (2009)

22. Blakeslee, S.: Lost on Earth: Wealth of Data Found in Space. New York Times, 20 March 1990. C1

23. Corti, L., Backhouse, G.: Acquiring qualitative data for secondary analysis. In: Forum: Qualitative Social Research, vol. 6, no. 2 (2005)

24. Humphrey, C.: Preserving research data: a time for action. In: Proceedings of the Preservation of Electronic Records: New Knowledge and Decision-making: Postprints of a Conference – Symposium 2003, pp. 83–89. Canadian Conservation Institute, Ottawa (2004)

25. Johnson, P.: Who you gonna call? Technicalities **10**(4), 6–8 (1990)

26. Sanin, C., et al.: Decisional DNA: a multi-technology shareable knowledge structure for decisional experience. Neurocomputing **88**, 42–53 (2012)

27. ROS Core Components (2016). http://www.ros.org/core-components/

28. Birman, K.P.: The process group approach to reliable distributed computing. Commun. ACM **36**(12), 36–53 (1993)

29. Mishra, S., Peterson, L.L., Schlichting, R.D.: Consul: a communication substrate for fault-tolerant distributed programs. Department of Computer Science, The University of Arizona, TR 91-32, November 1991

30. Oki, B., Pfluegl, M., Siegel, A., Skeen, D.: The information bus - an architecture for extensible distributed systems. In: Operating Systems Review, vol. 27, no. 5, pp. 58–68, December 1993

31. Powell, D. (Guest ed.): Group communication. Commun. ACM, vol. 39, no. 4, April 1996

32. Skeen, D.: Vitria's Publish-Subscribe Architecture: Publish-Subscribe Overview (1998). http://www.vitria.com/

33. ROS Concepts. AaronMR (2014). http://wiki.ros.org/ROS/Concepts

34. Brézillon, P., Pomerol, J.C.: Contextual knowledge and proceduralized context. In: Proceedings of the AAAI-99 Workshop on Modeling Context in AI Applications, Orlando, Florida, USA, July. AAAI Technical Report (1999)

Assessing Individuals Learning's Impairments from a Social Entropic Perspective

José Neves[1(✉)] , Filipa Ferraz[2] , Almeida Dias[3] ,
António Capita[3,4] , Liliana Ávidos[3] , Nuno Maia[2] ,
Joana Machado[5] , Victor Alves[1] , Jorge Ribeiro[6] ,
and Henrique Vicente[1,7]

[1] Centro Algoritmi, Universidade do Minho, Braga, Portugal
{jneves,valves}@di.uminho.pt
[2] Departamento de Informática, Escola de Engenharia, Universidade do Minho,
Braga, Portugal
filipatferraz@gmail.com, nuno.maia@mundiservicos.pt
[3] CESPU, Instituto Universitário de Ciências da Saúde, Gandra, Portugal
a.almeida.dias@gmail.com,
antoniojorgecapita@gmail.com,
liliana.avidos@ipsn.cespu.pt
[4] Instituto Superior Técnico Militar, Luanda, Angola
[5] Farmácia de Lamações, Braga, Portugal
joana.mmachado@gmail.com
[6] Escola Superior de Tecnologia e Gestão, ARC4DigiT – Applied Research
Center for Digital Transformation, Instituto Politécnico de Viana do Castelo,
Viana do Castelo, Portugal
jribeiro@estg.ipvc.pt
[7] Departamento de Química, Escola de Ciências e Tecnologia,
Centro de Química de Évora, Universidade de Évora, Évora, Portugal
hvicente@uevora.pt

Abstract. Individuals with *Learning Impairments* (*LI*) may have not only language problems as reading, spelling and writing, but also difficulties in terms of their relationship with the society, i.e., may have glitches not only with language but also with social handiness. On the other hand, the dimension that most contributes to the process of acceptance or integration of an individual in a learning environment or in the society in general, is *Emotional Intelligence*, i.e., emotion is the foundation for creativity, passion, optimism, drive, and transformation. Motivation is a synonym for enthusiasm, initiative, and persistence. The technical skills have passed to the bottom, since they denote the rational that sustains an individual's involvement in such processes. Indeed, among other things, this is the reason why this work focuses on *LI* and its various manifestations and how it may affect, evaluate and treat the natural development of a human being and the environment (society) in which him/her is immersed.

Keywords: Learning Impairment · Emotional Intelligence ·
Learning Impairment Entropic based Social Machine · Logic Programming ·
Knowledge Representation and Argumentation · Artificial Neural Networks

© Springer Nature Switzerland AG 2019
N. T. Nguyen et al. (Eds.): ACIIDS 2019, LNAI 11431, pp. 62–73, 2019.
https://doi.org/10.1007/978-3-030-14799-0_6

1 Introduction

Social Entropy (SE) may be understood as the force that not only strives to dissolve the higher or medium institutions of any society, namely by reducing the individuals learning impairments to its component forms, i.e., it is not like a force that only appears when the law ceases to work. Rather, it is humans violating the social norms of predictability in terms of how society functions and affecting in some way the economic predictability of society (i.e., causing damage which costs resources and time to repair). Being left unchecked, it may act as a destructive force that may cause lawlessness, riots, unpredictability, and a social concern. Our societies are dependent on what people are learning and following, either looking at the norms in terms of getting up, getting to work, getting to learn, or not stealing or vandalizing property or hurting others. Indeed, while the underlying cause of *Learning Impairments (LI)* remains a mystery, experts agree that it is neurologically-based, meaning that it results from differences in the way the brain is wired and processes information. Although *LI* cannot be outgrown, it can be compensated and remediated. The better one understands the nature of individuals' learning difficulties, the greater the chances of helping them achieve scientific and social success. Thus, considering previous studies on this subject [1, 2], a computational tool was engendered as an *Artificial Neural Network (ANN)* [3] that will stand for a *Learning Impairment Entropic based Social Machine (LIESM)*, which will be in the front line to help individuals to be seen as full elements of the society, i.e., *LIESM* is concerned with the scientific study, design, analysis, and applications of computational tools that learn concepts, envisage models and behaviors. One's model of Emotional Intelligence is a process framework for accessing and using emotions effectively. Unlike other theoretical models, this is an actionable, practical, simple process that facilitates performance. Last but not least, a brief description of one's approach to *Knowledge Representation and Argumentation (KRA)* is set out in the next section, followed by a case focusing on the screening of the individuals learning's impairments, its implications on a proper sociality, and the *LIESM* outlook. Finally, conclusions are drawn and directions for future work are outlined.

2 Fundamentals

2.1 Knowledge Representation and Argumentation

We would argue that most societies are characterized by *Social Entropy (SE)*, i.e., their context is introduced as the entropic potential of the energy that is transferred to them. It considers the overall transfer from the energy in its initial and finite states, starting as primary energy, and ending when it becomes part of the internal energy of the ambient. Therefore, a function or predicate term will stand for a set of different elements, namely its *Entropic State Range (ESR)*, *ESR's Quality-of-Information (QoI_{ESR})* and *Degree-of-Confidence (DoC_{ESR})*, and the *Vagueness Entropic Range (VER)*, *VER's Quality-of-Information (QoI_{VER})* and *Degree-of-Confidence (DoC_{VER})*. *VER* sets a measure to *Predicate's Vagueness*, i.e., it is used in the sense that it is not known which are the terms that make a predicate extension, it may or not may consider information in the form, viz.

$$exception_{p1},\ \cdots\ ,exception_{pj}\ (0\le j\le k),\ being\ k\ an\ interger\ number$$

as it is shown below. The predicates' extensions that induce the discussion's cosmos are now given in the form [4], viz.

$$
\begin{aligned}
&\{\\
&\quad \neg p \leftarrow not\,p, not\,exception_p\\
&\quad p \leftarrow p_1,\cdots,p_n,\,not\,q_1,\cdots,\,not\,q_m\\
&\quad ?(p_1,\cdots,p_n,\,not\,q_1,\cdots,\,not\,q_m)(n,m\ge 0)\\
&\quad exception_{p_1},\cdots,exception_{p_j}(0\le j\le k), being\ k\ an\ integer\ number\\
&\}::entropic\ state'score
\end{aligned}
$$

where the p_s and q_s are makings of the kind, viz.

$$
\bigwedge_{1\le i\le n} predicate_i - clause_j\Big(\Big(ESR_{x_1}, QoI_{ESR_{x_1}}, DoC_{ESR_{x_1}}, VER_{x_1}, QoI_{VER_{x_1}},
$$
$$
DoC_{VER_{x_1}}\Big),\cdots,\Big(ESR_{x_m}, QoI_{ESR_{x_m}}, DoC_{ESR_{x_m}}, VER_{x_m}, QoI_{VER_{x_m}}, DoC_{VER_{x_m}}\Big)\Big)
$$
$$
:: ESR_{ESR_j} :: QoI_{ESR_j} :: DoC_{ESR_j} :: VER_{VER_j} :: QoI_{VER_j} :: DoC_{VER_j}
$$

n, m, and Λ symbolize the number of $predicate_i$'s terms, its arguments' cardinality, and logical juxtaposition, respectively. Items like ESR_j, QoI_{ESR_j}, DoC_{ESR_j}, VER_j, QoI_{VER_j} and DoC_{VER_j} allow for the understanding of how a complex whole (i.e., in this case that of a data item) is broken down into its parts or elements [5].

2.2   Evaluating a Qualitative Data Item in Terms of Its Quantitative Complement

As a sample, let us consider 3 (three) questions on a particular topic (here fixed as *Issue's Data*), with a choice limited to 3 (three) alternatives, namely *low · medium · high*, which are divided into 3 (three) segments that set a circle of area 1 (one) (Fig. 1). The *ESR* stands for a measure of the unavailable energy in a closed thermodynamic system, i.e., a process of degradation, running down or a trend to disorder, and is given by dark colored areas in Fig. 1. Areas with a gray color indicate a relaxation in the central values, i.e., the *VER* and corresponding energy values may or may not have been spent.

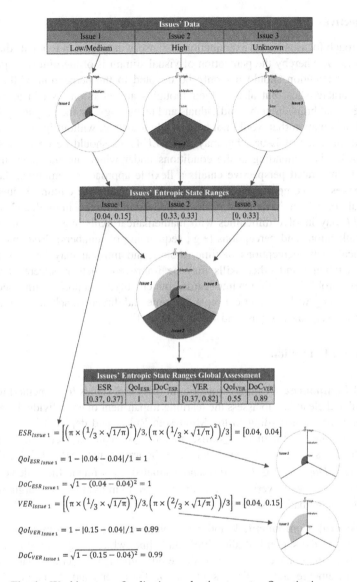

Fig. 1. Working on a *Qualitative* evaluation to get a *Quantitative* one.

3   Case Study

The case study was selected to reflect a situation where making eligibility determinations are challenging. It progresses through the *Specific Learning Disabilities (SLD)* identification process and likewise correlates to the contents of the abstract and introduction section, plus the content of the section that follows, providing the context for the activities and procedures that will occur next [6].

3.1 Objectives

One's approach focuses on the examination of sensory data from individuals diagnosed with specific *LI*, whereby the perception of visual stimuli is of the utmost importance, i.e., a special attention should not only be devoted to the perception of the brain's physical characteristics, but also to their thoughts and emotions, set in terms of an environment that brings together individuals and technology, interacting and producing deeds and information that would not be possible to extract without interaction among all. Indeed, in terms of issues that may affect *LI*, focus should be on economic and social policies, i.e., attending at the conditions under which an individual lives and learns. This integrated perspective entails a flexible approach to emotional behavior, i.e., the expressive comportment, the action tendencies and the cultural influence and feelings, all of which have their place in understanding the individuals' sensitive changes. *LI* may involve difficulties with mathematical skills (e.g., addition, subtraction, multiplication) and perceptions (e.g., sequencing of numbers). Remembrance of mathematical facts, perceptions of time, money and musical may be also induced. However, language and other skills may be improved, such as severe positioning capabilities, problems with reading cards, punctuality, grappling with mechanical processes, dealing with abstract concepts like time and direction, schedules, the course of time or the sequence of past and future events (Fig. 2).

3.2 Feature Extraction

Technical Component. Feature extraction, in this case, stands for a method to obtain meaningful and clear data to assess the learning impairment of an individual, and based on a criteria list set according to the Minnesota Rules, and depicted in terms of the relations/predicates mathematical_skills, memory_of_material_facts and language, whose extensions are given below in terms of the arguments ESR_j, QoI_{ESR_j}, DoC_{ESR_j}, VER_j, QoI_{VER_j}, DoC_{VER_j} ($j \in 1 \ldots 100$) and evaluated according to the scale very easy · easy · not so easy · hard · very hard, and exemplified to individual #1 to argument time of predicate mathematical_skills (Fig. 3).

The Behavioral Component. Emotional development action, communication and understanding is the support of adult life's guideline, which progress according to the peculiarities of the context. In this unstable environment, it is certain that a series of conflicts will emerge and require a solution. On the other hand, social emotions and feelings are the regulators of the process of involvement and participation of people in the face of organizational challenges. As Damasio shapes, our lives must be regulated not only by our desires and feelings, but also by our concern for the desires and feelings of others [7]. It is this apprehension to reach a point of balance that leads us to consider the dimension of *Affections*, i.e., *Emotions* and *Feelings*, which are here set and evaluated in terms of *Envy* and *Indignation* (Fig. 4).

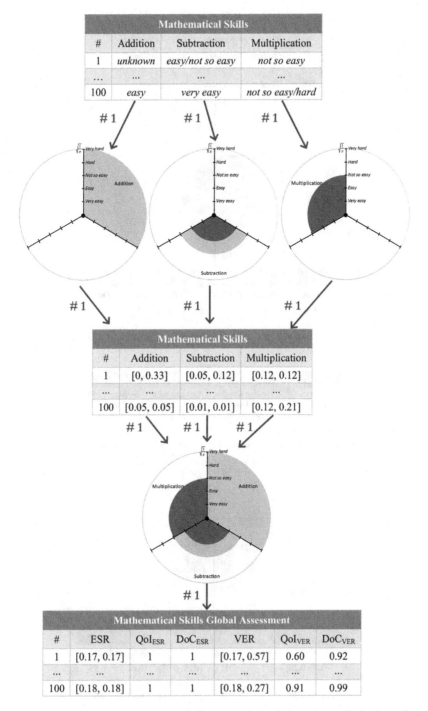

Fig. 2. A knowledge-based fragment of the extension of the relational database for the *Individuals Learning Impairment Assessment.*

Memory of Material Facts			
#	Time	Money	Musical Concepts
1	*not so easy*	*unknown*	*not so easy/hard*
...	...	...	...
100	*hard*	*very hard*	*unknown*

#1   #1   #1

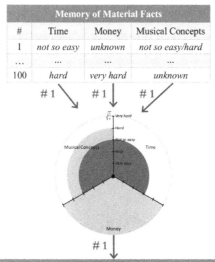

#1

Memory of Material Facts Global Assessment					
#	ESR	QoI$_{ESR}$	DoC$_{ESR}$	VER	QoI$_{VER}$	DoC$_{VER}$
1	[0.24, 0.24]	1	1	[0.24, 0.67]	0.57	0.90
...	...	...	...	...	...	...
100	[0.55, 0.55]	1	1	[0.55, 0.88]	0.67	0.94

Language			
#	Sense of Direction	Trouble Reading Maps	Dealing with Schedules	Sequence of Past and Future Events
1	*easy/not so easy*	*not so easy/hard*	*hard*	*unknown*
...	...	...	...	...
100	*not so easy/hard*	*easy*	*very easy*	*very hard*

#1   #1   #1   #1

#1

Language Global Assessment					
#	ESR	QoI$_{ESR}$	DoC$_{ESR}$	VER	QoI$_{VER}$	DoC$_{VER}$
1	[0.29, 0.29]	1	1	[0.29, 0.67]	0.62	0.95
...	...	...	...	...	...	...
100	[0.39, 0.39]	1	1	[0.39, 0.46]	0.93	0.99

Fig. 2. (*continued*)

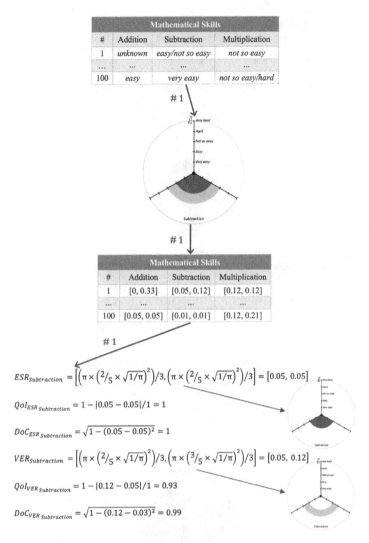

Fig. 3. Evaluation of ESR_j, QoI_{ESR}, DoC_{ESR}, VER_j, QoI_{VER}, DoC_{VER} parameters to *individual #1* with respect to the argument *Subtraction* of the relation *Mathematical Skills*.

4 Computational Model

The data from Figs. 2 and 4 is used to train the *LIESM* (the data set's cardinality is 100 (one hundred)), with 25% of them being used to assess its performance. The *LIESM* is now defined as an *ANN*, the topology of which is given in Fig. 5 [3]. We show no comparison with other work to display the contrast effectiveness of our approach once we are spatially limited, but not only this issue will be considered in the upcoming work, but it is also our purpose to expand the experimental sample.

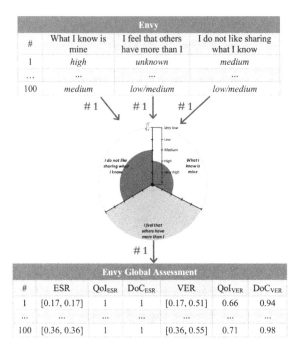

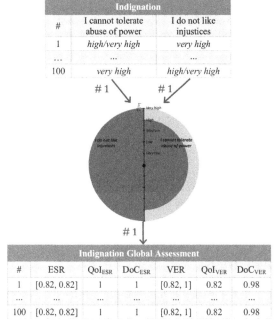

Fig. 4. A knowledge-based fragment of the extension of the relational database that set the *Behavioral Component*.

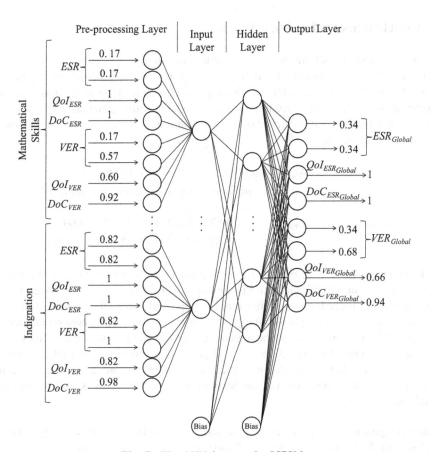

Fig. 5. The *ANN* that sets the *LIESM*.

In the preprocessing layer it was used as an activation function the linear one (i.e., the activation is proportional to the input), and in the remaining layers the sigmoid (i.e., the output of the activation function will be in the range [0, 1]). The *confusion matrix* with respect to the *ANN* presented in Fig. 5 is shown in Table 1. A reading from Table 1 shows that the *ANN* model correctly classifies between 88 to 94 cases from a total of 100 cases with an accuracy in the range [0.88, 0.94], suggesting that the performance of the model is up to standard.

Table 1. The *Confusion Matrix* with respect to the proposed model.

Output	Model output	
	True (1)	False (0)
True (1)	[70, 72]	[4, 7]
False (0)	[2, 5]	[18, 22]

5   Conclusions and Future Work

A syntheses and characterization of a feasible methodology for problem solving that allows for the classification of an individual's *LI* set as a *Learning Impairment Entropic based Social Machine* (*LIESM*), which may also be understood as a *Many-Valued Empirical Social Machine*, once it returns an individual learning impairment assessing in terms of a *Truth Value Valuation* of a person entropic state, i.e., given an individual record in the form, viz.

$$I_{individual}L_{earning}I_{mpairment}A_{ssessing}(Mathematical\ Skills, Memory\ of$$
$$Material\ Facts, Language, Envy, Indignition) \cong$$

$$I_{individual}L_{earning}I_{mpairment}A_{ssessing}([0.17, 0.17], 1, 1, [0.17, 0.57], 0.60,$$
$$0.92, \cdots, [0.82, 0.82], 1, 1, [0.82, 1], 0.82, 0.98,)$$

as input to the *LIESM*, it returns an assignment of truth values to *ESR*, *VER* and respective *QoIs* and *DoCs* that unfolds the individual's *LI* entropic state under a predefined context. Indeed, one's approach not only proved successful in such a task, but also explains workings on *Qualitative and Social Data*. Considering how social factors may put an individual at risk, we have gone further and yield results on data from a set of immaterial variables that glimpse social perception. Focusing on such attributes, which may be indicative of dissimilarities in the cognition arena, we were able to quantify a degree of disorder, i.e., the level of entropy. Future work will follow this trend [8].

Acknowledgments. This work has been supported by COMPETE: POCI-01-0145-FEDER-007043 and FCT – Fundação para a Ciência e Tecnologia within the Project Scope: UID/CEC/00319/2013.

References

1. Ferraz, F., Neves, J.: A brief look into dyscalculia and supportive tools. In: Proceedings of the 5th IEEE International Conference on E-Health and Bioengineering (EHB 2015), pp. 1–4. IEEE Edition, Los Alamitos (2015)
2. Ferraz, F., Neves, J., Alves, V., Vicente, H.: Dyscalculia: a behavioural vision. In: Ntalianis, K., Croitoru, A. (eds.) APSAC 2017. LNEE, vol. 489, pp. 199–206. Springer, Cham (2019). https://doi.org/10.1007/978-3-319-75605-9_28
3. Fernández-Delgado, M., Cernadas, E., Barro, S., Ribeiro, J., Neves, J.: Direct Kernel Perceptron (DKP): ultra-fast kernel ELM-based classification with non-iterative closed-form weight calculation. J. Neural Netw. **50**, 60–71 (2014)
4. Neves, J.: A logic interpreter to handle time and negation in logic databases. In: Muller, R., Pottmyer, J. (eds.) Proceedings of the 1984 Annual Conference of the ACM on the 5th Generation Challenge, pp. 50–54. ACM, New York (1984)

5. Neves, J., et al.: Predicative vagueness in lung metastases in soft tissue sarcoma screening. In: Groza, A., Prasath, R. (eds.) MIKE 2018. LNCS (LNAI), vol. 11308, pp. 80–89. Springer, Cham (2018). https://doi.org/10.1007/978-3-030-05918-7_8
6. Weinberg, V.: Determining the Eligibility of Students with Specific Learning Disabilities – A Technical Manual. Minnesota Department of Education, Minnesota (2006)
7. Damásio, A.: Looking for Spinoza: Joy, Sorrow and the Feeling Brain. A Harvest Book Harcourt Inc., London (2003)
8. Fernandes, B., Vicente, H., Ribeiro, J., Analide, C., Neves, J.: Evolutionary computation on road safety. In: Cos Juez, F., et al. (eds.) Hybrid Artificial Intelligent Systems. Lecture Notes in Artificial Intelligence, vol. 10870, pp. 647–657. Springer International Publishing, Cham (2018). https://doi.org/10.1007/978-3-319-92639-1_54

Text Processing and Information Retrieval

Robust Web Data Extraction Based on Unsupervised Visual Validation

Benoit Potvin[(✉)] and Roger Villemaire[(✉)]

Department of Computer Science, Université du Québec à Montréal,
Montréal H3C 3P8, Canada
benoit.potvin@protonmail.com, villemaire.roger@uqam.ca

Abstract. Visual validation is the process of validating sets of extracted entities by means of visual information. The main advantage of visual validation is to make use of visual information for web information extraction without impacting on the robustness of extractors. In this paper, we show that unsupervised visual validation can be used to create robust web data extractors. More precisely, we evaluate the performance of visual validation on a corpus of visually heterogeneous documents. The selected extraction task consists in extracting the price, name, description, and SKU of unspecified products from unseen documents. Our corpus contains 1000 various products from 100 different sources, which we render public. Results also show that visual validation improves web data extraction even when the extractor is trained with visual features.

Keywords: Visual validation · Robustness · Isolation forest ·
Web information extraction · Classifiers

1 Introduction

Information Extraction (IE) is the task of extracting structured information from unstructured documents that were intended for human usage. IE is hence paramount to leverage the vast amount of information available in digital form. This is particularly true in the context of the *World Wide Web* (WWW), where an ever growing number of documents—mostly web pages—offer a broad spectrum of information whose ubiquitous availability defines the Information Age. Extracting information from web documents, known as *Web Information Extraction* (WIE), is also central to bridging the gap between the vastly unstructured web mostly experienced today, and the objective of a truly *Semantic Web* [2] that would allow much easier knowledge extraction.

The WWW is also nowadays the preferred medium for businesses to advertise services and goods, allowing would-be customers to compare products and prices. Business web pages are therefore designed to appeal to customers, emphasizing for instance, availability of new products or special discounts and pricing. As appealing as the page layout and presentation can be from a marketing point-of-view, this usually makes it even more difficult to devise methods that

© Springer Nature Switzerland AG 2019
N. T. Nguyen et al. (Eds.): ACIIDS 2019, LNAI 11431, pp. 77–89, 2019.
https://doi.org/10.1007/978-3-030-14799-0_7

automatically extract factual information such as the product name, sale price, description, and the Stock Keeping Unit (SKU). Information extraction methods must hence do their best to leverage the web page's structure, with its elements, their position and the way they are rendered (e.g. fonts and colors), in order to extract meaningful data values.

One can distinguish two broad approaches to web data extraction [7]. One can first leverage the semi-structured nature of HTML documents, for instance the Document Object Model (DOM) tree. Alternatively one can use Machine Learning (ML) methods, and train a classifier to predict whether an element belongs to a class of interest. In the former case, DOM-based (or tree-based) methods offer excellent extraction performance on previously seen templates but cannot extract information from novel documents [21]. In the latter case, ML methods usually require large manually labelled training sets, whose production is time-consuming and error-prone. Furthermore, the selection of discriminant features is a laborious task. In practice, ML-based methods are often website-dependent as generalizing websites absent from the training set usually sharply degrades performance [7,23]. Therefore extraction on novel documents, regardless of the selected approach, is highly challenging.

In this paper, we show that one can create a robust extractor for product information extraction by combining supervised classification with unsupervised methods. We consider the task of extracting a specific set of data types, i.e. the name, sale price, description, and SKU of unspecified products for sale from unseen web pages. In other words, we aim at creating a robust product information extractor that can extract a specific set of information for any product from any web page.

In fact, we introduce a general method to extend the range of supervised classification and create robust extractors. While a classifier's performance will sharply degrade as one move further away from the types of entities present in the training set, we show that Visual Validation, an unsupervised learning method, can restore classification performance. In the context of WIE, this means that one can rely on *flexible* features in order to extract information from as many documents as possible. The opposite scenario would be to use very specific features in order to obtain outstanding results on a limited set of documents or a specific task but extremely poor results on general tasks and large corpora. Visual Validation therefore enables to create extremely robust extractors with minimal effort in features selection and training set annotation, while relying on unsupervised methods to assure generalization to unseen entities. The proposed method uses visual information of web pages without being conducive to layout dependencies, which for instance is not the case when training classifiers with visual features. We will show that Visual Validation improves extraction results even if the original classifier has been trained with visual features.

2   Related Work

Web data extraction relies on information patterns (textual, structural or visual) in order to extract relevant entities [5]. The success of extraction therefore depends on the presence of identified patterns in analyzed documents [17,23]. In the literature, we often encounter the terms "language-dependent" [22], "template-dependent" [21], or "layout-dependent" [11] to describe methods that respectively use textual, structural or visual patterns.

A major challenge for WIE is to create robust extractors that lead to good performance. Robustness is the degree of insensitivity of an extractor to web page modifications, including changes in the syntax, template, formatting and layout [16]. In other words, it is the ability of an extractor to rely on discriminant patterns that are unlikely to change, whether a document is modified or extraction is performed on novel documents.

When designing an extractor, one will have to find the right balance between the performance and the robustness of the extractor. For instance, one can easily create an extractor that will obtain perfect results on few documents but also extremely poor results on novel documents. We refer to this tradeoff as the performance-robustness tradeoff [17].

Robustness has been extensively studied in the literature with a special focus on DOM-based methods [7]. These methods rely on structural patterns, i.e. regularities across the DOM tree (template-dependent). As a consequence, DOM-based methods lead to excellent performance at the expense of robustness [21]. Several techniques have been suggested in order to automatically adapt the relevant regularities into the new DOM tree [6]. This however only concerns the maintenance of extractors and do not make these methods applicable to unseen documents.

Visual information plays an important role in the definition of web documents. The use of discriminant visual regularities for WIE can improve extraction results [1,4,9,14]. However, these improvements are often obtained at the expense of robustness [19]. In fact, visual regularities are rarely expected to be consistent across all web documents. Therefore, visual information is either exploited in sets of documents that have visual similarities [1], or in specific tasks that rely on an object with similar visual cues across all documents, such as a table [8], a text block [12], etc. In the first case, the extraction process relies on a set of visual regularities and robustness is therefore impacted. In the second case, most visual information is ignored.

Visual Validation, an unsupervised learning method that rely on the visual information of extracted entities, has recently been used to improve information extraction without impacting on the robustness of extractors [19]. The method has only been used on documents with visual similarities. In this paper, we evaluate the performance of Visual Validation on a corpus of heterogeneous documents and show the positive impact of the method for robust extraction. This is, to our knowledge, the first example of a method where visual information is broadly used for robust web information extraction.

3   Background

3.1   Web Data Extraction as a Classification Problem

In Supervised ML a dataset of entities labelled by their correct class is used to devise a *classifier* in order to predict the class of previously unseen entities. A web data extraction task can be seen as a classification problem where a classifier is trained to predict the classes of entities that one wants to extract [20]. In this setting, entities can be tokens that are generated from the textual content (common in NLP tasks), DOM nodes or any other relevant entity. The DOM is a W3C recommendation that allows programs and scripts to dynamically access and update the content, structure, and style of documents. Since it is the most common approach to manipulate web documents, we will solely consider entities that are DOM nodes. Training is hence done on a dataset of DOM nodes labeled with their correct class and the objective is to extract relevant nodes that belong to the classes of interest in each document.

XPath is the W3C recommended and preferred tool to address nodes of the DOM and has been largely used in WIE systems and web annotation tools. Each node of a document can be located through its XPath expression and its textual content, style properties, and position can then be retrieved (e.g. with a headless browser). Based on a set of discriminant features, a vector of values describes each node. The classifier will hence map each vector to its corresponding class. The training set is formed of labeled web documents, which is a set of nodes' vectors labeled with the class to which the node belongs.

In order to predict multiple classes one can either use a specific (binary) classifier for each class or a single (multinomial) classifier predicting one of many classes. For instance, in order to predict product name, sale price, description and SKU one can have four classifiers, one for each class. On the other hand, a multinomial classifier takes an entity and returns the class, which would be in this case either "product name", "sale price", "description", "SKU" or "None of these classes".

A web document can contain several thousand DOM nodes. Extraction is done by having the classifier process each node in order to extract those that are relevant. This method is usually used in settings where one intends to extract many nodes. For instance, a content extractor should identify all nodes containing a part of the main content, and the objective is to extract as many nodes of the main content as possible. In our case, we extract exactly four data values, hence four nodes, from each page. We therefore simply rely on the classifier's underlying real-valued output [24] and select the node with the highest output for each of the four classes "product name", "sale price", "description" and "SKU".

3.2   Visual Validation

Visual Validation (VV) is the process of validating sets of extracted entities by means of visual information [19]. It aims at identifying false positive entities from those returned by a web data extractor. VV consists in three steps. First, visual information of extracted entities is obtained. A common approach is to use XPath expressions to locate each entity and obtain its visual characteristics. Secondly, visual information is used to identify visual outliers in the set of extracted entities. Finally, visual outliers are discarded.

VV assumes that visual outliers in a set of extracted entities are false positives. A web data extractor is designed to extract a specific type of information and resulting sets therefore contain similar entities, at the exception of false positives. In visually-rich documents such as web pages and PDFs, it is common that similar entities have visual similarities. Discriminant visual patterns are often used to identify relevant entities and facilitate web data extraction, in the same manner that visual information helps human users to navigate across a web page and understand its content [4,7,9,14]. VV however uses visual information in a specific way. VV does not aim at identifying visual patterns across true positive entities but only assumes that such patterns exist in order to state that visual outliers, i.e. entities that visually differ from the norm, are false positives.

Visual outliers are furthermore point anomalies, i.e. single anomalous instances that differ from the norm. Point anomalies have been extensively studied in the domain of anomaly detection. In order to determine point anomalies, one usually uses unsupervised anomaly detection methods.

Isolation Forest (iForest) [15] is an unsupervised anomaly detection algorithm that makes use of the concept of "isolation", i.e. the process of separating an instance from all other instances. The rationale behind iForest is that anomalies are rare items that are different, and therefore they are more susceptible to isolation than normal instances. iForest consists of two steps. In the first step, random decision trees (called iTrees) are constructed by recursively partitioning sub-samples of the given dataset until instances are isolated or a specific tree height has been reached. In the evaluation step, all instances of the dataset are passed through each iTrees, and anomaly scores are computed based on the average path length. Anomalies are instances with shorter path lengths (as they are easier to isolate), and the average tree height is sufficient for anomaly detection. This enables iForest to use sub-sampling to an extent that is not feasible with other methods. iForest achieves high detection performance with small sub-sampling sizes. iForest can hence process large datasets in linear time with low memory requirements. iForest is therefore particularly well suited for web data extraction tasks.

4 Methodology

Our method goes as follows Fig. 1. We first train a classifier to extract product information from unseen web pages. Secondly, processing the whole dataset, node by node, we select for each page the nodes with the highest output for each of the four classes "product name", "sale price", "description" and "SKU". Finally, we use VV to filter out visual outliers from the set of all selected nodes.

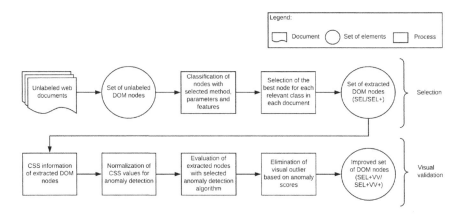

Fig. 1. The successive steps of the proposed method

4.1 Selection

In the second step of our approach, a name, price, description, and SKU are selected for each page. Note that at this step precision, recall, and F_1 score are the same, as precision and recall are equal. This is the case as four nodes are selected for each document (true and false positive), while there is exactly the same number of relevant elements (true positives and false negatives).

4.2 Visual Validation

Note however that some nodes selected in the second step will be removed in the last step, leaving some pages with missing data values. This is consistent with the fact that these missing values have been removed by VV since they are outliers, and we hence expect them to be erroneous.

Although this is not the case in this paper's experiments, data values are often absent of web pages. For example, not all product web pages have a SKU. Contrary to DOM-based methods, a classifier will fill all data slots, which can lead to a considerable amount of false positive entities. VV could help to mitigate this problem by eliminating false positive entities.

Finally, note that VV filters out elements selected in the second step of our method. The number of retrieved true positives can hence only decrease, and recall cannot improve. In our experiments, we will hence investigate the possibility of increasing precision and F_1 score. Since recall cannot improve, increasing F_1 score means that the increase in precision compensated for the decrease in recall.

5   Experimental Setup

We train the classifier on a set of shallow features. Shallow features have previously been used to create domain- and language-independent classifiers [13]. The selected shallow features for this task are: (1) the string length, (2) the number of digits in the string, (3) the number of whitespace characters in the string, (4) the number of line breaks in the string, (5) the number of currency symbols in the string, (6) the number of hash symbols in the string and (7) the XPath expression depth level (i.e. the number of slashes). For simplicity, the same set of features is used for all classes.

5.1 Dataset

To our knowledge, there is no available corpus of web documents where visual information, as it is rendered on the user's screen is available i.e. that includes CSS, JavaScript and other third party files that can change the appearance of a web page. Therefore, we annotated our own corpus that one can access online.[1] The dataset contains 1000 labeled web pages from 100 e-commerce websites. For each website, ten products have been randomly selected. The corpus includes a large selection of products from several countries and in different currencies. Examples of products are: clothes, musical instruments, bicycles, car parts, food, furniture, cosmetics, art pieces, games, tools, sport goods, computers, etc. On each page the same four entities have been identified: the name, price, description and SKU of the main product for sale.

To these 4000 *positive* labeled entities, i.e. 1000 entities for each of the four classes (one per page), we add 4000 *negative* entities corresponding to the "None of these classes" class. As each page contains a unique node corresponding to the four classes of interest, any other node can randomly be chosen for the fifth class "None of these classes". The annotation set hence includes 8000 labeled entities: 4000 positives cases and 4000 negative cases.

Websites were found using Google search engine with keywords such as "on sale", "buy online", etc. All web pages are in English.

5.2 Selection

At the second step, we select the DOM nodes that have the best scores for the classifier, one for each class in each document. When multiple DOM nodes have the same high score, a node is selected randomly. Precision is computed. We will refer to this result as SEL. As explained in Sect. 4.1, precision, recall, and F_1 score are equal in this case.

5.3 Visual Validation

We use iForest for VV, developing a single anomaly model for all classes. We evaluated the case where a model is learned for each class, e.g. by evaluating price entities with a model trained only with prices. However, both approaches lead to similar results, and we therefore selected the simpler approach. For VV,

---

[1] Files can be downloaded at this link: https://drive.google.com/drive/folders/1GYU6 ZgZOXsNq4-F7o8v3qjr47DcLvK3.

we furthermore used the following visual properties: the position, font-color, font-size, and font-weight.

DOM nodes are hence sorted in ascending order according to their computed anomaly scores. For iForest a low score denotes an anomalous instance. Visual outliers are successively deleted and precision, recall, and F_1 score are computed. We will refer to these results as the SEL+VV results.

5.4 Tools and Implementation

Web pages have been annotated with Pundit Annotator [10]. Pundit can be used through a Chrome extension, and annotations are exported in the JSON-LD format. Labeled elements are located through XPointer, and their CSS characteristics and position are extracted with Headless Chrome Node API Puppeteer.

Classifiers' implementations are those from scikit-learn Python library [18], as well as iForest anomaly detection algorithm. In order to obtain values in the range [0,1] the "predict_proba" function from scikit-learn is used. If not indicated, all parameters are those by default.

6   Experimental Evaluation

Using 10-fold cross-validation on our labeled dataset, we readily identified that the most promising classifiers for the selection step are Random Forest (RF) [3] and Multi-Layer Perceptron (MLP) classifiers.[2]

In this section we first compare SEL and SEL+VV results with these two types of classifiers (for selection) and iForest for VV (Fig. 2 and Table 1). Finally, we retrain the classifiers using visual features and then compare the same values, which we will identify by SEL+ and SEL+VV+ to distinguish them from the previous results (Fig. 3 and Table 2).

6.1 Training Classifiers

In order to get fair results on the classifiers' ability to extract information from unseen documents we divide our dataset in four sets where documents from a same website are always gathered in the same set (and absent from other sets). This way, we make sure that documents from a same website will not be at the same time in the training and test datasets, which could drastically improve results. In our case where 1000 documents have been annotated, we obtain four sets of documents, each containing 250 documents from 25 websites. There are four instances of each classifier (MLP and RF), each instance being trained on three distinct sets in order to evaluate it on the fourth one containing only novel documents. Overall results for a classifier are the average of the results obtained

---

[2] Tested classifiers for this task were: a Gaussian Naive Bayes classifier, a k-nearest neighbor classifier, a multi-class SVM classifier (one-versus-one), and the two selected classifiers.

with each instance, making sure that the documents in one set are not easier to extract than documents in other sets.

In this setting, we trained seven classifiers, i.e. a random forest (RF) classifier and six multi-layer perceptron (MLP) classifiers with respectively one, three and ten hidden layers (100 neurons) with "tanh" and "ReLU" activation functions (i.e. 1 hidden layer with ReLU, 1 hidden layer with tanh, 3 hidden layers with Relu, 3 hidden layers with tanh, etc.).

Therefore, there are 28 instances of classifiers, i.e. seven classifiers trained in four instances. Each classifier predicts the classes of all DOM nodes contained in its test set.

MLP classifiers had similar performance across all versions, independently of the number of hidden layers, the number of neurons on each layer, or the selected activation function. For simplicity, we only exhibit the results obtained with one hidden layer (100 neurons) MLP using "tanh" activation function.

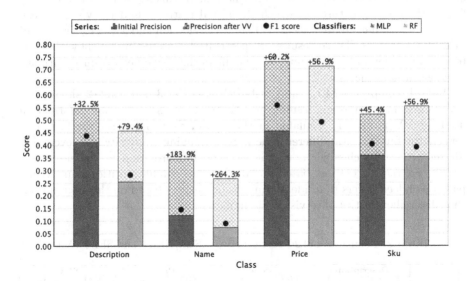

Fig. 2. SEL (solid bars) and SEL+VV (crosshatched bars) results

6.2 Comparing SEL and SEL+VV

Initial results (SEL) for both classifiers (MLP and RF) have been improved with VV (SEL+VV) as shown in Fig. 2. Solid bars denote the initial classifiers' results (SEL) while the crosshatched areas show improvements obtained with VV (SEL+VV). In the case of the SEL results, as explained in Sect. 4.1, precision, recall, and F_1 score are equal. Black dots denote the F_1 scores for SEL+VV. The number on top of each bar shows the percentage by which the precision is increased (SEL to SEL+VV). F_1 percentage increases are presented in Table 1.

Table 1. Comparison of F_1 scores between SEL and SEL+VV

	Description		Name		Price		SKU	
	MLP	RF	MLP	RF	MLP	RF	MLP	RF
SEL	0.4118	0.2543	0.1214	0.0732	0.4554	0.4133	0.3580	0.3526
SEL+VV	0.4357	0.2807	0.1429	0.0873	0.5574	0.4891	0.4017	0.3892
	(+5.81%)	(+10.38%)	(+17.67%)	(+19.23%)	(+22.34%)	(+18.34%)	(+12.19%)	(+10.39%)

We note a consistent improvement both in precision and F_1 score from SEL to SEL+VV, attesting to the improvement provided by VV.

6.3   Comparing SEL+ and SEL+VV+ and Classification to VV

The results of Sect. 6.2 were obtained with unsupervised VV on classifiers that have not been trained with visual information. In this section, we train classifiers with a set of features to which the visual features used for VV has been added, and again compare selection alone with selection and VV, i.e., SEL+ and SEL+VV+ results. These results are particularly interesting as they allow us to evaluate whether VV can improve classifiers that have been trained with the same visual information.

Indeed, comparing SEL+ and SEL+VV+ in Fig. 3 and Table 2 shows that VV consistently improves results, even if the same set of visual information used for VV is already available in the classifier's training set. In fact, this reflects the fundamental difference between leveraging *similarity* and leveraging *dissimilarity*. The classifier does not use visual information in the same way as VV. Extracting entities based on the fact that similar entities share visual patterns indeed differs from identifying visual outliers based on the assumptions that dissimilar entities have visual dissimilarities.

Table 2. Comparison of F_1 scores between SEL, SEL+ and SEL+VV+

	Description		Name		Price		SKU	
	MLP	RF	MLP	RF	MLP	RF	MLP	RF
SEL	0.4118	0.2543	0.1214	0.0732	0.4554	0.4133	0.3580	0.3526
SEL+	**0.2623**	0.3184	0.2553	0.6306	**0.3429**	0.7302	**0.1596**	0.4435
	(-36.30%)	(+25.18%)	(+110.19%)	(+761.57%)	(-24.71%)	(+76.69%)	(-55.42%)	(+25.78%)
SEL+VV+	0.2869	0.3364	0.2965	0.6660	0.3747	0.7745	0.1799	0.4661
	(+9.38%)	(+5.66%)	(+16.14%)	(+5.61%)	(+9.27%)	(+6.07%)	(+12.71%)	(+5.10%)

Comparing SEL+ to SEL in Table 2, we note a large improvement in F_1 scores for RF and a large degradation for MLP, except in the case of the name class. In fact, for RF, SEL+ always outperform SEL+VV, while for MLP this is the case only for the name class. Therefore, using more features to train the classifier can lead to large performance improvement, but this can sharply depend on the type of classifier.

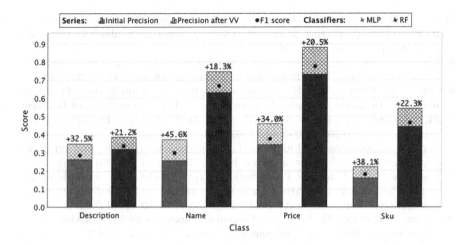

Fig. 3. SEL+ and SEL+VV+ results

7 Conclusion

In this paper, we showed that unsupervised visual validation can improve web data extraction from heterogeneous documents. First, unsupervised VV can be combined with supervised classifiers that have been trained with shallow features in order to obtain robust extractors. This strategy allows to reduce the workload that is usually required to create supervised classifiers with acceptable performance. For instance, in our experiments the same features are used for all classes, and textual information is only analyzed quantitatively. Secondly, we compared the results obtained by using the same set of visual information for VV and for training the classifier. We hence determined that VV can still improve extraction results in this setting, which shows that selection, and VV use visual information in a fundamentally different way. Finally, we showed that iForest can effectively be used to perform VV.

Acknowledgments. The authors gratefully acknowledge the financial support of the Natural Sciences and Engineering Research Council of Canada (NSERC).

References

1. Apostolova, E., Pourashraf, P., Sack, J.: Digital leafleting: extracting structured data from multimedia online flyers. In: Proceedings of the 2015 Conference of the North American Chapter of the Association for Computational Linguistics: Human Language Technologies, pp. 283–292 (2015)
2. Berners-Lee, T., Hendler, J., Lassila, O., et al.: The semantic web. Sci. Am. **284**(5), 28–37 (2001)
3. Breiman, L.: Random forests. Mach. Learn. **45**(1), 5–32 (2001)

4. Burget, R., Rudolfova, I.: Web page element classification based on visual features. In: First Asian Conference on Intelligent Information and Database Systems, ACI-IDS 2009, pp. 67–72. IEEE (2009)
5. Chang, C.H., Kayed, M., Girgis, M.R., Shaalan, K.F.: A survey of web information extraction systems. IEEE Trans. Knowl. Data Eng. **18**(10), 1411–1428 (2006)
6. Ferrara, E., Baumgartner, R.: Automatic wrapper adaptation by tree edit distance matching. In: Hatzilygeroudis, I., Prentzas, J. (eds.) Combinations of Intelligent Methods and Applications. SIST, vol. 8, pp. 41–54. Springer, Heidelberg (2011). https://doi.org/10.1007/978-3-642-19618-8_3
7. Ferrara, E., De Meo, P., Fiumara, G., Baumgartner, R.: Web data extraction, applications and techniques: a survey. Knowl.-Based Syst. **70**, 301–323 (2014)
8. Gatterbauer, W., Bohunsky, P., Herzog, M., Krüpl, B., Pollak, B.: Towards domain-independent information extraction from web tables. In: Proceedings of the 16th International Conference on World Wide Web, pp. 71–80. ACM (2007)
9. Gogar, T., Hubacek, O., Sedivy, J.: Deep neural networks for web page information extraction. In: Iliadis, L., Maglogiannis, I. (eds.) AIAI 2016. IAICT, vol. 475, pp. 154–163. Springer, Cham (2016). https://doi.org/10.1007/978-3-319-44944-9_14
10. Grassi, M., Morbidoni, C., Nucci, M., Fonda, S., Ledda, G.: Pundit: semantically structured annotations for web contents and digital libraries. In: SDA, pp. 49–60 (2012)
11. Han, H., Noro, T., Tokuda, T.: An automatic web news article contents extraction system based on RSS feeds. J. Web Eng. **8**(3), 268 (2009)
12. Kang, J., Choi, J.: Detecting informative web page blocks for efficient information extraction using visual block segmentation. In: International Symposium on Information Technology Convergence, ISITC 2007, pp. 306–310. IEEE (2007)
13. Kohlschütter, C., Fankhauser, P., Nejdl, W.: Boilerplate detection using shallow text features. In: Proceedings of the Third ACM International Conference on Web Search and Data Mining, pp. 441–450. ACM (2010)
14. Krüpl-Sypien, B., Fayzrakhmanov, R.R., Holzinger, W., Panzenböck, M., Baumgartner, R.: A versatile model for web page representation, information extraction and content re-packaging. In: Proceedings of the 11th ACM Symposium on Document Engineering, pp. 129–138. ACM (2011)
15. Liu, F.T., Ting, K.M., Zhou, Z.H.: Isolation forest. In: Eighth IEEE International Conference on Data Mining, ICDM 2008, pp. 413–422. IEEE (2008)
16. Liu, L., Özsu, M.T.: Encyclopedia of Database Systems, vol. 6. Springer, New York (2009)
17. Parameswaran, A., Dalvi, N., Garcia-Molina, H., Rastogi, R.: Optimal schemesfor robust web extraction. Proc. VLDB Conf. **4**(11)(2011)
18. Pedregosa, F., et al.: Scikit-learn: machine learning in python. J. Mach. Learn. Res. **12**(Oct), 2825–2830 (2011)
19. Potvin, B., Villemaire, R.: When different is wrong: visual unsupervised validation for web information extraction. In: Perner, P. (ed.) MLDM 2018. LNCS (LNAI), vol. 10935, pp. 132–146. Springer, Cham (2018). https://doi.org/10.1007/978-3-319-96133-0_10
20. Tang, J., Hong, M., Zhang, D.L., Li, J.: Information extraction: methodologies and applications. In: Emerging Technologies of Text Mining: Techniques and Applications, pp. 1–33. IGI Global (2008)
21. Wang, J., et al.: Can we learn a template-independent wrapper for news article extraction from a single training site? In: Proceedings of the 15th ACM SIGKDD International Conference on Knowledge Discovery and Data Mining, pp. 1345–1354. ACM (2009)

22. Wang, R.C., Cohen, W.W.: Language-independent set expansion of named entities using the web. In: ICDM, pp. 342–350. IEEE (2007)
23. Weninger, T., Palacios, R., Crescenzi, V., Gottron, T., Merialdo, P.: Web content extraction: a metaanalysis of its past and thoughts on its future. ACM SIGKDD Explor. Newsl. **17**(2), 17–23 (2016)
24. Witten, I.H., Frank, E., Hall, M.A., Pal, C.J.: Data Mining: Practical Machine-learning Tools and Techniques. Morgan Kaufmann, Burlington (2016)

A Character-Level Deep Lifelong Learning Model for Named Entity Recognition in Vietnamese Text

Ngoc-Vu Nguyen[1,2], Thi-Lan Nguyen[1], Cam-Van Nguyen Thi[1],
Mai-Vu Tran[1], and Quang-Thuy Ha[1(✉)]

[1] University of Engineering and Technology (UET),
Vietnam National University, Hanoi (VNU), Hanoi, Vietnam
{14020249, vanntc, vutm, thuyhq}@vnu.edu.vn
[2] Department of Information Technology, MoNRE,
144, Xuan Thuy, Cau Giay, Hanoi, Vietnam
nnvu@monre.gov.vn

Abstract. Lifelong Machine Learning (LML) is a continuous learning process, in which the knowledge learned from previous tasks is accumulated in the knowledge base, then the knowledge will be used to support future learning tasks, for which it may be only a few of samples exists. However, there is a little of studies on LML based on deep neural networks for Named Entity Recognition (NER), especial in Vietnamese. We propose DeepLML-NER model, a lifelong learning model based on using deep learning methods with a CRFs layer, for NER in Vietnamese text. DeepLML-NER includes an algorithm to extract the knowledge of "prefix-features" of Named Entities in previous domains. Then the model uses the knowledge stored in the knowledge base to solve a new NER task. The effect of the model was demonstrated by in-domain and cross-domain experiments, achieving promising results.

Keywords: Deep LML · Named Entity Recognition (NER) ·
Deep Lifelong Learning · Deep LML for NER in Vietnamese text

1 Introduction

Named Entity Recognition (NER) in textual documents has recently become a highly challenging topic with a great attention. In the few recent years, some neural network models have been proposed for sequence labeling and NER such as Chiu and Nichols [3], Huang et al. [6], Lample et al. [8], Ma and Hovy [10], and these models are competitive with traditional models.

Lifelong machine learning (LML) is an advanced learning paradigm (Thrun and Mitchell [19], Silver et al. [18], Bendale and Boult [1], Chen and Liu [2]). In LML, the learning process is continuous, the knowledge learned from previous tasks is accumulated in the knowledge base, then the knowledge are used to support future learning tasks. So far, there have been studies that combine LML with deep learning in different tasks such as human action detection (Parisi et al. [11]), image recognition (Rusu et al.

© Springer Nature Switzerland AG 2019
N. T. Nguyen et al. (Eds.): ACIIDS 2019, LNAI 11431, pp. 90–102, 2019.
https://doi.org/10.1007/978-3-030-14799-0_8

[14]). However, LML models with deep learning methods for sequence labeling are limited, especial for Vietnamese NER.

This paper presents a Vietnamese NER Deep LML model based on using a Bi-LSTM network combined with CRFs (DeepLML-NER model). This paper has two main contributions: (i) propose an algorithm for extracting the knowledge of "prefix-features" of NEs in Vietnamese domains, (ii) propose a Deep Lifelong Learning Model for NER in Vietnamese Text. DeepLML-NER can perform well on the current domain by storing and transferring "prefix-features" knowledge simultaneously from the previous task to the new one. Experiments were designed on both in-domain and cross-domain data to evaluate the effectiveness of the model.

The rest of this paper is organized as follows. In next section, we brief outline fundamental concept of Bidirectional Long Short-term Memory (Bi-LSTM) and the hybrid architecture which combines Bi-LSTM with a CRFs layer for NER task. Section 3 describes the proposed Deep Lifelong Learning Model for Vietnamese NER, which includes an algorithm for extracting the knowledge of "prefix-features". Section 4 shows experiments on sixteen Vietnamese domains. Comparisons the performance of our Vietnamese NER model and other Vietnamese NER models have been shown. Some related works are introduced in Sect. 5. In the last section, we present conclusions and future work.

2 Bidirectional Long Short-Term Memory (Bi-LSTM) with a CRFs Layer

In this section, we brief outline fundamental concept of Bidirectional Long Short-term Memory (Bi-LSTM) and the hybrid architecture, which combines Bi-LSTM with a CRFs layer for NER task.

2.1 Bidirectional LSTM (Bi-LSTM)

LSTM is a well-known variant of the Recurrent Neural Networks (RNN), which has been originally introduced as a solution to overcome the problem of vanishing and exploding gradient and hence allows deeper networks to perform well in practice (Hochreiter and Schmidhuber [5]). This idea was implemented in LSTM cell by creating internal memory state which is simply added to the processed input in order to reduce the multiplicative effect of small gradient.

2.2 Bi-LSTM+CRF

For NERT, traditional LSTM models have a big disadvantage due they do not using the current suffix of the sentence then the Bidirectional LSTM (Schuster and Paliwal [16]) is designed to address the disadvantage. Figure 1 illustrates the architecture of Bidirectional LSTM (Bi-LSTM) networks. The Bi-LSTM contains two single LSTM networks including forward and backward. For each input word, the forward LSTM learns the effect of the prefix of the word and the backward LSTM learns the effect of the suffix of the word.

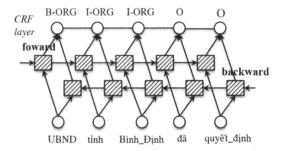

Fig. 1. Bi-LSTM+CRF architecture for an example Vietnamese sentence input

In the Bi-LSTM models, the final label on the output is independent of the softmax function. It means that the final tagging of a token does not depend on the label of words around it. Thus, with the advantage of allowing the context to be labeled in the CRFs (Lafferty et al. [7]), adding a CRFs layer to an LSTM or BI-LSTM model will allow the model to learn the sequence label in the best way, thereby maximizing the accuracy of the model.

3   The Proposed Character-Level Deep Lifelong Learning Model

Prefix patterns are words or phrases usually stands in front of NEs, for example, the word Công_ty (company) usually stands in front of the name of an organization (ORG label), the words Ông (Mr.), Bà (Mrs.) usually stands in front of a person entity (PER), etc. Using a prefix list for labels, we can extract more features for each word by calculating the correlation between the word and the list of prefixes. Besides, we apply the scalable prefix list - an important feature that the model can apply for lifelong learning.

Let K be a reliable prefix extracted from previous tasks using a model M, where a Bi-LSTM+CRF is used. The model M is trained based on using the training dataset D^t. Initially, the set K is the set K^t (the set of all reliable prefixes of the training data set D^t). Suppose M deals with more tasks and more reliable prefixes are extracted, so that the size of K is larger. When dealing with the D_{n+1} task, K allows for extra prefix extraction, model M can yield better results for the new task.

3.1   Training Model

The proposed Character-level Deep Lifelong Learning Model named DeepLML-NER includes three phases (see Fig. 2): (1) Preprocessing, (2) Feature extraction, and (3) Training model using Bi-LSTM+CRF.

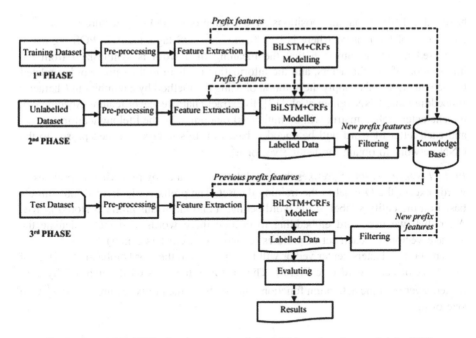

Fig. 2. DeepLML-NER: the character-level deep lifelong learning model for NER

Preprocessing

We ran word segmentation on the input data set, added POS tag for each word using VNCoreNLP[1] and converted data to IOB2 format (Ramshaw and Marcus [15]). Since the aim of this model is to detect person, location, organization names and the miscellaneous label, there are total of seven possible tags for a token: B-PER, I-PER, B-LOC, I-LOC, B-ORG, I-ORG, O.

Feature Extraction

Assume the sentence include n words $S = [w_1, w_2, \ldots, w_n]$. In this phase, the feature of each word w_i will be extracted and represented as a vector x_i.

Word Embedding-Based and POS Tag Feature: We apply pre-trained word vector models for non-English languages named Pre-trained Word2vec[2]. This model was trained on a Wikipedia repository of 74 million Vietnamese records, of which the word embedding matrix is 10,087 words, each word is a 100-dimensional vector. We also randomly initializes a 100-dimensional vector in order to represent the words contained in the dataset but not in the embedded matrix. Each word w_i in the input sentence will be searched in the embedded matrix to obtain a feature vector w_i^w.

Additional feature (POS tag) was transformed into one-hot vectors. A set with m POS tag will be encoded through an $m * m$ identity matrix (only the diagonals matrix

---

[1] https://github.com/vncorenlp/VnCoreNLP.

[2] https://github.com/Kyubyong/wordvectors.

have value 1, the remaining positions have a value of 0). POS tag feature vector w_i^p is the ith row of the matrix, where is j $\in [0; m - 1]$ POS tag's index in POS tag list.

Based on the assumption that the meaning of a word is synthesized from the information of the characters, and the information of them in the same word is related to each other. We also apply the solution of using the method by a combining Character Embedding and CNN (Pham et al. [13]) to extract character-level feature. Specifically, the character embed matrix will be initialized randomly for the lifelong learning model and each letter of word will be encoded based on this matrix and then passed to the CNN network to obtain the feature vector w_i^C.

Prefix Feature: A set of prefixes that we define in advance by retrieving all prefixes of words except the O label in the training dataset. The prefix set is divided into subsets based on the entity's label of the word behind it and all of this prefix is stored in KB. After that, we use word embedding to encode these words, and then calculate the average vectors on each set of prefixes (called prefix clusters center).

The prefix clusters center vector will be appended to the word embedded vectors of the prefix of each word (see Fig. 3). The vector is then passed through a fully connected layer with the activation function *tanh* to obtain the prefix feature vector w_i^f with size of d_f.

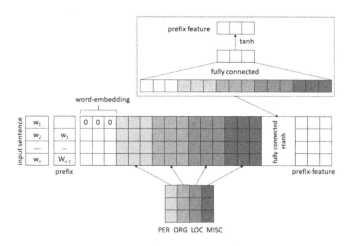

Fig. 3. Prefix feature representation

Feature Vector Representation: Finally, all extracted features will be put into a single vector for each word i^{th}

$$x_i = x_i^w \oplus x_i^p \oplus x_i^c \oplus x_i^f \tag{2}$$

Training Model Using Bi-LSTM+CRF

We extracts and presents training data in the form of characteristic vectors as described above. The training dataset will be divided into batches to be processed in batches. The model will calculate the loss and measurement (will be presented in the model evaluation phase) in batch, the end of each batch will update the weight once. To update the weights of the model after each batch, we uses the Adaptive Moment Estimation (ADAM) optimization function.

Training the Bi-LSTM+CRF neural network with training data sets will obtain the model used for extraction and inference in the next phase.

3.2 Lifelong Prefix Feature Extraction

In DeepLML-NER, a lifelong feature extraction algorithm for aspect mining proposed by Shu et al. [17] is applied to extract lifelong feature, but instead of using the general dependency feature, DeepLML-NER uses the prefix feature.

Algorithm 1 Lifelong Feature Extraction

1: $K_p \leftarrow \phi$
2: **loop**:
3:     $F \leftarrow FeatureGeneration(D_{n+1}, K)$
4:     $E_{n+1} \leftarrow ApplyModel(M, F)$
5:     $S \leftarrow S \cup \{E_{n+1}\}$
6:     $K_{n+1} \leftarrow Frequent-Prefixs-Mining(S, \lambda)$
7:     **if** $K_p = K_{n+1}$ **then**
8:         **break**
9:     **else**
10:         $K \leftarrow K^t \cup K_{n+1}$
11:         $K_p \leftarrow K_{n+1}$
12:         $S \leftarrow S - \{E_{n+1}\}$
13:     **end if**
14: **end loop**

Fig. 4. The proposed character-level deep lifelong learning algorithm

In this phase, the first step is to collect data from Dantri[3], a Vietnamese online newspaper. The data is then passed through the VnCoreNLP tool to preprocessing and extracting the characteristic. Model M trained in the previous phase is used to recognize the entity and extract prefixes from new tasks. M is unchanged and the data from new tasks is unlabeled. All the results from the new domain are stored in the past information store S. After M has extracted the prefix with n domains, and when faced with the $(n+1)th$ task, model M uses trusted prefixes $K_{(n+1)}$ originate from S and $K^t (K = K^t \cup K_{(n+1)})$. The prefixes in K^t from the dataset are trusted because they are

---

[3] http://dantri.com.vn/.

manually labeled. It is not possible to use all of the prefixes obtained from older tasks as reliable data because the labeling model may have errors. However, if the prefix appears multiple times, the prefix is likely to be more accurate.

Algorithm 1 (see Fig. 4) describes the lifelong feature extraction of our proposed model. Detailed steps in the algorithm are presented as follows:

- Generating features F on the data D_{n+1} and applies to model M to produce a set of entities $E_{(n+1)}$ (Line 3).
- $E_{(n+1)}$ is added to S which is the past information store. From S, we mine a set of frequent prefix $K_{(n+1)}$ using threshold λ.
- If $K_{(n+1)}$ is the same as K_p from the previous iteration that means no prefix is found, then the loop will stop.
- Else: some additional reliable prefix are found. M can be labeled more accurately in the next iteration. Lines 10 and 11 update the two sets for the next iteration.

3.3   Evaluating Model

In this phase, testing data are passed through the BiLSTM+CRF model and uses the KB derived from previous task to extract the features. Testing data are also preprocessed by VnCoreNLP tool then extract the same features as the above phase but use the additional KB to extract prefix feature. Then, model filters prefixes occurring frequently with threshold λ, and evaluates the accuracy of the measurements described below.

Evaluation Method

We use the same training datasets and test datasets on the same models. Domain datasets are combined for training and testing in two ways: in-domain and cross-domain. Assume that the lifelong feature extraction phase was performed on unlabeled domain datasets with thresholds λ.

Cross-Domain Experiments: Combining 9 domains for training and testing on 10 different domains (not used in training phase) and obtained 10 results. This evaluation method desires to have a training model for effective use in different domains, thus saving the effort of manually assigning labels.

In-Domain Experiments: Training and testing on 9 same domains and got 10 results. In addition, in order to evaluate the improvement over the model when solving the new task after taking advantage of the knowledge learned from the old tasks, we comparing the Deep LML model with two unused models the knowledge of the old tasks.

In addition, in order to evaluate the improvement of the model when solving the new task after taking advantage of the knowledge learned from the previous task, we compare the Deep LML model with two models, which do not use knowledge from the old task:

- CRFs model: this model only use features including two features $F = (w_i, w_i^p)$.
- Bi-LSTM + CRF model: this model use features $F = \left(w_i, w_i^p, w_i^c, w_i^f \right)$ which is similar to Deep LML but does not use prior knowledge.

For CRFs, we use CRFSuit[4] in order to fast implement this model.

Evaluation Metric

The model is evaluated with Precision (P), Recall (R) and F_1 score (F_1), where

$$F_1 = \frac{2 * P * R}{P + R}$$

In particular, Precision is the percentage named entities found by the model that are correct, Recall is the percentage of named entities present in the data sets that are found. The average Precision, Recall and F_1 scores, which reflect the quality of the model, are the weighted (based on the number of tokens belong to each class) average of precision, recall and F_1 of each class.

4 Experimental Results

4.1 Datasets

VLSP 2018 NER Data

Experiments were conducted on the VLSP 2018 dataset for NER. This dataset are provided by the Vietnamese Language and Speech Community, which are collected from electronic newspapers published on webs including 10 domains.

Four types of NEs are compatible with the descriptions in the CoNLL Shared Task 2003[5], i.e. PER (person), ORG (organization), LOC(locations) and MISC (miscellaneous). The main aim of the VLSP 2018 - Evaluation Campaign for NER is to evaluate of ability of recognizing that entities. Input data followed MUC-6 NER task[6] format: only tags to identify named entities chunks are provided, without word segmentation and other information (POS tag, chunk tag, etc.). This dataset are divided into train, development and test set. The statistic summarization of the given data set is described in Table 1.

Table 1. The statistic of VLSP 2018 NER dataset

Dataset	PER	ORG	LOC	MISC	#total of NEs
Train	4,600	5,587	6,289	743	17,219
Develop	492	723	795	63	2,073
Test	1,883	2,126	2,377	178	6,564
Total	6,978	8,436	9,461	984	25,856

---

[4] https://sklearn-crfsuite.readthedocs.io/en/latest/.

[5] https://www.clips.uantwerpen.be/conll2003/ner/.

[6] https://cs.nyu.edu/cs/faculty/grishman/muc6.html.

Table 2 compares the entity of intersections between domains (the probability of an entity in X of domain A is also the entity in X of domain B) in the VLSP 2018 dataset. It can be seen that the number of entities crossing between domains is low (highest is 15%). Therefore, when cross-domain testing is performed, the meaning of learning from the old tasks for the new task will be seen.

Table 2. Percentage of entities intersecting between domains in the VLSP2018 dataset

Domain	Life	Entertainment	Education	Science/Technology	Economy	Law	World	Sport	Culture	Society
Life	–	0.14	0.13	0.10	0.14	0.14	0.09	0.11	0.15	0.13
Entertainment	0.05	–	0.05	0.05	0.07	0.04	0.04	0.06	0.08	0.05
Education	0.06	0.07	–	0.06	0.12	0.11	0.05	0.06	0.12	0.12
Science/Technology	0.06	0.08	0.07	–	0.13	0.06	0.12	0.08	0.12	0.10
Economy	0.04	0.06	0.08	0.07	–	0.08	0.06	0.05	0.09	0.11
Law	0.05	0.04	0.08	0.04	0.09	–	0.04	0.03	0.08	0.10
World	0.03	0.05	0.04	0.08	0.08	0.04	–	0.05	0.09	0.07
Sport	0.03	0.05	0.04	0.04	0.05	0.03	0.04	–	0.05	0.04
Culture	0.03	0.05	0.06	0.05	0.07	0.06	0.05	0.04	–	0.08
Society	0.04	0.05	0.08	0.05	0.11	0.08	0.05	0.04	0.10	–

Dantri Data

The second dataset is an unassigned dataset was collected from 1,600 articles in 16 areas on Dantri. It contains 246,586 sentences corresponding to 6,682,201 tokens. The statistical of each domain was described in Table 3.

Table 3. The statistic of **Dantri** dataset

Domain	#number of tokens	#number of sentences
Strange story	350,450	15,169
Entertainment	271,666	11,086
Education	680,809	25,331
Business	483,219	15,795
Young Lifestyle	309,252	10,910
Auto and moto	480,321	14,680
Law	462,295	16,003
Health	475,327	17,885
Strength	427,959	14,374
Event	404,959	14,480
Charity	180,972	6,746
World	401,711	14,664
Sports	402,051	17,215
Love	514,916	23,872
Culture	433,822	15,947
Society	402,472	13,463
Summary	**6,682,201**	**246,586**

4.2 Experiment Setting

Overall Setting: We set the dimension of word embedding and prefix feature size are 100. Since the data set is relatively large, we select the batch size of 40 and the unit number in the LSTM layer is 100. Number of CNN filters is 30 and CNN window size is 3. For the parameters of the optimization function Adam, we set the parameters of the library that we use by default, i.e. *learningrate* = $0:01$; $\beta_1 = 0:9$; $\beta_2 = 0:999$; $\epsilon = 10^{-7}$. Threshold λ is set equal to 2.

Cross-*Domain Experiments:* Each domain dataset is chosen as the testing dataset and the combination of 9 other domain datasets is chosen as the training dataset and we obtained 10 results. This evaluation method desires to have a training model for effective use in different domains, thus saving the effort of manually assigning labels.

In-Domain Experiments: Training dataset and testing dataset were two disjoint sets of each domain datasets among 10 domain datasets and we got 10 results. In addition, in order to evaluate the improvement over the model when solving the new task after taking advantage of the knowledge learned from the old tasks, we comparing the Deep LML model with two unused models the knowledge of the old tasks.

4.3 Experiment Results and Analysis

Table 4 shows the detailed experimental results of the proposed model, where the symbol $-X$ means all the domains except X. We found that for both experimental strategies, the results of the BiLSTM+CRF model is better than CRFs and Deep LML model LML results are higher than the other two models.

Cross-Domain Result: Each -X domain in the Training column means that X is not used for the training phase. Each X domain in the Testing column means that X is used for the evaluation phase. Looking at the result's table, we can see that the Deep LML is better than the other two models with an average F1 scoring of 69.45%, while the conventional CRFs model is only 62.53% and that score in the model Bi-LSTM+CRF did not use prior knowledge just reached 66.63%.

In-Domain Result: Each -X domain in the Training column and the Testing column means that all 9 domains except X are used for training and evaluation. We found that Deep LML gave better results than CRFs and Bi-LSTM+CRF with a F1 measurement of 70.51%. However, the percentage was not significantly higher. Nevertheless, this result is reasonable because most of the labeled data in the training phase are the same domain of the evaluation phase.

5 Related Works

There are three works on Vietnamese NER in VLSP 2018 (Fifth International Workshop on Vietnamese Language and Speech Processing in conjunction with CICLing 2018). All of three works focus to extract four kinds of entities of PER, LOC, ORG, and MIS in the whole domain dataset of VLSP 2018.

Table 4. Cross-Domain and In-Domain results

Cross-Domain

Training	Testing	CRF			Bi-LSTM+CRF			Deep LML		
		P(%)	R(%)	F_1(%)	P(%)	R(%)	F_1(%)	P(%)	R(%)	F_1(%)
- Life	Life	75.71	67.05	70.36	65.82	67.89	66.84	70.41	77.50	73.80
- Entertainment	Entertainment	64.00	53.96	55.73	63.86	70.00	66.79	68.38	68.68	68.53
- Eduction	Education	70.83	63.42	66.27	78.36	76.14	77.23	80.12	75.69	77.84
- Sci/Tech	Sci/Tech	60.18	62.80	57.89	66.47	53.74	59.43	67.63	53.92	60.00
- Economy	Economy	74.38	64.53	67.12	69.63	69.38	69.50	68.76	72.88	70.76
- Law	Law	83.47	75.78	78.92	83.73	79.05	81.32	84.91	87.10	85.99
- International	International	50.00	56.55	59.08	67.02	62.20	64.52	66.11	63.41	64.73
-Sports	Sports	62.62	37.47	42.54	39.39	49.79	43.99	43.93	61.43	51.23
- Culture	Culture	62.53	49.55	53.17	61.18	62.75	61.95	61.95	63.79	62.85
- Society	Society	82.38	68.57	74.23	75.47	73.97	74.71	77.02	80.52	78.73
Average		68.61	59.97	**62.53**	67.09	66.49	**66.63**	68.92	70.49	**69.45**

In-Domain

Training	Testing	CRF			Bi-LSTM+CRF			Deep LML		
		P(%)	R(%)	F_1(%)	P(%)	R(%)	F_1(%)	P(%)	R(%)	F_1(%)
Life	Life	70.64	63.88	66.34	65.69	70.95	68.22	70.02	71.67	70.84
Entertainment	Entertainment	71.46	63.84	66.79	66.57	69.25	67.89	69.53	69.34	69.44
Education	Education	69.82	66.36	67.23	68.96	68.28	68.62	69.53	69.81	69.67
Sci/Tech	Sci/Tech	70.74	64.58	66.70	68.41	71.80	70.07	70.30	71.09	70.69
Economy	Economy	70.78	61.75	65.02	68.20	70.99	69.57	68.07	72.36	70.15
Law	Law	69.06	61.21	64.30	66.47	69.07	67.75	66.85	71.73	69.20
International	International]	72.03	63.61	66.84	69.39	71.29	70.33	68.97	72.30	70.60
Sports	Sports	72.18	66.15	68.56	70.48	75.20	72.77	72.75	74.48	73.60
Culture	Culture	73.70	66.17	69.07	70.35	70.55	70.45	70.19	74.23	72.16
Society	Society	69.34	62.19	64.73	67.05	71.34	69.13	67.06	72.64	69.74
Average		70.98	63.97	66.56	68.16	70.87	**69.48**	69.33	71.97	70.61

Dong [4] introduces three experimental systems for Vietnamese NER: a CRF-based system (CRF-based approach), an 1-layer LSTM (LSTM-based approach for each type of entity), and a 3-layers LSTM (joint model approach). In F1 measure, the CRF-based system has a performance of 0.61, the 1-layer LSTM has a performance of 0.54, and the 3-layers LSTM has a performance of 0.52.

Luong and Pham [9] propose a system combinated Bidirectional Long Short-Term Memory and Conditional Random Field. Firstly, a FastText algorithm is applied for pretraining word embedding, then a BiLSTM is used to build a part of word embedding from the characters. Secondly, new word embeddings are fed to BiLSTM and CRF to predict the NER tags. This system achieved an overall F1 scores of 0.74 (Level 1) and 0.68 (Nested) on the standard VLSP NER Test set.

Pham [12] proposes a feature-based NER model which combines word, word-shape features, Brown-cluster-based features, and word-embedding-based features. The NER model focus to extract of three levels of entities: (i) level-1 entities are entities that do not contain other entities inside them, (i) level-2 entities are entities contain only level-1 entities inside them, and (iii) level-3 entities are entities that contain at least one level-2 entity and may contain some level-1 entities. The system achieved an overall F1 scores of 0.735 at all levels.

Our work does not consider the whole domain dataset but ten domain datasets in of VLSP 2018 that is why the work focus to propose a Vietnamese NER Deep LML model based on using a Bi-LSTM network combined with CRFs.

6   Conclusions and Future Work

This paper proposed a lifelong learning model using a combination of Bi-LSTM and CRFs (Deep LML-NER model). The model to retain the "prefix pattern" knowledge gained from the old task and utilize that knowledge to improve the learning speed and performance of the new task solving.

We also conducted experiments on ten domain datasets form the VLSP 2018 dataset. Experimental results have been shown to be satisfactory with mean cross-domain and in-domain F1 measurements achieving 69.45% and 70.51% respectively. The results show that lifelong learning methods can improve the performance of the model based on these prior knowledge.

Four kinds of entities of PER, ORG, LOC, and MISC do not have specific characteristics of ten domains, so other kinds of entities should be considered in the future. Moreover, kinds of knowledge and solutions for finding knowledge from previous NER tasks and using the knowledge for future NER tasks should be studied.

Acknowledgments. This work was supported in part by Grant TNMT.2017.09.02.

References

1. Bendale, A., Boult, T.E.: Towards open world recognition. In: IEEE Conference on Computer Vision and Pattern Recognition, pp. 1893–1902 (2015)
2. Chen, Z., Liu, B.: Lifelong Machine Learning. Synthesis Lectures on Artificial Intelligence and Machine Learning, 2nd edn. (2018)
3. Chiu, J.P.C., Nichols, E.: Named entity recognition with bidirectional LSTM-CNNs. CoRR, abs/1511.08308 (2015)
4. Dong, N.T.: An investigation of Vietnamese nested entity recognition models. In: VLSP 2018 Workshop, pp. 14–16 (2018)
5. Hochreiter, S., Schmidhuber, J.: Long short-term memory. Neural Comput. **9**(8), 1735–1780 (1997)
6. Huang, Z., Xu, W., Yu, K.: Bidirectional LSTM-CRF models for sequence tagging. CoRR, abs/1508.01991 (2015)
7. Lafferty, J.D., McCallum, A., Pereira, F.C.N.: Conditional random fields: probabilistic models for segmenting and labeling sequence data. In: ICML 2001, pp. 282–289 (2001)
8. Lample, G., Ballesteros, M., Subramanian, S., Kawakami, K., Dyer, C.: Neural architectures for named entity recognition. CoRR, abs/1603.01360 (2016)
9. Luong, V.-T., Phan, L.K.: Zaner: Vietnamese named entity recognition at VLSP 2018 evaluation campaign. In: VLSP 2018 Workshop, pp. 10–13 (2018)
10. Ma, X., Hovy, E.H.: End-to-end sequence labeling via bi-directional LSTM-CNNS-CRFs. CoRR, abs/1603.01354 (2016)
11. Parisi, G.I., Tani, J., Weber, C., Wermter, S.: Lifelong learning of human actions with deep neural network self-organization. Neural Netw. **96**, 137–149 (2017)
12. Pham, Q.N.M.: A feature-based model for nested named-entity recognition at VLSP-2018 NER evaluation campaign. In: VLSP 2018 Workshop, pp. 5–9 (2018)
13. Pham, T.-H., Pham, X.-K., Nguyen, T.-A., Le-Hong, P.: NNVLP: A neural network-based Vietnamese language processing toolkit. arXiv preprint arXiv:1708.07241 (2017)
14. Rusu, A.A., et al.: Progressive neural networks. CoRR, abs/1606.04671 (2016)
15. Ramshaw, L.A., Marcus, M.: Text chunking using transformation-based learning. In: VLC@ACL (1995)
16. Schuster, M., Paliwal, K.K.: Bidirectional recurrent neural networks. Trans. Sig. Proc. **45** (11), 2673–2681 (1997)
17. Shu, L., Xu, H., Liu, B.: Lifelong learning CRF for supervised aspect extraction. CoRR, abs/1705.00251 (2017)
18. Silver, D.L., Yang, Q., Li, L.: Lifelong machine learning systems beyond learning algorithms. In: AAAI Spring Symposium Lifelong Machine Learning (2013)
19. Thrun, S., Mitchell, T.M.: Lifelong robot learning. Robot. Auton. Syst. **15**(1–2), 25–46 (1995)

Identification of Conclusive Association Entities by Biomedical Association Mining

Rey-Long Liu[✉]

Department of Medical Informatics, Tzu Chi University, Hualien, Taiwan
rlliutcu@mail.tcu.edu.tw

Abstract. *Conclusive association entities* (CAEs) in the title and the abstract of an article *a* are those biomedical entities (e.g., genes, diseases, and chemicals) that are *specific* targets on which *conclusive* findings about their associations are reported in *a*. Identification of the CAEs is essential for the analysis of conclusive associations, which is a task routinely conducted by many biomedical scientists. However, CAE identification is challenging, as it is difficult to identify the *specific* entities and then estimate how *conclusive* the findings on the entities are. In this paper we present an association mining technique to improve CAE identification. The technique is based on a hypothesis: two candidate entities in an article are likely to be CAEs of the article if a strong association between them is mined from a collection of articles. Experimental results show that, by integrating the technique with representative keyword identification indicators, CAE identification can be significantly improved. The results are of technical and practical significance to the indexing, curation, and exploration of conclusive associations reported in biomedical literature.

Keywords: Biomedical literature · Conclusive association entity · Association mining

1 Introduction

Conclusive association entities (CAEs) in an article *a* are those biomedical entities (e.g., genes, diseases, and chemicals) that are *specific* targets on which *conclusive* findings about their associations are reported in *a*. Given candidate entities in the title and the abstract of an article, identification of the CAEs is essential to the analysis of conclusive findings on specific entities, which is a task routinely conducted by many biomedical scientists. For example, CTD (Comparative Toxicogenomics Database),[1] GHR (Genetic Home Reference),[2] and OMIM (Online Mendelian Inheritance in

---

[1] Update information of CTD can be found at http://ctdbase.org/help/faq/;jsessionid=92111C8A6B218E4B2513C3B0BEE7E63F?p=6422623 (accessed, September 2018).

[2] A large number of biomedical scientists join the curation tasks of GHR, see http://ghr.nlm.nih.gov/ExpertReviewers (accessed, September 2018).

© Springer Nature Switzerland AG 2019
N. T. Nguyen et al. (Eds.): ACIIDS 2019, LNAI 11431, pp. 103–114, 2019.
https://doi.org/10.1007/978-3-030-14799-0_9

Human)[3] recruit many biomedical scientists to frequently update entity associations based on those articles with conclusive findings on the associations.

However, it is challenging to identify the CAEs, as it is difficult to identify the *specific* entities and then estimate how *conclusive* the findings on the entities are. An article often mentions many entities in its title and abstract, but (1) many of them are not specific targets (e.g., those entities that are not, but their derivatives are specific targets), (2) some of them are not the targets on which the reported findings are conclusive enough (e.g., those entities whose associations may not exist under certain conditions), and (3) some entities may be related to the background of the article, rather than the main finding concluded in the article. Therefore, only a small number of the entities can be CAEs of the article.

One way to tackle the challenge is to build complete and scalable knowledge and intelligent semantic processing techniques to determine whether an entity is a *specific* target on which the reported findings are *conclusive enough*. However, it is both difficult and costly to build such knowledge and techniques. In this paper we aim at developing an association mining technique to improve CAE identification, without needing to build such complicated knowledge and techniques. The technique is thus named AMICAE (association mining for identification of CAEs).

AMICAE is developed based on a hypothesis that two candidate entities in an article are likely to be CAEs of the article if a strong association between them is mined from a collection of articles. More specifically, it is based on the following observations: (1) given two entities e_1 and e_2 that are *potential* CAEs in an article, there may be an association $<e_1, e_2>$ between them, (2) associations derived based on the potential CAEs in literature may be used to infer possible associations (e.g., given $<e_1, e_2>$ and $<e_2, e_3>$ that are derived based on CAEs, an inferred association may be $<e_1, e_3>$), and (3) if two candidate entities in an article a are involved in an inferred association (e.g., e_1 and e_3 are candidate entities in a, and $<e_1, e_3>$ is an inferred association), they are likely to be CAEs of a.

Results of the paper are of technical significance. With AMICAE, the system may be more intelligent in identifying CAEs in *an article* based on the *potential* CAEs identified in *a collection of articles*. Therefore, CAE identification for the article is actually improved by *CAE-based association mining* on the collection of articles. Results of the paper are also of practical significance to the indexing, curation, and exploration of conclusive associations reported in biomedical literature.

2   Related Work

Several types of related studies are discussed to clarify the contributions of the paper: improving CAE identification by CAE-based association mining.

---

[3] OMIM updates association information on a daily basis, see http://www.omim.org/about (accessed, September 2018).

2.1 Article Indexing by Biomedical Terms

CAE identification is different from biomedical article indexing, which aims at labeling the articles with a controlled vocabulary such as MeSH (Medical Subject Heading),[4] as done by many previous studies (e.g., techniques presented in the BioASQ workshop [20] and the Medical Text Indexer tool [14]). The indexing task can be seen as a classification task, as it actually classifies an article into certain categories denoted by terms (MeSH terms). Instead of classifying an article to certain categories, CAE identification investigated in the paper aims at *prioritizing* candidate entities in the title and the abstract of the article, so that CAEs of the article can be ranked high for biomedical scientists (e.g., experts of CTD, GHR, and OMIM) to study how biomedical associations were published in literature as *conclusive* findings.

2.2 Extraction of Biomedical Entity Associations

Many previous studies aimed at extracting (recognizing) biomedical associations mentioned in articles (e.g., [8, 10, 19, 23]). The association extraction task is different from CAE identification as well, because an association that happens to be mentioned in an article is *not* necessarily the *conclusive* finding of the article. The association may not be conclusive enough (e.g., associations may not exist under certain conditions), or only related to the background of the article, rather than the main finding concluded in the article. Entities in an association extracted from an article are thus not necessarily CAEs of the article.

2.3 Estimation of Entity-Article Relatedness

Many previous studies developed techniques to estimate the relatedness between entities (terms) and articles. They can be used for various purposes such as keyword extraction and article retrieval. These techniques can be used for CAE identification as well, because an entity (term) that is related to the main contents of an article is likely to be a CAE of the article.

However, none of the previous techniques estimated the entity-article relatedness by CAE-based association mining on the titles and abstracts of articles. For example, in the biomedical domain, the MetaMap indexing tool (MMI) indexes (labels) articles with MeSH terms [14]. It only worked on MeSH terms in an article, with the depth of each term in the MeSH tree as a critical factor to rank MeSH terms [2]. Instead of only working on MeSH terms, the proposed technique AMICAE can work on those entities in other ontologies (e.g., OMIM and the Entrez-Gene database, which are considered by curators of CTD [22]), as it works by CAE-based association mining without relying on the structure of any specific ontology.

As another example, consider BioCreative, which defined a task of ranking important genes in a full-text article [1]. Many strategies were thus developed to work on full texts of the articles (e.g., preferring those genes in the abstract, figure legends,

---

[4] MeSH (available at https://www.ncbi.nlm.nih.gov/mesh) is a controlled vocabulary for indexing biomedical articles.

table captions, or certain sections of the article [1]), and hence they cannot work well when only titles and abstracts are available. Our technique AMICAE can be applicable to more articles, as it works on titles and abstracts, which are more commonly available than full texts.

Many entity-article relatedness indicators were developed for keyword extraction (from articles) and article retrieval (by search engines) as well. Typical indicators include (1) *frequency-based* indicators (those terms that appear frequently in the article are likely to be key terms, e.g. [2]), (2) *locality-based* indicator (those terms appearing in certain parts of the article, such as titles and first sentences, may be key terms, e.g. [2, 16, 21, 22]), (3) *co-occurrence-based* indicators (those terms that co-occur with many terms in the article are likely to be key terms, e.g. [9, 12, 13, 17, 18]), and (4) integration of frequency-based and *rareness-based* indicators (those terms that appear frequently in the article but rarely in other articles are likely to be key terms, e.g. [3, 12, 18]). When compared with these typical types of indicators, our technique AMICAE is the first technique that employs CAE-based association mining to refine entity-article relatedness estimation. We will show that, by integrating AMICAE with these indicators, CAE identification can be significantly improved.

2.4   Prediction of New Associations by Existing Associations

Previous studies have also found that associations between entities were helpful for *predicting* new possible associations that deserve further analysis by biomedical researchers [4, 5, 15]. When compared with these studies, AMICAE is novel in two ways: (1) AMICAE aims at *identification* of CAEs that have already been published in each individual article (rather than predicting associations that have not yet been reported in literature), and technically (2) AMICAE improves CAE identification by the associations that are inferred by *potential* CAEs in literature (i.e., given two entities e_1 and e_2 that are potential CAEs in an article, the association $<e_1, e_2>$ may be used to inferred more associations to improve CAE identification).

3   Association Mining to Improve CAE Identification

As noted above, AMICAE woks by hypothesizing that two candidate entities in an article are likely to be CAEs of the article if an association between them is mined from a set of articles. There are thus three challenges in developing AMICAE: (C1) identifying potential CAEs of each article in the given set of articles, (C2) based on the potential CAEs, inferring possible entity associations, and (C3) based on the possible entity associations, refining the strength of each candidate entity of being a CAE of a new article.

To tackle Challenge C1, given a candidate entity e in the title and the abstract of an article a, AMICAE estimates its strength of being a potential CAE of a ($PotentialS_{e,a}$). The strength is estimated by Eq. 1, where $TF(e,a)$ is the number of occurrences of e in a (i.e., term frequency of e in a). It is actually estimated by the well-known TFIDF (term frequency × inverse document frequency) technique, which was found to be one of the best way to estimate term-article relatedness for various purposes [3, 12, 18].

Based on the strength estimation, AMICAE selects top-2 entities in each article as the *potential* CAEs of the article.

$$PotentialS_{e,a} = TF(e,a) \times Log_2 \frac{1 + Total\ number\ of\ articles}{1 + Number\ of\ articles\ mentioning\ e} \quad (1)$$

For Challenge C2, AMICAE constructs a network of *potential* associations based on the potential CAEs, and accordingly infers *indirect* associations. Two entities e_1 and e_2 that are potential CAEs of an article is said to have a *potential* association (see the solid lines in Fig. 1). Strength of the association is estimated by the number of articles having e_1 and e_2 as CAEs (i.e., a potential association is more reliable if more articles have the two entities as CAEs). Based on the potential associations, *indirect* associations can be inferred. An indirect association $<e_1, e_3>$ is inferred if both $<e_1, e_2>$ and $<e_2, e_3>$ are potential associations (see the dashed lines in Fig. 1). Strength of an indirect association between e_1 and e_3 ($IndirectS_{e1,e3}$) is set to the smaller one of the strengths of $<e_1, e_2>$ and $<e_2, e_3>$ (i.e., an indirect association is more reliable if both its corresponding potential associations have larger strengths).

For Challenge C3, AMICAE employs the indirect associations to refine CAE identification. Given a test article a whose CAEs are to be identified, AMICAE estimates how each candidate entity in a indirectly associates with each other. For a candidate entity e in a, Eq. 2 is employed to estimate its strength of correlating to other candidate entities (*CorStrength*), which is the sum of the logarithm strengths of the indirect associations between e and other candidate entities in a. Therefore, an entity that has strong indirect associations with other entities in a is likely to be a CAE of a.

$$CorStrength_{e,a} = \sum_{\substack{x\ is\ a\ candidate\ entity\ in\ a; \\ <e,x>\ is\ an\ indirect\ association}} Log_2(1 + IndirectS_{e,x}) \quad (2)$$

Note that *CorStrength* of an entity e in an article a is estimated based on *association mining* of a set of articles, rather than those indicators that measure how e appears in a (e.g., frequency and position of e in a, as well as how e co-occurs with other entities in a), which were routinely employed by many entity-article relatedness estimation techniques (Ref. Sect. 2.3). Actually, these indicators are helpful for CAE identification as well, as they directly reflect how articles mention the individual entities. Therefore, *CorStrength* and these indicators can provide different kinds of helpful information that should be integrated to achieve better performance in CAE identification. AMICAE employs a learning-based integration approach implemented by RankingSVM [7], which is one of the best techniques routinely used to integrate multiple indicators by SVM (Support Vector Machine) to achieve better ranking (e.g., [11]). We employ SVM^{rank} to implement RankingSVM.[5]

---

[5] SVM^{rank} is available at http://www.cs.cornell.edu/people/tj/svm_light/svm_rank.html.

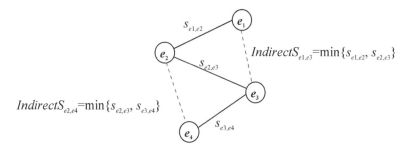

Fig. 1. Inferring possible associations to improve CAE identification: *indirect* associations (see the dashed lines) are inferred from *potential* associations (see the solid lines) between potential CAEs in articles (given two entities e_1 and e_2 that are potential CAEs in an article, there is a potential association $<e_1, e_2>$ between them). The strength of an indirect association $<e_1, e_3>$ is estimated based on the strengths of its corresponding potential associations $<e_1, e_2>$ and $<e_2, e_3>$. Strengths of indirect associations (*IndirectS*) are employed to improve identification of CAEs in new articles.

4 Empirical Evaluation

4.1 Basic Experimental Settings

Experimental data is collected from CTD,[6] which employed a controlled vocabulary to curate biomedical articles whose main focuses were on associations between entities in the vocabulary (including chemicals, genes, and diseases [4]). The vocabulary is "customized" for the curation purpose of CTD (e.g., entities for species-specific entities are added, while some entities not considered by CTD are removed[7]). Therefore, for our evaluation purpose, candidate entities in each article are identified based on the vocabulary of CTD. Other potential entities not in the vocabulary are beyond the scope of consideration in the evaluation, as they are not considered by CTD. This experimental setting is thus helpful for evaluating how systems perform in identifying those CAEs that are identified by domain experts of CTD.

More specifically, we randomly collect about 50% of all articles in CTD (60,507 articles) with their CAEs appearing in their titles or abstracts. For each article a, we collect all associations that are annotated with a. All the involving entities of the associations are thus CAEs of a (i.e., the golden standard for a), because CTD experts think that a reports conclusive findings on these entities. As noted above, candidate entities in each article need to be identified based on the vocabulary of CTD. We thus map candidate entities in each article to their IDs by a dictionary-based name normalization approach, which was employed by many previous studies as well (e.g., [6,

---

[6] In one of our previous projects (ID: MOST 105-2221-E-320-004), we ever employed articles from CTD as experimental data.

[7] More information about the customized vocabulary is available at http://ctdbase.org/help/faq/; jsessionid=92111C8A6B218E4B2513C3B0BEE7E63F?p=6422623, http://ctdbase.org/help/geneDetailHelp.jsp, and http://ctdbase.org/help/diseaseDetailHelp.jsp (accessed, May 2017)

15]). To further fit the approach to our evaluation purpose, given an article a, all terms that CTD experts have selected as CAEs of an article a are first mapped to their corresponding entity IDs, as the existence of the entities in a has been confirmed by the CTD experts. Other terms are then identified by checking whether official symbols or names of entities in the vocabulary appear in a; and if no, synonyms of entities are checked. Moreover, authors of articles often employ their own abbreviations (or symbols) to represent an entity. These abbreviations are mapped to their corresponding entity IDs as well.

As noted in Sect. 3, AMICAE estimates *CorStrength* (Ref. Eq. 2) by association mining, and SVM^{rank} is employed as a learning-based strategy to integrate *CorStrength* and other statistical indicators. We thus require training data, and hence we randomly select 80% of the 60,507 articles as training data, and the other articles as test data.

4.2   Baselines and Evaluation Criteria

As noted in Sect. 2.3, many statistical indicators were developed for keyword extraction and article retrieval. We thus employ them as baseline indicators for CAE identification. We investigate their performance, as well as how their performance may be improved when AMICAE is integrated with them. These baseline indicators include (1) *TF* (term frequency), (2) *IDF* (inverse document frequency), (3) *TITLE*, (4) *AbstractLoc* (abstract location), and (5) *CoOcc* (term co-occurrence). They are respectively defined in Eqs. 3, 4, 5, 6 and 7, where e is a candidate entity in a given article a, $S_e(a)$ is the set of sentences (in a) mentioning e, and $S_{e \cap x}(a)$ is the set of sentences (in a) mentioning both e and another entity x (i.e., set of sentences that e and x co-occur).

$$TF(e, a) = \textit{Number of occurrences of } e \textit{ in } a \tag{3}$$

$$IDF(e) = Log_2 \frac{1 + \textit{Total number of articles}}{1 + \textit{Number of articles mentioning } e} \tag{4}$$

$$TITLE(e, a) = \begin{cases} 1, & \text{if } e \text{ appears in the title of } a \\ 0, & \text{otherwise.} \end{cases} \tag{5}$$

$$AbstractLoc(e, a) = \begin{cases} 1, & \text{if } e \text{ is in the first and the last two sentences in abstract of } a \\ 0, & \text{otherwise.} \end{cases} \tag{6}$$

$$CoOcc(e, a) = \sum_{x \in a, x \neq e} \frac{|S_{e \cap x}(a)|}{|S_e(a)|} \tag{7}$$

The five indicators are representative indicators with potential application to CAE identification. *TF* and *IDF* were routinely employed as essential indicators for keyword extraction and article retrieval [2, 3, 12, 18]. *TITLE* and *AbstractLoc* are typical locality-based indicators employed by many previous techniques (e.g., [2, 16, 21, 22]), based on the observation that those terms that appear in certain parts of an article (e.g., titles, as well as the first and the last sentences) may be key terms of the article. *CoOcc* is a co-occurrence-based indicator, which was considered in many previous studies as well (e.g. [9, 12, 13, 17, 18]), based on the expectation that those terms that co-occur with many terms in an article are likely to be key terms of the article.

Note that, as we are evaluating possible contributions of AMICAE to CAE identification, we are particularly concerned with whether integration of the five representative indicators and *CorStrength* (produced by AMICAE, Ref. Eq. 2) can achieve significantly better performance in CAE identification (recall that the integration is realized by SVM^{rank}, Ref. Sect. 4.1).

To measure performance in CAE identification, we employ three evaluation criteria. The first criterion is *mean average precision* (MAP), which is defined by Eq. 8, where |A| is the number of test articles in the experiment, k_i is number of entities that are believed (by the experts) to be CAEs of the ith article, and $Seen_i(j)$ is the number of entities whose ranks are higher than or equal to that of the jth CAE for the ith article. Therefore, $AP(i)$ is actually the average precision (AP) for the ith article. It is the average of the precision when each CAE is output in the ranked list. Given an article, if a ranker can rank higher those CAEs in the article, AP for the article will be higher. MAP is simply the average of the AP values for all test articles.

$$MAP = \frac{\sum_{i=1}^{|A|} AP(i)}{|A|}, \quad AP(i) = \frac{\sum_{j=1}^{k_i} \frac{j}{Seen_i(j)}}{k_i} \tag{8}$$

The second criterion is *average precision at top-X* (Average P@X), which is simply the average of the P@X values in all test articles (see Eq. 9). Equation 10 defines P@X, which is the precision when top-X entities are shown to the readers. Therefore, when X is set to a small value, P@X measures how a system ranks CAEs very high. In the experiments, we set X to 1, 2, and 3.

$$\text{Average P@X} = \frac{\sum_{i=1}^{|A|} P@X(i)}{|A|} \tag{9}$$

$$P@X(i) = \frac{\text{Number of top-X entities that are CAEs in the } i^{th} \text{ article}}{X} \tag{10}$$

The third evaluation criterion is *%P@X>0*, which is the percentage of the test articles whose CAEs are ranked at top positions (X = 1, 2 and 3). It can be a good measure to indicate whether a system successfully identifies CAEs for a large portion of the test articles.

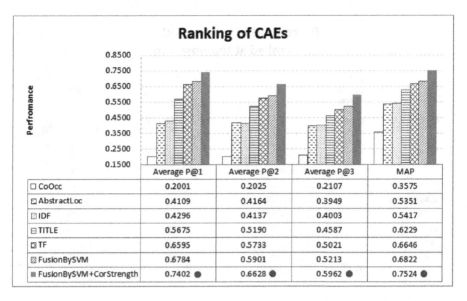

Fig. 2. Results in average P@X and MAP: Integrating all the five baseline indicators by SVM can achieve better performance than each individual indicator (see *FusionBySVM*). When *CorStrength* is added (i.e., there are six indicators integrated by SVM, see *FusionBySVM +CorStrength*), the performance can be further improved significantly (a dot on a system indicates that performance difference between the system and others is statistically significant). AMICAE can thus provide another kind of helpful information to improve CAE identification.

4.3   Results

Figure 2 shows how the systems perform in average P@X and MAP. *TF* performs better than all individual indicators in CAE identification, however, by integrating all the indicators with SVM (see *FusionBySVM*), performance can be further improved. Therefore, these indictors can actually provide different kind of information that can be used to improve CAE identification.

It is interesting to note that, by integrating *CorStrength* (produced by AMICAE) and these indicators (see *FusionBySVM+CorStrength*), system performance can be further improved. To verify whether the performance differences are *statistically significant*, we conduct significance tests by two-sided and paired t-tests with 99% confidence level. The results show that *FusionBySVM+CorStrength* performs significantly better than all the indicators and their fused version *FusionBySVM*, indicating that AMICAE can provide an additional kind of helpful information to improve CAE identification significantly.

Figure 3 presents results in %P@X>0, which measures how the systems rank CAEs at top positions for a large portion of the test articles. Again, with the information provided by AMICAE, *FusionBySVM+CorStrength* successfully identifies CAEs for the largest percentage of the test articles. Over 95% of the test articles can have CAEs ranked at top-3 positions by *FusionBySVM+CorStrength*. The results are of practical

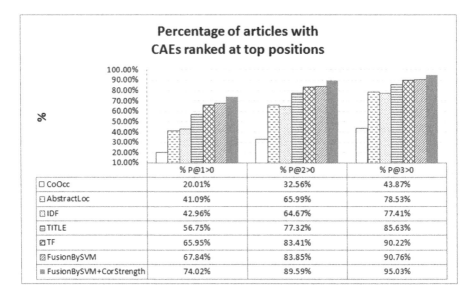

Fig. 3. Results in *%P@X>0*: When integrating *CorStrength* and the five baseline indicators (see *FusionBySVM+CorStrength*), larger percentage of test articles can have their CAEs ranked at top positions (top-1 to top-3). The results are of practical significance, as AMICAE can help systems to successfully identify CAEs for more articles.

significance to the exploration and curation of conclusive evidence reported in biomedical literature.

5   Conclusion and Future Extension

Identification of CAEs in the title and the abstract of an article is challenging, as it is difficult to identify those *specific* entities on which research findings reported in the article is *conclusive enough*. We present a novel technique AMICAE that can provide helpful information to improve CAE identification. AMICAE works by assuming that two candidate entities in an article are likely to be CAEs of the article if a strong association between them is mined from a collection of articles. We thus present how the possible associations can be mined without relying on complicated and costly domain knowledge, and how the possible associations can be used to refine the ranking of candidate entities so that CAEs of test articles can be ranked high. Experimental results show that, by integrating AMICAE and representative keyword identification indicators, CAE identification can be improved significantly. The results are of both technical and practical significance to the analysis of conclusive evidence reported in biomedical literature.

AMICAE can be extended in several ways. It improves CAE identification by mining *indirect* associations, which are inferred by those *potential* associations derived by selecting top-2 entities in each article as potential CAEs. Although we have shown that this way works well, selecting the entities more intelligently (e.g. thresholding on the strengths of being CAEs in each article) is an interesting extension to AMICAE.

Another interesting extension is to relax the constraint for inferring the indirect (possible) associations. As noted in Sect. 3, an indirect association $<e_1, e_3>$ is inferred only if there exist two potential associations $<e_1, e_x>$ and $<e_x, e_3>$. The constraint may be relaxed by considering more indirect relationships: inferring $<e_1, e_3>$ even when there exist three associations $<e_1, e_x>$, $<e_x, e_y>$, and $<e_y, e_3>$ (i.e., association between e_1 and e_3 is indirectly inferred through two intermediate entities e_x and e_y, rather a single entity e_x). Obviously, an association inferred in this less-constrained way tends to have a smaller strength. It is interesting to investigate the possible contributions of relaxing the constraint.

Acknowledgment. This research was supported by Ministry of Science and Technology, Taiwan (grant ID: MOST 107-2221-E-320-004).

References

1. Arighi, C.N., et al.: BioCreative III interactive task: an overview. BMC Bioinform. **12** (Suppl. 8), S4 (2011)
2. Aronson, A.R.: The MMI Ranking Function (1997). https://ii.nlm.nih.gov/MTI/Details/mmi.shtml. Accessed May 2018
3. Boyack, K.W., et al.: Clustering more than two million biomedical publications: comparing the accuracies of nine text-based similarity approaches. PLoS ONE **6**(3), e18029 (2011)
4. Davis, A.P., et al.: The comparative toxicogenomics database: update 2017. Nucleic Acids Res. **45**(Database issue), D972–D978 (2017)
5. Frijters, R., van Vugt, M., Smeets, R., van Schaik, R., de Vlieg, J., Alkema, W.: Literature mining for the discovery of hidden connections between drugs, genes diseases. PLoS Comput. Biol. **6**(9), e1000943 (2010). https://doi.org/10.1371/journal.pcbi.1000943
6. Heo, G.E., Kang, K.Y., Song, M.: A flexible text mining system for entity and relation extraction in PubMed. In: Proceedings of DTMBIO 2015 (2015)
7. Joachims, T.: Optimizing search engines using clickthrough data. In: Proceedings of ACM SIGKDD, Edmonton, Alberta, Canada, pp. 133–142 (2002)
8. Kim, J., So, S, Lee, H.J., Park, J.C., Kim, J.J., Lee, H.: DigSee: disease gene search engine with evidence sentences (version cancer). Nucleic Acids Res. **41**(Web Server issue), W510–W517 (2013). https://doi.org/10.1093/nar/gkt531
9. Kwon, K., Choi, C.H., Lee, J., Jeong, J., Cho, W.S.: A graph based representative keywords extraction model from news articles. In: Proceedings of the 2015 International Conference on Big Data Applications and Services, pp. 30–36 (2015)
10. Li, L., Liu, S., Qin, M., Wang, Y., Huang, D.: Extracting biomedical event with dual decomposition integrating word embeddings. IEEE/ACM Trans. Comput. Biol. Bioinform. **13**(4), 669–677 (2016)
11. Liu, R.-L., Huang, Y.-C.: Ranker enhancement for proximity-based ranking of biomedical texts. J. Am. Soc. Inf. Sci. Technol. **62**(12), 2479–2495 (2011)
12. Matsuo, Y., Ishizuka, M.: Keyword extraction from a single document using word co-occurrence statistical information. Int. J. Artif. Intell. Tools **13**(01), 157–169 (2004)
13. Mihalcea, R., Tarau, P.: TextRank: bringing order into texts. In: Proceedings of the Conference on Empirical Methods in Natural Language Processing (2004)

14. Mork, J., Aronson, A., Demner-Fushman, D.: 12 years on - Is the NLM medical text indexer still useful and relevant? J. Biomed. Semant. **8**, 8 (2017)
15. Özgür, A., Vu, T., Erkan, G., Radev, D.R.: Identifying gene-disease associations using centrality on a literature mined gene-interaction network. Bioinformatics **24**(13), i277–i285 (2008)
16. PubMed: Algorithm for finding best matching citations in PubMed. https://www.ncbi.nlm. nih.gov/books/NBK3827/#pubmedhelp.Algorithm_for_finding_best_ma.　　　Accessed September 2018
17. Shah, P.K., Perez-Iratxeta, C., Bork, P., Andrade, M.A.: Information extraction from full text scientific articles: where are the keywords? BMC Bioinform. **4**, 20 (2003)
18. Thomas, J.R., Bharti, S.K., Babu, K.S.: Automatic keyword extraction for text summarization in e-Newspapers. In: Proceedings of ICIA-16 (2016)
19. Thuy Phan, T.T., Ohkawa, T.: Protein-protein interaction extraction with feature selection by evaluating contribution levels of groups consisting of related features. BMC Bioinform. **17** (Suppl 7), 246 (2016)
20. Tsatsaronis, G., et al.: An overview of the BIOASQ large-scale biomedical semantic indexing and question answering competition. BMC Bioinform. **16**, 138 (2015)
21. Tudor, C.O., Schmidt, C.J., Vijay-Shanker, K.: eGIFT: mining gene information from the literature. BMC Bioinform. **11**, 418 (2010)
22. Wiegers, T.C., Davis, A.P., Cohen, K.B., Hirschman, L., Mattingly, C.J.: Text mining and manual curation of chemical-gene-disease networks for the comparative toxicogenomics database (CTD). BMC Bioinform. **10**, 326 (2009)
23. Žitnik, S., Žitnik, M., Zupan, B., Bajec, M.: Sieve-based relation extraction of gene regulatory networks from biological literature. BMC Bioinform. **16**(Suppl. 16), S1 (2015)

A Data Preprocessing Method to Classify and Summarize Aspect-Based Opinions Using Deep Learning

Duy Nguyen Ngoc[1,2(✉)], Tuoi Phan Thi[1], and Phuc Do[3]

[1] Ho Chi Minh City University of Technology,
VNU-HCM, Ho Chi Minh City, Vietnam
{1680475, tuoi}@hcmut.edu.vn
[2] Posts and Telecommunications Institute of Technology,
Ho Chi Minh City, Vietnam
duynn@ptithcm.edu.vn
[3] University of Information Technology, VNU-HCM,
Ho Chi Minh City, Vietnam
phucdo@uit.edu.vn

Abstract. Opinion summarization is based on aspect analyses of products, events or topics, which is a very interesting topic in natural language processing. Opinions are often expressed in various different ways in regards to objects. Therefore, it is important to express the characteristics of a product, event or topic in a final summary compiled by an automatic summarizing system. This paper proposes a method for conducting data preprocessing on the sentence level of a text using Convolutional Neural Networks. The corpus includes Vietnamese opinions on cars collected from social networking sites, forums, online newspapers and the websites of automobile dealers. The data processing phase will standardize terms for aspects that occur in opinion expressing aspects of the product. These aspects are used by manufacturers. Similarly, the standardization will be performed for both positive and negative terms used in opinions. The sentiment terms in the opinions will be replaced by standardized sentiment terms expressing the same sentiment polarities as those being replaced. This standardization is performed with the support of a semantic and sentiment ontology which has a tree hierarchy in the case of cars. This ontology ensures that the semantics and sentiment of the original opinion are not changed. The experimental results of the paper show that the proposed method gives better results than using no data preprocessing method for deep learning.

Keywords: Corpus · Deep learning · Classification · CNN ·
Convolution Neural Network · Sentiment · Summarization

1 Introduction

The number of opinions about products, events or other problems in the world has been increasing due to the development of the Internet environment. Hence, there is a large demand for the exploitation of information resources. Using manual methods to analyze and evaluate opinions about products takes a serious amount of work.

© Springer Nature Switzerland AG 2019
N. T. Nguyen et al. (Eds.): ACIIDS 2019, LNAI 11431, pp. 115–127, 2019.
https://doi.org/10.1007/978-3-030-14799-0_10

There are many ways to evaluate a product, an event or a problem based on their aspects. Since personal opinions about each object are so diverse, opinion summarization will be invalid if it does not determine which aspects of the product the opinions are referring to and only evaluates the product in a general manner. The methods for classifying and summarizing aspect-based opinions have been studied since the early 2000s [8]. These methods include the Support Vector Machine (SVM), Latent Semantic Analysis (LSA), and Probabilistic Latent Lemantic Analysis (PLSA) methods [9], all of which produce good results. The SVM method [9] provides an accuracy of 85.4%, and its recall reaches 83.63%. Recently, the methodology of deep learning for the natural language processing field has emerged as a highly effective and accurate machine learning method which has generated great interest among scientists. Many works have applied this method and achieved very good results [1–3]. Most of the works have precision and recall scores of over 80%, with the precision in [3] reaching 91%.

The basic deep learning model for sentiment classification suggested by [7] is displayed in Fig. 1. The Word Embedding layer is the weight matrix for the data in the corpus. An enhancement of the efficiency of the deep learning system can be achieved by enhancing algorithms and configuring deep learning systems (the Convolution, Pooling and Fully Connected layers) or by improving the quality of the data for the Word Embedding layer. This matrix is built by applying statistical methods to the lexical terms in the corpus [10], which makes it possible to determine the correlation between the words in the corpus. As a result, language characteristics do not have important roles in this model. However, human emotions are often expressed very subtly based on the linguistic characteristics. If a method could be found for representing the linguistic features in a data set for a deep learning system, then the system would become more efficient.

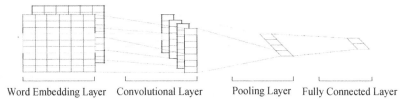

Word Embedding Layer   Convolutional Layer   Pooling Layer   Fully Connected Layer

Fig. 1. Basic Convolutional Neural Network model

This paper proposes a method for representing language characteristics and storing them in an ontology with the structure of a semantic and sentiment vocabulary hierarchical tree (an SSVHT ontology). In order to examine the characteristics of an SSVHT, consider, for example, the semantic components of the SSVHT in terms of the aspects of a car (the object). There are two types of aspects of a car: the standard and equivalent standard aspects. The standard aspects are words or phrases that indicate the technical characteristics the manufacturer used to introduce the car. An equivalent standard aspect is a word or phrase that the user uses to describe the car. The sentiment component, in comparison, is what a user uses to express opinions while evaluating aspects of the car. There is a relationship between the semantic and sentiment components which represents the characteristic expression of the specific topic. The extraction of sentiment and semantic characteristics in opinions is done with the support of an SSVHT.

The rest of the paper is organized as follows: Sect. 2 reviews the related works. Section 3 introduces our proposed data pre-processing for deep learning, and Sect. 4 discusses the configuration of the deep learning system and the experimental results. Finally, Sect. 5 presents the conclusion of the paper.

2   Related Works

When considering the trend in using deep learning for natural language processing, the problems of opinion classification and summarization have been studied and great results have been achieved. The data set used for training in deep learning systems is users' opinions about products on e-commerce sites such as amazon.com [2] or Yelp.com [6] or twitter [11]. All of the data is in English. Other languages, such as Chinese [1], have been tested. These projects usually process the data corpus using the word2vec tool to create the input matrix for the deep learning system.

A deep learning model is often used as a Convolution Neural Network - CNN [1–3]. The combination of different CNNs has been tested, such as in [2], which combines CNNs in a vertical (cascade) with a horizontal (multitask) for different tasks, and [3] combined a Recurrent Neural Network (RNN) with a CNN in their deep learning systems. The results obtained from the experiments on natural language processing by deep learning systems are very good. Wu et al. in [2] obtained precision and recall F-measure scores of over 80%, the highest of which was 84.1%. Djanush et al. in [3] achieved a precision of 91%, a recall of 86.2% and an F1 of 88.5%.

All of the deep learning systems which have been applied in natural language processing have had good results, although they have no special processing designed for data corpora. Thus, it can be said that the results obtained from a deep learning system applied in natural language processing depend on the configuration of this system. The configuration could lead the system to soon reach a limit if no other solution enhances its effect, especially its effect in the data preprocessing stage.

With a standard and rich corpus in English, the deep learning systems can achieve high accuracy without data preprocessing, as seen in [1–3, 6]. However, in other languages which have poor and limited data corpora, such as Vietnamese, data preprocessing is an important enhancement for all deep learning systems. Therefore, this paper proposes a method for Vietnamese data preprocessing. This method also processes Vietnamese characteristics, which are quite different from English characteristics.

3   Proposed Method

The input data matrix for the Word Embedded layer is the important data for a deep learning system. Without the feature processing phases of the word2vec method, the Word Embedding layer is implemented on a statistical basis. There are not many expressions based on linguistic elements in word2vec, but word2vec is based on correlations between words in the statistical corpus. Some works that use deep learning methods in natural language processing have achieved very good results, but the objects processed in these works are data in some popular domains, such as phones, computers, and so forth. Data from other domains requiring a representation of linguistic elements have yet to be included in experiments involving deep learning models.

Consequently, the paper proposes a method for data preprocessing which is based on the meaning of words to build a corpus for the objects (cars) in Vietnamese for a deep learning system. Based on the experimental results, this paper shows the advantages as well as the disadvantages of the proposed method in comparison with other deep learning systems lacking preprocessed data. The data preprocessing model used in our proposed method is shown in Fig. 2.

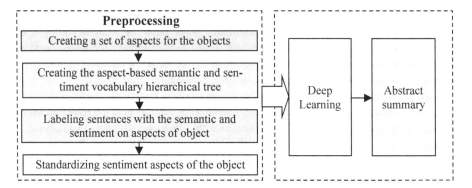

Fig. 2.   Data preprocessing for a deep learning system

3.1   Data Preprocessing Model

The data preprocessing model shown in Fig. 2 demonstrates four tasks. The first task builds aspects in terms of the object. The second task builds an aspect-based SSVHT. The third task involves labeling the semantic and sentiment aspects. The fourth task is standardizing the sentiment aspects of the object.

Creating a Set of Aspects for the Objects. The aspects are the technical features of the product that are introduced to consumers by the manufacturer. Also, these aspects are subject to customer reviews. Each aspect is represented in two manners, one is the name used by the manufacturer and the other is the name used in a customer review on the Internet. For example, some car's aspects are determined below:

- Engine (machine, transmission system, gas, and so on).
- Interior (seats, steering wheel, handle assist, and so forth).

Creating the Aspect-Based SSVHT. An aspect determining a car's specifications is a noun that may be replaced by a user's word having the same, or approximately the same, meaning as the aspect. The set of aspects is determined by two bases.

The Word or Phrase in a Specification Which is Determined by the Manufacturer. There are a lot factors involving an object, for example, a car, that customers are interested in and that manufacturers want to introduce to their customers. There are many specifications related to the technology involved in a car which the manufacturer groups into specification categories with specific group names. Each manufacturer has a way to name these groups. For example, engine, safety, interior, exterior, and so forth. This paper combines these group names to produce a list of standard aspects.

The Specification Word or Phrase That is Used by Users. Users can comment about a car in terms of aspects that are identified by the manufacturer, i.e., by using standard aspects or terms. Users can also use words or word phrases that have the same or approximately the same meanings as the standard terms, called user's terms. For example, a user may use the word "machine" instead of "engine" or "drive" instead of "operate." These user's terms are identified through five following steps:

- Step 1: As a part of speech tagging, finding the noun or phrase which referred to as a term. These terms refer to car specifications in the corpus.
- Step 2: Arranging these terms found in step 1 according to standard words.
- Step 3: Selecting all terms that occur in the corpus two or more times.
- Step 4: Selecting the terms from the term set found in step 3 that have the same meaning as the standard term referring to each aspect. The set of these selected terms is synonymic with the referred to aspect.
- Step 5: Using the word2vec tool to enrich the set of terms found in 4 step. From the results of word2vec, find terms related closely to each sentiment aspect. The found terms can refer to the aspect.

Each noun as a name of an aspect has an amount of adjectives, adverbs, and verbs which modify this noun and also are sentiment aspect of opinions. For example, the search for sentiment terms by the word2vec tool is performed as follows.

- Starting with the aspect *"động cơ engine"* the sentiment terms *"mạnh potent"* *"bốc explosive"* and *"cùi bắp terrible"* are detected by word2vec. The sentiment term *"cùi bắp terrible"* and the aspect term *"máy engine"* are used in the sentence:

  "Máy của chiếc Xe này phải nói là cực kỳ **cùi bắp"** (*The **engine** of this car can be said to be extremely **terrible***)

- In the next step, word2vec can use the sentiment term *"cùi bắp terrible"* to search any other terms which have a high correlation with *"cùi bắp terrible."* Assume that the terms "táp lô dashboard," *"vô lăng steering wheel"* *"hiện đại morden"* and so on are detected. Now, the *"táp lô dashboard"* term indicates the aspect *"nội thất interior."* The terms *"taplo dashboard"* and *"bảng điều khiển control panel"* have the same meaning as *"táp lô dashboard,"* but they occur more often than *"táp lô dashboard"* in the corpus. However, due to the word2vec tool, the term "táp lô dashboard" is detected even though it occurs rarely.

For example, the *"cùi bắp terrible"* and *"táp lô dashboard"* terms are used in the sentence: "Cái táp lô nhìn quá cùi bắp" (*This dashboard looks too terrible*)

An ontology with the structure of an SSVHT is presented in Fig. 3.

Sentiment terms are simple words that usually have a positive or negative polarity on the aspect level. These sentiment terms often occur in car reviews on websites are called center words, such as đắt expensive, chát austere, bốc explosive, hầm hố heady, and so forth.

For example:

- Interior: sang $_{luxury}$/tệ $_{dud}$, chắc chắn $_{firm}$/ọp ẹp $_{rickety}$, and so on.
- Exterior (front and rear fog lamps, wheels, remote keyless entry, and so on).

When the above center words are combined with sentiment adverbs to form new sentiment terms that increase or decrease the emotional level in comparison with center words, these sentiment adverbs are called complements. For example, **quá** đắt (*too expensive*) and **cực** sang (*extremely luxury*) are complements. The adverbs are also divided into five levels: **intensifier, booster, diminished, minimizer** and **modifier** [4].

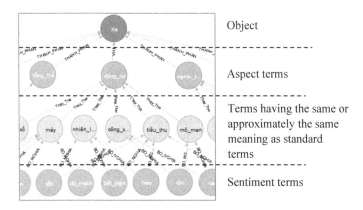

Fig. 3. Semantic and sentiment vocabulary hierarchical tree

The different positions of the center word in comparison with its complementary positions in the phrase determine different levels of positive or negative polarities.

For example: **cực kỳ** sang (*extremely luxurious*) > **quá** sang (*too luxurious*) > sang (*luxurious*) > **khá** sang (*rather luxurious*) > **cũng** sang (*seemingly luxurious*) > **không** sang (*not luxurious*).

Some adverbs representing the degree at each sentiment level are presented in Table 1.

Table 1. Some Vietnamese adverbs by degree with their scalings

Intensifier	Booster	Diminished	Minimizer	Modifier
cực kỳ$_{extremely}$	rất $_{very}$	khá $_{rather}$	hơi $_{a\ bit}$	không $_{no}$
cực $_{strongly}$	quá $_{too}$	hiếm $_{scarcely}$	cũng $_{seemingly}$	chẳng $_{no}$
vô cùng $_{utmost}$	lắm $_{much}$	từng $_{already}$		chả $_{no}$

Labeling Sentences with the Semantic and Sentiment on Aspects of Object. In order to maintain high data quality for training, the labeling is performed manually. Labeling is divided into two stages. In the first stage, the sentence is labeled for a specific aspect. In the second stage, the sentence is labeled for a specific sentiment polarity as positive, negative or neutral. The labeling process removes Vietnamese sentences without accents, without specific topics, and without sentiments. Note that sentences in the first stage that are tagged for the aspects shown in Table 3.

Standardizing Sentiment Aspects of the Object. The sentences in the corpus may be rewritten using a simple sentence structured, as seen below:

$$N + A \tag{1}$$

$$N + R + A \tag{2}$$

$$N + A + R \tag{3}$$

Where: N: noun representing an aspect
R: adverb or adverbs
A: sentiment complements (adjectives or verbs).

The algorithm standardizing a sentence using the patterns (1), (2) and (3) is shown in Fig. 4.

Input: *SSVHT* – Semantic and sentiment vocabulary hierarchical tree ontology,
 S – Sentence labeled with sentiment polarity
Output: *SS* – sentences have shortened
Begin
 for i = 0 **to** *number_of_leafs_SSVHT* **do** L = *leaf_of_treef_SSVHT[i]*
  if exists *L* **in** *S* **then**
   for j = 0 **to** *Nodes_are_not_leaf_SSVHT*
    if *T[j] is_father_of_L* **then**
     if *T[j] is_not_child_node* **then** SS = *T[j]* + L
     else SS = *father_of_T[j]* + L
 return *SS*;
end

Fig. 4. Algorithm standardizing sentences based on aspects

This algorithm applies (1), (2) and (3) and the SSVHT ontology, as follows:

(i) **Replacing pronouns and synonyms of aspects with standard aspects.**
For example: "Nhìn là muốn xúc rồi" (*I want to scoop (it means buy) it immediately at the first sight.*) → "Tổng thể hấp dẫn" (*Overall, it is attractive*). The word "*sight*" does not refer to any specification aspects of the car, but it expresses a general evaluation of the car, so the algorithm replaces "*I want to scoop (it means buy) it immediately at the first sight.*" with "*Overall, it is attractive*", where "*sight*" is mapped to the "*overall*" aspect (Table 3). The word "*scoop*" is a slang word that expresses the desire to buy a car. In the context of the sentence above, it expresses the desire to buy this car.

(ii) **Replacing the sentiment term with the appropriate sentiment standard term**.
For example: "Đạp ga rất sướng" (*Pressing the gas is very pleasurable*) → "Máy rất mạnh" (*Machine is very strong*) → "Động cơ rất mạnh" (*Engine is very strong*). The term "gas" is used to indicate the power control of a motor. This term has the direct father node "machine" in the *SSVHT* ontology, and the word "machine" has the appropriate meaning as the aspect "engine."
The term "very pleasurable" expresses the excitement of the user, meaning that he or she wants to control the car's motor. The term "very pleasurable" has the same sentiment polarity as the term "very strong."

(iii) **Any sentence with multiple aspects is reviewed and separated into simple sentences, in such a way that each sentence has only one aspect**.
For example 1: "Ngoại thất đẹp, động cơ quá được, có điều giá quá chát." (*Nice exterior, engine is very much ok, but the price is too high*).
"Ngoại thất đẹp" (*Nice exterior*) → "Ngoại thất đẹp" (*Nice exterior*).
"Động cơ quá được" (*engine is very much ok*) → "Máy mạnh" (*strong engine*).
"Giá quá chát" (*price is too high*) → "Giá quá đắt" (*such an expensive price*).
When a sentence has more than one aspect, but all aspects have the same sentiment term, then the sentence is separated into simple sentences in such a way that each sentence has only one aspect with the same sentiment term.
Example 2: "Nội và ngoại thất đều ngon" (*Interior and exterior are good*).
"Nội thất ngon" (*Good interior*) → "Nội thất ngon" (*Good interior*).
"Ngoại thất ngon" (*Good exterior*) → "Ngoại thất ngon" (*Good exterior*).
Thanks to the algorithm in Fig. 4 and the SSVHT ontology, the aspect terms are standardized, so that the number of lexical items in the corpus is reduced. In such a way, the corpus obtains the items corresponding to semantic and sentiment aspects in the specific domain (in this case, the car domain). Due to the data preprocessing model, the speed for processing the deep learning model may be improved.

3.2   Opinion Summarization

After classifying sentiments via the Convolution Neural Network system, the reviews are summarized with the support of SSVHT. The paper uses an abstract method to summarize opinions. The positive and negative opinions are summarized, but neutral ones are not. The sentiment sentences may be rewritten with standard aspects in the cases in which the sentences contain terms in the spoken language with which the standard aspects are synonymous. The rewritten sentences keep the same sentiment and meaning as the originals by one of three formulas at (1), (2), (3). This paper used ROUGE [5] to evaluate the result of this phase. The summarizing model is presented in Fig. 5.

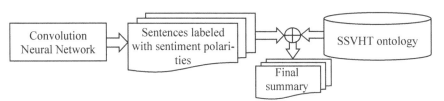

Fig. 5. Aspect-based opinion summarization model

4   Experimental Results

4.1   Corpus

The reviews are collected from the websites for commenting on cars, forums for cars, and the websites of the companies trading in cars. These comments are data processed via the method presented in Sect. 3. The organization of the corpus after preprocessing in the previous section is presented in Table 2. A car sample is a brand in a particular version. For example, the Honda Accord version 2018.

The number of sentences labeled with neutral polarity is much higher than the number of sentences labeled with either positive or negative polarity. This fact can affect the experimental results.

Table 2.   The organization of the corpus

Features of corpus	Quantity
Sample of cars	93
Opinions	2,097
Training sentences	4,013
Testing sentences	816
Sentiment labels	3 (positive, neutral, negative)
Sentences labeled with positive polarity	1,503
Sentences labeled with neutral polarity	1,977
Sentences labeled with negative polarity	1,349
Aspect term standard (the term that the manufacturer used)	10
Aspect term	101
Sentiment term	1,578

Table 3.   Number of sentences labeled with aspects

Standard aspect	Number of sentences
Động cơ Engine	549
Giá bán Price	762
Vận hành Transmission	393
An toàn Safety	419
Ngoại thất Exterior	536
Nội thất Interior	519
Tiện nghi Convenience	571
Kích thước Dimension	270
Trọng lượng Weight	221
Tổng thể Overall	589

In addition to the standard aspects of the car, we add the "overall" aspect. This aspect is added due to the fact there are many vague opinions about cars which do not refer to any particular aspect. This kind of opinion occurs when a customer glances at the object. The "overall" aspect presents basic features, such as vehicle color, model, styling and other basic aesthetics of vehicle design.

4.2   Proposed System Configuration

Word2vec. This paper sets the general configuration of the word2vec tool to process different preprocessed data sets in order to evaluate the effect of our propose method. This configuration is presented in Table 4.

Table 4. The configuration of the word2vec tool

Specification	Value
Size	300
Window	10
min_count	2
Workers	10
Algorithm	CBOW

The experimental results of the stages, such as detecting aspects and sentiment classifications, are better when the paper uses word2vec with the Continuous Bag of Words (CBOW) algorithm in comparison with word2vec using the Skip-gram algorithm.

The Convolution Neural Network for Opinion Classification. This paper sets a general configuration of CNN to process different preprocessed data sets in the opinion stage in order to evaluate the effect of our propose method. This configuration is presented in Table 5. Factor L2 is chosen manually through the testing process.

Table 5. The configuration of the Convolution Neural Network

Specification	Value
Word Embedding size	300
Filter size	3, 4, 5
Dropout	0.5
Batch size	64
L2 weight decay	0.0012

Opinion Summarization. The result for this phase of system is in abstract summarization. The summarization method is presented in Sect. 3.2.

4.3 Experiments

Sentiment Classification. The sentences with "positive" or "negative" polarities are easy to identify, but "neutral" sentiment sentences are not. Therefore, the paper performs sentiment classifications on different training data sets in order to evaluate the effect of the proposed method. In addition, two basic data sets, the raw set and the preprocessed one (shown in Sect. 3), as well as the positive and negative data sets and the set with all three sentiment polarities are experimented upon. The experimental results are presented in Tables 6, 7, 8 and Fig. 6.

Table 6. The experimental results for the sentiment classification of the positive data set

Data	F1 (%)	Precision (%)	Recall (%)
Three layers are not standardized	58.26	73.33	48.32
Three layers are standardized	61.81	74.90	52.62
Two layers are not standardized	73.91	73.33	74.50
Two layers are standardized	77.01	76.86	77.17

Table 7. The experimental results for the sentiment classification of the negative data set

Data	F1 (%)	Precision (%)	Recall (%)
Three layers are not standardized	54.88	71.18	44.66
Three layers are standardized	58.86	73.80	48.99
Two layers are not standardized	71.43	72.05	70.82
Two layers are standardized	74.51	74.67	74.35

Table 8. The experimental results for the sentiment classification of the neutral data set

Data	F1 (%)	Precision (%)	Recall (%)
Three layers are not standardized	59.64	59.64	59.64
Three layers are standardized	65.09	66.27	63.95

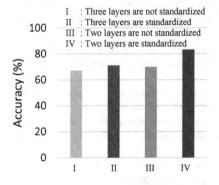

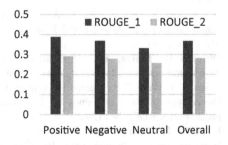

Fig. 6. The accuracy of the experimental results.

Fig. 7. The experimental results of summarization

The experimental results on sentiment classification in Tables 6, 7, 8 and Fig. 6 are not so good, but the significance of the proposed method lies in:

– Results on standardized data are better than on non- standardized data in all experiments. The experimental results of the two testing data sets are very different from each other. The accuracy of three non-standard data layers is less than the accuracy of two standardized data layers by approximately 16%.
– As mentioned in Sect. 4.1, the data set with the neutral sentiment polarity lessens the system quality. However, the number of positive and negative sentences in the data set (two data layers) is less than the number of sentences in the data set with all three sentiment polarities (three layers), but the experimental results on two layers are better than on three layers.

Summarization. The experimental results of summarization are presented in Fig. 7.

The data in Fig. 7 shows that the summarizing method used in the paper is convincing. The SSVHT ontology has improved the accuracy of the results.

5   Conclusion

This paper has proposed a method for preprocessing and creating a training data set from Vietnamese reviews of cars. This method, which employs an SVHT ontology, is then applied in a deep learning system to improve the system performance. The experimental results show that this method has a certain advantage. The paper also shows that the results obtained on data sets processed by the our method always have higher accuracy than those obtained when not using this method.

Acknowledgments. This paper was supported by research project TNCS_KHMT_2017_06, funded by the Ho Chi Minh City University of Technology, VNU-HCM.

References

1. Li, Q., Jin, Z., Wang, C., Zeng, D.D.: Mining opinion summarizations using convolutional neural networks in Chinese microblogging systems. Knowl.-Based Syst. **107**, 289–300 (2016)
2. Wu, H., Gu, Y., Sun, S., Gu, X.: Aspect-based opinion summarization with convolutional neural networks. In: International Joint Conference Neural Networks, IJCNN, pp. 3157–3163. IEEE (2016)
3. Dhanush, D., Thakur, A.K., Diwakar, N.P.: Aspect-based sentiment summarization with deep neural networks. Int. J. Eng. Res. Technol. **5**(5), 371–375 (2016)
4. Thien, K.T., Tuoi, T.P.: Computing sentiment scores of verb phrases for Vietnamese. In: Proceedings of the Conference on Computational Linguistics and Speech Processing, pp. 204–213. ROCLING (2016)
5. Lin, C.Y.: Rouge: a package for automatic evaluation of summaries. In: Proceedings of the Workshop on Text Summarization Branches Out, pp. 74–81 (2004)
6. Xu, L., Lin, J., Wang, L., Yin, C., Wang, J.: Deep convolutional neural network based approach for aspect-based sentiment analysis. Adv. Sci. Technol. Lett. **143**, 199–204 (2017)

7. Kim, Y.: Convolutional neural networks for sentence classification. In: Proceedings of the 2014 Conference on Empirical Methods in Natural Language Processing, EMNLP, pp. 1746–1751 (2014)
8. Hu, M., Liu, B.: Mining opinion features in customer reviews. Assoc. Adv. Artif. Intell. 4(4), 755–760 (2004)
9. Liang, L.C., Hoar, H.W., Hoang, L.C., Chi, L.G., Emery, J.: Movie rating and review summarization in mobile environment. IEEE Trans. Syst. Man Cybern.-Part C: Appl. Rev. 42, 397–406 (2012)
10. Xin, R.: word2vec parameter learning explained, in arXiv preprint arXiv:1411.2738 (2016)
11. Liao, C., Wang, J., Yu, R., Sato, K., Cheng, Z.: CNN for situations understanding based on sentiment analysis of Twitter data. Procedia Comput. Sci. 111, 376–381 (2017)

Word Mover's Distance
for Agglomerative Short Text Clustering

Nigel Franciscus[✉], Xuguang Ren, Junhu Wang, and Bela Stantic

Institute for Integrated and Intelligent Systems, Brisbane, QLD, Australia
{n.franciscus,x.ren,j.wang,b.stantic}@griffith.edu.au

Abstract. In the era of information overload, text clustering plays an important part in the analysis processing pipeline. Partitioning high-quality texts into unseen categories tremendously helps applications in information retrieval, databases, and business intelligence domains. Short texts from social media environment such as tweets, however, remain difficult to interpret due to the broad aspects of contexts. Traditional text similarity approaches only rely on the lexical matching while ignoring the semantic meaning of words. Recent advances in distributional semantic space have opened an alternative approach in utilizing high-quality word embeddings to aid the interpretation of text semantics. In this paper, we investigate the word mover's distance metrics to automatically cluster short text using the word semantic information. We utilize the agglomerative strategy as the clustering method to efficiently group texts based on their similarity. The experiment indicates the word mover's distance outperformed other standard metrics in the short text clustering task.

Keywords: Word mover's distance · Text clustering · Short text · Social media

1 Introduction

The abundance of textual data particularly in the social media domain creates a high demand for efficient and accurate information processing. In many cases, news often appears on social media simultaneously or right before they appear in the traditional news media. Social media also plays a pivotal role in the rapid propagation of information including factoid and non-factoid information (rumors). Thus, it is evidence to provide an efficient tool to enable the interested readers to rapidly detect recent breaking news in social media streams. Text processing pipeline often involves three subtasks: selection, clustering, and re-rank or summarization of tweets [16]. In this work, we address the task of tweet clustering as one of the central subtasks required in the overall pipeline.

Traditional approaches to clustering textual data are associated with the construction of a document-term matrix (e.g. TF-IDF), which represents each document as a bag-of-words. While these methods still play an important role, they suffer from data sparsity due to the nature of the short texts. Post preprocessing is likely to generate the average length of less than ten per document. One way

© Springer Nature Switzerland AG 2019
N. T. Nguyen et al. (Eds.): ACIIDS 2019, LNAI 11431, pp. 128–139, 2019.
https://doi.org/10.1007/978-3-030-14799-0_11

to overcome sparseness in the document-term matrix is to drop all the infrequent terms and consider only the terms that appear frequently across the collection. While this procedure temporarily alleviates the problem, most tweets will be ignored. Consequently, they cannot influence the clustering outcomes which hinders the low frequency but important information. Other popular techniques such as pLSI [6] and LDA [1] are able to capture the probabilistic representation of topics in the document, however, they often fail when the documents are sparse. Moreover, these techniques cannot identify the distance between two documents if the documents are previously unseen.

Word Embeddings introduce a solution by projecting words in a semantic space based on their relationship with each other using a shallow neural network [3,11,15]. Word embeddings represent each word as a vector in a dense semantic space which can capture the semantic relationships between words. For example,

- Entity analogies, vec(Berlin) - vec(Germany) + vec(Portugal) $\approx vec$(Lisbon).
- Syntactic analogies, vec(quick) + vec(quickly) $\approx vec$(slow) + vec(slowly).

Recently, many high-quality pre-trained word embeddings from Wikipedia or GoogleNews corpus are available such as Google's word2vec[1], Stanford's GloVe[2], and Facebook's fastText[3].

Intuitively, representation of meaning is a better approach to measure text similarity - rather than relying on the traditional syntactical or lexical matching. Word embeddings are often used to measure how similar a text compare to others, for example "good" and "better" or "sick" and "unwell" should be relatively similar. Several approaches such as paragraph vector [9], soft-cosine [14], and earth mover's distance [8] have shown promising results to capture the similarity of text semantics. Based on the recent findings in the semantic similarity, in this paper, we consider the similarity distance to group similar tweets using the hierarchical ward linkage approach for short text clustering.

Contribution. To be specific, we present an agglomerative scheme to cluster tweets by utilizing word mover's distance and word embedding results. We opt to use word mover's distance as it generates a better cluster representation in terms of tight and loose balance. By using the bottom-up hierarchical approach, we can compute the similarity score for each tweet as the intermediate results which further enables the distance-based clustering. In the experiments, we compare the word mover's distance with other state-of-the-art distance metrics and measure both the quantitative and qualitative evaluation of the cluster outputs.

Organization. The rest of the paper is organized as follows: in Sect. 2, we present some related work; in Sect. 3, we present the details of calculating word mover's distance; in Sect. 4, we explain the clustering process; in Sect. 5, we provide the experiment results; and finally in Sect. 6 we conclude the findings and indicate the future work.

---

[1] https://code.google.com/archive/p/word2vec/.
[2] https://nlp.stanford.edu/projects/glove/.
[3] https://fasttext.cc/.

2   Related Work

We categorized the related work into three sections, the mapping of raw tweets into embeddings level, choosing the clustering distances, and to utilize the embeddings for clustering task.

Tweet Representation. Recently, researchers have explored the use of deep learning to generate the vector embeddings for tweets. The tweet embeddings also known Tweet2Vec. Tweet2Vec itself has two versions, the first used the Bidirectional Gated Recurrent Unit (Bi-GRU) encoder to predict the hashtags of given tweets [4]. The second version proposed the Convolutional Neural Network with Long Short-Term Memory (CNN-LSTM) encoder-decoder for tweet semantic similarity and tweet sentiment categorization tasks [17]. Both versions are using character-level to learn the tweet representation. While the noise and the idiosyncratic nature in social media make character-level more sensible, it is often harder and takes longer to train than the word-level representation. Our work investigates the quality of results of both levels.

Similarity Distance. The core of the clustering is to measure the distance between points. The objective of the previous work is to group similar text together which can be classified as text similarity. For example, Liu et al. employed the traditional bag-of-words with cosine similarity to group Facebook's post comments [10], Kenter and De Rijke utilized word embeddings results (e.g. word2vec) to measure the cosine similarity between the terms and then applied a modified BM25 metrics [7]. Our work examines the recently proposed word mover's distance [8] to go beyond the cosine and Euclidean distance in measuring the text similarity.

Tweet Clustering. Since document matrix does not generate a good representation of tweets, researchers try to take advantage of the vector embeddings. Vakulenko et al. extended the character-level Tweet2Vec to generate cluster using the fastcluster hierarchical clustering method with distance-threshold selection [16]. Another similar work adopted the Hierarchical DBSCAN [2] to cluster the tweet embeddings for tweets conversation mining [18]. The density-based clustering also uses pure Euclidean distance as the base metrics to measure the similarity. Our work is similar to this work in generating the clusters. However, we employ the word mover's distance metrics instead of straightforward euclidean distance.

3   Word Mover's Distance

An unsupervised learning approach, Word Mover's Distance (WMD) was introduced as an alternative in measuring the similarity of two texts [8]. The method originated from the Earth Mover's transportation problem that calculates dissimilarity as the minimum distance between the two objects [12]. The WMD takes advantage of the word embeddings results between words as the Euclidean distance. More precisely, the transportation problem is calculated using the following constraint,

$$\min_{\mathbf{T} \geq 0} \sum_{i,j=1}^{n} \mathbf{T}_{ij} c(i,j)$$

$$\text{subject to} : \sum_{j=1}^{n} \mathbf{T}_{ij} = d_i \; \forall i \in \{1, ..., n\} \quad (1)$$

$$\sum_{i=1}^{n} \mathbf{T}_{ij} = d_{j}' \; \forall j \in \{1, ..., n\}$$

with $c(i,j)$ as the approximate euclidean distance $\|\mathbf{x}_i - \mathbf{x}_j\|_2$ between the two words in the embedding space. The travel cost between two words becomes the natural building block of distance between texts. For example, let $\mathbf{d}$ and $\mathbf{d}'$ be the representation of the two texts with each word i in $\mathbf{d}$ to be transformed into any word in $\mathbf{d}'$. $\mathbf{T} \in R^{n \times n}$ is a sparse flow matrix where $\mathbf{d}'$ represents how much of i in $\mathbf{d}$ travels to word j in $\mathbf{d}'$. The outgoing flow from word i must equal to d_i to enable the transformation of $\mathbf{d}$ entirely into $\mathbf{d}'$, e.g. $\sum_j \mathbf{T}_{ij} = d_i$. Subsequently, the amount of incoming flow to word j must match d_j', e.g. $\sum_i \mathbf{T}_{ij} = d_j'$.

As an illustration, below is the example of the two text snippets,

1. The President is having dinner in Jakarta.
2. The Prime Minister eats lunch in Sydney.

After removing the stop-words, the important words can be taken from the word embedding space (Fig. 1). Note that if two words often co-occur together, then it is common to treat them as a phrase, e.g. Prime Minister. Words that have the *closest embedding distance* are coupled together to fill the outgoing flow, for example, (Prime Minister - President) or (Jakarta-Sydney).

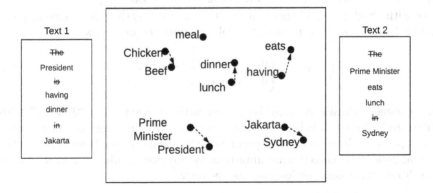

Fig. 1. Mapping words from the embedding space

It is intuitive when all words are line up with each other, however, when the length of two sentences is different, WMD follows the transportation optimization problems. For example, given two texts with a different number of words,

1. The President is having dinner in Jakarta.
2. The Treasurer landed in Sydney.

Figure 2 shows a detailed illustration for calculating the flow for each word pair. After the stop-words removal, every unique word $w \in N$ will be given a flow of $\frac{1}{N}$. Thus, in text one and text two each word will have a flow of $\frac{1}{4}$ and $\frac{1}{3}$ respectively. Since text two > text one, text two will have an excess flow towards text one. In this instance, the remaining flow will be distributed evenly to the word which does not have a pair. Note that the remaining word (dinner) will be distributed to every other word in the shorter text with the flow weight of $e.g.$ $\mathbf{T}_{ij} = \frac{1}{3} - \frac{1}{4}$.

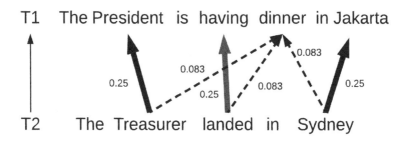

Fig. 2. The flow between two texts with different numbers of words

The main problem with WMD is the computational complexity of $\mathcal{O}(n^3 \log n)$ [12] where n denotes the number of unique words in the documents. The cost of the optimum flow lies on the assigning a unique word pair with the closest Euclidean distance in the word embeddings space. For example, in the previous illustration, regardless of the position, the word "Jakarta" has the closest distance with "Sydney", and therefore must be linked together. A relaxed approximation is only to trial a random way of assigning the unique pairs.

$$\mathbf{T}_{ij}^* = \begin{cases} d_i, & \text{if } j = \operatorname{argmin}_j c(i,j) \\ 0, & \text{otherwise} \end{cases} \tag{2}$$

For example, "President" may be paired with "Sydney" and "Jakarta" with "Treasurer". This will bring the complexity to $\mathcal{O}(n^2)$ as suggested in the original paper [8]. Based on the experiment, the computation time between the original wmd and relaxed wmd approximation is not significant once the word embeddings have been loaded into the memory.

4   Clustering Method

In this section, we give an overview of the clustering approach which can be divided into several interconnecting segments: (i) *Initial preprocessing*, where we collect the raw tweet and store them into MongoDB. (ii) *WMD Clustering*, where we compute the dissimilarity of two tweets to determine the clustering distance and merge similar tweets into clusters. (iii) *Refining Cluster Selection*, where we further improve the quality of clusters. Figure 3 depicts the overview of bottom-up clustering approach. For the word embeddings, we choose Google News corpus with 3 million unique vocabularies[4].

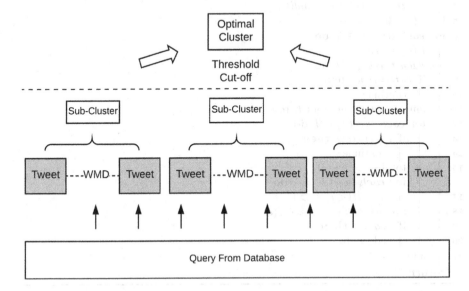

Fig. 3. Agglomerative structure of clustering overview

4.1   Initial Preprocessing

An efficient storage architecture is essential to manage raw data and handling preprocessing. The initial preprocessing is primarily built based on the precomputing architecture inside the NoSQL database [5], where we partition whole data into filtered seed key terms to classify the collection based on topics. Then, to maintain the word consistency with the global word embedding, we apply recurrent neural network spelling corrector based on the news corpus. Any non-ASCII characters and non-English words will be removed. Finally, we apply tokenization with stop-words removal to build the dictionary.

4.2   WMD Clustering

In this subsection, we present our algorithm of clustering using the wmd metrics, as shown in Algorithm 1. Algorithm 1 takes a list of preprocessed tweets T and

---

[4] https://code.google.com/archive/p/word2vec/.

Algorithm 1. WMDCLUSTERING

Input: a collection of preprocessed tweets T, threshold ϕ
Output: a collection of clusters C
1 HashMap $H \leftarrow \emptyset$
2 **for** *each* tweet $t_i \in T$ **do**
3 $\quad$ **for** *each* tweet $t_j \in T$ **do**
4 $\quad\quad$ | $H(< t_i, t_j >) \leftarrow wmd(t_i, t_j)$
5 Flag $F \leftarrow \emptyset$
6 **for** *each* tweet $t_i \in T$ **do**
7 $\quad$ | $F(t_i) \leftarrow false$
8 **for** *each* tweet $t_i \in T$ **do**
9 $\quad$ **if** $F(t_i)$=*true* **then**
10 $\quad\quad$ | continue
11 $\quad$ init a cluster c and put t_i to c
12 $\quad$ **for** *each* tweet $t_j \in T$ **do**
13 $\quad\quad$ **if** $F(t_j)$=*true* **then**
14 $\quad\quad\quad$ | continue
15 $\quad\quad$ total score $s \leftarrow 0$
16 $\quad\quad$ **for** *each* tweet t_z *in* c **do**
17 $\quad\quad\quad$ | $s \leftarrow H(< t_j, t_z >)$
18 $\quad\quad$ average $avg \leftarrow s/sizeOf(c)$
19 $\quad\quad$ **if** $avg < \phi$ **then**
20 $\quad\quad\quad$ | add t_j to c
21 $\quad$ add c to C
22 **return** C

---

a threshold ϕ as input and returns the clusters C where the elements in each cluster are similar. In Line 1 to 4, we use a hashmap to calculate and cache the wmd score for each pair of tweets. The wmd computation is usually fast enough once the word embeddings have been loaded into the memory.

We use a flag to indicate whether a tweet has been assigned to any cluster (Line 5 to 7). After that, for each tweet t_i in T which does not belong to any cluster, we initialize a new cluster c for it (Line 8 to 11). Then for each other unassigned tweet t_j, we calculate the average wmd score between t_j with every member of cluster c (Line 15 to 18). If the average score is less than the threshold, we add t_j to cluster c. (Line 19, 20). Algorithm 1 ensures the average wmd distance between the members of a cluster is lower than a given threshold to maintain the relatedness of tweets. The complexity of Algorithm 1 is $\mathcal{O}(n^2)$ where n is the number of tweets in the collection T.

4.3 Refining Cluster Selection

In this subsection, we present an approach for refining the clusters generated by Algorithm 1 to further improve the quality of each cluster. The refining algorithm is given in Algorithm 2 which takes the output clusters from Algorithm 1, a lower wmd threshold ϕ and a cluster size threshold ψ as input.

Algorithm 2. CLUSTERREFINING

Input: a collection of clusters C, a lower threshold ϕ , cluster size threshold ψ
Output: a refined collection of clusters C'

1 reuse the HashMap H from Algorithm 1
2 **for** *each cluster* $c_i \in C$ **do**
3   **if** $sizeOf(c_i) < \psi$ **then**
4   continue
5   **for** $c'_i \in wmdClustering(c_i, \phi)$ **do**
6   **if** $sizeOf(c'_i) > \psi$ **then**
7   add c'_i to C'
8 **return** C'

There are two key-ideas in Algorithm 2, (1) using the same process of Algorithm 1 to further split each of the old clusters with a lower wmd threshold (Line 5). (2) using a cluster size threshold to remove clusters with few members (Line 3,4 and 6,7). Note that we can reuse the hashmap from Algorithm 1 without re-calculating it in Line 5. It is important to mention that Algorithm 2 can be recursively called to refine the generated clusters more than once, depending on the quality of the actual outputs.

5 Experiments

We used Twitter datasets based on the sample streaming which occurs in a continuous timeline. The average length after preprocessing is approximately 10 tokens for each tweet. We vary the number of documents to test the ability of clusters generation. The initial approximation has a complexity of $\mathcal{O}(n^2)$ since we have to compare tweets against each other. We evaluate the clustering result based on both quantitative and qualitative measurements.

5.1 Cluster Evaluation

As an illustration, we take the sample of a thousand tweets as the benchmark comparison. We select word mover's distance, soft-cosine distance and pure cosine distance (without the word embeddings). We measure the silhouette score based on the distance threshold for each metrics. We choose 0.2 as the minimum and 0.8 as the maximum distance since closer to 1 will harm the cluster quality. Note that since there is no ground truth labels or distances, we quantify both intra and inter-cluster using the respective metrics.

Figure 4 highlights the average of clusters generated based on the similarity distance threshold. Intuitively, a lower threshold will generate smaller clusters as texts tend to group and as the opposite, higher threshold will generate a larger number of clusters due to the tight bound of similarity. Subsequently, the refined clusters will also reflect from the original results. Note that we set the

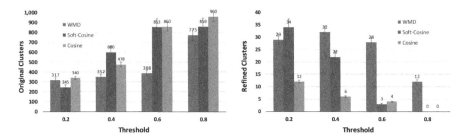

Fig. 4. Number of clusters generated   **Fig. 5.** Number of refined clusters

cluster threshold to 3 to compensate small dataset. However, it is interesting to observe that wmd tend to consistently cluster text within an appropriate balance (Fig. 5).

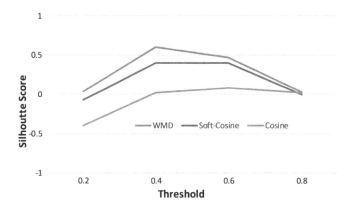

Fig. 6. Average Silhouette score for distance metrics

Silhouette score has been the gold standard to measure the cluster validity index [13]. It is an unsupervised intrinsic evaluation metric that measures the quality of the produced clusters without the ground truth labels. In particular, Silhouette score is measured as $\frac{(b-a)}{max(a,b)}$ where a is the mean intra-cluster distance and b is the mean of nearest-cluster (minimum) distance for each sample. Figure 6 represents the clusters quality indicated by the silhouette score. In general, choosing between lower and middle bounds generates a better cluster representation. In contrast, upper bound tends to generate a stricter similarity matching which impairs the overall cluster quality.

5.2   Qualitative Analysis

To measure the quality aspect of clusters produced, we manually evaluate the cluster based on the topics or categories. Specifically, we obtain sample tweets

Table 1. WMD sample results separated by categories

Sample cluster	Evaluation
Commonwealth games 2018 Gold Coast **queens relay** Commonwealth Games: Historic Gold Coast Games **set to begin** **Queen's Baton Relay** live on Channel7 as we **count down** to...	Correct
Miller **wins** Commonwealth Games **gold** in the hammer Athletics... Commonwealth Games: Wales' David Phelps **wins** 50m prone **gold** Flora Duffy **Wins Gold** At Commonwealth Games! England's Learmonth **wins** triathlon **silver** at Commonwealth Games...	Correct
The latest The kenyan banker Daily! #gc2018 #monitorupdates **The latest** The Melanie Ryding Daily! #cycling **The latest** The Rhino.Neill Motorsport Daily! Thanks to	Incorrect

during the 2018 Commonwealth Games. Then, we use the sample from the refined cluster selection. Tables 1 and 2 illustrate some of the manual evaluation from wmd clusters and soft-cosine respectively. Relevant topics are highlighted in bold. Note that, we do not include the cosine metrics since it tends to generate poor results.

Table 2. Soft-Cosine sample results separated by categories

Sample cluster	Evaluation
Commonwealth games 2018 Gold Coast **queens relay** Commonwealth Games: Historic Gold Coast Games **set to begin**	Correct
Follow highlights of **Commonwealth Games** 2018 opening ceremony... What's the purpose of these **Commonwealth Games**? What are the **Commonwealth** Video **games**? **Commonwealth Games** 2018 Live	Incorrect
Congratulation #MirabaiChanu for getting **India's** first #Gold medal... **Congratulations** #Gururaja on winning the first medal for **India**... **Congratulations** Gururaja for winning **India's** first #Silver...	Correct

Based on the manual evaluation, we observe that wmd tends to produce a better generalization compared to soft-cosine. This is particularly reflected from the way wmd considers the length of texts (normalized bag-of-words) into the calculation. For example, in the first row of Table 1, the context is referring to the opening of the games. Wmd is able to generalize the "count down" while soft-cosine tends to be stricter and completely ignores that relevant tweet. The second row also indicates a good generalization of "win gold" and "win silver". One of the notable observations are both techniques seem to generalize words that co-occur in the same position. For example, on the third row on both tables, phrase "The latest" and "Congratulations" which occur at the beginning of the sentence will likely to be grouped. Although for wmd, it is not particularly accurate. This is most likely due to the limitation of word embeddings to distinguish between semantic and syntactic similarity in strings [16].

6 Conclusion

We presented the agglomerative scheme to cluster short text such as tweets. Within the scheme, we investigated the text similarity distance, in particular the word mover's distance to generate clusters. Overall, word mover's distance produced better clusters due to better generalization. Based on our scheme, we were able to detect similar texts and cluster them into a relevant group. Through the practical implementation and evaluation in our experiments and by using real-world dataset, we showed the effectiveness of the clustering process, especially for the short text. Some future works including the improvement in handling social media tokens by training a high-quality spelling corrector to mitigate out of vocabulary and by training the word embeddings based on tweets. We also plan to improve the word mover's distance to better handle word position in the short texts.

References

1. Blei, D.M., Ng, A.Y., Jordan, M.I.: Latent dirichlet allocation. J. Mach. Learn. Res. **3**(Jan), 993–1022 (2003)
2. Campello, R.J.G.B., Moulavi, D., Sander, J.: Density-based clustering based on hierarchical density estimates. In: Pei, J., Tseng, V.S., Cao, L., Motoda, H., Xu, G. (eds.) PAKDD 2013. LNCS (LNAI), vol. 7819, pp. 160–172. Springer, Heidelberg (2013). https://doi.org/10.1007/978-3-642-37456-2_14
3. Collobert, R., Weston, J., Bottou, L., Karlen, M., Kavukcuoglu, K., Kuksa, P.: Natural language processing (almost) from scratch. J. Mach. Learn. Res. **12**(Aug), 2493–2537 (2011)
4. Dhingra, B., Zhou, Z., Fitzpatrick, D., Muehl, M., Cohen, W.W.: Tweet2vec: character-based distributed representations for social media. In: The 54th Annual Meeting of the Association for Computational Linguistics, p. 269 (2016)
5. Franciscus, N., Ren, X., Stantic, B.: Answering temporal analytic queries over big data based on precomputing architecture. In: Nguyen, N.T., Tojo, S., Nguyen, L.M., Trawiński, B. (eds.) ACIIDS 2017. LNCS (LNAI), vol. 10191, pp. 281–290. Springer, Cham (2017). https://doi.org/10.1007/978-3-319-54472-4_27
6. Hofmann, T.: Probabilistic latent semantic analysis. In: Proceedings of the Fifteenth Conference on Uncertainty in Artificial Intelligence, pp. 289–296. Morgan Kaufmann Publishers Inc. (1999)
7. Kenter, T., De Rijke, M.: Short text similarity with word embeddings. In: Proceedings of the 24th ACM International on Conference on Information and Knowledge Management, pp. 1411–1420. ACM (2015)
8. Kusner, M., Sun, Y., Kolkin, N., Weinberger, K.: From word embeddings to document distances. In: International Conference on Machine Learning, pp. 957–966 (2015)

9. Le, Q., Mikolov, T.: Distributed representations of sentences and documents. In: International Conference on Machine Learning, pp. 1188–1196 (2014)
10. Liu, C.Y., Chen, M.S., Tseng, C.Y.: Incrests: towards real-time incremental short text summarization on comment streams from social network services. IEEE Trans. Knowl. Data Eng. **27**(11), 2986–3000 (2015)
11. Mnih, A., Hinton, G.E.: A scalable hierarchical distributed language model. In: Advances in Neural Information Processing Systems, pp. 1081–1088 (2009)
12. Pele, O., Werman, M.: Fast and robust earth mover's distances. In: ICCV, vol. 9, pp. 460–467 (2009)
13. Rousseeuw, P.J.: Silhouettes: a graphical aid to the interpretation and validation of cluster analysis. J. Comput. Appl. Math. **20**, 53–65 (1987)
14. Sidorov, G., Gelbukh, A., Gómez-Adorno, H., Pinto, D.: Soft similarity and soft cosine measure: similarity of features in vector space model. Computación y Sistemas **18**(3), 491–504 (2014)
15. Turian, J., Ratinov, L., Bengio, Y.: Word representations: a simple and general method for semi-supervised learning. In: Proceedings of the 48th Annual Meeting of the Association for Computational Linguistics, pp. 384–394. Association for Computational Linguistics (2010)
16. Vakulenko, S., Nixon, L., Lupu, M.: Character-based neural embeddings for tweet clustering. In: Proceedings of the Fifth International Workshop on Natural Language Processing for Social Media, pp. 36–44 (2017)
17. Vosoughi, S., Vijayaraghavan, P., Roy, D.: Tweet2vec: learning tweet embeddings using character-level CNN-LSTM encoder-decoder. In: Proceedings of the 39th International ACM SIGIR Conference on Research and Development in Information Retrieval, pp. 1041–1044. ACM (2016)
18. Vosoughi, S., Vijayaraghavan, P., Yuan, A., Roy, D.: Mapping twitter conversation landscapes. In: Proceedings of the Eleventh International Conference on Web and Social Media, ICWSM, 15–18 May 2017, pp. 684–687 (2017)

Improving Semantic Relation Extraction System with Compositional Dependency Unit on Enriched Shortest Dependency Path

Duy-Cat Can[✉], Hoang-Quynh Le, and Quang-Thuy Ha

Faculty of Information Technology, University of Engineering and Technology,
Vietnam National University Hanoi, Hanoi, Vietnam
{catcd,lhquynh,thuyhq}@vnu.edu.vn

Abstract. Experimental performance on the task of relation extraction/classification has generally improved using deep neural network architectures. In which, data representation has been proven to be one of the most influential factors to the model's performance but still has many limitations. In this work, we take advantage of compressed information in the shortest dependency path (SDP) between two corresponding entities to classify the relation between them. We propose (i) a compositional embedding that combines several dominant linguistic as well as architectural features and (ii) dependency tree normalization techniques for generating rich representations for both words and dependency relations in the SDP. We also present a Convolutional Neural Network (CNN) model to process the proposed SDP enriched representation. Experimental results for both general and biomedical data demonstrate the effectiveness of compositional embedding, dependency tree normalization technique as well as the suitability of the CNN model.

Keywords: Relation extraction · Dependency unit ·
Shortest dependency path · Convolutional neural network

1 Introduction

Relation extraction (RE) is an important task of natural language processing (NLP). It plays an essential role in knowledge extraction tasks from information extraction [17], question answering [13], medical and biomedical informatics [4] to improving the access to scientific literature [5], etc. The relation extraction task can be defined as the task of identifying the semantic relations between two entities e_1 and e_2 in a given sentence S to a pre-defined relation type [5].

Many deep neural network (DNN) architectures are introduced to learn a robust feature set from unstructured data [15], which have been proved effective, but, often suffer from irrelevant information, especially when the distance

© Springer Nature Switzerland AG 2019
N. T. Nguyen et al. (Eds.): ACIIDS 2019, LNAI 11431, pp. 140–152, 2019.
https://doi.org/10.1007/978-3-030-14799-0_12

between two entities is too long. Previous researches have illustrated the effectiveness of the shortest dependency path between entities for relation extraction [4]. We, therefore, propose a model that using convolution neural network (CNN) [9] to learn more robust relation representation through the SDP.

The on-trending researches demonstrated that machine learn a language better by using a deep understanding of words. The better representation of data may help machine learning models understanding data better. Word representation has been studied for a long time, several approaches to embed a word into an informative vector has been proposed [1,11], especially with the development of deep learning. Up to now, enriching word representation is still attracting the interest of the research community; in most cases, sophisticated design is required [7]. Meanwhile, the problem of representing the dependency between words is still an open problem. In our knowledge, most previous researches often used a simple way to represent them, or even ignore them in the SDP [18].

Considering these problems as motivation to improve, in this paper, we present a compositional embedding that takes advantage of several dominant linguistic and architectural features. These compositional embedding then are processed within a dependency unit manner to represent the SDPs.

The main contributions of our work can be concluded as:

1. We introduce a enriched representation of SDP that utilizes a major part of linguistic and architectural features by using compositional embedding.
2. We investigate the effectiveness of dependency tree normalizing before generating the SDP.
3. We propose a deep neural architecture which processes the above enriched SDP effectively; we also further investigate the contributions of model components and features to the final performance that provide a useful insight into some aspects of our approach for future research.

2   Related Work

Relation extraction has been widely studied in the NLP community for many years. There has been a variety of computational models applied to this problem, and supervised methods have shown to be the most effective approach. Generally, these methods can be divided into two categories: feature engineering-based methods and deep learning-based methods.

With feature-based methods, researchers concentrate on extracting a rich feature set. The typical studies are of Le et al. [8] and Rink et al. [14], in which variety of handcrafted features that capture the, semantic and syntactic information are fed to an SVM classifier to extract the relations of the nominals. However, these methods suffer from the problem of selecting a suitable feature set for each particular data that requires tremendous human labor.

In the last decade, deep learning methods have made significant improvement and produced the state-of-the-art result in relation extraction. These methods usually utilize the word embeddings with various DNN architectures to learn the features without prior knowledge. Socher et al. [15] proposed a Recursive

Neural Network (mvRNN) on tree structure to determine the relations between nominals. Study of Zhou et al. [21] presents an ensemble model using DNN with syntactic and semantic information. Some other studies use all words in sentence with position feature [20] to extract the relations within it.

In recent years, many studies attempt other possibilities by using dependency tree-based methods. Panyam et al. [12] exploit graph kernels using constituency parse tree and dependency parse tree of a sentence. The SDP also receives more and more attention on relation extraction researches. CNN models (Xu et al. [18]) are among the earliest approaches applied on SDP. Xu et al. [19] rebuilt an Recurrent Neural Network (RNN) with Long Short-Term Memory (LSTM) unit on the dependency path between two marked entities to utilize sequential information of sentences. Various of improvements have been suggested to boost the performance of RE models, such as negative sampling [18], exploring subtrees go along with SDP's node [10], voting schema and combining several deep neural networks [7].

3   Enriched Shortest Dependency Path

3.1   Dependency Tree and Shortest Dependency Path

The dependency tree of a sentence is a tree-structural representation, in which each token is represented as a node and each token-token dependency is represented as a directed edge. The original dependency tree provides the full grammatical information of a sentence, but some of this information may be not useful for the relation extraction problem, even bring noises.

The Shortest Dependency Path (SDP) is the shortest sequence go from a starting token to the ending token in the dependency tree. Because the SDP represents the concise information between two entities [3], we suppose that the SDP contain necessary information to shows their relationship.

3.2   Dependency Tree Normalization

In this work, we applied two techniques to normalize the dependency tree, in order to reduce noise as well as enrich information in the SDP extracted from the dependency tree (see Fig. 1 for example).

Preposition Normalization: We collapse the *"pobj"* dependency (object of preposition) with the predecessor dependency (*e.g.*, *"prep"*, *"acl"*, etc) into a single dependency, and cut the preposition off from the SDP.

Conjunction Normalization: Base on the assumption that two tokens that linked by a conjunction dependency *"conj"* should have the same semantical and grammatical roles; we then add a skip-edges to ensure that these conjuncted tokens have same dependencies with other tokens.

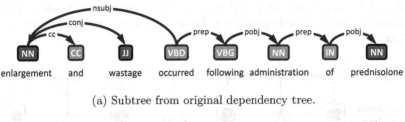

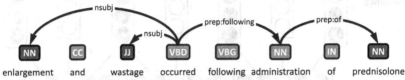

(a) Subtree from original dependency tree.

(b) Subtree from normalized dependency tree.

Fig. 1. Example of normalized dependency tree.

3.3   The Dependency Unit on the SDP

According to the study of [7], a pair of a token and its ancestor has the difference in meaning when they are linked by a different dependency relation. We make use of this structure and represent the SDP as a sequence of substructures like "$t_a \xleftarrow{r_{ab}} t_b$", in which t_a and t_b are token and its ancestor respectively; r_{ab} is the dependency relation between them. This substructure refers to the Dependency Unit (DU) as described in Fig. 2.

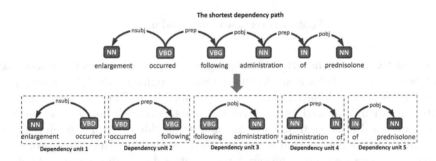

Fig. 2. Dependency units on the SDP.

4   Proposed Model

We design our cduCNN model to learn the features on the sequence of DUs that consist of both token and dependency information. Figure 3 depicts the overall architecture of our proposed model. The model mainly consists of three components: compositional embeddings layer, convolution phase, and a softmax classifier.

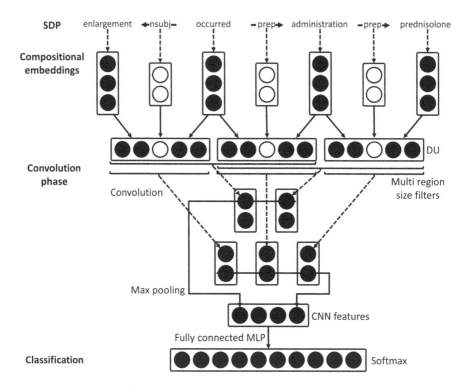

Fig. 3. An overview of proposed model.

Given the dependency tree of a sentence as input, we extract the shortest path between two entities from the tree, pass it through an embedding generation layer for token embeddings and dependency embeddings. These two embeddings matrix are then composed into dependency units. A convolution layer is applied to capture local features from each unit and its neighbors. A max pooling layer thereafter gathers information from these features combines these features into a global feature vector, and a softmax layer is followed to perform a $(K+1)$-class classification. This final $(K+1)$-class distribution indicates the probability of each relation respectively. The details of each layer are described below.

4.1   Compositional Embeddings

In the embeddings layer, each component of the SDP (*i.e.*, token or dependency) is transformed into a vector $w_e \in \mathbb{R}^d$, where d is the desired embedding dimension. In order to capture more features along the SDP, we compositionally represent the token and dependency on SDP with various type of information.

Dependency Embeddings: The dependency directions are proved effective for the relation extraction task [18]. However, treated the dependency relations with opposite directions as two separated relations can induce that two vectors of the same relation are disparate. We represent dependency relation dep_i as a vector that is the concatenation of dependency type and dependency direction. The concatenated vector is then transform into a final representation d_i of dependency relation as follow:

$$d_i = \tanh\left(W_d\left[d_i^{typ} \oplus d_i^{dir}\right] + b_d\right) \tag{1}$$

where $d^{typ} \in \mathbb{R}^{d^{dtyp}}$ represents the dependency relation type among 62 labels; and $d^{dir} \in \mathbb{R}^{d^{ddir}}$ is the direction of the dependency relation, *i.e.* from left-to-right or vice versa on the SDP.

Token Embeddings: For token representation, we take advantage of five types of information, including:

- **Pre-trained fastText embeddings** [1]: which learned the word representation based on its external context, therefore allows words that often appear in similar context to have similar representations. Each token in the input SDP is transformed into a vector t_i^w by looking up the embedding matrix $W_w^e \in \mathbb{R}^{d^w \times |V^w|}$, where V^w is a vocabulary of all words we consider.
- **Character-based embeddings**: CNN is an effective approach to learn the character-level representations that offer the information about word morphology and shape (like the prefix or suffix of word). Given a token composed of n characters $c_1, c_2, ..., c_n$, we first represent each character c_i by an embedding r_i using a look-up table $W_c^e \in \mathbb{R}^{d^{char} \times |V^c|}$, where V^c is the alphabet. A deep CNN with various window sizes is applied on the sequence $\{r_1, r_2, ..., r_n\}$ to capture the character features. A pooling layer is followed to produce the final character embedding t_i^c.
- **Position embeddings**: To extract the semantic relation, the structure features (*e.g.*, the SDP between nominals) do not have sufficient information. The SDP is lack of in-sentence location information that the informative words are usually close to the target entities. We make use of position embeddings to keep track of how close each SDP token is to the target entities on the original sentence. We first create a 2-dimensional vector $[d_i^{e1}, d_i^{e2}]$ for each token that is combination of relative distances from current token to two entities. Then, we obtain the position embedding t_i^p as follow:

$$t_i^p = \tanh\left(W_p[d_i^{e1}, d_i^{e2}] + b_p\right) \tag{2}$$

- **POS tag embeddings**: A token may have more than one meaning representing by its grammatical tag such as noun, verb, adjective, adverb, etc. To address this problem, we use the part-of-speech (POS) tag information in the token representation. We randomly initialize the embeddings matrix $W_t^e \in \mathbb{R}^{d^t \times 56}$ for 56 OntoNotes v5.0 of the Penn Treebank POS tags. Each POS tag is then represented as a corresponding vector t_i^t.

– **WordNet embeddings**: WordNet is a large lexical database containing the set of the cognitive synonyms (synsets). Each synset represents a distinct concept of a group and has a coarse-grained POS tag (*i.e.*, nouns, verbs, adjectives or adverbs). Synsets are interlinked by their conceptual-semantic and lexical meanings. For this paper, we heuristically select 45 F1-children of the WordNet root which can represent the super-senses of all synsets. The WordNet embedding t_i^n of a token is in form of a sparse vector that figure out which sets the token belongs to.

Finally, we concatenate the word embedding, character-based embedding, position embedding, POS tag embedding, and WordNet embedding of each token into a vector, and transform it into the final token embedding as follow:

$$t_i = \tanh\left(W_t[t_i^w \oplus t_i^c \oplus t_i^p \oplus t_i^t \oplus t_i^n] + b_t\right) \tag{3}$$

4.2   CNN with Dependency Unit

Our CNN receives the sequence of DUs $[u_1, u_2, ..., u_n]$ as the input, in which two token embeddings t_i, t_{i+1} and dependency relation d_i are concatenate into a d-dimensional vector u_i. Formally, we have:

$$u_i = t_i \oplus d_i \oplus t_{i+1} \tag{4}$$

In general, let the vector $u_{i:i+j}$ refer to the concatenation of $[u_i, u_{i+1}, ..., u_{i+j}]$. A convolution operation with region size r applies a filter $w_c \in \mathbb{R}^{rd}$ on a window of r successive units to capture a local feature. We apply this filter to all possible window on the SDP $[u_{1:r}, u_{2:r+1}, ..., u_{n-r+1:n}]$ to produce convolved feature map. For example, a feature map $c^r \in \mathbb{R}^{n-r+1}$ is generated from a SDP of n DUs by:

$$c^r = \left\{\tanh(w_c u_{i:i+r-1} + b_c)\right\}_{i=1}^{n-r+1} \tag{5}$$

We then gather the most important features from the feature map, which have the highest values by applying a max pooling [2] layer. This idea of pooling can naturally deal with variable sentence lengths since we take only the maximum value $\hat{c} = \max(c^r)$ as the feature to this particular filter.

Our model manipulates multiple filters with varying region sizes $(1-3)$ to obtain a feature vector f which take advantage from wide ranges of n-gram features that can boost relation extraction performance.

4.3   Classification

The features from the penultimate layer are then fed into a fully connected multi-layer perceptron network (MLP). The output h_n of the last hidden layer is the higher abstraction-level features, which is then fed to a softmax classifier to predict a $(K+1)$-class distribution over labels $\hat{y}$:

$$\hat{y} = \text{softmax}\left(W_y h_n + b_y\right) \tag{6}$$

4.4   Objective Function and Learning Method

The proposed cduCNN relation classification model can be stated as a parameter tuple θ. The $(K + 1)$-class distribution $\hat{y}$ predicted by the softmax classifier denotes the probability that SDP is of relation R. We compute the the penalized cross-entropy, and further define the training objective for a data sample as:

$$L(\theta) = - \sum_{i=0}^{K} y_i \log \hat{y}_i + \lambda \|\theta\|^2 \qquad (7)$$

where $y \in \{0, 1\}^{(K+1)}$ indicating the one-hot vector represented the target label, and λ is a regularization coefficient. To compute the model parameters θ, we minimize $L(\theta)$ by applying mini-batch gradient descent (GD) with Adam optimizer [6] in our experiments. θ is randomly initialized and is updated via back-propagation through neural network structures.

Table 1. System's performance on SemEval-2010 Task 8 dataset

Model	Feature set	F1
SVM [14]	Lexical features, dependency parse, hypernym, NGrams, PropBank, FanmeNet, NomLex-Plus, TextRunner	82.2
CNN [20]	Word embeddings	69.7
	+ Lexical features, WordNet, position feature	82.7
mvRNN [15]	Word embeddings	79.1
	+ WordNet, NER, POS tag	82.4
SDP-LSTM [19]	Word embeddings	82.4
	+ WordNet, GR, POS tag	83.7
depLCNN [18]	Word embeddings	81.9
	+ WordNet, word around nominals	83.7
	+ Negative sampling	85.6
Baseline	Word embeddings	83.4
	+ DU	83.7
cduCNN (our model)	Compositional Embedding, DU	84.7
	+ Normalize conjunction	85.1
	+ Ensemble	86.1
	+ Normalize object of a preposition	80.6

5   Experimental Evaluation

5.1   Dataset

Our model was evaluated on two different datasets: SemEval-2010 Task 8 for general domain relation extraction and BioCreative V CDR for chemical-induced disease relation extraction in biomedical scientific abstracts.

The SemEval-2010 Task 8 [5] contains 10, 717 annotated relation classification examples and is separated into two subsets: 8, 000 instances for training and 2, 717 for testing. We randomly split 10 percents of the training data for validation. There are 9 directed relations and one undirected Other class.

The BioCreative V CDR task corpus [16] (BC5 corpus) consists of three datasets, called training, development and testing set. Each dataset has 500 PubMed abstracts, in which each abstract contains human annotated chemicals, diseases entities, and their abstract-level chemical-induced disease relations.

In the experiments, we fine-tune our model on training (and development) set(s) and report the results on the testing set, which is kept secret with the model. We conduct the training and testing process 20 times and calculate the averaged results. For evaluation, the predicted labels were compared to the golden annotated data using standard precision (P), recall (R), and F1 score metrics.

5.2   Experimental Results and Discussion

System's Performance: Table 1 summarizes the performances of our model and comparative models. For a fair comparison with other researches, we implemented a baseline model, in which we interleave the word embeddings and dependency type embeddings for the input of CNN. It yields higher F1 than competitors which are feature-based or DNN-based with information from pre-trained Word embeddings only. With the improvement of 0.3% when applying DU on the baseline model, our model achieves the better result than the remaining comparative DNN approaches which utilized full sentence and position feature without the advanced information selection methods (*e.g.*, attention mechanism). This result is also equivalent to other SDP-based methods.

The results also demonstrate the effectiveness of using compositional embedding that brings an improvement of 1.0% in F1. Our cduCNN model yields an F1-score of 84.7%, outperforms other comparative models, except depLCNN model with data augmented strategy, by a large margin. However, the ensemble strategy by majority voting on the results of 20 runs drives our model to achieve a better result than the augmented depLCNN model.

It is worth to note that we have also conducted two techniques to normalize the dependency tree. Unfortunately, the results did not meet our expectations, with only 0.4% improvement of conjunction normalization. Normalizing the object of preposition even degrades the performance of the model with 4.1% of F1 reduction. A possible reason is that the preposition itself represent the relation on SDP, such as *"scars from stitches"* shows Cause-Effect relation while *"clip about crime"* shows Message-Topic relation. With the cut-off of prepositions, the SDP is lack of information to predict the relation.

Contribution of Components on Enriched SDP: Figure 4 shows the changes in F1 when ablating each component and information source from the cduCNN model. The F1 reductions illustrate the contributions of all proposals

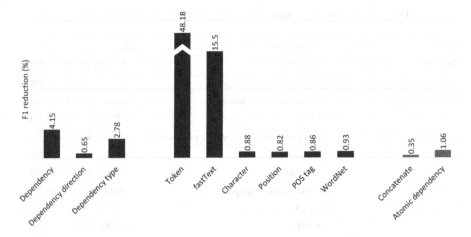

Fig. 4. Contribution of each component. The black columns indicate the kick-out of components. The grey columns indicate the alternative methods of embedding.

to the final result. However, the important levels are varied among different components and information sources. Both dependency and token embeddings have a great influence on the model performance. Token embedding plays the leading role, eliminating it will reduce the F1 by 48.18%. However, dependency embedding is also an essential component to have the good results. Removing fast-Text embedding, dependency embedding and dependency type make significant changes of 15.5% 4.15% and 2.78% respectively. The use of other components brings a quite small improvement.

An interesting observation comes from the interior of dependency and token embeddings. The impact of kicking the whole component out is much higher than the total impact of kicking each minor component out. This proves that the combination of constituent parts is thoroughly utilized by our compositional embedding structure.

Another experiment on using alternative methods of embedding also proves the minor improvement of compositional embedding. The result lightly reduces when we concatenate the embedding elements directly without transforming into a final vector or treat two divergent directional relations as to atomic relations.

Model's Adaptation to Other Domain: Table 2 shows our results on biomedical BioCreative V CDR corpus compared to some related researches. Our model outperforms the traditional SVM model using rich feature set without additional data and the hybrid DNN model with position feature. The average result is lower than ASM model using dependency graph. However, the conjunction normalization and ensemble technique can boost our F1-score 1.15%.

We further apply the post processing rules on the predictions of the model to improve the recall and achieve the best result among competing models with

Table 2. System's performance on BioCreative V CDR dataset

Model	Feature set	P	R	F1
BioCreative benchmarks	Average result*	47.09	42.61	43.37
	Rank no. 1 result*	55.67	58.44	57.03
UET-CAM [8]	SVM, rich feature set	53.41	49.91	51.60
	+ silverCID corpus	57.63	60.23	58.90
hybridDNN [21]	Syntactic feature, word embeddings	62.15	47.28	53.70
	+ Context	62.39	47.47	53.92
	+ Position	62.86	47.47	54.09
ASM [12]	Dependency graph	49.00	67.40	56.80
Baseline	Word embeddings	60.25	49.37	54.27
	+ DU	60.33	50.36	54.90
cduCNN (our model)	Compositional Embedding, DU	57.24	55.27	56.24
	+ Normalize conjunction	56.95	56.14	56.54
	+ Ensemble	58.74	56.10	57.39
	+ Post processing	52.09	70.09	59.75
	+ Normalize object of a preposition	56.66	55.94	56.30

*Results are provided by the BioCreative V.

59.75%. The results also highlight out the limitation of our model about cross-sentence relation. We leave this issue for our future works.

6   Conclusion

In this paper, we have presented a neural relation extraction architecture with the compositional representation of the SDP. The proposed model is capable of utilizing the dominant linguistic and architectural features, such as word embeddings, character embeddings, position feature, WordNet and part-of-speech tag.

The experiments on SemEval-2010 Task 8 and BioCreative V CDR datasets showed that our model achieves promising result when compared with other comparative models. We also investigated and verified the rationality and contributions of each model's constituent parts, features, and additional techniques. The result also demonstrated the adaptability of our model on classifying many types of relation in different domains.

Our limitation of cross-sentence relations extraction is highlighted since it resulted in low performance on the BioCreative V CDR corpus compared to state-of-the-art results which handled this problem significantly. Moreover, the SDP between two nominals may be lack of supported information, raising the motivation to take advantages of more informative feature for augmenting the SDP. We aim to address these problems in our future works.

References

1. Bojanowski, P., Grave, E., Joulin, A., Mikolov, T.: Enriching word vectors with subword information. Trans. Assoc. Comput. Linguist. **5**, 135–146 (2017)
2. Boureau, Y.L., Ponce, J., LeCun, Y.: A theoretical analysis of feature pooling in visual recognition. In: Proceedings of the 27th International Conference on Machine Learning (ICML 2010), pp. 111–118 (2010)
3. Bunescu, R.C., Mooney, R.J.: A shortest path dependency kernel for relation extraction. In: Proceedings of the Conference on Human Language Technology and Empirical Methods in Natural Language Processing, pp. 724–731. Association for Computational Linguistics (2005)
4. Ching, T., et al.: Opportunities and obstacles for deep learning in biology and medicine. J. R. Soc. Interface **15**(141), 20170387 (2018)
5. Hendrickx, I., et al.: Semeval-2010 task 8: multi-way classification of semantic relations between pairs of nominals. In: Proceedings of the Workshop on Semantic Evaluations, pp. 94–99 (2009)
6. Kingma, D.P., Ba, J.: Adam: a method for stochastic optimization. arXiv preprint arXiv:1412.6980 (2014)
7. Le, H.Q., Can, D.C., Vu, S.T., Dang, T.H., Pilehvar, M.T., Collier, N.: Large-scale exploration of neural relation classification architectures. In: Proceedings of the 2018 Conference on Empirical Methods in Natural Language Processing, pp. 2266–2277 (2018)
8. Le, H.Q., Tran, M.V., Dang, T.H., Ha, Q.T., Collier, N.: Sieve-based coreference resolution enhances semi-supervised learning model for chemical-induced disease relation extraction. Database **2016** (2016). https://doi.org/10.1093/database/baw102. ISSN: 1758-0463
9. LeCun, Y., Bottou, L., Bengio, Y., Haffner, P.: Gradient-based learning applied to document recognition. Proc. IEEE **86**(11), 2278–2324 (1998)
10. Liu, Y., Wei, F., Li, S., Ji, H., Zhou, M., Houfeng, W.: A dependency-based neural network for relation classification. In: Proceedings of the 53rd Annual Meeting of the Association for Computational Linguistics and the 7th International Joint Conference on Natural Language Processing (Volume 2: Short Papers), vol. 2, pp. 285–290 (2015)
11. Mikolov, T., Sutskever, I., Chen, K., Corrado, G.S., Dean, J.: Distributed representations of words and phrases and their compositionality. In: Advances in Neural Information Processing Systems, pp. 3111–3119 (2013)
12. Panyam, N.C., Verspoor, K., Cohn, T., Ramamohanarao, K.: Exploiting graph kernels for high performance biomedical relation extraction. J. Biomed. Semant. **9**(1), 7 (2018)
13. Rajpurkar, P., Zhang, J., Lopyrev, K., Liang, P.: Squad: 100,000+ questions for machine comprehension of text. In: Proceedings of the 2016 Conference on Empirical Methods in Natural Language Processing, pp. 2383–2392 (2016)
14. Rink, B., Harabagiu, S.: UTD: classifying semantic relations by combining lexical and semantic resources. In: Proceedings of the 5th International Workshop on Semantic Evaluation, pp. 256–259. Association for Computational Linguistics (2010)
15. Socher, R., Huval, B., Manning, C.D., Ng, A.Y.: Semantic compositionality through recursive matrix-vector spaces. In: Proceedings of the 2012 Joint Conference on Empirical Methods in Natural Language Processing and Computational Natural Language Learning, pp. 1201–1211. ACL (2012)

16. Wei, C.H., et al.: Overview of the BioCreative V chemical disease relation (CDR) task. In: Proceedings of the Fifth BioCreative Challenge Evaluation Workshop, pp. 154–166 (2015)
17. Wu, F., Weld, D.S.: Open information extraction using Wikipedia. In: Proceedings of the 48th Annual Meeting of the Association for Computational Linguistics, pp. 118–127. Association for Computational Linguistics (2010)
18. Xu, K., Feng, Y., Huang, S., Zhao, D.: Semantic relation classification via convolutional neural networks with simple negative sampling. In: Proceedings of the 2015 Conference on Empirical Methods in Natural Language Processing, pp. 536–540 (2015)
19. Xu, Y., Mou, L., Li, G., Chen, Y., Peng, H., Jin, Z.: Classifying relations via long short term memory networks along shortest dependency paths. In: Proceedings of the 2015 Conference on Empirical Methods in Natural Language Processing, pp. 1785–1794 (2015)
20. Zeng, D., Liu, K., Lai, S., Zhou, G., Zhao, J.: Relation classification via convolutional deep neural network. In: Proceedings of the 25th International Conference on Computational Linguistics: Technical Papers, pp. 2335–2344 (2014)
21. Zhou, H., Deng, H., Chen, L., Yang, Y., Jia, C., Huang, D.: Exploiting syntactic and semantics information for chemical–disease relation extraction. Database **2016** (2016). https://doi.org/10.1093/database/baw048. ISSN: 1758-0463

An Adversarial Learning and Canonical Correlation Analysis Based Cross-Modal Retrieval Model

Thi-Hong Vuong[1(✉)], Thanh-Huyen Pham[1,2], Tri-Thanh Nguyen[1],
and Quang-Thuy Ha[1]

[1] Vietnam National University, Hanoi (VNU),
VNU-University of Engineering and Technology (UET),
No. 144, Xuan Thuy, Cau Giay, Hanoi, Vietnam
{hongvt57,ntthanh,thuyhq}@vnu.edu.vn
[2] Ha Long University, Quang Ninh, Vietnam
phamthanhhuyen@daihochalong.edu.vn

Abstract. The key of cross-modal retrieval approaches is to find a maximally correlated subspace among multiple datasets. This paper introduces a novel Adversarial Learning and Canonical Correlation Analysis based Cross-Modal Retrieval (ALCCA-CMR) model. For each modality, the ALCCA phase finds an effective common subspace and calculates the similarity by canonical correlation analysis embedding for cross-modal retrieval. We demonstrate an application of ALCCA-CMR model implemented for the dataset of two modalities. Experimental results on real music data show the efficacy of the proposed method in comparison with other existing ones.

Keywords: Cross-modal retrieval · Adversarial learning ·
Canonical correlation analysis

1 Introduction

Due to the explosion multimodal data, i.e., the data describing the same events or topics such as texts, audio, images, and videos, cross-modal retrieval (CMR) has drawn much attention. In order to optimally benefit from the source of multimodal data and make maximal use of the developing multimedia technology, automated mechanisms need to set up a similarity link from one multimedia item to another if parallel datasets is semantically correlated. Constructing a joint representation invariant across different modalities is of significant importance in many multimedia applications. Previous studies have focused mainly on single modality scenarios [2,7,11]. However, these techniques mainly use metadata such as keywords, tags or associated descriptions to calculate similarity other than content-based information. In this study, we use content-based multimodal data for cross-modal retrieval [5,13,14,18]. The various approaches have been proposed to deal with this problem, which can be roughly divided into two

© Springer Nature Switzerland AG 2019
N. T. Nguyen et al. (Eds.): ACIIDS 2019, LNAI 11431, pp. 153–164, 2019.
https://doi.org/10.1007/978-3-030-14799-0_13

categories as [16]: real-value representation learning [13,14,18] and binary representation learning [5,17,22]. The approach in this paper focuses on in the category of real-value representation.

The features of multimodal data have inconsistent distribution and representation, therefore a modality gap needs to be bridged in a certain way to access the semantic similarity of items across modalities. A common approach to bridge the modality gap is representation learning. The goal is to find projections of data items from different modalities into a common feature representation subspace in which the similarity between them can be assessed directly. Recently, the study have focused on maximization the cross-modal pairwise item correlation or item classification accuracy like canonical correlation analysis (CCA) [10,19,20]. However, the existing approaches fail to explicitly address the statistical aspect of the transformed features of multimodal data, the similarity between their distributions must be measured in a certain way. The practical challenge is the difficulty in obtaining well-matches across datasets that are essential for data-driven learning such as deep learning [12,15,18].

We focus on the real-value approach for the supervised representation learning by the adversarial learning and CCA for cross-modal retrieval. The adversarial learning, which explored deep neural network (DNN) by a system of two neural networks contesting with each other, was inspired by the effectiveness in image applications [6,14,21]. CCA is a statistical technique that extracted correlation between two datasets [3,4,9]. It capitalizes on the knowledge that the different modalities represent different sets of descriptors for characterizing the same object. On the one hand, CCA and DNN can be combined together for deep representations in computer vision, e.g., Deep-CCA (DCCA) method [1] trained an end to end system between DNN and CCA. In this paper, we propose a ALCCA-CMR method for cross-modal retrieval. We evaluated the proposed approach on a music dataset and the experimental results show that it significantly outperforms the state-of-the-art approach in cross-modal retrieval. The rest of this paper is organized as follows. Section 2 shows the detail of ALCCA-CMR method and we evaluate it in Sect. 3. Section 4 describes the related work. Finally, Sect. 5 concludes the paper and gives some future work.

2   ALCCA-CMR Model

2.1   Problem Formulation

The ALCCA-CMR contains two sub-problems: ALCCA and CMR. The ALCCA phase finds an common subspace effectively by adversarial learning and CCA. Then, CMR phase retrieve cross-modal based on the common subspace.

In ALCCA phase, the inputs are the feature matrices of two modalities such as $\mathbf{A} = \{a_1, \ldots, a_n\}$, $\mathbf{T} = \{t_1, \ldots, t_n\}$ and a label matrix $\mathbf{L} = \{l_1, \ldots, l_n\}$, where n is the number of samples; $\mathbf{A}$ and $\mathbf{T}$ are audio and lyrics text datasets, correspondingly. The output is an ALCCA model which finds an common subspace S for mapping cross-modal. In S, the similarity of different points reflects the semantic closeness between their corresponding original inputs. We assume that

f_A and f_T can take $\mathbf{A}$ and $\mathbf{T}$ in $S = \{S_A, S_T\}$ such as $S_A = f_A(\mathbf{A}; \theta_A)$ and $S_T = f_T(\mathbf{T}; \theta_T)$. We have two mappings $f_A(\mathbf{a}; \theta_A)$ and $f_T(\mathbf{t}; \theta_T)$ that transform audio and lyrics text features into d dimensional vector s_A and s_T with $s_A^i = f_A(\mathbf{a}_i; \theta_A)$ and $s_T^i = f_T(\mathbf{t}_i; \theta_T)$.

In CMR phase, given an input as an audio/lyric query, the output is a list of relevant audio and lyrics by cosine similarity measure.

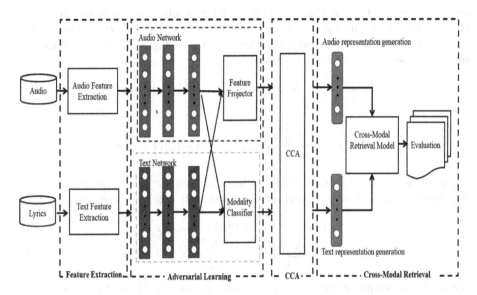

Fig. 1. The general flowchart of the proposed method. Given parallel audio and lyrics datasets, the feature extraction phase extracts audio features and lyrics text features. For each modality, ALCCA seeks an effective common subspace in the adversarial learning phase and calculates their similarity by CCA embedding for CMR

2.2   Proposed Framework

The process of cross-modal retrieval is showed in Fig. 1. The feature extraction phase extracts audio feature and lyrics text feature. The ALCCA phase tries to generate a common subspace for supervised multimodal data. Adversarial learning is the interplay between feature projector and modality classifier D with parameter θ_D, conducted as a minimax game. The feature projector and classifier trained under the adversarial leaning. Audio and lyrics features first pass through respective transformation f_A and f_T. The goal of modality classifier is to maximize its prediction precision given a transformed feature vector. Whereas, the feature projector are trained to generate modality invariant features minimizing the classifier's prediction precision. Then, transformed features are calculated their similarity by CCA function. The CMR implements cross-modal retrieval and evaluates the performance.

2.3   Adversarial Learning and CCA

Adversarial Learning. We based on the adversarial learning the same as [14] to implement for audio and lyrics text. In the adversarial learning, *feature projector* trained to generate modality invariant features to maximize the modality classifier error while *modality classifier* is trained to minimize its error.

Feature Projector. The goal of feature projector implements the process of modality-invariant embedding of audio and lyrics into a common subspace. In the feature projector, we use embedding loss L_{emb} that is formulated as the combination of the intra-modal discrimination loss L_{imd} and the inter-modal invariant loss L_{imi} with regularization L_{reg}.

$$L_{emd}(\theta_A, \theta_T, \theta_{imd}) = \alpha.L_{imi} + \beta.L_{imd} + L_{reg} \tag{1}$$

$$L_{imd}(\theta_{imd}) = -\frac{1}{n}\sum_{i=1}^{n}(m_i.(log\hat{p}_i(a_i) + log(1 - \hat{p}_i(t_i)) \tag{2}$$

where m_i is the ground-truth modality label of each instance, expressed as one-hot vector, $\hat{p}$ is probability distribution of semantic categories per item.

$$L_{imi}(\theta_A, \theta_T) = L_{imi}(\theta_A) + L_{imi}(\theta_T). \tag{3}$$

$$= \sum_{i,j,k} l2(a_i, t_j) + \sum_{i,j,k} l2(t_i, a_j)) \tag{4}$$

where the hyper-parameters α and β control the contributions of the two terms. All the distances between the feature mapping $f_A(A; \theta_A)$ and $f_T(T; \theta_T)$ per couple (audio, text) item pair (a_i, t_i) were used l2 norm.

$$L_{reg} = \sum_{l=1}^{L}(||W_a^l||F + ||W_t^l||F) \tag{5}$$

where F denotes the Frobenius norm and W_a, W_t represent the layer-wise parameters of deep neural networks.

Modality Classifier. is a modality classifier D with parameter θ_D which actives as discriminator. The cross-entropy loss of modality classification is calculated as Eq. 6.

$$L_{adv}(\theta_D) = -\frac{1}{n}\sum_{i=1}^{n}(m_i.(logD(a_i; \theta_D) + log(1 - D(t_i; \theta_D))). \tag{6}$$

Optimization. The optimization goals of the two objective functions are opposite, the process runs as minimax game [6] as follow:

$$\hat{\theta_A}, \hat{\theta_T}, \hat{\theta_{imd}} = \underset{(\theta_A, \theta_T, \theta_{imd})}{argmin} (L_{emd}(\theta_A, \theta_T, \theta_{imd}) - L_{adv}(\hat{\theta_D})). \tag{7}$$

$$\hat{\theta_D} = argmax_{(\theta_D)}(L_{emd}(\hat{\theta_A}, \hat{\theta_T}, \hat{\theta_{imd}}) - L_{adv}(\theta_D)). \tag{8}$$

As in [14], minimax optimization was performed efficiently by incorporating Gradient Reversal Layer (GRL). If GRL is added before the first layer of the modality classifier, we update the model parameters using following rules

$$\theta_A \leftarrow \theta_V - \mu.\nabla_{\theta_A}(L_{emb} - L_{adv}), \tag{9}$$

$$\theta_T \leftarrow \theta_T - \mu.\nabla_{\theta_T}(L_{emb} - L_{adv}), \tag{10}$$

$$\theta_{imd} \leftarrow \theta_{imd} - \mu.\nabla_{\theta_{imd}}(L_{emb} - L_{adv}), \tag{11}$$

$$\theta_D \leftarrow \theta_D + \mu.\nabla_{\theta_{imd}}(L_{emb} - L_{adv}). \tag{12}$$

where μ is learning rate. The results of the adversarial learning learn representation in common subspace: $f_A(A; \theta_A)$ and $f_T(T; \theta_T)$.

The pseudo-code of the proposed method using ALCCA for cross-modal retrieval is shown in Algorithm 1.

Algorithm 1. Pseudocode of ALCCA

1: **procedure** ALCCA(A, T)
2:     Extract audio and lyrics features $F_A \leftarrow A, F_T \leftarrow T$
3:     **for** each epoch **do**
4:         Randomly divide F_A, F_T to batches
5:         **for** each batch of audio and lyrics **do**
6:             **for** each pair (a, t) **do**
7:                 Compute representations f_A and f_T
8:                 **for** k steps **do**
9:                     Update parameters θ_A as Eq. 9
10:                    Update parameters θ_T as Eq. 10
11:                    Update parameters θ_{imd} as Eq. 11
12:                Update parameters θ_D as Eq. 12
13:                learned representation in $S=(f_A, f_T)$
14:                $x \leftarrow f_A(a)$
15:                $y \leftarrow$ by $f_T(t)$
16:            Get converted batch (X, Y) from (x, y)
17:            Apply CCA on (X, Y) to compute w_X, w_Y as Eq. 13
18:            Compute number of canonical components

CCA. CCA is used to maximally correlated between two multi-dimension variables $X \in \mathbb{R}^{p \times n}$ and $Y \in \mathbb{R}^{q \times n}$. Where n is the number of samples, p and q are the number of features of X and Y, respectively. When a linear projection is performed, CCA tries to find two canonical weights w_x and w_y, so that the correlation between the linear projections $w_x X^T$ and $w_y Y^T$ is maximized. The correlation coefficient ρ is defined as

$$
\rho = \underset{(w_x, w_y)}{argmax}\, corr(w_x^T x, w_y^T y)
$$

$$
= \underset{(w_x, w_y)}{argmax} \frac{w_x^T C_{xy} w_y}{\sqrt{w_x^T C_{xx} w_x \cdot w_y^T C_{yy} w_y}}. \tag{13}
$$

where C_{xy} is the cross-covariance matrix of X and Y; C_{xx} and C_{yy} are covariance matrices of X and Y, respectively. CCA obtains two directional basis vectors w_x and w_y such that the correlation between $X^T w_x$ and $Y^T w_y$ is maximum. Regularized CCA (RCCA) [4] is an improved version of CCA which used a ridge regression optimization scheme to prevent over-fitting in case of insufficient training data. However, RCCA computation is very expensive because of this regularization process. We use CCA and CCA variants to calculate the similarity between audio and lyrics in the common subspace with a number of canonical components for cross-modal retrieval. In the subspace, we use CCA with the number of components from 10 to 100.

2.4   Cross-Modal Retrieval

Evaluation Metric. In the retrieval evaluation, we use the standard evaluation measures used in most prior work on cross-modal retrieval [20], i.e., mean reciprocal rank 1 (MRR1) and recall@N. Because there is only one relevant audio or lyrics, MRR1 is able to show the rank of the result. MRR1 is defined by Eq. 14

$$
MRR1 = \frac{1}{N_q} \sum_{i=1}^{N_q} \frac{1}{rank_i(1)}, \tag{14}
$$

where N_q is the number of the queries and $rank_i(1)$ corresponds to the rank of the relevant item in the i-th query. We also evaluate recall@N to see how often the relevant item is included in the top of the ranked list. Let S_q be the set of its relevant items ($|S_q| = 1$) in the database for a given query and the system outputs a ranked list K_q ($|K_q| = N$). Then, recall@N is computed by Eq. 15 which is averaged over all queries.

$$
recall@N = \frac{|S_q \bigcap K_q|}{|S_q|} \tag{15}
$$

3 Experiments

3.1 Experimental Setup

We implemented the proposed method on a music dataset and compared with the methods mentioned in [20]. First, the music dataset consists of 10,000 pairs of audio and lyrics with 20 most frequent mood categories *aggressive, angry, bittersweet, calm, depressing, dreamy, fun, gay, happy, heavy, intense, melancholy, playful, quiet, quirky, sad, sentimental, sleepy, soothing, sweet.*

Audio Feature Extraction. The audio signal is represented as a spectrogram. We mainly focus on mel-frequency cepstral coefficients (MFCCs). For each audio signal, a slice of 30 s was resampled to 22,050 Hz with a single channel. Each audio was extracted 20 MFCC sequences and 161 frames for each MFCC.

Lyrics Text Feature Extraction. From the sequence of words in the lyrics, textual feature is computed, more specifically, by a pre-trained Doc2vec [8] model, generating a 300-dimensional feature for each song.

Implementation Details. We deployed our proposed method as follow: the adversarial learning, with three-layer feed-forward neural networks activated by *tanh* function, nonlinearly projected the raw audio and lyrics text features into a common subspace, i.e., ($A \rightarrow 1000 \rightarrow 200$ for audio modality and $T \rightarrow 200 \rightarrow 200$ for lyric modality). With modality classifier, we stuck to the three fully connected layers ($f \rightarrow 50 \rightarrow 2$). We used the same parameters in [14] with batch size is set to 100 and the training takes 200 epochs for the proposed method.

After having learned the representation in common subspace, we used CCA function to calculate their similarity for cross-modal retrieval. Here, we evaluated the impact of the number of CCA components, which affects the performance of both the baseline methods and the proposed methods. The number of CCA components is adjusted from 10 to 100.

Comparison with the Baseline Methods. We compared our proposed method against all the methods used in [20], such as PretrainCNN-CCA, Spotify-DCCA, PretrainCNN-DCCA, JointTrain-DCCA on the same dataset. This comparison can verify the effectiveness of our proposed adversarial and correlation learning for cross-modal retrieval.

3.2 Experimental Results

In the CMR phase, we use 20% data to evaluate the performance of the ALCCA when using audio or lyrics as queries. We apply 5 fold cross-validation on multi-modal data.

There are two kinds of MRR1 measures to evaluate the effectiveness as [20]: instance-level MRR1 and category-level MRR1. Instance-level MRR1 is to retrieve items of different datasets without label. Category-level MRR1 is to retrieve multi-modal data with label. I-MRR1-A, C-MRR1-A are instance-level

MRR1 and category-level when using audio as queries. I-MRR1-L, C-MRR1-L are instance-level MRR1 when using lyrics as queries.

Proposed Method Results. The proposed method results with 5 fold cross-validation on the dataset with MRR1, R@1 and R@5 measures when using audio or lyrics as queries.

Table 1. Performance cross-modal retrieval of the propose method

#CCA	I-MRR1-A	I-MRR1-L	C-MRR1-A	C-MRR1-L	R@1-A	R@1-L	R@5-A	R@5-L
10	0.08	0.081	0.213	0.212	0.045	0.047	0.100	0.099
20	0.200	0.200	0.305	0.305	0.137	0.136	0.251	0.253
30	0.300	0.300	0.387	0.387	0.224	0.224	0.371	0.376
40	0.370	0.366	0.448	0.445	0.288	0.284	0.454	0.447
50	0.415	0.411	0.488	0.484	0.335	0.327	0.498	0.496
60	0.439	0.436	0.506	0.506	0.358	0.354	0.523	0.519
70	0.453	0.449	0.519	0.517	0.371	0.367	0.539	0.535
80	0.456	0.452	0.521	0.519	0.373	0.370	0.540	0.536
90	0.447	0.444	0.515	0.513	0.365	0.362	0.531	0.529
100	0.427	0.425	0.497	0.497	0.349	0.346	0.507	0.505

In Table 1, the performance of the cross-modal retrieval overall measures is approximate between using audio and lyrics as queries, which demonstrates that the cross-modal common subspace is useful for both audio and lyrics retrieval. When the number of CCA components increases from 10 to 40, the performance also significantly increases from 10% to 30%. After that, there is a slight increase from 30% to 40% when the number of CCA components gets more than 40. The category-level MRR1 and recall@5 are higher and more stable than another measures.

Comparison with the Baseline Methods. The ALCCA-CMR model performance is more effective than that of the baseline methods on the same music dataset, and overall measures when using audio/lyrics as queries.

The Fig. 2 demonstrates that our proposed method significantly outperforms PretrainCNN-CCA, DCCA, PretrainCNN-DCCA and JointTrainDCCA on the instance-level MRR1 measure when the number of components gets more than 30. The results of the proposed method are high and stable about 40% while the results are about 25% with JointTrainDCCA, 20% with PretrainCNN-DCCA, about 15% with DCCA, about 10% with PretrainCNN-CCA.

The results in Fig. 3 show that the our proposed method is better than PretrainCNN-CCA, DCCA, PretraiinCNN-DCCA and JointTrainDCCA on the category-level MRR1 measure when the number of component gets more than 30. The results of the proposed method are high from 40% to 50% while the results are about 35% with JointTrainDCCA, 32% with PretrainCNN-DCCA, about 25% with DCCA, about 20% with PretrainCNN-CCA.

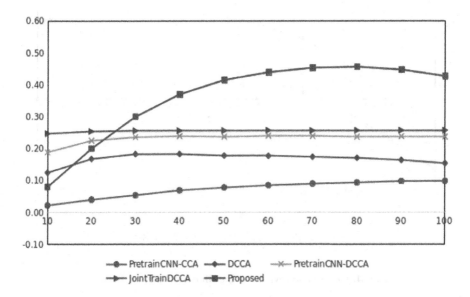

Fig. 2. Comparison with the baseline methods on instance-level MRR1

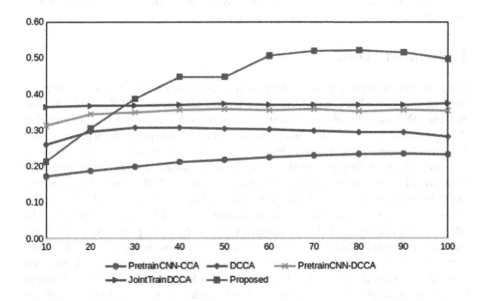

Fig. 3. Comparison with the baseline methods on category-level MRR1

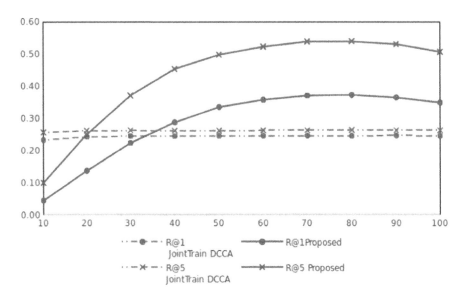

Fig. 4. Comparison with the baseline methods on recall

The results in Fig. 4 show that the our proposed method is more effective than JointTrainDCCA on the recall@1 and recall@5 when the number of components gets more than 40. The results of the proposed method are high from 40% to 50% with R@5 and about 35% with R@1. While the results of JointTrainDCCA are stable about 25% on R@1 and R@5 measures.

4   Related Work

With the rapid development of deep neural network models, DNN has increasingly been deployed in the cross-modal retrieval context as well [5,14,15,18]. The existing DNN-based cross multimedia retrieval models mainly focus on ensuring the pairwise similarity of the item pairs in a common subspace which multi-modal data can be compared directly. However, a common representation learned in this way fails to fully preserve the underlying cross-modal semantic structure in data. In [14], the adversarial cross-modal retrieval (ACMR) method used adversarial learning which was proposed by Goodfellow et al. [6] in GAN for image generation, as regularization into cross-modal retrieval for image and text. The adversarial learning used the correlation through features projections and regularized their distributions on modality classifier. Through the joint exploitation of two processes in [14] as a minimax game, the underlying cross-modal semantic structure of bimodal data is better preserved when this data is projected into the common subspace. The adversarial approach learned an effective subspace representation for image and text retrieval.

CCA has many characteristics that make it suitable for analysis of real-world experimental data. First, CCA does not require that the datasets to have the same dimensionality. Second, CCA can be used with more than two datasets simultaneously. Third, CCA does not presuppose the directionality of the relationship between datasets. Fourth, CCA characterizes relationships between datasets in an interpretable way. This is in contrast to correlation methods that merely quantify the similarity between datasets. In recent years, deep learning and CCA are used to fuse heterogeneous data such as pixel values of images and text [18], audio and images [3]. Regularized CCA (RCCA) is an advance version of CCA, which used a ridge regression optimization scheme [4,9] in the presence of insufficient training data to prevent over-fitting.

The approach proposed in this paper focuses on real-value approach for music retrieval. We combine supervised representation learning by the adversarial learning and CCA for audio and lyrics retrieval. Our approach is inspired by the effectiveness of the adversarial learning for image applications [6,14,21]. On the one hand, CCA and DNN combined together to deep representations in computer vision, like DCCA method [1]. Furthermore, our approach is motivated in music instead of focus on image applications.

5   Conclusion and Future Work

The paper proposed a ALCCA-CMR model for cross-modal retrieval. Our approach is inspired by the effectiveness of the deep learning and CCA for the supervised multi-modal data. The ALCCA phase seeks an effective common subspace in the adversarial learning phase and calculates their similarity by CCA embedding for CMR phase. The results demonstrated that our method is more effective than the baseline methods for both using audio and lyrics as queries. In the future, we will improve cross-modal retrieval accuracy by CCA variants and retrieval time.

References

1. Andrew, G., Arora, R., Bilmes, J., Livescu, K.: Deep canonical correlation analysis. In: International Conference on Machine Learning, pp. 1247–1255 (2013)
2. Boutell, M., Luo, J.: Photo classification by integrating image content and camera metadata. In: 2004 Proceedings of the 17th International Conference on Pattern Recognition, ICPR 2004, vol. 4, pp. 901–904. IEEE (2004)
3. Chaudhuri, K., Kakade, S.M., Livescu, K., Sridharan, K.: Multi-view clustering via canonical correlation analysis. In: Proceedings of the 26th Annual International Conference on Machine Learning, pp. 129–136. ACM (2009)
4. De Bie, T., De Moor, B.: On the regularization of canonical correlation analysis. In: International Symposium on ICA and BSS, pp. 785–790 (2003)
5. Feng, F., Li, R., Wang, X.: Deep correspondence restricted boltzmann machine for cross-modal retrieval. Neurocomputing **154**, 50–60 (2015)
6. Goodfellow, I., et al.: Generative adversarial nets. In: Advances in Neural Information Processing Systems, pp. 2672–2680 (2014)
7. Hu, X., Downie, J.S., Ehmann, A.F.: Lyric text mining in music mood classification. Am. Music **183**(5,049), 2–209 (2009)

8. Le, Q., Mikolov, T.: Distributed representations of sentences and documents. In: International Conference on Machine Learning, pp. 1188–1196 (2014)
9. Mandal, A., Maji, P.: Regularization and shrinkage in rough set based canonical correlation analysis. In: Polkowski, L., et al. (eds.) IJCRS 2017. LNCS (LNAI), vol. 10313, pp. 432–446. Springer, Cham (2017). https://doi.org/10.1007/978-3-319-60837-2_36
10. Mandal, A., Maji, P.: FaRoC: fast and robust supervised canonical correlation analysis for multimodal omics data. IEEE Trans. Cybern. **48**(4), 1229–1241 (2018)
11. McAuley, J., Leskovec, J.: Image labeling on a network: using social-network metadata for image classification. In: Fitzgibbon, A., Lazebnik, S., Perona, P., Sato, Y., Schmid, C. (eds.) ECCV 2012. LNCS, vol. 7575, pp. 828–841. Springer, Heidelberg (2012). https://doi.org/10.1007/978-3-642-33765-9_59
12. Ngiam, J., Khosla, A., Kim, M., Nam, J., Lee, H., Ng, A.Y.: Multimodal deep learning. In: Proceedings of the 28th International Conference on Machine Learning (ICML 2011), pp. 689–696 (2011)
13. Peng, Y., Huang, X., Qi, J.: Cross-media shared representation by hierarchical learning with multiple deep networks. In: IJCAI, pp. 3846–3853 (2016)
14. Wang, B., Yang, Y., Xu, X., Hanjalic, A., Shen, H.T.: Adversarial cross-modal retrieval. In: Proceedings of the 2017 ACM on Multimedia Conference, pp. 154–162. ACM (2017)
15. Wang, K., He, R., Wang, W., Wang, L., Tan, T.: Learning coupled feature spaces for cross-modal matching. In: Proceedings of the IEEE International Conference on Computer Vision, pp. 2088–2095 (2013)
16. Wang, K., Yin, Q., Wang, W., Wu, S., Wang, L.: A comprehensive survey on cross-modal retrieval. arXiv preprint arXiv:1607.06215 (2016)
17. Xia, R., Pan, Y., Lai, H., Liu, C., Yan, S.: Supervised hashing for image retrieval via image representation learning. In: AAAI, vol. 1, p. 2 (2014)
18. Yan, F., Mikolajczyk, K.: Deep correlation for matching images and text. In: Proceedings of the IEEE Conference on Computer Vision and Pattern Recognition, pp. 3441–3450 (2015)
19. Yao, T., Mei, T., Ngo, C.W.: Learning query and image similarities with ranking canonical correlation analysis. In: Proceedings of the IEEE International Conference on Computer Vision, pp. 28–36 (2015)
20. Yu, Y., Tang, S., Raposo, F., Chen, L.: Deep cross-modal correlation learning for audio and lyrics in music retrieval. arXiv preprint arXiv:1711.08976 (2017)
21. Zhang, H., et al.: StackGAN: text to photo-realistic image synthesis with stacked generative adversarial networks. arXiv preprint (2017)
22. Zhang, J., Peng, Y., Yuan, M.: Unsupervised generative adversarial cross-modal hashing. arXiv preprint arXiv:1712.00358 (2017)

Event Prediction Based on Causality Reasoning

Lei Lei$^{(\boxtimes)}$, Xuguang Ren, Nigel Franciscus, Junhu Wang, and Bela Stantic

Institute for Integrated and Intelligent Systems, Brisbane, QLD, Australia
{l.lei6,x.ren,n.franciscus,j.wang,b.stantic}@griffith.edu.au

Abstract. Event prediction is a challenging task which is controversial and highly debatable in the area of text mining. Previous works in utilizing causality reasoning for event prediction have shown a promising outcome in recent years. Many causality-based event prediction systems have been designed where various models, methods, and algorithms are used respectively to (1) extract events from texts, (2) construct causality relationships among events and (3) predict future events based on the causality relationships. In this paper, we first present a brief survey to systematically discuss the existing literature and highlight a common framework of the causality based event prediction systems. We discovered that most systems are not context-aware, which in reality leads to a poor performance. Based on that, we present a revised framework for causality-based event prediction, where we consider the context-related information. Our preliminary experiments suggest that the revised framework improved the overall quality of predicted events.

Keywords: Event prediction · Causality reasoning

1 Introduction

The problem of event prediction is to forecast future events that can be caused by a given event. In many situations, its importance reflects an effective decision making where one can determine the potential risks and outcomes. For example, if an earthquake just occurred nearby the ocean, it is easy to predict that tsunami may happen due to the earthquake. With this causality ((*earthquake, sea*) → (*tsunami*)) in mind, the authority may take precautions for pre-emptive evacuation. However, the real-world environments are much more complex where the causality relationships are normally hidden and difficult to justify. In the past, event prediction in the complex environment was not practical due to the limitation of their perceptive capabilities. Fortunately, with the help of the World Wide Web and advantages of data mining, an intelligent agent may perform efficient event prediction and make smarter decisions [24, 25].

The problem of event prediction is debatable in the area of text mining and natural language processing (NLP). In recent years, many efforts have been devoted to event prediction which includes [5, 12, 16, 19, 20, 24–26, 33]. Amongst

© Springer Nature Switzerland AG 2019
N. T. Nguyen et al. (Eds.): ACIIDS 2019, LNAI 11431, pp. 165–176, 2019.
https://doi.org/10.1007/978-3-030-14799-0_14

them, the causality based approaches [12, 24, 33] have achieved some exciting progress. In [33], the causality based approach achieved higher precision than several state-of-the-art link prediction models.

Causality-based approaches inference future events based on causalities among the previous events. Causality between events is the relationship between one event (cause) and a second event (effect), where the second event is the consequence of the first [21]. The philosophy that supports the causality-based event prediction approaches to perform well lies in its higher logic: causality provides the basis for explaining how things have happened and for predicting how the outcomes would be when their causes have changed [1].

Although the idea of causality based event prediction is simple, there are many challenging tasks.

(1) **Event Detection.** The events are hidden in the massive web information, normally in text formats such as tweets and online news. The challenge is how to effectively filter out the texts without events and detect the events from those containing events? Event detection has been widely studied in information retrieval [3, 15, 32]. A comprehensive survey related to event detection in twitter can be found in [6].

(2) **Event Representation.** In traditional event detection, events are normally represented by the *bag of words* model [2, 6]. However, in order to power up the underlying event prediction, we may need to organize the words in a specific manner, such as named entity [18], object-relation triplet [24] Verbs-Nouns [33] etc. The challenge here is how to select the right model to most effectively represent the events for a system aiming for event prediction?

(3) **Causality Extraction.** Causality relationships from natural language texts is a popular and challenging problem in the artificial intelligence area. And it has been extensively studied in a wide range of disciplines, including Psychology [31], Philosophy [30], and Linguistics [8]. A comprehensive survey of causality extraction from natural language texts can be found in [10]. The challenge here is how to extract the causality of events from natural language texts?

(4) **Causality Rule Construction.** There may be millions of specific events in the world while many of them are sharing similar patterns. It is not reliable to predict an event based on a causality of two particular events since one can argue that the respective causality is unique and cannot be applied to other cases [24, 33]. In contrast, it is more reasonable to predict an event based on a cluster of similar events (the idea of data mining of big data). Therefore, the challenge here is how to cluster the causalities and generalize some pervasive causality rules which can be used for the task of event prediction?

(5) **Event Matching and Prediction.** Although not specifically designed for event prediction, there are existing methods and models proposed for event detection, representation, causality extraction, and rule generalization. We may borrow their techniques when designing the system of event prediction. However, given a query event, how to predict event based on the extracted

causality relationships? This is the core challenge for the research of event prediction based on causality reasoning.

Although the approaches in [24, 33] achieved some success, they are still facing some critical problems.

Given a query event e, the existing approaches only search for similar historical events of e and identify corresponding results from related causalities. They are not considering the underlying contexts of e. For example, a football match will be held in a place at several nights of each month. If it rains on the football night, there will be a traffic jam. However, only the causality between the rain and the traffic jam is directly mentioned in the media. Therefore, we might only have a causal rule $(rain) \rightarrow (traffic\ jam)$. Simply based on this rule, it is inaccurate to predict traffic jams for any rain event in most other days. In this example, a football match is a context which should be considered when predicting any specific rain event.

Contribution. In this paper, we study the problem of event prediction based on causality reasoning. To the best of our knowledge, there is no existing survey for this particular topic. Our first contribution is a brief survey that systematically discusses several state-of-the-art approaches, emphasizing the strategies of how they solved the above challenges. We highlight a common framework that they normally follow. The second contribution of our work is a revised framework which considers the context information. We define two types of contexts: *event-related* and *factor-related*, for which we designed two relativeness functions, respectively. The third contribution is a preliminary experiment to evaluate our revised framework.

Organization. The rest of the paper is organized as follows: in Sect. 2, we present a preliminary to explain some key terms; in Sect. 3, we present the survey; in Sect. 4, we present our revised framework; in Sect. 5, we provide the experimental study; and finally in Sect. 6 we conclude the paper and indicate some future work.

2   Preliminary

Event Prediction and New Event Detection. Event prediction is different from the traditional New Event Detection (NED). NED is to detect the *first* story on a topic of interest, where a topic is defined as "a seminal event or activity, along with directly related events and activities" [2]. Event prediction focuses on the prediction of future events which have not yet happened. In contrast, NED focuses on detecting the first article discussing recently occurring events.

Causality Patterns. Causality patterns are explicit expressions to indicate causal relationships. They can be classified into the following [10]:

(1) *Causality links* linking clauses or sentences [4], which include (1) adverbial links, e.g. so, hence, therefore, (2) prepositional links, e.g. because of,

on account of, (3) subordination, e.g. because, as, since, and (4) clause-integrated links, e.g. that's why, the result was.

(2) *Causative verbs* are transitive verbs whose meanings include a causal element [29].

(3) *Resultative constructions* are sentences in which the object of a verb is followed by a phrase describing the state of the object as a result of the action denoted by the verb [28].

(4) *If-Then* conditionals often indicate that the antecedent causes the consequent.

(5) *Causation adverbs and adjectives* have causal element in their meanings [11].

Social Media and Microblogs. Social Media and Microblogs have evolved to become a source of various kinds of information. This is due to the nature of social media and microblogs where users post real-time messages about their opinions on a variety of topics and discuss current issues. Twitter is the most popular microblog service due to the availability of dataset.

3   Survey of Causality Based Event Prediction

In this section, we present a survey for several state-of-the-art approaches of event prediction based on causality. We first highlight a common framework that are followed by them. Then we discuss their strategies of tackling the challenges (2), (4) and (5) in Sect. 1. Event detection and causality extraction has been extensively studied in the literature [6,10]. We omit the discussion of the strategies of challenges (1) and (3) since the key ideas are borrowed from previous research more or less.

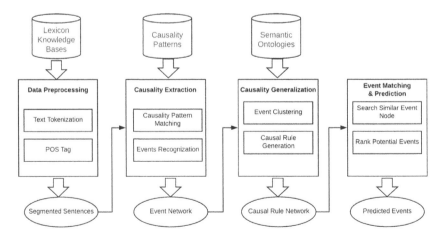

Fig. 1. Overall framework

3.1 Framework

The framework is as shown in Fig. 1. The framework is a pipeline consisting of four components:

(1) *Data pre-processing*, where the texts will be sanitized and tokenized. A POS tagging sub-process is also conducted to attach a label to each word in the sentence. The output of this component is a list of segmented sentences. And in this component, a lexicon knowledge is normally needed, such as WordNet [22] and VerbNet [27].
(2) *Causality Extraction*, where the sentences will be filtered out if not containing any causality patterns. Then for each sentence containing two causal related events, the events will be identified and finally used as nodes in the event network.
(3) *Causality Generalization*, based on the event network from the last component, this component groups the event nodes into clusters where the events in the same cluster are similar to each other. Semantic ontologies are normally needed in order to compute the similarities of events.
(4) *Event Matching and Prediction*, this component accepts event queries from the user and search the causality rule network. Based on the similar nodes and its corresponding causal rule, it ranks the potential events. After that, top-K possible events will be sent back to the user.

3.2 Event Representation

Now we first discuss the event representation which corresponds to Challenge-2. The format to represent events determines the process of constructing the event network in Component-2 of the framework.

Note that different approaches are using different formats to represent the events. And it is difficult to claim which format is more expressive than others. However, researchers tend to design the event format using a bottom-up manner, where the final event format is designed to suit algorithms of their event comparing. Since to decide whether two events are similar, a proper comparison scheme is the core of the event prediction system. There are four common formats to represent events in the literature:

(1) **Noun Phrases.** Has been used to represent events in [9,14,16,17], where each element is mapped to a syntax-driven noun phrase.
(2) **Triplet.** The triplet format is originally designed in [18] where given a set of (physical or abstract) objects O, an event triplet is in a format of $[O_i, P, t]$ where $O_i \subseteq O$ is a set of objects, P is a relation (or a property) over objects and t is a time interval.
(3) **Triplet with Roles.** [24] extends the format of Triplet by adding roles to the property relation and results in an ordered set $e = <P, O_1, \ldots, O_4, t>$. Each event consists of a temporal action or state that the event's objects exhibit (P), one ore more actors O_1 that conducted the action, one or more objects O_2 on which the action was performed, one or more instruments O_3

the action was conducted with, and one or more locations O_4 of the event. For example, given an event "The U.S army destroyed a warehouse in Iraq with explosives", which occurred on October 2004, is modelled as: Destroy (Action); U.S Army (Actor); warehouse (Object); explosives (Instrument); Iraq (Location); October 2004 (Time).

(4) **Verbs and Nouns.** Focusing on news headlines where the text information is brief and with a large fraction of constituents omitted. A *Verbs & Nouns* is used in [33]. Formally, an event e is represented as $\{W_i | W_i \in Verbs \cup Nouns\}$ in which $Verbs$ and $Nouns$ are the sets of verb and nouns in headlines, respectively. In this format, the order of W_i is consistent with their original order in the text.

3.3   Causality Rule Construction

Now we discuss the techniques used for causal rule construction which is Challenge-4 and corresponds to the Component-3 in the framework.

The key idea of causality rule construction is to group similar events into clusters where each cluster can be marked with a general type. For example, the causal events "(earthquake happened in Japan) → (22 people died)" and "(earthquake happened in South Korea) → (45 people died)" are similar and can be generalized together, where we can get a causal rule "(earthquake happened in [place]) → (people died)".

The methods of causality rule construction of [24,33] are state-of-the-art. Their key strategies are quite similar: generalizing over objects and actions [24] and generalizing words and verbs [33], where the objects and words, actions and verbs are refereeing the same information of the event.

(1) **Generalizing over objects and actions.** [24] treats the event network as a semantic network $G = (V, E)$, where the nodes $V = O$ are the objects in the universe, and the labels on the edges are relations such as isA, partOf, and capitalOf. The labels of the nodes and edges are generated by using the Linked Data ontology [7]. A similarity function between two event nodes is defined based on the *minimal generalization path*. The HAC hierarchical clustering algorithm [13] is used as the clustering method of [24]. A structure "abstract tree" is used to represent the generalized causal rules.

(2) **Generalizing words and verbs.** Comparing with [24], the generalizing process of [33] is more straightforward. Given a specific event represented by a set of *Nouns* and *Verbs*, [33] generalizes each noun to its hypernym in WordNet (e.g. the noun "chips" is generalized to "dishes"), and each verb to its class in VerbNet (e.g. "kill" is generalized to "murder-42.1"). A structure "hierarchical causal network" is used to represent the generalized causal rules.

3.4   Event Prediction

Now we discuss the event prediction which corresponds to the challenge-5 and whose key process is given in the framework of Fig. 1.

The event prediction of [24] consists of two main steps: (1) propagation of the event in the abstract tree retrieving similar nodes that match the new event. (2) application of the similar node rule on the event to produce the effect of the event. [24] ranks the potential events by the similarity defined by the *minimal generalization path*. In contrast, [33] embed event causality network into a continuous vector space, which simplifies the event manipulation and still preserves the cause-effect relationship. An energy function is designed to rank potential events where true cause-effect pairs are assumed to have low energies.

However, we noticed that most existing approaches have not considered the context information when predicting the events. This significantly decreases the prediction quality when dealing with some complex scenarios where the hidden cause-effect information is not mentioned explicitly. This problem can be demonstrated by the football night example in Sect. 1. Context-features were considered in [16] when extracting the causality relationships. However, [16] does not use any context information when predicting events based on extracted causality relationships.

The query event given by a user is usually in the format of short text which does not contain enough context information. A naive strategy is to use a static ontology to attach some historical information to the query event. However, this cannot cover any context information of real time at which point when the query event happened.

In this paper, we revise the framework in Fig. 1 by considering real-time context information when predicting the events. We propose the revised framework in the next section.

4 Context-Aware Event Prediction

Before we present the detailed steps of our event prediction, we first study two types of context information as follows:

(1) **Event-Related Context.** Upon receiving an event e from users, we consider the related events that happened close to e both temporally and spatially. For example, the event of *football match* happened on the same day and in the same city as the user's query event e is a potential context that may affect the prediction of e.

Formally, we have the following definition:

Definition 1. *Given an event happened at time T and in location L, there are some other events E happened within the time period of $T \pm \Delta_T$ and the location range $L \pm \Delta_L$, where Δ_T is a time window and Δ_L is a geolocation window. We call E_r the event-related context of e if for any $e' \in E_r$, $REL_{event}(e, e') > \tau$, where $E_r \subseteq E$, REL_{event} is a relativeness function, τ is a relativeness threshold.*

We can use the techniques of new event detection (NED) from [3,6] to get the collection of events which satisfy the temporal and spatial conditions of $T \pm \Delta_T$

and $L \pm \Delta_L$. However, the next question is to define the relativeness function REL_{event}. The similarity is different from relativeness where the former requires the events to follow a specific pattern (e.g., semantic pattern) while the latter requires the events to have some context-oriented interconnection (e.g., involve some common people). We propose a relativeness function based on the co-occurrence of the entities of events.

Definition 2. *Let $G = (V, E)$ be a event network extracted from a text collection C, each node n in V represents the entities of an event, each label $e = (n_1, n_2)$ in E records the number of co-occurrences of the entities of nodes n_1 and n_2 in C. Then we have*

$$REL_{event}(e, e') = \frac{\sum_{e \in P} Label(e)}{|P|^\rho} \tag{1}$$

where P is the shortest path from any entity of e to any entity of e', and ρ is a parameter to tune the score.

Intuitively, the above definition tries to find the shortest entity path that connects two events. The relativeness score is the co-occurrence frequencies of entities on the path divided the path length to the power of ρ. If two events are sharing some common entities, their relativeness score is high. If the relativeness between an event with query event e is below a threshold τ, this event will not be considered as context of e.

(2) **Factor-Related Context.** Factor-related context is the real time status of some factors that may affect the event prediction. The factors are represented by *Nouns* and the statuses are represented by *Adjectives*. For example, *rain* is a factor and *heavy* is a status, *rain is heavy* is a factor context of traffic events.

Formally, we have the following definition:

Definition 3. *Given an event e happened at time T, its factor-related context is defined as a set of triplets $F = \{< N, A, T >\}$ where N is a Noun representing a factor, and A is an Adjective representing the status of N at time T.*

Given a user's query event e, we search the causality network for similar events with e. Considering the factor-related context, we require the target events to have a high relativeness in terms of their factor-related contexts. We propose a relativeness definition for two events with the consideration of factor-related contexts.

Definition 4. *Given two events e, e' with their factor-related contexts F_e and $F_{e'}$, the relativeness of e, e' regarding factor-related contexts is defined as*

$$REL_{factor} = \frac{|\{f | f \in (F_e \cap F_{e'}), f(e) = f(e')|\}}{|F_e \cup F_{e'}|} \tag{2}$$

The definition of REL_{factor} is derived from the Jaccard similarity [23] function. If more factors are in the same status when two events happened, their REL_{factor} will be higher.

With the above two context definitions ready, now we present the revised framework, as shown in Fig. 2. Since the main difference is the component of event matching & prediction, here we omit other parts of the framework.

Before we search the similar event nodes, we add two sub-processes to get the information of event-related context and factor-related context respectively. Both sub-processes requires data sources of real time. Hence, we can choose such as Microblogs, such as Twitter to construct the factor-related contexts.

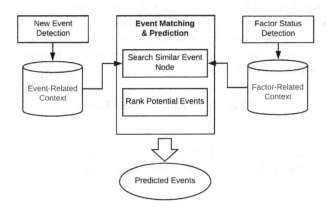

Fig. 2. Context-aware event prediction

5   Experiments

In this section, we present the results of our preliminary experimental study.

5.1   Environment Settings

For the construction of event network and causality rule network, we used the New York Times corpus which contains over 1.8 million articles published between January 1, 1987 and June 19, 2007. We extracted events and causality rules following the approach of [24].

For the query events, we used a dataset from Kaggle[1], which contains 143,000 articles. We extracted 1000 cause-effect event pairs within the year of 2016 and 2017. For each cause event, there may be more than one effect events caused by it. We use the most frequently mentioned one as effect event. We use those

---

[1] https://www.kaggle.com/snapcrack/all-the-news.

1000 cause-effect event pairs as the standard to evaluate the precision of event prediction.

For each query event, we used the above Kaggle dataset to construct the event-related context and we used Twitter dataset to construct the factor-related context for the query events. The Twitter dataset is downloaded through the Twitter API[2].

We implemented the system using Java 1.8. We loaded the raw news articles into MongoDB[3]. We used Neo4j[4] to store the event network and causality rules.

5.2   Results

For each cause-effect event pair, we feed the cause event to the system. If the predicted event fall into the top-K predictions of the system, we call it a K-hit. For a particular K, the accuracy is measure by the proportion of K-hits. We compared the accuracy our approach with [24] by varying K from 1 to 10. The results are as shown in Table 1.

Table 1. Accuracy comparison

K-hits	1	2	3	4	5	6	7	8	9	10
EP	35.2%	36.0%	40.1%	43.5%	46.7%	51.7%	56.8%	61.9%	65.3%	71.1%
Con-EP	37.3%	37.4%	42.9%	46.9%	51.0%	54.3%	60.3%	62.5%	65.5%	71.3%

In Table 1, we use **EP** to denote the event prediction without considering the context information, and use **Con-EP** to denote our context aware event prediction. As we can see the results, the precision of our method is consistently better than without context information. Both the methods reached more than 70% in the top-10 hit.

We also compared the response time of our context-aware system with existing approaches. Our approach is only slightly slower due to the construction of context information and the relativeness computation.

6   Conclusion

In this paper, we studied the problem of event prediction based on causality reasoning. We presented a survey to systematically discuss the state-of-the-art solutions of this problem. To the best of knowledge, there is no existing survey covering this topic. We noticed the shortages of current approaches when dealing with user event without enough context information. Two types of context information are formally defined. We also designed the relativeness functions to compare events more effectively. We revised the overall event prediction framework

---

[2] https://developer.twitter.com/en/docs.html.
[3] https://www.mongodb.com/.
[4] https://neo4j.com/.

by adding the context information. Finally, a experimental study was conducted and our event-aware method has shown superiority over existing approach which has not considered context information.

Since the source codes of most event prediction systems based on causality are not publicly available. One of our future tasks is to implement several more state-of-the-art systems so as to comprehensively compare the performance of our approach with them. We also noticed that there is no benchmark evaluation datasets in this area. Another future work from us is to build a benchmark data for the sake of fair comparison.

References

1. Aliferis, C.F., Statnikov, A., Tsamardinos, I., Mani, S., Koutsoukos, X.D.: Local causal and Markov blanket induction for causal discovery and feature selection for classification part I: algorithms and empirical evaluation. J. Mach. Learn. Res. **11**(Jan), 171–234 (2010)
2. Allan, J., Carbonell, J., Doddington, G., Yamron, J., Yang, Y., et al.: Topic detection and tracking pilot study: final report. In: Proceedings of the DARPA Broadcast News Transcription and Understanding Workshop, vol. 1998, pp. 194–218. Citeseer (1998)
3. Allan, J., Papka, R., Lavrenko, V.: On-line new event detection and tracking. In: ACM SIGIR Forum, vol. 51, pp. 185–193. ACM (2017)
4. Altenberg, B.: Causal linking in spoken and written English. Studia Linguistica **38**(1), 20–69 (1984)
5. Amodeo, G., Blanco, R., Brefeld, U.: Hybrid models for future event prediction. In: Proceedings of the 20th ACM International Conference on Information and Knowledge Management, pp. 1981–1984. ACM (2011)
6. Atefeh, F., Khreich, W.: A survey of techniques for event detection in Twitter. Comput. Intell. **31**(1), 132–164 (2015)
7. Bizer, C., Heath, T., Berners-Lee, T.: Linked data: the story so far. In: Semantic Services, Interoperability and Web Applications: Emerging Concepts, pp. 205–227. IGI Global (2011)
8. Blank, A., Koch, P.: Historical Semantics and Cognition, vol. 13. Walter de Gruyter, Berlin (2013)
9. Chan, K., Lam, W.: Extracting causation knowledge from natural language texts. Int. J. Intell. Syst. **20**(3), 327–358 (2005)
10. Cheong, H., Shu, L.: Automatic extraction of causally related functions from natural-language text for biomimetic design. In: International Design Engineering Technical Conferences and Computers and Information in Engineering Conference, ASME 2012, pp. 373–382. American Society of Mechanical Engineers (2012)
11. Cresswell, M.J.: Adverbs of causation. In: Cresswell, M.J. (ed.) Adverbial Modification. SLAP, vol. 28, pp. 173–192. Springer, Dordrecht (1981). https://doi.org/10.1007/978-94-009-5414-4_7
12. Dami, S., Barforoush, A.A., Shirazi, H.: News events prediction using Markov logic networks. J. Inf. Sci. **44**(1), 91–109 (2018)
13. Eisen, M.B., Spellman, P.T., Brown, P.O., Botstein, D.: Cluster analysis and display of genome-wide expression patterns. Proc. National Acad. Sci. **95**(25), 14863–14868 (1998)

14. Girju, R., Moldovan, D.I., et al.: Text mining for causal relations. In: FLAIRS Conference, pp. 360–364 (2002)
15. Guralnik, V., Srivastava, J.: Event detection from time series data. In: Proceedings of the Fifth ACM SIGKDD International Conference on Knowledge Discovery and Data Mining, pp. 33–42. ACM (1999)
16. Hashimoto, C., et al.: Toward future scenario generation: extracting event causality exploiting semantic relation, context, and association features. In: Proceedings of the 52nd Annual Meeting of the Association for Computational Linguistics (Volume 1: Long Papers), vol. 1, pp. 987–997 (2014)
17. Khoo, C.S., Chan, S., Niu, Y.: Extracting causal knowledge from a medical database using graphical patterns. In: Proceedings of the 38th Annual Meeting on Association for Computational Linguistics, pp. 336–343. Association for Computational Linguistics (2000)
18. Kim, J.: Supervenience and Mind: Selected Philosophical Essays. Cambridge University Press, Cambridge (1993)
19. Letham, B., Rudin, C., Madigan, D.: Sequential event prediction. Mach. Learn. **93**(2–3), 357–380 (2013)
20. Liben-Nowell, D., Kleinberg, J.: The link-prediction problem for social networks. J. Am. Soc. Inf. Sci. Technol. **58**(7), 1019–1031 (2007)
21. Mackie, J.L.: The Cement of the Universe: A Study of Causation. Oxford University Press, Oxford (1974)
22. Miller, G.A.: WordNet: a lexical database for English. Commun. ACM **38**(11), 39–41 (1995)
23. Niwattanakul, S., Singthongchai, J., Naenudorn, E., Wanapu, S.: Using of Jaccard coefficient for keywords similarity. In: Proceedings of the International MultiConference of Engineers and Computer Scientists, vol. 1 (2013)
24. Radinsky, K., Davidovich, S., Markovitch, S.: Learning causality for news events prediction. In: Proceedings of the 21st International Conference on World Wide Web, pp. 909–918. ACM (2012)
25. Radinsky, K., Davidovich, S., Markovitch, S.: Learning to predict from textual data. J. Artif. Intell. Res. **45**, 641–684 (2012)
26. Radinsky, K., Horvitz, E.: Mining the web to predict future events. In: Proceedings of the Sixth ACM International Conference on Web Search and Data Mining, pp. 255–264. ACM (2013)
27. Schuler, K.K.: VerbNet: a broad-coverage, comprehensive verb lexicon (2005)
28. Simpson, J.: Resultatives. Indiana University Linguistics Club (1983)
29. Thomson, J.J.: Verbs of action. Synthese **72**(1), 103–122 (1987)
30. White, P.A.: Ideas about causation in philosophy and psychology. Psychol. Bull. **108**(1), 3 (1990)
31. Wolff, P., Song, G.: Models of causation and the semantics of causal verbs. Cogn. Psychol. **47**(3), 276–332 (2003)
32. Yang, Y., Pierce, T., Carbonell, J.: A study of retrospective and on-line event detection. In: Proceedings of the 21st Annual International ACM SIGIR Conference on Research and Development in Information Retrieval, pp. 28–36. ACM (1998)
33. Zhao, S., et al.: Constructing and embedding abstract event causality networks from text snippets. In: Proceedings of the Tenth ACM International Conference on Web Search and Data Mining, pp. 335–344. ACM (2017)

A Method for Detecting and Analyzing the Sentiment of Tweets Containing Conditional Sentences

Huyen Trang Phan[1][iD], Ngoc Thanh Nguyen[2][iD], Van Cuong Tran[3][iD], and Dosam Hwang[1(✉)][iD]

[1] Department of Computer Engineering, Yeungnam University, Gyeongsan, Republic of Korea
huyentrangtin@gmail.com, dosamhwang@gmail.com
[2] Faculty of Computer Science and Management, Wroclaw University of Science and Technology, Wroclaw, Poland
Ngoc-Thanh.Nguyen@pwr.edu.pl
[3] Faculty of Engineering and Information Technology, Quang Binh University, Dong Hoi, Vietnam
vancuongqbuni@gmail.com

Abstract. Society is developing daily, and consequently, the population is more interested in public opinion. Surveys are frequently organized for detecting the attitude as well as the belief of the community in situations and their opinion about the measures or products. Users particularly express their feelings through comments posted on social networks, such as Twitter. Tweet sentiment analysis is a process that automatically detects personal information from the public emotion of the users about the events or products related to them from published tweets. Many studies have solved the sentiment analysis problem with high accuracy for the general tweets. However, these previous studies did not consider or dealt with low performance in case of tweets containing conditional sentences. In this study, we focus on solving the detection and sentiment analysis problem of a specific tweet type that includes conditional sentences. The results show that the proposed method achieves high performance in both the tasks.

Keywords: Sentiment analysis · Conditional sentence · Conditional sentence detection

1 Introduction

With the development of society, the population increasingly want to acquire more information and there is a large amount of information available for sharing. Social networks are one of the available channels that can satisfy the needs of users. Twitter is one of the most well-known platforms. Twitter has attracted numerous users who post any type of information on any topic of their interest.

© Springer Nature Switzerland AG 2019
N. T. Nguyen et al. (Eds.): ACIIDS 2019, LNAI 11431, pp. 177–188, 2019.
https://doi.org/10.1007/978-3-030-14799-0_15

The information posted on Twitter frequently contains opinions on products, services, and celebrities and details of relevant events. Twitter, with over 319 million monthly active users[1], has now become important for individuals as well as organizations who have an interest in maintaining and enhancing their power and reputation in political, social, and economic areas. Tweet sentiment analysis (TSA) solves the problem of analyzing the content of tweets posted on Twitter in terms of user's expressed sentiments. TSA provides these organizations with the ability to survey the opinions and sentiments of Twitter's users in real time.

Numerous methods have been published to solve the TSA problem. The recent works on analyzing the user sentiment can be divided into three types of approaches [17]. The first approach relies on well-known machine learning algorithms [3,6]. This approach has some limitations as a large number of labeled training tweets need for the supervised methods while the unsupervised techniques have to seek unlabeled training tweets. The second approach employs lexicon-based methods [1,2,8]. This approach depends on searching the sentiment lexicon. The limitations of lexicon-based methods are very time consuming and cannot be used alone; they have to be combined with other automated methods as a final check for avoiding errors. The third approach is a hybrid method which combines both the above approaches [15,18]. This method is widely used and plays a pivotal role in the majority of the sentiment analysis methods.

In general, the previous methods focused on analyzing the sentiment of tweets without considering tweets contain different types of sentences, for example, tweets containing conditional sentences (CSs) and sarcastic sentences. In the paper [16], the authors argued that a divide-and-conquer approach was necessary, i.e. a study focusing on each type of sentence could perform sentiment analysis more accurately. To understand clearly the limitations of the existing methods, we analyze the following examples:

(i)   *"If Hue city has a beautiful view, I will go there."* This example does not express any sentiment of the writer, although *"beautiful"* is a positive sentiment word. However, If the word *"If"* is removed, the tweet expresses a positive sentiment, which was the result produced by most of the existing methods.

(ii) *"If the Ho Chi Minh city is not nice, go to great Da Lat city."* This example expresses a positive sentiment for Da Lat city but does not exhibit any sentiment of the writer for *"Ho Chi Minh city."* However, if we do not consider the word *"If"*, then the above example expresses a negative sentiment for *"Ho Chi Minh city"* and a positive sentiment for *"Da Lat city"*. The previous methods usually produce neutral sentiment result.

(iii) *"If you like the beach with quiet, do not go to Cua Lo beach."* This example expresses a negative sentiment for the *"Cua Lo beach."* However, most of the methods produced the same result of the sentence having a positive sentiment because they only focused on the word *"like"* without considering the impact of the word *"If"*.

---

[1] https://en.wikipedia.org/wiki/Twitter.

The above examples revealed that tweets containing CSs have various effects on the accuracy of the TSA methods. This motivated us to propose a method for analyzing the sentiment of tweets containing CSs. The proposed method consists of the following steps: First, a set of features related to syntactic, lexical, semantic, and sentiment of the words is extracted. Second, tweets containing CSs are filtered. Finally, the sentiment of each tweet containing CSs is analyzed.

The difference between this method and the previous studies is that the former focuses on classifying the tweets containing CSs and analyzing their sentiment using the Multilayer Perceptron (MLP) model [10] and the Convolutional Neural Network (CNN) model [12]. Thus, the main contribution of this proposal provides a detection and sentiment analysis approach based on a divide-and-conquer strategy applied to tweets containing CSs.

The remainder of the paper is organized as follows. In Sect. 2, we summarize the work related to sentiment analysis approaches. The research problem is described in Sect. 3 and the proposed method is presented in Sect. 4. The experimental results and evaluations are shown in Sect. 5. The conclusion and future works are discussed in Sect. 6.

2   Related Works

A lot of studies regarding the tweets sentiment analysis are published. In this section, we analyze some specific studies to know the motivation for our proposal.

Bakliwalm et al. [3] used the SVM classifier for addressing TSA. In that study, the SVM classifier was trained on 11 features by considering the following steps: (i) Different preprocessing techniques were employed sequentially to quantify their effectiveness. (ii) The accuracy of the classifier was increased by implementing actions such as spelling correction, stemming, and stop-word removal. (iii) The authors assessed their approach by using two different datasets, namely, the Stanford [9] and the Mejaj [5] datasets.

Dong et al. [6] proposed an adaptive recursive neural network (AdaRNN) for entity-level TSA. In this method, a dependency tree was used to identify the words which were related to a topic and propagate the sentiment from the sentiment words associated with the topic. AdaRNN was assessed using a manually annotated dataset consisting of 6248 training and 692 testing tweets and obtained an F_1 score of 65.9%.

In the paper [2], the authors presented a sentiment analysis method applying a multistage hybrid scheme with a unified framework. This method aims to improve the performance of Twitter-based sentiment analysis systems by combining 4 classifiers (slang classifier, emoticon classifier, SentiWordNet classifier, and improved domain-specific classifier). This approach has the advantages as: Classification results are better than the baseline methods and the framework is generalized can classify tweets in any domain.

Asghar et al. [1] proposed a lexicon-based method for classifying user sentiments of online communities. They combined different lexicons, including SWN and user-defined dictionaries, with different classifiers (SWN-based classifier, negation classifier, and domain-specific classifier). An advantage of this app-roach, in addition to the relationships between the words, was the extraction of domain-specific words to resolve the issue of domain dependency. The significant limitations of the approach of Asghar et al. included a lack of dealing with CSs, slang and sarcastic sentences, which if incorporated, could result in enhanced sentiment classification.

Ghiassi et al. [7] presented a novel hybrid method by combining a dynamic artificial neural network with n-gram. Two classifiers: SVM and dynamic archi-tecture for artificial neural networks (DAN2) were built by using emoticons and tweets which contained the word *"love"* or *"hate"* or their synonyms as the fea-tures. This approach was tested on a collection of tweets crawled utilizing the subject, *"Justin Bieber"*. The results show that DAN2 outperformed SVM.

Reference [16] focused on studying the types of CSs, which have some unique characteristics requiring distinct handling. This study was performed from both linguistic and computational perspectives. In linguistic research, the authors focused on canonical tense patterns. In the computational study, SVM models were built to automatically predict whether the opinions on the topics were positive, negative, or neutral. The experimental results showed the effectiveness of the method.

In general, the *precision*, *recall*, and F_1 score exhibit that the previous meth-ods can produce results with a high accuracy and be used to classify tweets in any domain. However, the previous approaches do not mention and often yield inaccurate classification results for tweets containing CSs. (For example, *"@osamahaj5: b'RT @dontphonecallme: Do not buy this Samsung phone If your Nokia phone is good..."* is labeled positive by previous systems while the result is actually negative). Moreover, the existing methods do not provide an automatic approach for detecting tweets containing CSs.

3   Research Problems

Tweets containing CSs [14] are the tweets that describe implications or hypothet-ical situations and their consequence. Tweets containing CSs typically contain two clauses: the condition clause and consequence clause. These two clauses are dependent on each other. Their relationship has a significant impact on deter-mining whether a tweet has a positive or negative sentiment.

To address the limitations of previous methods, the main question for the research is as follows: How can analyze the sentiment of tweets containing conditional sentences? This question is partitioned into the two following sub-questions:

The first question: How to detect tweets containing CSs? The answer to this question is whether a tweet contains CSs. For instance, *"@tradefx2012: b'yup .. trump already knows whats in all of this .. its our time now"* is not tweet containing CSs, but *"@taekim07670251: b'RT @waytocum: If you think you have the right to treat people with small accounts like they dont matter just bc you have bunch of follower"* is tweet containing CSs.

The second question: What is the sentiment in tweets containing CSs? The answer to this question is the sentiment of tweets containing conditional sentences be positive, negative, or neutral.

4   The Proposed Method

In this section, we present a method to detect tweets containing CSs and analyze the sentiment of tweets containing CSs. The workflow of the method is shown in Fig. 1. Our proposed method consists of two main steps: First, the features of tweets are extracted. Second, MLP model is used to classify tweets into two sets, in which, the first set includes tweets containing CSs, the second set consists of remain tweets. Finally, sentiment analysis is applied to each tweet containing CSs to identify whether the sentiment is positive, negative, or neutral based on a CNN model. The detail of the steps is explained in the next sections.

4.1   Features Selection

To distinguish between tweets containing CSs and those not, we have to select the features which are used to recognize the CSs. In this study, the information related to the lexical, syntactic, semantic, and polarity sentiment are employed as features.

Conditional Connective Words: In this case, we are interested in the conditional connective words, such as if, only if, and even if. Each conditional connective word appearing in a tweet becomes an entry in the feature vector with feature value of 1.

n-Grams: n-grams includes 1-gram, 2-grams, and 3-grams. Each type of n-grams is created based on the term frequency inverse document frequency (tf-idf) values. Each n-grams appearing in a tweet becomes an entry in the feature vector with the corresponding feature value of tf-idf.

Part-Of-Speech (POS) Tags of Words: POS tags of the words are used as features for indicating the words which are adverbs or conjunctions. These words are the conditional connective words existing in each tweet. The NLTK toolkit [4] is used to annotate the POS tags and implement tokenization for all the tweets. POS tags with their corresponding tf-idf values are the syntactic features and features value, respectively.

Word Embeddings: A significant challenge when handling with tweets is that the lexicons and syntactics used in a tweet are informal and much different from normal sentences. It is challenging to well-capture the properties related to the lexical and syntactic features. This problem is dealt by applying the word embeddings approach to compute tweet vector representations.

A word embedding is a type of word representation which allows words with similar meanings having the same representation, in which each word is represented as a real-valued vector in the vector space. Concerning English, there are

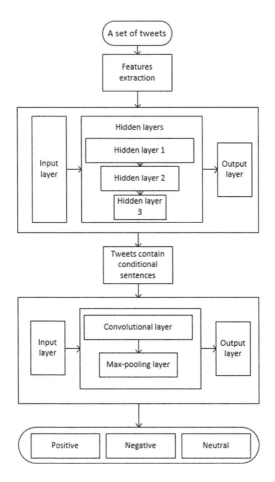

Fig. 1. Workflow of analyzing sentiment from tweets containing conditional sentences

three pre-trained models which can be used to learn a word embedding from text data such as Word2vec[2], Glove[3], and Levy Word2vec on Dependencies[4].

In this work, the glove model is applied because it is an extension to the Word2vec model. It is an approach for combining both the global statistics of matrix factorization techniques such as latent semantic analysis and social context-based learning in Word2vec. Moreover, rather than using a window to define the local context, the glove model constructs an explicit word-context or a co-occurrence word matrix using statistics across the entire text corpus. Therefore, in general, the model achieves the result in better word embeddings.

---

[2] https://code.google.com/p/word2vec/.
[3] http://nlp.stanford.edu/projects/glove/.
[4] https://levyomer.wordpress.com/2014/04/25/dependency-based-word-embeddings/.

Negation Words: Negation words can reverse the polarity of sentiment words. Example, if a negation word appears before a positive word then the sentiment of the tweet is changed to negative and vice versa. A list of negation words is created to determine them in tweets. The feature is extracted using a window of 1 to 3 words before a sentiment word and search for these kinds of words.

Sentiment Words: Sentiment words have a significant role in determining the sentiment polarity of tweets. Sentiment words are used as a feature for indicating the location of the sentiment words in each tweet. The sentiment dictionaries provided by Hu and Liu [11] are employed to determine the positive and negative words in tweets.

4.2 Detecting Tweets Containing the Conditional Sentence

In this study, the MLP model is used for detecting tweets containing CSs. The detection model is built with an architecture consisting of three layers, namely, an input layer, a hidden layers, and an output layer. In the detection model, the input layer is a feature vector, which is a combination of the features represented in each tweet. There are three hidden layers; for each hidden layer, the activation function is a scaled exponential linear units (SELU) [13]. This is a light new activation function appearing from 2017 and gives better performance compared with other activation functions [13]. Moreover, the activation function used in the output layer is the Softmax function. In addition, an early stopping technique is employed to avoid overfitting. This technique can stop the algorithm before the value of the loss function becomes extremely small. The detail of the hyperparameters of the MLP model is presented in Table 1.

Table 1. Hyperparameters for MLP model

Hyperparameters	Values
# hidden layer	3
# neural of hidden layer	1000, 500, 200
L_2 regularization	0.0001
Learn rate	0.001
# epochs	100
Early stop	40
# batch size	500
# k-fold	15

4.3 Sentiment Analysis of Tweets Containing Conditional Sentence

In the proposed method, we focus on classifying the sentiments in each tweet as positive, negative, or neutral. The CNN model is employed to implement this task. The architecture of this model includes layers such as an input layer, a convolutional layer, a max-pooling layer, and an output layer. The input layer is the layer representation of a tweet in low-dimensional vectors based on the word embedding technique. The convolutional layer implements convolutions over the vector using multiple filter sizes. Because the convolutional layer uses filters of different size, it implies that each convolution gives a tensor of an unusual shape. Therefore, the max-pooling layer iterates and merges the tensors into one significant feature vector using the max function. For the output layer, the Softmax function is used to classify the results. The detail of the hyperparameters of the CNN model is listed in Table 2.

Table 2. Hyperparameters for CNN model

Hyperparameters	Values
Dimensionality of character embedding	300
Comma-separated filter sizes	5, 5, 5
Dropout keep probability	0.1
L_2 regularization	0.0001
# epochs	300
# batch size	100

5   Experiment

In this section, we attempt to prove that the proposed approach give a better performance than the other methods by considering the following two hypotheses:

Hypothesis 1: The proposed method can identify whether a tweet contains CSs with a high accuracy.
Hypothesis 2: The proposed method can be used to determine the sentiment of tweets containing CSs as a positive, negative, or neutral sentiment.

5.1   Data Acquisition

In this study, the Python package Tweepy[5] is used to collect 5158 tweets and we only deal with tweets written in English. The elements in tweets such as punctuation marks, re-tweet symbols, URLs, hashtags, and query term are extracted and removed. Next, a describing text replaces an emoji icon in tweets by using the Python emoji package[6]. In addition, tweets are informal in which the users

---

[5] https://pypi.org/project/tweepy/.
[6] https://pypi.org/project/emoji/.

can use the acronyms as well as make spelling mistake. These can affect the accuracy of the result. Therefore, the Python-based Aspell library[7] is employed to implement spell correction. Then, the tweets are divided and stored into two separate database files to use for the experiment as follows:

- For the detection case: the training data consists of 4373 tweets (2190 tweets containing CSs, 2183 tweets not containing CSs), the testing data includes 785 tweets (653 tweets containing CSs, 132 tweets not containing CSs).
- For the sentiment analysis case: the training data includes 2190 tweets (686 positive tweets, 679 negative tweets, 825 neutral tweets), the testing data has 653 tweets (197 positive tweets, 203 negative tweets, 253 neutral tweets).

5.2   Evaluation Results

Metrics used to assess the proposed method include *precision, recall, F_1,* and *accuracy.* The values of *precision, recall, F_1,* and *accuracy* are computed as follows:

$$Precision = \frac{TP}{TP + FP} \tag{1}$$

$$Recall = \frac{TP}{TP + FN} \tag{2}$$

$$F_1 = 2 \times \frac{Precision \times Recall}{Precision + Recall} \tag{3}$$

$$Accuracy = \frac{TP + TN}{TP + TN + FP + FN} \tag{4}$$

where, *TP* (True Positive) represents the number of exactly classified items, *FP* (False Positive) is the number of misclassified items, *TN* (True Negative) is the number of exactly classified non-items, *FN* (False Negative) is the number of misclassified non-items.

5.3   Results and Discussion

The performance of the proposed is assessed from the results of the two subtasks, namely, detection of the tweets containing CSs and analysis the sentiment of each tweet containing CSs. The performance of the two steps is evaluated by comparing with other methods.

a. Results of the detection step
Seventy of the total 785 tweets in the test dataset contain CSs. This accounts for approximately 7.65% of the total number of tweets. This value is small, but it impacts significantly the accuracy of the sentiment analysis methods. Table 3 presents the performance of the detection step in terms of the four metrics.

---

[7] https://pypi.org/project/aspell-python-py2/.

Table 3. Results of the detection step

Metrics	Proposed method
Precision	0.83
Recall	0.85
F_1	0.84
Accuracy	0.83

b. Results of the sentiment analysis step

After implementing the sentiment analysis step on the set of tweets containing conditional sentences, we obtain 157 positive tweets, 181 negative tweets, and 191 neutral tweets. Therefore, there are 40 positive tweets, 22 negative tweets, and 62 neutral tweets are misclassified.

From these results, Eqs. (1, 2, 3 and 4) are applied to compute *precision, recall, F_1,* and *accuracy* of the method. Table 4 presents the performance of the sentiment analysis step in terms of the four metrics.

Table 4. Results of the sentiment analysis step

	Precision	Recall	F_1	Accuracy
Positive	0.83	0.80	0.81	0.89
Negative	0.75	0.89	0.82	0.87
Neutral	0.85	0.76	0.80	0.86
Average	0.81	0.82	0.81	0.87

c. The accuracy compared to other methods

The accuracy of the different methods is compared in Table 5. For a fair comparison, the methods are implemented on the same dataset and parameters. We find that the proposed method can be used to detect as ell as analyze the sentiment of tweets containing CSs.

Table 5. Accuracy of the methods

Methods	Detection step	Sentiment analysis step
Naive bayes	0.76	0.71
Logistic regression	0.83	0.74
Support vector machine	0.76	0.68
K nearest neighbors	0.73	0.51
Proposed method	0.83	0.87

According to Table 5, the proposed method obtains better results than the other methods. This study also overcomes other disadvantages as mentioned in Sect. 1. Why does the proposed method achieve better results than other methods? In this paper, the information related to the lexical, syntactic, semantic, and polarity sentiment of words are extracted and used as features. Besides, we considered affecting of conditional connective words and negation words to both tasks. In addition, algorithms used in the proposed method are state-of-art algorithms for detecting and analyzing sentiment. The results again confirm that tweets containing CSs have a significant impact on the accuracy of the sentiment analysis methods. If the CSs is not considered, the other methods could give higher performance than the proposed method, but when considering CSs, the proposed method yields better results than baseline methods.

6   Conclusion and Future Work

This work proposed a method for detecting tweets containing conditional sentences and analyzing the sentiment of these tweets by combining the MLP and CNN models. First, the proposed method extracted the features related to the lexical, semantic, syntactic, and sentiment of the words. Second, tweets containing conditional sentences were detected to analyze as positive, negative, or neutral. The experiment analysis revealed that the proposed method improved a significant performance.

There are some possible limitations of the proposed approach, namely the method only considered the tweets containing CSs without considering the impact of slangs and emoticons. In addition, the influence of the sarcastic tweets on the sentiment analysis was not considered. In the future, we plan to analyze in more detail the information related to the sentiment of tweets containing CSs, such as objects of sentiments and sentiment of each object in entire tweets.

Acknowledgment. This research was supported by Basic Science Research Program through the National Research Foundation of Korea (NRF) funded by the Ministry of Science, ICT & Future Planning (2017R1A2B4009410).

References

1. Asghar, M.Z., Khan, A., Ahmad, S., Qasim, M., Khan, I.A.: Lexicon-enhanced sentiment analysis framework using rule-based classification scheme. PLoS One **12**(2), e0171649 (2017)
2. Asghar, M.Z., Kundi, F.M., Ahmad, S., Khan, A., Khan, F.: TSAF: Twitter sentiment analysis framework using a hybrid classification scheme. Expert Syst. **35**(1) (2018). https://doi.org/10.1111/exsy.12233
3. Bakliwal, A., Arora, P., Madhappan, S., Kapre, N., Singh, M., Varma, V.: Mining sentiments from tweets. In: Proceedings of the 3rd Workshop in Computational Approaches to Subjectivity and Sentiment Analysis, pp. 11–18 (2012)
4. Bird, S., Loper, E.: NLTK: the natural language toolkit. In: Proceedings of the ACL 2004 on Interactive Poster and Demonstration Sessions, p. 31. Association for Computational Linguistics (2004)

5. Bora, N.N.: Summarizing public opinions in tweets. Int. J. Comput. Linguist. Appl. **3**(1), 41–55 (2012)
6. Dong, L., Wei, F., Tan, C., Tang, D., Zhou, M., Xu, K.: Adaptive recursive neural network for target-dependent Twitter sentiment classification. In: Proceedings of the 52nd Annual Meeting of the Association for Computational Linguistics (Short Papers), vol. 2, pp. 49–54 (2014)
7. Ghiassi, M., Skinner, J., Zimbra, D.: Twitter brand sentiment analysis: a hybrid system using n-gram analysis and dynamic artificial neural network. Expert Syst. Appl. **40**(16), 6266–6282 (2013). https://doi.org/10.1016/j.eswa.2013.05.057
8. Gilbert, C.H.E.: VADER: a parsimonious rule-based model for sentiment analysis of social media text. In: Eighth International Conference on Weblogs and Social Media (ICWSM 2014) (2014). http://comp.social.gatech.edu/papers/icwsm14. vader.hutto.pdf. Accessed 20 April 2016
9. Go, A., Bhayani, R., Huang, L.: Twitter sentiment classification using distant supervision. CS224N Project Report, Stanford 1(12) (2009)
10. Hornik, K., Stinchcombe, M., White, H.: Multilayer feedforward networks are universal approximators. Neural Netw. **2**(5), 359–366 (1989)
11. Hu, M., Liu, B.: Mining and summarizing customer reviews. In: Proceedings of the Tenth ACM SIGKDD International Conference on Knowledge Discovery and Data Mining, pp. 168–177. ACM (2004)
12. Kim, Y.: Convolutional neural networks for sentence classification. arXiv preprint arXiv:1408.5882 (2014)
13. Klambauer, G., Unterthiner, T., Mayr, A., Hochreiter, S.: Self-normalizing neural networks. In: Advances in Neural Information Processing Systems, pp. 971–980 (2017)
14. Liu, B.: Sentiment analysis and opinion mining. Synth. Lect. Hum. Lang. Technol. **5**(1), 1–167 (2012)
15. Mohammad, S.M., Kiritchenko, S., Zhu, X.: NRC-Canada: building the state-of-the-art in sentiment analysis of tweets. arXiv preprint arXiv:1308.6242 (2013)
16. Narayanan, R., Liu, B., Choudhary, A.: Sentiment analysis of conditional sentences. In: Proceedings of the 2009 Conference on Empirical Methods in Natural Language Processing, vol. 1, pp. 180–189. Association for Computational Linguistics (2009)
17. Thakkar, H., Patel, D.: Approaches for sentiment analysis on Twitter: a state-of-art study. arXiv preprint arXiv:1512.01043 (2015)
18. Zhang, L., Ghosh, R., Dekhil, M., Hsu, M., Liu, B.: Combining lexicon-based and learning-based methods for twitter sentiment analysis. HP Laboratories, Technical Report HPL-2011 89 (2011)

Aggregating Web Search Results

Marek Kopel$^{(\boxtimes)}$ ⓘ and Maksim Buben

Faculty of Computer Science and Management,
Wroclaw University of Science and Technology,
wybrzeze Wyspiańskiego 27, 50-370 Wroclaw, Poland
`marek.kopel@pwr.edu.pl`

Abstract. In this paper a method for aggregating Web search results is proposed. The aggregator results are compared with the results of most popular search engines: Google, Bing and Yandex. There are 3 stages of the comparison, one for each of the languages: English, Polish and Russian. The quality of the aggregator search results is tested based on user preferences and measured with normalized discounted cumulative gain (nDCG).

Keywords: Aggregation · Metasearch · SERP · Search engine ·
Google · Bing · Yandex

1 Introduction

Is Google always giving the best answer? The short answer is: no. And if you came here from a link bait we could stop here. But if you want to get into the subtle details and cases when the Search Giant is actually losing to a meta search engine, then we came up with a study. Fifteen users assessing 30 results sets (SERPs) from three search engines plus aggregation in three localizations (languages) and one normalized measure to find 'the best search results' winner.

1.1 Search Engine Results Page

Search Engine Results Page (SERP) is one of the pages displayed by a search engine in response to a user query, the query which should express user information needs. The main component of the SERP is the listing of results that are returned by the search engine in response to a keyword query - they are the organic search results. The other component are sponsored search results (i.e., advertisements), as discussed in [6].

The results are traditionally ranked by relevance to the query, but sponsored results have higher priority, appearing usually at the top of the SERP. Each organic search result displayed on the SERP includes a title, a link to the actual page and a snippet of text on that page, usually showing the matched keywords. For sponsored results, the advertiser chooses what to display.

© Springer Nature Switzerland AG 2019
N. T. Nguyen et al. (Eds.): ACIIDS 2019, LNAI 11431, pp. 189–197, 2019.
https://doi.org/10.1007/978-3-030-14799-0_16

Since the number of results for a query is usually a high number, search engine would display ten results per SERP, showing the most relevant first. With each next result page the relevance of displayed results will be lower and so more unlikely to be viewed by user. In fact, "the best place to hide a dead body is page two of Google" as discussed in [1]. So in our study we only focus on the first SERP for each engine.

1.2   Metasearch - The Aggregation of SERPs

Since SERP for different engines for the same query may differ, the question arises "which is better?". And maybe some results from one engine SERP and not present in another engine SERP can still be valuable to the user, as claimed in [4]. To deal with this problem, the aggregation of SERP or, so called, "metasearch" came to be, which is "searching the search engines". In effect the result of metasearch is a composition of best results from different engines - in our case: from Google, Bing and Yandex - called aggregation. And so the engine doing the metasearch and delivering the aggregation shall be called the aggregator.

1.3   Related Works

The metasearch engines are about as old as the search engines themselves. The first most known was Metacrawler. While competing with each other, the regular Web search engines also incorporated techniques of indexing competitor SERPs, making them partially metasearch engines. The aggregation was also a part of meta data and Semantic Web concept, which, in case of Google, resulted with patents like [8]. On the other hand, the "no aggregation" version of search also became a thing and so a patents like [2] were needed.

From newer works in the field of Web search aggregation, beside the trend of incorporating aggregation into an actual Web search engine, like in [5], other trends concern technical aspects like aggregation optimization, i.e. in [12] or new takes on Semantic Web - the promised intelligent Web 3.0, as in [10].

2   Method and Experiment

The main question in this study is the following: Will the aggregated results be more relevant than the results of individual search engines? So first we must choose the aggregation method.

2.1   Aggregation

A classical algorithm described in [9] was chosen as the aggregation algorithm. It is based on document position in various search engines results. For individual search engines, a specific value is assigned with each search result position. If search result A is on first position - it gets 10 points, when on second place

- 9 points and so on. The 10th position document gets 1 point and all next documents get 0 points. Then points for each document from different search engines are summed up [1] to give the new ranking value and from that a new position in aggregated result list.

2.2 Measures

In order to assess the quality of search results, the idea of Discounting cumulative gain (DCG) is used as in [3,7] and further discussed in [11].

The idea of DCG is that the lower the position in the rating of a given document, the less useful it becomes for the user, because it is less likely that it will be viewed. Therefore highly relevant documents appearing lower in a search result list get relevance value reduced logarithmically proportional to the position of the result. It is computed as in 1.

$$DCG = \sum_{r=1}^{R} \frac{Gain(result@r))}{log_b(r+1))} \tag{1}$$

where

R : length of the result list
r : current iteration over

Two assumptions are made in using DCG:

1. Highly relevant documents are more useful when appearing earlier in a search engine result list (have higher ranks).
2. Highly relevant documents are more useful than marginally relevant documents, which are in turn more useful than irrelevant documents.

Finally, a normalized version of DCG is used called Normalized discounted cumulative gain, or NDCG, and computed as in 2.

$$nDCG = \frac{DCG}{idealDCG} \tag{2}$$

The ideal DCG is the value calculated for a version of SERP where all results are shown in descending ordered by their relevance.

2.3 Queries

The quality (relevance) evaluation is made on the basis of user ratings for individual search engines: Google, Bing and Yandex, as well as for an aggregator with its own version of SERP. The search queries are in three natural languages: English, Polish and Russian. Each query asks about the objective, actual state of the world witch should allow user to easily assess relevance of each result document for each query. It was authors intend to avoid issues subjective and in question, leaving the concept of 'fake news' aside. English version of queries is presented below:

1. canada macroeconomic indicators 2000 2010
2. adam mickiewicz date of arrest
3. albert einstein nobel prize for what
4. bmw x6 m50d horsepower
5. elon musk age
6. where did the battle of grunwald take place
7. kim jong un education
8. kim jong il birthplace
9. gdp poland 1991
10. population of mozambique 2015

Russian and Polish version of queries was similar, buy sometimes modified to better match the localization. Queries in Russian run with localization set to Russia:

1. макроэкономические показатели Канады 2000 2010
2. дата рождения адама мицкевича
3. альберт эйнштейн нобелевскую премию за что получил
4. бмв x6 m50d сколько лошадей
5. элон маск возраст
6. дата битвы под грюнвальдом
7. ким чен ын образование
8. ким чен ир место рождения
9. ввп польши 1991
10. мозамбик население 2015

Queries in Polish run in Polish localization:

1. wskaźniki makroekonomiczne kanada 2000 2010
2. adam mickiewicz data aresztowania
3. albert einstein nagroda nobla za co
4. bmw x6 m50d ile koni
5. elon musk wiek
6. data bitwy pod grunwaldem
7. kim dzong un wykształcenie
8. kim jong il miejsce urodzenia
9. pkb polski 1991
10. populacja mozambique 2015

All of the queries were run in November 2017.

2.4   Relevance Assessment

In the study the assessment team consisted of 15 people: SEO (Search Engine Optimization) specialists, specialists in the field of PPC (Pay Per Click), developers and Internet marketing specialists. Each of the experts received paper and electronic versions of a questionnaire to fill after assessing each query (see Fig. 1). Each of the experts was asked to assess results for 2 queries. In total, each of the experts gave feedback for up to 60 documents. The scale used in the assessment is presented in Table 1.

Table 1. Evaluation scale for relevance assessment.

Evaluation of relevance	Points	Description
Vital	10	This is the highest rating that a search result can get. A vital evaluation will most likely receive an official query page
Useful	7	This is the second most important result of the page rating. This evaluation is given to the document (website) where you can find information not only exactly matching the query, but also having an additional information value
Relevant	5	The document responds to the request, but does not have the added value of information as in the Useful evaluation
Not relevant	2	The rating is given to pages that do not answer the query, but are somehow related to the intention of the user
Off-topic	1	This is the lowest positive rating that a page that is completely incompatible with the query will receive
Foreign language	0	This rating will be assigned to the website if the target language of the query does not match the language of the page
Didn't load	0	Such evaluation will be given to the Web page, returning the 404 error, the "page not found" error, the "product not found" error, the "server time out" error, the 403 forbidden error, if authorization etc. is required
Unratable	0	Rating awarded if the site can not be evaluated

Table 2. nDCG for queries in English.

Question	Google	Bing	Yandex	Aggregator
1	0.967094751	0.933049507	0.319023198	0.872017982
2	0.825572726	0.63057043	0.992312373	0.897789659
3	0.845176096	0.80557925	0.880899522	0.966518375
4	0.917222156	0.819354823	0.485457396	0.881289017
5	0.732028955	0.942458084	0.83698606	0.944755662
6	0.969939234	0.613330561	0.845732126	0.977435009
7	0.970130578	0.820087533	0.833992932	0.899617805
8	0.829549721	0.827767526	0.600463836	0.933800969
9	0.926333895	0.921587575	0.795387436	0.943238286
10	0.922871274	0.831829117	0.799463912	0.902135097
Avg	0.890591939	0.814561441	0.738971879	**0.921859786**

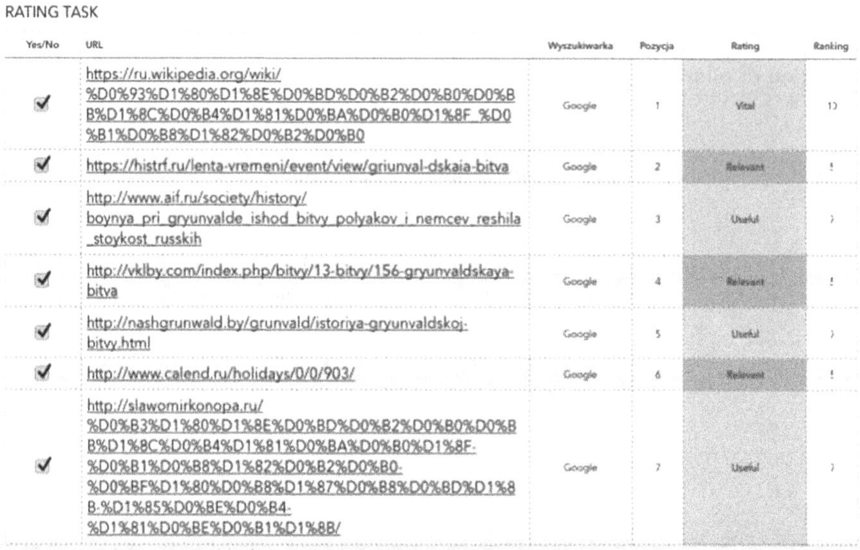

RATING TASK

Yes/No	URL		Wyszukiwarka	Pozycja	Rating	Ranking
✔	https://ru.wikipedia.org/wiki/ %D0%93%D1%80%D1%8E%D0%BD%D0%B2%D0%B0%D0%B B%D1%8C%D0%B4%D1%81%D0%BA%D0%B0%D1%8F_%D0 %B1%D0%B8%D1%82%D0%B2%D0%B0		Google	1	Vital	13
✔	https://histrf.ru/lenta-vremeni/event/view/griunval-dskaia-bitva		Google	2	Relevant	5
✔	http://www.aif.ru/society/history/ boynya_pri_gryunvalde_ishod_bitvy_polyakov_i_nemcev_reshila _stoykost_russkih		Google	3	Useful	3
✔	http://vklby.com/index.php/bitvy/13-bitvy/156-gryunvaldskaya-bitva		Google	4	Relevant	5
✔	http://nashgrunwald.by/grunvald/istoriya-gryunvaldskoj-bitvy.html		Google	5	Useful	3
✔	http://www.calend.ru/holidays/0/0/903/		Google	6	Relevant	5
✔	http://slawomirkonopa.ru/ %D0%B3%D1%80%D1%8E%D0%BD%D0%B2%D0%B0%D0%B B%D1%8C%D0%B4%D1%81%D0%BA%D0%B0%D1%8F- %D0%B1%D0%B8%D1%82%D0%B2%D0%B0- %D0%BF%D1%80%D0%B8%D1%87%D0%B8%D0%BD%D1%8 B-%D1%85%D0%BE%D0%B4- %D1%81%D0%BE%D0%B1%D1%8B/		Google	7	Useful	3

Fig. 1. A filled-in questionnaire for a query rating task. The fields with coloured background show the evaluation for each Web page appearing in a SERP according to scale presented in Table 1.

Table 3. nDCG for queries in Russian

Question	Google	Bing	Yandex	Aggregator
1	0.915611808	0.661764405	0.910845706	0.938635335
2	0.939000348	0.878427736	0.995483776	0.993653734
3	0.922173622	0.679071893	0.795189759	0.917404545
4	0.907464398	0.692446399	0.88300339	0.946523138
5	1	0.899495632	0.936980207	1
6	0.955935878	0.552977671	0.985100413	0.913026778
7	0.65589738	0.653882911	0.841142961	0.831979759
8	0.816423491	0.715936491	0.973641206	0.863869369
9	0.889243271	0.81315996	0.8336009	0.930861319
10	0.795391596	0.638138456	0.963668729	0.911012941
Avg	0.879714179	0.718530155	0.911865705	**0.924696692**

3   Results

Tables 2, 3 and 4 present the values of nDGC calculated for each query in 3 corresponding groups, in 3 languages: EN, RU and PL, based on points from the evaluation of relevance as shown in Table 1. The evaluation was done as described earlier using the questionnaire presented in Fig. 1.

Table 4. nDCG for queries in Polish

Question	Google	Bing	Yandex	Aggregator
1	0.851186383	0.4250813	0.246175938	0.606393201
2	0.967461449	0.580303852	0.461774159	0.70852278
3	0.984657242	0.932006786	0.67545195	0.891474674
4	0.940951605	0.827707235	0	0.755485232
5	0.856691478	0.899498975	0.482404919	0.955516751
6	1	0.813415391	0.773281909	0.907261943
7	0.86867189	0.822351389	0.808398771	0.993251859
8	0.630514225	0.499452346	0.484181547	0.662664269
9	0.790982483	0.780794773	0.785013688	0.69820307
10	0.697925427	0.69019418	0.436889463	0.883287352
Avg	**0.858904218**	0.727080623	0.515357234	0.806206113

Table 5. Average nDCG for queries in 3 languages

Language	Google	Bing	Yandex	Aggregator
English	0.890591939	0.814561441	0.738971879	0.921859786
Russian	0.879714179	0.718530155	0.911865705	0.924696692
Polish	0.858904218	0.727080623	0.515357234	0.806206113
Avg	0.876403445	0.75339074	0.722064939	**0.884254197**

The results in Table 5 show that the average nDCG for aggregator is slightly higher than the measures of the three individual search engines. The best result among classic search engines was scored by Google, followed by Bing, and then Yandex.

3.1 Statistical Significance

The Wilcoxon signed-rank test was used as a statistical test. This is a non-parametric statistical hypothesis test used to compare two samples (SERP pairs) to assess whether their population mean ranks differ.

For the Aggregator-Bing and Aggregator-Yandex pair, the level of asymptotic significance of p is less than 0.05 ($p < 0.05$), which means that the data of the data pair is statistically different from each other, which in our case means increase in the quality of the results. The null hypothesis for these cases should be denied.

For the pair Aggregator-Google p(1-tail) = 0.2358 and p(2-tail) = 0.4715, which is more than 0.05, which in turn means that these results are statistically similar, i.e. the differences between the samples are not statistically significant. From this we can conclude that the quality of the search results of the aggregator is not worse than the quality of the search for each of the classic search engines individually (Google, Bing, Yandex) and better than the two search engines (Bing, Yandex).

4   Conclusions

The experiment outcome shows that in some cases the compilation (or aggregation) of SERPs from different engines make user more happy (or informed) than SERPs from a single engine (read "Google"). Therefore using an aggregator of query results for queries about the current state is justified and can improve the quality of search results, which in turn increases user satisfaction. The use of the aggregator increases the search range, which in turn makes the use of meta-search engines advantageous in those segments of the Internet, where there is no dominance of one of the search engines.

Examples of such countries may be: China, Russia and the Czech Republic. The markets of these countries represent the greatest potential for increasing search quality when using search aggregators.

The quality of the search results in both aggregator and individual search engines based on 30 queries of the actual state was investigated. This sample is too small at the scale of search engines, therefore, if there could be more people involved (Search Quality Raters), it is necessary to conduct such research on a larger sample of queries and SERPs.

References

1. Brorsson, M., Lindhom, H.: The best place to hide a dead body is page 2 on Google search results (2016)
2. Colgrove, C., Martin, G., Campanini, J.: Simple web search. U.S. Patent 8,868,537, 21 October 2014
3. Croft, W.B., Metzler, D., Strohman, T.: Search Engines: Information Retrieval in Practice, vol. 283. Addison-Wesley Reading, Boston (2010)
4. Glover, E.J., Lawrence, S., Birmingham, W.P., Giles, C.L.: Architecture of a metasearch engine that supports user information needs. In: Proceedings of the Eighth International Conference on Information and Knowledge Management, pp. 210–216. ACM (1999)
5. Ishii, H., Tempo, R.: The PageRank problem, multiagent consensus, and web aggregation: a systems and control viewpoint. IEEE Control Syst. **34**(3), 34–53 (2014)
6. Jansen, B.J., Spink, A.: Investigating customer click through behaviour with integrated sponsored and nonsponsored results. Int. J. Internet Mark. Advertising **5**(1), 74 (2009)
7. Järvelin, K., Kekäläinen, J.: Cumulated gain-based evaluation of IR techniques. ACM Trans. Inf. Syst. (TOIS) **20**(4), 422–446 (2002)
8. Newman, E., Lockett, J.: Dynamic aggregation and display of contextually relevant content. U.S. Patent 7,917,840, 29 March 2011
9. Patel, B., Shah, D.: Ranking algorithm for meta search engine. IJAERS Int. J. Adv. Eng. Res. Stud. **2**(1), 39–40 (2012)

10. Sherkhonov, E., Cuenca Grau, B., Kharlamov, E., Kostylev, E.V.: Semantic faceted search with aggregation and recursion. In: d'Amato, C., et al. (eds.) ISWC 2017. LNCS, vol. 10587, pp. 594–610. Springer, Cham (2017). https://doi.org/10.1007/978-3-319-68288-4_35

11. Wang, Y., Wang, L., Li, Y., He, D., Chen, W., Liu, T.Y.: A theoretical analysis of NDCG ranking measures. In: Proceedings of the 26th Annual Conference on Learning Theory (COLT 2013), vol. 8 (2013)

12. Yun, J.M., He, Y., Elnikety, S., Ren, S.: Optimal aggregation policy for reducing tail latency of web search. In: Proceedings of the 38th International ACM SIGIR Conference on Research and Development in Information Retrieval, pp. 63–72. ACM (2015)

Cross-Lingual Korean Speech-to-Text Summarization

HyoJeon Yoon[1], Dinh Tuyen Hoang[1], Ngoc Thanh Nguyen[2],
and Dosam Hwang[1(✉)]

[1] Department of Computer Engineering, Yeungnam University,
Gyeongsan, South Korea
hjyoon314@ynu.ac.kr, hoangdinhtuyen@gmail.com, dosamhwang@gmail.com
[2] Faculty of Computer Science and Management,
Wroclaw University of Science and Technology, Wrocław, Poland
Ngoc-Thanh.Nguyen@pwr.edu.pl

Abstract. The development of a cross-lingual text summarization of a language differing from that of the source document has been a challenge in recent years. This paper describes a summarization system built to auto-translate Korean speech into an English summary text. Recent studies have discussed two separate tasks in this area, namely, obtaining the analysis information from one of the two languages by providing early or late translation approaches. The early translation tries to translate the original documents into the target language, and then summarizes the results by considering the information of the translated texts, whereas the late translation approach attempts to summarize the original documents and then translate them into the target language. We propose a method for automatically converting Korean speech into an English summary text. The Korean transcript is segmented and analyzed for sentence clustering. A word-graph is then used to compress and generate a unique, concise, and informative compression. Experiments prove that our method achieves better accuracy in comparison with other methods.

Keywords: Cross-lingual · Text summarization · Speech to text

1 Introduction

With the explosive growth of the Internet, the amount of information available is quickly increasing. The amount of audio data is also increasing, leading to information overload and impacting users when gathering information. Owing to the availability of big data, manual data analysis has become a difficult and time-consuming process. Auto-speech summarization, which is used to extract

---

D. T. Hoang—The author contributed equally first author to this work.

© Springer Nature Switzerland AG 2019
N. T. Nguyen et al. (Eds.): ACIIDS 2019, LNAI 11431, pp. 198–206, 2019.
https://doi.org/10.1007/978-3-030-14799-0_17

relevant information and remove redundant or incorrect information, has become a good method for transcribing spontaneous speech.

Automatic cross-lingual document summarization is related to the summary of an original document in another language [1,4]. In general, this type of task can be divided into two sub-tasks: machine translation and text summarization. However, for each sub-task, the result has errors, and setting one task as an input of the other may accumulate such errors and reduce the quality of the cross-lingual document summarization [3,22].

Some previous systems have been built for solving the problem of cross-lingual document summarization, [5,12,24], including SUMMARIST [10], which extracts sentences from documents in a different language and translates the results. Conversely, Newsblaster [6] proceeds with the translation first and then selects appropriate sentences. However, these approaches are not without their drawbacks, and are dependent on the results of machine translation systems. Other researchers [18,25] have attempted to analyze the information of a given document in both (the source and target) languages when considering the specifications and determining the most appropriate sentences. Recent studies have confirmed that analyzing the information of a document in both (the source and target) languages provides better results. Inspired by these ideas, we propose a framework for auto-translating Korean speech into an English summary text. First, we use Google Translate for automatically segmenting a Korean transcript and translating it into English. Next, we calculate the sentence clustering in both Korean and English. A compression method is proposed for creating a short and unique summary. Finally, a cross-lingual text summarization is generated using a word-graph to obtain the most relevant information without redundancy.

The remainder of this paper is organized as follows. In the next section, related studies are reviewed to present general knowledge regarding automatic cross-lingual text summarization. The details of the proposed method are then presented in Sect. 3. In Sect. 4, we provide the experiment results. Finally, some concluding remarks and areas of future work are presented in Sect. 5.

2   Related Work

With advances in storage and networking, an enormous amount of information has been made accessible to an ever-increasing portion of the world population, with audio data experiencing an explosive growth. The problems of cross-lingual and speech-to-text summarization have attracted the attention of researchers over the years [18,23,25]. Unlike text summarization, in which a document is segmented into sentences, speech-to-text has encountered a number of difficulties, with the absence of sentence boundaries complicating the recognized text. In addition, the speech recognition application may not recognize certain words, or may confuse them. However, recent works have confirmed that text summarization techniques can be applied to speech-to-text systems. For example, in [19], the authors built a system for speech-to-text summarization using phrase extraction. They used an automatic speech recognizer for recorded speech, which they transcribed into text. A syntactic analyzer module was then used to identify phrases in the recognized text. A summary test was generated based on the phrases and term frequency–inverse document frequency [21].

Fig. 1. Early translation

Pontes et al. [18] built a framework for French-to-English cross-lingual transcript summarization. They segmented a French transcript and analyzed the information in both the source and target languages to determine the saliency of the sentences. A multi-sentence compression method was then used to concentrate and enhance the semantics of the sentences. Hori et al. [9] presented an automatic speech summarization method that uses the summarization score to indicate words extracted from a transcribed speech. A dynamic programming method is applied for the extraction stage. The authors showed a good performance for both English and Japanese news speech. Cross-lingual summarization is a relative field of text summarization, and has been undertaken by adjusting existing extraction-based multi-document systems for treating English documents. The source documents delivered are written in several languages, and automatic multi-lingual summarization includes creating a summary in the target language. Two traditional approaches for resolving this problem are (1) machine translating the generated summaries, as shown in Fig. 2, and (2) translating the documents before the sentence extraction phase, as shown in Fig. 1 [20]. However, such approaches have a downside as they completely depend on the performance of the machine translation system. It is more useful to try and combine multi-sentence and sentence compression methods to generate a communicative cross-lingual text summarization [19, 20].

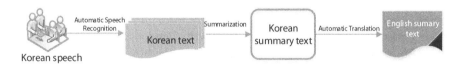

Fig. 2. Late translation

With regard to Korean text summarization, Kim et al. [11] developed a Korean text summarization system using the aggregate similarity. Each document was considered as a weighted graph, which is called a text relationship map, where each node represents a vector of nouns in a sentence. Two nodes are

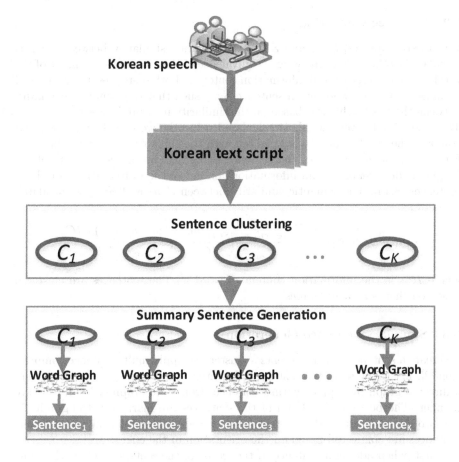

Fig. 3. Workflow of our proposed system

connected when two sentences are semantically related, and are considered an edge, and the similarity between a pair of sentences is considered as a weight of the graph (Fig. 3).

3 Cross-Lingual Speech-to Text Summarization Method

3.1 Speech-to-Text Translation

There are certain tools that can help us record a Korean transcript from audio data. We checked some of these tools and used Google Translate[1] for automatically recording a Korean transcript and translating it into English owing to its availability and high-level of performance. Because of the absence of sentence boundaries, we used a method proposed by Narayan et al. [16] to extract the sentences.

---

[1] https://github.com/ssut/py-googletrans.

3.2   Sentence Clustering

Given two sentences, Sen_1 and Sen_2, the semantic similarity between them is calculated as the ratio of the weight equivalence of the information content of the words in Sen_1 to the overall information content in both sentences. For each word w_i in Sen_1, we find word w_j in sentence Sen_2 such that the semantic similarity between them is the highest based on the similarity measure between the words. The similarity between words is calculated using the word embedding method [15], and the word similarity is weighted with the similar word content in Sen_1. Finally, for the similarity of the sentence semantics, we divide the sum of the weighted equivalences for all information content of the words included in both sentences [8,14]. The semantic similarity between Sen_1 and Sen_2 is calculated as follows:

$$Sim(Sen_1, Sen_2) = \frac{\sum_{w_i \in Sen_1, w_j \in Sen_2} max(Sen_w(w_i, w_j)) * IC_{w_i}}{\sum_{w_i \in Sen_1} IC_{w_i} + \sum_{w_j \in Sen_2} IC_{w_j}} \tag{1}$$

where IC_w is the information content of word w. The sentences are clustered based on their similarity values.

3.3   Summary Sentence Generation

To create a sentence that represents a cluster, we apply multi-sentence compression [2]. Let C_i be a cluster containing k sentences, $S = s_1, s_2, ..., s_k$. We create a directed graph by supplementing sentences from S to graph G in an iterative manner. The beginning and end of the sentences are marked in the graph [7]. The words are considered as vertices of the graph. The adjacent words in the sentences are connected based on the orientation of the edges.

First, when adding a sentence, if the words of the sentences follow the same word or the same parts-of-speech tagging, they are mapped to the same button in the graph. The order of precedence used to create the graph is as follows: words that do not appear in the graph, content words that may display multiple mappings or words that appear more than once in a sentence, and stop words. To avoid incomplete sentences, the minimum path length (in words) is set to we eight. For each cluster, we choose a maximum of 100 randomly selected paths to reduce the time complexity and select the best path among them. We use the method in [2] to find the best path.

Let $P_i^{C_j}$ be a path in cluster C_j. Let K be the number of paths in cluster C_j, where $K = min\{|C_j|, 100\}$. The best paths (Bf) of the clusters are computed by combining the informativeness $IF(P_i^{C_j})$ and the quality of linguistic $QL(P_i^{C_j})$ as follows (Fig. 4):

$$Bf(P_1^{C_1}, ..., P_K^{C_N}) = \sum_{j=1}^{N} \sum_{i=1}^{K} \frac{1}{F(P_i^{C_j})}.IF(P_i^{C_j}).QL(P_i^{C_j}).P_i^{C_j} \tag{2}$$

where $F(P_i^{C_j})$ is the number of tokens in a path.

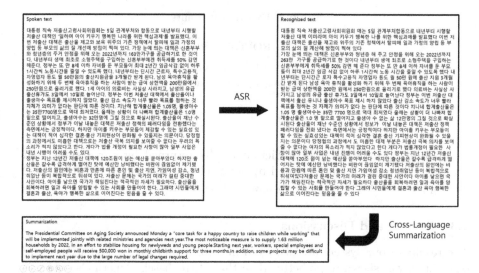

Fig. 4. Example of Korean speech-to-text summarization

4   Experiment

4.1   Dataset

There are no standard datasets available for evaluating a Korean speech-to-text summarization. Because the most common method for evaluating the informativeness of an automatic summarization is to compare it with human-made model summaries, we have to create the summary on our own. We collected 200 articles from a Korean news source. We then asked five students in our lab to summarize the articles in both Korean and English as the ground truth. The Korean students spoke while the Google Translate API was used to record the Korean transcript and translate it into English.

4.2   Evaluation

To evaluate the proposed method, we compared the prediction results with our ground truth. We used the Recall-Oriented Understudy for Gisting Evaluation (ROUGE) method proposed by Lin et al. [13] to calculate the differences between the distribution of words in the prediction results and the ground truth. The prediction summary and the ground truth were split into n-grams to compute their intersection. The common ROUGE-n is ROUGE-1, ROUGE-2, which are unigrams and bigrams. The value of ROUGE-n is computed as follows:

$$ROUGE - n = \frac{\sum_{n-grams} \in \{Sum_{Pre} \cap Sum_{Gt}\}}{\sum_{n-grams} \in Sum_{Gt}} \tag{3}$$

Table 1. Experiment results

System	ROUGE-1	ROUGE-2
Our system	0.4550	0.1011
Baseline 1	0.4205	0.0945
Baseline 2	0.3912	0.0898

Here, Sum_{Pre} indicates the prediction results returned by our system, and Sum_{Gt} is the ground truth created by our students. ROUGE-1 and ROUGE-2 were applied to evaluate the performance of our system.

4.3 Results and Discussions

The results of our system are displayed in Table 1 and Fig. 5. The accuracy of our system was shown to be better than that of the baseline methods, which is understandable because early translation and late translation methods are normally completely dependent on the performance of machine translation systems. In addition, the baseline methods have other problems such as speech recognition errors, no sentence boundaries, a wide range of sentence sizes, and an uneven distribution of information. To choose the most appropriate sentences for our proposed method, we considered the text in the target languages. We proceeded with the clustering of similar sentences to analyze the subjects of the document independently. A word-graph was then built to compress clusters into sentences. We selected the most relevant sentences and reduced the level of redundancy.

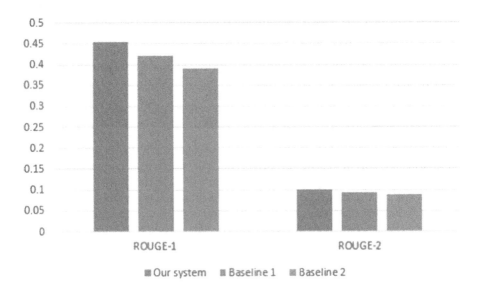

Fig. 5. Accuracy comparison between our system and other methods

5 Conclusion and Future Work

In this study, we proposed a method for auto-translating Korean speech into an English summary text. First, we used Google Translate for automatically segmenting a Korean transcript and translating it into English. We then calculated the sentence relevance used for sentence clustering. A compression method was built to create a short and unique summary. Finally, a cross-lingual text summarization was generated by merging the most relevant sentences generated through a word graph without redundancy. The experiment results demonstrate that our proposed method achieves a higher level of accuracy in comparison to other baseline methods.

As future work, we plan to consider the grammaticality of the summarization. In addition, the translation and quality of the transcripts can be improved using a deep neural network and integrating data from external resources [17].

Acknowledgment. This research was funded by the Basic Science Research Program through the National Research Foundation (NRF) of Korea, funded by the Ministry of Science, ICT, and Future Planning (2017R1A2B4009410).

References

1. Aggarwal, C.C., Zhai, C.: Mining Text Data. Springer, New York (2012). https://doi.org/10.1007/978-1-4614-3223-4
2. Banerjee, S., Mitra, P., Sugiyama, K.: Multi-document abstractive summarization using ILP based multi-sentence compression. In: IJCAI, pp. 1208–1214 (2015)
3. Bednár, P.: Cross-language dependency parsing using part-of-speech patterns. In: Sojka, P., Horák, A., Kopeček, I., Pala, K. (eds.) TSD 2016. LNCS (LNAI), vol. 9924, pp. 117–124. Springer, Cham (2016). https://doi.org/10.1007/978-3-319-45510-5_14
4. Best, C.T.: A direct realist view of cross-language speech perception. In: Speech Perception and Linguistic Experience: Issues in Cross-language Research, pp. 171–204 (1995). Chap. 6
5. Durgunoğlu, A.Y., Nagy, W.E., Hancin-Bhatt, B.J.: Cross-language transfer of phonological awareness. J. Educ. Psychol. **85**(3), 453 (1993)
6. Evans, D.K., Klavans, J.L., McKeown, K.R.: Columbia newsblaster: multilingual news summarization on the web. In: Demonstration Papers at HLT-NAACL 2004, pp. 1–4. Association for Computational Linguistics (2004)
7. Filippova, K.: Multi-sentence compression: finding shortest paths in word graphs. In: Proceedings of the 23rd International Conference on Computational Linguistics, pp. 322–330. Association for Computational Linguistics (2010)
8. Hoang, D.T., Tran, V.C., Nguyen, V.D., Nguyen, N.T., Hwang, D.: Improving academic event recommendation using research similarity and interaction strength between authors. Cybern. Syst. **48**(3), 210–230 (2017)
9. Hori, C., Furui, S., Malkin, R., Yu, H., Waibel, A.: Automatic speech summarization applied to English broadcast news speech. In: 2002 IEEE International Conference on Acoustics, Speech, and Signal Processing (ICASSP), vol. 1, pp. I-9. IEEE (2002)

10. Hovy, E., Lin, C.Y.: Automated text summarization and the SUMMARIST system. In: Proceedings of a Workshop, Baltimore, Maryland, 13–15 October 1998, pp. 197–214. Association for Computational Linguistics (1998)

11. Kim, J.H., Kim, J.H., Hwang, D.: Korean text summarization using an aggregate similarity. In: Proceedings of the Fifth International Workshop on Information Retrieval with Asian Languages, pp. 111–118. ACM (2000)

12. Kuhl, P.K., et al.: Cross-language analysis of phonetic units in language addressed to infants. Science **277**(5326), 684–686 (1997)

13. Lin, C.Y.: Rouge: a package for automatic evaluation of summaries. In: Text Summarization Branches Out (2004)

14. Liu, X.Y., Zhou, Y.M., Zheng, R.S.: Measuring semantic similarity within sentences. In: 2008 International Conference on Machine Learning and Cybernetics, vol. 5, pp. 2558–2562. IEEE (2008)

15. Mikolov, T., Sutskever, I., Chen, K., Corrado, G.S., Dean, J.: Distributed representations of words and phrases and their compositionality. In: Advances in Neural Information Processing Systems, pp. 3111–3119 (2013)

16. Narayan, S., et al.: Document modeling with external attention for sentence extraction. In: Proceedings of the 56th Annual Meeting of the Association for Computational Linguistics (Volume 1: Long Papers), vol. 1, pp. 2020–2030 (2018)

17. Nguyen, N.T.: Advanced Methods for Inconsistent Knowledge Management. Springer, London (2008). https://doi.org/10.1007/978-1-84628-889-0

18. Linhares Pontes, E., González-Gallardo, C.-E., Torres-Moreno, J.-M., Huet, S.: Cross-lingual speech-to-text summarization. In: Choroś, K., Kopel, M., Kukla, E., Siemiński, A. (eds.) MISSI 2018. AISC, vol. 833, pp. 385–395. Springer, Cham (2019). https://doi.org/10.1007/978-3-319-98678-4_39

19. Rott, M., Červa, P.: Speech-to-text summarization using automatic phrase extraction from recognized text. In: Sojka, P., Horák, A., Kopeček, I., Pala, K. (eds.) TSD 2016. LNCS (LNAI), vol. 9924, pp. 101–108. Springer, Cham (2016). https://doi.org/10.1007/978-3-319-45510-5_12

20. Torres-Moreno, J.M.: Automatic Text Summarization. Wiley, Hoboken (2014)

21. Vanderwende, L., Suzuki, H., Brockett, C., Nenkova, A.: Beyond sumbasic: task-focused summarization with sentence simplification and lexical expansion. Inf. Process. Manag. **43**(6), 1606–1618 (2007)

22. Wan, X.: Co-training for cross-lingual sentiment classification. In: Proceedings of the Joint Conference of the 47th Annual Meeting of the ACL and the 4th International Joint Conference on Natural Language Processing of the AFNLP, vol. 1, pp. 235–243. Association for Computational Linguistics (2009)

23. Wan, X., Luo, F., Sun, X., Huang, S., Yao, J.G.: Cross-language document summarization via extraction and ranking of multiple summaries. Knowl. Inf. Syst. 1–19 (2018)

24. Werker, J.F., Tees, R.C.: Cross-language speech perception: evidence for perceptual reorganization during the first year of life. Infant Behav. Dev. **7**(1), 49–63 (1984)

25. Zhang, J., Zhou, Y., Zong, C.: Abstractive cross-language summarization via translation model enhanced predicate argument structure fusing. IEEE/ACM Trans. Audio Speech Lang. Process. **24**(10), 1842–1853 (2016)

On Some Approach to Evaluation in Personalized Document Retrieval Systems

Bernadetta Maleszka$^{(\boxtimes)}$ (iD)

Faculty of Computer Science and Management,
Wroclaw University of Science and Technology,
Wybrzeze Wyspianskiego 27, 50-370 Wroclaw, Poland
Bernadetta.Maleszka@pwr.edu.pl

Abstract. Due to the information overload in the Internet, it is a hard task to obtain relevant information. New techniques and sophisticated methods are developed to improve efficiency of the searching process. In our research, we focus on a Personalized Document Retrieval System which allows to adjust relevance of searched documents. Based on user data, usage data and social connections between users, it determines up-to-date user profile and recommends better documents. In the work we analyze a methodology for experimental evaluations in simulated environment.

Keywords: Ontology-based user profile · Knowledge integration ·
Experimental evaluations · "Cold-start" problem

1   Introduction

Nowadays, there exists many information sources and newer and newer methods for searching and combining information are developed. At one time a user had to look for the information, whereas in the present days user is overloaded by information. His or her task is rather to select most relevant document out of a set of results. User satisfaction can be measured in terms of document relevance or in time that user spends using the retrieval system.

The main objective of personalization systems is to alleviate a problem of information overload and to avoid "cold-start" problem. In many approaches to personalization systems one can find the following types: content-based, collaborative filtering, hybrid and context-aware systems [12]. Recommender algorithms are very popular in the Internet services: bookshops, music or movies. They improve profits by few percent of income. Nowadays, it is popular to based our choice of product on the comments of other users or friends.

In this paper we present the idea of Personalized Document Retrieval System (Fig. 1). Effectiveness of personalization system depends on gathered information about users' activities – relevant and irrelevant documents. Important aspect of

© Springer Nature Switzerland AG 2019
N. T. Nguyen et al. (Eds.): ACIIDS 2019, LNAI 11431, pp. 207–216, 2019.
https://doi.org/10.1007/978-3-030-14799-0_18

the system is a collaborative part - the system can find users with similar interests and recommend them similar documents (not necessarily the same documents). The system is still being developed. Some issues has been presented in previous papers: method for user modeling (selecting proper profile structure, building and updating user profile) [7], information filtering, clustering users group [5], and detecting and analyzing evolution of the single users as well as a group of users.

In this paper we propose a methodology to perform experimental evaluations in simulated environment when real usage data is not available.

The further part of the paper is structured in the following way. Section 2 presents overview in related works. In Sect. 3 we describe in details a model of Personalized Document Retrieval System. Section 4 contains a method for determining a representative of group of users. A methodology and result of performed experimental evaluations are presented in Sect. 5. Section 6 contains summary and an idea of future works.

2 Related Works

Traditional recommendation systems can be divided to content-based information and collaborative filtering [12]. The first way takes into account user activities, preferences or some additional information given by the user, sometimes not in a direct way, e.g. browser cookies. Collaborative filtering approach focuses on the activities of user groups and finds users with similar interests – the information can be obtained from some social networks.

Both approaches have advantages and disadvantages: content-based solution can analyze only single user relevant and irrelevant documents and collaborative filtering methods presents the same recommendations (documents) for different users – the assumption is that the same documents are relevant for users with similar interests. In this approach one should consider "outliers" – users with different interests than other users in the system.

A hybrid systems that combine both approaches eliminate the disadvantages and enhance the advantages of both solutions.

2.1 Ontology-Based User Profile

Information about documents and users can be stored in different type of structure. According to [2] and [1], domain ontologies can be used to enhance retrieval process with semantics. It can be used for better understanding of users' queries or for expending user query to disambiguate the meaning.

Ramkumar et al. [11] claim that ontology-based profile allows to combine syntax and semantic search.

They show the following approaches to information retrieval due to the degree of ontology usage:

– ontology driven semantic search – domain knowledge is presented in a form of an ontology; using it improves results relevance;

- ontology-based information retrieval – ontology is used to annotate documents in semi-automated way;
- semantic approach to personalized retrieval – an ontology-based profile is developed; it is enriched by interests score for concepts.

In our work we assume a domain ontology – it is created based on a librarian system of Wroclaw University of Science and Technology [13]. The aim is to develop semantic approach: each concept in the ontology has a value that reflects user's interest in the concept. Bayesian network approach is used to describe user interests between the concepts.

2.2 Distance Between Bayesian Networks

We can differentiate two levels of comparing Bayesian networks: structure of the graph and conditional probability tables for the fixed structure of the graph. Jongh et al. [4] presented methods for the first level. If the structures of the networks are the same, it is sufficient to compare CPTs. If the networks have different graphs (relations between nodes) it is necessary to find proper nodes to compare its CPTs.

Bayesian networks have very important property at the structural level: the graph in directed but without loops. Fenz et al. [3] claim that occurring loops in the structure precludes possibility to calculate CPTs.

Let us consider two classical ways to calculate the distance between two Bayesian networks: on the first level – structural one, it is easy to focus on differences between CPTs.

$$dist_{BN}(CPT_{BN1}(t), CPT_{BN2}(t)) = \frac{1}{2^k} \sum_{i=1}^{k} |P(t|pa_1(t)) - P(t|pa_2(t))| \qquad (1)$$

where $pa_1(t)$ is a set of parents of term t in the first profile, $k = \#\{pa(t)\}$ is number of parents of node t, $CPT_{BN}(t)$ is a conditional probability table for term t in network BN and $P(t|pa(t))$ is a conditional probability of t given its parents $pa(t)$.

To compare Bayesian networks with different structure, one should take into account the source information. In the case of our system, source data are sets D_r and D_{ir} of relevant and irrelevant documents appropriately.

Let us consider the following formula (2) at the data level:

$$dist_{set}(A, B) = \frac{\#(A \cap B)}{\#(A \cup B)} \qquad (2)$$

where A and B are sets of documents based on which the profiles were created.

Both measures are normalized and satisfy conditions of distance measures.

In our system we assumed that the structure of the profiles is fixed – we can add a new concept to the profile if a user is interested in documents with this concept.

3   Personalized Document Retrieval Model

In this section we present a Personalized Document Retrieval System, developed in some previous articles [5,7].

The Personalized Document Retrieval System consists of three main modules (Fig. 1): a clustering module (cluster user profiles), a representative module (for each group obtained in the first module, here a representative profile is determined), and an adaptation module (reacts to changes of user interests). When a new user joins the system, he or she is classified into a group based on rough classification method and a representative profile is recommended to him or her. Based on users current activities (usage data and social information), his or her profile is changed. Such an approach allows us to alleviate "cold-start" problem, as the first profile is not empty.

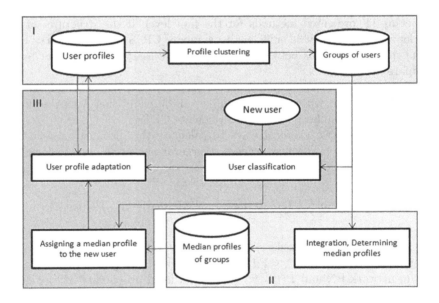

Fig. 1. Schema of Personalized Document Retrieval System.

In our current work, we focus on the second module correlated with determining a representative profile of a group of similar users.

3.1   Domain Ontology

A model of domain ontology is a basis for both: document and user profile.

Let us assume that a domain ontology is a triple [10], [6]:

$$O = (C, R, I) \tag{3}$$

where C is a finite set of concepts, R is a finite set of relations between concepts $R = \{r_1, r_2, ..., r_n\}$, $n \in N$ and $r_i \subset C \times C$ for $i \in \{1, n\}$ and I is a finite set of instances.

In the presented approach a concept is a set of main index term and its synonyms. We define two kinds of relations: "is-a" which reflects to specification – generalization and "is-correlated" which reflects to "see also". Relation "is-a" is directed and can be considered as a basis for Bayesian network.

An example of the term is presented in Table 1 based on the librarian system of the Main Library and Scientific Information Centre in Wroclaw University of Science and Technology [13]. It has almost 200000 documents which are described using about 6000 terms.

Table 1. A concept of "Databases" in domain ontology.

Main term	Databases
Broader terms	Computer Science, Information retrieval systems
Narrower terms	Data – analysis, Data format, Files, Relative databases, Fuzzy databases, Temporal databases, Data warehouses, WWW
See also	Information sources

We assume that in a library we have n_d documents:

$$D = \{d_i : i = 1, 2, \ldots, n_d\} \tag{4}$$

where n_d represents a number of documents and d_i represents i^{th} document – it is an instance of domain ontology.

3.2 User Profile

In this paper we build user profile based on user activities – documents that were relevant and irrelevant for the user. It is based on domain ontology and enriched by conditional probabilities tables (CPT) for each node.

The definition of user profile is as follows:

$$UP = (O, \{CPT_c : c = 1, 2, \ldots, n_c\}) \tag{5}$$

where O is the domain ontology, CPT_k is a conditional probability table for concept c and n_c is a number of concepts in domain ontology.

User profile is built using data-driven model of building Bayesian network [8] for known structure and full observability. The main idea is to find the maximum likelihood estimates of the parameters of conditional probability distribution for each node.

4   Algorithm for Determining Representative Profile for a Group of Users

This section contains detailed description of determining a median profile of the group of profiles.

In this section we try to solve the following problem:

Calculate a representative profile UP_r for a given set of m ontology-based users profiles: $\{UP_j : j \in \{1, 2, \ldots, m\}\}$ and conditional probability tables for each concept in profiles.

We assume that profile of each user in the group is based on the domain ontology. It means that set of concepts in a single profile is a subset of all concepts from the domain ontology. Also, a set of relations in single profile is a subset of relations set in the domain ontology. In our approach we consider only "is-a" relation because we need directed relation to build Bayesian network (and to avoid cycles in the network).

Each node in the domain ontology can have one or more narrower concepts (children) and also one or more broader concepts (parents) – this means that it is not a hierarchical structure (some branches can be a hierarchy but the whole domain ontology has a structure of graph). Such a structure can be a basis for a Bayesian network.

Combining profiles of a set of user can be divided into two levels: structural level and probability level. At the first level the task is to merge structures. It is rather simple task as each profile is a consistent sub-graph of the domain ontology graph. As a result we should consider all nodes and relations between them that have occurred in input profiles.

The second level is to determine conditional probability tables for each concept in a result Bayesian network. To calculate values of each probabilities we can consider existing methods for determining a mean value of conditional probabilities. According to [9] we can calculate median value or arithmetical average value of probability for each concepts to satisfy appropriately O_1 and O_2 criteria. It is important to check if the result probabilities satisfy conditions for CPTs.

Developed algorithm for calculating a representative profile is presented below (Algorithm 1).

Algorithm 1: Determining a representative profile for a group of users.

Input: A set $\{UP_j : j \in \{1, 2, \ldots, m\}\}$ of m user profiles.
Output: A representative profile UP_r.
  1. Obtain the structure of the result profile $T_r = \cup_{j=1}^{m} T_{UP(j)}$.
  2. Prepare CPT for each node based on the "is-a" relation.
  3. Calculate a conditional probability for each concept (average or median).
  4. Determine the result profile as $UP_r = (O, \{CPT_k : k = 1, 2, \ldots, n_t\})$, where n_t is a terms' number in the set T_r.

---

5   Experimental Evaluations

In this section we present a methodology of performed evaluations and result of simulations. Unfortunately, we need to simulate the process of document retrieval as performing experimental evaluations based on real data is time-consuming.

5.1   Methodology of Simulations

The overall idea of the simulations is presented in Fig. 2. Let us consider a set of m users profiles that are based on the domain ontology. To calculate profile for each user we use sets of relevant and irrelevant documents (positive and negative examples for Bayesian network creating procedure). The representative profile is determined using method described in Algorithm 1.

To judge the accuracy of developed profile it is necessary to know the real profile of the whole group. It can be calculated based on the sum of sets of relevant and irrelevant documents taken from all the users in the group. To compare these two profiles we need a quality condition which is described in the next subsection.

5.2   Quality Condition

To check the effectiveness of proposed method for determining the representative profile it is necessary to consider two profiles created in a different ways:

1. UP_{base} is a baseline profile – it is calculated as Bayesian network using a methods presented in [8]. The basis for the creating procedure is a sum of relevant and irrelevant documents sets of all the users in the group.
2. UP_r is a representative profile calculated based on profiles of each member of the group using Algorithm 1.

Let us define a *quality condition*:
  The distance between representative profile UP_r and baseline profile UP_{base} should be smaller than assumed threshold ϵ.

$$dist(UP_r, UP_{base}) < \epsilon \qquad (6)$$

In our previous work [7] we have confirm a desired property of the algorithm - more users in the group we have, smaller ϵ we can use. Our current experimental evaluation are performed to check if the diameter of the group has any impact on the result.

5.3   Obtained Results

The aim of the simulations was to check if the characteristic of the group of users (diameter of the group) has an impact on the quality condition.

The diameter $dist_{set}$ of the group of profiles is calculated using formula 2 for each pair of relevant and irrelevant documents taken from all the users in the

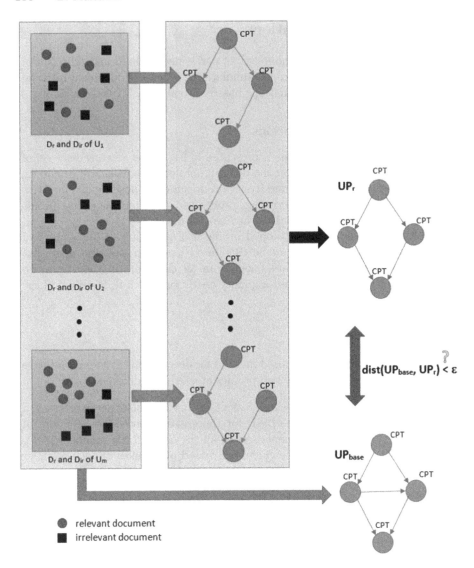

Fig. 2. An overall idea of experimental evaluations methodology.

group. The result is obtained as a maximum distance of calculated distances. We have performed simulations for the following values of distances:

$$dist_{set} \in \{0.2; 0.3; 0.4; 0.5\}.$$

Obtained distances between representative user profiles and baseline profiles (in each of 4 cases) are gathered in Table 2 and in Fig. 3.

Table 2. Results of performed simulations.

$dist_{set}$	0.2	0.3	0.4	0.5
$dist_{BN}$	0.0019	0.0032	0.0052	0.0078

The presented result have confirmed that the bigger diameter of the profile group is, the greater distance between representative and baseline profiles is. We can also observe that the dependency can not be a linear one.

Obtained results are important as they can be used in the future works. If the quality condition is no more satisfied (calculated distance between representative and baseline profile is greater than assumed threshold ϵ), it can be a reason to recalculate the groups in the first module of the whole system (clustering module in Fig. 1). It can also mean that many new users have been classified into this group but the dynamic of the group as a whole goes into different directions.

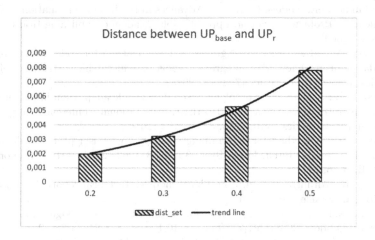

Fig. 3. Distances between UP_{base} and UP_r for different values of documents sets.

6 Summary

In this paper we have presented a method for checking the accuracy of the method for determining a representative profile of the group of users in the Personalized Document Retrieval System. We have considered a Bayesian network as a basis for user profile. The task was to analyze the influence of group characteristic on the method's accuracy. It was shown that more variety of input profiles determines less accurate representative profile.

Obtained result will be used in the future works: if the quality condition is not satisfied, the users in the system should be re-clustered as the user interests can change with time. Other conditions or criteria should be defined to analyze the dynamic of the whole system.

Acknowledgments. This research was partially supported by the Polish Ministry of Science and Higher Education.

References

1. Aknouche, R., Asfari, O., Bentayeb, F., Boussaid, O.: Integrating query context and user context in an information retrieval model based on expanded language modeling. In: Quirchmayr, G., Basl, J., You, I., Xu, L., Weippl, E. (eds.) CD-ARES 2012. LNCS, vol. 7465, pp. 244–258. Springer, Heidelberg (2012). https://doi.org/10.1007/978-3-642-32498-7_19
2. Al-Nazer, A., Helmy, T., Al-Mulhem, M.: User's profile ontology-based semantic framework for personalized food and nutrition recommendation. Procedia Comput. Sci. **32**, 101–108 (2014)
3. Fenz, S.: An ontology-based approach for constructing Bayesian networks. Data Knowl. Eng. **73**, 73–88 (2012)
4. Jongh, M., Druzdzel, M.J.: A comparison of structural distance measures for causal Bayesian network models. In: Recent Advances in Intelligent Information Systems, Challenging Problems of Science, pp. 443–456. Academic Publishing House EXIT, Warsaw (2009)
5. Maleszka, M., Mianowska, B., Nguyen, N.T.: A method for collaborative recommendation using knowledge integration tools and hierarchical structure of user profiles. Knowl.-Based Syst. **47**, 1–13 (2013)
6. Maleszka, B.: A method for ontology-based user profile adaptation in personalized document retrieval systems. In: 2016 IEEE International Conference on Systems, Man, and Cybernetics (SMC), pp. 3187–3192 (2016)
7. Maleszka, B.: A method for determining ontology-based user profile in document retrieval system. J. Intell. Fuzzy Syst. **32**, 1253–1263 (2017). https://doi.org/10.3233/JIFS-169124
8. Murphy, K.: An introduction to graphical models. Technical report, University of California, Berkeley, May 2001
9. Nguyen, N.T.: Advanced Methods for Inconsistent Knowledge Management. Springer, London (2008). https://doi.org/10.1007/978-1-84628-889-0
10. Pietranik, M., Nguyen, N.T.: A multi-attribute based framework for ontology aligning. Neurocomputing **146**, 276–290 (2014)
11. Ramkumar, A.S., Poorna, B.: Ontology based semantic search: an introduction and a survey of current approaches. In: 2014 International Conference on Intelligent Computing Applications. IEEE (2014). https://doi.org/10.1109/ICICA.2014.82
12. Ricci, F., Rokach, L., Shapira, B., Kantor, P.B. (eds.): Recommender Systems Handbook. Springer, Boston (2011). https://doi.org/10.1007/978-0-387-85820-3
13. Main Library and Scientific Information Centre in Wroclaw University of Science and Technology (2018). http://aleph.bg.pwr.wroc.pl/

Machine Learning and Data Mining

A Class-Cluster k-Nearest Neighbors Method for Temporal In-Trouble Student Identification

Chau Vo$^{(\boxtimes)}$ and Hua Phung Nguyen$^{(\boxtimes)}$

Ho Chi Minh City University of Technology, Vietnam National University,
Ho Chi Minh City, Vietnam
{chauvtn, nhphung}@hcmut.edu.vn

Abstract. Temporal in-trouble student identification is a classification task at the program level that predicts a final study status of a current student at the end of his/her study time using the data gathered from the students in the past. Moreover, this task focuses on correct predictions for the in-trouble students whose predicted labels are at the lowest performance level. Educational datasets in this task have many challenging characteristics such as multiple classes, overlapping, and imbalance. Simultaneously handling these characteristics has not yet been investigated in educational data mining. For the existing general-purpose works, their methods are not straightforwardly applicable to the educational datasets. Therefore, in this paper, a novel method is proposed as an effective solution to the previously defined task. Combining the traditional k-nearest neighbors and clustering ensemble methods, our method is designed with three new features: relax the number k of the nearest neighbors, use a set of the cluster-based neighbors newly generated by partitioning the subspace of each class, and set four new criteria to decide a final class label rather than the majority voting scheme. As a result, it is a new lazy learning method able to provide correct predictions of more instances belonging to a positive minority class. In an empirical evaluation, higher Accuracy, Recall, and F-measure confirmed the effectiveness of our method as compared to some popular methods on our two real educational datasets and the benchmarking "Iris" dataset.

Keywords: Student classification · Educational data mining ·
k-nearest neighbors · Clustering ensemble · Fisher's discriminant ratio ·
Data imbalance

1 Introduction

In educational data mining, data of the students in the past has been widely used to support the mining tasks for current students. Student classification at the program level is a well-defined task to predict a final study status of each current student at the end of a program using only the educational data observed in a few first years of his/her study and the data of other students in the past. The students with predicted labels at the lowest performance level are called in-trouble students, forming a student group of our interest. Therefore, a student classification task with a focus on those in-trouble students is called an in-trouble student identification task. When temporal connections

© Springer Nature Switzerland AG 2019
N. T. Nguyen et al. (Eds.): ACIIDS 2019, LNAI 11431, pp. 219–230, 2019.
https://doi.org/10.1007/978-3-030-14799-0_19

between the past data and the current ones are examined, an in-trouble student iden-tification task is called a temporal one. This task is more practical in the real world; but more challenging than the previous ones because in-trouble students often belong to a minority class, also sharing other characteristics with normal students as they might fail some courses and succeed in the others at some various performance levels. Such cases lead to overlapping and imbalance in educational datasets. In our work, a new solution is investigated for the temporal task.

Nowadays, several works in [2, 9, 10, 12, 15, 17, 18] were dedicated to student classification and performance prediction for different purposes with many various approaches. [2, 10, 15, 17, 18] followed a supervised learning approach. [2, 10, 17, 18] utilized the existing models while [15] proposed a new grammar-based genetic pro-gramming method for predicting student failure at school. [2] used C4.5, k-nearest neighbors, support vector machines, PART rule learner, and Naïve Bayes; [10] used support vector machines with different kernel functions and bagging and boosting schemes; [17] used statistical models, decision tree models, rule induction models, fuzzy rule induction models, and neural networks; and [18] used C4.5, Naïve Bayes, k-nearest neighbors, and NB trees. In addition, [9] dealt with data shortage by means of multiple data sources for predicting student performance with more effectiveness. [12] considered a semi-supervised learning approach to enhance the training dataset over the time and obtain better classifiers. However, only [15] handled data imbalance while the others did not. In addition, no temporal connection between the training dataset and new one is considered in these works. Therefore, it is an open question if the achieved methods could be generalized for new students. Besides, overlapping has not yet been resolved in connection with imbalance. So, our task along with its context has not yet been addressed in the educational data mining area.

Nevertheless, [4, 6, 11, 19] are among the most recent works for a classification task on imbalanced overlapping datasets. [6] was proposed specifically for datasets in smart home systems while the others were general-purpose for different datasets in several application domains. Unfortunately, no dataset in the educational domain was tested in their experiments. In addition, [4, 11, 19] required the data space to be divided into several kinds of various regions and then built a classifier for each region while [6] conducted cluster-based undersampling in the preprocessing phase. Such overall approaches are not straightforwardly applicable to our imbalanced datasets with mul-tiple high overlapping classes because only binary classification was supported in these works. For only data imbalance handling, SMOTE and its improved versions intro-duced in [7] can be investigated. Discussed in [13], not only imbalance but also other data characteristics like those in educational datasets must be overcome.

In this paper, we propose a novel method, named Class-Cluster-k-Nearest Neigh-bors (Class-Cluster-k-NN), for the temporal in-trouble student identification task. Our method is defined as a lazy learning algorithm by combining k-nearest neighbors (k-NN) and clustering ensemble. First, for each prediction of an unlabeled instance, it uses k-NN to select an initial set of the k nearest neighbors. The value for k is user-specified but required in a larger range as compared to the true optimal value for k in k-NN. This relaxation makes our method friendlier. Second, the initial set is refined to yield another set of the new neighbors. This refinement process is based on a resulting set of class-based clusters generated by the partitioning-based clustering ensemble method. Each cluster is then used to prepare a new neighbor. Using such cluster-based neighbors instead of the real instances makes the data space less dense with a more balanced contribution from each class. Finally, these new neighbors are used to predict a class label of the unlabeled instance according to four new criteria instead of the majority voting scheme. These criteria help avoiding random selection in case of more than one solution meeting the requirements. Designed with these new features, our method overcomes the challenging characteristics of educational datasets as well as the difficulty of *k*-NN in an instance-based learning approach.

Through experimental results, compared to several popular methods, our method has demonstrated its overall prediction capability with higher Accuracy values on our two real educational datasets and the benchmarking "Iris" dataset. In addition, it can provide correct predictions for more instances of a positive minority class with higher Recall and F-measure values for two real educational datasets. As a result, our method is confirmed to be an effective solution to our task in a more practical context. It helps identifying more in-trouble students who might fail to achieve a degree. Using the returned predictions, more support can be given to them and our teaching and learning activities can be changed appropriately for those students' ultimate success in study.

2   A Temporal In-Trouble Student Identification Task

In our work, we reconsider the in-trouble student identification (aka dropout prediction or student classification focusing on correctly predicting in-trouble students) task at the program level. Its more practical context is empirically examined with the temporal aspects of training and new (testing) datasets.

First, each student is characterized by means of the results of all the courses already studied and also those to be studied in the program. In a computational form, each student is represented by a vector X_i in a p-dimensional data space: $X_i = (v_{i,1}, v_{i,2}, ..., v_{i,p})$, where each element $v_{i,d}$ at a dimension d ($d = 1...p$) is a positive real number in $[0, 10]$ which is a grade of a subject. Besides, each student is monitored towards his/her ultimate success at the program level. Therefore, a final study status of each student is observed and denoted as a label: $Y_i = \{y_i\}$, where in our work, y_i is "graduating", "studying", or "study-stop". The in-trouble students of our interest are labeled "study-stop". After all, each student is associated with a tuple (X_i, Y_i).

For a supervised learning process of the task, we are given a training dataset L of n vectors which are labeled as defined above: $L = \{(X_1, Y_1), (X_2, Y_2), ..., (X_n, Y_n)\}$. This training dataset contains the data of the students with known final study status in the past. Let us denote T_{past} as the past academic years of these students.

The task input is an unlabeled dataset U of un vectors: $U = \{(X_1, NULL), (X_2, NULL), ..., (X_{un}, NULL)\}$. Each vector X_j in U represents a current student whose final study status Y_j is NULL, an unknown value required to be predicted. $T_{current}$ are the current academic years of these students. In practice, T_{past} must be less than $T_{current}$.

The output is a label set $\{Y_1, ..., Y_j, ..., Y_{un}\}$ predicted for $X_1, ..., X_j, ..., X_{un}$ in U, respectively, where each Y_j for $j = 1...un$ belongs to a given set of the final status labels.

Finally, given U in $T_{current}$ and L in T_{past}, a temporal in-trouble student identification task is a classification task to predict a final study status of each current student with a key concentration on in-trouble students labeled with "study-stop". Educational datasets are highly imbalanced with overlapping data between classes. These properties make our task on such datasets more challenging.

3   The Class-Cluster k-Nearest Neighbors Method

For the task defined in Sect. 2, our work proposes a novel method, named Class-Cluster-k-Nearest Neighbors (Class-Cluster-k-NN), designed in an instance-based learning paradigm. The traditional k-nearest neighbors (k-NN) method is its base algorithm for the simplicity and acceptable performance of k-NN on many existing datasets. k-NN is a typical lazy learning algorithm [5] well-known for being generalized for new instances. It has only one parameter k specified for the number of nearest neighbors used in prediction with the majority voting scheme. Nevertheless, how to determine an appropriate value for k is difficult for each prediction. If k is small, prediction might be affected by noises. By contrast, if k is large, it is hard to make a correct prediction with many instances from different classes. In our work, the proposed method handles this difficulty of k-NN once utilizing it in the educational domain.

On the other hand, overlapping and imbalance are among the important data characteristics influencing the effectiveness of a learning process. Dealing with overlapping, [4, 6, 11, 19] have been proposed recently. In [4, 11, 19], overlapped and non-overlapped regions were determined and treated differently while in [6], clustering-based undersampling was used. The first approach is sophisticated but not scalable for larger datasets while the second one is not suitable for smaller datasets. Besides, only binary classification was examined in these works. Different from these works, our work addresses multiclass classification on overlapping imbalanced educational datasets. As a result, their achieved solutions are not directly adapted to the datasets in our context from the theoretical perspectives. Therefore, our method is defined with the following new features to overcome data challenges and the difficulty of k-NN.

First, our method relaxes the number of the nearest neighbors, k. Instead of asking its users for an exact value of k, this method requires a larger value as an upper bound of k. This manner is much easier, even if they specify the size of the training dataset.

Second, refinement is offered for the initial set of the k nearest neighbors. Because there might exist noises, leading to some not-good nearest neighbors, reorganization of this initial set must be done for a better set of the true nearest neighbors.

Last but not least, prediction based on the majority voting scheme does not always work well when there are more than one answer meeting the requirements. Therefore, our method sets more criteria and their priorities to avoid such a situation.

3.1   Method Descriptions

Our method is designed with three main phases: 1. Selection of the k nearest neighbors, 2. Refinement of the initial set of the k nearest neighbors, 3. Prediction.

In phase 1, an initial set of the k nearest neighbors is prepared for each unlabeled instance, X_j in U for $j = 1...un$, using the traditional k-NN method. As a result, only the most similar instances to X_j are considered for its label prediction.

In phase 2, refinement of the initial set of the k nearest neighbors is conducted to yield another set of the true nearest neighbors that deal with overlapping and imbalanced data. This process is done for each group of the instances belonging to the same class. In each class-based group, clustering is performed to partition its data space into many dense regions. Each region contains only the instances that are the most similar to each other. A partitioning method like k-means is preferred to separate these dense regions. A spherical shape is desired because with a large number of varying-size clusters, their spherical shapes can cover the whole space. In order to obtain such a large enough number of varying-size clusters, a clustering ensemble is used.

A new neighbor is then prepared from each cluster with a class label which is the common class label of all the instances belonging to this cluster. Each neighbor has a representative which is the center of its cluster. It also has a set of members each of which is from the initial set of the k nearest neighbors. For majority classes, a cluster contains more similar instances than that for minority classes. This leads to the reduction in quantity for the majority class when such a cluster-based neighbor is used instead of the real instances. Using cluster-based neighbors helps us balance the number of the neighbors for each class and thus, its contribution to the learning process. Using these neighbors also helps us resolve the overlapping issue. When representatives of these neighbors are used, the distribution of the initial set is now replaced by the distribution of the representatives. The separation between the representatives is higher than that between the real instances. This increases a chance of making the space less dense and a chance of making overlapping data reduced.

In phase 3, using the new set of the neighbors, our method derives a class label for an unlabeled instance X_j according to several prioritized criteria rather than the majority voting scheme. The similarity between X_j and a neighbor is first examined by considering an average distance between X_j and its members. If the same, a distance between X_j and its center is then examined. If the same, a minimum distance between X_j and its members is then examined. Otherwise, its size, i.e. the number of the members, is examined. The rationale behind these prioritized criteria is to avoid the following cases: a class label associated with a larger set of the nearest instances is not always true and random selection often happens once the same requirements are met.

Its pseudo code is sketched phase by phase to predict a class label for each X_j in U as follows. In this description, $d()$ is a distance measure, k-$nn()$ is the traditional k-NN method, *class-based-separation*() is a method that groups the instances in a given set into at most $|Y|$ groups each of which contains the instances of the same class, *clustering*() is a partitioning method, and *cluster-based-neighbor-preparation*() is a method that prepares a new neighbor from a cluster of all the instances belonging to the same class. In our case, Euclidean is used for $d()$ and k-means [14] for *clustering*().

Input: X_j in U which is an instance in an unlabeled dataset U, L which is a given training dataset, k which is the user-specified number of the nearest neighbors
Output: Y_j which is predicted for X_j
Process:

Phase 1. Selection of the k nearest neighbors

//knnSet is a result set of the k nearest neighbors using $d()$.
(1). knnSet $\leftarrow$ k-nn(L, X_j, k)

Phase 2. Refinement of the initial set of the k nearest neighbors

//nearestSet is a set of the new nearest neighbors after this Refinement phase.
(2). nearestSet $\leftarrow$ $\emptyset$
//Y is a given set of the class labels in the domain.
(3). knnClassSet $\leftarrow$ *class-based-separation*(knnSet, Y)
(4). for each set CS of a class Y_{CS} in knnClassSet
    //kCluster is the number of clusters in the clustering process.
(5).   for each kCluster from 2 to |CS|-1 /*|CS|-1 can be user-specified if desired*/
(6).   clusters $\leftarrow$ *clustering*(CS, kCluster)
(7).   for each cluster C in clusters
        //nearestNeighbor is a new neighbor from a cluster with class Y_{CS}
(8).   nearestNeighbor $\leftarrow$ *cluster-based-nearest-preparation*(C, Y_{CS})
(9).   nearestSet $\leftarrow$ nearestSet $\cup$ {nearestNeighbor}
(10).   end for
(11). end for
(12). end for

Phase 3. Prediction

(13). for each neighbor NS in nearestSet
    //$X_i \in$ NS
(14).   averageDistance $\leftarrow$ $\Sigma_{i=1..|NS|}d(X_i,X_j)/|NS|$
    //C is the center of the neighbor NS
(15).   centerDistance $\leftarrow$ $d(C, X_j)$
(16).   minDistance $\leftarrow$ $min_{i=1..|NS|}(d(X_i,X_j))$
(17). end for
(18). Y_j $\leftarrow$ $argmin_{NS\in nearestSet}$(averageDistance$_{NS}$)
(19). if there exist more than one neighbor with the same averageDistance value
    /*nearestSet1 is a subset of nearestSet including the neighbors with the same averageDistance value*/
(20). Y_j $\leftarrow$ $argmin_{NS\in nearestSet1}$ (centerDistance$_{NS}$)
(21)   if there exist more than one neighbor with the same centerDistance value
      /*nearestSet2 is a subset of nearestSet1 including the neighbors with the same centerDistance value*/
(22). Y_j $\leftarrow$ $argmin_{NS\in nearestSet2}$ (minDistance$_{NS}$)
(23).   if there exist more than one neighbor with the same minDistance value
      /*nearestSet3 is a subset of nearestSet2 including the neighbors with the same minDistance value*/
(24). Y_j $\leftarrow$ $argmax_{NS\in nearestSet3}$ (|NS|)
(25).   end if
(26).   end if
(27). end if

3.2   Method Characteristics

Class-Cluster-k-NN is the first method that combines k-NN and a clustering ensemble for multiclass classification on educational datasets. This method aims at more predictions of the instances in the minority class which is also a positive class. When we deploy this method for the task, temporal aspects of our datasets are considered.

Our method has only one parameter k for a large number of the nearest neighbors. Its users feel free to specify a large number for k and the refinement process in our method will do the rest to yield a better set of the new cluster-based neighbors.

Besides, our method can generate more groups of some similar instances of the same class by means of a clustering ensemble. Using cluster-based neighbors instead of the real instances helps considering data overlapping regions and data imbalance.

For comparison, our method is different from the more recent works in [4, 6, 11, 19] as follows. In [4], an overlapping dataset is split into overlapping and non-overlapping regions. Each region is then clustered into k clusters. Each cluster in the training dataset is then used to build a support vector machine model. Each cluster in the testing dataset is then compared with other clusters in the training dataset to find the closest cluster and the corresponding model is used for classification. Our work is more practical and general because no explicit overlapping or non-overlapping region has to be discovered in training or new (testing) datasets. When we split the dataset into regions, the number of instances is just small in each region. Therefore, a statistical learning model built on such a region might not be effective for the task. In addition, a clustering ensemble is used in our method. Above all, a multiclass classification task can be supported by our method directly while not by [4].

In [6], clustering-based undersampling is used to remove the overlapping classes and the existing supervised learning algorithms are then applied on the resulting dataset. In their approach, a clustering method is conducted to obtain several clusters. Each cluster is examined for a good mix with respect to a threshold τ by removing the instances of the majority class. It is not easy to determine an appropriate threshold and select the instances of the majority classes to be removed on a multiclass dataset like those in our work. This is because in [6], only binary datasets were supported.

In [11], an overlap-sensitive margin classifier is proposed by combining a modified fuzzy support vector machine and k-NN. First, it defines a weighting scheme to assign a weight for each instance using the DEC and k-NN algorithms. Using those weights, the modified fuzzy support vector machine is then constructed for an optimal hyperplane to separate the space into soft-overlap and hard-overlap regions. The instances in the soft-overlap regions are classified by the decision boundary of the proposed method while those in the hard-overlap regions by the 1-NN algorithm. For new instances, a prediction is decided by combining overlap-sensitive margin and 1-NN classifiers. Like the previous ones, [11] supported only binary classification. In addition, their method is quite close to our method, which is also based on k-NN. The main difference is that our method does not distinguish subregions in the data space because there are more than two classes and our data overlapping is very much. Instead, our method determines non-overlapping regions in terms of class-based clusters. Furthermore, our method is more practical than the one in [11] because it is defined for multiclass datasets, has only one parameter k, and examines the data space more finely with a large number of varying-size clusters from a clustering ensemble.

In [19], a soft-hybrid algorithm is proposed for binary classification. The data space is first divided into obviously non-overlapped, borderline, and overlapped regions. For each region, a classifier is then constructed. The RBFN model is built in the obviously non-overlapped region, the dDBSCAN model in the borderline region, and the modified kernel learning model in the overlapped region. As compared to their approach with three learning algorithms required several parameters, our method is also more practical for multiclass classification with only one parameter k.

In short, [4, 6, 11, 19] achieved some positive results in some domains, not including the educational domain. Turning their solutions to the one on educational datasets is not feasible. The first reason is related to data characteristics such as the number of classes, imbalance, and overlapping while the second one to method characteristics such as the number of parameters, complexity, and testing applicability in our temporal context. Therefore, our method has been defined as a new solution to this task.

4   An Empirical Evaluation

In this section, several experiments were conducted on two real datasets ("Year 2007" and "Year 2008") and the benchmarking dataset ("Iris"). The two real datasets described in Table 1 were collected from the grades of 43 Computer Science courses of the third-year regular undergraduate students at Faculty of Computer Science and Engineering, Ho Chi Minh City University of Technology, Vietnam National University – Ho Chi Minh City [1]. Each student is characterized by 43 attributes corresponding to those 43 courses. Three classes ("Graduating", "Studying", and "Study-Stop") are defined for the final study status. "Study-stop" is a positive class of our interest. The Iris[1] data set is also used for its popularity. It has 150 instances characterized by 4 attributes, labeled with 3 classes, and 50 instances per class.

Table 1. Data descriptions

Dataset	Role	Data in years	Number of instances	"Graduating"		"Studying"		"Study-stop"	
				#	%	#	%	#	%
Year 2007	Training	2005–2006	689	378	54.86	266	38.61	45	6.53
	Testing	2007	317	167	52.68	88	27.76	62	19.56
Year 2008	Training	2005–2007	1006	545	54.17	354	35.19	107	10.64
	Testing	2008	311	168	54.02	77	24.76	66	21.22

For our educational datasets, overlapping between the classes is also examined using Fisher's discriminant ratio [8, 13]. We modified the original overlapping degree from two classes to three classes by using their maximum and average values. The smaller overlapping degree values show the more inter-class overlapping in a dataset. The maximum and average overlapping degree values for "Year 2007" are 0.81 and

---

[1] UCI Machine Learning Repository [http://archive.ics.uci.edu/ml].

0.59, and those for "Year 2008" 0.84 and 0.5, respectively. Those for "Iris" are 50.94 and 29.04, respectively. The overlapping degree values of "Iris" are much larger than those of our educational datasets. It is thus figured out that obtaining an effective classifier from our educational datasets is harder and it is much more challenging to identify the "study-stop" students who correspond to the instances of a minority class.

In addition, we use only data of the students in the past to build a classifier for prediction on the more current students so that our method can be evaluated in a more practical context for temporal in-trouble student identification than the existing works. In particular, the holdout scheme is used instead of the k-fold cross validation scheme. The generalization of the method to new data can be then examined. With the data gathered in 2005–2008, we defined two datasets "Year 2007" and "Year 2008" for the prediction models applied on the 2007 data using the 2005–2006 data for training and on the 2008 data using the 2005–2007 data for training, respectively. This results in a smaller training dataset for "Year 2007" as compared to that for "Year 2008". Therefore, such data shortage has an impact on the effectiveness of our prediction models.

For method comparison, we used the well-known methods in the Weka[2] library:

- k-NN: the traditional k-nearest neighbors method [5] with different values of k in [1, 15] with a gap 1 and the Euclidean metric.
- C4.5: a J48 decision tree model [16] used as a baseline method for classification.
- Neural Network: a Multilayer Perceptron model with default parameter settings.
- Random Forest: a random forest model [3] which is an ensemble model of 300 random trees with the default number of random features.
- Support Vector Machine (Poly): an SMO support vector machine model with the polynomial kernel and different values of C in {500, 1000, 1500, 2000, 2500}.
- Support Vector Machine (RBF): an SMO support vector machine model with the Radial Basis Function kernel using different values of gamma in [0.5, 0.95] with a gap 0.05 and different values of C in {500, 1000, 1500, 2000, 2500}.

For our method, Class-Cluster-k-NN is executed with the Euclidean metric for every distance calculation. The number k of the nearest neighbors is calculated as a ratio r to $N/Class\#$ where r varies from 1 to $N/Class\#$, N is the number of instances in a training dataset, and $Class\#$ is the number of distinct class values. This value is an upper bound of a true value for the number of the nearest neighbors in traditional k-NN.

For an overall evaluation, Accuracy is used in the range [0, 100] and for a specific evaluation on predicting the instances of the positive class, Recall, Precision, and F-measure in the range [0, 1]. The higher values imply the better model. In all the experiments, the best results were recorded in Tables 2 and 3. For comparison, the best results for each measure are presented in bold while the second best ones in italic.

In addition, we raise two questions in this empirical evaluation as follows:

- Is our method more effective than the existing methods on some given datasets?
- Is our method an effective solution to temporal in-trouble student identification?

---

[2] Weka 3 [http://www.cs.waikato.ac.nz/ml/weka].

Table 2. Accuracy values from different methods

Dataset	k-NN	C4.5	Neural network	Random forest	Support vector machine (Poly)	Support vector machine (RBF)	Class-cluster-k-NN
Year 2007	56.15	58.36	58.68	58.68	59.31	*59.62*	**61.20**
Year 2008	60.13	54.34	61.41	*62.38*	58.20	61.74	**65.92**
Iris	*96.67*	92.67	94.67	94.67	96.00	94.67	**97.33**

Table 3. Experimental results for In-Trouble student identification

Dataset	Method	Accuracy	Recall	Precision	F-measure
Year 2007	k-NN	56.15	0.08	0.42	0.14
Year 2007	C4.5	58.36	*0.18*	*0.65*	*0.28*
Year 2007	Neural network	58.68	0.13	0.62	0.21
Year 2007	Random forest	58.68	0.03	**0.67**	0.06
Year 2007	Support vector machine (Poly)	59.31	0.05	0.38	0.09
Year 2007	Support vector machine (RBF)	*59.62*	0.10	0.60	0.17
Year 2007	Class-cluster-k-NN	**61.20**	**0.24**	*0.65*	**0.35**
Year 2008	k-NN	60.13	0.36	0.65	0.47
Year 2008	C4.5	54.34	0.41	0.48	0.44
Year 2008	Neural network	61.41	0.23	0.54	0.32
Year 2008	Random forest	*62.38*	0.24	**0.89**	0.38
Year 2008	Support vector machine (Poly)	58.20	0.18	0.63	0.28
Year 2008	Support vector machine (RBF)	61.74	*0.44*	*0.67*	*0.53*
Year 2008	Class-cluster-k-NN	**65.92**	**0.56**	*0.67*	**0.61**

For the first question, we have got a "yes" answer with higher accuracy values from our method in Table 2. Compared to the second best results, on Iris, accuracy of our method is about 0.66% higher than that of k-NN; on "Year 2007", about 0.58% Support Vector Machine (RBF); and on "Year 2008", about 3.54% Random Forest. This shows that our method outperforms the others. Such results stem from the capability of our method in examining and selecting the neighbors in the form of clusters. The elaborated criteria for a final class decision instead of the majority voting scheme in k-NN also have a good contribution to the performance of the resulting classifier.

For the second question, we focus on not only overall correct predictions but also correction predictions specific for the positive class. In our case, the positive class is also a minority class. Besides Accuracy, we thus observe Recall, Precision, and F-measure of the positive class. For "Year 2007", it is realized that Recall values of the existing methods are very low. When balanced with Precision values, F-measure values are a bit higher. This implies that just a few instances of the positive class can be correctly predicted. Different from the existing methods, our method can achieve much higher Recall and F-measure values. Meanwhile, it maintains higher Accuracy values. In particular, it has about 6% and 7% higher Recall and F-measure values than the second best returned by C4.5. Similarly, it also obtained better Recall and F-measure values on "Year 2008" with about 12% and 8% higher than the second best by Support Vector Machine (RBF), respectively. This confirms the appropriateness of the design features of our method. Looking at each target instance, our method can examine a group of instances for its true class. In a dataset with high overlapping, considering only individual instances for prediction is not a good choice as a chance of a mis-classification is higher than considering a group of similar instances. Besides, more criteria were added to finalize the nearest group for the target instance. These help avoiding random selections among the nearest groups with the same shortest distance in k-NN. Such results give us a "yes" answer to the second question.

In short, the effectiveness of our method is confirmed in this empirical evaluation.

5   Conclusions

In this paper, we have defined a novel Class-Cluster k-Nearest Neighbors method as a solution to temporal in-trouble student identification. The method takes into account overlapping and imbalance in a training dataset so that a more effective classifier can be built in an instance-based learning manner. In our method, k-nearest neighbors is redefined with a clustering ensemble. In particular, instead of individual instances for the k nearest neighbors, many groups of the most similar instances are formed by means of a clustering process and then examined in the 1-nearest neighbor method. Besides, the majority voting scheme is extended with several more elaborated criteria to avoid random selections of the nearest neighbors with the same shortest distances. Our method also relaxes the strict specification of an appropriate value for k, the desired number of the nearest neighbors. Those design features make our method more effective than the existing methods on two real educational datasets and the bench-marking "Iris" one. Temporal characteristics of the educational datasets are also con-sidered in our empirical evaluation. Better experimental results for Accuracy, Recall, and F-measure in the experiments have reflected the effectiveness of our method.

The proposed method is still in its infancy. In the future, other issues such as sparseness need to be tackled. Sparse data has been studied in our recent works while putting them altogether in this task has not yet been considered. In addition, feature learning is of our interest so that a feature space can be effectively established for the proposed method. Finally, a more practical effective model will be included in the educational decision support system for advancing learning and teaching activities.

Acknowledgments. This research is funded by Vietnam National University Ho Chi Minh City, Vietnam, under grant number C2017-20-18.

References

1. Academic Affairs Office: Ho Chi Minh City University of Technology, Vietnam. http://www.aao.hcmut.edu.vn. Accessed 29 June 2017
2. Bayer, J., Bydzovska, H., Geryk, J., Obsivac, T., Popelinsky, L.: Predicting drop-out from social behaviour of students. In: Proceedings of the 5th International Conference on Educational Data Mining, pp. 103–109 (2012)
3. Breiman, L.: Random forests. Mach. Learn. **45**(1), 5–32 (2001)
4. Chujai, P., Chomboon, K., Chaiyakhan, K., Kerdprasop, K., Kerdprasop, N.: A cluster based classification of imbalanced data with overlapping regions between classes. In: Proceedings of the International Multi-Conference of Engineers and Computer Scientists I, pp. 1–6 (2017)
5. Cover, T., Hart, P.: Nearest neighbor pattern classification. IEEE Trans. Inf. Theory **13**, 21–27 (1967)
6. Das, B., Krishnan, N.C., Cook, D.J.: Handling class overlap and imbalance to detect prompt situations in smart homes. In: Proceedings of the 2013 IEEE 13th International Conference on Data Mining Workshops, pp. 1–8 (2013)
7. Fernández, A., García, S., Herrera, F., Chawla, N.V.: SMOTE for learning from imbalanced data: progress and challenges, marking the 15-year anniversary. J. Artif. Intell. Res. **61**, 863–905 (2018)
8. Ho, T., Basu, M.: Complexity measures of supervised classification problems. IEEE Trans. Pattern Anal. Mach. Intell. **24**, 289–300 (2002)
9. Koprinska, I., Stretton, J., Yacef, K.: Predicting student performance from multiple data sources. Artif. Intell. Educ. **9112**, 678–681 (2015)
10. Kravvaris, D., Kermanidis, K.L., Thanou, E.: Success is hidden in the students' data. Artif. Intell. Appl. Innov. **382**, 401–410 (2012)
11. Lee, H.K., Kim, S.B.: An overlap-sensitive margin classifier for imbalanced and overlapping data. Expert Syst. Appl. **98**, 72–83 (2018)
12. Livieris, I.E., Drakopoulou, K., Tampakas, V.T., Mikropoulos, T.A., Pintelas, P.: Predicting secondary school students' performance utilizing a semi-supervised learning approach. J. Educ. Comput. Res. (2018)
13. López, V., Fernández, A., Moreno-Torres, J.G., Herrera, F.: Analysis of preprocessing vs. cost-sensitive learning for imbalanced classification. Open problems on intrinsic data characteristics. Expert Syst. Appl. **39**, 6585–6608 (2012)
14. MacQueen, J.: Some methods for classification and analysis of multivariate observations. In: Proceedings of the 5th Berkeley Symposium on Mathematical Statistics Probability, vol. 1, pp. 281–297 (1967)
15. Márquez-Vera, C., Cano, A., Romero, C., Ventura, S.: Predicting student failure at school using genetic programming and different data mining approaches with high dimensional and imbalanced data. Appl. Intell. **38**, 315–330 (2013)
16. Quinlan, J.R.: C4.5: Programs for Machine Learning. Morgan Kaufmann, Burlington (1993)
17. Romero, C., Espejo, P.G., Zafra, A., Romero, J.R., Ventura, S.: Web usage mining for predicting final marks of students that use Moodle courses. Comput. Appl. Eng. Educ. **21**, 135–146 (2013)
18. Taruna, S., Pandey, M.: An empirical analysis of classification techniques for predicting academic performance. In: Proceedings of the IEEE International Advance Computing Conference, pp. 523–528 (2014)
19. Vorraboot, P., Rasmequan, S., Chinnasarn, K.: Improving classification rate constrained to imbalanced data between overlapped and non-overlapped regions by hybrid algorithms. Neurocomputing **152**, 429–443 (2015)

Multidimensional Permutation Entropy
for Constrained Motif Discovery

Yomna Rayan[1](✉)(iD), Yasser Mohammad[1,2](✉)(iD), and Samia A. Ali[1](iD)

[1] Assiut University, Asyut, Egypt
{yomna_rayan,Samya.hassan}@eng.au.edu.eg, y.mohammad@aist.go.jp
[2] AIST, Tokyo, Japan

Abstract. Constrained motif discovery was proposed as an unsupervised method for efficiently discovering interesting recurrent patterns in time-series. The de-facto standard way to calculate the required constraint on motif occurrence locations is change point discovery. This paper proposes the use of time-series complexity for finding the constraint and shows that the proposed approach can achieve higher accuracy in localizing motif occurrences and approximately the same accuracy for discovering different motifs at three times the speed of change point discovery. Moreover, the paper proposes a new extension of the permutation entropy for estimating time-series complexity to multidimensional time-series and shows that the proposed extension outperforms the state-of-the-art multi-dimensional permutation entropy approach both in speed and usability as a motif discovery constraint.

1 Introduction

Multidimensional time series data get generated from all kinds of industrial and scientific activities at an ever increasing rate. Finding interesting patterns in this data for useful information is a daunting task that is attracting more attention in the age of big data. Several information discovery methods rely on the ability to discover approximately recurring short patterns in long time-series. These patterns are called *motifs*.

The history of motif discovery (MD) is rich and dates back to the 1980s [17]. The problem started in the bioinformatics research community focusing on discovering recurrent patterns in RNA, DNA and protein sequences. Since the 1990s, data mining researchers started to shift their attention to motif discovery in real-valued time-series and several formulations of the problems and solutions to these formulations were and are still being proposed.

One of the earliest approaches was to discretize the time-series (usually but not necessarily using SAX [4,6]) then apply a discrete motif discovery algorithm from the bioinformatics literature. The most widely used such algorithm is Projections [4]. The main idea of the algorithm is to select several random hash functions and use them to hash the input sequences. Occurrences of the hidden motif are expected to hash frequently to the same value (called *bucket*) with a

© Springer Nature Switzerland AG 2019
N. T. Nguyen et al. (Eds.): ACIIDS 2019, LNAI 11431, pp. 231–243, 2019.
https://doi.org/10.1007/978-3-030-14799-0_20

small proportion of background noise. An extension of Projections-based MD was proposed by Tanaka [22] that uses minimum description length (MDL) and PCA to handle multidimensional time-series and find only statistically significant ones.

Another popular approach (used for example by the MK algorithm [20]) is to find efficiently the most similar pair of subsequences of a given length in a single-dimensional time-series using the triangular inequality for pruning of unneeded calculations. Several extensions were proposed that can discover multiple pairs [16], multiple motif lengths [11] and full motif enumeration of scale invariant motifs [12,13] and [19].

One problem with both of the aforementioned approaches is that they do not scale well with very long time series (in the length of millions of points). Stochastic motif discovery does not suffer from this problem because its computational requirements can be adjusted to match available computational resources. Stochastic motif discovery algorithms sample subsequences from the time-series and compare them using some predefined distance function searching for recurrent patterns with small distances. The simplest of these algorithms was proposed by Catalano et al. [2] and it samples subsequences randomly (i.e. from a uniform distribution over subsequence start points) then uses the distances between short overlapping parts of these subsequences to discover if two full occurrences of the same motif exist in these sampled subsequences. This approach is only viable when motifs appear frequently in the time-series that random sampling can have a chance of sampling two complete occurrences.

Constrained Motif Discovery (CMD) is a special case of Stochastic MD which utilizes some constraint to bias the sampling process toward parts of the time-series in which there is higher probability of finding a motif occurrence. This can be achieved using domain knowledge but, more interestingly, it can be achieved by utilizing a change point discovery (CPD) algorithm. This use of CPD for seeding MD is based on the assumption that when motif occurrences do not overlap or immediately follow each other, a change in the generation process must – by definition – happen at the beginning and end of each occurrence and CPD can be used to find these locations and bias the search process of MD. The toolbox has three Constrained MD functions [17]. Variants of this approach include MCFull and MCInc [9], Distance Graph Relaxation [7], GSteX [8] and shift–density based approach [14].

In this paper, we propose a new approach for finding the constraint used by constrained MD algorithms using analysis of the complexity of time-series sequences. The advantage of relying on complexity measures instead of change point scores for constructing the constraint is that it automatically finds motifs that have *interesting/complex* structure avoiding the common problem of finding trivial motifs faced by CPD based algorithms. Moreover, this paper will show that the proposed method is faster to execute.

Bandt and Pompe proposed Permutation Entropy (PE) as a complexity measure for time series [1], which is similar to Lyapunov exponents and is applicable to any type of time-series. One of the main limitations of the PE algorithm in its original form is its inability to handle multidimensional time-series.

Several approaches have been proposed for a multivariate/multidimensional version of PE. Multivariate Permutation Entropy (MPE), focuses exclusively on the order of values at different channels at the same time step [5].

Multivariate Multi-Scale Permutation Entropy (MMSPE), applies PE in a matrix taking both orders of a single channel at different times and of different variables and simply combines the frequencies in the whole matrix. Moreover MMSPE uses a straightforward multi-scaling through averaging [18]. This method is the most related to the proposed approach and we use it as the baseline approach when evaluating the proposed method.

We propose an extension of PE that is more efficient than MMSPE and MPE, as it takes in consideration time-series values at different channels at different time steps.

The rest of this paper is organized as follows: Sect. 2 gives a mathematical definition for Permutation Entropy. Section 3 details our proposed algorithm which is evaluated in Sect. 4. Then the paper is concluded.

2 Permutation Entropy

Consider given a single dimensional time series x_t for $t = 1, 2, ..., T$, and its time-delay embedding representation $(X_t^n)_i = \{x_i, x_{i+\tau}, ..., x_{i+(n+1)\tau}\}$ for $i = 1, 2, ..., T - (n-1)\tau$, where n is the embedding dimension or permutation order and τ is a time delay that represents the time difference between the sample points for each segment of the $n!$ possible ordinal patterns.

To compute the PE of the time series, there will be $n!$ possible ordinal patterns $\pi_{j=1}^{n!}$. For each single motif π_j the relative frequency will be:

$$p(\pi_j^{n,\tau}) = \frac{\#\{k|0 \leq k \leq T - (n-1)\tau, (x_{k+1}, ..., x_{k+(n-1)\tau}) \text{ has type } \pi_j^{n,\tau}\}}{T - (n-1)\tau})$$

(1)

Then the permutation entropy (PE) of embedding dimension $n \geq 2$ and time-lag τ is:

$$H(n, \tau) = \sum_{j=1}^{n!} p(\pi_j^{n,\tau}) \log_2 p(\pi_j^{n,\tau})$$

(2)

Using Eqs. (1), and (2) it is trivial to show that $0 \leq H(n, \tau) \leq \log_2 n!$ [1]. For a completely random time-series: $H(n)$ will equal $\log_2(n!)$ (which is the upper bound). For a very regular time series (e.g. a constant value time-series), $H(n)$ will equal zero.

PE depends on two-parameters: n the embedding dimension, and τ the time-lag [21]. Figure 1 shows an example of PE calculation for $n = 3$, and $\tau = 1$, and 2.

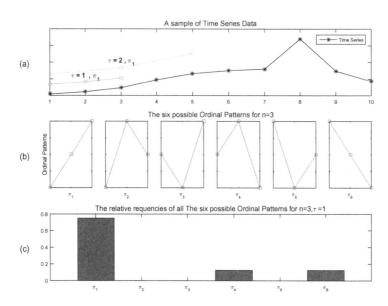

Fig. 1. Example describing the PE calculation. (a) Sample of time series data with extracting ordinal pattern π_1 for different value of τ, $(\tau = 1, 2)$ and $n = 3$. (b) The six possible Ordinal Patterns for $n = 3$. (c) The relative frequencies for all the six possible Ordinal Patterns for $n = 3$.

3   Multidimensional Permutation Entropy Variant

Given a multidimensional time series $\mathbf{x}_t \in \mathbb{R}^N$ for $t = 1, 2, 3, ..., T$, where N is the dimensionality of the time series. To order $\mathbf{x}_t$ points based on their values [1], we propose using the distance between all time series' points and a reference point (q). The rest of the PE calculation is the same as in Sect. 2. This technique can intuitively be understood as giving an ordering of time-series points at some single-dimensional projection specified by the reference point q converting the problem into a single-dimensional evaluation.

Any distance metric can be used for this ordering. In this paper we evaluate the following three variants:

1. Euclidean distance (see Eq. (3)), using the first point in time series $\mathbf{x}_1$ as (q) a reference point.

$$d(\mathbf{x}, \mathbf{q}) = \sqrt{(x_1 - q_1)^2 + (x_2 - q_2)^2 + (x_3 - q_3)^2 + ... + (x_N - q_N)^2} \quad (3)$$

2. Manhattan distance (see Eq. (4)), using the first point in time series $\mathbf{x}_1$ as (q) a reference point.

$$d(\mathbf{x}, \mathbf{q}) = |(x_1 - q_1)| + |(x_2 - q_2)| + |(x_3 - q_3)| + ... + |(x_N - q_N)| \quad (4)$$

3. Normalized distance which uses Eq. (3), but employs the zero point **0** as the reference point.

The resulting single dimension time series after distance calculation represents the rank order of the successive time series points which can be easily used as input to PE for complexity estimation. This conversion of a multi-dimensional PE problem into a standard single dimensional PE problem by utilizing a distance measure is called MPE hereafter (Multidimensional-PE).

4 Evaluation

The main contribution of this paper is applying multi-dimensional complexity analysis to calculate the constraint used in constrained motif discovery. This section reports two experiments using a real-world dataset to evaluate the proposed variance against state-of-the-art Singular Spectrum Analysis (SSA) based change point discovery (called CPD hereafter) [10], and MMSPE [18] as an example of the state-of-the-art in multidimensional complexity measures. Because CPD is a single dimensional algorithm, we apply Principal Component Analysis (PCA) to the time-series before applying CPD as suggested in [10]. This gives a total of five algorithms.

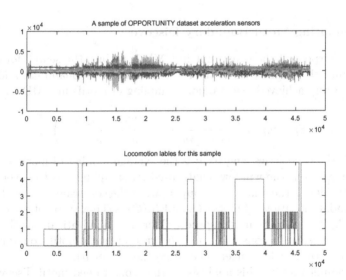

Fig. 2. OPPORTUNITY Dataset

4.1 Dataset

The OPPORTUNITY Dataset for Human Activity Recognition [3] contains the readings of motion sensors recorded while users executed typical daily activities. These sensors divide to:

- Body-worn sensors: 7 inertial measurement units, 12 3D acceleration sensors, 4 3D localization information.
- Object sensors: 12 objects with 3D acceleration and 2D rate of turn.
- Ambient sensors: 13 switches and 8 3D acceleration sensors.

In this paper, only the body-worn sensors are used. These sensors give a 36-dimensional time-series. One run from each user was used for hyper-parameter estimation and the remaining five runs were used for evaluation. The same hyper-parameters were used for all PE based algorithms including the proposed variants. Sensor data was annotated for different levels of activities (low level, medium level and high level). We consider only the mode of locomotion annotation which has five classes labelled as $[Stand = 1, Walk = 2, Sit = 3, Lie = 4]$. These classes define five different *motifs* that repeat in the time series but in slightly different ways corresponding to the definition of approximately recurring motifs.

Figure 2 shows a sample from OPPORTUNITY acceleration sensors data, and their locomotion labels. Missing data was ignored.

4.2 Evaluating Motif Boundary Discovery

The main purpose of the constraint in CMD is to bias the search for motifs to areas around the boundaries of motifs. As shown by Mohammad and Nishida [9], the speedup achieved over random sampling of motifs in CMD is:

$$Speedup = \left(\frac{n}{m \log_2 m} \times \frac{D\left(C||G\right) + H\left(C\right)}{w - \bar{l} + 1} \right)^2, \tag{5}$$

where n is the time-series length, m is the average number of motif occurrences for each motif, w is the window length used for sampling, l is the average motif length, $D\left(C||G\right)$ is the relative entropy between the constraint C and the ground truth boundary locations of motifs G, and $H\left(C\right)$ is the entropy of the constraint. From Eq. 5, it is clear that the speedup increases quadratically with increased similarity between the constraint and the ground truth motif boundaries. This effect is also stronger for longer time-series (i.e. larger n).

One complication to this analysis, is that constrained motif discovery algorithms do not require the constraint to have a high score exactly at the motif occurrence boundary but *near* it. This means that direct calculation of relative entropy is too conservative for estimating the performance of CMD algorithms given any constraint.

A more appropriate measure of the quality of a constraint (i.e. its *similarity* to ground truth motif occurrence boundaries) that takes this *acceptable shift* in motif discovery into account is the Equal Sampling Rate (ESR) [10], which is defined as the following:

$$ESR\,(p,q,n,s) = \sum_{t=1}^{T} q\,(t) \sum_{j=\max(1,t-n)}^{j=\min(T,t+n)} p\,(j)\,s\,(j-t), \qquad (6)$$

where $p(t)$ is the *true* motif occurrence boundary distribution (normalized to sum to one) and $q(t)$ is the constraint (normalized to sum to one). $p\,(t)$ and $p\,(t)$ are defined on the range $1 \leq t \leq T$, n is a positive integer specifying *acceptable shift* in detection of the motif boundary, $s(i)$ is a scaling function.

The main idea behind ESR is to treat the change scores as pdfs (Probability Density Function) and finding the probability that a random sample from this pdf will be within a given allowed delay from a true change point. This notion of sampling equality to the true distribution is directly measuring the appropriateness of the algorithm for use in Constrained Motif Discovery [9], but it also corresponds to our subjective sense that if a better motif occurrence boundary discovery algorithm is one that gives high scores *near* true motif occurrence boundaries and low scores otherwise. The main difference between this approach and relative entropy is that it incorporates the concept of a neighbourhood of acceptable shift.

The ESR measure is a non-symmetric measure of similarity. This allows us to use it to estimate the precision (specificity) of the constraint as well as its recall (sensitivity) by evaluating $ESR\,(C,G)$ and $ESR\,(G,C)$ respectively. The MC^2 toolbox [15] was used for these evaluations. Figure 3 shows the estimated precision and recall of the five algorithms with two window sizes $w = 200$ and $w = 400$, and CPD used after applying Principle Component Analysis (PCA) on the acceleration sensors data for dimensionality reduction. For the CPD algorithm we use the parameter settings recommended by Mohammad and Nishida [10]: $m =$ the width of the dynamic window, and $k =$ the number of windows to consider ($m = \frac{m}{4}, k = \frac{3m}{4}$). For PE based algorithms, we set $n = 5$ and $\tau = 1$. The values for these parameters were determined by trial-and-error on a single session from a single user to avoid over-fitting. According to [21] recommended values for n and τ are $\tau = 1$ and $3 \leq n \leq 7$ which agree with the results of our hyper-parameter tuning.

To get a general performance evaluation independent of the acceptable shift, the Aggregated ESR (AESR) [10] was also calculated for each algorithm. Figure 4a shows these results. It is clear that the proposed three variants significantly outperform MMSPE and CPD both in terms of AESR and speed.

To assess the statistical significance of these results, we applied factorial t-test with the conservative Bonferroni's multiple comparisons. For $F_{0.5}$, the only statistically significant results were Normalized MPE vs. MMSPE ($p = 0.002$ and $t = 3.35$), Euclidean MPE vs. MMSPE ($p = 0.003$ and $t = 3.12$), Manhattan MPE vs. MMSPE ($p = 0.0049$ and $t = 3.01$) and MMSPE vs. CPD ($p = 0.003$ and $t = -3.2$) confirming that the proposed variants are better than MMSPE which is worse than CPD for the evaluated task. Differences in speed showed the same pattern of significance but will not be reported due to lack of space.

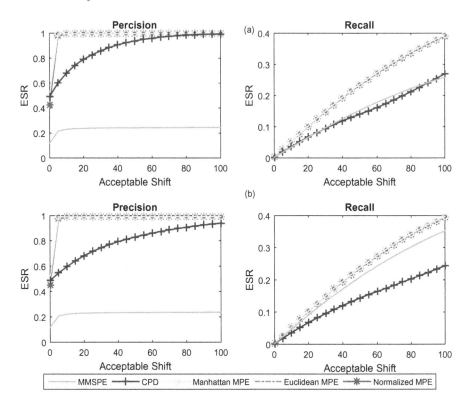

Fig. 3. Motif occurrence boundary discovery for the OPPORTUNITY Dataset using ESR (a) for window size equal = 200 points, (b) for window size equal = 400 points.

The main limitation of this evaluation methodology is that it compares discovery of *motif occurrences* instead of *motifs*. In some applications, it may not be important to discover all occurrences of a motif but only representative samples. The next experiment evaluates the proposed variants in terms of motif discovery instead of motif occurrence discovery.

4.3 Evaluating Motif Discovery Results

In this experiment, we compared the actual motif discovery performance of the GSteX Constrained Motif Discovery algorithm [8] when the constraint is calculated using the proposed variants against both CPD and MMSPE. The same hyper-parameters as in the first experiment were used. The GSteX algorithm expects a discrete set of *probable* motif locations instead of a real-valued score. A simple local-maxima finding algorithm was used to find these probable locations for all algorithms as recommended by [8].

In this experiment, we consider a discovered motif a true-positive if it corresponds to occurrences of one of the five ground-truth motifs (locomotion activities), and a false-positive if it does not. A ground truth motifs for which no

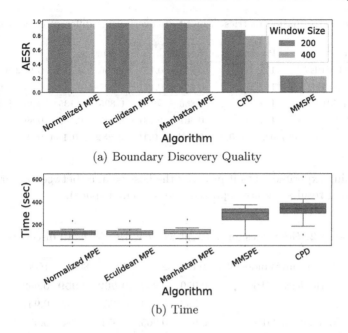

(a) Boundary Discovery Quality

(b) Time

Fig. 4. Comparing motif occurrence boundary discovery using aggregated ESR

occurrences were covered by any discovered motifs is considered a false-negative and true-negatives are found as the time-steps that do not correspond to any of the aforementioned three categories as multiples of average motif occurrence lengths. Because most applications of motif discovery put more emphasis on sensitivity, we calculate the $F_{0.5}$ measure as an overall quality assessment (Eq. 7) as well as accuracy.

$$\frac{(1 + \alpha^2)Precision \times Recall}{(\alpha^2 \times Precision) + Recall}, \alpha = 0.5. \tag{7}$$

Table 1 summarizes the results of this experiment for window size 400 (Varying the window size from 200 to 400 did not affect the results and is not reported due to lack of space). It is clear that the proposed variants consistently outperform MMSPE at all evaluation criteria while achieving comparable results to CPD.

4.4 Sensitivity Analysis

In this experiment, we evaluate the effect of hyperparameters on the quality of motif discovery (using the same evaluation methodology).

Table 2 shows the $F_{0.5}$ results with variable window sizes between 100 and 1600. It is clear that for all algorithms, varying the window size did not affect the performance significantly. Based on that a window size of 400 was used in

Table 1. Evaluation metrics of motif discovery using different constraints

Algorithm	Precision	Recall	Accuracy	$F_{0.5}$
Normalized MPE	1 ± 0	0.833 ± 0.15	0.856 ± 0.15	$\mathbf{0.962 \pm 0.043}$
Eculidean MPE	1 ± 0	0.822 ± 0.141	0.837 ± 0.141	0.957 ± 0.041
Manhattan MPE	1 ± 0	0.856 ± 0.135	0.825 ± 0.135	0.954 ± 0.039
MMSPE	1 ± 0	0.667 ± 0.181	0.675 ± 0.182	0.896 ± 0.072
CPD (m = 100 k = 300)	0.991 ± 0.039	$\mathbf{0.889 \pm 0.157}$	$\mathbf{0.882 \pm 0.154}$	0.962 ± 0.051

the remaining experiments as it provides the best balance between accuracy and speed (timing results are not reported due to lack of space).

Table 2. $F_{0.5}$ values for different window sizes for motif discovery

Algorithm/WindowSize	100	200	400	800	1600
Normalized MPE	0.959	0.962	**0.962**	**0.959**	0.962
Eculidean MPE	0.952	0.958	0.957	0.957	0.957
Manhattan MPE	0.95	0.965	0.954	0.954	0.957
MMSPE	0.901	0.897	0.896	0.9	0.903
CPD	**0.971**	**0.974**	**0.962**	0.958	**0.98**

Table 3 shows $F_{0.5}$ values of the proposed variants at $\tau = 1$ for different n values while Table 4 shows the same metric for different τ values at $n = 3$. There was no statistically significant differences due to varying either of these parameters.

Table 3. Effect of n on the performance of the propose variants ($F_{0.5}$)

Proposed variant	$n = 3$	$n = 4$	$n = 5$	$n = 6$	$n = 7$	$n = 8$	$n = 9$	$n = 10$
Normalized MPE	0.962	0.962	0.962	0.959	0.959	0.959	0.952	0.958
Eculidean MPE	0.967	0.954	0.957	0.966	0.955	0.955	0.959	0.959
Manhattan MPE	0.962	0.954	0.954	0.978	0.937	0.952	0.963	0.962

Table 4. Effect of τ on the performance of proposed variants at $n = 3$

Proposed variant	Value of τ									
	1	5	10	20	30	40	50	60	70	80	90
Normalized	0.962	0.962	0.962	0.959	0.962	0.962	0.962	0.962	0.962	0.959	0.962
Eculidean	0.967	0.935	0.981	0.963	0.961	0.94	0.961	0.959	0.95	0.95	0.944
Manhattan	0.962	0.975	0.966	0.946	0.966	0.966	0.966	0.951	0.962	0.952	0.944

4.5 Discussion

Taken together, the results of the first two experiments reported in this section show that the proposed variants provide a better constraint for motif discovery than either change point discovery or MMSPE in terms of occurrence boundary discovery (Fig. 3 and 4a) and outperforms MMSPE in motif discovery (Table 1). This superior performance is achieved despite the fact that the proposed variants are two times faster than MMSPE and three times faster than CPD.

One limitation of the proposed method for complexity analysis in general is that it relies directly on the concept of *entropy*. It can be argued that high entropy signals as well as low entropy signals are both *simple* which is not captured by the proposed method. This limitation, nevertheless, does not affect our goal of finding an effective constraint for motif discovery. In the future, we will consider using more rigorously defined complexity techniques based on minimum description length. Another limitation of the proposed approach is the dependence on some form of Euclidean or Manhattan distance metric which is not always meaningful for comparing time series (consider for example sound signals). In the future, the use of other metrics will be explored and the proposed method will be applied to other types of signals to assess its generality.

5 Conclusion

In this paper, we proposed a new approach for calculating time series complexity using Permutation Entropy for multi-dimensional time-series. The proposed complexity measure was shown to provide a better way to detect motif occurrence boundaries than standard change point discovery in constrained motif discovery applications. Evaluation results established the superiority of the proposed variants to a state-of-the-art time-series complexity measure in a real-world evaluation dataset. Moreover, the paper shows that the proposed method is robust to variations in its hyper-parameters.

References

1. Bandt, C., Pompe, B.: Permutation entropy: a natural complexity measure for time series. Phys. Rev. Lett. **88**(17), 174102 (2002). https://doi.org/10.1103/PhysRevLett.88.174102
2. Catalano, J., Armstrong, T., Oates, T.: Discovering patterns in real-valued time series. In: Fürnkranz, J., Scheffer, T., Spiliopoulou, M. (eds.) PKDD 2006. LNCS (LNAI), vol. 4213, pp. 462–469. Springer, Heidelberg (2006). https://doi.org/10.1007/11871637_44
3. Chavarriaga, R., et al.: The opportunity challenge: a benchmark database for on-body sensor-based activity recognition. Pattern Recogn. Lett. **34**(15), 2033–2042 (2013). https://doi.org/10.1016/j.patrec.2012.12.014

4. Chiu, B., Keogh, E., Lonardi, S.: Probabilistic discovery of time series motifs. In: 9th ACM SIGKDD International Conference on Knowledge Discovery and Data Mining, KDD 2003, pp. 493–498. ACM, New York (2003).https://doi.org/10.1145/956750.956808

5. He, S., Sun, K., Wang, H.: Multivariate permutation entropy and its application for complexity analysis of chaotic systems. Phys. A Stat. Mech. Appl. **461**, 812–823 (2016)

6. Lin, J., Keogh, E., Wei, L., Lonardi, S.: Experiencing sax: a novel symbolic representation of time series. Data Min. Knowl. Disc. **15**(2), 107–144 (2007)

7. Mohammad, Y., Nishida, T.: Learning interaction protocols using augmented Baysian networks applied to guided navigation. In: IEEE/RSJ International Conference on Intelligent Robots and Systems, IROS 2010, pp. 4119–4126. IEEE, October 2010. https://doi.org/10.1109/IROS.2010.5651719

8. Mohammad, Y., Ohmoto, Y., Nishida, T.: G-SteX: greedy stem extension for free-length constrained motif discovery. In: Jiang, H., Ding, W., Ali, M., Wu, X. (eds.) IEA/AIE 2012. LNCS (LNAI), vol. 7345, pp. 417–426. Springer, Heidelberg (2012). https://doi.org/10.1007/978-3-642-31087-4_44

9. Mohammad, Y., Nishida, T.: Constrained motif discovery in time series. New Gener. Comput. **27**(4), 319–346 (2009)

10. Mohammad, Y., Nishida, T.: On comparing SSA-based change point discovery algorithms. In: IEEE/SICE International Symposium on System Integration, SII 2011, pp. 938–945 (2011)

11. Mohammad, Y., Nishida, T.: Exact discovery of length-range motifs. In: Nguyen, N.T., Attachoo, B., Trawiński, B., Somboonviwat, K. (eds.) ACIIDS 2014. LNCS (LNAI), vol. 8398, pp. 23–32. Springer, Cham (2014). https://doi.org/10.1007/978-3-319-05458-2_3

12. Mohammad, Y., Nishida, T.: Scale invariant multi-length motif discovery. In: Ali, M., Pan, J.-S., Chen, S.-M., Horng, M.-F. (eds.) IEA/AIE 2014. LNCS (LNAI), vol. 8482, pp. 417–426. Springer, Cham (2014). https://doi.org/10.1007/978-3-319-07467-2_44

13. Mohammad, Y., Nishida, T.: Exact multi-length scale and mean invariant motif discovery. Appl. Intell. **44**, 1–18 (2015)

14. Mohammad, Y., Nishida, T.: Shift density estimation based approximately recurring motif discovery. Appl. Intell. **42**(1), 112–134 (2015)

15. Mohammad, Y., Nishida, T.: MC^2: an integrated toolbox for change, causality and motif discovery. In: Fujita, H., Ali, M., Selamat, A., Sasaki, J., Kurematsu, M. (eds.) IEA/AIE 2016. LNCS (LNAI), vol. 9799, pp. 128–141. Springer, Cham (2016). https://doi.org/10.1007/978-3-319-42007-3_12

16. Mohammad, Y., Nishida, T.: Unsupervised discovery of basic human actions from activity recording datasets. In: IEEE/SICE International Symposium on System Integration, SII 2012, pp. 402–409. IEEE (2012)

17. Mohammad, Y., Ohmoto, Y., Nishida, T.: CPMD: a matlab toolbox for change point and constrained motif discovery. In: Jiang, H., Ding, W., Ali, M., Wu, X. (eds.) IEA/AIE 2012. LNCS (LNAI), vol. 7345, pp. 114–123. Springer, Heidelberg (2012). https://doi.org/10.1007/978-3-642-31087-4_13

18. Morabito, F.C., Labate, D., Foresta, F.L., Bramanti, A., Morabito, G., Palamara, I.: Multivariate multi-scale permutation entropy for complexity analysis of Alzheimer's disease EEG. Entropy **14**(7), 1186–1202 (2012)

19. Mueen, A.: Enumeration of time series motifs of all lengths. In: 2013 IEEE 13th International Conference on Data Mining (ICDM). IEEE (2013)

20. Mueen, A., Keogh, E., Zhu, Q., Cash, S., Westover, B.: Exact discovery of time series motifs. In: SIAM International Conference on Data Mining, SDM 2009, pp. 473–484 (2009)
21. Riedl, M., Müller, A., Wessel, N.: Practical considerations of permutation entropy. Eur. Phys. J. Spec. Top. **222**(2), 249–262 (2013)
22. Tanaka, Y., Iwamoto, K., Uehara, K.: Discovery of time-series motif from multi-dimensional data based on MDL principle. Mach. Learn. **58**(2/3), 269–300 (2005)

A LSTM Approach for Sales Forecasting of Goods with Short-Term Demands in E-Commerce

Yu-Sen Shih[1(✉)] and Min-Huei Lin[2(✉)]

[1] Department of Information and Finance Management,
National Taipei University of Technology, Taipei, Taiwan
yusen.shih@gmail.com
[2] Department of Information Management, Aletheia University,
New Taipei City, Taiwan
au4052@au.edu.tw

Abstract. This study proposed a model to forecast short-term goods demand in E-commerce context. The model integrated LSTM approach with sentiment analysis of consumers' comments. In the training stage, the sales figures and comments crawled from "taobao.com" were preprocessed, and the sentiment rating of comments were analyzed for "positive", "negative" and confidence. The LSTM model was trained to learn the prediction of future value according to the time-series sequence of sales and sentiment rating of comments. Due to the characteristics of short-term goods, there are not enough history data to evaluate cyclic and periodic variation, so the decision makers have to react to market conditions and take appropriate actions as soon as possible. It also suggested that to adjust the weight of sentiment rating appropriately could further improve the forecasting accuracy. The study fulfilled the goal for supporting them to make use of minimal trading data to achieve maximal predictive accuracy. The results demonstrated that the proposed LSTM approach performed high-level accuracy for sales forecasting of goods with short-term demands.

Keywords: LSTM · Short-term goods demands · Sales forecasting ·
Sentiment analysis

1 Introduction

Firms use product sales forecasting as a foundation to estimate sales revenue and make decisions regarding production, operation and marketing strategies [11, 13]. "The forecast is always wrong", is a common saying among those professional in the supply chain industry. No matter how good the predictive model is, one is never going to achieve 100% accuracy or even a number, which is close to the figure [6].

The products with short time period demand are categorized as short life cycle products, and so the historical sales or other related information is available only for short duration of time such that traditional time series forecasting models do not work well [8]. It will cost highly if the sales prediction is not accurate whether how strong the distributor network of company is. Sale forecasting is necessary, because the stock does

© Springer Nature Switzerland AG 2019
N. T. Nguyen et al. (Eds.): ACIIDS 2019, LNAI 11431, pp. 244–256, 2019.
https://doi.org/10.1007/978-3-030-14799-0_21

not meet the demand of customer, it will lose patron, and pay the unreasonable price due to excessive inventory. These problems will influence the company's operating and working capital.

Because there is no entire cycle data can refer to the characteristics of short-term commodities, so future sales changes are difficult to speculate. There is not enormous historical data for short-term goods referred to forecast future sales, therefore in the process of forecasting, it must be careful and be cautious, and the reaction to sudden huge change must be faster. While the ups and downs of sales does not pose a threat, once the huge fluctuations in sales are observed to continue to decline, it is said that the Fashion Cycle may soon end, and appropriate disposal should be made quickly.

In order to have high accuracy and efficient short-term products sales forecasting, the study have conducted experiments through combinations of various training time-series length and window size. The difference in accuracy was compared and tested experimentally, including the effect of time-series length, window size, and sentimental factor's weight on accuracy. The differences in sentiment analysis for LSTM sales predictions were also compared.

This study proposed effective short-term sales forecasting methods and verifications, and was sufficient to cope with dramatic market changes. Moreover, the first sales forecasting model used LSTM model combined with sentiment analysis to highlight the innovation and importance of this research. The common companies will have the opportunity to become market leaders, and to achieve the cost savings by constantly sales forecasting.

2 Literature Review

2.1 Sales Prediction and Sentiments of Comments

With the rapid development of economic globalization and science and technology, the life cycle of products is getting shorter and shorter, and consumer demand is increasingly personalized and diversified. This market background makes sales forecasting an indispensable part of enterprise management [14]. The traditional methods of sales forecasting can be divided into two categories, one is the qualitative analysis method. These methods are based on the wisdom of experts with extensive experience and extensive wisdom, professional subjective analysis and judgment; second, quantitative analysis, such methods are through historical data, mathematical methods are used to analyze historical data to make predictions. Lots of previous researches on the issues of quantitative sales forecasting were about various industries and used various intelligent computing methodologies for constructing forecasting models, such as linear statistical models, nonlinear time-series models, and until recently machine learning techniques for time-series predictions.

Sentiments analysis is one of the main categories of natural language processing, that also include text reading, speech recognition, Chinese automatic word segmentation, part-of-speech tagging, machine translation, automatic summarization, and so on. Sentiments analysis is especially important in the marketing field [1]. Opinion exploration, also known as sentiment analysis, is designed to analyze people's subjective views (e.g., opinions, emotions, ratings, etc.) on certain entities of interest (e.g., products, services, organizations, events, etc.). The collected comments or text materials will be aggregated according to the attributes of the goods or services, become structured customer knowledge, help the company understand the opinions and needs of consumers, and enable the company to make informed and correct decisions [5, 9, 10, 12, 15]. Using "too satisfied, white plus velvet slightly fat, good cloth" as an example, the opinions of the users derived are about the color of the clothes and the material of the clothes, and are negative and positive respectively.

Fan, Huang, & Chen [3] designed an algorithm for applying digital word of mouth to sales forecasting, and proved that the proposed method outperformed several traditional sales forecasting methods. Fan, Che, & Chen [4] developed a novel method for product sales forecasting that combined the Bass/Norton model and sentiment analysis while using historical sales data and online review data. The computational results revealed that the combination of the Bass/Norton model and sentiment analysis provided higher forecasting accuracy than the standard Bass/Norton model and some other sales forecasting models.

To sum up, the study proposed a LSTM approach in modeling the sales prediction with sentiments of consumers' comments for goods with short-term demands in E-commerce context, and investigated the sensitivity of comment sentiments to sales prediction.

2.2   LSTM-RNN in Time-Series Sales Forecasting for Short-Life Cycle Products

Chniti, Bakir, & Zaher [2] provided a robust forecasting model to predict phone prices on European markets using Long Short Term Memory neural network (LSTM) and Support Vector Regression (SVR). They conducted a comparison study of time series forecasting models for these two techniques. After studying and comparing several univariate models, SVR and LSTM neural network appear to be the most accurate ones.

Yu, Wang, Strandhagen, & Wang [16] tested a LSTM recurrent neural network (RNN) on 45 weeks point of sale (POS) data of 66 products without considering the impact of seasonality and promotions. One fourth of products had a relatively low forecasting error, which validated the feasibility of the LSTM network to some degree.

A good deal of research has already proved that LSTM is able to predict time serial data accurately and effectively. However, nearly no research could accurately predict short-term demands for short-life cycle products. Thus, the study presents a LSTM approach combined with sentiment analysis in accurately modeling the sales prediction for products with short-term demand, and expects to predict short-term demands most accurately with the least time-series and the optimal weight of comment sentiments.

3 Methodology

3.1 Research Design

To challenge the accurate sales forecasting of products with short time period demand, this study integrated the strategy with task-technology fitness and data with information features needed to achieve the goal of the sales forecasting system, as the research framework depicted in Fig. 1.

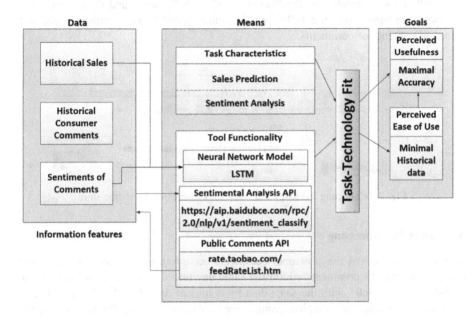

Fig. 1. The research framework

Based on the Integrated TAM/TTF (technology acceptance model/task-technology fit model) model, the goal of the study is to gain the ease of use and usefulness of the sales forecasting model to raise the acceptance of the sales forecasting service. The ease of use means that the business may use minimal historical data, manual efforts of data preparation, and computing resources, and then the usefulness is the capability of making the accuracy of sales forecasting maximal to support the informed decision-making. To achieve the goal, the study analyzed and match the task characteristic of sales prediction with sentiments of comments and the available functionality of technological tools, such as the neural network model and various APIs used to build predictive model, crawl comments and perform sentiment analysis.

The study implemented these research instruments by Python3 with different APIs and the TensorFlow backend on the Jupyter notebook environment of Google Colaboratory. The procedure for predicting sales of products with short time period demand in this study was stated as the following sections.

3.2   Data Collection

The study collected both the historical sales and online comments data from online store, and the product selected for experiments was the "Gift box multi-taste Gift chocolate" sold in Dove official flagship store on the "taobao.com". The sales data collected through public webpage was acquired from the chrome-extension developed by Huidianshang, which is a professional e-commerce seller integrated service provider affiliated to the brand of Xi'an QianFan Electronic Commerce Co. Ltd. as shown in Fig. 2. The consumer comments on products were captured through requesting the open API service provided by the "taobao.com" with Python3, and there were 5018 comments gathered with the following attributes: member id, comment date, and complete contents of comments.

Fig. 2.   Collecting the sales and comments data from the online store

3.3   Data Preprocessing

The study reorganized the consumer comments according to the comment date, and concatenated all the comments within the same day into one article, then joined these well-organized articles with the sales data by comment date. Besides the quantitative sales data, the other information feature used to co-predict the future sales is the qualitative comment data. The study extracted the quantitative sentiments of comments from the qualitative comment data, and the attributes included the sentiment ratings of "positive", "negative" and confidence. The sentiment analysis was proceeded through requesting the sentiment analysis API on the "ai.baidu.com" platform (Fig. 3).

Fig. 3.   The example of performing sentiment analysis

3.4 Model Architecture

The model shown in Fig. 4 was the neural network structure for sales the study used for sales forecasting in the research experiment.

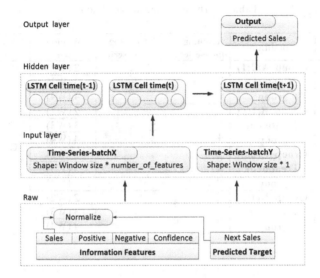

Fig. 4. The model architecture for time-series sales forecasting

During the model-training phase, the input layer received training time-series data together with a predicted target. The training data included the normalized sales and the positive, negative, and confidence features generated from data preprocessing, and the predicted target was the normalized next sales. The shape of Time-Series-batchX depended on the window size and number of features used for sales prediction. These well-formatted time-series was fed into the hidden layer of the neural network in the forward pass. The hidden layer consisted of Basic LSTM recurrent neural network cells in TensorFlow, inside the LSTM cell are LSTM units and tanh is the default activation function of the inner states. The loss function used the MSE and the optimizer was the Adam.

3.5 Experiments

To investigate rigorously the minimal extent of dataset for training the model to learn the maximal predictive accuracy of future sales, the study designed twelve combinations of various training time-series length and window size into twelve groups. As shown in Table 1, the Group_A(2:1) contained three control groups and the Group_B (1:1), Group_C(2:3), Group_D(1:2) were totally of nine treatment groups. The study designated the traditional partition of time-series, i.e., two-thirds for training and one-third for testing, as control groups. Besides, the treatment groups included three conditions, such as half for training and half for testing, two over five for training and three over five for testing, one-third for training and two-thirds for testing. Every group was trained and validated in both with-sentiment and without-sentiment conditions.

Furthermore, the calibration and adjustment of comment sentiments rating was conducted for some groups to explore the impacts of comment sentiments on predictive accuracy under various time-period length training datasets used.

Table 1. The experimental conditions

	Condition	Training time-series	:	Testing time-series	Window size	Sentiment Ratings	
1	Group_A(2:1)				2 (W2)		
2	Control	$\dfrac{2}{3}$		$\dfrac{1}{3}$	3 (W3)		
3	group				4 (W4)		
4	Group_B(1:1)				2 (W2)		
5	Treatment	$\dfrac{2}{4}$		$\dfrac{2}{4}$	3 (W3)		
6	group				4 (W4)	O	×
7	Group_C(2:3)				2 (W2)		
8	Treatment	$\dfrac{2}{5}$		$\dfrac{3}{5}$	3 (W3)		
9	group				4 (W4)		
10	Group_D(1:2)				2 (W2)		
11	Treatment	$\dfrac{2}{6}$		$\dfrac{4}{6}$	3 (W3)		
12	group				4 (W4)		

3.6   Evaluation

The MAPE is the most commonly used for measuring sales forecasting performance [7]. To evaluate the sales forecasting accuracy of the proposed predictive approach on various experimental conditions, the study used the MAPE (Mean Absolute Percentage Error) against the testing time-series.

$$MAPE = \frac{1}{N} \sum_{i=1}^{n} \left| \frac{e_i}{\hat{y}_i} \right| \times 100$$

N is the number of predicted sales, $\hat{y}_i$ denotes the predicted sale and the e_i denotes the error between actual sale and the predicted sale.

In addition, to verify the forecasting performance of every treatment group, the study compared it respectively with every control group by *T-test*, in which the sales errors of every two testing time-series were examined.

4   Results and Discussions

The aim of the study was to challenge the accurate sales forecasting of products with short time period demand, and to investigate the impact of training time-series length, window size and sentiment weight calibration on maximal predictive accuracy and minimal training dataset. The study used Python3 to implement the whole system on the free Google Colaboratory Jupyter notebook environment based on the research framework developed in Sect. 3. As stated in the "3.2 data collection", this study collected the time-series with thirty-six days available from July 17 2018 to August 20 2018 describing sales and comments related to the target product for the experiment and explanation.

4.1 Model Fitting and Comparison Within- and Between- Group

Analysis of Sales Predicting Performance Within a Group

The MAPE against the testing time-series for all the groups was summarized in Table 2 and the sales forecasting results of every group are shown in Fig. 5(a) and (b). In Fig. 5(a) and (b), the green line represented the actual sales, the yellow line was predicted sales by training time-series, and the red line was predicted sales by using testing time-series. As expected, the longer time-series used for model training the lower predictive errors, like the Group_A(2:1) and Group_B(1:1) performed. In addition, there existed some extreme predictive performance, like that the Group_A (2:1)_W4 with sentiment owned especially high predictive accuracy, but the MAPEs of both Group_C(2:3)_W3 and Group_C(2:3)_W4 with sentiment exceeded 10%.

The closer the MAPE is to zero, the better the sales forecasting is. The experiment results showed that MAPEs of most groups were less than 10%, and indicated that the high accuracy have been achieved.

Table 2. The experiment results – MAPE against the testing time-series

	Experimental conditions	MAPE with sentiment	MAPE without sentiment
Control groups	Group_A(2:1)_W2	5.32890099	5.742631349
	Group_A(2:1)_W3	5.48861293	6.872902678
	Group_A(2:1)_W4	2.96368197	10.71652315
Treatment groups	Group_B(1:1)_W2	4.92182926	6.397893914
	Group_B(1:1)_W3	6.64304135	7.537821816
	Group_B(1:1)_W4	5.92298457	6.46368591
	Group_C(2:3)_W2	7.69228619	6.109069937
	Group_C(2:3)_W3	11.1110589	7.68926864
	Group_C(2:3)_W4	10.1537683	8.439346306
	Group_D(1:2)_W2	7.54477657	5.800122257
	Group_D(1:2)_W3	8.94626736	7.140886152
	Group_D(1:2)_W4 .	9.89742348	8.339694597

Analysis of Sales Forecasting Performance Between Groups

As stated in the "3.5 Experiments", to investigate the differences of forecasting performance between every pair of the control group and the treatment group, the study conducted the *T-test* on testing time-series. Every *T-test* received the data contained two columns, error and group, and all the data rows from the two groups consisted of the absolute values of actual sales subtracted from predicted sales as error, and the group name as group, as depicted in Fig. 5(c), and the results of the *T-test* were shown in Table 3.

Except that the Group_A(2:1)_W4 was the special case, the differences of *T-test* between any pairs of control group and treatment group were not significant. It meant that the study used less time-series for training learn the sales prediction as well as the traditional one that used much more time-series for training. The result supported that the sales forecasting approach proposed by this study outperformed the traditional ones.

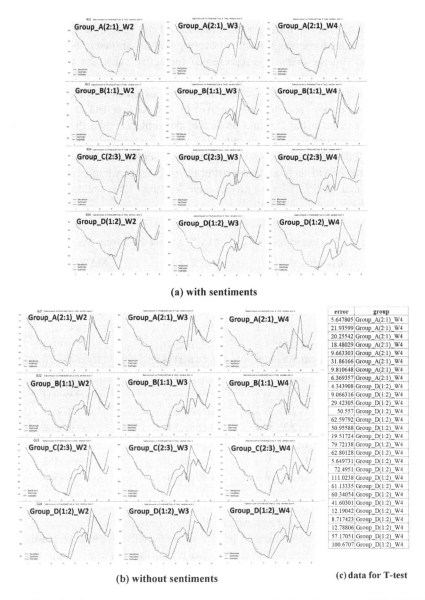

(a) with sentiments

(b) without sentiments   (c) data for T-test

Fig. 5. The sales forecasting results on all the training and testing time-series (W2, W3, W4 denote the window size 2, 3, 4 respectively) (Color figure online)

4.2   Impacts of Comment Sentiment Ratings on Sales Forecasting

In Fig. 6, the line chart showed the MAPEs summarized in Table 2, the green line represented prediction errors under with sentiments condition and the red line was the ones under without sentiments condition. Starting from the Group_A(2:1)_W2, the MAPEs of with sentiments condition was less than the ones of without sentiments condition until Group_B(1:1)_W4, the green line and red line intersected and reversal happened. It should be able to interpret that sentiments could increase accuracy while using the longer time-series for training. Toward to the right of x-axial, the shorter time-series for training and the sentiments let the error increase.

Table 3. The *T-test* results – significances against the testing time-series

Condition	Group_B (1:1) W2	Group_B (1:1) W3	Group_B (1:1) W4	Group_C (2:3) W2	Group_C (2:3) W3	Group_C (2:3) W4	Group_D (1:2) W2	Group_D (1:2) W3	Group_D (1:2) W4
Group_A (2:1) W2	0.572	0.756	0.677	0.582	0.732	0.573	0.445	0.485	0.361
Group_A (2:1) W3	0.477	0.931	0.848	0.734	0.888	0.738	0.584	0.631	0.517
Group_A (2:1) W4	0.125	*0.014	*0.013	*0.023	*0.032	*0.013	*0.018	*0.018	*0.004

Note: p < .05 (W2, W3, W4 denote the window size 2, 3, 4 respectively)

Therefore, the study further experimented by lowering weight of sentiment impacts to raise predictive accuracy for the groups in Group_C(2:3) and Group_D(1:2).

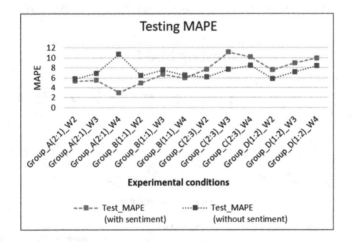

Fig. 6. The Testing MAPEs for sales forecasting with/without sentiments (Color figure online)

4.3   Model Calibration for Comment Sentiments Ratings on Shorter Time-Series for Training

In the study, the sentiment rating of comments were analyzed for "positive", "negative" and confidence. The study claimed that the sentiment credibility became not so trustworthy while the time-series for training became shorter. Therefore, the study calibrated the model and reduced the weights of confidence, the results of model calibration was listed in Table 4. While the weight of sentiment confidence decreased 40%, the accuracy increased (as the blue line in Fig. 7 with the confidence * 0.6), the blue line stayed stable between the green and red lines, even more some performed better than the red line did, like the Group_C(2:3)_W4 and the Group_D(1:2)_W4. Through the experiments of model calibration, it showed that the results were consistent with the proposed claim.

Table 4. The calibration and adjustment - MAPE against the testing time-series

Experimental conditions	MAPE with sentiment	MAPE without sentiment	With sentiment confidence*0.6
Group_A(2:1)_W2	5.32890099	5.742631349	–
Group_A(2:1)_W3	5.48861293	6.872902678	–
Group_A(2:1)_W4	2.96368197	10.71652315	–
Group_B(1:1)_W2	4.92182926	6.397893914	–
Group_B(1:1)_W3	6.64304135	7.537821816	–
Group_B(1:1)_W4	5.92298457	6.46368591	–
Group_C(2:3)_W2	7.69228619	6.109069937	7.857451565
Group_C(2:3)_W3	11.1110589	7.68926864	8.765309932
Group_C(2:3)_W4	10.1537683	8.439346306	6.407624807
Group_D(1:2)_W2	7.54477657	5.800122257	7.024713039
Group_D(1:2)_W3	8.94626736	7.140886152	8.459406192
Group_D(1:2)_W4	9.89742348	8.339694597	7.788219483

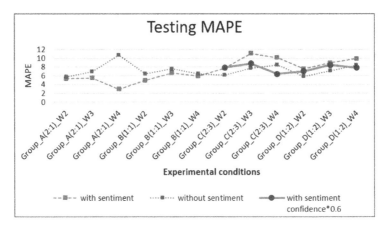

Fig. 7. The forecasting accuracy resulted from performing adjustments the weight of sentiment ratings (Color figure online)

5 Conclusions

People use Artificial Intelligence to support themselves solving the tasks or problems in the future, and increase the opportunities to make the personalized service real. The study proposes a LSTM approach in modeling the sales forecasting with sentiments of consumers' comments for goods with short time-period demand in E-commerce context. By the research design and rigorous experiments, the study successful challenged the accurate sales forecasting of products with short time period demand to achieve the goal of using minimal historical data, manual efforts of data preparation, and computing resources, and making the accuracy of sales forecasting maximal. Namely, the study used AI and computing intelligence to achieve the goal of ease of use and usefulness based on both the fitness of task and technology, and the well-organized information features. In the future, the study will investigate various categories of stores and commodity to confirm the generalization of the proposed sales forecasting model. Moreover, it may be a valuable application of deep learning in management issues that using deep learning to discover the potential commercial opportunities base on consumers' comments.

References

1. Chen, A.Y.: Using the text mining and sentiment analysis technology to develop the store commodity evaluation module. Master's thesis of Graduate Institute of Information Management, 48 p. National Taipei University, Taipei (2017)
2. Chniti, G., Bakir, H., Zaher, H.: E-commerce time series forecasting using LSTM neural network and support vector regression. In: Proceedings of the International Conference on Big Data and Internet of Thing - BDIOT2017, pp. 80–84. ACM (2017)
3. Fan, Y.N., Huang, H.W., Chen, C.C.: A solution for sales forecasts of fashion products based on electronic word-of-mouth. J. Inf. Manag. **19**, 27–50 (2012)
4. Fan, Z.P., Che, Y.J., Chen, Z.Y.: Product sales forecasting using online reviews and historical sales data: a method combining the Bass model and sentiment analysis. J. Bus. Res. **74**, 90–100 (2017). https://doi.org/10.1016/j.jbusres.2017.01.010
5. Feldman, R.: Techniques and applications for sentiment analysis: the main applications and challenges of one of the hottest research areas in computer science. Commun. ACM **56**(4), 82–89 (2013). https://doi.org/10.1145/2436256.2436274
6. Goyal, A., Kumar, R., Kulkarni, S., Krishnamurthy, S., Vartak, M.: A solution to forecast demand using long short-term memory recurrent neural networks for time series forecasting. In: 2018 Midwest Decision Sciences Institute Conference, pp. 1–18 (2018)
7. Kahn, K.B.: Benchmarking sales forecasting performance measure. J. Bus. Forecast. Methods Syst. **17**(4), 19–23 (1998)
8. Kadam, S., Apte, M.D.: A survey on short life cycle time series forecasting. Int. J. Appl. Innov. Eng. Manag. **4**, 445–449 (2015)
9. Liu, B.: Sentiment analysis and subjectivity. Handb. Nat. Lang. Process. **2**, 627–666 (2010)
10. Liu, B.: Sentiment Analysis and Opinion Mining. [Electronic Resource]. Morgan & Claypool, San Rafael (2012)
11. Marshall, P., Dockendorff, M., Ibáñez, S.: A forecasting system for movie attendance. J. Bus. Res. **66**(10), 1800–1806 (2013)

12. Pang, B., Lee, L.: Opinion mining and sentiment analysis. Found. Trends Inf. Retrieval **2**(1–2), 1–135 (2008). https://doi.org/10.1561/1500000011
13. Shi, X., Li, F., Bigdeli, A.Z.: An examination of NPD models in the context of business models. J. Bus. Res. **69**(7), 2541–2550 (2016)
14. Shu, L.Z.: Research on sales forecasting methods. Mod. Mark. **7**, 80 (2011). https://doi.org/10.3969/j.issn.1009-2994.2011.07.051
15. Yang, C.S., Xie, P.Y., Shih, H.P.: Mining consumer knowledge from social media: development of an opinion mining technique. NTU Manag. Rev. **27**, 1–28 (2017). https://doi.org/10.6226/NTUMR.2017.JUN.F104-008
16. Yu, Q., Wang, K., Strandhagen, J.O., Wang, Y.: Application of long short-term memory neural network to sales forecasting in retail—a case study. In: Wang, K., Wang, Y., Strandhagen, J., Yu, T. (eds.) IWAMA 2017, pp. 11–17. Springer, Singapore (2018). https://doi.org/10.1007/978-981-10-5768-7_2

Detection of Rare Elements in Investigation of Medical Problems

Piotr Kulczycki[1,2]($\boxtimes$) and Damian Kruszewski[1]

[1] Systems Research Institute, Centre of Information Technology for Data
Analysis Methods, Polish Academy of Sciences, Warsaw, Poland
kulczycki@ibspan.waw.pl, damian.kruszewski@o2.pl
[2] Faculty of Physics and Applied Computer Science,
Division for Information Technology and Systems Research,
AGH University of Science and Technology, Kraków, Poland
kulczycki@agh.edu.pl

Abstract. The task of detecting atypical (rare) elements is of major significance in the field of medical problems and its conditions seem to be specific in practice. Such elements, mostly concerned with pathology, are very different in nature and their set is often small in size with a low level of representativeness. A frequency approach was applied in the presented research, which, in conjunction with nonparametric methods, enabled the detection of atypical elements – in the case of distributions with many modes – also located between them, and not only lying on the peripheries of the population. Within the framework of the procedure investigated here, the database is artificially extended, which significantly improves the quality of results. The presented method has been successfully used for two medical problems: biochemical blood tests and the influence of hemoglobin levels on mortality.

Keywords: Detection · Atypical element · Rare element ·
Frequency approach · Nonparametric methods · Medical applications

1 Introduction

Atypical elements can naturally be treated as considerably differing from remaining elements of data [1, 2]. Depending on the nature of a given application problem, such a general notion is most precisely concerned as lying far from the rest of the population (outliers) or occurring most infrequently (rare elements). However, this does not exhaust all the possibilities – in specific applications, one can treat e.g. elements which (singularly) have too large an influence over research results as being atypical.

In this paper, the second of the above approaches, regarding rare elements, will be applied due to it being universal, illustrative and offering considerable analytical possibilities. In particular, it allows the identification of atypical elements lying not only on the fringes of a population but also in the multimodal distribution located internally, between particular modes. Moreover, by using nonparametric methods it is possible to explore the distributions of any shapes, including those which consist of several separate (non-coherent) components. The description of the nonparametric kernel estimators is presented in Sect. 2.

© Springer Nature Switzerland AG 2019
N. T. Nguyen et al. (Eds.): ACIIDS 2019, LNAI 11431, pp. 257–268, 2019.
https://doi.org/10.1007/978-3-030-14799-0_22

In medical applications, atypical elements mostly represent pathologies, in practice having a very diverse, unknown *a priori* nature [19]. Often they have varying characteristics, because they represent people who are, unfortunately, living shorter lives. The initial version of the procedure is introduced within Sect. 3 in an unsupervised form, based on the data set from the entire population and consisting of both atypical and typical elements.

By their very nature, atypical elements comprise a small group. The concept of extension of the data set, presented in the next Sect. 4, enables improvements in the estimation of the quantities used, and in consequence, in the results of the detection procedure.

In Sect. 5, the presented material is applied to two practical tasks in the medical domain: the identification of biochemical blood test results lying outside the standard range, and the influence of hemoglobin levels on mortality.

The material investigated and described in this paper is given in a ready-to-use form, while its illustrative nature facilitates the creation of potential individual modifications.

2   Mathematical Preliminaries: Kernel Estimators

The characteristics describing the data set under research will be established in this paper by the kernel estimators methodology. This distribution-free concept does not require any preliminary knowledge about a specific type of distribution occurring in the tested reality. Detailed description of this method appears in the classic books [7, 20]. The publications [8–10, 14] also present sample applications for data analysis problems.

Consider the n-dimensional continuous random variable X, the distribution of which possesses the density X. Its kernel estimator $\hat{f} : \mathbb{R}^n \to [0, \infty)$, obtained on the basis of the acquired m-element random sample x_i for $i = 1, 2, \ldots, m$, can be defined in a basic form as

$$\hat{f}(x) = \frac{1}{mh^n} \sum_{i=1}^{m} K\left(\frac{x - x_i}{h}\right), \tag{1}$$

where $m \in \mathbb{N}\backslash\{0\}$, the parameter $h > 0$ takes the name of a smoothing parameter, and the measurable function $K : \mathbb{R}^n \to [0, \infty)$ fulfilling the condition $\int_{\mathbb{R}^n} \hat{f}(x)\mathrm{d}x = 1$, having a weak global maximum in zero and symmetrical with respect to this point, is called a kernel.

The form of the kernel is unimportant from a statistical perspective, thanks to which the user can primarily consider desired features of the estimator achieved or computational properties, appropriate for the specific task under research; for details, see the monographs [7 – Sect. 3.1.3; 20 – Sects. 2.7 and 4.5]. The normal (Gauss) kernel

$$K(x) = \frac{1}{\sqrt{2\pi}} \exp\left(-\frac{x^2}{2}\right) \tag{2}$$

and the uniform kernel

$$K(x) = \begin{cases} \frac{1}{2} & \text{dla } x \in [-1, 1] \\ 0 & \text{dla } x \notin [-1, 1] \end{cases} \tag{3}$$

are applied in the following for the one-dimensional case (i.e. when $n = 1$). The former is usually treated as basic. The support of the latter is bounded and it assumes only two values, which will successfully used further in this investigation. However, a so-called product kernel will be applied for the multidimensional case. Its idea relies on separately treated particular variables, and in this situation the multidimensional kernel is given as a product of n one-dimensional kernels defined for particular coordinates. Thus, kernel estimator (1) can be described in the following form:

$$\hat{f}(x) = \frac{1}{mh_1h_2\ldots h_n} \sum_{i=1}^{m} K_1\left(\frac{x_1 - x_{i,1}}{h_1}\right) K_2\left(\frac{x_2 - x_{i,2}}{h_2}\right) \ldots K_n\left(\frac{x_n - x_{i,n}}{h_n}\right), \tag{4}$$

where K_j means one-dimensional kernels, e.g. normal (2) or uniform (3), while h_j designate smoothing parameters for successive coordinates $j = 1, 2, \ldots, n$, and denoting for coordinates

$$x = \begin{bmatrix} x_1 \\ x_2 \\ \vdots \\ x_n \end{bmatrix} \quad \text{and} \quad x_i = \begin{bmatrix} x_{i,1} \\ x_{i,2} \\ \vdots \\ x_{i,n} \end{bmatrix} \quad \text{for} \quad i = 1, 2, \ldots, m. \tag{5}$$

Such established kernels fulfill all conditions required by procedures applied in the following.

The value of the smoothing parameter is, however, important from the point of view of the estimation quality. Auspiciously, useful algorithms are investigated for calculating it from the random sample. In the investigations presented here, the essential method [7 – Sect. 3.1.5; 20 – Sect. 3.2.1] is used, based on the following formula:

$$h_j = \left(\frac{8\sqrt{\pi} \, W(K_j)}{3} \frac{1}{U(K_j)^2} \frac{1}{m}\right)^{1/5} \hat{\sigma}_j \quad \text{for} \quad j = 1, 2, \ldots, n, \tag{6}$$

where $W(K_j) = \int_{-\infty}^{\infty} K_j(x)^2 \, dx$ and $U(K_j) = \int_{-\infty}^{\infty} x^2 K_j(x) \, dx$, while $\hat{\sigma}_j$ denotes the estimator of a standard deviation for the j-th coordinate:

$$\hat{\sigma}_j = \sqrt{\frac{1}{m-1} \sum_{i=1}^{m} (x_{i,j})^2 - \frac{1}{m(m-1)} \left(\sum_{i=1}^{m} x_{i,j}\right)^2} \quad \text{for} \quad j = 1, 2, \ldots, n. \tag{7}$$

The verification results prove that although this is simple and fast, it turn out to be sufficiently precise for the presented procedure. The values of the functionals W and U appearing in equality (6) are respectively

$$W(K_j) = \frac{1}{2\sqrt{\pi}}, \quad U(K_j) = 1 \tag{8}$$

for normal kernel (2), and

$$W(K_j) = \frac{1}{2}, \quad U(K_j) = \frac{1}{3} \tag{9}$$

in the case of uniform (3). For individual applications, the more complex plug-in method [7 – Sect. 3.1.5; 20 – Sect. 3.6.1] can be used in its one-dimensional form, for the n-dimensional case applied sequentially n times for each coordinate.

Various generalizations and adjustments of the basic kernel estimator form introduced above can be proposed, matching its properties to the conditioning of reality under investigation. It should however be mentioned that they complicate formulas and interpretations, which causes the problem to be not so clear and convenient for possible users. For additional information about the kernel estimators methodology, see the books [7, 20].

3 Procedure for Atypical Elements Detection

The basic concept for the atypical elements detection procedure, presented in this paper, originates from the significance test introduced in the work [15]. It is worth emphasizing that thanks to using nonparametric methods, there is no need to assume any concrete type of distribution for the tested data.

Consider the set consisting of elements representative for the examined population

$$x_1, x_2, \ldots, x_m \in \mathbb{R}^n. \tag{10}$$

Treating the above elements as realizations of the random variable X, one can obtain the kernel estimator $\hat{f}$ of its distribution density, based on material described in Sect. 2, using here normal kernel (2). Next, let the set of the following values

$$\hat{f}_{-1}(x_1), \hat{f}_{-2}(x_2), \ldots, \hat{f}_{-m}(x_m) \tag{11}$$

be given, where $\hat{f}_{-i}$ denotes the kernel estimator $\hat{f}$ achieved with the exclusion of the i-th element $(i = 1, 2, \ldots, m)$. Note that independently of the random variable X dimension, the set (11) values are one-dimensional (real).

Assume the number

$$r \in (0,1) \tag{12}$$

defining the sensitivity of the designed procedure for atypical elements detection. This number will decide on the share of atypical elements in the whole population, i.e. the ratio of atypical elements number to the atypical and typical elements in sum. For practical tasks

$$r = 0.01, 0.05, 0.1 \tag{13}$$

is the commonest, and the most frequently the second choice. In specific cases one can, however, use different values of this parameter, if it is advantageous for an individual problem.

Treat set (11) as realizations of a one-dimensional (real) random variable and obtain the quantile of the order r estimator. The positional concept presented in the publications [6, 18] will be used, modified to the following formula:

$$\hat{q}_r = \begin{cases} z_1 & \text{for} \quad mr < 0.5 \\ (0.5 + i - mr)z_i + (0.5 - i + mr)z_{i+1} & \text{for} \quad mr \geq 0.5 \end{cases}, \tag{14}$$

where $i = [mr + 0.5]$, while $[d]$ denotes an integral part of the number $d \in \mathbb{R}$, and z_i is the i-th value in size of set (11) after being sorted; thus

$$\{z_1, z_2, \ldots, z_m\} = \{\hat{f}_{-1}(x_1), \hat{f}_{-2}(x_2), \ldots, \hat{f}_{-m}(x_m)\} \tag{15}$$

with $z_1 \leq z_2 \leq \ldots \leq z_m$. This type of the quantile estimator ensures that its value belongs to the random variable support. In the problem investigated here it implies that the condition $\hat{q}_r > 0$ is fulfilled, since normal kernel (2) applied here has positive values.

In the general case, there are no specific indications on the selection of sorting algorithm [3], which is used to obtain set (15). Although, in interpreting definition (14) with condition (13) note that only the $i + 1$ smallest values of the set $\{z_1, z_2, \ldots, z_m\}$, i.e. about 1–10% of its elements, should be sorted. Therefore, a natural algorithm can be applied, where one in turn finds the $i + 1$ smallest elements of the set $\{z_1, z_2, \ldots, z_m\}$ avoiding those already found.

Summarizing: if for the examined element $\tilde{x} \in \mathbb{R}^n$, the condition $\hat{f}(\tilde{x}) \leq \hat{q}_r$ is met, then it needs to be recognized as atypical; in the opposite case $\hat{f}(\tilde{x}) > \hat{q}_r$ it should be considered as typical. Note that the proportion between typical and atypical elements, defined by the parameter r is maintained for the properly established estimators $\hat{f}$ and $\hat{q}_r$.

Thanks to properties of kernel estimators, the investigated procedure, in the multidimensional case, inferences also on the basis of relations between particular coordinates of a tested element but not only on the values of separate coordinates.

4 Extension of Population Pattern

Although the algorithm described in the preceding section looks theoretically complete, for the values r commonly used in practice (see formula (13)) and when the size m is relatively small, the quantile $\hat{q}_r$ estimation can be unacceptably inexact, because of the very small number of elements z_i less than the estimated value. To counteract this, additional elements originating from a distribution characterizing set (10), representing the whole population, can be added to it. For this purpose, Neumann's elimination concept [4] will be applied. This enables a random numbers sequence to be generated, having a distribution with a support bounded to the interval $[a, b]$, where $a < b$, as its density f is bounded by the positive number c, i.e.

$$f(x) \leq c \text{ for every } x \in [a, b]. \tag{16}$$

In the multidimensional case, the interval $[a, b]$ generalizes to the n-dimensional cuboid $[a_1, b_1] \times [a_2, b_2] \times \ldots \times [a_n, b_n]$, while $a_j < b_j$ for $j = 1, 2, \ldots, n$.

First the one-dimensional case is considered. Let two pseudorandom numbers u and v be generated from a uniform over the intervals $[a, b]$ and $[0, c]$ distributions, respectively. If the condition

$$v \leq f(u) \tag{17}$$

is fulfilled, then the value u is accepted as the sought random variable realization from the distribution with the density f, i.e.

$$x = u. \tag{18}$$

In the other case, the above steps (17)–(18) should be repeated, till the wanted record of pseudorandom numbers has been collected.

In this concept, the density f will be established using the kernel estimators method, presented in Sect. 2. To achieve such an estimator $\hat{f}$, the uniform kernel is used, which enables effective determining of the support boundaries a and b, and also the value c occurring in inequality (16). Namely:

$$a = \min_{i=1,2,\ldots,m} x_i - h \tag{19}$$

$$b = \max_{i=1,2,\ldots,m} x_i + h \tag{20}$$

and

$$c = \max_{i=1,2,\ldots,m} \left\{ \hat{f}(x_i - h), \hat{f}(x_i + h) \right\}. \tag{21}$$

Note that when the uniform kernel is employed, a kernel estimator maximum occurs on the border of some kernel, which justifies formula (21). Thanks to the proper choice of the kernel form, calculation of parameters (19)–(21) needs little effort, since the maxima are sought in a finite set.

The multidimensional case will be considered now. Neumann's elimination procedure is analogous to the earlier presented. The borders of the n-dimensional cuboid $[a_1, b_1] \times [a_2, b_2] \times \ldots \times [a_n, b_n]$ can be obtained based on formulas (19)–(20) for each coordinate separately. The maximum of the kernel estimator $\hat{f}$ is placed on one of the kernels corners, so

$$c = \max_{i=1,2,\ldots,m} \left\{ \hat{f} \left(\begin{bmatrix} x_{i,1} \pm h_1 \\ x_{i,2} \pm h_2 \\ \vdots \\ x_{i,n} \pm h_n \end{bmatrix} \right) \right\} \quad \text{for all } \pm \text{ combinations.} \quad (22)$$

Note that the number of such combinations is finite. Having the above parameters, n specific coordinates of the pseudorandom vector u and the proper number v can be generated analogously to the procedure presented for the one-dimensional case, and finally inequality (17) checked.

The empirical verification proves that for set (10) enlarging in this way, the algorithm for atypical elements detection worked out here enables hold the ratio of such elements with respect to the entire population, with an accuracy acceptable in practice.

5   Empirical Verification

The concept presented in this paper has been comprehensively tested using synthetic, illustrative data and two major coronaries should be highlighted here.

The consequence of the extension of data set (10), described in Sect. 4, shows that an improvement of results does not only rely on an increase in its size. Namely, even if the number of generated elements m^* is equal to the initial size m, the results are more precise and stable. The procedure investigated in Sect. 4 acts here in a similar manner as a filter.

The increasing of the generated sample size m^* is effective only to a certain degree, above which results do not improve further – information included in set (10) has already been fully used and further progress is impossible.

An empirical verification of the investigated procedure used for two medical problems is presented below.

5.1   Biochemical Blood Tests

Results of research concerning biochemical tests of blood in the subject of plasma component analysis or more precisely electrolytes glucose, potassium and sodium concentration, have been carried out using data from the *National Health and Nutrition Examination Survey*, conducted in the USA in the years 2007–08 [16]. Generally the results of clinical interpretation of this laboratory research comprises comparisons with reference ranges. In order to determine how much a specific result differs from the norm, for evaluation of a case intensity, the standard terminology of the *National Cancer Institute* [17] was used. The five-level scale 0, 1, 2, 3, 4 was used. Level 0 means that the result is consistent with the laboratory norm, level 1 represents the lowest diffusion but not pathological, whereas the higher levels 2, 3, 4 denotes degrees of full-symptomic disease, with the possibility of tendency to worsen the functioning of a patient.

The one-dimensional cases representing subsequently glucose, potassium and sodium, as well as their two-dimensional combinations were subject to emanation. Table 1 shows the proportions of elements from the database under consideration, mapped to levels 0–4 using standard reference research. It was used to determine the parameter r values, defining the atypical elements identification procedure sensitivity. For particular factors, this parameter value equals the proportions in levels 1–4 sum; it has been denoted in the second column of Table 2.

The atypical elements detection procedure was used with set (10) enlarged to the size $m^* = 10,000$. Table 3 presents the results of mapping to the particular level of illness advancement. Thus, observations from levels 2, 3, 4 were treated as atypical elements, while belonging to level 1, not being pathological, which concerned more or less half of the patients. Approximately 5% of elements from level 0 were included in level 1. It should be underlined here that the parameter r can be modified due to preferences of a misclassification of atypical elements as typical and *vice versa* risk, according to individual conditions of the investigated problem.

Table 1. Proportions of subsequent levels 0–4 [11].

Electrolyte	Level 0	Level 1	Level 2	Level 3	Level 4
Glucose	0.757	0.106	0.095	0.041	0.001
Potassium	0.891	0.058	0.034	0.014	0.003
Sodium	0.872	0.109	0.011	0.007	0.001
Combination of glucose and potassium	0.681	0.139	0.121	0.055	0.004
Combination of glucose and sodium	0.663	0.183	0.105	0.047	0.003
Combination of potassium and sodium	0.787	0.147	0.041	0.021	0.004

Table 2. Obtained proportions of atypical elements [11].

Electrolyte	r	Share of identified atypical elements	Error
Glucose	0.243	0.250	2.9%
Potassium	0.109	0.105	3.7%
Sodium	0.128	0.129	0.8%
Combination of glucose and potassium	0.319	0.309	3.1%
Combination of glucose and sodium	0.337	0.333	1.2%
Combination of potassium and sodium	0.213	0.213	0.0%

Table 3. Obtained proportions of atypical elements, accounting for subsequent levels [11].

Electrolyte	Level 0	Level 1	Level 2	Level 3	Level 4
Glucose	0.040 (5.2%)	0.070 (68.5%)	0.097 (100.0%)	0.043 (100.0%)	0.001 (100.0%)
Potassium	0.031 (3.4%)	0.024 (44.0%)	0.032 (100.0%)	0.014 (100.0%)	0.005 (100.0%)
Sodium	0.038 (4.4%)	0.080 (66.1%)	0.006 (100.0%)	0.004 (100.0%)	0.001 (100.0%)
Combination of glucose and potassium	0.049 (7.3%)	0.080 (54.6%)	0.125 (99.8%)	0.052 (100.0%)	0.003 (100.0%)
Combination of glucose and sodium	0.050 (7.8%)	0.117 (61.2%)	0.115 (100.0%)	0.050 (100.0%)	0.001 (100.0%)
Combination of potassium and sodium	0.054 (7.1%)	0.092 (56.1%)	0.046 (100.0%)	0.019 (100.0%)	0.002 (100.0%)

5.2 The Influence of Hemoglobin Levels on Mortality

In pharmacological tasks, a dependence between factors and resultant phenomena is frequently described by a convex function – both too small and too large values of a factor cause an increasing intensity of the generated phenomena. Such a dependence can be observed between the level of hemoglobin and mortality rates, or to be more precise: mortality resulting from various medical conditions, in particular cardiac arrest and circulatory illnesses. In this section, the data originating from *The National Health and Nutrition Examination Survey* 2009–10, regarding hemoglobin levels, and also *The National Center for Health Statistics*, providing information concerning the mortality of previously examined patients, including time and cause of death [16]. The main reasons for death in the observed population were malignant cancer, heart disease, chronic lower respiratory diseases, vascular cerebral illnesses, and diabetes. During the analyzed period of observation, there were 102 recorded deaths from 5,919 registered patients. Detailed results are presented in Table 4.

Table 4. Number of living patients at the end of the research period or causes of death.

Status at the end of research period or cause of death	Number	Percentage
Living	5,817	98.28
Heart disease	25	0.42
Malignant cancer	31	0.52
Chronic lower respiratory diseases	8	0.14
Vascular cerebral illnesses	4	0.07
Diabetes	3	0.05
Influenza and pneumonia	5	0.08
Others	26	0.44

According to the content in Sect. 4, the procedure with a population sample extended to $m^* = 10,000$ was used for the analysis. It was assumed $r = 0.017$, which results from the percentage share of deceased persons (see Table 4). Special attention should be paid to the strongly unbalanced proportion of survivors to deceased persons, and also the significant heterogeneity in causes of death, which are superior to unsupervised methods, based on data sets representing the whole population. The investigated procedure enabled the division of the data into atypical and typical elements. Here, 108 specimens were recognized as being atypical and 5,811 as typical (precision 7.3% according to the parameter r value assumed).

For the interpretation of the obtained results, the time factor should be considered by analysis if, in the group identified as atypical, the global survival time in the considered period was shorter than for typical elements. For the purpose of estimating the risk of death as well as the confidence interval for groups of atypical and typical elements, an analysis of time of survival was performed. To achieve this, the time from the begining of observation until death is treated as a random variable. In the case of loss of contact with a patient, his survival will be defined for the moment of the last contact with him [5].

The values of the surviving function in a given moment of time (the product of conditional probabilities of surviving through subsequent time periods) are shown using the product Kaplan-Meier estimator in Fig. 1. Note that in any moment of the observation period, the probability of surviving in the group of atypical elements is smaller than for the typical group; the curves do not cross. The probability that a person survives longer than 25 months equals 98% in the group with typical values of hemoglobin, while in the group of atypical elements, it is only 91%. Characteristic steps' occur in the moments of death for any patient under observation.

The results of tests comparing the surviving functions in groups of atypical and typical elements, shown in Table 5, confirm the statistical significance of the difference between groups under an analysis.

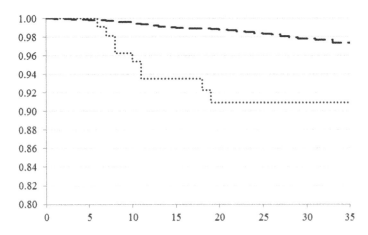

Fig. 1. Probability of surviving subsequent months in atypical (dotted line) and typical (dashed line) groups; product Kaplan-Meier estimator.

Table 5. Tests comparing surviving functions in groups of atypical and typical elements.

Test	Value of test statistic	p-value
Log Rank	27.6	<0.0001
Wilcoxon	37.8	<0.0001
-2Log(LR)	14.6	0.0001

6 Final Comments

One should finally underline that although the method presented in this paper has been illustrated using medical data, its nature is universal and can be applied in numerous different problems of contemporary science and practice including engineering, sociology, management, natural sciences as well as others.

In the work [12] the method designed to represent the result in fuzzy and intuitionistic fuzzy forms is described. In turn, in the publication [13] an unsupervised version of the atypical elements detection is transposed to a convenient supervised form with equalized patterns, which allows the application of numerous and diverse classification apparatuses, with a method that is appropriately well-matched for each specific applicational task.

Acknowledgments. The work was supported in parts by the Systems Research Institute of the Polish Academy of Sciences in Warsaw, and the Faculty of Physics and Applied Computer Science of the AGH University of Science and Technology in Cracow, Poland.

References

1. Aggarwal, C.C.: Outlier Analysis. Springer, New York (2013). https://doi.org/10.1007/978-1-4614-6396-2
2. Barnett, V., Lewis, T.: Outliers in Statistical Data. Wiley, New York (1994)
3. Canaan, C., Garai, M.S., Daya, M.: Popular sorting algorithms. World Appl. Program. **1**, 62–71 (2011)
4. Gentle, J.E.: Random Number Generation and Monte Carlo Methods. Springer, New York (2003). https://doi.org/10.1007/b97336
5. Hosmer, D.W., Lemeshow, S.: Applied Survival Analysis: Regression Modelling of Time to Event Data. Wiley, New York (1999)
6. Kulczycki, P.: Wykrywanie uszkodzeń w systemach zautomatyzowanych metodami statystycznymi. Alfa, Warsaw (1998)
7. Kulczycki, P.: Estymatory jądrowe w analizie systemowej. WNT, Warsaw (2005)
8. Kulczycki, P., Charytanowicz, M.: An algorithm for conditional multidimensional parameter identification with asymmetric and correlated losses of under- and overestimations. J. Stat. Comput. Simul. **86**, 1032–1055 (2016)
9. Kulczycki, P., Charytanowicz, M., Kowalski, P.A., Łukasik, S.: The complete gradient clustering algorithm: properties in practical applications. J. Appl. Stat. **39**, 1211–1224 (2012)
10. Kulczycki, P., Kowalski, P.A.: A complete algorithm for the reduction of pattern data in the classification of interval information. Int. J. Comput. Methods **13**, Paper ID: 1650018 (2016)

11. Kulczycki, P., Kruszewski, D.: Identification of atypical elements by transforming task to supervised form with fuzzy and intuitionistic fuzzy evaluations. Appl. Comput. **60**, 623–633 (2017)
12. Kulczycki, P., Kruszewski, D.: Detection of atypical elements with fuzzy and intuitionistic fuzzy evaluations. In: Mitkowski, W., Kacprzyk, J., Oprzędkiewicz, K., Skruch, P. (eds.) KKA 2017. AISC, vol. 577, pp. 774–786. Springer, Cham (2017). https://doi.org/10.1007/978-3-319-60699-6_75
13. Kulczycki, P., Kruszewski, D.: Detection of atypical elements by transforming task to supervised form. In: Shankar, B.U., Ghosh, K., Mandal, D.P., Ray, S.S., Zhang, D., Pal, S.K. (eds.) PReMI 2017. LNCS, vol. 10597, pp. 458–466. Springer, Cham (2017). https://doi.org/10.1007/978-3-319-69900-4_58
14. Kulczycki, P., Łukasik, S.: An algorithm for reducing dimension and size of sample for data exploration procedures. Int. J. Appl. Math. Comput. Sci. **24**, 133–149 (2014)
15. Kulczycki, P., Prochot, C.: Identyfikacja stanów nietypowych za pomocą estymatorów jądrowych. In: Bubnicki, Z., Hryniewicz, O., Kulikowski, R. (eds.) Metody i techniki analizy informacji i wspomagania decyzji, pp. 57–62. EXIT, Warsaw (2002)
16. National Health and Nutrition Examination Survey. http://www.cdc.gov/nchs/nhanes.htm/. Accessed 10 May 2016
17. National Cancer Institute. http://ctep.cancer.gov/. Accessed 10 May 2016
18. Parrish, R.: Comparison of quantile estimators in normal sampling. Biometrics **46**, 247–257 (1990)
19. Piros, P., et al.: An overview of myocardial infarction registries and results from the Hungarian myocardial infarction registry. In: Fujita, H., Selamat, A., Omatu, S. (eds.) New Trends in Intelligent Software Methodologies, Tools and Techniques, pp. 312–320. IOS Press, Amsterdam (2017)
20. Wand, M., Jones, M.: Kernel Smoothing. Chapman and Hall, London (1995)

Cardiac Murmur Effects on Automatic Segmentation of ECG Signals for Biometric Identification: Preliminary Study

C. Duque-Mejía[1,5], M. A. Becerra[1,5(✉)], C. Zapata-Hernández[1,5],
C. Mejia-Arboleda[2,5], A. E. Castro-Ospina[2,5], E. Delgado-Trejos[6],
Diego H. Peluffo-Ordóñez[4,5], P. Rosero-Montalvo[3,5],
and Javier Revelo-Fuelagán[4,5]

[1] Institución Universitaria Pascual Bravo, Medellín, Colombia
migb2b@gmail.com
[2] Instituto Tecnológico Metropolitano, Medellín, Colombia
[3] Universidad Técnica del Norte-Ecuador, Ibarra, Ecuador
[4] Universidad de Nariño, Pasto, Colombia
[5] SDAS Research Group, Yachay Tech, Urcuquí, Ecuador
[6] Quality, Metrology and Production Research Group,
Instituto Tecnológico Metropolitano ITM, Medellín, Colombia
http://www.sdas-group.com

Abstract. Biometric identification or authentication is a pattern recognition process, which is carried out acquiring different measures of human beings to distinguish them. Fingerprint and eye iris are the most known and used biometric techniques; nevertheless, also they are the most vulnerable to counterfeiting. Consequently, nowadays research has been focused on physiological signals and behavioral traits for biometric identification because these allow not only the authentication but also determine that the subject is alive. Electrocardiographic signals (ECG-S) have been studied for biometric identification demonstrating their capability. Taking into account that some pathologies are detected using ECG-S, these can affect the results of biometric identification; nonetheless, some diseases such as cardiac murmurs are not detected by ECG-S, but they can distort their morphology. Therefore, these signals must be analyzed considering different pathologies. In this paper, a biometric study was carried out from 40 subjects (20 with cardiac murmurs and 20 without cardiac affections). First, the ECG-S were preprocessed and segmented using the fast method for detecting T waves with annotation of P and T waves, then feature extraction was carried out using discrete wavelet transform (DWT), maximal overlap DWT, cepstral coefficients, and statistical measures. Then, rough set and relief F algorithms were applied to datasets (pathological and normal signals) for attribute reduction. Finally, multiple classifiers and combinations of them were tested. The results of the segmentation were analyzed achieving low results for signals affected by cardiac murmurs. On the other hand, according to the cardiac murmur effects analyzed, the performance of the classifiers in

© Springer Nature Switzerland AG 2019
N. T. Nguyen et al. (Eds.): ACIIDS 2019, LNAI 11431, pp. 269–279, 2019.
https://doi.org/10.1007/978-3-030-14799-0_23

cascade shown the best accuracy for human identification from ECG-S, minimizing the impact of variability generated on ECG-S by cardiac murmurs diseases.

Keywords: Automatic segmentation · Biometric · Heart murmur · Electrocardiographic signal · Pattern recognition

1 Introduction

Currently identifying people based on traits and physiological characteristics is no longer a matter of large companies, governments or banks, and has come to have a direct impact on the daily lives of people. An example of this are the mobile devices, which with technological advancement have developed tools to use biometric identification (BI) as a parameter additional security, health institutions and even universities where they have been putting aside the methods of identification based on barcodes or passwords, and they have been dabbling in the use of physiological traits for identification of individuals. Fields of application of BI are varied, and range from the identification of disease with the diagnostic, even the automation of identification processes, in a world where information is just as valuable as the money itself and in the wrong hands can endanger the financial and judicial integrity of individuals and corporations. BI based on fingerprints is a desirable method because each person has different morphological characteristics [22] providing a higher degree of safety compared to conventional identification methods (codes and barcodes) but remains vulnerable to counterfeiting [18,20]. The fingerprint has rapidly gained strength in different fields but just as quickly have developed a series of techniques to circumvent this identification system, which falsifies different that acquire of this trace materials illegally [29]; a similar situation is presented by the biometric identification based on the iris.

In order to replace the deficiencies of traditional biometrics [5,11,17], the Electrocardiographic signals (ECG-S) have been applied for BI [4,6,22,24,28, 32,34]. The ECG-S compared with traditional biometrics, gets a more reliable biometric solution and satisfies the necessary conditions of universality, uniqueness, permanence, avoidance [25], and life detection [8] required by BI systems. It also has several important features such as the spatial location of the heart, heart size among others. Moreover, the ECG signals can be acquired more easily every day, using minimal contact on the skin through fingers [15]. In Table 2 is summarized the studies of ECG signals for human identification considering techniques for features extraction, classifiers and number of subjects analyzed. Most of the studies have applied the wavelet transform for characterization and classifiers based on support vector machine. Besides, it is reported high variability respect to the number of subjects considered in the experiments and the accuracies achieved, reaching results 100% by some studies. Nonetheless, there are not reported works that analyze the cardiac murmur effects on BI systems except for our previous paper [2], but without considered the automatic segmentation.

Table 1. Studies of ECG for human identification

Cite	Year	Database size	Feature extraction	Clasiffiers	Accuracy (%)
[16]	2017	18	DCT, window removal method, autocorrelation and scaling	SVM, LDA, NN	100
[26]	2017	–	DCT, transformation of Haar	SVM, k-NN, MLP, GMM-UBM	94.9
[33]	2017	85	DCT	SVM, red network	96.6, 97.6
[31]	2017	18	FROS	BPNN	91.76
[25]	2017	20	R-peak, position normalization	SVM, ANN, k-NN	93.709
[21]	2017	10	Statistical measures	SVM	100
[9]	2017	88	SPCA, wavelet coefficient, QRS wave, complete beat	RPROP, BP, SVM, k-NN	93.2
[7]	2016	89	DWT, GF, PCA, Euclidean	SVM, Network	95
[13]	2016	47	PCA, LDA, KPCA	C-SVC, SVM-OAA	98.29
[14]	2016	52	DWT, AC	SVM-OAA	88.41
[27]	2016	50	Wavelet transform	SVM	97
[12]	2017	27	BEMD analysis of ECG-S and frequency distributions	k-NN	99.5
[1]	2015	20	Fractional system of commensurate order	k-NN	100
[30]	2017	184	Wavelet coefficient	Random forest, wavelet distance	99.52

The ECG-S are very applied for heart diagnosis, and some biometric systems based on ECG-S have analyzed some cardiac diseases to enhance its yield and reliability [30]. Nonetheless, some diseases such as cardiac murmurs (CMs) are not detected by ECG-S, but their morphology is distorted. CMs are common heart failures which are studied using phonocardiography and echocardiography [3] for supporting the diagnosis of different abnormality functionalities of the heart such as left atrial abnormality in mitral stenosis, ventricular hypertrophy in aortic stenosis, among others (Table 1).

Taking this into account, we consider essential to analyze the effects of valvular conditions on ECG signals and their consequences on the performance of biometric systems through all process stages. In this work is presented a study on 40 subjects for BI from ECG-S investigating the effects of cardiac diseases generated by cardiac murmurs on the automatic segmentation of ECG signals, which has not been analyzed according to the literature review, and a second stage of the previous work which was done using manual segmentation.

In this work, four main stages were accomplished after collecting the database as follows: Preprocessing standardization and automatic segmentation then, feature extraction. Then, five features selection algorithms were applied. In the last stage, five classifiers and six mixtures of them were validated using 10-cross-validation. Besides, k-Nearest Neighbor (k-NN) and Linear Discriminant Classifier (LDC) were tested in cascade, which achieved the best result 74.68%. The principal contribution of this paper is the study of cardiac murmur effects on automatic segmentation and consequently on the performance of human identification from ECG signals.

2   Experimental Setup

The proposed procedure in this work is shown in Fig. 1. This follows the previous study discussed in [2], and the same database of ECG-S was used. The signals were standardized and then, were segmented using the fast method for detecting T waves with annotation of P and T waves (FMDT-APT). Then, a set of attributes were calculated from ECG-S using multiple techniques based on Linear Frequency Cepstral coefficients (LFCC), linear and non-linear measures, Discrete Wavelet Transform (DWT), and Maximal overlap DWT (MODWT). Attribute reduction was carried out applying relief F algorithm and four algorithms based on Rough Set (RS) theory as follows: RS using Neighbor Distance, RS using Entropy estimation, Fuzzy RS using Neighbor Distance, and FRS using Entropy estimation, which are denoted by (RS-N), (RS-E),(FRS-N), and (FRS-E) respectively. The results given by all algorithms were analyzed using a histogram for selecting the relevant attributes. Finally, six classification methods were tested as follows: Support Vector Machines (SVM), Quadratic Bayes Normal Classifier (QDC), Optimization of the Parzen classifier (PZ), k-NN, and LDC, alongside k-NN-LDC classifiers in cascade. Moreover, six different mixture among them were performed.

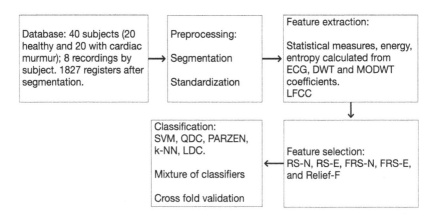

Fig. 1. Experimental procedure

2.1   Database

A set of 320 ECG-S acquired from 40 subjects were applied in this study The number of ECG-S were balanced for healthy and unhealthy subjects. The unhealthy subjects were diagnosed with different types of CMs. Forty subjects were selected from a database of 143 subjects (55 healthy and 88 unhealthy). The selection was performed considering the completeness of the number of signals per subject. The database is widely depicted in our previous work [2] (Fig. 2).

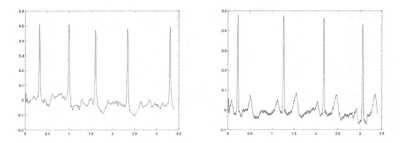

Fig. 2. Normal and pathological non-segmented ECG-S

2.2 Preprocessing

The ECG signals were standardized among $[-1, 1]$. Then, the segmentation process was carried out following the FMDT-APT depicted in [10]. First, to eliminate baseline noise and power line interference a second order Butterworth bandpass filter was used with a lower cutoff frequency of 0.5 Hz and high cutoff 60 Hz, keeping the P wave, QRS complex (10 Hz–40 Hz), P (8 Hz–10 Hz) and T wave (5 Hz–8 Hz). The signals were segmented considering two heartbeats. In Fig. 3 is shown a normal ECG signal and three pathological signals where evidence changes of the signal respect to normal signal but they were correctly segmented. While in Fig. 4 are shown two signals morphologically affected by cardiac murmurs and the adverse effects generated in the segmentation.

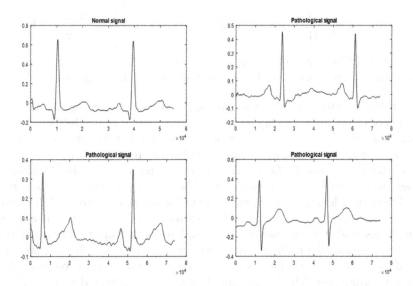

Fig. 3. Normal and pathological ECG signals

274   C. Duque-Mejía et al.

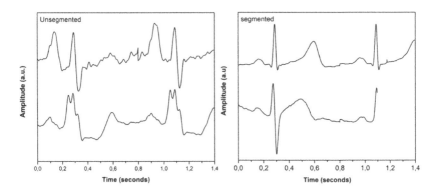

Fig. 4. Non-segmented and segmented ECG signals

R Peak Detection. The peak of QRS complex is the most prominent peak, this is detected with the reference point, to identify the P wave and T wave. The detection of R point is based in (Thomks algorithm) due to its computational simplicity. We used differentiated ECG signal to get slope information of QRS complex, a squaring function to achieve a non-linear amplification and decision rule system that contemplates the minimum distance between R peaks and minimum amplitude of R peaks.

QRS Elimination. Generally, the P wave and the T wave are of low amplitude, to easily identify these waves, the QRS complex is removed using moving average.

$$QRS_{cut}[n] = \frac{1}{W_1} \left(x \left[n - \frac{W_1 - 1}{2} \right] \dots + x[n] + x \left[n + \frac{W_1 - 1}{2} \right] \right) \quad (1)$$

where $x[n]$ is ECG signal and W_1 represents the windows size of approximately the duration of QRS complex. The value assigned for W_1 is 120 ms.

P and T Wave Detection. The detection of P wave and T wave are based on the detection of peaks using moving average and duration of peaks. Initially moving average is applied to detect of peaks of P wave and T wave and then through of average length of waves, it is confirmed that the peaks are within the range for each wave. In some ECG signals which have pathologies, the P waves are smaller than normal, in these cases, only the duration of the peaks is considered. The normal length of the P wave and QT segment wave are 80 ± 30 ms and 400 ± 40 ms, respectively. The average size windows size for P wave is approximately 80 ms, and average windows size for QT segment is approximately 400 ms. For the case where the peaks are not detected were used search windows for P wave of 0.3 ms and 0.55 ms for T wave. These rules were designed based on from an expert cardiologist. The P wave and T wave will be detected simultaneously; the windows size will be adjusted in the relation of P wave duration.

$$P/T_{wave}[n] = \frac{1}{W_2} \left(y \left[n - \frac{W_2 - 1}{2} \right] \dots + y[n] + y \left[n + \frac{W_2 - 1}{2} \right] \right) \quad (2)$$

where $y[n] = QRS_{cut}[n]$ and W_1 is equal to 55 ms which is half windows width of P wave. A smaller windows size of the healthy subject was selected for detecting P and T wave in arrhythmia cases.

2.3 Feature Extraction

The attributes were calculated from segmented ECG-S in order to analyze the effects of the segmentation. A set of 140 attributes were obtained using statistical measures applied on ECG-S, and the coefficients given by DWT and MODWT cite Barzegar2018. The DWT and MODWT were applied using db10 of 4 levels. Besides, 39 attributes were obtained by calculating the LFCC from ECG-S based on delta coefficients, delta-delta coefficients, and the logarithm of the energy, using 24 Hamming shaped filters and sliding hamming windows (50% overlap).

Table 2. Measures calculated from ECG signals - MODWT and DWT

Renyi entropy	Root mean square	Standard deviation	Entropy	Variance	Power	Energy
Shannon entropy	Log energy entropy	Covariance	Kurtosis	Min/Max	Mean	

2.4 Feature Selection

The methodology presented in [23] was applied for setting the parameters of the algorithms RS-N, RS-E, FRS-N, and FRS-E algorithms to determine the relevant features. Also, the Relief F algorithm was applied. The results obtained by all algorithms of attribute selection were analyzed using a histogram to select the more relevant features. This process was applied for three datasets: (i) Attributes calculated from segmented unhealthy ECG-S, (ii) Attributes calculated from segmented healthy ECG-S, and (iii) Attributes calculated from all segmented ECG-S.

2.5 Classification

SVM, QDC, PZ, k-NN and LDC classifiers were tested. Besides, six combinations of all classifiers were tested, and k-NN and LDC classifiers were tested in cascade. The mixture of classifiers used in this work are the following: Product (Prod), Mean combiner, Median combiner, Maximum combiner (Max), Minimum combiner (Min), and majority voting combiner (Vm). These methods are explained in [19].

3   Results and Discussion

The results of the identification rate of classifiers are shown in Table 3. All classi-
fication systems were tested with the healthy ECG-S (HS) and unhealthy ECG-S
(US), and All ECG-S (HS and US). The best result 74.68% was achieved by the
k-NN and LDC classifiers in cascade for the US dataset, and 60.38% for all signals
(HS and US datasets together). Individually, the best results were obtained by
PZ and k-NN classification systems with a mean of 57.51% and 58.98% respectively. On the other hand, the combination of the five classifiers achieved the
best result using majority vote. Nonetheless, comparing respect to a previous
work carried out using manual segmentation, the results of this work shown that
the effects of cardiac murmur affect profoundly to the performance for human
identification based on automatically segmented ECG signals (see Table 4).

Table 3. Accuracy(%) of classification systems

Signals	SVM	QDC	PZ	k-NN	LDC	k-NN–LDC	Prod	Mean	Median	Max	Min	Vm
US	24.05	34.18	70.89	70.89	63.29	74.68	37.97	34.18	63.29	34.18	37.97	62.15
HS	32.5	42.5	53.75	51.25	45.00	65.00	32.50	42.5	53.75	51.25	45.00	53.23
ALL	27.50	40.23	52.3	50.4	42.76	60.38	43.54	37.98	47.51	54.12	49.87	52.15

Table 4. Comparison between this work (automatic segmentation) and previous work
(manual segmentation)

	Manual Segm. (previous work [2])	Automatic Segm. (this work)
Accuracy (%)	91.19	60.38

4   Conclusions

In this work, an analysis of cardiac murmur effects on automatic segmentation
of ECG-S applied to biometric identification was presented. PZ classifier shows
the best individual global performance of 70.89%, but using k-NN and LDC
classifiers in cascade increased the accuracy to 74.68% for pathological ECG-S.
Based on the results achieved with the pathological ECG-S vs. normal ECG-S
for human identification the best results were obtained with the pathological
signals with a difference of 9.8%. However, using all ECG-S, the difference was
14.3%, which evidence the adverse effects of the cardiac murmur on the global
performance of the biometric system based on ECG-S. The best results of the
unhealthy ECG-S is explained considering that cardiac murmurs generate different effects on the dynamic of the heart causing changes in the electrical responses
which are captured by ECG-S. Consequently, taking into account that this is not
a generality of the human beings, the cardiac murmurs can facilitate the identification of the subjects. On the other hand, considering the results achieved in the
previous work is evidenced that the cardiac murmurs generate adverse effects on

automatic segmentation of ECG-S affecting the human identification from ECG-S. Therefore, new strategies for segmentation should be proposed or consider no apply segmentation, and analyze longer ECG-S for this type of BI. The global low accuracy respect to studies reported in the literature can be attributed to the database, the quality of the automatic segmentation, and due to the signals were acquired in two-phases: post-expiratory and post-inspiratory apnea. On the other hand, the extracted features can differ from other works despite that in this work were considered a lot type of features. As future work, we propose to include a public database, compare multiples algorithms of segmentation, extract new features, and increase the number of registers, and individuals to enhance the discernibility and generality. Besides, unveil the apnea phases effects on biometric systems based on ECG-S.

Acknowledgment. The authors acknowledge to the research project titled "Desarrollo de una metodología de visualización interactiva y eficaz de información en Big Data" supported by Agreement No. 180 November 1st, 2016 by VIPRI from Universidad de Nariño. Besides, acknowledge to SDAS Research Group and to the project P17202 supported by Instituto Tecnologico Metropolitano ITM of Medellin.

References

1. Assadi, I., Charef, A., Mentouri, F., El-bey, R.A.: QRS Complex Based Human Identification, pp. 248–252 (2015)
2. Becerra, M.A., et al.: Exploratory study of the effects of cardiac murmurs on electrocardiographic-signal-based biometric systems. In: Yin, H., Camacho, D., Novais, P., Tallón-Ballesteros, A.J. (eds.) IDEAL 2018. LNCS, vol. 11314, pp. 410–418. Springer, Cham (2018). https://doi.org/10.1007/978-3-030-03493-1_43
3. Becerra, M.A., Orrego, D.A., Delgado-Trejos, E.: Adaptive neuro-fuzzy inference system for acoustic analysis of 4-channel phonocardiograms using empirical mode decomposition. In: 2013 35th Annual International Conference of the IEEE Engineering in Medicine and Biology Society (EMBC), pp. 969–972. IEEE, July 2013. https://doi.org/10.1109/EMBC.2013.6609664
4. Belgacem, N., Fournier, R., Nait-Ali, A., Bereksi-Reguig, F.: A novel biometric authentication approach using ECG and EMG signals. J. Med. Eng. Technol. **39**(4), 226–238 (2015). https://doi.org/10.3109/03091902.2015.1021429
5. Bhatnagar, S.: Cooperative biometric multimodal approach for identification. In: Satapathy, S.C., Joshi, A. (eds.) ICTIS 2017. SIST, vol. 83, pp. 13–19. Springer, Cham (2018). https://doi.org/10.1007/978-3-319-63673-3_2
6. Bugdol, M.D., Mitas, A.W.: Multimodal biometric system combining ECG and sound signals. Pattern Recogn. Lett. **38**, 107–112 (2014). https://doi.org/10.1016/J.PATREC.2013.11.014
7. Chun, S.Y.: Single Pulse ECG-based Small Scale User Authentication using Guided Filtering
8. Dar, M.N., Akram, M.U., Shaukat, A., Khan, M.A.: ECG based biometric identification for population with normal and cardiac anomalies using hybrid HRV and DWT features. In: 2015 5th International Conference on IT Convergence and Security, ICITCS 2015 - Proceedings (2015). https://doi.org/10.1109/ICITCS.2015.7292977

9. Duffy, V.G. (ed.): DHM 2017. LNCS, vol. 10287. Springer, Cham (2017). https://doi.org/10.1007/978-3-319-58466-9

10. Elgendi, M., Eskofier, B., Abbott, D.: Fast T wave detection calibrated by clinical knowledge with annotation of P and T waves. Sensors **15**(7), 17693–17714 (2015). https://doi.org/10.3390/s150717693

11. Elhoseny, M., Essa, E., Elkhateb, A., Hassanien, A.E., Hamad, A.: Cascade multimodal biometric system using fingerprint and iris patterns. In: Hassanien, A.E., Shaalan, K., Gaber, T., Tolba, M.F. (eds.) AISI 2017. AISC, vol. 639, pp. 590–599. Springer, Cham (2018). https://doi.org/10.1007/978-3-319-64861-3_55

12. Ferdinando, H., Seppänen, T., Alasaarela, E.: Bivariate Empirical Mode Decomposition for ECG - based Biometric Identification with Emotional Data, pp. 450–453 (2017)

13. Hejazi, M., Al-Haddad, S.A., Hashim, S.J., Aziz, A.F.A., Singh, Y.P.: Feature level fusion for biometric verification with two-lead ECG signals. In: Proceeding - 2016 IEEE 12th International Colloquium on Signal Processing and its Applications, CSPA 2016, pp. 54–59, March 2016. https://doi.org/10.1109/CSPA.2016.7515803

14. Hejazi, M., Al-Haddad, S.A., Singh, Y.P., Hashim, S.J., Abdul Aziz, A.F.: ECG biometric authentication based on non-fiducial approach using kernel methods. Digital Signal Process.: Rev. J. **52**, 72–86 (2016). https://doi.org/10.1016/j.dsp.2016.02.008

15. da Silva, H.P., Carreiras, C., Lourenco, A., Fred, A., das Neves, R.C., Ferreira, R.: Off-the-person electrocardiography: performance assessment and clinical correlation. Health Technol. **4**, 309–318 (2015)

16. Jung, W.H., Lee, S.G.: ECG identification based on non-fiducial feature extraction using window removal method. Appl. Sci. **7**(12), 1205 (2017). https://doi.org/10.3390/app7111205. http://www.mdpi.com/2076-3417/7/11/1205

17. Kanchan, T., Krishan, K.: Loss of fingerprints: forensic implications. Egypt. J. Forensic Sci. **8**(1), 19 (2018). https://doi.org/10.1186/s41935-018-0051-0

18. Martinez-Diaz, M., Fierrez, J., Galbally, J., Ortega-Garcia, J.: An evaluation of indirect attacks and countermeasures in fingerprint verification systems. Pattern Recognit. Lett. **32**(12), 1643–1651 (2011). https://doi.org/10.1016/J.PATREC.2011.04.005

19. Moreno-Revelo, M., Ortega-Adarme, M., Peluffo-Ordoñez, D.H., Alvarez-Uribe, K.C., Becerra, M.A.: Comparison among physiological signals for biometric identification. In: Yin, H., Gao, Y., Chen, S., Wen, Y., Cai, G., Gu, T., Du, J., Tallón-Ballesteros, A.J., Zhang, M. (eds.) IDEAL 2017. LNCS, vol. 10585, pp. 436–443. Springer, Cham (2017). https://doi.org/10.1007/978-3-319-68935-7_47

20. Murillo-Escobar, M., Cruz-Hernández, C., Abundiz-Pérez, F., López-Gutiérrez, R.: A robust embedded biometric authentication system based on fingerprint and chaotic encryption. Expert Syst. Appl. **42**(21), 8198–8211 (2015). https://doi.org/10.1016/j.eswa.2015.06.035

21. Nawal, M., Sharma, M.K., Bundele, M.M.: Design and implementation of human identification through physical activity aware 12 lead ECG. In: 2016 International Conference on Recent Advances and Innovations in Engineering, ICRAIE 2016 (2017). https://doi.org/10.1109/ICRAIE.2016.7939536

22. Odinaka, I., Lai, P.H., Kaplan, A.D., O'Sullivan, J.A., Sirevaag, E.J., Rohrbaugh, J.W.: ECG biometric recognition: a comparative analysis (2012). https://doi.org/10.1109/TIFS.2012.2215324
23. Orrego, D., Becerra, M., Delgado-Trejos, E.: Dimensionality reduction based on fuzzy rough sets oriented to ischemia detection. In: Proceedings of the Annual International Conference of the IEEE Engineering in Medicine and Biology Society, EMBS (2012). https://doi.org/10.1109/EMBC.2012.6347186
24. Pal, S., Mitra, M.: Increasing the accuracy of ECG based biometric analysis by data modelling. Measurement 45(7), 1927–1932 (2012). https://doi.org/10.1016/J.MEASUREMENT.2012.03.005
25. Patro, K., Kumar, P.: Machine learning classification approaches for biometric recognition system using ECG signals. Eng. Sci. Technol. Rev. 10(6), 1–8 (2017). https://doi.org/10.25103/jestr.106.01
26. Pinto, J., Cardoso, J., Lourenço, A., Carreiras, C.: Towards a continuous biometric system based on ECG signals acquired on the steering wheel. Sensors 17(10), 2228 (2017). https://doi.org/10.3390/s17102228. http://www.mdpi.com/1424-8220/17/10/2228
27. Ramli, D.A., Hooi, M.Y., Chee, K.J.: Development of heartbeat detection kit for biometric authentication system. Procedia Comput. Sci. 96, 305–314 (2016). https://doi.org/10.1016/j.procs.2016.08.143
28. Sidek, K.A., Khalil, I., Jelinek, H.F.: ECG biometric with abnormal cardiac conditions in remote monitoring system. IEEE Trans. Syst. Man Cybern.: Syst. 44(11), 1498–1509 (2014). https://doi.org/10.1109/TSMC.2014.2336842
29. Song, W., Kim, T., Kim, H.C., Choi, J.H., Kong, H.J., Lee, S.R.: A finger-vein verification system using mean curvature. Pattern Recognit. Lett. 32(11), 1541–1547 (2011). https://doi.org/10.1016/J.PATREC.2011.04.021
30. Tan, R., Perkowski, M.: Toward improving electrocardiogram (ECG) biometric verification using mobile sensors: a two-stage classifier approach. Sensors (Switzerland) 17(2) (2017). https://doi.org/10.3390/s17020410
31. Tseng, K.K., Lee, D., Hurst, W., Lin, F.Y., Ip, W.H.: Frequency rank order statistic with unknown neural network for ECG identification system. In: Proceedings - 4th International Conference on Enterprise Systems: Advances in Enterprise Systems, ES 2016, pp. 160–167 (2017). https://doi.org/10.1109/ES.2016.27
32. Wahabi, S., Pouryayevali, S., Hari, S., Hatzinakos, D.: On evaluating ECG biometric systems: session-dependence and body posture. IEEE Trans. Inf. Forensics Secur. 9(11), 2002–2013 (2014). https://doi.org/10.1109/TIFS.2014.2360430
33. Zhang, Y., Wu, J.: Practical human authentication method based on piecewise corrected Electrocardiogram. In: Proceedings of the IEEE International Conference on Software Engineering and Service Sciences, ICSESS, vol. 61571268, pp. 300–303 (2017). https://doi.org/10.1109/ICSESS.2016.7883071
34. Zhao, Z., Yang, L.: ECG identification based on matching pursuit. In: 2011 4th International Conference on Biomedical Engineering and Informatics (BMEI), pp. 721–724. IEEE, October 2011. https://doi.org/10.1109/BMEI.2011.6098470

An Approach to Estimation of Residential Housing Type Based on the Analysis of Parked Cars

Marcin Kutrzyński[1] , Zbigniew Telec[1] ,
Bogdan Trawiński[1(✉)] , and Hien Cao Dac[2]

[1] Faculty of Computer Science and Management,
Wrocław University of Science and Technology, Wrocław, Poland
{marcin.kutrzynski,zbigniew.telec,
bogdan.trawinski}@pwr.edu.pl
[2] Nguyen Tat Thanh University, Ho Chi Minh City, Vietnam
cdhien@ntt.edu.vn

Abstract. A method for prediction of residential housing types based on an analysis of the number of cars parked near buildings in consideration is proposed in the paper. The source of data constitute satellite or aerial images of a given residential area where cars and building can be identified. The machine learning models are build based on the distribution of car parked in the area. The resulting classification models allow for distinguishing between low-rise, mid-rise and high-rise housing. The effectiveness of the method was proved using aerial images of three residential districts of a big city in Poland and the WEKA data mining system.

Keywords: Residential housing · Satellite images · Aerial images · Machine learning · Classification algorithms

1 Introduction

Public geo-information systems surpass in terms of precision and completeness official government or commercial solutions in some regions [1–3]. Open access and quality of datasets make them useful to visualize urban areas [4], constitute a road network base for navigation systems [5], construct video games [6], or perform urban analysis [7]. They contain precise information about the position, shape, usage, and cultural context of objects, however they often lack data related to height [4, 8].

There are several techniques for acquiring information about the building height in a large area. The most accurate, i.e. geodetic measurements and the LiDAR method, are expensive, rarely performed and publicly available results are very often outdated [9]. Moreover, license issues often arise due to the fact that proprietary data cannot be used in open systems.

The data from the SRTM system (Shuttle Radar Topography Mission) are old and not accurate [10]. Methods based on automatic visual analysis of satellite imagery have also limitations. Another method for determining the height is volumetric shadow analysis of monoscopic imagery. If the exact date and time when satellite images were

© Springer Nature Switzerland AG 2019
N. T. Nguyen et al. (Eds.): ACIIDS 2019, LNAI 11431, pp. 280–289, 2019.
https://doi.org/10.1007/978-3-030-14799-0_24

taken is known (i.e. the angle of sunlight), the height can be computed from trigonometric equations [11, 12]. Simplicity is an advantage in this case, however this method contains a number of limitations: it cannot be applied in the mountain area or to objects whose walls are not vertical [11].

Another precise method is designed to determine the building height based on multi-view imagery [13, 14]. Taking several shots of an object from different positions at a known angle, makes it possible to accurately outline the base of a building. Knowing the base, the length of the wall in the projection and the angle of shooting, the height can be precisely calculated. Unfortunately, this method also has its limitations. It requires to take a several shots of the subject at acute angle to the ground. In high density areas it is often impossible to do this.

All methods of satellite imagery analysis encounter issues related to object detection [11, 15]. The shadow is often confused with a road, because they both have similar spectra. Precision of these methods could be very diverse.

In this paper the authors propose a new method for prediction of the residential housing type based on an analysis of the number of cars parked near houses. The cars can be identified in satellite or aerial images of a given residential area. The idea behind the method is that the height of a residential building could be estimated based on the density of cars located around that building. The larger number of cars the higher the building. A machine learning models are built on the basis of the distribution of cars parked in the residential area. The classification models can be used to distinguish between low-rise, mid-rise and high-rise housing.

So far, the authors have built and investigated single and ensemble regression models to aid in real estate appraisal using machine learning algorithms included in the KEEL and WEKA data mining systems [16–18]. In this work WEKA classification algorithms such as support vector machine, k-nearest neighbours, decision trees, bagging, boosting, and random forest are employed [19].

2 Method for Estimation of Residential Housing Type

The method proposed in this paper assumes that the users have satellite or aerial imagery data of housing areas at their disposal. Moreover they are able to identify address points of residential buildings and parked cars in the images. The task is to construct classification models which can be utilized to distinguish between diverse types of residential housing. It depends on local standards and requirements which buildings can be classified into low-rise, mid-rise, and high-rise housing types. In our case the low-rise housing includes single-family housing, i.e. buildings up to 2 storeys high. In turn, multiple family houses, i.e. buildings up to five storeys high and without a lift, belong to the mid-rise housing. Finally, the high-rise housing comprises buildings with lifts which are higher than five storeys.

The method for the preparation of input data to build classification models is depicted in Fig. 1. This figure illustrates the technique for collecting data of parked car distribution for one address point of a building. In Fig. 1 the red squares stand for parked cars and blue icons for address points of buildings. Input data vector of each building is composed of numbers of cars identified in successive concentric annuli. The address point of a given building is a common central point of all concentric annuli.

All annuli have the same width r. So that the input vector $Build_j$ of a building j, to be more precise, - an address point j, can be expressed by means of Formula 1:

$$Build_j = \left(car(a_1), car(a_2), \ldots, car(a_n), class(b_j)\right) \tag{1}$$

where $car(a_i)$ stands for the number of cars identified within the $annulus_i$, n means the number of annuli considered, and $class(b_j)$ denotes the class representing the type of residential housing of the building b_j.

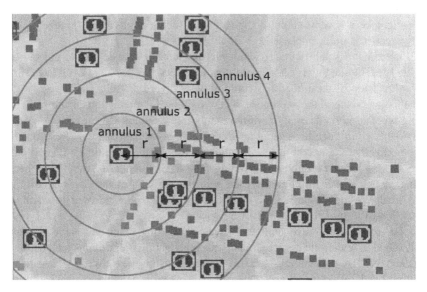

Fig. 1. Illustration of input data preparation for one address point of a building. (Color figure online)

Six machine learning algorithms designed for classification problems included in the WEKA data mining system are employed to generate predictive models [19]. They are:

SVM – sequential minimal optimization for support vector classification (*Class SMO*),
KNN – k-nearest-neighbours classifier (*Class IBk*),
ADA – boosting a classifier using the AdaBoostM1 method (*Class AdaBoostM1*),
BAG – bagging a classifier to reduce variance (*Class Bagging*),
TRE – C4.5 decision tree learner (*Class J48*),
FOR – constructs a forest of random trees (*Class RandomForest*).

3   Setup of Evaluation Experiments

The goal of the evaluation experiment was to examine classification accuracy of the proposed method using real-world data of images of three residential districts of a big city in Poland. The classification accuracy was defined as the ratio between the number of correctly classified buildings to the total number of classified buildings.

The aerial images of individual districts are shown in Figs. 2, 3, and 4. Due to space limitations the scales of individual images are different. The address points were denoted with blue colour, the cars parked were indicated with red colour and areas of low-rise, mid-rise, and high-rise housing types were marked with green, yellow and red background colours respectively. The District 1 (Fig. 2) is the biggest one, where all three classes are balanced and have clear boundaries. The District 2 has few instances. The classes are balanced and high-rise and mid-rise buildings are mixed. In turn, the high-rise housing borders on low-rise ones. On the other hand, the District 3 does not contain practically the high-rise housing.

Table 1. Statistics of buildings and cars identified in individual districts

Area	Low-rise (Class 1)	Mid-rise (Class 2)	High-rise (Class 3)	Total no of address points	Total no of cars identified
District 1	254	344	237	835	4680
District 2	52	64	61	177	1861
District 3	226	144	12	382	1766

The statistics of buildings and cars identified in aerial images of individual districts is presented in Table 1. The largest number of objects was recognized for District 1. In turn, District 1 and District 2 are characterized by the balanced numbers of objects within individual classes.

Fig. 2. Location of residential buildings and parked cars in District 1 (Color figure online)

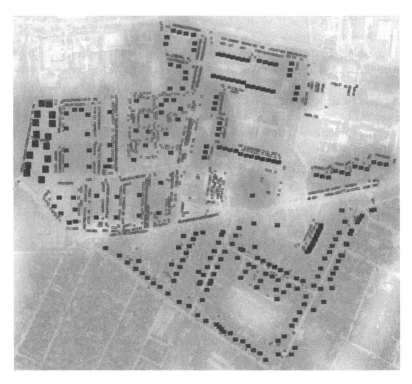

Fig. 3. Location of residential buildings and parked cars in District 2 (Color figure online)

Fig. 4. Location of residential buildings and parked cars in District 3 (Color figure online)

Pre-processing stage of experiments was performed in order to find values of hyper parameters of WEKA machine learning algorithms which provide the best value of classification accuracy. As a result, the following values were utilized in the evaluation experiments:

SVM (*class SMO*) – calibration model: a simple linear regression model (*class SimpleLinearRegression*), kernel: Pearson VII function-based universal kernel (*class PUK*), Omega parameter: 0.6, Sigma parameter: 1.2,

KNN (*Class IBk*) – number of nearest neighbours: 6, the nearest neighbour search algorithm used *LinearNNSearch*, distance function used: *ChebyshevDistance*,

ADA (*class AdaBoostM1*) – number of iterations: 20, base classifier: *J48*, confidence threshold for pruning: 0.2, minimum number of instances per leaf: 20,

BAG (*class Bagging*) – number of iterations: 50, size of each bag: 100% of the training set size, random number seed: 1, base classifier: *J48*, confidence threshold for pruning: 0.25, minimum number of instances per leaf: 2,

TRE (*class J48*) – confidence threshold for pruning: 0.15, minimum number of instances per leaf: 20,

FOR (*class RandomForest*) – number of iterations: 100, size of each bag: 100% of the training set size, base classifier: *REPTree*, minimum numeric class variance proportion of train variance for split: 0.001, minimum number of instances per leaf: 1, seed for random number generator: 1.

4   Analysis of Experimental Results

Two runs of experiments were executed. The first run aimed at assessing the accuracy of the proposed method in categorizing the types of residential housing. Therefore, 10-fold cross-validation approach was employed to build classification models built over aerial imagery data. The results are illustrated in Fig. 5, where the x axis shows the number of annuli taken into account while the percent accuracy appears on the y axis. The number of annuli ranged from 5 to 30 and width of all annuli was constant and equal to 10 m. The general trend could be seen in Fig. 5, namely the larger number of annuli considered the better accuracy. Moreover, the support vector machine models revealed the best performance. In turn, the lowest accuracy provided the C4.5 decision trees and AdaBoostM1 classifiers generated using these decision trees. The accuracy of the models ranged from about 60% for five annuli to above 85% for 30 annuli. The latter can be regarded as a very good performance. The investigation of the models with the split of the data from all districts into training and test sets in the proportion 70:30 showed that these models exhibited similarly good performance as the 10cv models (compare the charts titled **Districts 1 + 2 + 3 – 10cv** and **Districts 1 + 2 + 3 – split 70:30**).

The goal of the second run of experiment was to check whether it is practical to build a classifier over training data coming from one district and then use this classifier to estimate the types of housing in another district. The results are shown in Fig. 6, where horizontal and vertical axes show the same data as in Fig. 5. The accuracy of the models was lower when compared with the 10cv approach. The best results were obtained for the lower number of annuli between 10 and 15. The accuracy varied between 60% and 80%, and for training over data from District 2 and 3 was even below 60%.

Fig. 5. Results of experiments with 10-fold cross-validation and split 70:30

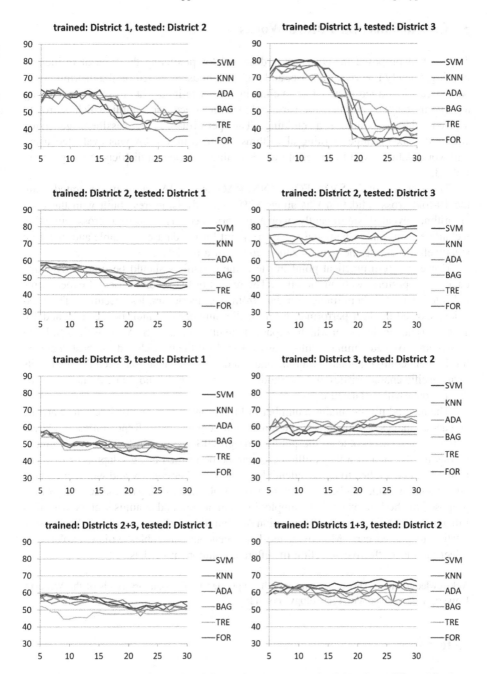

Fig. 6. Results of experiments with training and test sets comprised data from different districts

5 Conclusions and Future Works

The method proposed in the paper is designed for predicting the type of residential housing, namely low-rise, mid-rise, and high-rise housing, based on the distribution of parked cars identified in aerial or satellite images. The input data to machine learning algorithms contain numbers of cars included in concentric annuli which surround a given address point of a building. The performance of classifiers generated employing six machine learning algorithms included in WEKA was evaluated using real-world data derived from aerial images of three residential districts in a big city in Poland.

The accuracy of the SVM, FOR, KNN, BAG, and ADA classifiers generated using the 10-fold cross-validation was above 80%. Only decision tress built with the C4.5 algorithm revealed significantly worse performance. The 10-fold cross-validation technique trains and tests models on every point from a dataset available and therefore suits well measuring the generalization ability of a model.

However, from practical point of view, the classifiers should be trained over one dataset and perform well on the other. The results of the second series of experiments indicated that classifiers trained on data from one district are less accurate. The classifiers showed the best performance for smaller number of annuli between 10 and 15. There was no single model which surpassed the other models. SVM, FOR, KNN, BAG classifiers provided similar results, however each of them produced the best accuracy for different configurations of training and test data. This means that each district has its own specific characteristics and the models for practical use should be trained over the diverse data with balanced classes.

The method itself contains a number of limitations, but is relatively cheap. Satellite and aerial images of sufficient resolution are publicly available. Repositories of address points are also accessible to the public through GIS services in the Internet. Computations for the area of a half million city can be performed using a laptop or PC. The method can be applied to any residential area in any other country and will work where other methods could not be used. Future work is planned to increase the accuracy of the proposed method for example by employing semi annuli and annulus sectors instead of full annuli and adjust them to the location of a building and position of an address point relative to parked cars. Moreover, the heterogeneous ensembles will be explored to select the most effective selection of different component models.

Acknowledgments. This paper was partially supported by the statutory funds of the Wrocław University of Science and Technology, Poland.

References

1. Haklay, M.: Beyond good enough? Spatial data quality and OpenStreetMap data. In: Proceedings of the State of the Map 2009, Amsterdam, The Netherlands (2009)
2. Haklay, M.: How good is OpenStreetMap information? A comparative study of OpenStreetMap and Ordnance Survey datasets for London and the rest of England. Environ. Plan. B: Plan. Des. **37**(4), 682–703 (2010). https://doi.org/10.1068/b35097

3. Zhou, Q.: Exploring the relationship between density and completeness of urban building data in OpenStreetMap for quality estimation. Int. J. Geogr. Inf. Sci. **32**(2), 257–281 (2018). https://doi.org/10.1080/13658816.2017.1395883
4. Over, M., Schilling, A., Neubauer, S., Zipf, A.: Generating web-based 3D City Models from OpenStreetMap: the current situation in Germany. Comput. Environ. Urban Syst. **34**, 496–507 (2010). https://doi.org/10.1016/j.compenvurbsys.2010.05.001
5. Ding, R., et al.: Detecting the urban traffic network structure dynamics through the growth and analysis of multi-layer networks. Physica A: Stat. Mech. Appl. **503**, 800–817 (2018). https://doi.org/10.1016/j.physa.2018.02.059
6. Games - OpenStreetMap Wiki. https://wiki.openstreetmap.org/wiki/Games. Accessed 30 Dec 2018
7. Helbich, M., Amelunxen, Ch., Neis, P., Zipf, E.: Comparative spatial analysis of positional accuracy of OpenStreetMap and proprietary geodata. In: Jekel, T., et al. (eds.) GI_Forum 2012: Geovisualisation, Society and Learning, pp. 24–33 (2012)
8. Feng, Q., Zhai, Z., Gaihong, D.: Building height estimation using Google Earth. Energy Build. **118**, 123–132 (2016). https://doi.org/10.1016/j.enbuild.2016.02.044
9. Rottensteiner, F., Briese, C.: a new method for building extraction in urban areas from high-resolution LIDAR data. Int. Arch. Photogram. Remote Sens. Spat. Inf. Sci. **34**(3/A), 295–301 (2002)
10. Karwel, A., Ewiak, I.: Assessment of usefulness of SRTM data for generation of DEM on the territory of Poland. Prace Instytutu Geodezji i Kartografii **52**(110), 75–87 (2006). (in Polish)
11. Taeyoon, L., Taejung, K.: Automatic building height extraction by volumetric shadow analysis of monoscopic imagery. Int. J. Remote Sens. **34**(16), 5834–5850 (2013). https://doi.org/10.1080/01431161.2013.796434
12. Biljecki, F., Ledoux, H., Stoter, J.: Generating 3D city models without elevation data. Comput. Environ. Urban Syst. **64**, 1–18 (2017). https://doi.org/10.1016/j.compenvurbsys.2017.01.001
13. Fradkin, M., Maître, H., Roux, M.: Building detection from multiple aerial images in dense urban areas. Comput. Vis. Image Underst. **82**(3), 181–207 (2001). https://doi.org/10.1006/cviu.2001.0917
14. Cheng, L., Gong, J., Li, M., Liu, Y.: 3D building model reconstruction from multi-view aerial imagery and LIDAR data. Photogram. Eng. Remote Sens. **77**(2), 125–139 (2011)
15. Ghanea, M., Moallem, P., Momeni, M.: Building extraction from high-resolution satellite images in urban areas: recent methods and strategies against significant challenges. Int. J. Remote Sens. **37**(21), 5234–5248 (2016). https://doi.org/10.1080/01431161.2016.1230287
16. Lasota, T., Telec, Z., Trawiński, B., Trawiński, K.: A multi-agent system to assist with real estate appraisals using bagging ensembles. In: Nguyen, N.T., Kowalczyk, R., Chen, S.-M. (eds.) ICCCI 2009. LNAI, vol. 5796, pp. 813–824. Springer, Heidelberg (2009). https://doi.org/10.1007/978-3-642-04441-0_71
17. Lasota, T., Łuczak, T., Trawiński, B.: Investigation of random subspace and random forest methods applied to property valuation data. In: Jędrzejowicz, P., Nguyen, N.T., Hoang, K. (eds.) ICCCI 2011. LNCS, vol. 6922, pp. 142–151. Springer, Heidelberg (2011). https://doi.org/10.1007/978-3-642-23935-9_14
18. Lasota, T., Łuczak, T., Trawiński, B.: Investigation of rotation forest method applied to property price prediction. In: Rutkowski, L., Korytkowski, M., Scherer, R., Tadeusiewicz, R., Zadeh, L.A., Zurada, J.M. (eds.) ICAISC 2012. LNCS, vol. 7267, pp. 403–411. Springer, Heidelberg (2012). https://doi.org/10.1007/978-3-642-29347-4_47
19. Witten, I.H., Frank, E., Hall, M.A., Pal, C.J.: Data Mining: Practical Machine Learning Tools and Techniques, 4th edn. Morgan Kaufmann, Burlington (2016)

Household Electric Load Pattern Consumption Enhanced Simulation by Random Behavior

Alabbas Alhaj Ali$^{(\boxtimes)}$, Doina Logofătu, Prachi Agrawal, and Sreshtha Roy

Department of Computer Science and Engineering,
Frankfurt University of Applied Sciences, 60318 Frankfurt, Germany
{alabbasa,sresthar}@stud.fra-uas.de, logofatu@fb2.fra-uas.de

Abstract. The demand for electricity is increasing exponentially and thus, the concern for energy conservation becomes important. The daily consumption of electricity by each family needs to be calculated which in turn would help to estimate the weekly, monthly and yearly electric consumption for a particular unit. The electricity consumed by each family depends upon various factors. The occupancy model of the family needs additionally to be considered. After studying the this model, one can predict at which hours of the day the load consumed is maximum and at which hours of the day it is minimum. Studying the load profiles of each family, the supplier of the electricity can estimate the consumption charge supply policy accordingly. While studying the load profile, we need to take into consideration various appliances and their demand behavior. In this paper, we summarize influential factors of house electrical consumption, the occupancy of the members of the house, and the electrical demand for lighting. It also explains various types of appliances usually employed in a house and their categorization based on behavior and how they contribute to the total load profile of a household.

Keywords: Electric load pattern · Electric demand ·
Software engineering · Simulation · Random behaviour

1 Introduction

As industrialization and modernization are growing at a very fast rate, it has become very important to study the demands and supplies of each and every aspect of our life and one such very important and soon to be one of the most demanded and carrying a threat of being depleted is electricity. We know that electricity plays a very necessary and important role in our day to day life. Thus, electricity is more of a necessity in our life. The demand for electricity has grown exponentially not only in the industrial and commercial areas but also has seen a sudden growth in the residential sectors. Therefore, it is of great importance to have the demand fulfilled in an efficient manner and at the same time make use of electricity in a proper way.

© Springer Nature Switzerland AG 2019
N. T. Nguyen et al. (Eds.): ACIIDS 2019, LNAI 11431, pp. 290–302, 2019.
https://doi.org/10.1007/978-3-030-14799-0_25

It is our responsibility to save such valuable resources otherwise will be a big problem. So, to effectively have the supply of the electricity and also for the planning and managing of the supply and the rates of the electricity per unit time it is very important to have a proper energy consumption statistics.

There is no proper statistics about the demand and supply of electricity to date. And this paper will briefly explain, what are the factors that need to be considered while statistically analyzing the use of electricity in any household.

There are two main motivations that have led us to study the electric load pattern consumption. The first and foremost is, electric load pattern consumption helps in the forecasting the electric use which in turn plays a vital role in the modern power systems. The second important motivation to study the electric load pattern consumption is, to make sure that the smart power grids that are designed to calculate the electric consumption can be used more effectively.

2   Related Works

The work of paper [1] propose a cross analysis of some existing methods capable of building up a residential electric load curve. In this paper the author reviewed and build a classification of all the works related to simulating the residential electric load curve. According to the authors the methods have been divided into two categories top-down and bottom-up approaches.

Many research papers have conducted various techniques to analyze the factors that influence the consumption. In the paper [8], the author had collected data from households as a part of the Energy Follow-Up Survey (EFUS) commissioned by the Department of Energy and Climate Change Householders by an interview survey. Another paper [4], the data analyzed was provided by the Bureau of Energy (BOE). This paper also compares the sectors other than household sector, but for our paper, we would require only the results from the household sector. In another paper [9], the author only discusses about the factors influencing the energy consumption in the power plants. Since, the factors are more or less the same as the household, we have used this paper just to finalize our conclusions for this topic. So, the above papers, forms the basis of our discussion further in this chapter. The list below summarize the influencing factors:

- **Impact of Socio Demographic factors.** The size of a household [2,3,5,10], Number of occupants [8], And the age of occupants [8].
- **Impact of Appliances and Lighting.** Number of appliances [3,13], And type of appliances [8].
- **Impact of Weather Conditions.** Temperature and Climate play other important factor in consumption [6,7].

The paper [12] propose a solidly different approach to simulation the electricity consumption by applying Artificial neural networks (ANNs) to prediction of the energy consumption of a passive solar system. Unfortunately, the method of prediction is constrained by having a massive amount of data about the consumption values in relation to all input variable which is not valid case in non-controllable environment. In this paper we try to introduce a general simulation bottom-up model by separation the simulation core form factors that influence the consumption like occupancy, appliances, lighting and Other fracture like Socio Demographic factors.

3 Load Curves

It is a graphical representation of the electric load consumption of any appliance with respect to time. i.e. it represents the power that is consumed by a appliance or group of appliances in a particular time frame.

There are various factors that influence the electric consumption of the house which then impacts the load curve:

– **User of electricity.** Who is the user of the electricity has a great influence on its consumption. Example: Residential, industrial, agriculture, etc.
– **Time of the day.** The load consumption curve has some peaks during the day. Also the consumption is different on the weekdays and the weekends.
– **Environmental factors.** Electric consumption also depends on the ambient temperature. Usage of heating, cooling and ventilation depends completely on the climate.

The appliances load curves sums up to construct the load profile of the house with small error can be ignored. The different information we get by studying a load curve are:

– The peak in the load curve represents the maximum power that is consumed at a specific time.
– The total area under the curve can be used to estimate the average load i.e. average power consumed in a certain amount of time.

$$AverageLoad(kW) = \frac{AreaUnderTheLoadCurve(kWh)}{no.ofhours(h)} \tag{1}$$

4 Occupancy Model

The availability of members at different hours of the day and the activities carried out by them during the course of the day results in the electric consumption per day which sums up to give us the total electric consumption of a household. The occupancy model of any household used in our simulator depends on various factors such as:

– Number of the members in the house,
– Number of working and non-working members in the house,
– Number of hours the members in the house are active or inactive,
– Age of members in the house.

Each factor mentioned plays a vital role in the electric consumption of each family.

5 Appliance Operational Modes and Activity Ontology

Activity ontology means the relation between the activities performed by the members in the house and the appliances that are associated with that activity. Every activity that is carried out in the house accompanies with it use of some of the appliances.

Appliance that is used in an household has some operational modes associated with it. Operational mode of any appliance is the detail about the working of that particular appliance like for example, the start time of the appliance, the end time of the appliance usage, the total duration for which appliance was used, the power that is withdrawn by the appliance per unit time, etc.

5.1 Classification of Appliances

According to Smart Homes and their users by Hargreaves and Wilson, a particular activity can have one or even more than one appliance associated with it. At a given time, any given appliance can definitely, possibly or indirectly indicate that a particular activity is taking place. Thus, depending on their behaviour, the appliances are classified into three types [14]:

- **Marker Appliances.** These are the appliances which when used, indicates that the activity is definitely happening.
- **Auxiliary Appliances.** These type of appliances when used, indicate possibility of the activity taking place.
- **Associated Appliances.** There are appliances which are marker for certain activity and are indirectly linked to some other activity. Such appliances are called associated appliances.

There are certain appliances which are on even when there is no activity performed or if the person is sleeping. Such appliances come under Type-III i.e. the appliances which exhibit infinite steps and continuous loop range of power, e.g. Refrigerator, HVAC appliances, etc. So this appliances cannot be categorized into any of the above classification. Thus, the appliances which are switched on continuously have no dependency on the activities that are performed.

Appliances Usage Frequency: We conducted a survey for few families and asked them about the number of days they use some of this appliances and based on statistical analyses of this surveys we extract a frequency table which gives us an inside of how are some of the common appliances used.

We have broadly classified the appliances into three categories namely,

- **Daily:** This means that, those appliances were used almost everyday of the week including the weekends.
- **Often:** The frequency of use of the appliances is nearly 4–5 days a week is termed to be used often.
- **Rarely:** The appliances which are hardly used once or twice in a week or two weeks are said to have rare frequency.

Table 1. Activity ontology: relations between the activities and the reflected appliances

Activity type	Appliances used
Morning course	Electric shower
Getting ready	Hair straightener
	Hair dryer
Cleaning home	Vacuum cleaner
Cooking	Microwave
	Oven
	Induction plates
	Electric cooker
	Microwave
Coffee making	Coffee maker
	Kettle
Toasting bread	Toaster
Dish washing	Dish washer
Washing clothes	Washing machine
	Clothes dryer
Ironing clothes	Iron
Watching TV	Television
	Set top box
Watching movie	DVD player
	CD player
Computing	Desktop
	Laptop
	Monitor
Printing	Printer
	Scanner
Telephoning	Mobile charger
	Landline telephone
Heating and ventilation	Electric space heating
	Storage heater

5.2   Classification of Activities

We have made few assumption of what can be different activity types and what are generally the appliances that are associated with that particular activity type. And also, when you go through this section you will see that, we have made assumptions regarding the activity inter-dependencies i.e. which activities can be performed or in general are performed together.

All the assumptions are based on the statistical analyses of this surveys we conducted. Table 1 shows few of the activity types and the appliances associating those activities which used in designing our simulator. In the Table 1, you can observe that we have tried segregating the activities to very basic level so as to make sure that while designing the simulator for this we can precisely chose the activity and the appliances and to have less error factor.

It is very important to consider if multiple activities are carried out at a time. For our project we have made certain assumptions about which activities are often not performed simultaneously and which activities can be performed at a time. Figure 1 show the interdependent between activities.

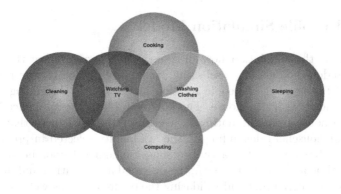

Fig. 1. Activities inter-dependency graph, the activities which can performed simultaneously.

5.3   Period of Activity

Each activity that is performed during the day has some start time and end time. We have taken into consideration few of the activities and with the help of the survey we have approximated the times at which those particular activities are generally performed. But, as we know that, not everything is done at the same time everyday. So we have considered respective deviation and the mean of time of our activity performance and considered it to be a standard normal distribution.

Table 2 shows few examples of the daily activities and what is the normal start, end and standard deviation. This table is constricted by conducting a survey and in the in our simulation this numbers are automatically calculated from the data input of the activities.

Table 2. Examples of appliances start and end time, with there relevant standard deviation. This table is constricted by conducting a survey

Activity	Start time	End time	Standard deviation
Cooking	20:00 h	22:00 h	30 min
Washing	10:30 h	11:30 h	2 h
Cleaning	12:00 h	13:30 h	1 h
Entertainment	19:00 h	22:00 h	3 h
Computing	20:30 h	21:30 h	20 h

6 Load Profile Simulation Model

Figure 2 show the simulation flow diagram. The diagram show the steps for creating random household be selecting availability randomly (the presence of people and the number of people residing in the house) depending upon the number of members in the house. The next step select the appliances that are used in the house from the database. We select the appliances that are used in particular household depending on the availability selected in previous step. In the next step we give the activities some random behavior be shifting the there start time and end time depending on standard deviation and mean time. The next step is correcting and validating the entered data as well as validating the data taken from the database and making sure that the activities have no overlapping or some appliances has been chosen more than once. The last two steps are simulating the total load profile of the house or total community load profile. This two steps will be explained in more details.

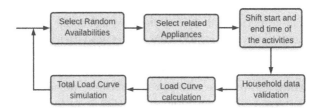

Fig. 2. General household consumption profile simulation flow diagram

6.1 Bottom-Up Approach for Modeling

Figure 3 illustrates the Bottom-Up approach for model used in our project to calculate the load profile of a household and generate a simulation. The simulator starts by creating the occupancy and active occupancy arrays. The occupancy array represent the number of people in the household in any time of the day.

While, the active occupancy array represent the number or active people in the household (people which are not sleeping). Active occupancy array is used as input for to lighting simulator. Lighting simulator is used to study the load consumed by the household due to the lamps present in the house and depending upon the active hours of the member, the simulator decides whether the light is used or not. The lighting simulator has been implemented with help of the research paper of Domestic lighting demand model by Richardson, Ian Thomson, Murray [11]. And the data for lighting has been collected from 100 houses with different bulbs set in this paper. After the lighting is simulated the simulator enter the curve generation loop which will be responsible of calculated the electric load for each time of the day. Along with the output of the lighting simulator, the other input that is given to the load curve generation loop is the set of active appliances that needs to be defined for comparing the activities that are performed at particular time/moment of the day along with that the operational modes of the particular appliances needs to be taken into account and should be continuously effective till the activity is carried out. The load curve generation loop then makes use of active occupancy model together with measured lighting data which is a dynamic inputs as well as the appliance data.

Finally, the household load at each moment and the load curves of the appliances which are switched ON continuously (e.g. refrigerator) are all summed up and the final load profile of the households taken into consideration is simulated.

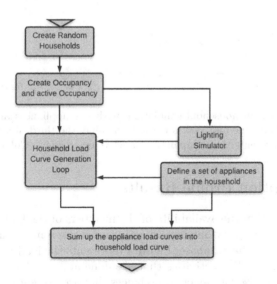

Fig. 3. Bottom-up approach for modeling of the household consumption profile simulator

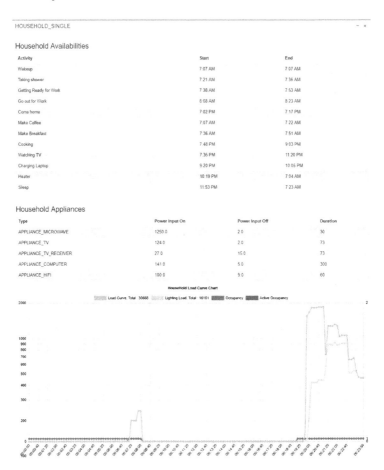

Fig. 4. A single person household simulation, with five appliances and 12 activities, x-axes represent time line and y-axes is the load value, the third dimension with pink and red colour represents the occupancy and active occupancy. (Color figure online)

7 Consumption Profile Results

During the simulation the availability of the members of the house and the time of the day when they carry out certain activities is taken into consideration and the data is stored in the database. Figure 4 represents the load profile of a house with a single person. The activities that are done and at what time of the day are considered and also the general appliances that are present in a single person house is taken into consideration. We can see in Fig. 4 that the load consumed has two peaks one in the morning and other at night. The load curve has a small peak in the morning due to the morning course the person carries out but is comparatively very less when we see the load consumption after 19:00 h. This is because we have assumed the single person to be working and we consider that the person is out of house from 8:00 am in the morning till 7:00 pm in the

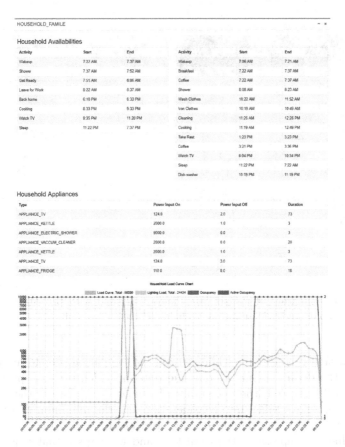

Fig. 5. A family household simulation, with two resident seven appliances, x-axes represent time line and y-axes is the load value, the third dimension with pink and red colour represents the occupancy and active occupancy. (Color figure online)

evening. The load profile is highest in the evening from 7:00 pm to 11:00 pm. i.e. because the person has his multiple activities such as cooking, watching TV, computing, etc carried out at those time.

The lighting simulation is seen by the yellow line in the simulation graph and we can see that the light is used only when the person is home in the evening. Thus, cumulatively we can say that in a single member house with a working member, we have maximum load consumption at the evenings. Figure 5 represents the load profile of a family house. We see that there are peaks in the load curve at particular hours of the day when all the members are active and are at home. The occupancy is considered to be maximum in the hours when all the members are at home that is in the morning before 8:00 am and after 6:00 pm in the evening.

Figure 6 show the total simulation of the number of sex households selected with the percentage of singles and families represents the total load profile of the

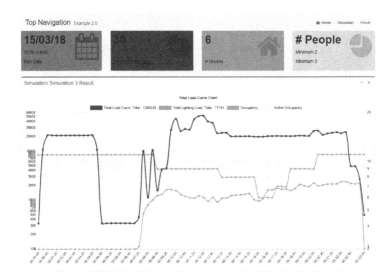

Fig. 6. Total simulation of the six households generated with probability of 0.3 for singles households and 0.7 families households. (Color figure online)

community that is considered while starting the simulation. The total simulation represents the details about the day and date on which the load profile of the houses is to be studied, the percentage of singles in the house, the number of houses that we have taken into consideration and the minimum and maximum people that the user have assumed while starting the simulation.

The blue line represents the total load and we can see that it is minimum in the mornings when the members are sleeping and night after 10:40 pm when people generally start going to sleep. The pink line explains us the occupancy which can be seen maximum from night 8:00 pm to morning 8:00 am when generally all the members are home. Active hours of the members is explained by the red line which is minimum when working people are generally out of home. Yellow line with lighting simulation can be seen constantly increasing over the course of the day. But again, the lighting depends a lot on the weather but we have not considered weather in our simulation.

8   Conclusions

This work has discussed a lot about the factors that influence the electric load pattern of residential areas and a simulator is designed and developed to generate the load profiles of the households. There are multiple factors that impact are day-to-day activities which in turn has a larger impact on the electricity consumption. Electricity is a very important resource for mankind and is one resource which will soon deplete if not taken care and used efficiently. To make use of this important resource efficiently its necessary to plan and have proper statistics as well as to plan the rate of per unit electricity at different hours of the day. So, all the factors and parameters discussed in this paper play important role in studying the electric load consumption pattern. We have considered a lot of parameters which are somehow important while designing the load simulator of any house for a day, month or year. But, we could not take all the situations and factors in actually implementation as of the time and the complexity. So, having a simulator with all the factors considered in it like the age of the members in the house, the income of family, weather conditions is what is the future scope of this work.

References

1. Grandjean, A., Adnot, J., Binet, G.: A review and an analysis of the residential electric load curve models. Renew. Sustain. Energy Rev. **41**, 6539–6565 (2012)
2. Bartusch, C., Odlare, M., Wallin, F., Wester, L.: Exploring variance in residential electricity consumption: household features and building properties. Appl. Energy **92**, 637–643 (2012). http://www.sciencedirect.com/science/article/pii/S030626191100256X
3. Bedir, M., Hasselaar, E., Itard, L.: Determinants of electricity consumption in Dutch dwellings. Energy Build. **58**, 194–207 (2013). http://www.sciencedirect.com/science/article/pii/S0378778812005257
4. Chen, Y.T.: The factors affecting electricity consumption and the consumption characteristics in the residential sector-a case example of taiwan. Appl. Energy **9**, 1484 (2017)
5. Cramer, J.C., et al.: Social and engineering determinants and their equity implications in residential electricity use. Energy **10**(12), 1283–1291 (1985). http://www.sciencedirect.com/science/article/pii/0360544285901392
6. Grandjean, A., Binet, G., Bieret, J., Adnot, J., Duplessis, B.: A functional analysis of electrical load curve modelling for some households specific electricity end-uses. Appl. Energy **hal-00770135**, 24 p. (2011). https://hal-mines-paristech.archives-ouvertes.fr/hal-00770135
7. Hobby, J.D., Tucci, G.H.: Analysis of the residential, commercial and industrial electricity consumption. In: 2011 IEEE PES Innovative Smart Grid Technologies, pp. 1–7. IEEE, November 2011
8. Huebner, G., Shipworth, D., Hamilton, I., Chalabi, Z., Oreszczyn, T.: Understanding electricity consumption: a comparative contribution of building factors, sociodemographics, appliances, behaviours and attitudes. Appl. Energy **177**, 692–702 (2016). http://www.sciencedirect.com/science/article/pii/S0306261916305360
9. Jones, H.B., Lee, S.R.: Factors influencing energy consumption and costs in broiler processing plants in the south. J. Agric. Appl. Econ. **10**(2), 63–68 (1978). https://doi.org/10.1017/S0081305200014382
10. Ndiaye, D., Gabriel, K.: Principal component analysis of the electricity consumption in residential dwellings. Energy Build. **43**(2), 446–453 (2011). http://www.sciencedirect.com/science/article/pii/S0378778810003592
11. Richardson, I., Thomson, M., Infield, D., Delahunty, A.: Domestic lighting: a high-resolution energy demand model. Appl. Energy **41**, 781–789 (2009)
12. Kalogirou, S.A., Bojic, M.: Artificial neural networks for the prediction of the energy consumption of a passive solar building. Energy **41**, 479–491 (2000). https://doi.org/10.1016/S0360-5442(99)00086-9

13. Wiesmann, D., Azevedo, I.L., Ferrão, P., Fernández, J.E.: Residential electricity consumption in Portugal: findings from top-down and bottom-up models. Energy Policy **39**(5), 2772–2779 (2011). http://www.sciencedirect.com/science/article/pii/S030142151100139X
14. Wilson, C., Hargreaves, T., Hauxwell-Baldwin, R.: Smart homes and their users: a systematic analysis and key challenges. Pers. Ubiquit. Comput. **19**(2), 463–476 (2015). https://doi.org/10.1007/s00779-014-0813-0

Anomaly Detection Procedures in a Real World Dataset by Using Deep-Learning Approaches

Alabbas Alhaj Ali[1]($\boxtimes$), Abdul Rasheeq[1], Doina Logofătu[1], and Costin Bădică[2]

[1] Department of Computer Science and Engineering,
Frankfurt University of Applied Sciences, 60318 Frankfurt am Main, Germany
{alabbasa,rasheeq}@stud.fra-uas.de, logofatu@fb2.fra-uas.de
[2] Department of Computer Sciences and Information Technology,
University of Craiova, 200285 Craiova, Romania
costin.badica@software.ucv.ro

Abstract. Water covers 71% of the Earth's surface and is vital for all known forms of life. Quality of drinking water is very important. The concentration of major chemical elements under the desirable limit is good for health but an increase in the concentration of the element above the desirable limit may cause adverse effects on human health. Major problems being faced by the world population are due to the presence of excess fluoride, sulfate, chloride, nitrate, and sodium in water. In this paper, we address the problem of changes in the drinking water quality and the crucial task for public water companies to monitor the quality of water. Requirements for drinking water quality monitoring change frequently, e.g., due to contamination by civilization itself or in the supply and distribution network. The proposed methods are *K-Nearest Neighbour Algorithm (KNN)* and *Classification Neural Network based on Logistic Regression* for obtaining an appropriate solution in an adequate period of time. Also, the paper compares of the result between the proposed methods and other methods applied in previous work. All experiments are carried out using data gathered from Thüringer Fernwasserversorgung (TFW) water company.

Keywords: Drinking water quality · Data analysis ·
Neural network · Logistic regression · K-Nearest Neighbour (KNN)

1 Introduction

While many parts of the world face major challenges due to limited freshwater availability, a significant amount of the limited freshwater resources in the world are contaminated by pollutants from industry, farming, energy generation, and other activities. Due to water contamination and improper conduct of water treatment, a lot of people suffer from chronic water born diseases and this can even cause death if the patient is not subjected to proper treatment on time.

© Springer Nature Switzerland AG 2019
N. T. Nguyen et al. (Eds.): ACIIDS 2019, LNAI 11431, pp. 303–314, 2019.
https://doi.org/10.1007/978-3-030-14799-0_26

The goal of the work is to develop capable procedures for detection in time series of drinking water composition data. Precise detection of changes in water quality is a crucial task for public water companies and an urgent requirement for a timely reaction to these changes. An adequate and accurate alarm system that allows for early recognition of all kinds of changes is a basic requirement for the provision of clean and safe drinking water [1].

KNN is a non-parametric learning algorithm [3]. It's a method based on which the data are separated into various classes in a database to predict the classification of a new dataset [3,6]. In this paper we will used KNN algorithm to evaluate the F1 score of different parameters on real word water dataset. Different values of parameters have been tried on each algorithm to ensure that the experimental results faithfully reflect the performance of the algorithms. Also the paper evaluate the significance of data visualization and data preprocessing results of ANN Multi Layer Neural. Finally comparing the evaluation of KNN in contradiction with ANN.

2   Problem Statement and Motivation

In this paper, we address the problem of changes in the drinking water quality and the crucial task for public water companies to monitor the quality of water. This is exactly an NP-complete set cover problem [7]. For the monitoring of the water quality, the Thüringer Fernwasserversorgung performs measurements at significant points throughout the whole water distribution system, in particular at the outflow of the waterworks and the inflow and outflow of the water towers. For this purpose, a part of the water is bypassed through a sensor system where the most important water quality indicators are measured. The data has been measured at different stations near the outflow of a waterworks. The data provided consists of one time series of water quality data [1]. The data contains one time series denoting the water quality and operational data. Given is the amount of chlorine dioxide in the water, its pH value, the redox potential, its electric conductivity and the turbidity of the water [1]. These values are the water quality indicators, any changes here are considered as events. The flow rate and the temperature of the water are considered as operational data, changes in these values may indicate changes in the related quality values but are not considered as events themselves. Table 1 gives an overview of the data provided. Where *EVENT* is a boolean value "TRUE" denotes significant change in the quality of water while "FALSE" denotes non-significant change in the quality of water.

As mentioned above in the context of data to be considered as an event or not, we can place this problem under Classification [2] Predictive Problem of two classes (Event or not an Event). We are going to denote these two classes with *TRUE or FALSE*. In the other hand in KNN Algorithm, the class probability estimation is based on majority voting.

Table 1. Description of the given time series data from Thüringer Fernwasserversorgung water company [1].

Column name	Description
Time	Time of measurement
Tp	The temperature of the water, given in $\circ C$
Cl	Amount of chlorine dioxide in the water, given in mg/L (MS1)
pH	PH value of the water
Redox	Redox potential, given in mV
Leit	Electric conductivity of the water, given in $\mu S/cm$
Trueb	Turbidity of the water, given in NTU
Cl_2	Amount of chlorine dioxide in the water, given in $mg/L(MS2)$
Fm	Flow rate at water line 1, given in $m3/h$
Fm_2	Flow rate at water line 2, given in $m3/h$
EVENT	Marker if this entry should be considered as a remarkable change resp. event, given in Boolean

3   Related Work

In previous contributions were used LDA *(Latent Dirichlet Allocation)*, SVM *(Support Vector Machine)*, ANN *(Artificial Neural Networks)* and LR *(Logistic Regression)* to find the best F1 score for the classification problem [4]. These literature have been using Boruta Random Forest and Recursive Feature Selection with Random Forest as feature selection algorithms.

In [11] the authors provide an overview of a big data set imputation by applying several different algorithms under a missing completely at random (MCAR) assumption. The evaluation criteria here was the Root Mean Squared Error (RMSE) and the experimental results show that the Random Forest method can be very useful for missing values imputation.

A general data driven approach to construct an automated online event detection system is proposed in [12]. The approach uses various tree ensemble models, especially gradient boosting methods overcome challenges in time series data imbalanced class and collinearity and provide satisfactory predictive performance.

Even the best classification algorithm selected to perfectly fit a given data set was not efficient enough without Data Visualization and Data Preprocessing. Table 2 shows the result of previous work done on detecting an anomaly in the quality of the water [4]. We can observe from the results that the best F1 score is gained using Logistic Regression Algorithm but we still have the prediction failure of more than 40% of the cases. While logistic regression is more sensitive to outliers and logistic loss function do not go to zero even if the point is classified sufficiently confidently we expect SVM to perform marginally better than logistic regression, but as our dataset is imbalanced data the SVM typically we could change the mis-classification penalty per class. The mis-classification penalty for the minority class is chosen to be larger than that of the majority class [10].

Table 2. F1 score of previous work on anomaly detection of water quality [4].

Alg	TP	FP	TN	FN	RESULT
LogRe	795	187	120407	945	0.58
LDA	603	17427	103167	1136	0.06
SVM	487	9707	111599	541	0.08
ANN	242	24392	96914	786	0.01

In this paper, our goal is to present improvements of the logistic regression algorithm for anomaly detection in water quality by applying it in Multi-Level Neural Network and by doing data visualization and data pre-processing.

4   Data Preprocessing

Data preprocessing is an essential step in data analysis. Any attempt to train a prediction or classification algorithm without a real understanding of the nature of data is doomed to failure. Data preprocessing include visualization the data, selecting variables, removing outliers and data scaling.

4.1   Variable Selection

In the statistical analysis the selection of independent (predictor) variables in a regression model that might influence the outcome variable is an important task. According to the problem description, the influencing variables are (Tp, Cl, pH, Redox, Leit, Trueb, Cl_2, EVENT) the number of variables are small and each of then has a significant role in the EVENT trigger. While Cl and Cl_2 are two variables which describe the amount of chlorine dioxide in the water given in mg/L and the each play the same role EVENT being triggered. So to reduce the imbalance effect of them decided to merge these two variable result and take the average value as one variable. To transform the problem into a classification one we encode the *EVENT* variables to two variables *TRUE* and *FALSE*.

The dataset is time series data but the event triggering is only passed on the current value of the sensor which make it possible to sample the data to train the neural network is randomly with 60% of the input data. The rest of the data is used for validating the neural network model.

4.2   Removing Outliers

The dataset contains a parentage of NAs values. This values should be removed. Also, we have some extreme values. All the NAs and extreme values should be

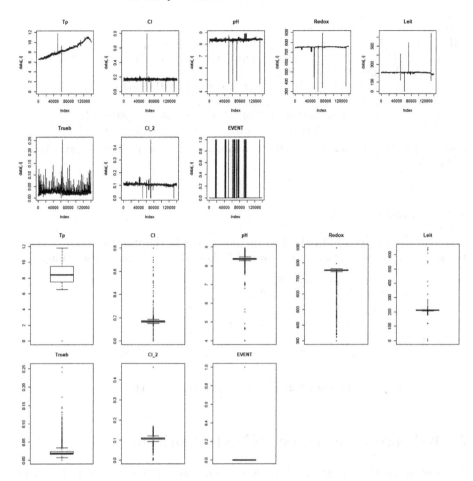

Fig. 1. Data plots and Box-plot of variables Tp, Cl, pH, Redox, Leit, Trueb, Cl_2 and EVENT.

handled carefully before training any prediction module. Figure 1 shows a plot of the data set after removing NAs values. We can easily detect which variables have the most extreme values. We find out that removing outliers from (Cl_avg, Redox, Leit and Trueb) variables should be applied before fitting the our models.

The outliers for Cl_avg and Leit are defined as 1.5 * IQR, where IQR *The Inter Quartile Range* is the difference between 75th and 25th quartiles. Outliers for Trueb are the values less than 0.004 and more than 0.062. And outliers for Redox are the values less than 700 and more than 800. Threshold values for Trueb and Redox are driven from data visualization and problem definition provided [4].

```
1 data <- data[complete.cases(data),]
2 qnt_Cl <- boxplot.stats(data$Cl_avg)$stats
3 H_Cl <- 1.5 * IQR(data$Cl_avg, na.rm = TRUE)
```

```
qnt_Leit <- boxplot.stats(data$Leit)$stats
H_Leit <- 1.5 * IQR(data$Leit, na.rm = TRUE)
data <- data[(data$Cl_avg > (qnt_Cl[1] - H_Cl) & data$Cl_
    avg < (qnt_Cl[5] + H_Cl)),];
data <- data[(data$Leit > (qnt_Leit[1] - H_Leit) & data$
    Leit < (qnt_Leit[5] + H_Leit)),];
data <- data[(data$Redox > 700 & data$Redox < 800),];
data <- data[(data$Trueb > 0.004 & data$Trueb < 0.062),];
```

Listing 1.1. R code to remove outliers from data variables (**Cl**, **Redox**, **Leit**, **Trueb** and **Cl_2**)

4.3   Scaling the Data

Scale the data helps to standardize the predictors in the interval [0-1]. It helps the model to converge faster. The code 1.2 show R code of our data scaling.

```
max = apply(data[,1:6], 2,
function(x) max(x, na.rm=TRUE));
min = apply(data[,1:6], 2,
function(x) min(x, na.rm=TRUE));
data[,1:6] = as.data.frame(
scale(data[,1:6], center = min, scale = max - min)
);
```

Listing 1.2. R code to scale the data set

5   K-Nearest Neighbour (KNN) Algorithm

The output of KNN is the optimal K of each training data-set [8,9]. K is the number of the fewest nearest neighbors a training data-set has to identify to get its correct class label. KNN uses standard Euclidean distance [3] to measure the difference or similarity between training and test instance. In our experiment, the value of K may be from 1 to 371 (approx. square root of the total number of observations).

When several of the K-nearest neighbors share the same category, then the per-neighbor weights of that category are added together, and the resulting weighted sum is used as the likelihood score of candidate categories. A ranked list is obtained for the test document. By thresholding on these scores, binary category assignments are obtained [6].

Figure 2 shows an example of modeling result of the KNN model for category TRUE. The number of centroids for a category depends on the dataset given for training.

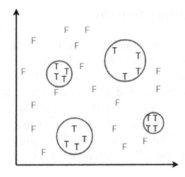

Fig. 2. An example representation of category "TRUE". Classification of TRUE for K = 4 based on K-Nearest Neighbour Algorithm.

5.1 Experiment and Evaluation

We have conducted experiments on the training data provided by Thüringer Fernwasserversorgung (TFW) [1]. Experimental results reported in this section are based on the so-called F1 - value, the harmonic mean of precision(P) and recall(R).

In plot Fig. 3a below, we observed that the majority of *TRUE* and *FALSE* are overlapping at a certain area of the plot. Whereas in Fig. 3b, the overlapping of points is comparatively less. Hence, the prediction based on the majority voting can be done efficiently.

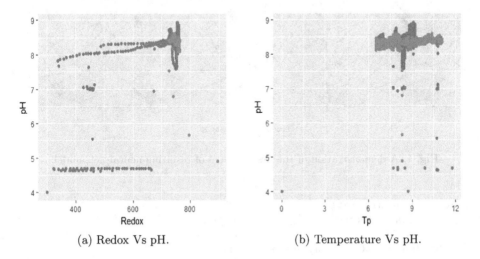

(a) Redox Vs pH.   (b) Temperature Vs pH.

Fig. 3. The overlapping between the majority of *TRUE* and *FALSE*, comparison of pH of water with redox potential and temperature.

6 Multi Layer Neural Network

Neural networks are one of the most powerful existing learning algorithms. A multi-layer neural network has one or more hidden layers along with the input and output layers, each layer contains several nodes that interconnect with each other by weight links. The number of node in the input layer will be the number of attributes in the data-set, nodes in the output layer will be the number of classes given in the data set. Each node in the network consists of a summation function of the weight and the basis value. This summation is applied to the activation function of the node. The activation function defines its output [5].

6.1 Training the Neural Net *(Fit a Model)*

We created a neural network model which has 1 input layer with 6 nodes *(one for each variable)*, 7 hidden layers *(the first 6 layers has 6 nodes each the last hidden layer has 4 nodes)* and one output layer with two node representing prediction results. Figure 4 shows the resulting neural net. The listing 1 shows the use of liberally "NEURALNET" to fit our neural network model.

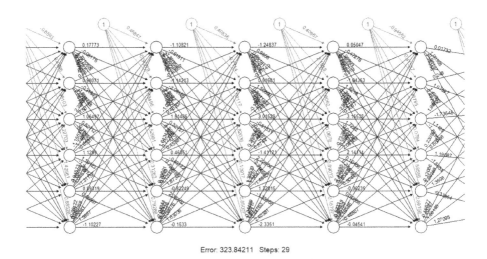

Error: 323.84211 Steps: 29

Fig. 4. A demonstration of the first 5 layers of resulting neural net model.

```
neNet = neuralnet(
  formula = form,
  data = data,
  hidden = c(6, 6, 6, 6, 6, 6, 4),
  act.fct = "logistic",
  linear.output = FALSE,
  lifesign = "minimal")
```

Listing 1.3. Fit the neural net model use of liberally "NEURALNET" in R code

Figure 4 demonstrate the first 5 layer of the result neuralnet.

7 Validation and Results

In our experiments, we used Multi-layer Neural Network and KNN Algorithm methods to evaluate the F1 score of different algorithms on the above data-set. Different values of parameters have been tried on each algorithm to ensure that the experimental results faithfully reflect the performance of the algorithms. The value of K for KNN algorithm includes 5, 10, 15, 25, 35, 45, 50, 55, 65, 80, 90, 100, 150, 200, 371. Normally, the value of K is the square root of the number of data items in the data set. We have 138521 data items in the data set and the square root is 371 (approx.). The experimental results for KNN method are listed below in Fig. 5a and b. The "RESULT" column corresponds to the "F1 Value". We can compare the F1 value achieved by different approaches.

```
> Recall<-tp/(tp+fn)
> Precision<-tp/(tp+fp)
> F1value<- 2*((Recall*Precision)/(Recall+Precision))
> Accuracy<- (tp+tn)/(tn+fp+tp+fn)
>
>
> Recall
[1] 0.9582851
> Precision
[1] 0.9957857
> F1value
[1] 0.9766755
> Accuracy
[1] 0.9994297
>
```

```
> completeResult
      SUBMISSION  TP FP      TN   FN   RESULT
1 ABDUL RASHEEQ 659 71 136724 1067 0.536645
>
```

(a) Evaluation-KNN Algorithm.   (b) Evaluation logistic regression model.

Fig. 5. KNN method experimental results in compare to logistic regression model applied to same reprocessing functions.

7.1 Cross Validation and Prediction

We have evaluated our neural network method by performing *k-fold cross validation*. Using the evergreen 10 fold cross validation with train and test sets splits the data into 90% training set and 10% test set. The result of cross-validation shows 98.8% accuracy in prediction.

After executing the algorithm for many times, we observed that training time for the neural network is conveniently fast enough. For 90% of the cases it was less than 30 s, but it also exceeded 14 min in 10% of the cases.

```
1    pr.nn <- compute(neNet, dataset)
2    # Extract results
3    pr.nn_ <- pr.nn\$net.result
4    if(pr.nn_[1] > pr.nn_[2]){ return (FALSE);}
5    return (TRUE);
```

Listing 1.4. R code for computing prediction of events

```
Cell Contents
|-----------------------|
|                     N |
|         N / Row Total |
|         N / Col Total |
|       N / Table Total |
|-----------------------|
```

Total Observations in Table: 138521

```
                          | trainingData_test_pred
trainingData_test_labels  |    FALSE  |     TRUE  | Row Total |
--------------------------|-----------|-----------|-----------|
                   FALSE  |   136787  |        8  |   136795  |
                          |    1.000  |    0.000  |    0.988  |
                          |    0.999  |    0.005  |           |
                          |    0.987  |    0.000  |           |
--------------------------|-----------|-----------|-----------|
                    TRUE  |       73  |     1653  |     1726  |
                          |    0.042  |    0.958  |    0.012  |
                          |    0.001  |    0.995  |           |
                          |    0.001  |    0.012  |           |
--------------------------|-----------|-----------|-----------|
            Column Total  |   136860  |     1661  |   138521  |
                          |    0.988  |    0.012  |           |
--------------------------|-----------|-----------|-----------|
```

Fig. 6. Cross table evaluation - KNN algorithm.

Figure 6 shows the cross evaluation of the KNN classification.

In prediction function, we reselect the variables. Any incomplete case is considered as *FALSE* production. Any outliers are considered as *TRUE* production. Otherwise, the case is scaled and then the model classifies the case and then we predict an event to be *TRUE* or *FALSE* according to max class probability result.

7.2   Results

We ran our detection algorithm over the data set provided by Thüringer Fernwasserversorgung water company [1]. We observed the following results for F1 score on data set from 2018 in Table 3 below. Where Tables 4 and 5 show the F1 score of running the algorithms on the two different data-set set from 2017. We observe an considerable enhancement in detecting the change in the drinking water quality. For dataset 2 from 2018 [1], we achieved F1 score of 0.851 and 0.976 by using Neural Network and KNN algorithm respectively. Considering the FN, the best result was achieved by KNN Algorithm. F1 score varies for different FN values.

Table 3. F1 score results on data set from 2018.

Model	TP	FP	TN	FN	F1 score
Neural network	1574	403	137437	152	0.851
KNN algorithm	**1653**	**8**	**136787**	**73**	**0.976**
Logistic regression	659	71	136724	1067	0.536

Table 4. F1 score results on data set one from 2017.

Model	TP	FP	TN	FN	F1 score
Neural network	951	87	120507	789	0.684
KNN algorithm	**967**	**6**	**121291**	**77**	**0.954**
Logistic regression	578	53	121021	822	0.534

Table 5. F1 score results on data set two from 2017.

Model	TP	FP	TN	FN	F1 score
Neural network	941	1	241899	1827	0.507
KNN algorithm	**969**	7	**243605**	**87**	**0.975**
Logistic regression	578	55	243208	827	0.537

8 Conclusion and Future Work

In this paper, we ran an investigation of a major approach used for estimation and prediction in monitoring drinking water quality under the framework of real-time data based learning - Neural Network with Logistic Regression and KNN algorithm model. The resolution of the model is depending on the data set preprocessing functionality. Visualization and understanding the data resolve in a good definition of outlines is a cornerstone of data preprocessing. By using the Neural Network with Logistic Regression, we achieved better results in detecting the anomaly in drinking water quality. All experimental results show that our proposed KNN model-based approach outperforms the Neural network with Logistic regression model approach in terms of F1 score. The FN value has a large effect on the F1 score. Each model has its limitations and advantages. Further investigation is required on how to improve the accuracy of data that falls marginally outside the regions of representatives.

Also we discussed the advantages of using neural network over traditional machine learning algorithm like LDA *(Latent Dirichlet Allocation)*, SVM *(Support Vector Machine)*, ANN *(Artificial Neural Networks)* and LR *(Logistic Regression)*. Enhancing the model may be possible by cooperating with other proficiency in the field of water quality to come within a clear definition of what to be considered an event and the boundaries of the variables.

References

1. Rehbach, F., Moritz, S., Chandrasekaran, S., Rebolledo, M., Friese, M., Bartz-Beielstein, T.: Industrial challenge: monitoring of drinking-water quality. In: GECCO (2018)
2. Baobao, W., Jinsheng, M., Minru, S.: An enhancement of K-Nearest Neighbor algorithm using information gain and extension relativity. In: Proceedings of International Conference on Condition Monitoring and Diagnosis, CMD 2008, April 2008

3. Bo, S., Junping, D., Tian, G.: Study on the improvement of K-Nearest-Neighbor algorithm. In: Proceedings of International Conference on Artificial Intelligence and Computational Intelligence, AICI 2009, vol. 4. IEEE Computer Society, November 2009

4. Muharemi, F., Logofătu, D., Andersson, C., Leon, F.: Approaches to building a detection model for water quality: a case study. In: Sieminski, A., Kozierkiewicz, A., Nunez, M., Ha, Q.T. (eds.) Modern Approaches for Intelligent Information and Database Systems. SCI, vol. 769, pp. 173–183. Springer, Cham (2018). https://doi.org/10.1007/978-3-319-76081-0_15

5. Bishop, C.M.: Neural Networks for Pattern Recognition. Oxford University Press, Oxford

6. Yang, Y., Liu, X.: A re-examination of text categorization methods. In: Proceedings of 22nd ACM International Conference on Research and Development in Information Retrieval, SIGIR 1999 (1999)

7. Zeng, D., Gu, L., Lian, L., Guo, S., Hu, J.: On cost-efficient sensor placement for contaminant detection in water distribution systems. IEEE Trans. Ind. Inform. **12**(6), 2177–2185 (2016). ieeexplore.ieee.org/document/7470468/. Accessed 14 June 2018

8. Xiao, X., Ding, H.: Enhancement of K-Nearest Neighbor algorithm based on weighted entropy of attribute value. In: Proceedings of 5th International Conference on BioMedical Engineering and Informatics, BMEI 2012. IEEE Press, October 2012. https://ieeexplore.ieee.org/document/6513101/. Accessed 06 June 2018

9. Jiang, L., Cai, Z., Wang, D., Jiang, S.: Survey of improving K-Nearest-Neighbor for classification. In: Proceedings of 4th International Conference on Fuzzy Systems and Knowledge Discovery, FSKD 2007, vol. 1, August 2007

10. Osuna, E., Freund, R., Girosi, F.: Support vector machines: training and applications. Technical report AIM-1602 (1997)

11. Muharemi, F., Logofătu, D., Leon, F.: Review on general techniques and packages for data imputation in R on a real world dataset. In: Nguyen, N.T., Pimenidis, E., Khan, Z., Trawiński, B. (eds.) ICCCI 2018. LNCS, vol. 11056, pp. 386–395. Springer, Cham (2018). https://doi.org/10.1007/978-3-319-98446-9_36

12. Nguyen, M., Logofătu, D.: Applying tree ensemble to detect anomalies in real-world water composition dataset. In: Yin, H., Camacho, D., Novais, P., Tallón-Ballesteros, A.J. (eds.) IDEAL 2018. LNCS, vol. 11314, pp. 429–438. Springer, Cham (2018). https://doi.org/10.1007/978-3-030-03493-1_45

Developing a General Video Game AI Controller Based on an Evolutionary Approach

Kristiyan Balabanov$^{(\boxtimes)}$ and Doina Logofătu

Department of Computer Science and Engineering,
Frankfurt University of Applied Sciences,
Nibelungenplatz 1, 60318 Frankfurt am Main, Germany
balabano@stud.fra-uas.de, logofatu@fb2.fra-uas.de

Abstract. The field of general intelligence is one, where humans can still easily outperform machines. In the context of our work we describe it as the ability to learn an activity, like playing a game, without any prior knowledge of goals and rules. The agent has to learn by doing/playing and examining the consequences of its actions. Many traditional techniques in reinforcement learning, such as SARSA and Q-Learning, can provide a good solution to this category of problems. In our paper, however, we propose an alternative method based on evolutionary algorithms to overcome the extensive computing for all state-action pairs needed in traditional approaches. We have evaluated various parent selection algorithms and two different fitness functions. "The General Video Game AI Competition" (GVGAI), where contestants submit a playing agent programmed with some learning algorithm to be tested against unknown games, has been used as a benchmark for the performance of our implementation.

Keywords: Artificial intelligence · General intelligence ·
Planning agents · Video game controllers

1 Introduction

Embodying some form of human-like intelligence in computers has been a hot topic for a long time, but still remains a major challenge. Even though the advancements in the field are remarkable in comparison to previous years, modern artificial intelligence (AI) cannot compete with the general intelligence of the average human. Games are an excellent benchmark for testing AI. Numerous video games employ AI to introduce agents against which human players can play. Nevertheless, in most cases the agents possess a so called specific intelligence. In other words their intelligence has been optimized with knowledge of the respective game's rules and best strategies. In a different context, however, the agents would be helpless. Humans on the other hand are much more flexible, and excel in adapting and learning, especially by observation. The General

© Springer Nature Switzerland AG 2019
N. T. Nguyen et al. (Eds.): ACIIDS 2019, LNAI 11431, pp. 315–326, 2019.
https://doi.org/10.1007/978-3-030-14799-0_27

Video Game AI Competition[1] offers programmers to compete against each other by implementing agents, that learn to play completely unknown games by only observing their environment.

A submitted AI agent has to play various video games without prior knowledge about game specifics, such as rules, strategies or long-term resp. short-term goals. The only information the agent can use during the course of the game for its decision-making is the state of its surroundings. In other words the agent has to learn "on the fly" how to best interact with the environment by relying only on its "sight", which, however, covers the entire game field. Games are created using a Video Game Description Language (VGDL) implemented in Python [8], and each offers a unique game experience. The type and count of objects and actions available in each game, as well as the interactions needed to reach a win condition can differ significantly. An agent cannot be "hardcoded" to expect certain information in each game, but rather has to invoke a more general approach when analyzing its surroundings, e.g. moving to resource objects (if any) would probably be beneficial in most games. All games are considered stochastic, i.e. the same controller could perform differently when playing the same level of a game multiple times.

2   Proposed Approach

Many of the top ranked participants in previous years of the contest have used traditional reinforcement learning techniques, such as SARSA and Q-Learning, complemented with various metaheuristics. This approach has proven to be reliable in terms of delivering satisfactory results. It was our desire, however, to implement a controller based on an evolutionary algorithm, which is more innovative, but lacks a proven track record of success, and compare it with the well established and reliable reinforcement learning. Moreover, the amount of objects, actions and possible interactions in the games results in a considerable number of state-action pairs to be evaluated, making SARSA and Q-Learning either too inefficient to be employed in a real-time scenario without an offline learning phase or heavily dependent on metaheuristics.

An Evolutionary Algorithm (EA) is a broad term for any metaheuristic inspired by evolution in nature, hence the name, and more precisely - *natural selection*. Such techniques have gained significant importance in the recent years, especially in the domain of *optimization*, since they can be quite rewarding as an efficient alternative to traditional deterministic methods, more so in very complex and intractable problem scenarios. The main advantage of EAs is that they can deliver a satisfactory solution in a reasonable time, even if the solution space is very big. The basic idea covers the following aspects:

- A system consists of a population of individuals
- Individuals vary in their characteristics (phenotype and genotype) and can reproduce

---

[1] http://www.gvgai.net/.

- There is a strong correlation between the ability to reproduce and an individual's characteristics
- Each iteration a new (better) generation of individuals is created from the old one, which is then fully or partially replaced.

The corresponding terminology used in computer science and engineering is slightly different. An individual is a *candidate solution* to the problem taken from the *solution space*, and it is represented using some *encoding scheme*, e.g. a string of zeros and ones. An individual's probability to reproduce is expressed as its *fitness*, i.e. how well does the solution solve the given problem, calculated with a *fitness function*. New individuals are created from the old ones by utilizing *genetic operators*, which are very similar to their real world equivalents. The idea behind the fitness function is to decrease the population diversity by eliminating poor candidate solutions, but thus improve the overall quality of the candidate solution space (called *survival of the fittest* or *natural selection* in the field of biology). The fitness function is a key element in evolution-based algorithms as it must maintain a proper balance between exploration for new potentially better solutions (population diversity) and exploitation of already known good solutions (population quality). The dynamics of the environment must be considered when modeling the fitness function - a rapidly changing environment would benefit more from exploration than exploitation as known good solutions quickly become obsolete, and vice versa in a static environment, where any found knowledge can be dependably used in the long-term [5].

Even though, this is a simplified abstraction of actual evolution as it happens in nature, it has proven extremely useful in scenarios where: (1) the solution space is too big to fully traverse; (2) a suboptimal solution is sufficient; (3) the solution must be obtained in reasonable time. Since the games are unknown, a specific strategy cannot be implemented. As already mentioned, there are many possible states for the agent to consider in a game. In other words, the given problem, although computationally solvable with finite resources in theory, can be seen as analytically intractable in practice (*solution space is too big*). In addition, many games have multiple win conditions based on the achieved score for instance. An optimal solution can win the game with maximum score, whereas a suboptimal one would just win it. The most significant criteria when controllers submitted to the competition are evaluated is the number of games won (*a suboptimal solution is sufficient*), followed by the achieved score and the time steps needed to win. Finally, each game is played in a maximum of 2000 discrete steps of 40 ms each, meaning that the controller has only limited time to make decisions (a solution is needed fast). It is for these reasons that we think a purely heuristics approach, such as an evolutionary algorithm, can prove quite beneficial. A comprehensive description of EA basics and application are covered in [3,4,6].

3   Implementation Details

We used the framework provided by the competition organizers together with the standard Java class libraries. A local testing environment was created to automate the tests conducted with the controllers. The first step when creating the evocontroller (evolutionary-algorithm-based controller) was to map the context of the given problem to the building blocks of an evolutionary algorithm as explained in Sect. 2, i.e. to design the agent's structure and elements in such a way that it incorporates the aforementioned algorithm. In that matter of thought the following mapping was done:

- **Individual** (candidate solution encoding) - a finite permutation of the agent's available actions for a given game (e.g. shoot, move up, move right etc.)
- **Population** (candidate solution set) - multiple action permutations of a fixed length that represent different strategies to play a given game (e.g. shoot, move up, then shoot again or just keep shooting)
- **Genetic modifiers** (recombination/mutation) - produce a new permutation by combining the elements of several others, i.e. recombination, or randomly alter one or more elements of an existing one, i.e. mutation
- **Fitness** (solution adequateness) - use a provided *forward model* to simulate the game progress when the agent uses a given action permutation and evaluate the reached state.

The central idea of the evocontroller is to initialize a population of candidate solutions at the beginning of the game and let it evolve during each game step, choosing the first element in the most fit solution's permutation at the end of the game step as the agent's next action. The evolving process is split into discrete iterations with each iteration producing a new generation of candidate solutions from the old one. An iteration consists of the following steps: (1) create an empty population with individuals of size s; (2) select n parents from the previous population with probability based on their fitness; (3) while there is enough time and the maximum population size has not been reached create a new child by applying crossover to the selected parents and mutating each element in the

---

Algorithm 1. evocontroller evolution iteration pseudocode

procedure ITERATE(Population $\mathcal{E}$)
   create empty population $\mathcal{E}'$
   select n fit parents $\mathcal{P} \subseteq \mathcal{E}$
   while time available AND children needed **do**
      new child $c \leftarrow$ CROSSOVER($\mathcal{P}$)
      $c \leftarrow$ MUTATE(c, m)
      CALCFITNESS(c)
      $\mathcal{E}' \leftarrow \mathcal{E}' \cup c$
   end while
   SORT($\mathcal{E}'$)
end procedure

Algorithm 2. evocontroller calcfitness pseudocode

---

procedure CALCFITNESS(Individual c, State s)
    state $s' \leftarrow$ COPY(s)
    individual score $f_i \leftarrow 0$
    for $i = 0$ **to** $SimulationDepth$ **do**
        if game not over **then**
            ADVANCE(s', action in c at position i)
            state score $f_s \leftarrow$ EVALUATESTATE(s')
            $f_i \leftarrow f_i + f_s \cdot \epsilon^i$
        else
            return
        end if
    end for
    fitness of $c \leftarrow f_i$
end procedure

---

child's permutation with probability m; calculate the newly created child's fitness and insert it into the new population; (4) sort the new population according to the individuals' fitness. The respective pseudocode of the iteration procedure can be seen in Algorithm 1.

When calculating the fitness of a candidate solution the evocontroller agent makes full use of the provided forward model to simulate the potential changes in the game state after a given action has been executed. The respective procedure's pseudocode is shown in Algorithm 2. It copies the current game state and sequentially executes every action in the candidate solution's permutation and adds the resulting state score, i.e. an evaluation of the consequences, multiplied by a factor $\epsilon < 1$ to account for the stochastic nature of the games. In our experiments ϵ was set to 0.95 and was gradually decreased during the evaluation of an individual due to the increasing uncertainty whether the simulated sequence of events would actually happen.

When deciding which parameters of the evolutionary model to change, two key factors were considered: (1) the game constraints and more specifically the maximum time per step; (2) the maximum number of steps per game. The first did not allow for a large population or deep simulation using the forward model, whereas the latter meant that the population had to evolve fast. Moreover, most of the benchmark games contained a large number of objects and interactions per game step providing for a very dynamic environment in which knowledge obtained by the agent could quickly become obsolete. This meant that emphasis had to be put on exploration rather than exploitation, i.e. making use of high mutation rates, various parent selection schemes, combining multiple parents when creating a child, short agent vision and others.

3.1 Mutation Rate

The rather limited number of individuals per population posed danger of high inbreeding that would gradually reduce the population diversity and prematurely converge to a suboptimal solution [7]. Thus, a high mutation rate of 20% was used to ensure sufficient exploration of the solution space, which meant that on average 2 elements in each new individual's permutation would be changed when it is created. A mutated element is guaranteed to be replaced with a different one, i.e. a move action would never be mutated to a move action.

3.2   Crossover

The crossover operation used, which is the process of combining the parents' genes to create a child, was an (n)-multi-parent crossover - each child gene is randomly selected from one of its n parents, where n was fixed to 3 for all experiments.

3.3   Parent Selection

To further cope with the possible inbreeding issue we concentrated on experimenting with the parent selection process, which is a part of the recombination operation together with the crossover. We employed 3 different selection techniques: *roulette-wheel*, *stochastic universal sampling* and *sigma scaling*. Each exhibits a different selection pressure defined as the ratio of the probability to select the most fit individual and the probability to select an average individual. Selection pressure can be used quite effectively to implement exploration indirectly. In high selection pressure environments more fit individuals are more likely to be selected as parents, therefore passing down their genes, i.e. obtained good knowledge is saved. On the other hand applying lower pressure during the selection process gives less fit individuals better chance to compete, effectively promoting exploration of suboptimal paths [1].

Roulette Wheel (RW). The selection probability $P(x_i)$ of an individual x_i is proportional to the ratio of its calculated fitness $f(x_i)$ relative to the summed fitness of the entire population of size N (refer to Eq. 1). RW applies relatively high selection pressure as very fit individuals are very likely to be chosen as parents and vice versa with very unfit ones [5].

$$P(x_i) = \frac{f(x_i)}{\sum_{j=0}^{N} f(x_j)} \tag{1}$$

Stochastic Universal Sampling (SUS). This is a variation of RW that increases the probability of fit individuals to be selected in scenarios where multiple parents are used to create a child. With RW an individual x_i with fitness $f(x_i) = a$ has the probability $P(x_i) = a$ to be selected or $\bar{P}(x_i) = (1 - a)$ not to be selected. If 4 parents are needed and the selection algorithm is executed 4 times, then $\bar{P}(x_i) = (1 - a)^4$. It is clear that in multi-parent scenarios the possibility to lose the most fit individual rapidly decreases, but it, nevertheless, remains greater than 0. In SUS [2] the selection algorithm chooses n parents with a single execution. If we align all individuals in a row and their summed fitness is f_{total}, then the algorithm need only generate a random number within the range $[0, \frac{f_{total}}{n}]$ in order the determine the first parent. The remaining $n-1$ parents are selected at intervals $\frac{1}{n}$ from it. With SUS an individual x_i with fitness $f(x_i) = a$ would have a probability of not being selected $\bar{P}(x_i) = \max(0, 1 - n \cdot a)$, where the $max()$ operation always returns the greater of the arguments passed to it (need in order to avoid working with negative probabilities). Using proof by induction one can easily verify the validity of Eq. 2, i.e. the probability for an individual to not be selected when multiple parents are needed to create a child is lower in SUS than in RW as long as the probability for it to be selected is greater than 0. Thus, SUS applies higher selection pressure.

$$(1 - a)^n > 1 - n \cdot a, \quad \forall a \in (0, 1], n > 1 \tag{2}$$

Sigma Scaling (SS). The fitness $f(x_i)$ of an individual x_i is normalized relative to the entire population's standard deviation. The resulting scaled fitness $f'(x_i)$ is calculated with Eq. 3.

$$f'(x_i) = \begin{cases} \max\left(1 + \frac{(f(x_i) - \bar{f})}{2 \cdot \sigma}, \epsilon\right) & \text{if } \sigma \neq 0 \\ 1 & \text{if } \sigma = 0 \end{cases}, \tag{3}$$

where $\bar{f}$ is the mean of the fitness values, σ is the standard deviation of the fitness values, and ϵ is a minimum scaled fitness greater or equal to 0 defined by the user. The idea of SS is to even out the selection probabilities of widely-differing fitness values, while spreading out the selection probabilities of closely-spaced fitness values. This essentially means lower selection pressure when compared to RW and SUS.

3.4   Fitness Function

Two different fitness functions were designed and tested for the purposes of this work. Each evaluates the game state at the discrete simulation steps based on given environment features and sums the resulting scores, which are then multiplied with a factor $\epsilon = 0.95$ to account for the stochastic nature of the benchmark games. Equation 4 shows the formula used to calculate the fitness of an individual x_i.

$$f(x_i) = \epsilon \cdot \sum_{i=0}^{S} s_i \tag{4}$$

S represents the depth of the simulation (the length of the solution vector), i.e. the length of a candidate solution's action permutation, and s_i is the score of the i-th game state reached during the simulation. Both fitness functions differ mainly in the environment features they consider when calculating the respective state score.

Score-Based Fitness. A very simple approach that uses the currently achieved in-game score g and whether the game has been won or not to evaluate a game state.

$$s_i = \begin{cases} +100000000 & \text{if game won} \\ -100000000 & \text{if game lost} \\ g & \text{otherwise} \end{cases} \qquad (5)$$

Object- and Score-Based Fitness. This is the score-based fitness formula that also considers common knowledge about video games. This common knowledge consists of assumptions based on our own experience when playing video games such as:

- moving towards objects in the game is mostly a good strategy
- moving towards portals in the game is mostly a good strategy
- the number of objects in the current game state must be as low as possible
- the number of gathered resources must be as large as possible.

Using these assumptions we extended the expression in Eq. 5 with the parameters d_o (distance to the closest object in the game relative to the agent), d_p (distance to the closest portal in the game relative to the agent), O (number of objects currently present) and R (number of resources currently gathered by the agent). The modified formula can be seen in Eq. 6.

$$s_i = \begin{cases} +100000000 & \text{if game won} \\ -100000000 & \text{if game lost} \\ g - 1000 \cdot d_o - 1000 \cdot O - 100 \cdot d_p + 100 \cdot R & \text{otherwise} \end{cases} \qquad (6)$$

4 Statistical Tests and Experimental Results

The evocontroller agent was tested against a total of 20 benchmark games requiring different strategies to win. Multiple runs per game were executed in order to collect statistically significant data. The controller's performance is measured as the chance to win a given game.

Roulette Wheel Selection. Figure 1 shows the win rate of the evocontroller for each game in the benchmark list. The results are quite self-explanatory: the agent managed to play 4 of the games quite well with a win chance of 70% or higher (even reaching a 100% in *surround*), 3 averagely good (about 50% win chance), and another 3 relatively poorly (it won approx. every fourth game). In the remaining 10 games the evocontroller displayed unsatisfactory performance.

Stochastic Universal Sampling Selection. The win rate of the evocontroller when parents are selected according to SUS, displayed in Fig. 1, is identical to the win rate when using RW. The small variations in the values can be explained with the stochastic nature of the video games.

Sigma Scaling Selection. The *sigma scaling* method to select parents managed to improve the win rate in the games *aliens* and *defender* slightly, effectively increasing the number of games the agent can play quite well from 4 to 5.

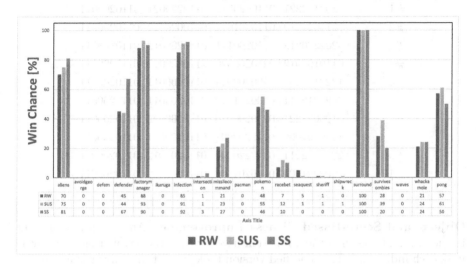

	aliens	avoidgeo rge	defem	defender	factorym anager	ikaruga	infection	intersecti on	missileco mmand	pacman	pokemo n	racebet	seaquest	sheriff	shipwrec k	surround	survivez ombies	waves	whacka mole	pong
RW	70	0	0	45	88	0	85	1	21	0	48	7	5	1	0	100	28	0	21	57
SUS	75	0	0	44	93	0	91	1	23	0	55	12	1	1	1	100	39	0	24	61
SS	81	0	0	67	90	0	92	3	27	0	46	10	0	0	0	100	20	0	24	50

Axis Title

■ RW   ■ SUS   ■ SS

Fig. 1. The agent's win rate for every game in the benchmark list after playing it multiple times using various selection algorithms with *score-based* fitness. Mutation rate set to 20%.

The Importance of Mutation. The recombination operation's exploration capabilities are highly proportional to population size. A big population is generally less prone to inbreeding issues as there are many individuals to chose parents from. Mutation, however, is significantly better and faster when exploring for new solutions, and it does not require a large gene pool to work with like recombination.

To prove this statement a series of tests with a mutation rate of 0% were conducted to analyze the effect of the reduced exploration on the agent's performance. SUS was used as the parent selection algorithm. The extremely small size of the population combined with the total lack of mutation yielded a rapid decrease in population diversity. Table 1 shows the population's individuals of different generations starting with the initial randomly created one. It can be seen that all individuals are gradually replaced by clones that do not necessarily represent an optimal solution to the problem at hand, i.e. the algorithm prematurely converges to a suboptimal solution with no way of further exploring better alternatives. As it can be seen from Table 1 as soon as the 7th generation the entire population consists of clones. Increasing the number of parents used during crossover had a negligible effect: with 3, 6, 9 and 10 parents a complete clone take-over was reached at the 7th, 10th, 12th and 19th generation respectively.

Table 1. The impact of inbreeding on the population's diversity.

Individual	Generation			
	1st	3rd	5th	7th
#1	0202122221	0110020012	0110220021	0110200011
#2	1211022210	0110201012	0110020022	0110200011
#3	2202220212	0110200111	0110200012	0110200011
#4	0110100012	0210200020	0110000012	0110200011
#5	1222111202	0100000022	0110000011	0110200011
#6	2021212211	0100201110	0110200011	0110200011
#7	0100021221	0210200010	0110000022	0110200011
#8	1120101000	0110200021	0110200011	0110200011
#9	2221102211	0210220000	0110200012	0110200011
#10	0202221100	0110201000	0110000012	0110200011

Object- and Score-Based Fitness Improvement. Another series of tests included modifying the fitness function used by the evocontroller agent to evaluate each individual. The modified version took note of the current game score, the amount of resources gathered by the agent so far, the distance to the closest object and portal in the current game state, and the number of objects present. The first two had to be maximized to improve the fitness and vice versa with the last three. This approach did achieve a much lower game endscore when compared to the previous ones, while exhibiting no significant change in the timing

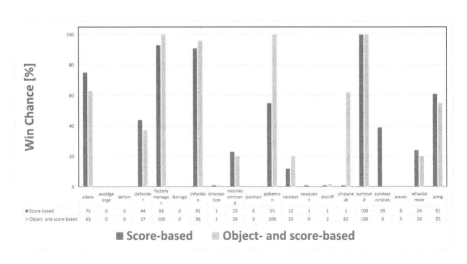

Fig. 2. Comparison of the agent's win rate for every game in the benchmark list after playing it multiple times using *stochastic universal sampling* selection with *score-based* versus *object- and score-based* fitness. Mutation rate set to 20%.

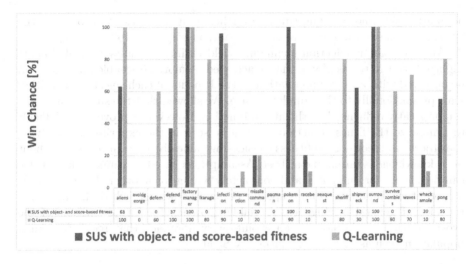

Fig. 3. Comparison of the agent's win rate for every game in the benchmark list after playing it multiple times using *stochastic universal sampling* selection with *object- and score-based* fitness and mutation rate set to 20% versus using modified *Q-Learning* with discount factor 0.35.

performance. There was, however, an interesting difference in the win rate of the agent shown in Fig. 2. It managed to play the games *pokemon* and *shipwreck* considerably better, increasing its win chance from approx. 50% and 0% to a staggering 100% for both. On the other hand it won none of the *survivezombies* games compared to winning every fourth game on average when using the simple score-based fitness function. This is easily explained with the presumable nature of the aforementioned games: pokemons in movies, series and other games are commonly gathered, whereas one usually tries to run away from zombies. The modified fitness function rewards the agent when it moves towards the closest object, and punishes it otherwise. In other words the agent is stimulated to move towards objects in the game. As there is no indication to which objects can be considered enemies, the success of this approach depends solely on the type of the game.

5   Conclusion and Future Work

Considering that the evocontroller was able to win more than half the benchmark games at least 20% of the time (achieving approx. 100% in four of the games), it is safe to say that general artificial intelligence can be simulated to some extent with an evolutionary technique. However, when compared to a controller employing the more traditional Q-Learning modified to account for the limited learning time, which in our tests achieved a win chance greater that 50% in more than half the benchmark games, it is easy to see that the evocontroller's performance is average at best (see Fig. 3). This can be related to the limited population size and evolving time. Nevertheless, it manages to outperform the Q-Learning in several games, more noticeably in *shipwreck* (62 vs. 30%). All three parent selection techniques yielded similar results with SUS being slightly better in terms of win chance achieved. A more noticeable difference can be observed with a bigger population as the benefits of each selection algorithm scale proportionally to the population size. We showed that the success of an EA can be greatly influenced by the fitness function. However, it proved very difficult to generalize the fitness function. As can be seen in the experiments involving the object- and score-based fitness considering game objects' importance without prior knowledge is unreliable at best. Assumptions based on common knowledge can be correct for one game and completely wrong for another. In other words a good fitness function requires specific domain knowledge, which diverges from the idea of general intelligence. A viable approach for future versions of the evocontroller is to implement an evolutionary approach that requires a small population such as *(1 + 1)*-Evolutionary Strategy.

References

1. Back, T.: Selective pressure in evolutionary algorithms: a characterization of selection mechanisms. In: Proceedings of the First IEEE Conference on Evolutionary Computation. IEEE World Congress on Computational Intelligence, vol. 1, pp. 57–62, June 1994. https://doi.org/10.1109/ICEC.1994.350042
2. Baker, J.E.: Reducing bias and inefficiency in the selection algorithm. In: Proceedings of the Second International Conference on Genetic Algorithms on Genetic Algorithms and Their Application, pp. 14–21. L. Erlbaum Associates Inc., Hillsdale (1987). http://dl.acm.org/citation.cfm?id=42512.42515
3. Eiben, A.E., Smith, J.E.: Introduction to Evolutionary Computing. Natural Computing Series, 2nd edn. Springer, Heidelberg (2015). https://doi.org/10.1007/978-3-662-44874-8. http://d-nb.info/1055840826/04
4. Gerdes, I., Klawonn, F., Kruse, R.: Evolutionäre Algorithmen: genetische Algorithmen - Strategien und Optimierungsverfahren - Beispielanwendungen. Computational intelligence, Wiesbaden, 1. aufl. edn. (2004). http://scans.hebis.de/HEBCGI/show.pl?12308575_toc.pdf
5. Holland, J.H.: Adaptation in Natural and Artificial Systems: An Introductory Analysis with Applications to Biology, Control and Artificial Intelligence. MIT Press, Cambridge (1992)
6. Michalewicz, Z.: Genetic algorithms + data structures: = evolution programs: with 36 tables. Berlin [u.a.], 3, rev. and extended ed. edn. (1996). http://scans.hebis.de/HEBCGI/show.pl?04961015_toc.pdf
7. Ronald, S.: Duplicate genotypes in a genetic algorithm. In: 1998 IEEE International Conference on Evolutionary Computation Proceedings. IEEE World Congress on Computational Intelligence, Cat. No. 98TH8360, pp. 793–798, May 1998. https://doi.org/10.1109/ICEC.1998.700153
8. Schaul, T.: A video game description language for model-based or interactive learning. In: 2013 IEEE Conference on Computational Intelligence in Games, CIG, pp. 1–8, August 2013. https://doi.org/10.1109/CIG.2013.6633610

Separable Data Aggregation by Layers of Binary Classifiers

Leon Bobrowski[1,2(✉)] and Magdalena Topczewska[1]

[1] Faculty of Computer Science, Bialystok University of Technology,
Wiejska 45A Str., Bialystok, Poland
{l.bobrowski,m.topczewska}@pb.edu.pl
[2] Institute of Biocybernetics and Biomedical Engineering,
PAS, Trojdena 4 Str., Warsaw, Poland

Abstract. Aggregating layers can be designed from binary classifiers on the principle of preserving data sets separability. Formal neurons or logical elements are treated here as basic examples of binary classifiers. Learning data sets are composed of such feature vectors which are linked to particular categories (classes). Separability of the learning sets is preserved during transformation of feature vectors from these sets by a dipolar layer of binary classifiers. The dipolar layer separates all such pairs of feature vectors that have been linked to different classes and belong to different learning sets.

Keywords: Feature vectors · Binary classifiers · Separable data aggregation · Dipolar aggregation · Hierarchical networks · Deep learning

1 Introduction

Some methods of pattern recognition are aimed at problem of classifiers designing from learning data subsets [1]. Learning sets are composed from such feature vectors which have been assigned to particular categories (classes). A given class is represented by a number of exemplary feature vectors. It is assumed here that learning sets are separable. It means that each feature vector from the learning sets represents only one category. Classification rules designed on the basis of learning sets should have generalization property [2]. Such rules should well classify not only the majority of the learning feature vectors but also new vectors not included in the learning sets.

Multiclass decision rules can be designed from binary classifiers [3]. Formal neuron is an example of a binary classifier. Formal neurons can transform multidimensional feature vectors in the binary numbers equal to one or to zero. The number one at the output of a given formal neuron appears only when the weighted sum of the input features is greater than the threshold. Such feature vectors that give the number one at the output of the formal neuron are separated from the zero-giving vectors by means of some separating hyperplanes in the feature space. Formal neuron has been used in the Perceptron, the first model of neural plasticity and learning processes in human brain [2].

Formal neurons with only one input signal have been used in this work as the binary classifiers. Such binary classifiers have been named logical elements [3]. The

© Springer Nature Switzerland AG 2019
N. T. Nguyen et al. (Eds.): ACIIDS 2019, LNAI 11431, pp. 327–338, 2019.
https://doi.org/10.1007/978-3-030-14799-0_28

separating hyperplanes of logical elements are perpendicular to only one coordinate axis of the feature space and parallel to other coordinate axes.

In the second section the separable learning sets are described, while in the third the binary classifiers are presented. In the next section the binary classifiers forming layers are used to separable data aggregation. In the fifth and sixth sections the dipolar and the ranked strategies of such layers are formulated. The seventh section describes the use of layers of univariate binary classifiers to aggregate data, and in the eight section the results of data aggregation are presented.

2 Separable Learning Sets

Let us assume that m objects O_j $(j = 1, \ldots, m)$ are represented in a standard manner by feature vectors $\mathbf{x}_j = [x_{j,1}, \ldots, x_{j,n}]^T$ belonging to the n-dimensional feature space $F[n]$ ($\mathbf{x}_j \in F[n]$). Components $x_{j,i}$ of the j-th feature vector $\mathbf{x}_j$ can be standardized numerical results of measurements of n features $x_{j,i}$ $(i = 1, \ldots, n)$ of the j-th object O_j ($x_{j,i} \in \{0, 1\}$ or $x_{ji} \in R$). Feature vectors $\mathbf{x}_j$ $(j = 1, \ldots, m)$ can be represented as a cloud of points in the n-dimensional feature space $F[n]$. We can consider the feature space $F[n]$ equal to the n-dimensional real space R^n ($F[n] = R^n$).

We assume also that each of m objects O_j belongs only to one *category* (*class*) ω_k $(k = 1, \ldots, K)$. Such feature vectors $\mathbf{x}_j$ that represent objects O_j from one class ω_k are collected as the k-th *learning set* C_k:

$$C_k = \{\mathbf{x}_j : j \in J_k\}, \tag{1}$$

where J_k is the set of indices j of objects O_j assigned to the k-th the class (ω_k).

Feature vectors $\mathbf{x}_j$ (1) may be represented as the data matrix X of the $m \times n$ dimension:

$$X = [\mathbf{x}_1, \ldots, \mathbf{x}_m]^T. \tag{2}$$

The X data matrix is *sparse*, if many of elements $x_{j,i}$ is equal to zero ($x_{j,i} = 0$) [8].

Definition 1: The learning sets C_k (1) are *separable* in the feature space $F[n]$, if they are disjoined in this space ($C_k \cap C_{k'} = \varnothing$, if $k \neq k'$). This means that feature vectors $\mathbf{x}_j$ and $\mathbf{x}_{j'}$ belonging to different learning sets C_k and $C_{k'}$ cannot be equal:

$$(k \neq k') \Rightarrow (\forall j \in J_k) \text{ and } (\forall j' \in J_{k'}) \ \mathbf{x}_j \neq \mathbf{x}_{j'}, \tag{3}$$

where the inequality $\mathbf{x}_{j'} \neq \mathbf{x}_j$ of the vectors $\mathbf{x}_{j'} = [x_{j',1}, \ldots, x_{j',n}]^T$ and $\mathbf{x}_j = [x_{j,1}, \ldots, x_{j,n}]^T$ means that there exists at least one feature x_i which was measured in both objects $O_{j'}$ and O_j and gave different measurement results ($x_{j',i} \neq x_{j,i}$).

The assumption of the learning sets C_k (1) separability (3) is connected to some constraints in the structure of data sets. We are also considering the separation of the learning sets C_k (1) by the below hyperplanes $H(\mathbf{w}_k, \theta_k)$ in the feature space $F[n]$:

$$H(\mathbf{w_k}, \theta_k) = \{\mathbf{x} : \mathbf{w}_k^T \mathbf{x} = \theta_k\}. \tag{4}$$

where $\mathbf{w_k} = [w_{k,1}, \ldots, w_{k,n}]^T \in R^n$ is the weight vector, $\theta_k \in R^1$ is the threshold, and $\mathbf{w}_k^T \mathbf{x}$ is the inner product of the vectors $\mathbf{w_k}$ and $\mathbf{x}$.

Definition 2: The learning sets C_k (1) are *linearly separable* in the n-dimensional feature space $F[n]$ if each of the sets C_k can be fully separated from the sum of the remaining sets C_i by some hyperplane $H(\mathbf{w_k}, \theta_k)$ (4):

$$(\forall k \in \{1, \ldots, K\}) \ (\exists \mathbf{w_k}, \theta_k) \ (\forall \mathbf{x_j} \in C_k) \ \mathbf{w}_k^T \mathbf{x_j} > \theta_k$$
$$\text{and} \ (\forall \mathbf{x_j} C_i, \ i \neq k) \ \mathbf{w}_k^T \mathbf{x_j} < \theta_k. \tag{5}$$

In accordance with the inequalities (5), all the vectors $\mathbf{x_j}$ from the learning set C_k are situated on the positive side of the hyperplane $H(\mathbf{w_k}, \theta_k)$ (4) and all the vectors $\mathbf{x_j}$ from the remaining sets C_i are situated on the negative side of this hyperplane.

3 Binary Classifiers

A binary classifier transforms an input feature vectors $\mathbf{x} = [x_1, \ldots, x_n]^T$ into the output number r equal to one or to zero ($r = 1$ or $r = 0$). The formal neuron $FN(\mathbf{w}, \theta)$ is one of the most important example of binary classifier. The formal neuron $FN(\mathbf{w}, \theta)$ was used in the *Perceptron*, one of the first model of neural networks plasticity and learning processes in human brain (Rosenblatt, 1962).

The output $r = r(\mathbf{w}, \theta)$ of the formal neuron $FN(\mathbf{w}, \theta)$ is equal to one ($r(\mathbf{w}, \theta) = 1$) if and only if the weighted sum (linear combination) $\mathbf{w}^T \mathbf{x} = w_1 x_1 + \ldots + w_n x_n$ of n inputs $x_i (x_i \in R)$ is greater or equal to some threshold θ (Fig. 1). If this sum is less than the threshold θ, then the output r is equal to zero ($r(\mathbf{w}, \theta) = 0$). The decision rule $r_1(\mathbf{w}, \theta)$ of the formal neuron $FN(\mathbf{w}, \theta)$ depends on the $n + 1$ parameters w_i ($i = 1, \ldots, n$) and θ.

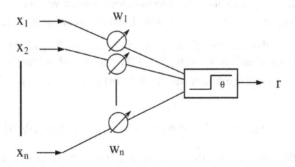

Fig. 1. A scheme of formal neuron $FN(\mathbf{w}, \theta)$

Each formal neuron $FN(\mathbf{w}, \theta)$ is linked to the hyperplane $H(\mathbf{w}, \theta)$ (4). The activation field $A_{FN}(\mathbf{w}, \theta)$ (6) of the formal neuron $FN(\mathbf{w}, \theta)$ is defined as the below half space:

$$A_{FN}(\mathbf{w}, \theta) = \{\mathbf{x} : r(\mathbf{w}, \theta) = 1\} = \{\mathbf{x} : \mathbf{w}^T\mathbf{x} \geq \theta\} \qquad (6)$$

The activation half space $A_{FN}(\mathbf{w}, \theta)$ contains such feature vector vectors $\mathbf{x} = [x_1, \ldots, x_n]^T$ which are situated on the positive side of the hyperplane $H(\mathbf{w}, \theta)$ (5).

Univariate binary classifiers (logical elements) $LE(\mathbf{w}_i, \theta_i)$ can be treated as formal neurons $FN(\mathbf{w}, \theta)$ with only one input x_i [3]. The activation field $A_{LE}(\mathbf{w}_i, \theta_i)$ (4.1) of the i-th logical element $LE(\mathbf{w}_i, \theta_i)$ depends on only two parameters w_i, θ_i and is defined as the below half space (Fig. 2(b)):

$$A_{LE}(\mathbf{w}_i, \theta_i) = \{\mathbf{x} : \mathbf{w}^T\mathbf{x} \geq \theta_i\} = \{\mathbf{x} : w_i x_i \geq \theta_i\} \qquad (7)$$

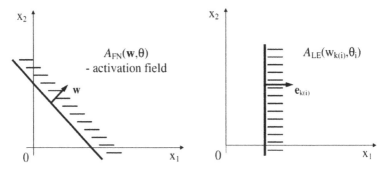

Fig. 2. An example of the activation field of the formal neuron $FN(\mathbf{w}, \theta)$ (a) and the i-th logical element $LE(\mathbf{w}_i, \theta_i)$ (b) in the two dimensional feature space F [2]

A layer of L binary classifiers transforms input feature vectors $\mathbf{x} = [x_1, \ldots, x_n]^T$ into the output vectors $\mathbf{r} = [r_1, \ldots, r_L]^T$ with L binary components r_i (Fig. 3).

Definition 3: The k-th *activation field* $A_k[(\mathbf{w}_1, \theta_1), \ldots, (\mathbf{w}_L, \theta_L)]$ of the layer of L formal neurons (binary classifiers) $FN(\mathbf{w}_i, \theta_i)$ is the set of such feature vectors $\mathbf{x}$ which give at the layer's output the binary vector (*pattern*) $\mathbf{r}_k = [r_{k,1}, \ldots, r_{k,L}]^T$, where $r_{k,i} \in \{0, 1\}$ (Fig. 4).

$$A_k[(\mathbf{w}_1, \theta_1), \ldots, (\mathbf{w}_L, \theta_L)] = \{\mathbf{x} : r(\mathbf{w}_1, \theta_1; \mathbf{x}) = r_{k,1}, \ldots, r_1(\mathbf{w}_L, \theta_L; \mathbf{x}) = r_{k,L}\} \qquad (8)$$

Each activation field $A_k[(\mathbf{w}_1, \theta_1), \ldots, (\mathbf{w}_L, \theta_L)]$ (6) of the layer of L formal neurons $FN(\mathbf{w}_i, \theta_i)$ is a convex polyhedron in the feature space $F[n]$ with the faces constituted by the hyperplanes $H(\mathbf{w}_i, \theta_i)$ (4).

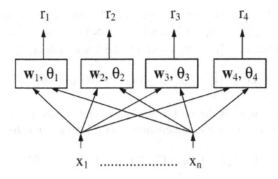

Fig. 3. The layer of four formal neurons $FN(\mathbf{w}_i, \theta_i)$ $(i = 1, 2, 3, 4)$

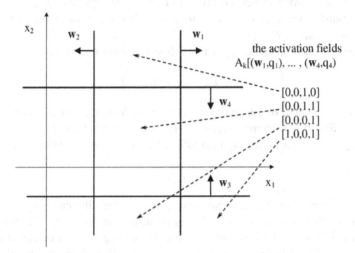

Fig. 4. An example of division of the feature space F [2] by the layer of $L = 4$ logical elements $LE(\mathbf{w}_i, \theta_i)$

4 Separable Data Aggregation by Layers of Binary Classifiers

Let us consider the layer of L binary classifiers designed for extraction of the k-th data subset C_k (1) from the data matrix X (2). For this purpose it is useful to consider the learning data subsets C_k and the *complementary subsets* C_k^c $(k = 1, \ldots, K)$:

$$C_k = \{\mathbf{x}_j : j \in J_k\}, \ and \ C_k^c = \{\mathbf{x}_j : j \in J_{k'}, \ where \ k' \neq k\}. \tag{9}$$

The k-th data subset $C_k = \{\mathbf{x}_j : j \in J_k\}$ (4) is composed of m_k feature vectors $\mathbf{x}_j = [x_{j,1}, \ldots, x_{j,n}]^T$ (1) and the complementary subset C_k^c is composed from the remaining $m - m_k$ vectors $\mathbf{x}_j$.

The layer of L binary classifiers aggregates m_k feature vectors $\mathbf{x}_j$ from the k-th data subset C_k (9) into m'_k ($m'_k < m_k$) different binary vectors $\mathbf{r}_{j'}$ (if $j' \neq j''$, then $\mathbf{r}_{j'} \neq \mathbf{r}_{j''}$). The transformed vectors $\mathbf{r}_{j'}$ ($\mathbf{r}_{j'} \neq \mathbf{0}$) constitute the below subset D_k:

$$D_k = \{\mathbf{r}_{j'} : (\exists \mathbf{x}_j \in C_k) \ \mathbf{r}_{j'} = f(W; \mathbf{x}_j) \neq \mathbf{0}\}, \tag{10}$$

where $W = [\mathbf{w}_1, \ldots, \mathbf{w}_{n'}]^T$ is the vector of n' parameters of the layer.

Feature vectors $\mathbf{x}_j$ from the complementary set C_k^c (9) are transformed into set D_k^c:

$$D_k^c = \{\mathbf{r}_{j'} : (\exists \mathbf{x}_j \in C_k^c(9)) \ \mathbf{r}_{j'} = f(W; \mathbf{x}_j) \neq \mathbf{0}\}. \tag{11}$$

Definition 4: The layer of L binary classifiers separably aggregates the k-th data subset C_k (1) if the transformed subsets D_k (10) and D_k^c (11) are separable (3) and the number m'_k of the transformed vectors $\mathbf{r}_{j'}$ ($\mathbf{r}_j \neq \mathbf{0}$) is less than the number m_k of the feature vectors $\mathbf{x}_j$ in the subset C_k (1).

The *aggregation coefficient* η_k of the k-th learning set C_k (1) has been defined as

$$\eta_k = (m_k - m'_k)/(m_k - 1), \tag{12}$$

where m_k is the number of the input vectors $\mathbf{x}_j$ in the learning set C_k (4), and m'_k is the number of different transformed vectors $\mathbf{r}_{j'}$ (9) in the subset D_k (10).

The aggregation coefficient η_k (12) fulfills the below normalization condition:

$$0 \leq \eta_k \leq 1. \tag{13}$$

It can be seen, that the minimal number m'_k of the different output vectors $\mathbf{r}_{j'}$ (9) from the k-th separable sublayer SL_k is equal to one $(m'_k = 1)$. The aggregation coefficient η_k (24) takes the maximal value equal to one $(\eta_k = 1)$ in this ideal situation. The maximal value of the number m_k' is equal to m_k. There is no aggregation in this case and the aggregation coefficient m'_k (24) is equal to zero $(\eta_k = 0)$.

5 Dipolar Strategy of Separable Layers Designing

The described strategy of designing layers of binary classifiers is aimed at separable aggregation of data sets C_k (9) and is based on the concept of the *mixed dipoles* and the *clear dipoles* [4].

Definition 5: Two feature vectors $\mathbf{x}_j$ and $\mathbf{x}_{j'}$ ($\mathbf{x}_{j'} \neq \mathbf{x}_j$) which belong to different sets $C_k(\mathbf{x}_j \in C_k)$ and $C_k^c(\mathbf{x}_{j'} \in C_k^c)$ (9) constitute the *mixed* dipole $\{\mathbf{x}_j, \mathbf{x}_{j'}\}$.

Definition 6: Two feature vectors $\mathbf{x}_j$ and $\mathbf{x}_{j'}$ which belong to the same set C_k or C_k^c (9) constitute the *clear* dipole $\{\mathbf{x}_j, \mathbf{x}_{j'}\}$.

Definition 7: The i-th formal neuron (binary classifier) $FN(\mathbf{w}_i, \theta_i)$ divides the dipole $\{\mathbf{x}_j, \mathbf{x}_{j'}\}$ if only one feature vector $\mathbf{x}_j$ or $\mathbf{x}_{j'}$ gives the output r_i of the binary classifier equal to one $(r(\mathbf{w}_i, \theta_i; \mathbf{x}_j) = 1$ or $r(\mathbf{w}_i, \theta_i; \mathbf{x}_{j'}) = 1$ (6)).

Remark: The layer of L binary classifiers is *separable* in respect to the k-th data set C_k (9) if and only if each *mixed* dipole $\{\mathbf{x}_j, \mathbf{x}_{j'}\}$ (Definition 5) is divided by at least one binary classifier of this layer [4].

The separable layer of binary classifiers can be enlarged gradually during the designing procedure. An additional classifier can be added to the designed layer during the k-th step of the procedure ($k = 1,..., K$). The below designing postulate can be applied during each step k of a given layer enlargement [4]:

Dipolar Designing Postulate
An additional binary classifier in the dipolar layer should divide the highest number of the yet undivided mixed dipoles $\{\mathbf{x}_j, \mathbf{x}_{j'}\}$ and the lowest number of the yet undivided clear dipoles.

Such mixed dipoles $\{\mathbf{x}_j, \mathbf{x}_{j'}\}$ which have been divided (Definition 7) by some binary classifier of the layer are removed from further considerations. The remaining, yet undivided mixed dipoles $\{\mathbf{x}_j, \mathbf{x}_{j'}\}$ are used in the next stage to find an additional binary classifier. The described scheme is repeated in successive steps l ($l = 1,...,$ L) until all mixed dipoles $\{\mathbf{x}_j, \mathbf{x}_{j'}\}$ are divided or the number of the undivided mixed dipoles it's small enough.

An example of two-dimensional data aggregation in the dipolar layer of binary classifiers can be found in [5].

6 Ranked Strategy of Separable Layers Designing

The ranked procedure of designing ranked layers is based on finding a sequence of the so called *admissible* binary classifiers.

Definition 8: The i-th formal neuron (binary classifier) $FN(\mathbf{w}_i, \theta_i)$ is admissible *in* respect to the k-th learning subset $C_k(i)$ during the i-th stage ($i = 1,..., L$) (9) if and only if some feature vectors $\mathbf{x}_j$ from only one of the subsets $C_k(i)$ or $C_k(i)^c$ (9) gives the output r_i of the binary classifier equal to one ($r_i = 1$):

$$(\forall i \in \{1,...,L\})\ (\exists\,\mathbf{w}_{k(i)}, \theta_{k(i)})\ (\exists \mathbf{x}_j \in C_k(i))\ \mathbf{w}_{k(i)}^T \mathbf{x}_j > \theta_{k(i)}$$
$$and\ (\forall \mathbf{x}_j \in C_k(i)^c)\ \mathbf{w}_{k(i)}^T \mathbf{x}_j < \theta_{k(i)} \tag{14}$$

Or

$$(\forall i \in \{1,...,L\})\ (\exists\,\mathbf{w}_{k(i)}, \theta_{k(i)})(\exists \mathbf{x}_j \in C_k(i)^c)\ \mathbf{w}_{k(i)}^T \mathbf{x}_j > \theta_{k(i)}$$
$$and\ (\forall \mathbf{x}_j \in C_k(i))\ \mathbf{w}_{k(i)}^T \mathbf{x}_j < \theta_{k(i)} \tag{15}$$

where the symbol $C_k(i)$ means the reduced learning set C_k (9) during the i-th stage.

During the next ($i + 1$)-th stage either the data subset $C_k(i)$ is reduced further by such feature vectors $\mathbf{x}_j$ which fulfill the inequality $\mathbf{w}_{k(i)}^T \mathbf{x}_j > \theta_{k(i)}$ (14) or the complementary subset C_k^c (9) is reduced by such feature vectors $\mathbf{x}_j$ which fulfill the inequality (15).

The parameters $(\mathbf{w}_{k(i)}, \theta_{k(i)})$ find during the admissible cuts (14) or (15) are used in definition of successive formal neurons $FN(\mathbf{w}_{k(i)}, \theta_{k(i)})$ the ranked layer ($i = 1,...,$ L) (Fig. 5).

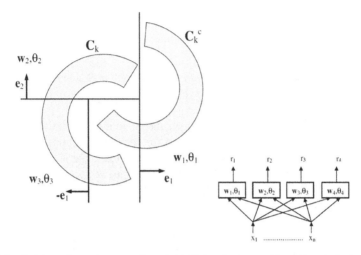

Fig. 5. The division of two data sets C_k and C_k^c (9) by the layer of four ($L = 4$) logical element $LE(\mathbf{w}_i, \theta_i)$ designed in accordance with the ranked strategy

The ranked layer of L formal neurons $FN(\mathbf{w}_{k(i)}, \theta_{k(i)})$ allows to transform the data sets C_k and C_k^c (9) into the linearly separable (5) sets D_k and D_k^c (9). The below theorem can be proved [3]:

Theorem: If the data sets C_k and C_k^c (9) are separable (3), then the transformed sets D_k and D_k^c obtained at the output of the ranked layer are linearly separable (5).

It can be seen that during successive ranked steps, the set of the mixed dipoles is reduced by such pairs $\{\mathbf{x}_j, \mathbf{x}_{j'}\}$, in which one of the feature vectors $\mathbf{x}_j$ or $\mathbf{x}_{j'}$ is contained in a given admissible cut (Definition 8). In the consequence, the ranked strategy can be treated as a special case of a more general dipolar strategy.

The ranked layer of L formal neurons $FN(\mathbf{w}_{k(i)}, \theta_{k(i)})$ allows transform the data sets C_k and C_k^c (5) into the linearly separable (5) sets D_k and D_k^c (9) composed of the transformed vectors $\mathbf{r}_{j'} = [r_{j',1}, \ldots, r_{j',L}]^T$ (11) with L binary components $r_{j'i}(r_{j'i} \in \{0, 1\})$. Additionally, the ranked layers allow for data aggregation. The number m_k' of different transformed vectors $\mathbf{r}_{j'}(if\ j' \neq j''$, *then* $\mathbf{r}_{j'} \neq \mathbf{r}_{j''})$ in the sets D_k can be much smaller than the number m_k of the feature vectors $\mathbf{x}_j$ in the learning set C_k (1).

7 Data Aggregation by Layers of Univariate Binary Classifiers

The separable layers can be designed both from multivariate formal neurons $FN(\mathbf{w}_k, \theta_k)$ as well as from univariate binary classifiers (logical elements) $LE(\mathbf{w}_{k,i}, \theta_k)$. Designing ranked or dipolar layers from multivariate formal neurons $FN(\mathbf{w}_k, \theta_k)$ requires specialized computational techniques. We are developing and using for this purpose machine learning algorithms based on minimization of the *convex and piecewise linear* (*CPL*) criterion functions [3]. Designing ranked or dipolar layers from univariate binary classifiers (logical elements) $LE(\mathbf{w}_{k,i}, \theta_k)$ is much easier.

The univariate binary classifiers $LE(\mathbf{w}_i, \theta_k)$ based on the i-th feature x_i and the l-th threshold θ_l ($\theta_l \in R^1$) can be characterized by the bellow decision rule $r_{i,l} = r_i(\theta_l; \mathbf{x}_j)$:

$$(\forall i \in \{1, \ldots, n\})\ (\forall l \in \{1, \ldots, L'\}) \tag{16}$$

$$r_{i,l} = r_i(\theta_l; \mathbf{x}_j) = \begin{array}{ll} 1 & \textit{if } \ w_i x_{j,i} \geq \theta_{i,l} \\ 0 & \textit{if } \ w_i x_{j,i} < \theta_{i,l} \end{array}$$

where $\theta_{i,l}$ is the value of the l-th threshold on the i-th axis x_i, $w_i = 1$ or $w_i = -1$, and $x_{j,i}$ is the i-th component of the j-th feature vector $\mathbf{x}_j = [x_{j,1}, \ldots, x_{j,n}]^T$ ($x_{j,i} \in R^1$ or $x_{j,i} \in \{0, 1\}$).

The learning set C_k (1) can be composed from a large number m_k of high dimensional feature vectors $\mathbf{x}_j = [x_{j,1}, \ldots, x_{j,n}]^T$. The univariate binary classifiers $LE(w_i, \theta_k)$ (18) are designed on the basis of the projection on the i-th axis x_i of the feature vectors $\mathbf{x}_j$ from the learning set C_k and the complementary set C_k^c (9).

We assume that the learning set C_k and the complementary set C_k^c (9) are separable (3). From the assumed separability (3) of the sets C_k and C_k^c (9) results that each mixed dipole $\{\mathbf{x}_j, \mathbf{x}_{j'}\}$ ($\mathbf{x}_j \in C_k$ and $\mathbf{x}_{j'} \in C_k^c$) can be divided by some univariate binary classifier $LE(w_i, \theta_k)$ (18) based on the i-th axis (feature) x_i. In result, by choosing an adequate k-th feature subset $F_k(n') = \{x_{i(1)}, \ldots, x_{i(n')}\}$ of n' features x_i we can assure the separability (3) of the entire sets C_k and C_k^c (9).

Problem of Feature Subsets Selection
Find such feature subset $F_k(n') \subset F(n)$ with the minimal number n' ($n' < n$) of features x_i which allow to assure separability (3) of the transformed sets D_k (10) and D_k^c (11).

8 Experimental Results

The Ionosphere data set from the UCI Machine Learning Repository [10] was chosen to present the results. It consists of 351 objects described by 34 continuous and integer features. The selected data subset C_1 (9) contained 126 points $\mathbf{x}_j$ (35.90%) and the complementary subset C_1^c contained 225 points $\mathbf{x}_j$ (64.10%). The number of the mixed dipoles $\{\mathbf{x}_j, \mathbf{x}_{j'}\}$ (14) was equal to 28350, and the number of clear dipoles was equal to 33075.

The mixed dipoles were divided by the separable layer of nine univariate classifiers $LE(w_{k,i}, \theta_k)$. In the first step using the attribute no. 5 16762 mixed dipoles were divided. Then the attribute no. 27 was chosen to separate 8372 mixed dipoles. Subsequently variable no. 4 with 1954 mixed dipoles, 28 with 684 dipoles; 10–312; 26–159; 6–61; 20–26; and finally attribute no. 16 was used to separate 12 mixed dipoles (Table 1).

Table 1. Results of the separable layer action

Lp	Attribute no.	Threshold	No. of divided mixed dipoles
1	5	0.418	16762
2	27	0.999	8372
3	4	−0.047	1954
4	28	0.0325	684
5	10	−0.023	312
6	26	−0.299	159
7	6	0.047	61
8	20	0.0001	26
9	16	0.199	12

All 126 points x_j were transformed into below 88 vectors $r_{j'} = \left[r_{j',1}, \ldots, r_{j',9}\right]^T$ (18) with nine binary components $r_{j',i}(j', i = 1, \ldots, 9)$. The aggregation coefficient (12) has a value $\eta_1 = 38/125$ in this example. Further, a noticeable benefit was achieved as a result of dimensionality reduction - instead of 34 attributes the number of binary components was only nine.

The second data set to present results is the Parkinson data set [9]. It contains biomedical voice measurements from 31 people among which 23 have Parkinson's disease. The original study was to use feature extraction methods while in this experiments the aggregation of data is checked to separate healthy people and those with PD. The selected data subset C_1 (PD) contained 147 points x_j (75.38%) and the complementary subset C_1^c contained 48 points x_j (24.62%). The number of the mixed dipoles $\{x_j, x_{j'}\}$ (14) was equal to 7056, and the number of clear dipoles was equal to 11859 (Table 2).

Table 2. Results of the separable layer action

Lp	Attribute no.	Threshold	No. of divided mixed dipoles
1	19	−6.318	4919
2	10	0.191	1304
3	2	206.452	532
4	20	0.213	195
5	5	0.000	68
6	4	0.0037	24
7	3	100.441	10
8	17	0.407	4

All 147 objects $\mathbf{x}_j$ were transformed into below 47 vectors $\mathbf{r}_{j'} = \left[r_{j',1}, \ldots, r_{j',8} \right]^T$ (18) with binary components $r_{j',i}(j', i = 1, \ldots, 8)$, thus the aggregation coefficient has a value $\eta_1 = 100/146 = 0.69$ in this example. This means that the aggregation is substantial, and additionally dimensionality reduction was achieved - instead of 22 attributes the number of binary components was only eight.

9   Conclusion Remarks

Separable layers of binary classifiers preserves separability (3) of the learning data subsets C_k and the complementary subsets $C_k^c (k = 1, \ldots, K)$. The layer of L binary classifiers separably aggregates the k-th data subset C_k (1) if the transformed subsets D_k (10) and D_k^c (11) are separable (3) and the number m_k' of the transformed vectors $\mathbf{r}_{j'}(\mathbf{r}_j \neq \mathbf{0})$ is less than the number m_k of the feature vectors $\mathbf{x}_j$ in the subset C_k (1).

Separable layers of binary classifiers have been designed by using formal neurons $FN(\mathbf{w}_i, \theta_i)$ or logical elements $LE(\mathbf{w}_i, \theta_k)$ (18). Formal neurons $FN(\mathbf{w}_i, \theta_i)$ can transform multidimensional feature vectors $\mathbf{x}_j$ into the binary numbers r_i. Logical elements (univariate binary classifiers) $LE(\mathbf{w}_i, \theta_k)$ (18) transform the numbers $x_{j,i}$ (the i-th components of the feature vectors $\mathbf{x}_j$ into the binary numbers r_i.

The dipolar and ranked strategies of the separable layers designing have been considered in this work. The ranked strategy can be treated as a special case of a more general dipolar strategy.

Separable layers of binary classifiers could be designed for the purpose of data aggregation. In this case, the number m_k' of the transformed vectors in the k-th set D_k (10) should be less then the number m_k of the feature vectors $\mathbf{x}_j$ in the learning data subsets C_k (9) $(m_k' < m_k)$.

Univariate binary classifiers used in separable layers are based on single features only. For this reason the proposed dipolar designing procedure is relatively low costly. Univariate binary classifiers could be particularly useful in processing of such data sets which are only partially structured. Many large data sets encountered in practice have such property [8]. An interesting possibility of using separable layers of univariate binary classifiers for feature selection has been demonstrated in this paper.

Acknowledgments. The presented study was supported by the grant S/WI/2/2013 from Bialystok University of Technology and funded from the resources for research by Polish Ministry of Science and Higher Education.

References

1. Hand, D., Smyth, P., Mannila, H.: Principles of Data Mining. MIT Press, Cambridge (2001)
2. Duda, O.R., Hart, P.E., Stork, D.G.: Pattern Classification. Wiley, Hoboken (2001)
3. Bobrowski, L.: Data mining based on convex and piecewise linear criterion functions. Technical University Białystok (2005). (in Polish)

4. Bobrowski, L.: Piecewise-linear classifiers, formal neurons and separability of the learning sets. In: Proceedings of ICPR 1996, pp. 224–228, 13th International Conference on Pattern Recognition, Vienna, Austria, 25–29 August 1996 (1996)

5. Bobrowski, L., Topczewska, M.: Dipolar data aggregation in the context of deep learning. In: Kůrková, V., Manolopoulos, Y., Hammer, B., Iliadis, L., Maglogiannis, I. (eds.) ICANN 2018. LNCS, vol. 11141, pp. 574–583. Springer, Cham (2018). https://doi.org/10.1007/978-3-030-01424-7_56

6. Bobrowski, L., Topczewska, M.: Linearizing layers of radial binary classifiers with movable centers. Pattern Anal. Appl. **18**(4), 771–781 (2015)

7. Arel, I., Rose, D.C., Karnowski, T.P.: Deep machine learning – a new frontier in artificial intelligence – a survey paper. IEEE Comput. Intell. Mag. (2013)

8. Little, M.A., McSharry, P.E., Roberts, S.J., Costello, D.A.E., Moroz, I.M.: Exploiting nonlinear recurrence and fractal scaling properties for voice disorder detection. BioMed. Eng. OnLine **6**, 23 (2007)

9. Wang, Z., et al.: Sparse Coding and its Applications in Computer Vision. World Scientific, Hackensack (2016)

10. Dua, D., Karra Taniskidou, E.: UCI Machine Learning Repository. University of California, School of Information and Computer Science, Irvine, CA (2017). http://archive.ics.uci.edu/ml

Adaptation and Recovery Stages for Case-Based Reasoning Systems Using Bayesian Estimation and Density Estimation with Nearest Neighbors

D. Bastidas Torres[1,2]([✉]), C. Piñeros Rodriguez[3], Diego H. Peluffo-Ordóñez[2,6],
X. Blanco Valencia[2], Javier Revelo-Fuelagán[3], M. A. Becerra[4],
A. E. Castro-Ospina[4], and Leandro L. Lorente-Leyva[5]

[1] Pontificia Universidad Javeriana, Cali, Colombia
daba89@live.com
[2] SDAS Research Group, Yachay Tech, Urcuquí, Ecuador
http://www.sdas-group.com/
[3] Universidad de Nariño, Pasto, Colombia
[4] Instituto Tecnológico Metropolitano, Medellín, Colombia
[5] Universidad Técnica del Norte, Ibarra, Ecuador
[6] Coorporación Universitaria Autónoma de Nariño, Pasto, Colombia

Abstract. When searching for better solutions that improve the medical diagnosis accuracy, Case-Based reasoning systems (CBR) arise as a good option. This article seeks to improve these systems through the use of parametric and non-parametric probability estimation methods, particularly, at their recovery and adaptation stages. To this end, a set of experiments are conducted with two essentially different, medical databases (Cardiotocography and Cleveland databases), in order to find good parametric and non-parametric estimators. The results are remarkable as a high accuracy rate is achieved when using explored approaches: Naive Bayes and Nearest Neighbors (K-NN) estimators. In addition, a decrease on the involved processing time is reached, which suggests that proposed estimators incorporated into the recovery and adaptation stage becomes suitable for CBR systems, especially when dealing with support for medical diagnosis applications.

Keywords: Case-based reasoning · Classification · Probability · Bayes · Parametric

1 Introduction

Nowadays, the necessity to search new and safer ways to present medical information has led to create a new form of thinking, by involving unconventional methods or strategies, making this process the most optimal possible. Case-based

This work is supported by Faculty of Engineering from Universidad de Nariño.

N. T. Nguyen et al. (Eds.): ACIIDS 2019, LNAI 11431, pp. 339–350, 2019.
https://doi.org/10.1007/978-3-030-14799-0_29

reasoning (CBR) is a problem solving approach that uses past experiences to tackle current problems [1]. The life cycle of a CBR-based system consists of four main phases: Firstly, to identify the current problem features and find similar ones in old cases (retrieve); Secondly, using the old case, or cases, found with its proper solution predict an approximate solution to the current problem (reuse/adaptation); Thirdly, evaluate the proposed solution (revise), and finally, if the solution is suitable, update the system to keep the case and the solution as valuable information for future references (retain) [2,3]. Lately the structure given by Case based Reasoning (CBR) systems have been used in many science fields, for example, engineering, business, medicine and many others. The conventional CBR (Aamodt methodology) [4,5] gives, as an answer, a single class or result, which in some cases, is enough as a proper response to a problem. However, this methodology in medical environments lacks information and generally, leads to inconclusive or incorrect diagnosis by the specialist. Therefore is necessary to give more elements, not only a single output, but the total relationship between a new case and the different classes within a group of data, as seen on [6]. This study shows the difficulty obtaining a classification model for every specific group of information, this could explain why many well-known methods for probability estimation (Parzen windows, K-NN among others) does not work very well with some particular data. This article analyzes a new way to implement a CBR system, presenting a probability estimation process with para- metric and no parametric estimators altogether, located in the adaptation and recovery stages within this system, showing as a result, not only a class, but the probability for a new case to belong in all the different classes in a group of data, displaying more complete and valuable information for an expert to evaluate.

This paper is organized as follows: Sect. 2 reviews some related works and outlines basics on CBR. Section 3 describes the operation of the proposed change to classification in a CBR system. Section 4 gathers some results and discussion. Acknowledgment section to Universidad de Nariño. Finally, conclusions and final remarks are presented in, Section 5.

2   Related Works and Background

There are some CBR related work based on medical diagnosis. An example [7] shows the comparison between 5 different techniques to discover a hepatic disease, the 5 techniques used were: Back propagation neural networks, regression trees, lineal regression, CBR with 10 nearest neighbors using Euclidian distance as measurement and a hybrid model CBR-ANN. Proving to be the hybrid combination, the most successful out of the 5 techniques. Another way of using CBR systems in the medicine field is medicine prescription as shown in [8], where a CBR system and a Bayesian classifier are implemented in a parallel formation. This structure recovers information from old cases related to new ones and by classification presents two medical prescriptions, based on specific features, then the two responses are compared using an IF-THEN clause, if the medicine is the same in both cases, the medicine is considered a solution, otherwise another procedure is used for the right amount of medicine, what kind of medicine and what dose is required for this new case. There is also a study [9] where a hybrid system between CBR and decisions trees is implemented for medical diagnosis, Different test were made using breast cancer and liver disease data. First a regression analysis is made to weigh different feature relevance, this process consist in transforming a similarity matrix into an equivalent one, with the final purpose of clustering this information. Once different groups are obtained, a decision tree is applied creating rules to know whether a case belongs to one group or another.

Another study, trends and current developments of CBR-based systems, analyzed articles published between 2004 and 2012, and concluded that CBR systems were applied in different medical scenarios to support not only diagnosis but classification and treatment as well [10]. Recently the CBR methodology is using probability-based techniques and statistical informatics, opening up promising opportunities to improve correct classification using large and complex data systems, in addition to determine or classify uncertain data [11]. Another conclusion of the study is that automatic adaptation is a weakness, especially in the medical field [10].

3 Proposed Methodology

This article proposal is focused on improvement of the adaptation and recovery stages within a CBR system, giving a more conclusive data representation and a more clear and cohesive medical diagnosis for the specialist using this type of systems. A new methodology is proposed in Fig. 1, where the adaptation and recovery stages are united and a probability estimation algorithm, with parametric and no parametric estimators, is implemented inside this new stage; the probability estimation refers to a measurement for a new case to belong to any of the defined classes in a group of data, for each class a percentage would be given as a result, this output is revised by a specialist, who will decide if:

1. Is correct or nor.
2. Is necessary to retain this information for future uses.
3. Would put it in quarantine.
4. Will be removed completely.

Another aspect that supports this methodology is the wide variety in nature of in- formation that this structure could handle. According to the nature of data, a single probability estimator could change its performance a lot, so different groups of records works better with parametric estimators like Naive Bayes or non-parametric like K-NN, proving in the end the malleability of this methodology.

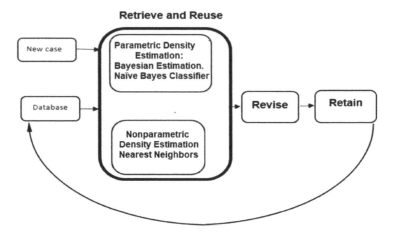

Fig. 1. Proposed methodology in which the adaptation and recovery stage of a CBR system is merged, in addition to using the parametric and non-parametric estimator.

3.1   Proposed Methodology Stages

– **Adaptation and recovery using parametric and non-parametric probability estimation:** The process of probability estimation within a CBR system is accomplished regardless of other classifying process before this stage. The estimators, initially, are trained with information related to the new case to solve (Knowledge base), this information was obtained through different experiences, which are the summary of a variety of medical examinations, scientifically research and professional background; then the new case is analyzed by two estimators and a percentage value of relationship for each class is given individually.
– **Review:** If the output, analyzed by an expert, is precise enough to be considered a solution, it would be given as a diagnosis. The results with ambiguous solutions would be considered again by the expert, who will decide if it is necessary to acquire more information (whether from the patient or another clinical check) or to be discarded completely.
– **Retain:** With a diagnosis already placed, the expert would determine if this new case and its respective solution would be added or discarded within the CBR system, taking into account its relevance as a measure. This way, new cases could benefit from this information as the system searches for similarities between old and new cases every time.

The foundation for this structure is its estimators, if the output is incorrect most of the time the system would not be reliable. Using estimators with low tendency to make mistakes is a priority in this type of systems, therefore the aim of this study is to determine the best parametric and no parametric estimators for two different heart diseases, angina pectoris called Cleveland database and anomalies in Cardiotocography tests.

4 Results and Discussion

4.1 Database

For the different tests and procedures in this work 2 medical databases were used from the machine learning repository UCI [12] listed below:

1. Cleveland database (heart disease), 303 cases, 5 classes and 75 features
2. Cardiotocography (heart disease), 2126 cases, 3 classes and 23 features.

This two databases were modified in size, by reducing its features without changing performance. For every additional feature, elements like execution time and error percentage would be affected negatively. More specifically a correlation strategy was used with the 'Best First' algorithm, as a result, features in Cleveland database were reduced from 75 to 7 and 23 to 10 in Cardiotocography database.

4.2 Methods

The methodology for choosing the best parametric and non-parametric estimator was realized as follows: a CBR simulation was made and different probability estimators were implemented within; then a group of cases with their respective classes were used for training the respective estimators and the remaining cases, without its classes, were used for testing. The output, which is related to the class of the test group, would be compared with the class in the original database of the same case, if the highest probability belongs to this class is considered a good estimation and is stored, if not, a specialist would revise the output and determine if is a good estimation or not. In this article and to verify stability, groups of 80% and 60% of the total database were used for training and 20% and 40% for test; boxplots are used to represent the success percentage of every estimator used in this article. To identify an appropriate procedure for parametric and no parametric estimators some algorithms generally mentioned for each type are tested, these procedures are

1. **Bayesian estimators:** Taking the most commonly known estimators we have:
   (a) *Linear Bayesian Normal Classifier or LDA (Bayes-Normal-1):* Known as Linear Discriminant Analysis (LDA), this classification algorithm, which makes use of qualitative features, is trained by two or more groups of information, with the later purpose of classifying accordingly new cases for each group of data used. Mathematically, by using Bayes theorem, the LDA method estimates the probability that a new observation or case, with a value given by a previous cycle, belongs to every single class known by the system [13]. Finally the new case is assigned to the class with the highest probability. This is the Bayes rule assuming normal distributions. All classes are assumed to have the same covariance matrix. The decision boundaries are thereby linear.

(b) *Quadratic Bayesian Normal Classifier or QDA (Bayes-Normal-2):* The Quadratic Discriminant Analysis, called QDA for short, is similar to LDA, with the only variation that every class in a QDA approach is related to a covariance matrix, and as result, the discriminating function is quadratic. Every new case is classified according to the highest probability, this probability is calculated for each class related to a group of information. The QDA method creates curved decision boundaries, therefore this method could be applied to non linear situations in general [14].

(c) *Naive Bayes:* The Bayesian classifier is based on the Bayes theorem, assuming independency between predictors [15].

2. **Non-parametric estimators:** The 3 more common non parametric estimators are:

(a) *Parzen Windows:* This method is based on probabilities, requiring a smoothing parameter, which improves during training for Gaussian distribution operations [16].

(b) *K-nearest neighbor or K-NN:* The principle behind the nearest neighbor procedures is finding a number or quantity of training samples near the new sample and then deduce the output variable [17].

4.3 Performance Measures

Performance value will help give the best parametric and non-parametric estimator, so boxplots show success percentage for every estimator. Only the class with the highest percentage would be noticed which will be taken as the output of the CBR.

4.4 Experiments

Experiments were done with databases with the feature selection algorithm implemented. First experiments each database were divided in 80% for training and 20% for test, a CBR simulation was run 100 times with random selected cases each time. Later groups of 60% and 40% were used for training and test the same amount of times.

1. **Experiment 1:** In this experiment parametric estimators like Naive Bayes, LDA and QDA were used. To verify performance the methodology explained before was used. Results can be seen in the next images:

Subfigure 2(a) shows a very good performance related to Naive Bayes using cardiotocography. The percentage of correct case diagnosis for both the 20% test group and the 40% test group were around 98%, this shows that this database has easily separable data and that they fit this estimator accordingly. For Subfig. 2(a), the estimator presents an acceptable result for the two different test groups in Cleveland. The accuracy percentages are under 55% for Naive Bayes, which could mean two things, the cases contained in this database are complex and difficult to separate or the estimator used does not fit well to the structure of the data. In regards to Subfigs. 3(a) and (b), which

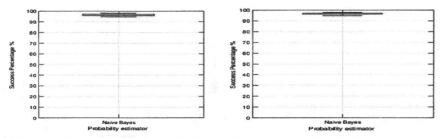

(a) *Boxplot for the Cardiography database. On the left is the success percentage result for the 20% test group and on the right is the success percentage result for the 40% test group, with the Naïve Bayes estimator.*

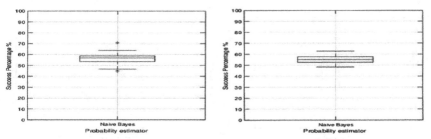

(b) *Boxplot for the Cleveland database. On the left is the success percentage result for the 20% test group and on the right is the success percentage result for the 40% test group, with the Naïve Bayes estimator.*

Fig. 2. Estimation of the Naïve Bayes parametric probability

correspond to the LDA classifier, a behavior similar to the results given by the Naive Bayes estimator are observed. In cardiotocography, both the 20% and 40% of data related to testing, showed a success rate of 98%. Conversely, for Cleveland, in both tests the success rate was about 5% lower than the previous one, this might be seen more clearly by the average of success percentage, it was about 50% lower in the two tests. The analysis of Fig. 3(c) and (d) shows the lowest results for both databases. For cardiotocography database the QDA estimator only succeed in approximately 50% of the time, below the previously studied estimators. In Cleveland database the behavior is similar, the average success percentage falls between 28% and 48% for the 20% and 40% cases respectively.

2. **Experiment 2:** This experiment uses non-parametric estimators K-NN and parzen windows. The same procedure was realized. Results are shown on figures:

   K-NN estimator presents a very good percentage with cardiotocography database. Subfigure 4(a), the results for the tests were approximately 98% of percentage of success, a value similar to the result of the parametric test (Naive Bayes), therefore, it is concluded that you can use this classifier or the Naive Bayes indifferently to obtain good results. For Cleveland database the average of success percentage is approximately 60% correct Fig. 4(b), showing a better performance than the parametric estimators. For the second test,

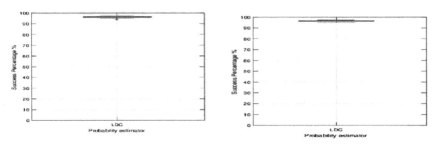

(a) *Boxplot for the Cardiography database. On the left is the success percentage result for the 20% test group and on the right is the success percentage result for the 40% test group, with the LDA estimator.*

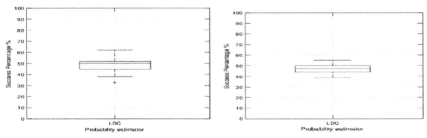

(b) *Boxplot for the Cleveland database. On the left is the success percentage result for the 20% test group and on the right is the success percentage result for the 40% test group, with the LDA estimator.*

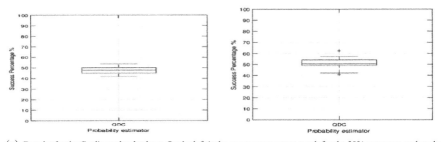

(c) *Boxplot for the Cardiography database. On the left is the success percentage result for the 20% test group and on the right is the success percentage result for the 40% test group, with the QDA estimator.*

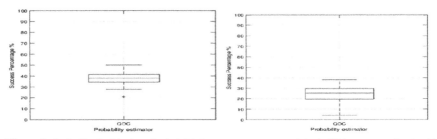

(d) *Boxplot for the Cleveland database. On the left is the success percentage result for the 20% test group and on the right is the success percentage result for the 40% test group, with the QDA estimator.*

Fig. 3. Estimation of LDA and QDA parametric probability

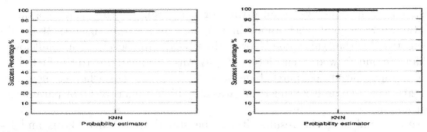

(a) *Boxplot for the Cardiography database. On the left is the success percentage result for the 20% test group and on the right is the success percentage result for the 40% test group, with the K-NN estimator.*

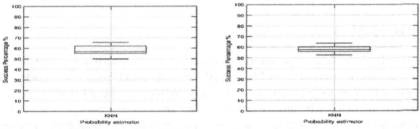

(b) *Boxplot for the Cleveland database. On the left is the success percentage result for the 20% test group and on the right is the success percentage result for the 40% test group, with the K-NN estimator.*

Fig. 4. Estimation of the non-parametric K-NN probability density

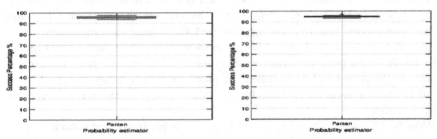

(a) *Boxplot for the Cardiography database. On the left is the success percentage result for the 20% test group and on the right is the success percentage result for the 40% test group, with the Parzen estimator.*

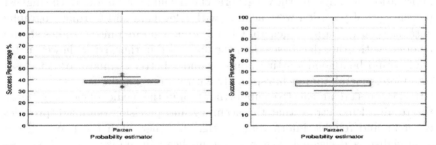

(b) *Boxplot for the Cleveland database. On the left is the success percentage result for the 20% test group and on the right is the success percentage result for the 40% test group, with the Parzen estimator.*

Fig. 5. Estimation of the non-parametric Parzen probability density

using Parzen Windows as an estimator, the results with cardiotocography database were good, with approximately 95% in average Fig. 5(a). It should be noticeable that this value is lower than K-NN. In Cleveland, a result lower than the one showed in cardiotocography was obtained, due to the fact that, on average, the success percentage was over 40% Fig. 5(b), whether the group of 20% was used or 40%; this result was only higher than the QDA parametric estimator.

Overall, analyzing the results, it can be stated that Naive Bayes and K-NN were the probability estimators with the highest success percentage or performance among the ones used, where Naive Bayes is representative for the parametric group and the K-NN estimator for the non-parametric group.

5   Conclusion

– In conclusion, adding a probability estimation process with a parametric and no parametric approach within a CBR system could be a good option to implement. A more complete a precise information to show, so an expert could give a more efficiently and correct diagnosis for each new patient or case that presents on a daily basis; also the probability estimation would give an expert a better understanding of the situation rather than a questionable result where a new case is difficult to classify. If the estimation process provides a high success rate, for example Cardiotocography with a success of 98%, the diagnosis for this particular disease would not require much attention by an expert; due to the high precision of the system. On the contrary, Cleveland database success rate did not surpass the 60% making the requirement for a specialist a necessity. Thus, this tool presented could be used as a support to medical diagnosis and not independently; although as more research is done in the field and higher values for precision and accuracy are reached an independent system for specific diseases could be used with little intervention by an expert.
– Probability estimation using Naive Bayes as a parametric approach and K-NN as a no parametric is better than others methods such as LDA, QDA and Parzen. Although is necessary to clarify that due to the context and type of information used in this study a general claim could not be stated, the highest percentages of precision were achieve with Cleveland and Cardiotocography databases only. If a more general CBR system is desired more studies should be done, with other databases than the ones used in this article, however this study could be a good starting point to find a better estimator or get closer to a more general approach, especially with heart related diseases.
– CBR systems are an effective supporting tool in the medical field. In general, the results demonstrate that the accuracy and precision obtained, provides coherent and reliable information so that together, with the knowledge of a specialist, a decision is achieved for medical diagnosis. Also the system shows versatility, different techniques could be implemented in the different stages, not only in the adaption or recovery. For example, this study combines the adaptation and recovery stages through probability estimation and

acknowledge them as one, other systems use them separately as it is common and others uses different types of measurements or algorithms like Euclidian, Mahalanobis or Manhattan distances. This shows that, depending on the application, the variation of stages within a CBR structure could lead to better results than a conventional approach.

Acknowledgment. The authors acknowledge to the research project 'Desarrollo de una metodología de visualización interactiva y eficaz de información en Big Data' supported by Agreement No. 180 November 1st, 2016 by VIPRI from Universidad de Nariño. Also, authors thank the valuable support given by the SDAS-Smart Data Analysis Systems Research Group (www.sdas-group.com).

References

1. Maher, M.L., Balachandran, M.B., Zhang, D.M.: Case-Based Reasoning in Design. Psychology Press, Abingdon (2014)
2. Pal, S.K., Shiu, S.C.: Foundations of Soft Case-Based Reasoning, vol. 8. Wiley, Hoboken (2004)
3. Craw, S., Aamodt, A.: Case based reasoning as a model for cognitive artificial intelligence. In: Cox, M.T., Funk, P., Begum, S. (eds.) ICCBR 2018. LNCS, vol. 11156, pp. 62–77. Springer, Cham (2018). https://doi.org/10.1007/978-3-030-01081-2_5
4. Aamodt, A., Plaza, E.: Case-based reasoning: foundational issues, methodological variations, and system approaches. AI Commun. **7**(1), 39–59 (1994)
5. Leake, D.B.: CBR in context: the present and future. In: Case-Based Reasoning, Experiences, Lessons & Future Directions, pp. 1–30 (1996)
6. Blanco Valencia, X., Bastidas Torres, D., Piñeros Rodriguez, C., Peluffo-Ordóñez, D.H., Becerra, M.A., Castro-Ospina, A.E.: Case-based reasoning systems for medical applications with improved adaptation and recovery stages. In: Rojas, I., Ortuño, F. (eds.) IWBBIO 2018. LNCS, vol. 10813, pp. 26–38. Springer, Cham (2018). https://doi.org/10.1007/978-3-319-78723-7_3
7. Kolodner, J.L.: Maintaining organization in a dynamic long-term memory. Cogn. Sci. **7**(4), 243–280 (1983)
8. Ting, S., Kwok, S.K., Tsang, A.H., Lee, W.: A hybrid knowledge-based approach to supporting the medical prescription for general practitioners: real case in a Hong Kong medical center. Knowl.-Based Syst. **24**(3), 444–456 (2011)
9. Fan, C.Y., Chang, P.C., Lin, J.J., Hsieh, J.: A hybrid model combining case-based reasoning and fuzzy decision tree for medical data classification. Appl. Soft Comput. **11**(1), 632–644 (2011)
10. Kitchenham, B.A.: Systematic review in software engineering: where we are and where we should be going. In: Proceedings of the 2nd International Workshop on Evidential Assessment of Software Technologies, pp. 1–2. ACM (2012)
11. Montani, S.: How to use contextual knowledge in medical case-based reasoning systems: a survey on very recent trends. Artif. Intell. Med. **51**(2), 125–131 (2011)
12. Lichman, M.: UCI machine learning repository (2013)
13. Gama, J.: A linear-Bayes classifier. In: Monard, M.C., Sichman, J.S. (eds.) IBERAMIA/SBIA 2000. LNCS, vol. 1952, pp. 269–279. Springer, Heidelberg (2000). https://doi.org/10.1007/3-540-44399-1_28
14. Srivastava, S., Gupta, M.R., Frigyik, B.A.: Bayesian quadratic discriminant analysis. J. Mach. Learn. Res. **8**(Jun), 1277–1305 (2007)

15. Suresh, K., Dillibabu, R.: Designing a machine learning based software risk assessment model using Naïve Bayes algorithm (2018)
16. Silverman, B.W.: Density Estimation for Statistics and Data Analysis. Routledge, Abingdon (2018)
17. Denoeux, T., Kanjanatarakul, O., Sriboonchitta, S.: EK-NNclus: a clustering procedure based on the evidential K-nearest neighbor rule. Knowl.-Based Syst. **88**, 57–69 (2015)

Feature Extraction Analysis for Emotion Recognition from ICEEMD of Multimodal Physiological Signals

J. F. Gómez-Lara[1]([✉]), O. A. Ordóñez-Bolaños[1], M. A. Becerra[2],
A. E. Castro-Ospina[3], C. Mejía-Arboleda[3], C. Duque-Mejía[3], J. Rodriguez[4],
Javier Revelo-Fuelagán[1], and Diego H. Peluffo-Ordóñez[5]

[1] Universidad de Nariño, San Juan de Pasto, Colombia
`jefersongomez008@gmail.com`
[2] Institución Universitaria Pascual Bravo, Medellín, Colombia
[3] Instituto Tecnológico Metropolitano, Medellín, Colombia
[4] Universidad Autónoma, Manizales, Colombia
[5] SDAS Research Group, Yachay Tech, Urcuquí, Ecuador
`https://sdas-group.com`

Abstract. The emotions identification is a very complex task due to depending on multiple variables individually and as a group. They are evaluated by different criteria such as arousal, valence, and dominance mainly. Several investigations have been focused on building prediction systems. Nevertheless, this is still an open research field. The main objective of this paper is the analysis of the Improved Complementary Ensemble Empirical Mode Decomposition (ICEEMD) for feature extraction from physiological signals for emotions prediction. Physiological signals and metadata of the DEAP database were used. First, the signals were preprocessed, then three decompositions were carried out using ICEEMD, Discrete Wavelet Transform (DWT), and Maximal overlap DWT. Feature extraction was carried out using Hermite coefficients, and multiple statistic measures from IMFs, coefficients DWT, and MODWT, and signals. Then, Relief F selection algorithms were applied to reducing the dimensionality of the feature space. Finally, Linear Discriminant Classifier (LDC) and K-NN cascade, and Random Forest classifiers were tested. The different decomposition techniques were compared, and the relevant signals and measures were established. The results demonstrated the capability of ICEEMD decomposition for emotions analysis from physiological signals.

Keywords: Emotion recognition ·
Improved complementary ensemble empirical mode decomposition ·
Multimodal · Physiological signals · Signal processing

1 Introduction

The emotions are a psycho-physiological induced process by the conscious or unconscious perception of a situation or object [20] and they can be classified

© Springer Nature Switzerland AG 2019
N. T. Nguyen et al. (Eds.): ACIIDS 2019, LNAI 11431, pp. 351–362, 2019.
https://doi.org/10.1007/978-3-030-14799-0_30

using discrete labels of motivation emotions, basic emotions, and social emotions [3]. Emotions are subjective experiences that make their study one of the most confusing and still open fields in psychology [15]. The need and importance of emotion recognition systems have increased for human-computer interface [30] and clinical applications [11]. However, identifying them is very complicated because they depend on multiple variables (individual variables, relationships, group and cultural relations) [31] and stimuli such as discussed in [9,10]. Besides, the emotional states are systematically interrelated and different methods had been applied for evaluating them. Russell et al. proposed a two-dimensional model in which emotions were given as coordinates denoting the degree of valence (the positive or negative quality of emotion), arousal (also called activation) goes from states like sleepy to excited and dominance correspond to the strength of the emotion [24]. However, other studies added other variables such as the pleasure [23] and the unpleasure [13], but the most commonly used model is the Circumplex Model of Affect (CMA), which only uses valence and arousal [24].

Multiple studies have been carried out to predict emotions, which are focused on methods of feature extraction and classification to improve the precision of these detection systems. Recently, multiple studies were discussed in [28] shown results among 60% and 95% mainly. In [31] were discussed a lot of studies of emotion identification which do not upper the 81.45% of accuracy applying as benchmark the recognized database named DEAP [20]. This database includes multiple physiological signals and analysis multimodal such as: EEG (electroencephalogram), GSR (resistance of the skin by positioning two electrodes on the distal phalanges of the middle and index fingers), Rb (respiration belt), skin temperature (ST), plethysmograph (PTG), electromyogram of Zygomaticus (zEMG) and Trapezius muscles (tEMG), and electrooculograms horizontal (hEOG) and vertical (vEOG). Since then, this database have been applied on multiple studies using different approaches or techniques such as: Deep Belief Networks [21], emotion modeling using fuzzy cognitive maps for regression [2], decision fusion for predicting the liking, valence, arousal and dominance [26], segment-to-response transformation method that uses an unsupervised generative model for generating a compact representation space which is applied to binary classification and regression recognition [36] among others. More recent studies increased the accuracy for identifying the emotions but limiting the number of emotions. In [6] applied multi-wavelet transform together with multiclass least squares support vector machine (SVM) for classifying four types of emotions achieving an accuracy of 91.04% using the Morlet kernel function. While [32] achieved an accuracy of 94.097% for classifying 4 emotions. Then in [22] implemented Discrete Wavelet Transform (DWT) to decompose into five frequency subbands the EEG signals and applied SVM and k-Nearest Neighbors (k-NN) for emotional states prediction achieving an accuracy of 84.05% for arousal and 86.75% for valence. However, the confidence of these systems is low yet for being considered an effective diagnostic support mechanism [1,29].

Other studies have been carried out with new databases [19,35]. In [5] was applied minimum-Redundancy-Maximum-Relevance (mRMR) for feature selection and SVM for classification based on two-dimension emotions model, which obtained levels of accuracy of 60.72% and 62.4% for excitation and valence respectively. Some algorithms are reported with the use of wavelet transforms to detect four emotional states: as calm, happy, fear and sad were classified with an accuracy of 90.9%, 63.63%, 90.90 and 100%, respectively [27]; SVM, artificial neural networks and K-NN stand out as the classifiers with the best performance in neuronal signals. Based on these studies, there is multiple analysis of physiological signals for predicting them. Nevertheless, this is an open research field, and new experiments with different techniques of features extraction, and classification should be carried out to improve the performance of these systems which can be still considered low.

On the other hand, EEG signals are nonlinear and non-stationary, for this reason, methods have been developed to cope with this kind of signals as the empirical mode decomposition (EMD) [4], this is a time domain processing method often used for decomposing in to the so-called intrinsic mode functions (IMF) [17]. This method has advantages of adaptability and flexibility [33]. EMD can only handle datasets single channel while MEMD already can handle multivariate data [14] and ensemble empirical mode decomposition (EEMD) allows the decomposition of non-stationary and non-linear data [12]. The EMD technique has been widely used for empirically decomposing, from which is carried out the feature extraction of this is performed by different methods. Several authors choose the best characteristics by statistically-significant features ($p \leq 0.05$) and the harmonics ratio for each IMF among others.

The main objective of this study is to analyze the capability of ICEEMD technique, together other feature extraction techniques, and classifiers to improve the accuracy (considering balanced labels) the emotion prediction from multimodal signals. In this study, an emotion prediction system is presented based on a multimodal analysis of physiological signals using Improved Complete Empirical Mode Decomposition (ICEEMD) which improves some problems of EEMD. DEAP database is used as a benchmark, and the results of ICEEMD are compared with DWT and MODWT. This work was developed as follows. First, the signals were preprocessed, and three different decompositions were carried out using ICEEMD, DWT, and MODWT. Feature extraction was carried out using Hermite coefficients, Linear frequency cepstral coefficients, Hilbert Huang transform, and multiple statistic measures calculated from IMFs, MODWT and DWT coefficients, and signals. Then, relief F algorithm was applied for reducing the dimensionality of the feature space. The most relevant features and signals were established. Finally, K-Nearest Neighbor (K-NN) and Linear Discriminant Classifier (LDC) in cascade, and Random Forest classifiers were applied for classifying four different categorical emotions besides valence and arousal. The results demonstrated the capability of ICEEMD decomposition for emotions analysis from physiological signals.

2   Experimental Setup

This study was carried out in 5 main stages as it is shown in Fig. 1. First, DEAP database was labeled as Low Arousal and Low Valence (LALV), High Arousal and Low Valence (HALV), Low Arousal and High Valence (LAHV), and High Arousal and High Valence (HAHV). Then, the EEG and peripheral signals were normalized between $[-1, 1]$ in pre-processing stage. In the third stage, three different decompositions were applied using ICEEMD, DWT, and MODWT [7]. These two last were applied using Daubechies 4 with a decomposition of 5 levels.

Besides, feature extraction was carried out using multiple techniques based on Linear Frequency Cepstral Coefficients (LFCC), Hermite Coefficients, Linear and Non-linear Statistical measures (statistical moments and entropies) were calculated from EEG signals and their decompositions. Reduction of the dimensionality of the feature space was performed relief F algorithm. Then for the problem of class imbalance [16], this study applied the synthetic minority sampling techniques (SMOTE). Finally, LDC and K-NN cascade, and Random Forest classifiers were tested.

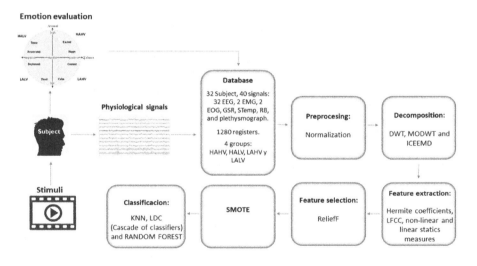

Fig. 1. Experimental procedure

2.1   Database

The DEAP database was used for this study. This has 1280 registers acquired from 40 trials applied to 32 subjects which were exposed to 40 videos to evoke emotions. 40 signals were acquired per trial correspond to 32 EEG signals acquired using the electrodes 10–20 system positioning as it is shown in Table 1. The last eight signals are named peripheral signals. The signals and their nomenclature are shown in Table 1.

Table 1. EEG signals based on 10–20 electrode placement system and 8 peripheral signals

EEG1 – Fp1	EEG2 – AF3	EEG3 – F3	EEG4 – F7	EEG5 – FC5	EEG6 – FC1
EEG7 – C3	EEG8 – T7	EEG9 – CP5	EEG10 – CP1	EEG11 – P3	EEG12 – P7
EEG13 – PO3	EEG14 – O1	EEG15 – OZ	EEG16 – PZ	EEG17 – FP2	EEG18 – AF4
EEG19 – Fz	EEG20 – F4	EEG21 – F8	EEG22 – FC6	EEG23 – FC2	EEG24 – CZ
EEG25 – C4	EEG26 – T8	EEG27 – CP6	EEG28 – CP2	EEG29 – P4	EEG30 – P8
EEG31 – PO4	EEG32 – O3	tEMG	zEMG	Rb	HEOG
VEOG	GSR	PTG	ST		

2.2 Decomposition

Improved Complete Mode Decomposition. The ICEEMD is developed for nonlinear and non-stationary signals, unlike the wavelet decomposition which is developed for non-linear but stationary signals. In this work, the physiological signals were decomposed using the algorithm ICEEMD presented in [12] and it was configured with the parameters shown in Table 2. In Fig. 2 are shown the results of decomposition of an EEG signal in 7 IMFs using ICEEMD. The three last IMFs contain low energy.

Table 2. Parameters of ICEEMD Algorithm

Signal noise relation	Noise standard deviation	Number of realizations	Max. iterations
Same SNR for all stages	0.2	50	150

The EMD technique decomposes a signal in multiple signals named IMFs. To be considered as an IMF, a signal must fulfill two conditions: *(i)* the number of extrema (maxima and minima) and the number of zero-crossings must be equal or differ at most by one; and *(ii)* the local mean, defined as the mean of the upper and lower envelopes, must be zero. Recently, Colominas et al. [12] proposed a noise-assisted data analysis method namely ICEEMD to address the mode mixing problem. The other advantages of this technique are: less residual noise and minimum reconstruction error in the modes and ensure completeness of the algorithm. ICEEMD is well suitable for biomedical signal processing [25]. IMF generation using ICEEMD is detailed as follows (Algorithm 1).

Notation used in algorithm: $E_l(\cdot) = lrh$ EMD mode, $M(\cdot)$ = local mean of the signal, $<\cdot> $ = averaging operator, W^j = realization of White Gaussian noise with zero mean and unit variance and X = input signal.

Algorithm 1. ICEEMD algorithm

1: Compute the local means of J realization $x^J = x + \beta_0 E_1(W^{(j)})$, $j = 1, 2, ..., J$ using EMD, to obtain first residue $r_1 \left\langle M(X^{(j)}) \right\rangle$.

2: At the first stage (l=1) compute the first IMF: $C_1 = x - r_1$.

3: For $l = 2, ..., L$ calculate r_l as $r_l = \left\langle M(r_{l-1} + \beta_{l-1} E_l(W^{(j)})) \right\rangle$.

4: Calculate the l-th mode as $C_l = r_{l-1} - r_l$.

5: Go to step 3 for next l.

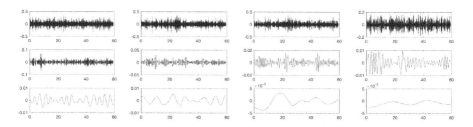

Fig. 2. EEG signal decomposition in 12 IMFs using ICEEMD

2.3   Feature Extraction

Six main features sets were extracted from each signal as follows: (i) 60 features were extracted of each IMF based on measures of Table 3; (ii) 40 features were obtained of multiple measures such as amplitude, angle, and frequency of results achieved applying Hilbert Huang Transform to IMF. (iii) 39 LFCC were calculated including delta coefficients, delta–delta coefficients, and log energy, using 24 Hamming shaped filters. (iv) 65 features were obtained using the measures shown in Table 3, which were calculated on coefficients of DWT, and 65 more were calculated on coefficients of MODWT. Finally, (v) 10 features were obtained using the measures shown in Table 3 applied to EEG signals directly. In total 179 features were extracted for each signal, and 7160 per trial.

Table 3. Measures calculated from EEG and peripheral signals - MODWT and DWT

Renyi entropy (REn)	Root mean square (RMS)	Standard deviation (Std)	Entropy (En)	Variance (Var)
Power (Pw)	Log energy entropy (LogEn)	Shannon entropy (ShEn)	Energy	Covariance (Cov)
Kurtosis (Kurt)	Min/Max (MM)	Mean		

2.4   Feature Selection and Classification

EEG signals are highly dimensional data that can contain many features with low relevance. The selection of the feature subset is a technique to reduce the space of a feature set removing irrelevant data. In this work was applied to the Relief F algorithm, which is a filter method that ponders each feature according to its relevance to each class. The weights of them are updated iteratively [34]. The selection of features was made taking as a criterion a sum of weights higher than 0.95. LDC and K-NN classifiers in cascade and Random Forest were applied to evaluate the capability of the selected features by the selectors, and they were normalized using z-score. The systems were validated using 10-folds cross-validation.

3   Results and Discussion

Table 4 shows the more relevant signals based on features obtained from ICEEMD and HHT-ICEEMD taking into account the sensitivity (Se), the specificity (Sp) and the accuracy (Acc) achieved for each channel taking into account individually and all together LALV, HALV, LAHV, and HAHV valence-arousal (V-A) spaces. In Table 5 is shown the Acc, Sp, and Se for DWT and MODWT, and in Table 6 is shown the Acc, Sp, and Se for Hermite coefficients and LFCC. The best results are achieved for Low Arousal Low Valence (LALV) regardless of the features. The best results for LALV was achieved by HHT-ICEEMD of EEG22 99.12% of accuracy; for HALV was achieved by LFCC of ST signal 88.10%; for LAHV was obtained by DWT of EEG8 signal 91.64% of accuracy, and for HAHV was obtained 90.72% of accuracy using features based on LFCC of EEG14. On the other hand, the best global performance for all classes 88.82% was achieved by Hermite coefficients of EEG31.

Table 4. Relevant signals based on features of ICEEMD and HHT-ICEEMD with KNN-LDC

Techn	ICEEMD					HHT-ICEEMD				
V-A	LALV	HALV	LAHV	HAHV	All	LALV	HALV	LAHV	HAHV	All
Signal	EEG30	EEG5	EEG15	EEG19	EEG6	EEG22	EEG31	EEG9	EEG13	EEG19-HEOG
Se	94.34	64.15	78.30	70.75	69.81	98.11	62.26	78.30	64.15	69.58
Sp	99.23	95.79	97.24	96.54	96.61	99.72	95.57	97.12	95.97	96.57
Acc	**97.50**	85.67	90.86	88.25	88.31	**99.12**	84.94	90.60	86.09	88.18

In Tables 7 and 8 is shown the average of the Acc, Sp, and Se of classification for each V-A space, using features calculated from ICEEMD, MODWT, DWT, Hermite coefficients, and LFCC applied to the best channels shown in Tables 4, 5, and 6. The best results were achieved by ICEEMD 97.92% for LALV, HHT-ICEEMD 85.22% and MODWT 85.63%for HALV, Hermite coefficients 89.31%

for LAHV, and HHT-ICEEMD 85.22% and MODWT 86.25% for HAHV. Finally, the best global result for all labels 88.56% was achieved by MODWT. In Table 9 is presented the Friedman test applied to classifications systems.

Table 5. Relevant signals based on features of DWT and MODWT with KNN-LDC

Techn	DWT					MODWT				
V-A	LALV	HALV	LAHV	HAHV	All	LALV	HALV	LAHV	HAHV	All
Signal	EEG15	HEOG	EEG8	EEG21	EEG21	EEG22	EEG9	EEG23	Rb	EEG4
Signal	–	–	–	EEG31	EEG32	EEG30		VEOG	–	EEG32
Se	95.28	62.26	81.13	65.09	69.58	94.34	64.15	77.36	64.15	69.58
Sp	99.30	95.59	97.41	96.07	96.55	99.15	95.51	96.81	95.75	96.57
Acc	**97.82**	84.98	**91.64**	86.45	88.14	**97.34**	85.04	89.80	85.58	88.19

Table 6. Relevant signals-features of Hermite coefficients and LFCC with KNN-LDC

Techn	Hermite coefficients					LFCC				
V-A	LALV	HALV	LAHV	HAHV	All	LALV	HALV	LAHV	HAHV	All
Signal	EEG4-12	EEG31	ST	EEG23	EEG31	EEG8	ST	tEMG	EEG14	EEG8
Signal	zEMG	–	–	GSR	–	–	–	–	–	–
Se	93.40	65.09	77.36	63.21	70.99	87.74	70.75	78.30	78.30	68.87
Sp	99.05	96.17	97.09	95.81	96.76	98.39	96.47	97.18	97.18	96.46
Acc	**96.98**	86.68	90.38	85.61	**88.82**	**94.73**	**88.10**	90.72	**90.72**	87.84

Table 7. Accuracy of KNN-LDC classifier using ICEEMD, HHT-ICEEMD, and MODWT

Valence-Arousal space	ICEEMD	HHT-ICEEMD	MODWT
Measures	Se-Sp-Acc	Se-Sp-Acc	Se-Sp-Acc
LALV	**95.28 - 99.36 - 97.92**	60.38 - 95.24 - 83.92	92.45 - 99.01 - 96.75
HALV	57.55 - 95.39 - 83.94	**64.15 - 95.60 - 85.22**	62.26 - 95.86 - 85.63
LAHV	72.64 - 96.77 - 89.02	69.81 - 96.15 - 87.26	62.26 - 95.87 - 85.63
HAHV	51.89 - 94.93 - 82.18	**64.15 - 95.59 - 85.22**	**64.15 - 96.03 - 86.25**
All	69.34 - 96.61 - 88.26	64.62 - 95.64 - 85.41	**70.28 - 96.69 - 88.56**

Table 10 illustrates a comparison of percentages between the result of this research with ICEEMD decomposition, one relevant analysis with the same database, which applied Sample entropy as a method of Feature extraction and SVM as classifier [18] and another reasearch with Pictorial stimuli of International Affective Picture System Database (IAPS), mean and covariance matrix as a method of Feature extraction and LDA and QDA as classifiers [8]. However, Despite that results are not comparable with other approaches, because

Table 8. Accuracy KNN-LDC classifier using DWT, Hermite coefficients, and LFCC

Valence-Arousal space	DWT	HERMITE	LFCC
Measures	Se-Sp-Acc	Se-Sp-Acc	Se-Sp-Acc
LALV	92.45 - 98.98 - 96.69	81.13 - 97.51 - 91.85	78.30 - 97.12
HALV	59.43 - 95.54 - 84.48	50.94 - 94.60 - 81.27	55.66 - 94.88 - 82.50
LAHV	69.81 - 96.48 - 87.97	**74.53 - 96.78 - 89.31**	71.70 - 96.40 - 88.08
HAHV	54.72 - 95.10 - 82.80	59.43 - 95.32 - 83.99	55.66 - 94.88 - 82.50
All	69.10 - 96.53 - 87.99	66.51 - 96.05 - 86.61	65.33 - 95.82 - 85.92

Table 9. p-value Friedman test - classification systems

Measure	DWT	HHT–ICEEMD	ICEEMD	LFCC	MODWT
p-value	2.32e−21	2.24e−22	3.84e−21	2.89e−13	3.63e−19

of our study analyzes each V-A space taking into account each physiological signal. In general, the results achieved can be considered upper per V-A space. However, taking all the features showed in the tables show above our approach showed similar results. Also, all results shown above were compared respect to the results of Random Forest classifier which were lowest. Consequently, these were not included in the tables. However, the results achieved the follows accuracy 92.97% for LALV using ICEEMD, 89.52% for HALV using DWT, 90.70% for LAHV using DWT, and 86.86% for HAHV using MODWT.

Table 10. Comparison of accuracy of classification among this work respect other works [8,18].

Valence-Arousal space	LALV	HALV	LAHV	HAHV
Xiang Jie et al.'s research [18]	71.16	80.43	71,16	80.43
Basu Saikat et al.'s research [8]	97.00	93.00	96.00	98.00
This Work	99.12	88.10	91.64	90.72

4   Conclusions

In this paper, a study of emotion classification from EEG and peripheral signals was presented. The capability of 40 different physiological signals for emotion classification was tested from different feature extraction techniques. ICEEMD decomposition shown the best performance followed by MODWT decomposition concerning the other decomposition techniques and features achieved directly from signals in the most cases. The results demonstrate the capability of each feature extraction technique and the physiological signals for emotion classification. The best signals for emotion classification were EEG8, EEG14, EEG22, and EEG31 based on the global results. However, we show the capability of some peripheral signals such as PTG, HEOG, VEOG, Rb, GSR, zEMG, and ST. This last is considered as the most relevant signal for HALV because of this achieved the best performance using LFCC, while the EEG signals were significant for the others labels.

As future work, we propose to test other classification systems using the features shown in this work, for improving the proposed system regarding discernibility and generality. Besides, we recommend to widen and adjust labels of emotions trying to increase the discernibility. Finally, a study that determines the relationship between emotion and subject considering contextual variables and wide the research including other databases.

Acknowledgment. The authors acknowledge to the research project "Desarrollo de una metodología de visualización interactiva y eficaz de información en Big Data" supported by Agreement No. 180 November 1st, 2016 by VIPRI from Universidad de Nariño.

References

1. Abadi, M.K., Subramanian, R., Kia, S.M., Avesani, P., Patras, I., Sebe, N.: DECAF: MEG-based multimodal database for decoding affective physiological responses. IEEE Trans. Affect. Comput. **6**(3), 209–222 (2015). https://doi.org/10.1109/TAFFC.2015.2392932
2. Akinci, H.M., Yesil, E.: Emotion modeling using fuzzy cognitive maps. In: 2013 IEEE 14th International Symposium on Computational Intelligence and Informatics (CINTI), pp. 49–55, November 2013. https://doi.org/10.1109/CINTI.2013.6705252
3. Al Mejrad, A.: Human emotions detection using brain wave signals: a challenging. Eur. J. Sci. Res. **44**(4), 640–659 (2010). https://www.scopus.com/inward/record.uri?eid=2-s2.0-79959391148&partnerID=40&md5=c98a158a7d5ed99b578c8d64210cf5b6, cited By 38
4. Alickovic, E., Kevric, J., Subasi, A.: Performance evaluation of empirical mode decomposition, discrete wavelet transform, and wavelet packed decomposition for automated epileptic seizure detection and prediction. Biomed. Sig. Process. Control **39**, 94–102 (2018). https://doi.org/10.1016/j.bspc.2017.07.022
5. Atkinson, J., Campos, D.: Improving BCI-based emotion recognition by combining EEG feature selection and kernel classifiers. Expert Syst. Appl. **47**, 35–41 (2016). https://doi.org/10.1016/j.eswa.2015.10.049. http://www.sciencedirect.com/science/article/pii/S0957417415007538
6. Bajaj, V., Pachori, R.B.: Human emotion classification from EEG signals using multiwavelet transform. In: 2014 International Conference on Medical Biometrics, pp. 125–130, May 2014. https://doi.org/10.1109/ICMB.2014.29
7. Barzegar, R., Asghari Moghaddam, A., Adamowski, J., Ozga-Zielinski, B.: Multistep water quality forecasting using a boosting ensemble multi-wavelet extreme learning machine model. Stoch. Env. Res. Risk Assess. **32**(3), 799–813 (2018). https://doi.org/10.1007/s00477-017-1394-z
8. Basu, S., et al.: Emotion recognition based on physiological signals using valence-arousal model. In: 2015 Third International Conference on Image Information Processing (ICIIP), pp. 50–55. IEEE (2015)

9. Becerra, M.A., et al.: Odor pleasantness classification from electroencephalographic signals and emotional states. In: Serrano C., J.E., Martínez-Santos, J.C. (eds.) CCC 2018. CCIS, vol. 885, pp. 128–138. Springer, Cham (2018). https://doi.org/10.1007/978-3-319-98998-3_10

10. Becerra, M.A., et al.: Electroencephalographic signals and emotional states for tactile pleasantness classification. In: Hernández Heredia, Y., Milián Núñez, V., Ruiz Shulcloper, J. (eds.) IWAIPR 2018. LNCS, vol. 11047, pp. 309–316. Springer, Cham (2018). https://doi.org/10.1007/978-3-030-01132-1_35

11. Bong, S.Z., Wan, K., Murugappan, M., Ibrahim, N.M., Rajamanickam, Y., Mohamad, K.: Implementation of wavelet packet transform and non linear analysis for emotion classification in stroke patient using brain signals. Biomed. Sig. Process. Control **36**, 102–112 (2017). https://doi.org/10.1016/j.bspc.2017.03.016. http://www.sciencedirect.com/science/article/pii/S1746809417300654

12. Colominas, M.A., Schlotthauer, G., Torres, M.E.: Improved complete ensemble EMD: a suitable tool for biomedical signal processing. Biomed. Sig. Process. Control **14**(1), 19–29 (2014). https://doi.org/10.1016/j.bspc.2014.06.009. http://dx.doi.org/10.1016/j.bspc.2014.06.009

13. Fontaine, J.R., Scherer, K.R., Roesch, E.B., Ellsworth, P.C.: The world of emotions is not two-dimensional. Psychol. Sci. **18**(12), 1050–1057 (2007). https://doi.org/10.1111/j.1467-9280.2007.02024.x. pMID: 18031411

14. Gaur, P., Pachori, R.B., Wang, H., Prasad, G.: A multi-class EEG-based BCI classification using multivariate empirical mode decomposition based filtering and Riemannian geometry. Expert Syst. Appl. **95**, 201–211 (2018). https://doi.org/10.1016/j.eswa.2017.11.007

15. Greco, A., Valenza, G., Lanata, A., Rota, G., Scilingo, E.P.: Electrodermal activity in bipolar patients during affective elicitation. IEEE J. Biomed. Health Inform. **18**(6), 1865–1873 (2014). https://doi.org/10.1109/JBHI.2014.2300940

16. Ha, T.M., Bunke, H.: Off-line, handwritten numeral recognition by perturbation method. IEEE Trans. Pattern Anal. Mach. Intell. **5**, 535–539 (1997)

17. Jia, J., Goparaju, B., Song, J., Zhang, R., Westover, M.B.: Automated identification of epileptic seizures in EEG signals based on phase space representation and statistical features in the CEEMD domain. Biomed. Sig. Process. Control **38**, 148–157 (2017). https://doi.org/10.1016/j.bspc.2017.05.015. http://linkinghub.elsevier.com/retrieve/pii/S1746809417301039

18. Jie, X., Cao, R., Li, L.: Emotion recognition based on the sample entropy of EEG. Bio-med. Mater. Eng. **24**(1), 1185–1192 (2014)

19. Khezri, M., Firoozabadi, M., Sharafat, A.R.: Reliable emotion recognition system based on dynamic adaptive fusion of forehead biopotentials and physiological signals. Comput. Methods Programs Biomed. **122**(2), 149–164 (2015). https://doi.org/10.1016/j.cmpb.2015.07.006. http://www.sciencedirect.com/science/article/pii/S0169260715001959

20. Koelstra, S., et al.: DEAP: a database for emotion analysis; using physiological signals. IEEE Trans. Affect. Comput. **3**(1), 18–31 (2012). https://doi.org/10.1109/T-AFFC.2011.15

21. Li, K., Li, X., Zhang, Y., Zhang, A.: Affective state recognition from EEG with deep belief networks. In: 2013 IEEE International Conference on Bioinformatics and Biomedicine, pp. 305–310, December 2013. https://doi.org/10.1109/BIBM.2013.6732507

22. Mohammadi, Z., Frounchi, J., Amiri, M.: Wavelet-based emotion recognition system using EEG signal. Neural Comput. Appl. **28**(8), 1985–1990 (2016). https://doi.org/10.1007/s00521-015-2149-8

23. Nicolaou, M.A., Gunes, H., Pantic, M.: Continuous prediction of spontaneous affect from multiple cues and modalities in valence-arousal space. IEEE Trans. Affect. Comput. **2**(2), 92–105 (2011). https://doi.org/10.1109/T-AFFC.2011.9

24. Posner, J., Russell, J.A., Peterson, B.S.: The circumplex model of affect: an integrative approach to affective neuroscience, cognitive development, and psychopathology. Dev. Psychopathol. **17**(3), 715–734 (2005). https://doi.org/10.1017/S0954579405050340

25. Rajesh, K.N., Dhuli, R.: Classification of imbalanced ECG beats using resampling techniques and AdaBoost ensemble classifier. Biomed. Sig. Process. Control **41**, 242–254 (2018). https://doi.org/10.1016/j.bspc.2017.12.004. http://dx.doi.org/10.1016/j.bspc.2017.12.004

26. Rozgić, V., Vitaladevuni, S.N., Prasad, R.: Robust EEG emotion classification using segment level decision fusion. In: 2013 IEEE International Conference on Acoustics, Speech and Signal Processing, pp. 1286–1290, May 2013. https://doi.org/10.1109/ICASSP.2013.6637858

27. Thejaswini, T., Ravikumar, K.M.: Detection of human emotions using features based on the mulitwavelet transform of EEG signals. Brain-Comput. Interfaces: Curr. Trends Appl. 119–122 (2018). https://books.google.com/books?id=2LUjBQAAQBAJ&pgis=1

28. Shu, L., et al.: A review of emotion recognition using physiological signals. Sensors **18**(7), 2074 (2018). https://doi.org/10.3390/s18072074

29. Soleymani, M., Asghari-Esfeden, S., Fu, Y., Pantic, M.: Analysis of EEG signals and facial expressions for continuous emotion detection. IEEE Trans. Affect. Comput. **7**(1), 17–28 (2016). https://doi.org/10.1109/TAFFC.2015.2436926

30. Thejaswini, S., Ravi Kumar, K.M., Rupali, S., Abijith, V.: EEG based emotion recognition using wavelets and neural networks classifier of emotion. J. Pers. Soc. Psychol. (2017). https://doi.org/10.1007/978-981-10-6698-6-10

31. Verma, G.K., Tiwary, U.S.: Multimodal fusion framework: a multiresolution approach for emotion classification and recognition from physiological signals. NeuroImage **102**, 162–172 (2014). https://doi.org/10.1016/j.neuroimage.2013.11.007. http://www.sciencedirect.com/science/article/pii/S1053811913010999, multimodal Data Fusion

32. Vijayan, A.E., Sen, D., Sudheer, A.P.: EEG-based emotion recognition using statistical measures and auto-regressive modeling. In: 2015 IEEE International Conference on Computational Intelligence Communication Technology, pp. 587–591, February 2015. https://doi.org/10.1109/CICT.2015.24

33. Yang, B., Zhang, T., Zhang, Y., Liu, W., Wang, J., Duan, K.: Removal of electrooculogram artifacts from electroencephalogram using canonical correlation analysis with ensemble empirical mode decomposition. Cogn. Comput. **9**(5), 626–633 (2017). https://doi.org/10.1007/s12559-017-9478-0

34. Zhang, Z., et al.: Modulation signal recognition based on information entropy and ensemble learning. Entropy **20**(3), 198 (2018)

35. Zheng, W.L., Guo, H.T., Lu, B.L.: Revealing critical channels and frequency bands for emotion recognition from EEG with deep belief network. In: 2015 7th International IEEE/EMBS Conference on Neural Engineering (NER), pp. 154–157, April 2015. https://doi.org/10.1109/NER.2015.7146583

36. Zhuang, X., Rozgić, V., Crystal, M.: Compact unsupervised EEG response representation for emotion recognition. In: IEEE-EMBS International Conference on Biomedical and Health Informatics (BHI), pp. 736–739, June 2014. https://doi.org/10.1109/BHI.2014.6864469

Using Fourier Series to Improve the Prediction Accuracy of Nonlinear Grey Bernoulli Model

Ngoc Thang Nguyen[1], Van Thanh Phan[2(✉)], and Zbigniew Malara[3]

[1] Tay Nguyen University, 567 Le Duan Street, Buon Ma Thuot City,
DakLak, Vietnam
ngthang67@yahoo.com
[2] Quang Binh University, 312 Ly Thuong Kiet Street,
Dong Hoi, Quang Binh, Vietnam
thanhkem2710@gmail.com
[3] Wroclaw University of Science and Technology,
27 Wybrzeże Wyspiańskiego Street, 50-370 Wrocław, Poland
zbigniew.malara@pwr.edu.pl

Abstract. In recent decades, the Nonlinear Grey Bernoulli model "NGBM (1, 1)" has been applied in various fields and achieved positive results. However, its prediction results may be inaccurate in different scenario. In order to expand the field of application and to improve the predict quality of NGBM (1, 1) model, this paper proposes an effective model (named as Fourier-NGBM (1, 1)). This model includes two main stages; first, we get the error values based on the actual data and predicted value of NGBM (1, 1). Then, we used Fourier series to filter out and to select the low- frequency their error values. To test the superior ability of the proposed model, the historical data of annual water consumption in Wuhan from 2005 to 2012 in He et al.' paper is used. Forecasted results proved that the performance of Fourier-NGBM (1, 1) model is better than three forecasting models which are GM (1, 1), NGBM (1, 1) and improved Grey-Regression model. In subsequent research, more methodologies can be used to reduce the residual error of NGBM (1, 1) model, such as Markov chain or different kinds of Fourier functions. Additionally, the proposed model can be applied in different industries with the fluctuation data and uncertain information.

Keywords: Fourier series · Nonlinear grey Bernoulli model · Modeling · Prediction accuracy

1 Introduction

Grey prediction model is the central part of grey system theory [1]. It is an efficient method in dealing with uncertain information and building the time series forecasting models with limited data. In the early 1980s, the Grey model GM (1, 1) was introduced by Deng based on the control theory [2, 3]. It was a commonly used than the other forecasting models in time series because this model is easy to simulate and need small sample sizes. It was constructed based on the first-order accumulation method and first-order differential equation. Fundamentally, it follows the exponential growth laws.

© Springer Nature Switzerland AG 2019
N. T. Nguyen et al. (Eds.): ACIIDS 2019, LNAI 11431, pp. 363–372, 2019.
https://doi.org/10.1007/978-3-030-14799-0_31

Based on the parameter value of a, we can make decision in the short-term, mid-term as well as long-term forecasting. Previous studies has been evidenced the advantages of grey forecasting model in dealing with the problem in uncertain information and the limited data [3–5]. Therefore, many scientists used this model to employ in different areas such as tourism demand forecasting [6, 7], cargo throughput forecasting or motor vehicle [8–10], gold or stock index price forecasting [11–13], the output value forecasting of high technology industry [14–17], energy demand forecasting [18–20], etc.

Recently, many scholars has been used different ways to expand the application areas and to improve the predict quality of grey models. Specifically, Lin et al. [21] and Wang et al. [22] provided a new ways to calculate the background values. Hsu [17] and Wang et al. [23] proposed a systematic approach to optimize the development coefficient or the grey input coefficient or both of two parameters. Some researchers focused on the residual error modification of GM (1, 1) model such as Hsu [15] and Wang et al. [24]. Additionally, many scientists proposed hybrid models based GM (1, 1) model. There are the grey econometric model [25], grey Markov model [26, 27], and the grey fuzzy model [21], etc. All of the aforementioned research concentrated on modification of background values or residual errors.

On the other hand, some scholars used the different equations to modify the GM (1, 1) model become the new ones models. For example, Chen [28] proposed the nonlinear grey Bernoulli model NGBM (1, 1) to forecast the unemployment rate. Guo et al. [9] proposed a grey Verhulst model to forecast the Cargo Throughput. Chen et al. [29] used NGBM (1, 1) model to predict the annual unemployment rates of ten countries. Zhou et al. [30] also applied proposed model to forecast the foreign exchange rates of Taiwan's major trading partners. Above real case studies evidenced that the NGBM (1, 1) shows significantly improved the efficient prediction of the original GM (1, 1) model.

With the flexible changing of the exponential parameter n in the NGBM (1, 1) forecasting model, this feature has made the model more suitable with the linear or nonlinear characteristics of real systems. Rely on this feature, Zhou et al. [30] provided the novel NGBM (1, 1) model to forecast the power load. Hsu [16] proposed the novel NGBM (1, 1) based genetic algorithm to forecast the output values of high tech industries in Taiwan. Chen et al. [31] proposed a Nash NGBM (1, 1) model based the Nash equilibrium concept to forecast the monthly stock indices in Taiwan. Wang, et al. [32] proposed optimized NGBM (1, 1) model to forecast the qualified discharge rate in 31 provinces in China. Pao et al. [33] proposed model named a NGBM-OP model to forecast three important indicators of clean energy economy in China which are the number of CO_2 emissions, energy consumption, and economic growth. The empirical results proved that the predict ability of proposed model has remarkably improved when comparison this model with GM (1, 1), ARIMA and NGBM (1, 1) model.

In fact, because the initial condition is a part of the predictive function, so, this is a main factor to affect the performance of grey prediction that reason why most of researchers concentrated on modification of the background values. This paper aims to reduce the residual error of NGBM (1, 1) model by modifying this sequence obtained from NGBM (1, 1) with Fourier series. Numerical example shows that the Fourier-NGBM (1, 1) model can significantly improved the prediction accuracy.

The remaining of this paper is organized as follows. Section 2 analyzed the related work. Section 3 briefly introduces the NGBM (1, 1) model and proposes the systematic approach for improve the prediction accuracy of NGBM (1, 1) by Fourier series. Section 4 proves the effectiveness of proposed model based on the numerical example in He and Tao [34] for water consumption forecasting. Finally, the paper concludes with some comments and suggestions.

2   Related Work

As mentioned in Sect. 1, the NGBM (1, 1) model expresses the superior ability in forecast when comparisons with GM (1, 1) and Grey Verhulst model. Because it does not require a specific number in the exponential value of n (n $\neq$ 1). So, it can be flexible determines the shape of the models' curve with the fluctuation sequence data. This characteristic has attracted many scientists' concern and prompted further research for improving the prediction accuracy of this model. For dealing with the solution of optimal the background interpolation value p, Zhou et al. has been used algorithm of particle swarm optimization [30]. To show the superior ability in forecasting of proposed model, the long-term power load during 1996 to 2007 is used for modeling. According to the MAPE indexes for in-sample and out-of sample, the outcome of simulation clearly indicated that the proposed model is get a better forecasted results than among four mentioned models which are the GM (1, 1), the optimal p value of GM (1, 1), the NGBM (1, 1) and proposed model. These results one again confirmed that the prediction accuracy of NGBM (1, 1) can be improved by used PSO algorithm.

Hsu [16] applied genetic algorithm to get the optimal values in NGBM (1, 1) model in order to enhance the performance prediction with small amount of data. Then, the author applied in forecast the economic trends in the integrated circuit industries in Taiwan during the period time 1990 to 2007. The empirical results indicated that the error values of proposed model in this study is lower than the traditional GM (1, 1) model and grey Verhulst model in term of MAPE index for in sample is 12.48% and out of-sample is 4.64%. Chen et al. [31] are based on the Nash equilibrium concept to propose a Nash NGBM (1, 1) model which is optimization of the exponential value n and the background interpolation value p in the NGBM (1, 1). For testing the proposed model adaptability, the monthly stock indices in Taiwan was used. The forecasting results show that the proposed model has actually improved the forecasting precision.

Wang et al. [32] proposed a novel NBGM (1, 1) model by optimization two parameters p (background interpolation value) and n (exponential value). To solve this problems, the authors used as LINGO software. In order to verify the efficiency of the novel NGBM (1, 1) model in environment with fluctuation data, the authors use a numerical example in Wen's book and the output value of electronics industry in Taiwan. The simulated results demonstrate that the forecast performance of the novel NBGM (1, 1) model is significantly improved compare to original NGBM (1, 1) model. In additional, authors also applied improved model to forecast the annual qualified discharge rate in 31 provinces in China during the period times from 2001 to 2011. Empirical results show that the proposed model is more effective and reliable in prediction with average *RPE index* (relative percentage error) less than 4% is 27 provinces is and more than 4% are 4 provinces.

Summary, all those previous studies focused on the modification of the background value and parameters by using the different mathematical algorithms. Based on the idea of Wang and Phan [24], the study presents a systematic approach I tries to enhance the performance of NGBM (1, 1) model by use the Fourier series.

3 Methodology

3.1 Nonlinear Grey Bernoulli Model "NGBM (1, 1)"

According to Chen et al. [29, 31], the NGBM (1, 1) is a first-order one-variable grey Bernoulli differential equation. For predictions related to nonlinear time series with the small sample, the forecasted results of NGBM (1, 1) is better than the original grey forecasting models such GM (1, 1), grey Verhulst.... The algorithm of NGBM (1, 1) is express as follows:

Step 1: Assume that we have the original time series data $X^{(0)}$

$$X^{(0)} = \left\{ x^{(0)}(t_i) \right\}, \ i = 1, 2, \ldots, m \tag{1}$$

Where $x^{(0)}(t_i)$ is the system output at time t_i, $x^{(0)}(t_i) \geq 0$
m is the total number of modeling data.

Step 2: To construct the $X^{(1)}$, we use the first-order accumulated generating operation (1-AGO) as follow equation

$$X^{(1)} = \left\{ x^{(1)}(t_i) \right\}, \ i = 1, 2, \ldots, m \tag{2}$$

$$\text{Where } x^{(1)}(t_k) = \sum_{i=1}^{k} x^{(0)}(t_i), \ k = 1, 2, \ldots, m \tag{3}$$

Step 3: $X^{(1)}$ is modeled by the Bernoulli differential equation:

$$\frac{dX^{(1)}}{dt} + aX^{(1)} = b \left[X^{(1)} \right]^n \tag{4}$$

Where a is called the developing coefficient
b is the grey input
n is any real number and $n \neq 1$.
Step 4: In order to estimate the parameter a and b, the Eq. (4) is approximated as:

$$\frac{\Delta X^{(1)}(t_k)}{\Delta t_k} + aX^{(1)}(t_k) = b \left[X^{(1)}(t_k) \right]^n \tag{5}$$

$$\text{Where } \Delta X^{(1)}(t_k) = x^{(1)}(t_k) - x^{(1)}(t_{k-1}) = x^{(0)}(t_k) \tag{6}$$

$$\Delta t_k = t_k - t_{k-1} \tag{7}$$

$$\text{let } \Delta t_k = 1, \text{ then } z^{(1)}(t_k) = px^{(1)}(t_k) + (1-p)x^{(1)}(t_{k-1}), \ k = 2,3,\ldots,n \tag{8}$$

To replace $X^{(1)}(t_k)$ in Eq. (5), we obtain

$$x^{(0)}(t_k) + az^{(1)}(t_k) = b\left[z^{(1)}(t_k)\right]^n, \ k = 2,3,\ldots,m \tag{9}$$

Where $z^{(1)}(t_k)$ is the background value,
p is production coefficient ($p \in [0,1]$), normally the p value set to 0.5.

Step 5: We use the ordinary least square method (OLS) to calculate the parameter a and b in Eq. (9) as follow:

$$\begin{bmatrix} a \\ b \end{bmatrix} = (B^T B)^{-1} B^T Y \tag{10}$$

$$\text{Where } B = \begin{bmatrix} -z^{(1)}(t_2) & (z^{(1)}(t_2))^n \\ -z^{(1)}(t_3) & (z^{(1)}(t_3))^n \\ \ldots & \ldots \\ -z^{(1)}(t_m) & (z^{(1)}(t_m))^n \end{bmatrix} \tag{11}$$

$$\text{And } Y = \left[x^{(0)}(t_2), x^{(0)}(t_3), \ldots, x^{(0)}(t_m))\right]^T \tag{12}$$

Step 6: Replace the parameter a and b into the Eq. (4), we can get the $\hat{x}^{(0)}(t_k)$ by equation below:

$$\hat{x}^{(1)}(t_k) = \left[\left(x^{(0)}(t_1)^{(1-n)} - \frac{b}{a}\right)e^{-a(1-n)(t_k-t_1)} + \frac{b}{a}\right]^{\frac{1}{1-n}}, \ n \neq 1, \ k = 1,2,3,\ldots \tag{13}$$

Step 7: To get the predicted value of $x^{(0)}(t_k)$, we adopted inverse accumulated generating operation (IAGO) as below function

$$\begin{cases} \hat{x}^{(0)}(t_1) = x^{(0)}(t_1) & \\ \hat{x}^{(0)}(t_k) = \hat{x}^{(1)}(t_k) - \hat{x}^{(1)}(t_{k-1}) & \end{cases}, \ k = 2,3,\ldots \quad \begin{matrix} (14) \\ (15) \end{matrix}$$

3.2 Residual Error Modification by Fourier Series

The Fourier series [6] was used to filter out and to select the low- frequency error values of the NGBM (1, 1) for improvement the performance in the prediction. The overall procedure as follow:

Step 1: Call ε is a residual error series. ε is defined by:

$$\varepsilon(k) = x(k) - \hat{x}(k), \; k = 2, 3, \ldots m \tag{16}$$

Step 2: Used the Fourier series for modifying the error values of NGBM (1, 1) as follow equation:

$$\hat{\varepsilon}(k) = \frac{1}{2}a_0 + \sum_{i=1}^{Z} \left[a_i \cos\left(\frac{2\pi i}{m-1}k\right) + b_i \sin\left(\frac{2\pi i}{m-1}k\right) \right], \; k = 1, 2, 3, \ldots, m \tag{17}$$

Where $Z = \left(\frac{m-1}{2}\right) - 1$ is the minimum deployment frequency and only take integer number [35], therefore, the residual error is estimate as follow:

$$\varepsilon = P \times C \tag{18}$$

Where

$$P = \begin{bmatrix} \frac{1}{2} & \cos\left(\frac{2\pi \times 1}{m-1} \times 2\right) \sin\left(\frac{2\pi \times 1}{m-1} \times 2\right) & \ldots & \cos\left(\frac{2\pi \times Z}{m-1} \times 2\right) \sin\left(\frac{2\pi \times Z}{m-1} \times 2\right) \\ \frac{1}{2} & \cos\left(\frac{2\pi \times 1}{m-1} \times 3\right) \sin\left(\frac{2\pi \times 1}{m-1} \times 3\right) & \ldots & \cos\left(\frac{2\pi \times Z}{m-1} \times 3\right) \sin\left(\frac{2\pi \times Z}{m-1} \times 3\right) \\ \ldots & \ldots & \ldots & \ldots \\ \frac{1}{2} & \cos\left(\frac{2\pi \times 1}{m-1} \times m\right) \sin\left(\frac{2\pi \times 1}{m-1} \times m\right) & \ldots & \cos\left(\frac{2\pi \times Z}{m-1} \times m\right) \sin\left(\frac{2\pi \times Z}{m-1} \times m\right) \end{bmatrix} \tag{19}$$

$$\text{And } C = [a_0, a_1, b_1, a_2, b_2, \ldots, a_Z, b_Z] \tag{20}$$

We use the OLS method to calculated the parameter a_0, a_1, b_1, a_2, $b_2, \ldots$, a_Z, b_Z by the equation below:

$$C = \left(P^T P\right)^{-1} P^T [\varepsilon]^T \tag{21}$$

To replace the parameters just estimated in Eq. (21) into Eq. (17), we can get the modified residual series $\hat{\varepsilon}$ as follows:

$$\hat{\varepsilon}(k) = \frac{1}{2}a_0 + \sum_{i=1}^{Z} \left[a_i \cos\left(\frac{2\pi i}{m-1}k\right) + b_i \sin\left(\frac{2\pi i}{m-1}k\right) \right] \tag{22}$$

Step 3: From the predicted series value $\hat{x}$ in Eq. (15) and the modification residual error $\hat{\varepsilon}$ in Eq. (22), the predicted series value of Fourier NGBM (1, 1) "$\hat{v}$" determined by:

$$\hat{v} = \{\hat{v}_1, \hat{v}_2, \hat{v}_3, \ldots, \hat{v}_k, \ldots, \hat{v}_m\} \tag{23}$$

$$\text{Where } \hat{v} = \begin{cases} \hat{v}_1 = \hat{x}_1 \\ \hat{v}_k = \hat{x}_k + \hat{\varepsilon}_k \end{cases} (k = 2, 3, \ldots, m) \tag{24}$$

3.3   Forecast Model Evaluation

In this paper, the Means Absolute Percentage Error (MAPE) index is used to evaluate the accuracy and reliability of predict models [35]. It is expressed as follows:

$$MAPE = \frac{1}{m} \sum_{k=2}^{m} \left| \frac{x^{(0)}(k) - \hat{x}^{(0)}(k)}{x^{(0)}(k)} \right| \times 100\% \tag{25}$$

Where $x^{(0)}(k)$ indicated the actual value while $\hat{x}^{(0)}(k)$ are the forecasted values in time period k. And m is the total number of predictions. According to Wang et al. [36, 37] the MAPE is divided into four levels forecast accuracy, level 1 is the excellent forecast (less than 1%), level 2 is a good forecast (from 1% to 5%), level 3 is a reasonable forecast (from 5% to 10%), and more than 10% is an inaccurate forecast.

4   Model Validation

To verify the improved of the Fourier-NGBM (1, 1) model in forecast, This paper take the historical data of annual water consumption in Wuhan from 2005 to 2012 in He et al. [34] to establish the model and then compare the performance among Fourier-NGBM (1, 1) and Coupling model of Grey system and Multivariate Linear Regression as well as original NBBM (1, 1).

According to the historical data of the annual water consumption demand from year 2005 to 2012 [see Table 1], this paper use the Microsoft Excel to simulate the mathematical algorithm of grey forecasting models. This software was provided by Microsoft Corporation. Excel software not only provides some basic functions but also provides two useful functions Mmult and Minverse to calculate the parameters in grey forecasting model, In particular, the function of Mmult (array 1, array 2) using for multiplying two relevant arrays matrices. The function of Minverse (array) use to calculate the inverse matrix. Moreover, Excel software was integrated the efficiency tool named Excel- solves for finding out the optimal values of parameters in forecasting models. For the sake of convenience, this paper just showed the detailed results of NGBM (1, 1) and Fourier-NGBM (1, 1) in Table 1. The other forecasted results are omitted here. Only the MAPE indexes of these models are listed in Table 2.

According to the average of MAPE index in Table 1, this results reveals that the residual error of Fourier-NGBM (1, 1) with n = −0.11 and p = 0.5 is lower than the original one. The average MAPE index decreased from 0.8% to 0%. This result also uses to compare with improved Grey-Regression Model in [34] as well as GM (1, 1). The overall MAPE indexes as well as the performance forecasting of these models was shown in Table 2.

The results from Table 2 show that the MAPE of the Fourier-NGBM (1, 1) is nearly 0%. This feature confirmed that the performance of Fourier-NGBM (1, 1) model is excellent. Comparison with three remaining models, the Fourier-NGBM (1, 1) shows the highest accuracy in forecast the total annual water consumption Demand in Wuhan.

Table 1. Forecasted results by NGBM (1, 1) and Fourier-NGBM (1, 1) models

Actual value		NGBM (1, 1) n = −0.11, p = 0.5		Fourier-NGBM (1, 1) n = −0.11, p = 0.5	
	$x^{(0)}(k)$	$\hat{x}^{(0)}(k)$	Error (%)	$\hat{x}^{(0)}(k)$	Error (%)
2005	367,204	367,204	0	367,204	0
2006	369,355	369,355	0	369,355	0
2007	363,883	364,331.6	0.1	363,883	0
2008	360,467	366,777.6	1.8	360,467	0
2009	377,944	372,695.6	1.4	377,944	0
2010	379,346	380,754.6	0.4	379,346	0
2011	397,345	390,359.2	1.8	397,345	0
2012	395,713	401,199.8	1.4	395,713	0
Average MAPE			0.8		0

Table 2. The accuracy indexes of forecasting model

Model	MAPE (%)	Prediction accuracy (%)	Performance
Original GM (1, 1)	1.163	98.837	Good
The proposed model in He and Tao [34]	1.156	98.84	Good
NGBM (1, 1)	0.8	99.2	Excellent
Fourier - NGBM (1, 1)	**0**	**100**	**Excellent**

5 Conclusion

In this paper, by using Fourier series aim to reduce the error values of NGBM (1, 1) model. We propose an effectiveness NGBM (1, 1) model named as Fourier-NGBM (1, 1). Through the simulation of the numerical example in He et al. [34] paper, this results indicated that the proposed model can significantly decrease the error of prediction. Comparison with three grey forecasting models, the Fourier-NGBM (1, 1) shows that the performance of Fourier-NGBM (1, 1) model is more accuracy in forecast with the MAPE less much than 1%.

The simulated results in this study one again confirmed that Fourier-NGBM (1, 1) can get more accurate prediction. This is a powerful tool for supporting the top managers or policymakers in making decisions. For future work, the authors will be used different methodologies to reduce the error of Fourier-NGBM (1, 1) model by optimization the parameters p and n in this model. Additionally, the authors can be applied this model to dealing with the complex issues in term of fluctuation data or uncertain information in different industries.

References

1. Deng, J.L.: Grey Prediction and Decision. Huazhong University of Science and Technology Press, Wuhan (2002). (in Chinese)
2. Deng, J.L.: Solution of grey differential equation for GM (1, 1|s, r) in matrix train. J. Grey Syst. **14**(1), 105–110 (2002)
3. Deng, J.L.: Control problems of grey systems. Syst. Control Lett. **5**, 288–294 (1982)
4. Yi, L., Liu, S.: A historical introduction to grey system theory. In: IEEE International Conference on System, Man and Cybernetics, pp. 2403–2408 (2004)
5. Liu, S., Forrest, J., Yang, Y.: A brief introduction to grey system theory. In: IEEE International Conference on Grey Systems and Intelligent Services, GSIS (2011)
6. Huang, Y.L., Lee, Y.H.: Accurately forecasting model for the stochastic volatility data in tourism demand. Modern Econ. **2**(5), 823–829 (2011)
7. Chu, F.L.: Forecasting tourism demand in Asian-Pacific countries. Annu. Tourism Res. **25**(3), 597–615 (1998)
8. Jiang, F., Lei, K.: Grey prediction of port cargo throughput based on GM (1, 1, a) model. Logist. Technol. **9**, 68–70 (2009)
9. Guo, Z.J., Song, X.Q., Ye, J.: A Verhulst model on time series error corrected for port cargo throughput forecasting. J. East. Asia Soc. Transp. Stud. **6**, 881–891 (2005)
10. Lu, I.J., Lewis, C., Lin, S.J.: The forecast of motor vehicle, energy demand and CO_2 emission from Taiwan's road transportation sector. Energy Policy **37**(8), 2952–2961 (2009)
11. Kayacan, E., Ulutas, B., Kaynak, O.: Grey system theory-based models in time series prediction. Expert Syst. Appl. **37**, 1784–1789 (2010)
12. Askari, M., Askari, H.: Time series grey system prediction-based models: gold price forecasting. Trends Appl. Sci. Res. **6**, 1287–1292 (2011)
13. Wang, Y.F.: Predicting stock price using fuzzy Grey prediction system. Expert Syst. Appl. **22**(1), 33–39 (2002)
14. Lin, C.T., Yang, S.Y.: Forecast of the output value of Taiwan's IC industry using the Grey forecasting model. Int. J. Comput. Appl. Technol. **19**(1), 23–27 (2004)
15. Hsu, L.C.: Applying the grey prediction model to the global integrated circuit industry. Technol. Forecast. Soc. Change **70**, 563–574 (2003)
16. Hsu, L.C.: A genetic algorithm based nonlinear grey Bernoulli model for output forecasting in integrated circuit industry. Expert Syst. Appl. **37**(6), 4318–4323 (2010)
17. Hsu, L.C.: Using improved grey forecasting models to forecast the output of opto-electronics industry. Expert Syst. Appl. **38**(11), 13879–13885 (2011)
18. Li, D.C., Chang, C.J., Chen, C.C., Chen, W.C.: Forecasting short-term electricity consumption using the adaptive grey-based approach-an Asian case. Omega **40**, 767–773 (2012)
19. Hsu, C.C., Chen, C.Y.: Application of improved grey prediction model for power demand forecasting. Energy Convers. Manag. **44**, 2241–2249 (2003)
20. Kang, J., Zhao, H.: Application of improved grey model in long-term load forecasting of power engineering. Syst. Eng. Proc. **3**, 85–91 (2012)
21. Lin, Y.H., Chiu, C.C., Lee, P.C.: Applying fuzzy grey modification model on inflow forecasting. Eng. Appl. Artif. Intell. **25**(4), 734–743 (2012)
22. Wang, Z.X., Dang, Y.G., Liu, S.F.: The optimization of background value in GM (1, 1) model. J. Grey Syst. **10**(2), 69–74 (2007)
23. Wang, C.H., Hsu, L.C.: Using genetic algorithms grey theory to forecast high technology industrial output. Appl. Math. Comput. **195**, 256–263 (2008)

24. Wang, C.N., Phan, V.T.: An enhancing the accurate of grey prediction for GDP growth rate in Vietnam. In: 2014 International Symposium on Computer, Consumer and Control, IS3C, pp. 1137–1139 (2014). https://doi.org/10.1109/is3c.2014.295

25. Liu, S.F., Lin, Y.: Grey Information Theory and Practical Applications. Springer, London (2006). https://doi.org/10.1007/1-84628-342-6

26. Dong, S., Chi, K., Zhang, Q.Y.: The application of a Grey Markov Model to forecasting annual maximum water levels at hydrological stations. J. Ocean Univ. China 11(1), 13–17 (2012)

27. Hsu, Y.T., Liu, M.C., Yeh, J., Hung, H.F.: Forecasting the turning time of stock market based on Markov-Fourier grey model. Expert Syst. Appl. 36(4), 8597–8603 (2009)

28. Chen, C.I.: Application of the novel nonlinear grey Bernoulli model for forecasting unemployment rate. Chaos Solitons Fractals 37(1), 278–287 (2008)

29. Chen, C.I., Chen, H.L., Chen, S.P.: Forecasting of foreign exchange rates of Taiwan's major trading partners by novel nonlinear Grey Bernoulli model NGBM (1, 1). Commun. Nonlinear Sci. Numer. Simul. 13(6), 1194–1204 (2008)

30. Zhou, J.Z., Fang, R.C., Li, Y.H.: Parameter optimization of nonlinear grey Bernoulli model using particle swarm optimization. Appl. Math. Comput. 207(2), 292–299 (2009)

31. Chen, C.I., Hsin, P.H., Wu, C.S.: Forecasting Taiwan's major stock indices by the Nash nonlinear grey Bernoulli model. Expert Syst. Appl. 37(12), 7557–7562 (2010)

32. Wang, Z.X., Hipel, K.W., Wang, Q.: An optimized NGBM (1, 1) model for forecasting the qualified discharge rate of industrial wastewater in China. Appl. Math. Model. 35(12), 5524–5532 (2011)

33. Pao, H.T., Fu, H.C., Tseng, C.L.: Forecasting of CO_2 emissions, energy consumption and economic growth in China using an improved grey model. Energy 40, 400–409 (2012)

34. He, F., Tao, T.: An improved coupling model of grey-system and multivariate linear regression for water consumption forecasting. Pol. J. Environ. Stud. 23(4), 1165–1174 (2014)

35. Makridakis, S.: Accuracy measures: theoretical and practical concerns. Int. J. Forecast. 9, 527–529 (1993)

36. Wang, C.N., Phan, V.T.: An improvement the accuracy of grey forecasting model for cargo throughput in international commercial ports of Kaohsiung Industrial and business forecasting methods. Int. J. Bus. Econ. Res. 1(3), 1–5 (2014)

37. Wang, C.N., Phan, V.T.: An improved nonlinear grey Bernoulli model combined with Fourier series. Math. Probl. Eng., Article ID 740272, 7 (2015)

Learning Hierarchical Weather Data Representation for Short-Term Weather Forecasting Using Autoencoder and Long Short-Term Memory Models

Yaya Heryadi[(⊠)]

Computer Science Department, BINUS Graduate Program – Doctor
of Computer Science, Bina Nusantara University, Jakarta, Indonesia
yayaheryadi@binus.edu

Abstract. Weather prediction task remains a challenging problem in computer vision field although some solutions have been used for many applications such as air/sea transportation. The increasing requirement toward safer human transportation that requires a robust weather prediction model has motivated the development of a vast number of weather prediction models. In the past decade, the advent of deep learning methods has opened up a new approach to weather prediction, mainly in two areas: automated learning hierarchical representation of weather data and robust weather prediction models. This paper presents a method for automatic feature extraction from weather time series data using Autoencoder model. The learned weather representation was used to train Long Short-term Memory model as a prediction model or regressor. Although it can be used to predict many other weather variables, in this study, the proposed model was tested to predict temperature, dew point, and humidity. The results show that the model performance measured by training and testing RMSE values are as follows. Predicting temperature: AE90-LSTM model (0.00003, 0.00010) and predicting dew point: AE199-LSTM model (0.00005, 0.00010). Interestingly, for predicting humidity 100LSTM model (0.00004, 0.00001) and AE100-LSTM model (0.00001, 0.00008) achieved almost similar performance.

Keywords: Autoencoder · Long Short-term memory · Weather prediction

1 Introduction

Weather prediction task is defined as *"the application of science and technology to predict the conditions of the atmosphere for a given location and time"* [1]. In the past, weather prediction models were mostly based on physical model of the atmosphere derived from Meteorology or Physics fields [2–4]. The drawbacks of this approach are mainly: high complexity models, requiring a large volume of variety datasets, high numeric computation complexity but inaccurate prediction for large periods of time due to the inability to handle unstable perturbations [5]. The advent of deep learning proposed by [6] has opened up a new approach to address weather prediction task. Following the popularity of deep learning method in the past decade, many efforts for predicting weather using deep

© Springer Nature Switzerland AG 2019
N. T. Nguyen et al. (Eds.): ACIIDS 2019, LNAI 11431, pp. 373–384, 2019.
https://doi.org/10.1007/978-3-030-14799-0_32

learning have been reported, for example, a study by [7] to predict temperature, humidity, pressure, dew point, wind speed, precipitation and visibility. Another study by [8, 9] used deep learning model for predicting wind speed/power.

Although many prediction models have been proposed, weather prediction remains a challenging task. Two main issues are extracting weather data feature and developing a robust prediction model. Although machine learning approach has shown a tremendous impact in solving many prediction tasks in many domains, it requires explicit feature extraction from raw input dataset. In many applications, this process becomes an additional workload and subjectivity on what features to be extracted from the raw dataset.

Due to the research interest in this research field, many weather prediction models have been proposed. However, to the best of our knowledge, most of the models used the raw weather data set as input to an end-to-end weather prediction process, without prior automated process for extracting weather features. Although both studies used Long Short-term Memory (LSTM) model as a weather prediction model, this study is different from the study reported by [7]. The main different was that this study proposed a combination of Autoencoder (AE) and LSTM model as a regression model. In the proposed model, the AE model part was intended to learn hierarchical representation of weather data. Finally, the LSTM model part was used as the regressor using hierarchical features generated by the first model.

2 Related Works

2.1 Deep Learning Approach to Weather Forecasting

The term machine learning was coined by Samuel [10]. It refers to algorithms to address particular tasks without explicitly programmed by making generalization of the pattern obtained in input dataset. Mitchell [11] further defined machine learning as algorithms that "*learn from experience E with respect to some class of tasks T and performance measure P, if its performance at tasks in T, as measured by P, improves with experience E.*" In the last ten years, there have been a vast number of studies using machine learning for weather classification [5]. However, the common issue in machine learning approach for weather classification as reported by [5] was feature selection. Inappropriate feature extracted from raw data might lead to low model performance.

Deep learning can be viewed as a particular class of machine learning or learning algorithms which "*... allows computational models that are composed of multiple processing layers to learn representations of data with multiple levels of abstraction*" [6]. The authors claimed that deep learning approach had several advantages over machine learning. While machine learning depends on manual feature extraction as input, deep learning generates features from raw input. With such a characteristic, many studies show some evidences that deep learning improved significantly the state-of-the-art solution in several areas including: high volume of images classification [12], speech recognition [13, 14] and many others.

Weather observation data contained many patterns (periodical) which are very difficult for human to accurately model the patterns. The ability of deep learning approach to produce hierarchical representation of weather data for weather modeling and analysis tasks had gained many research interest, for example, weather forecasting using LSTM model reported by [7].

2.2   Autoencoder Model

Autoencoder model is a neural network trained unsupervisedly to copy its input to its output [15]. The importance of this model, among others, is that the outputs of hidden layers can be interpreted as a hierarchical representation of the input. Therefore, the trained autoencoder model can be used for: (1) learning a hierarchical features from input dataset and (2) reconstructing input dataset from its hierarchical features [16].

As Fig. 1 indicates, in an autoencoder model structure, an encoding function $f(x, \theta_f)$ where: $\theta_f = \{W, b\}$ maps an input x to a hidden representation y using a affine transformation with a projection matrix W and a bias b. Sigmoid function $\sigma(x)$ is typically used as a squashing function

$$\sigma(x) = \frac{1}{1 + e^{-x}} \tag{1}$$

So that:

$$y = f(x, \theta_f) = \sigma(Wx + b) \tag{2}$$

Then a decoding function $g(y, \theta_g)$ where $\theta_g = \{W', b'\}$ maps the hidden representation y back to a reconstruction of input z. A decoding function can be either linear or nonlinear.

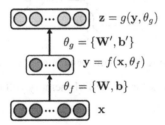

Fig. 1. The general structure of autoencoder model (Source: [17])

The autoencoder model parameters W, b, W' and b' were estimated using an unsupervised training algorithm by selecting the most optimum model from hypothesis space. Given an objective function $\mathcal{L}$ as follows.

$$\mathcal{L} = \underset{W,b,W',b'}{\mathrm{argmin}} \; \mathbb{E}\{\|x - z\|\}_2^2 \tag{3}$$

The most optimum model in the hypothesis space was selected as the model that minimizes the objective function $\mathcal{L}$ with respect to parameters W, b, W' and b' (see Eq. 3) where the term $x - z$ represents reconstruction error. In the autoencoder structure (see Fig. 1), the hidden layer y that was generated by the training algorithm can be viewed as a hierarchical feature that represents internal structure of the input x.

2.3   Long Short-Term Memory Model

Long Short-term Memory (LSTM) model was proposed by [18] and was initially designed to address the drawback of prominent Real-time Recurrent Learning [19] and backpropagation through time training algorithm [20, 21]. The weakness of the prominent training algorithms, according to [18], was a potential error explosion or diminishing due to iterative backward process of error signals.

The basic unit of a LSTM model structure is a complex unit called memory cell which comprises a node, input, output and forget gates which control input/output signals in the respective cell (see Fig. 2). The input gate function determines when the input signal to be processed, the forget gate controls when the output value from the previous step is significant enough to be remembered or forgotten, and output gate controls when should output the value. With such a structure, LSTM model was possible to learn long-term dependencies [18]. Many studies in the past decade showed LSTMs have achieved great success in addressing temporal sequence tasks such as weather prediction using rectifying variable [22] and speech recognition [23].

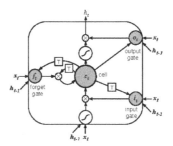

Fig. 2. Long Short-term memory cell (Source: [24])

Following [18], parameters for the LSTM memory cell can be computed using the following equations:

$$f_t = \sigma_g\left(W_f x_t + U_f h_{t-1} + b_f\right) \tag{4}$$

$$i_t = \sigma_g\left(W_i x_t + U_i h_{t-1} + b_i\right) \tag{5}$$

$$o_t = \sigma_g\left(W_o x_t + U_o h_{t-1} + b_o\right) \tag{6}$$

$$c_t = f_t \circ c_{t-1} + i_t \circ \sigma_c\left(W_c x_t + U_c h_{t-1} + b_c\right) \tag{7}$$

$$h_t = o_t \circ \sigma_h(c_t) \tag{8}$$

Where: f_t be forget gate's activation vector; i_t be input gate's activation vector; o_t be output gate's activation vector; W_f, W_i, W_o, U_f, U_i, U_o are weight matrices to be learned during model training; σ be activation function; $\circ$ be Hadarmard product or element-wise multiplication; and h_t be output of the LSTM memory cell.

The LSTM training algorithm was used to train LSTM model supervisedly. The most optimum model (hypothesis) was selected from hypothesis space that minimized the following objective function:

$$\mathcal{L}_2 = \underset{W_i,W_o,W_f,W_c,U_o,U_i,U_o,U_c}{\mathrm{argmin}} \mathbb{E}\{\|y_t - \hat{y}_t\|\}_2^2 \tag{9}$$

Where: y_t be actual weather data and $\hat{y}_t$ the predicted weather data.

3 Research Method

3.1 Dataset

Similar to the dataset used in the study reported by [22], the dataset for this observation comprised weather observation data from Weather Underground (https://www.wunderground.com/). The last access to this website was 19 March 2017. The location of weather observation for this research was the Hang Nadim Airport located in Batam, Riau Islands, Indonesia. The study location has a DMS coordinate of 1°07′9.00″ N 104°07′4.20″ E. This airport is very strategic for civil as well as military air transportation because it is located very near to the Indonesia boundary with Singapore and has relatively close distance to neighboring countries such as Thailand, Malaysia, Vietnam and others. The weather dataset for this study contains: 40,025 observation data including: temperature, dew point and humidity from 2014 to year 2016.

3.2 Data Preprocessing

In this study, weather data modeling and analysis were limited to temporal sequence data or time series with 30 min interval between two successive data. Several pre-processing steps were adopted including: missing data handling, rescaling and moving average. The missing observation was predicted using the data grand average. Data rescaling aimed to remove the effect of measurement unit by rescaling the data so that the data were in range [0, 1] using Eq. (10).

$$x_t = \frac{x_t - x_{min}}{x_{max} - x_{min}} \tag{10}$$

Simple moving average with time lag n was computed as the average of sequence data in n-length interval using Eq. (11). In this study $n = 9$ or about 4.5 h observation range.

$$x_t = \frac{1}{n}(x_t + x_{t-1} + \ldots + x_{t-n+1}) \tag{11}$$

The aim of simple moving average was to smooth out data by filtering out the "noise" from random fluctuations of weather data.

Following [25], the basic structure of weather data for training prediction model was represented in terms of frames. In this study, each frame comprised 100 overlapped sequence data of point observation or about 2 day observation range and every 100th data from the starting frame was used as the target data. Consider $x_1, x_2, ..., x_T$ is observation dataset of X weather variable with length T. The weather frame s_i with length 100 and prediction target t_i for $i = 1, 2, .., m$ is defined as:

$$s_i = (x_i, x_{i+1}, ..., x_{i+99}) \tag{12}$$

$$t_i = x_{i+100} \tag{13}$$

3.3   Model Training and Evaluation

The proposed weather prediction model in this study was a combined Autoencoder-LSTM model. In this model, Autoencoder (AE) model was used to copy its input to its output unsupervisedly. Then, the hidden layer of the trained AE model was used as the input feature for the successive LSTM model (see Fig. 3) (Tabl 1).

In the above table, the first and second hidden layers of the AE model were used as hierarchical representation of the input dataset (Tables 2 and 3).

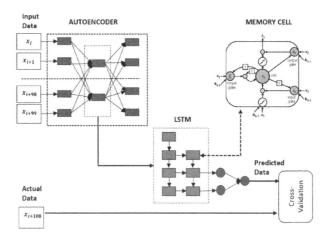

Fig. 3. The Autoencoder-LSTM (AE+LSTM) model

Table 1. The structure of autoencoder model

Layer (Type)	Output Shape	Parameter #
input_1 (InputLayer)	(None, 100)	0
dense_1 (Dense)	(None, 100)	10,100
dense_2 (Dense)	(None, 90)	9,090
dense_3 (Dense)	(None, 50)	4,550
dense_4 (Dense)	(None, 3)	153
dense_5 (Dense)	(None, 50)	200
dense_6 (Dense)	(None, 90)	4,590
dense_7 (Dense)	(None, 100)	9,100
dense_8 (Dense)	(None, 100)	10,100

Table 2. The structure of 100LSTM model

Layer (Type)	Output Shape	Parameter #
lstm_1 (LSTM)	(None, 1, 100)	41,600
lstm_2 (LSTM)	(None, 1, 90)	68,760
lstm_3 (LSTM)	(None, 1, 50)	28,200
time_distributed_1 (TimeDistributed)	(None, 1, 50)	2,550
flatten_1 (Flatten)	(None, 50)	0
dense_10 (Dense)	(None, 1)	51
activation_1 (Activation)	(None, 1)	0

Table 3. The Structure of 90LSTM model

Layer (Type)	Output Shape	Parameter #
lstm_1 (LSTM)	(None, 1, 90)	33,840
lstm_2 (LSTM)	(None, 1, 50)	28,200
time_distributed_1 (TimeDistributed)	(None, 1, 50)	2,550
flatten_1 (Flatten)	(None, 50)	0
dense_10 (Dense)	(None, 1)	51
activation_1 (Activation)	(None, 1)	0

4 Experiment Result and Discussion

4.1 Dataset

The proposed prediction model was tested using 40,025 observation data of the following variables: temperature, dew point and humidity. Data distribution of each variable is visualized in Figs. 4 and 5.

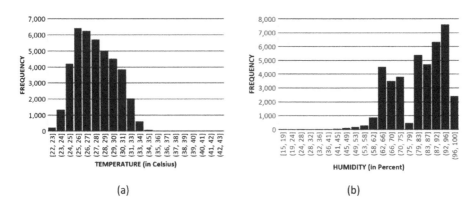

(a)   (b)

Fig. 4. Data Histogram of (a) temperature and (b) humidity

As can be seen from Fig. 5(a, b and c), the data distribution was not normally distributed. For that reason, these data were rescaled to standardize the value range into [0, 1].

Each variable of the dataset was then converted into some overlapped frames with length = 100 (equivalent to 2 day observation) using Eqs. 12 and 13. The data frame was represented as a matrix $X = [x_{i,j}]$ where $i = 1, 2, .., m$ and $j = 1, 2, .., 100$; the target data vector was represented as a vector $Y = [t_i]$. Prior to training model, the dataset was divided randomly into data for training the model (training dataset) and testing the model (testing dataset) with a proportion of 80:20.

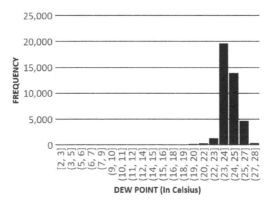

Fig. 5. Data Histogram of dew point

4.2   Model Evaluation

In this study, the trained models were evaluated using RMSE as performance metric. The training RMSE was computed as the average of deviation between predicted value and the actual data (training dataset), while the testing RMSE was computed as the average of deviation between predicted value and the actual data (testing dataset). In this study, the two prediction models were explored as follows.

(1) LSTM model: two models were explored namely 90LSTM and 100LSTM models. The number was associated with the number of memory cells in the first layer of the LSTM. The models were trained using Adam training algorithm [26] with the training dataset as input. Root mean squared error (RMSE) was used as the objective function. Next, the trained models were tested using the testing dataset.

(2) AE + LSTM model: two models were explored namely AE90-LSTM and AE100-LSTM models. The number was associated with the number of nodes in the AE hidden layer whose output was later used in the next training step as the learned representation from training dataset as input of LSTM model training. Hence, training process of both models was implemented in two steps. *First*, the Autoencoder was trained using Rmsprop training algorithm [27] with the training dataset as input. Root mean squared error (RMSE) was used as the objective function. Having been trained, the hidden layer of the Autoencoder model was treated as a coding of hierarchical representation of the input dataset. The outputs of the hidden layer were then forwarded to LSTM model as the input dataset. Next, the LSTM model was trained supervisedly. In particular, the LSTM model was trained using Adaptive Moment Estimation (Adam) training algorithm proposed by [26] which used mean squared error as the objective function. *Finally*, the trained models were tested using the testing dataset.

Model cross-validation was implemented using Leave-one-out technique in which the proportion of training dataset and testing dataset was 80:20 respectively. Training performance of the predictor models was measured using training and testing deviations which are summarized in Tables 4, 5 and 6 and Figs. 6 and 7.

Table 4. Performance comparison of temperature prediction model

RMSE	100LSTM	90LSTM	AE100-LSTM	AE90-LSTM
Training RMSE	0.00003	0.00005	0.00001	0.00001
Testing RMSE	0.00010	0.00040	0.00010	0.00001

Table 5. Performance comparison of dew point prediction model

RMSE	100LSTM	90LSTM	AE100-LSTM	AE90-LSTM
Training RMSE	0.00003	0.00005	0.00001	0.00001
Testing RMSE	0.00040	0.00010	0.00010	0.00020

Table 6. Performance comparison of humidity prediction model

RMSE	100LSTM	90LSTM	AE100-LSTM	AE90-LSTM
Training RMSE	0.00004	0.00006	0.00001	0.00002
Testing RMSE	0.00001	0.00040	0.00008	0.00020

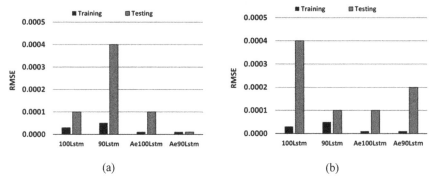

(a)   (b)

Fig. 6. Model Prediction Performance for (a) Temperature and (b) Dew Point

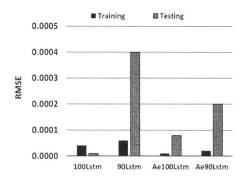

Fig. 7. Model Prediction Performance for Humidity

The predictor model (regressor) performance can be summarized as follows:

(1) Predicting temperature: AE90-LSTM model achieved 0.00003 Training RMSE and 0.00010 Testing RMSE.

(2) Predicting dew point: AE199-LSTM model achieved 0.00005 Training RMSE and 0.00010 Testing RMSE.

(3) Predicting humidity: 100LSTM model achieved 0.00004 Training RMSE and 0.00001 Testing RMSE; and AE100-LSTM model achieved 0.00001 Training RMSE and 0.00008 Testing RMSE. Both models have almost similar performance.

5   Conclusion

Weather prediction task has gained wide research interests resulted in a plethora of proposed prediction models. However, the task remains a challenging problem in computer vision research field. The advent of deep learning in the past decade has brought about a new approach to weather prediction mainly in learning hierarchical representation of weather data and developing robust prediction models.

Many previous reports showed some evidences that LSTM model achieved high accuracy for time series prediction. Motivated by previous studies on using LSTM for weather prediction, this study explored the role of autoencoder (AE) as feature extractor of weather variable to improve performance of LSTM as weather regression model. The empirical results from this study were obtained by comparing the performance of AE + LSTM and LSTM models to predict temperature, dew point and humidity time series data using RMSE as the performance metric. The main findings show that performance of the LSTM for weather prediction can be improved by combining it with Autoencoder model which was capable of learning hierarchical weather feature automatically from weather dataset. Exploring some variants of LSTM model for developing a robust weather predictor model is the future step of this research.

References

1. Wikipedia contributors: Weather forecasting. In: Wikipedia, The Free Encyclopedia. https://en.wikipedia.org/w/index.php?title=Weather_forecasting&oldid=868621761. Accessed 8 Nov 2018
2. Santhanam, M.S., Patra, P.K.: Statistics of atmospheric correlations. Phys. Rev. E **64**, 016102 (2001)
3. Kalnay, E.: Atmospheric Modeling, Data Assimilation and Predictability. Cambridge University Press, Cambridge (2003)
4. Dirmeyer, P.A., Schlosser, C.A., Brubaker, K.L.: Precipitation, recycling, and land memory: an integrated analysis. J. Hydrometeorol. **10**(1), 278–288 (2009)
5. Holmstrom, H., Liu, D., Vo, C.: Machine learning applied to weather forecasting. http://cs229.stanford.edu/proj2016/report/HolmstromLiuVo-MachineLearningAppliedToWeather Forecasting-report.pdf. Accessed 1 Dec 2018
6. LeCun, Y., Bengio, Y., Hinton, G.: Deep learning. Nature **521**(7553), 436 (2015)
7. Fente, D.N., Singh, D.K.: Weather forecasting using artificial neural network. In: IEEE 2018 Second International Conference on Inventive Communication and Computational Technologies (ICICCT), pp. 1757–1761 (2018)
8. Qureshi, A.S., Khan, A., Zameer, A., Usman, A.: Wind power prediction using deep neural network based meta regression and transfer learning. Appl. Soft Comput. **58**, 742–755 (2017)
9. Hu, Q., Zhang, R., Zhou, Y.: Transfer learning for short-term wind speed prediction with deep neural networks. Renew. Energy **85**, 83–95 (2016)
10. Samuel, A.L.: Some studies in machine learning using the game of checkers. IBM J. Res. Dev. **3**(3), 210–229 (1959)
11. Mitchell, T.M.: Machine Learning. McGraw-Hill, Boston (1997)

12. Krizhevsky, A., Sutskever, I., Hinton, G.E.: ImageNet classification with deep convolutional neural networks. In: Advances in Neural Information Processing Systems, pp. 1097–1105 (2012)
13. Hinton, G., et al.: Deep neural networks for acoustic modeling in speech recognition: the shared views of four research groups. IEEE Signal Process. Mag. **29**(6), 82–97 (2012)
14. Graves, A., Mohamed, A.R., Hinton, G.: Speech recognition with deep recurrent neural networks. In: 2013 IEEE International Conference on Acoustics, Speech and Signal Processing (ICASSP), pp. 6645–6649 (2013)
15. Goodfellow, I., Bengio, Y., Courville, A.: Deep learning, vol. 1. MIT press, Cambridge (2016)
16. Becker, S.: Unsupervised learning procedures for neural networks. Int. J. Neural Syst. **1 & 2**, 17–33 (1991)
17. Kang, Y., Lee, K.-T., Eun, J., Park, S.E., Choi, S.: Stacked denoising autoencoders for face pose normalization. In: Lee, M., Hirose, A., Hou, Z.-G., Kil, R.M. (eds.) ICONIP 2013. LNCS, vol. 8228, pp. 241–248. Springer, Heidelberg (2013). https://doi.org/10.1007/978-3-642-42051-1_31
18. Hochreiter, S., Schmidhuber, J.: Long short-term memory. Neural Comput. **9**, 1735–1780 (1997)
19. Robinson, A.J., Fallside, F.: The utility driven dynamic error propagation network. Technical report CUED/F-INFENG/TR.1, Cambridge University Engineering Department (1987)
20. Werbos, P.J.: Generalization of backpropagation with application to a recurrent gas market model. Neural Netw. **1**, 339–356 (1988)
21. Williams, R.J., Zipser, D.: Gradient-based learning algorithms for recurrent networks and their computational complexity. In: Back-Propagation: Theory, Architectures and Applications. Erlbaum, Hillsdale (1992)
22. Salman, A.G., Heryadi, Y., Abdurahman, E., Suparta, W.: Weather forecasting using merged long short-term memory model. Bull. Electr. Eng. Inf. **7**(3), 377–385 (2018)
23. Graves, A., Jaitly, N.: Towards end-to-end speech recognition with recurrent neural networks. In: JMLR Workshop and Conference Proceedings, vol. 32, no. 1, pp. 1764–1772 (2014)
24. Geiger, J.T., Zhang, Z., Weninger, F., Schuller, B., Rigoll, G.: Robust speech recognition using long short-term memory recurrent neural networks for hybrid acoustic modelling. In: Fifteenth annual conference of the international speech communication association (2014)
25. Basak, J., Sudarshan, A., Trivedi, D., Santhanam, M.S.: Weather data mining using independent component analysis. J. Mach. Learn. Res. **5**(Mar), 239–253 (2004)
26. Kingma, D.P., Ba, J.L.: Adam: a method for stochastic optimization. In: International Conference on Learning Representations, pp. 1–13 (2015)
27. Hinton, G., Srivastava, N., Swersky, K.: Neural networks for machine learning. https://www.cs.toronto.edu/~tijmen/csc321/slides/lecture_slides_lec6.pdf. Accessed 18 Nov 2018

Logo and Brand Recognition from Imbalanced Dataset Using MiniGoogLeNet and MiniVGGNet Models

Sarwo[✉], Yaya Heryadi, Widodo Budiharto, and Edi Abdurachman

Computer Science Department, BINUS Graduate Program – Doctor of Computer
Science, Bina Nusantara University, Jakarta, Indonesia
Sarwo.jowo@gmail.com,
{yayaheryadi,wbudiharto,ediA}@binus.edu

Abstract. Deep learning model tends to promote models with deep structure. Despite its high accuracy, the model was not practical when high computing power was not available. Thus, deep model with not-so-deep structure or less number of model parameters is needed for low capacity computer. Logo and brand recognition task is an important and challenging problem in computer vision with wide potential applications. The inherent challenge to address this task is not only due to the presence of logo in various direction and clutters as well as imbalanced dataset but also because of high computing workload when deep learning models were adopted. This paper presents empirical results of logo recognition method using MiniVGGNet and MiniGoogleNet models combined with augmentation technique to increase variation and number of samples. The results show that the proposed model combined with augmentation technique increased accuracy of model accuracies and fasten training convergence of both models.

Keywords: Logo detection · MiniVGGNet · MiniGoogLeNet · Augmentation technique

1 Introduction

Logo and brand recognition is an interesting computer vision problem with wide potential applications, including but not limited to ad placement, product placement validation and online brand management system. Although it is an easy problem for humans, it remains a challenging problem for computers due to several factors, such as viewpoint variation, scale variation, occlusion, deformation, illumination variation and background clutter. The wide availability of dataset and the advent of machine learning methods in the past ten years have motivated many researchers to adopt machine learning approach to ad-dress this problem.

A study reported by Iandola [1], for example, argued that logo recognition is vital for marketing in the digital era. Having been experienced by KFC, PEPSI, logo detection was shown to be a powerful and innovative promotion tool. Further research by [2] reported some evidences that logo recognition achieved successful results for addressing (i) identification and counterfeiting of products, and (ii) automatic monitoring of ad placements. The other research on logo detection has focused on:

© Springer Nature Switzerland AG 2019
N. T. Nguyen et al. (Eds.): ACIIDS 2019, LNAI 11431, pp. 385–393, 2019.
https://doi.org/10.1007/978-3-030-14799-0_33

(i) checking problem car ads by Psyllos [3], (ii) marketing brand tracking on Instagram, image searching, image classification [2, 4], and (iii) CAPTCHA Recognition [5].

Despite various studies focusing on object detection, to the best of our knowledge, few deal with automated logo and brand recognition. The challenges of addressing automated logo and brand recognition are mainly related to: (i) imbalanced logo dataset which made the classifier tend to bias the majority class, and (ii) high computation workload of deep learning model which made it impractical to train such a model when high computing power computer was not available. In this study, imbalanced issues in training dataset are addressed using augmentation technique. Meanwhile, the high computation work load was addressed by proposing a simplified GoogleNet and simplified VGGNet to speed up training process.

The rest of this paper is organized as follows. Section 2 describes related works. Next, Sect. 3 explains research method followed by results and discussion presented in Sect. 4. Finally, Sect. 5 concludes this paper.

2   Related Works

2.1   Deep Learning

Machine learning is a term which refers to a branch of computer science that studies algorithms which can learn from and make predictions on data [1]. Specifically, Mitchel [1997] defined machine learning algorithm as algorithm that *"learn[s] from experience E with respect to some class of tasks T and performance measure P, if its performance at tasks in T, as measured by P, improves with experience E."* Many studies reported successful application of machine learning in many areas [1, 2].

Despite many successful applications of machine learning, its performance depends heavily on the choice of features extracted from its input dataset. Unfortunately, feature extraction and selection is a challenging task. To address this issue, Hinton et al. [2–5] proposed deep neural networks which are able to learn a hierarchy of features from input data. The models are later called deep learning models. Hence, deep learning is a subset of machine learning. Unlike machine learning models, deep learning models integrate feature extraction with model training.

2.2   Logo Detection

In the past ten years, several works on logo detection have been reported. For example, research conducted by Han Su [10] used various methods that represented data using edge, color histogram, HOG, SIFT [11–17] where the research still employed prominent methods. The advent of deep learning methods and many successful results in object detection have motivated many researchers to adopt deep learning method for logo detection, for example, logo detection research which was reported by Zhang [18]. In this study, the logo recognition was used on TV where each TV station has different video types. Several other studies on logo detection using deep learning have been reported by [3, 19–21]. Much research on deep learning model that involved image processing and video processing used high performance computer hardware like GPU which was not practical when such computer platform was not available.

2.3   Augmentation Method

Augmentation method refers to a method for creating new image datasets from a given dataset using image processing techniques such as: translations, rotations, shears, and flips [6]. Similar to upsampling methods, the augmentation method can reduce model bias to the majority classes or model overfitting. Some previous studies showed evidences that augmentation method can increase classifier accuracy. Some previous studies in computer vision have used this technique as part of the proposed methods [6–11].

The augmentation technique typically comprises three main image operations. The first is shifting, an image operation which can be performed by moving element vector into a number element. Some previous studies have used this operation, such as research reported by Es Sabry et al. [5, 7] in which shift method was adopted for grayscale image encryption. A study by Allawi [8] used the shift method for image encryption based on linear feedback.

The second is scaling proposed by [9, 10], which is one of resampling methods. Some studies adopted scaling methods to increase or decrease the total number of pixels of image. Finally, rotation is a resampling technique that is done by rotating images into several positions. Several studies have used this model such as [11, 12].

3   Research Method

3.1   Dataset

Dataset for this research comprises 1,079 samples of logo images from Flicker-27 which consist of 27 classes. The class labels were Adidas, Apple, BMW, Citroen, Cocacola, DHL, Fedex, Ferrari, Ford, Google, HP, Heineken, Intel, McDonalds, Mini, NBC, Nike, Pepsi, Porsche, Puma, RedBull, Sprite, Starbucks, Texaco, Unicef, and Vodafone.

3.2   Augmentation Results

Augmented method was adopted to create synthetic images from original logo sample images. The augmentation techniques in this study include: shifting, scaling and rotation (see Fig. 2). The techniques produced 598,092 new logo images from 4,531 original sample images. The objective of augmentation method was to reduce imbalanced data in the input dataset by generating some sample datasets from minority class so that the discrepancy between majority and minority logo class was reduced. The total logo sample images resulted from upsampling process were then splitted randomly into training and testing dataset with proportion of 70:30.

3.3   Model Training

The deep models explored in this study were variant of GoogleNet [22] and VGG [23]. The proposed models were designed by modifying architecture of the former models mainly reducing the number of layers in order to reduce the number of model parameters. Thus, reducing computational workload of the training process. The resulted models were named: MiniGoogleNet and MiniVGGNet respectively. The two models can be viewed as a Convolutional Neural Network (CNN) as described in Fig. 1.

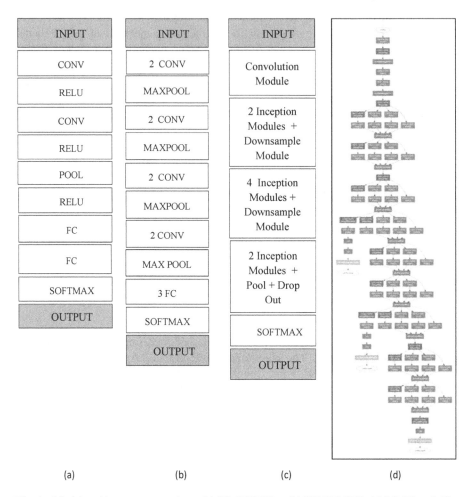

| (a) | (b) | (c) | (d) |

Fig. 1. Model architecture comparison: (a) MiniVGGNet, (b) VGG19 [23], (c) MiniGoogleNet, and (d) GoogleNet [22]

As can be seen from Fig. 1, by removing some convolutional layers, the proposed MiniVGGNet has a simpler architecture and less number of model parameters than the corresponding VGG model proposed by [23]. Similarly, by removing some inception modules in the GoogleNet proposed by [22], the proposed MiniGoogleNet has less number of parameters than the corresponding GoogleNet.

3.4   Logo Detection

Following Bianco et al. [6], logo detection method can be divided into two main parts: training and testing processes (see Figs. 2 and 3). Raw datasets for this study were:

Flicker-27 dataset which comprises 1,079 images from 27 classes. The combined dataset were used to train MiniVGGNet and MiniGoogleNet. These trained models were tested using the training datasets.

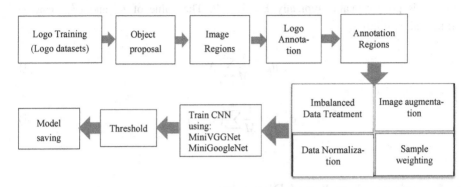

Fig. 2. Logo recognition training framework (Source: [6])

Figure 2 above explains the flow of training process. The first step starts with logo data acquisition. Next, object proposals [15] and image regions computation. Finally, the model is generated and saved. Whilst, the testing phase can be seen in Fig. 3. The testing stage is the stage where the model obtained from the training process will be tested. In this research, sample batch-based testing was carried out with the proportion of 30% of the total datasets.

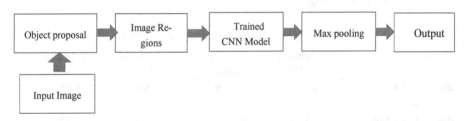

Fig. 3. Logo recognition testing framework (Source: [6]

Let $F_{X_{n,i}}$ is the nth layer of the model-2 and F_{X_n} of the model-1. Let X_n is the addition layer to the nth layer of previous model-1 [13–15]. The proposed function implemented in the nth layers of model-2 can be represented as:

$$F_{X_{n,i}} = F_{X_n} + X_i \tag{1}$$

Following [16, 17], loss function optimization in this experiments was based on mini batch. Consider $\hat{X}$ is mini batch loss function, calculation of the mini batch loss function can be represented as:

$$\hat{X} = \frac{X_i - \mu_\beta}{\sqrt{\sigma_{beta}^2 - \pounds}} \tag{2}$$

where $\pounds$ is positive value typically $\pounds = 1e - 7$. The value of μ_β and σ_{beta}^2 can be formulated as follows:

$$\mu_\beta = \frac{1}{M} \sum_{1=1}^{m} X_i \tag{3}$$

$$\sigma_{beta}^2 = \frac{1}{M} \sum_{1=1}^{m} (X_{i-\mu_\beta}) \tag{4}$$

4 Experiment Result and Discussion

The effect of image augmentation on the model accuracy of MiniVGGNet using Flicker-27 dataset is summarized in Table 1. As Table 1. indicates, augmentation technique affect in reducing training and validation loss measured by root mean square error (RMSE) as well as increasing training and validation accuracy. Interestingly, the accuracy of validation and training from MiniVGGNet with augmentation technique was more or less stable. In contrast, for MiniVGGNet without augmentation, the accuracy of validation tends to be higher than the accuracy of training which was typically a sign of overfitting.

Table 1. Training performance of MiniVGGNet

Model	Training RMSE	Training accuracy	Validation RMSE	Validation accuracy
MiniVGGNet without augmentation	0.0177	0.9946	0.1349	0.9820
MiniVGGNet with augmentation	0.0043	0.9992	0.0048	0.9996

The results of the experiment using MiniVGGNet with Augmentation method can be seen below. At this stage, there are four measurements namely: training RMSE, training accuracy, validation RMSE and validation accuracy (Fig. 4).

The results of experiments using MiniGoogleNet are presented in Table 2. As Table 2 shows, augmentation technique affects reducing training and validation loss as well as increasing training and validation accuracy.

In addition, as Tables 1 and 2. indicate, MiniGoogleNet tends to show higher training accuracy than MiniVGGNet. In addition, with augmentation, both models can achieve high training and validation accuracy and reduced model overfitting (Fig. 5).

The accuracy and loss of the testing phase using the MiniVGGNet and Mini-GoogleNet models are presented in Table 3.

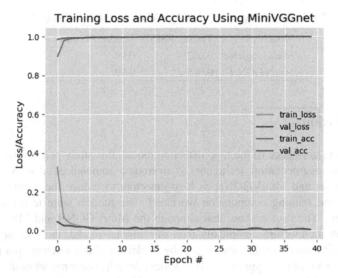

Fig. 4. Training loss (RMSE) of MiniVGGNet with augmentation technique

Table 2. Training performance of MiniGoogleNet

Model	Training RMSE	Training accuracy	Validation. RMSE	Validation accuracy
MiniGoogleNet without augmentation	0.0115	0.9953	0.1057	0.9820
MiniGoogleNet with augmentation	0.0019	0.9997	0.0020	0.9996

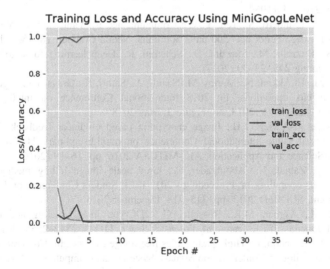

Fig. 5. Training loss (RMSE) of MiniGoogLeNet

Table 3. Testing accuracy of MiniVGGNet and MiniGooglenet

Model	Testing RMSE	Testing Accuracy
MiniGooglenet	0.002	0.99
MiniVGGNet	0.005	0.99

5 Conclusion

To improve performance of the logo detection model, this study presents the empirical results using augmentation technique to overcome imbalanced data and explores MiniGooglenet and MiniVGGNet as logo detectors to reduce the number of model parameters and training computation workload. The models were tested using three open datasets. The results show that although the MiniVGGNet and MiniGooglenet have simpler architecture complexity than the original model, the accuracy was not affected significantly. On the other hand, the model simplicity opened up a possibility to be improved further as poorman's (on-device) deep learning model or deep learning model designed for low capacity computers. Unfortunately, in terms of accuracy, the proposed models are still under performance of the following models: SSD [29], RPN [30, 31], YOLO [32] for object detection. Improving accuracy of the proposed models will become the next research agenda.

Acknowledgment. This research is partially supported by Binus IntelSys Research Interest Group.

References

1. Samuel, A.L.: Some studies in machine learning using the game of checkers. IBM J. Res. Dev. **3**(3), 210–229 (1959)
2. Hinton, G.E., Osindero, S., Teh, Y.-W.: Communicated by Yann Le Cun A fast learning algorithm for deep belief nets 500 units 500 units. Neural Comput. **18**, 1527–1554 (2006)
3. Bianco, S., Buzzelli, M., Mazzini, D., Schettini, R.: Deep learning for logo recognition. Neurocomputing **245**, 23–30 (2017)
4. Es-Sabry, M., El Akkad, N., Merras, M., Saaidi, A., Satori, K.: Grayscale image encryption using shift bits operations. In: 2018 International Conference on Intelligent System Computer Vision, ISCV 2018, no. Figure 3, May 2018
5. Allawi, S.T., Al-A'Meri, J.H.: Image encryption based on linear feedback shift register method. In: Al-Sadiq International Conference on multidisciplinary IT Communication Technical Science and Application, AIC-MITCSA 2016, pp. 16–19 (2016)
6. Wadi, S.M., Zainal, N., Abdulgader, A.: Grey scale image hiding method based on decomposition operation. In: Proceeding - 2013 IEEE Student Conference on Research and Development, SCOReD 2013, pp. 315–318, December 2015
7. Mohanty, M., Asghar, M.R., Russello, G.: 2DCrypt: image scaling and cropping in encrypted domains. IEEE Trans. Inf. Forensics Secur. **11**(11), 2542–2555 (2016)
8. Owen, C.B., Makedon, F.: High quality alias free image rotation. In: Conference Record of the Thirtieth Asilomar Conference on Signals, Systems and Computers, vol. 1, pp. 115–119 (1997)

9. Barnett, S.M., Zambrini, R.: Resolution in rotation measurements. J. Mod. Opt. **53**(5–6), 613–625 (2006)

10. Su, H., Zhu, X., Gong, S.: Deep Learning Logo Detection with Data Expansion by Synthesising Context. CoRR, vol. abs/1612.0 (2016)

11. Kalantidis, Y., Pueyo, L.G., Trevisiol, M., van Zwol, R., Avrithis, Y.: Scalable triangulation-based logo recognition (Flickr 27). In: Proceedings of the ACM International Conference on Multimedia Retrieval (ICMR 2011), pp. 20:1–20:7 (2011)

12. Romberg, S., Clara, S.: Scalable logo recognition in real-world Images categories and subject descriptors

13. Revaud, J., Schmid, C.: Correlation-Based Burstiness for Logo Retrieval Categories and Subject Descriptors: Keywords, pp. 965–968 (2012)

14. Revaud, J., et al.: DeepMatching: Hierarchical Deformable Dense Matching To cite this version: DeepMatching: Hierarchical Deformable Dense Matching (2015)

15. Romberg, S., Lienhart, R.: Bundle min-hashing for logo recognition. ACM (2013)

16. Boia, R., Bandrabur, A., Florea, C.: Local description using multi-scale complete rank transform for improved logo recognition, pp. 4–7 (2014)

17. Li, K., Chen, S., Su, S.: Logo detection with extendibility and discrimination (2013)

18. Zhang, Y., et al.: Deep learning for logo recognition. In: International Conference on Intelligent Systems Design and Applications, ISDA, vol. 245, no. 36, pp. 2051–2054 (2017)

19. Arivazhagan, N.: Logo Recognition. pp. 1–2

20. Su, H., Zhu, X., Gong, S.: Deep learning logo detection with data expansion by synthesising context. In: Proceedings of the 2017 IEEE Winter Conference Applications of Computer Vision (WACV 2017), pp. 530–539 (2017)

21. Pan, C., Yan, Z., Xu, X., Sun, M., Shao, J., Wu, D.: Learning architecture in video surveillance. pp. 123–126. IEEE (2013)

22. Szegedy, C., et al.: Going deeper with convolutions. In: Proceedings of the IEEE Computer Society Conference on Computer Vision and Pattern Recognition. pp. 1–9, 07–12 June 2015

23. Simonyan, K. Zisserman, A.: Very deep convolutional networks for large-scale image recognition. In: International Conference on Learning Representations, pp. 1–14 (2015)

24. He, Z.: Deep residual learning for image recognition. arXiv.org e-Print Arch **7**(3), 171–180 (2015)

25. He, K., Zhang, X., Ren, S., Sun, J.: Identity mappings in deep residual networks. In: Leibe, B., Matas, J., Sebe, N., Welling, M. (eds.) Computer Vision – ECCV 2016. ECCV 2016. LNCS, vol. 9908. Springer, Cham (2016). https://doi.org/10.1007/978-3-319-46493-0_38

26. Xie, S., Girshick, R., Dollár, P., Tu, Z., He, K.: Aggregated residual transformations for deep neural networks. In: Proceedings of the 30th IEEE Conference on Computer Vision and Pattern Recognition, CVPR 2017, January 2017, pp. 5987–5995 (2017)

27. Ioffe, S., Szegedy, C.: Batch normalization: accelerating deep network training by reducing internal covariate shift (2015)

28. Pyimagesearch: Deep Learning for computer Vision using Python. Book (2017)

29. Liu, W., et al.: SSD: single shot multibox detector. arXiv, pp. 1–15 (2016)

30. Girshick, R.: Fast R-CNN. arXiv.org e-Print Arch. (2015)

31. Ren, S., He, K., Girshick, R., Sun, J.: Faster R-CNN: towards real-time object detection with region proposal networks. Nips, pp. 1–10 (2015)

32. Impiombato, D., et al.: You only look once: unified, real-time object detection. Nucl. Instruments Methods Phys. Res. Sect. A Accel. Spectrometers, Detect. Assoc. Equip. **794**, 185–192 (2015)

Decision Support and Control Systems

Selected Problems of Controllability of Semilinear Fractional Systems-A Survey

Artur Babiarz[1]([⊠])[iD] and Jerzy Klamka[2][iD]

[1] Silesian University of Technology, Akademicka 2A, 44-100 Gliwice, Poland
`artur.babiarz@polsl.pl`
[2] Institute of Theoretical and Applied Informatics, Polish Academy of Sciences,
Bałtycka 5, 44-100 Gliwice, Poland
`jerzy.klamka@iitis.pl`

Abstract. The following article presents recent results of controllability problem of dynamical systems in infinite and finite-dimensional spaces. Roughly speaking, we describe selected controllability problems of fractional order systems, including approximate controllability and complete controllability.

Keywords: Fractional systems · Controllability ·
Fixed point theorem · Banach space

1 Introduction

Controllability is very important property of dynamical systems and it plays a crucial role in many control problems. The assumption that the control system is controllable performs fundamental establishment among other in optimal control, stabilizability and pole placement problem [26, 35]. In general controllability means that there exists a control function which steers the solution of the dynamical system from its initial state to a final state using a set of admissible controls, where the initial and final states may vary over the entire space. A standard approach to the controllability problems is to transform it into a fixed point problem for an appropriate operator in a functional space. There are many studies are related to the controllability problem. In [5, 6, 15, 20, 23, 24, 28, 30] authors used the theory of fractional calculus. A fixed point approach we can find in [3, 12, 17, 37].

Nowadays the fractional calculus and its applications in control theory are very popular and they became standard tool in designing of control systems. Moreover, the fractional calculus has become a powerful tool in modeling several complex phenomena in numerous seemingly diverse and widespread fields such as engineering, chemistry, mechanics, aerodynamics, physics, etc. [13, 14].

A lot of dynamical systems based on mathematical modelling of realistic models can be described as partial fractional differential or integrodifferential inclusions [4, 9, 32, 33]. A new approach to obtain the existence of mild solutions

© Springer Nature Switzerland AG 2019
N. T. Nguyen et al. (Eds.): ACIIDS 2019, LNAI 11431, pp. 397–407, 2019.
https://doi.org/10.1007/978-3-030-14799-0_34

and controllability results are presented in [41]. For this purpose they avoid hypotheses of compactness on the semigroup generated by the linear part and any conditions on the multivalued nonlinearity expressed in terms of measures of noncompactness. Author of [39] studied fractional evolution equations and inclusions and they presented application of obtained results in control theory. Many authors [7,8,21,31,40] investigated the existence of solutions for fractional semilinear differential or integrodifferential equations.

The special case of dynamical systems so-called the impulsive differential systems can be used to model processes which are subject to sudden changes and which cannot be described by the standard types of differential systems [22]. In [36] authors considered the controllability problem for impulsive differential and integrodifferential systems in Banach spaces. Articles [29] and [38] are devoted to the controllability of fractional evolution systems. The problem of controllability and optimal controls for functional differential systems has been extensively investigated in many articles [1,2].

In this paper we discuss selected problems of controllability for various types of fractional order systems. We present the latest results for finite and infinite-dimensional fractional nonlinear systems.

2   Basic Notations

In this section, we introduce some definitions. Let $(X, \| \cdot \|)$ be a Banach space, $J = [0, t_1]$, $\alpha \in (0, 1)$ and $f : J \to X$ be a given function.

Definition 1 [16]. *The Caputo fractional derivative of order α is given as follows*

$$^{C}D^{\alpha}f(t) = \frac{1}{\Gamma(1-\alpha)} \int_0^t \frac{f'(s)ds}{(t-s)^{\alpha}},$$

where: f is the function which has absolutely continuous derivative, Γ is the Gamma function and f' is the derivative of function f.

For completeness of presentation, below the definition of the measure noncompactness is shown. It is the generalization of the Schauder's theorem [18].

Definition 2. *Let $(X, \| \cdot \|)$ be a Banach space and E be a bounded subset of X. Then the measure noncompactness of the set E is defined as*

$$\mu(E) = \inf\{r > 0 : E \text{ can be covered by}$$
$$\text{a finite number of balls whose radii are smaller than } r\}.$$

Theorem 1 (Darbo fixed point theorem) [27]. *Let Q be a nonempty, bounded, convex and closed subset of the space X and let $F : Q \to Q$ be a continuous function such that*

$$\mu(F(S)) \leq k\mu(S),$$

for all nonempty subset S of Q, where $k \in [0, 1)$ is a constant. Then F has a fixed point in the set Q.

Theorem 2 (Schauder's fixed point theorem) [19]. *Let Q be a closed convex subset of X and $F : Q \to Q$ be a continuous function. Then F has at least one fixed point in the set Q.*

Theorem 3 (Krasnoselskii fixed point theorems) [11]. *Let X be a Banach space and Q be a bounded, closed, and convex subset of X. Let V_1, V_2 be maps of Q into X such that $V_1 x + V_2 y \in Q$ for every $x, y \in Q$. If V_1 is a contraction and V_2 is compact and continuous, then equation $V_1 x + V_2 x = x$ has a solution on Q.*

Moreover the relative and approximate controllability definitions are recall below.

Definition 3 [25]. *The dynamical system is said to be relative controllable on interval $[0, t_1]$, if for every initial function ϕ and every final state $x_1 \in \mathbb{R}^n$ there exists a control function u defined on $[0, t_1]$ such that the solution of dynamical system satisfies $x(t_1) = x_1$.*

Definition 4. *The dynamical system is said to be approximately controllable in time interval $[0, t_1]$, if for every desired final state x_1 and $\varepsilon > 0$ there exists a control function u such that the solution of dynamical system satisfies*

$$\|x(t_1) - x_1\| < \varepsilon.$$

3 Controllability of Semilinear Fractional Systems

In this section mathematical models of fractional systems with different delays in state and control will be presented. Let us introduce the following necessary notation:

- $0 < \alpha < 1$ is order of derivative,
- ϕ is a continuous function on $[-h, 0]$, $h \in [0, \infty)$, $\phi : [-h, 0] \to \mathbb{R}^n$,
- A, B are $n \times n$ dimensional matrices and C is $n \times m$ dimensional matrix,
- u is the control function $u : [-h, \infty) \to \mathbb{R}^m$,
- $\mathcal{L}$ is the Laplace transform,
- $X_\alpha(t) = \mathcal{L}^{-1} \Big[[s^\alpha \cdot I - A - Be^{-s}]^{-1} s^{\alpha-1} \Big](t)$,
- $X_{\alpha,\alpha}(t) = t^{t-\alpha} \int_0^t \frac{(t-s)^{\alpha-2}}{\Gamma(\alpha-1)} X_\alpha(s) ds$,
- $x_L(t; \phi) = X_\alpha(t)\phi(0) + \int_{-h}^0 (t-s-h)^{\alpha-1} X_{\alpha,\alpha}(t-s-h)\phi(s) ds$,
- $H(t, s)$ is an $n \times m$ matrix, continuous in t for fixed s, $H : J \times [-h, 0] \to \mathbb{R}^{n \times m}$,
- $\int_{-h}^0 d_s H(t, s)$ denotes the integrals in the Lebesgue-Stieltjes sense with respect to s.

The next definition will be used in investigation about controllability of semilinear dynamical systems.

Definition 5 [25]. *Assume that there exist positive real constants K and k with $0 \leq k < 1$ such that*

$$|f(t, x, y, z, u)| \leq K, \tag{1}$$

$$|f(t, x, y, z, u) - f(t, x, y, \overline{z}, u)| \leq k(|z - \overline{z}|) \tag{2}$$

for all $x, y, z, \overline{z} \in \mathbb{R}^n$ and $u \in \mathbb{R}^m$, where f is nonlinear function.

3.1   Fractional System with Distributed Delays in Control

The fractional delay dynamical system with distributed delays in control can be presented by the following equation:

$$
\begin{aligned}
{}^C D^\alpha x(t) &= Ax(t) + Bx(t - h) + \int_{-h}^{0} d_s H(t, s) u(t + s) + \\
&\quad + f\big(t, x(t), x(t - h), {}^C D^\alpha x(t), u(t)\big) \\
x(t) &= \phi(t), \quad t \in [-h, 0].
\end{aligned}
\tag{3}
$$

Using the well-known result of the unsymmetric Fubini theorem [10], the solution of (3) can be expressed by the following form:

$$x(t) = x_L(t; \phi)$$

$$+ \int_{-h}^{0} dH_\tau \int_{\tau}^{0} \big(t - (s - \tau)\big)^{\alpha - 1} X_{\alpha, \alpha}\big(t - (s - \tau)\big) H(s - \tau, \tau) u_0(s) ds$$

$$+ \int_{0}^{t} \int_{-h}^{0} \big(t - (s - \tau)\big)^{\alpha - 1} X_{\alpha, \alpha}\big(t - (s - \tau)\big) d_\tau H_t(s - \tau, \tau) u(s) ds$$

$$+ \int_{0}^{t} (t - s)^{\alpha - 1} X_{\alpha, \alpha}(t - s) f\big(s, x(s), x(s - h), {}^C D^\alpha x(s), u(s)\big) ds,$$

where

$$H_t(s, \tau) = \begin{cases} H(s, \tau), & s \leq t, \\ 0, & s > t, \end{cases}.$$

The below, we present the main result of relative controllability for the system (3).

Theorem 4. *Assume that the nonlinear function f satisfies the conditions (1) and (2) and suppose that the controllability Gramian*

$$W = \int_{0}^{t_1} S(t_1, s) S^*(t_1, s) ds$$

where:

$$S(t, s) = \int_{-h}^{0} \big(t - (s - \tau)\big)^{\alpha - 1} X_{\alpha, \alpha}\big(t - (s - \tau)\big) d_\tau H_t(s - \tau, \tau).$$

is positive definite. Then the nonlinear system (3) is relatively controllable on J.

In order to prove of Theorem 4 authors using Darbo fixed point theorem. Specifying the matrix function $H(t, s)$, it is possible to obtain systems with different lumped delays in control.

3.2   Fractional Systems with Multiple Delays in Control

In the same article [25] the authors focus on the implicit fractional delay dynamical system with time varying multiple delays in control given by the equation

$$^C D^\alpha x(t) = Ax(t) + Bx(t-h) + \sum_{i=0}^{M} C_i u(\sigma_i(t))$$

$$+ f\big(t, x(t), x(t-h), {}^C D^\alpha x(t), u(t)\big), \tag{4}$$

$$x(t) = \phi(t), \quad t \in [-h, 0],$$

where C_i for $i = 0, 1, \ldots, M$ are $n \times l$ matrices.

In order to obtain the main result we present some necessary hypotheses.

Hypothesis 1. *The functions $\sigma_i : J \to \mathbb{R}$, $i = 0, 1, \ldots, M$, are twice continuously differentiable and strictly increasing in J. Moreover $\sigma_i(t) \leq t$, $i = 0, 1, \ldots, M$, for $t \in J$.*

Hypothesis 2. *Introduce the time lead functions $r_i(t) : [\sigma_i(0), \sigma_i(t_1)] \to [0, t_1]$, $i = 0, 1, \ldots, M$, such that $r_i(\sigma_i(t)) = t$ for $t \in J$. Further, $\sigma_0(t) = t$ and for $t = t_1$ the following inequality holds:*

$$\sigma_M(t_1) \leq \cdots \leq \sigma_{l+1}(t_1) \leq 0 = \sigma_l(t_1)$$
$$< \sigma_{l-1}(t_1) = \cdots = \sigma_1(t_1) = \sigma_0(t_1) = t_1. \tag{5}$$

Hypothesis 3. *Given $\sigma > 0$, for functions $u : [-\sigma, t_1] \to \mathbb{R}^l$ and $t \in t_1$, we use the symbol u_t to denote the function on $[-\sigma, 0]$ defined by $u_t(s) = u(t+s)$ for $s \in [-\sigma, 0)$.*

Using (5), we can write solution of (4):

$$x(t) = x_L(t; \phi) + H(t)$$

$$+ \sum_{i=0}^{l} \int_0^t (t - r_i(s))^{\alpha-1} X_{\alpha,\alpha}(t - r_i(s)) C_i \dot{r}_i(s) u(s) ds \tag{6}$$

$$+ \int_0^t (t - s)^{\alpha-1} X_{\alpha,\alpha}(t - s) f\big(s, x(s), x(s-h), {}^C D^\alpha x(s), u(s)\big) ds,$$

where

$$H(t) = \sum_{i=0}^{l} \int_{\sigma_i(0)}^{0} (t - r_i(s))^{\alpha-1} X_{\alpha,\alpha}(t - r_i(s)) C_i \dot{r}_i(s) u_0(s) ds$$

$$+ \sum_{i=l+1}^{M} \int_{\sigma_i(0)}^{\sigma_i(t)} (t - r_i(s))^{\alpha-1} X_{\alpha,\alpha}(t - r_i(s)) C_i \dot{r}_i(s) u_0(s) ds.$$

The main theorem is given by the following form.

Theorem 5. *Assume that the Hypotheses 1–3 hold. Further assume that the nonlinear function satisfies the condition* (1) *and* (2) *and suppose that determinant of Gramian matrix*

$$W = \sum_{i=0}^{l} \int_0^{t_1} \Big(X_{\alpha,\alpha}(t_1 - r_i(s)) C_i \dot{r}_i(s) \Big) \Big(X_{\alpha,\alpha}(t_1 - r_i(s)) C_i \dot{r}_i(s) \Big)^T ds$$

is positive definite. Then the nonlinear system (4) *is relatively controllable on J.*

As before, the proof was obtained using Darbo fixed point theorem.

4   The Controllability of Nonlocal Nonlinear Fractional Systems

In this section, we present a recent results concerning nonlinear fractional system in infinite-dimensional space.

4.1   Approximate Controllability of Fractional Nonlocal Evolution Equation with Multiple Delays

Authors of [34] consider dynamical system described as follows:

$$\begin{aligned} {}^C D^\alpha x(t) &= Ax(t) + f\left(t, x(t), x(t - \tau_1), \ldots, x(t - \tau_n)\right) + Bu(t), \\ t &\in J = [0, T], \quad x(t) + g(x) = \varphi(t), \quad t \in [-b, 0], \end{aligned} \tag{7}$$

where ${}^C D^\alpha$ is Caputo fractional derivative of order $\alpha \in (0, 1)$, $T > 0$ is a constant, A generates a compact analytic semigroup $S(t)\,(t \geq 0)$ of uniformly bounded linear operator, $u \in L^2(J, X)$ is a control, X is a Banach space, X_α is the Banach space of $D(A^\alpha)$ with norm $\|x\|_\alpha := \|A^\alpha x\|$ for any $x \in D(A^\alpha)$, $B : X \to X_\alpha$ is a linear bounded operator, $\tau_1, \tau_2, \ldots, \tau_n$ are positive constants, $b = \max\{\tau_1, \tau_2, \ldots, \tau_n\}$, $\varphi : [-b, 0] \to X_\alpha$ is continuous, f and g are given functions.

To obtain the main results it should pose a some assumptions.

Hypothesis 4. *The function $f : J \times X_\alpha^{n+1} \to X$ is continuous and there exist positive constants $\beta_0, \beta_1, \ldots, \beta_n$ and $K \geq 0$ such that*

$$\|f(t, \nu_0, \nu_1, \ldots, \nu_n)\| \leq \sum_{i=0}^{n} \beta_i \|\nu_i\|_\alpha + K, \quad t \in J, \ (\nu_0, \nu_1, \ldots, \nu_n) \in X_\alpha^{n+1}.$$

Hypothesis 5. *The function $g : C\left([-b, T], X_\alpha\right) \to X_\alpha$ is continuous and there exists a constant $L \geq M$, $M \geq 1$, such that*

$$\|g(x) - g(y)\|_\alpha \leq \frac{\|x - y\|_C}{L + \|x - y\|_C}, \quad x, y \in C\left([-b, T], X_\alpha\right),$$

where $C\left([-b, T], X_\alpha\right)$ is the Banach space of all continuous X_α-valued functions on the interval $[-b, T]$ with norm $\|x\|_C = \max_{t \in [-b, T]} \|x(t)\|_\alpha$ for any $x \in C\left([-b, T], X_\alpha\right)$.

Hypothesis 6. *The function* $f : J \times X_\alpha^{n+1} \to X$ *is bounded.*

Hypothesis 7. *The linear fractional system given in the form:*

$$^C D^\alpha x(t) = Ax(t) + Bu(t), \quad t \in [0, T],$$
$$x(0) = x_0 \in X_\alpha$$

is approximately controllable.

Then, we can present the main results of [34].

Theorem 6. *Assume that the Hypotheses 4–7 hold. Then the fractional nonlocal control system* (7) *is approximately controllable.*

To prove obtained results author used fixed point theory (see [34]).

4.2 Approximate Controllability of Impulsive Nonlocal Nonlinear Fractional Systems

The controllability problem for impulsive nonlocal nonlinear fractional systems is discussed in [11]. That system is given in the following form:

$$^C D_\alpha x(t) = Ax(t) + f\left(t, x(t), (Wx)(t)\right) + Bu(t),$$
$$t \in (0, b] \setminus \{t_1, t_2, \ldots, t_m\},$$
$$x(0) + g(x) = x_0 \in X, \ \Delta x(t_i) = I_i\left(x(t_i^-)\right) + D\nu(t_i^-), \tag{8}$$
$$i = 1, 2, \ldots, m,$$

where $^C D_\alpha$ is the Caputo fractional derivative of order $\alpha \in (0, 1)$, the state $x(\cdot)$ takes its values in a Banach space X with norm $\| \cdot \|$, and $x_0 \in X$. Let $A : D(A) \subset X \to X$ be a sectorial operator of type (M, θ, α, μ) on X, $\mu \in \mathbb{R}$, $0 < \theta < \frac{\pi}{2}$, $M > 0$, $W : I \times I \times X \to X$ represents a Volterra-type operator such that $(Wx)(t) = \int_0^t h(t, s, x(s)) ds$, the control functions $u(\cdot)$ and $\nu(\cdot)$ are given in $L^2(I, U)$, U is a Banach space, B and D are bounded linear operators from U into X. Here, one has $I = [0, b]$, $0 = t_0 < t_1 < \cdots < t_m < t_{m+1} = b$, $I_i : X \to X$ are impulsive functions that characterize the jump of the solutions at impulse points t_i, the nonlinear term $f : I \times X \times X \to X$, the nonlocal function $g : PC(I, X) \to X$, where $PC(I, X)$ is the space of X-valued bounded functions on I with the uniform norm $\|x\|_{PC} = \sup\{\|x(t)\|, t \in I\}$ such that $x(t_i^+)$ exists for any $i = 0, \ldots, m$ and $x(t)$ is continuous on $(t_i, t_{i+1}]$, $t_0 = 0$ and $t_{m+1} = b$, $\Delta x(t_i) = x(t_i^+) - x(t_i^-)$, $x(t_i^+)$ and $x(t_i^-)$ are the right and left limits of x at the point t_i, respectively.

The main assumptions are formulated as follows.

Hypothesis 8. *The operators* $S_\alpha(t) = \frac{1}{2\pi i} \int_c e^{\lambda t} \lambda^{\alpha-1} R(\lambda^\alpha, A) d\lambda$ *and* $T_\alpha(t) = \frac{1}{2\pi i} \int_c e^{\lambda t} R(\lambda^\alpha, A) d\lambda$ *with* c *being a suitable path such that* $\lambda^\alpha \notin \mu + S_\theta$ *for* $\lambda \in c$, *generated by* A, *are bounded and compact, such that* $\sup_{t \in I} \|S_\alpha\| \le M$ *and* $\sup_{t \in I} \|T_\alpha\| \le M$.

Hypothesis 9. *The nonlinearity* $f : I \times X \times X \to X$ *is continuous and compact; there exist functions* $\mu_i \in L^\infty(I, \mathbb{R}^+)$, $i = 1, 2, 3$, *and positive constants* q_1 *and* q_2 *such that* $\|f(t, x, y)\| \le \mu_1(t) + \mu_2(t)\|x\| + \mu_3(t)\|y\|$ *and* $\|f(t, x, Wx) - f(t, y, Wy)\| = q_1\|x - y\| + q_2\|Wx - Wy\|$.

Hypothesis 10. *Function* $g : PC(I, X) \to X$ *is completely continuous and there exists a positive constant* β *such that* $\|g(x) - g(y)\| \le \beta\|x - y\|$, $x, y \in X$.

Hypothesis 11. *Associated with* $h : \Delta \times X \to X$, *there exists* $m(t, s) \in PC(\Delta, \mathbb{R}^+)$ *such that* $\|h(t, s, x(s))\| \le m(t, s)\|x\|$ *for each* $(t, s) \in \Delta$ *and* $x, y \in X$, *where* $\Delta = \{(t, s) \in \mathbb{R}^2 | t_i \le s, \, t \le t_{i+1}, \, i = 0, \ldots, m\}$.

Hypothesis 12. *For every* $x_1, x_2, x \in X$ *and* $t \in (t_i, t_{i+1}]$, $i = 1, \ldots, m$, I_i *are continuous and compact and there exist positive constants* d_i, e_i *such that*

$$\left\| I_i\left(x_1(t_i^-)\right) - I_i\left(x_2(t_i^-)\right) \right\| \le d_i \sup_{t \in (t_i, t_{i+1}]} \|x_1(t) - x_2(t)\|$$

and

$$\left\| I_i\left(x(t_i^-)\right) \right\| \le e_i \sup_{t \in (t_i, t_{i+1}]} \|x(t)\|.$$

Additionally, to formulate the main results it have to define a relevant operator:

$$\mathcal{R}(\lambda, \Psi_{t_{k-1}, i}^{t_k}) = \left(\lambda I + \Psi_{t_{k-1}, i}^{t_k}\right)^{-1}, \quad i = 1, 2$$

for $\lambda > 0$, where

$$\Psi_{t_{k-1}, 1}^{t_k} = \int_{t_{k-1}}^{t_k} T_\alpha(t_k - s) B B^* T_\alpha^*(t_k - s) ds, \quad k = 1, 2 \ldots, m + 1,$$

$$\Psi_{t_{k-1}, 2}^{t_k} = S_\alpha(t_k - t_{k-1}) D D^* S_\alpha^*(t_k - t_{k-1}), \quad k = 2, 3, \ldots, m + 1$$

are the controllability operators associated with the linear impulsive fractional control system:

$$^C D^\alpha x(t) = Ax(t) + Bu(t)$$
$$x(0) = x_0 \in X, \tag{9}$$
$$\Delta x(t_i) = D\nu(t_i^-), \quad i = 1, \ldots, m.$$

The main theorem of [11] is formulated as follows:

Theorem 7 [11]. *If Hypotheses 8-12 are satisfied and* $\lambda\mathcal{R}\left(\lambda, \Psi_{t_{k-1}, i}^{t_k}\right) \to 0$ *in the strong operator topology as* $\lambda \to 0^+$, $i = 1, 2$, *then the impulsive nonlocal fractional system* (8) *is approximately controllable on* $t \in [0, b] \setminus \{t_1, \ldots, t_m\}$.

For prove presented results authors of [11] used fractional calculus, sectorial operators and Krasnoselskii fixed point theorems.

5   Conclusions

After scrutinizing the selected articles in presented survey we observe a research methodology, which is used to solve the controllability problem. Below it is shown the methodology resulting from in-depth analysis of the papers concerning the controllability of semilinear and nonlinear fractional systems:

(a) showing a mathematical model of dynamical system;
(b) formulation the assumptions concerned dynamical systems;
(c) proof of solution existence of state-space equation using various types of fixed-point theorem or generally fixed-point technique;
(d) formulation theorem contains necessary conditions of controllability;
(e) proof of the above-mentioned theorem.

We also notice that the main role plays the assumption about Lipschitz continuity.

Acknowledgment. The research presented here was done by authors as parts of the projects funded by the National Science Centre in Poland granted according to decision UMO-2017/27/B/ST6/00145 (JK) and DEC-2015/19/D/ST7/03679 (AB).

References

1. Agarwal, R.P., de Andrade, B., Siracusa, G.: On fractional integro-differential equations with state-dependent delay. Comput. Math. Appl. **62**(3), 1143–1149 (2011)
2. de Andrade, B., Dos Santos, J.P.C.: Existence of solution for a fractional neutral integro-differential equation with unbounded delay. Electron. J. Differ. Equ. **2012**(90), 1–13 (2012)
3. Babiarz, A., Klamka, J., Niezabitowski, M.: Schauder's fixed-point theorem in approximate controllability problems. Int. J. Appl. Math. Comput. Sci. **26**(2), 263–275 (2016)
4. Bachelier, O., Dabkowski, P., Galkowski, K., Kummert, A.: Fractional and nd systems: a continuous case. Multidimension. Syst. Sig. Process. **23**(3), 329–347 (2012)
5. Balasubramaniam, P., Tamilalagan, P.: Approximate controllability of a class of fractional neutral stochastic integro-differential inclusions with infinite delay by using Mainardi's function. Appl. Math. Comput. **256**, 232–246 (2015)
6. Balasubramaniam, P., Vembarasan, V., Senthilkumar, T.: Approximate controllability of impulsive fractional integro-differential systems with nonlocal conditions in Hilbert space. Numer. Funct. Anal. Optim. **35**(2), 177–197 (2014)
7. Debbouche, A., Nieto, J.J.: Sobolev type fractional abstract evolution equations with nonlocal conditions and optimal multi-controls. Appl. Math. Comput. **245**, 74–85 (2014)
8. Debbouche, A., Torres, D.F.M.: Sobolev type fractional dynamic equations and optimal multi-integral controls with fractional nonlocal conditions. Fractional Calc. Appl. Anal. **18**(1), 95–121 (2015)
9. Debbouche, A., Torres, D.F.: Approximate controllability of fractional delay dynamic inclusions with nonlocal control conditions. Appl. Math. Comput. **243**, 161–175 (2014)

10. Fubini, G.: Sugli integrali multipli. Rendiconti Reale Accademia dei Lincei **5**, 608–614 (1907)
11. Guechi, S., Debbouche, A., Torres, D.F.M.: Approximate controllability of impulsive non-local non-linear fractional dynamical systems and optimal control. Miskolc Math. Notes **19**(1), 255–271 (2018)
12. Guendouzi, T., Farahi, S.: Approximate controllability of Sobolev-type fractional functional stochastic integro-differential systems. Boletín de la Sociedad Matemática Mexicana **21**(2), 289–308 (2015)
13. Herrmann, R.: Fractional Calculus: An Introduction for Physicists. World Scientific, Singapore (2011)
14. Kaczorek, T., Sajewski, L.: The Realization Problem for Positive and Fractional Systems, vol. 1. Springer, Cham (2014). https://doi.org/10.1007/978-3-319-04834-5
15. Karthikeyan, S., Balachandran, K., Sathya, M.: Controllability of nonlinear stochastic systems with multiple time-varying delays in control. Int. J. Appl. Math. Comput. Sci. **25**(2), 207–215 (2015)
16. Kilbas, A., Srivastava, H., Trujillo, J.: Theory and Applications of Fractional Differential Equations. Elsevier Science, Amsterdam (2006). No. 13
17. Klamka, J., Babiarz, A., Niezabitowski, M.: Banach fixed-point theorem in semilinear controllability problems-a survey. Bull. Pol. Acad. Sci. Tech. Sci. **64**(1), 21–35 (2016)
18. Klamka, J.: Schauder's fixed-point theorem in nonlinear controllability problems. Control Cybern. **29**, 153–165 (2000)
19. Kumlin, P.: Functional Analysis Notes: A Note on Fixed Point Theory. Chalmers & GU, Gothenburg (2004)
20. Liang, J., Yang, H.: Controllability of fractional integro-differential evolution equations with nonlocal conditions. Appl. Math. Comput. **254**, 20–29 (2015)
21. Liang, S., Mei, R.: Existence of mild solutions for fractional impulsive neutral evolution equations with nonlocal conditions. Adv. Differ. Equ. **2014**(1), 101 (2014)
22. Liu, Z., Li, X.: On the controllability of impulsive fractional evolution inclusions in Banach spaces. J. Optim. Theory Appl. **156**(1), 167–182 (2013)
23. Liu, Z., Li, X.: On the exact controllability of impulsive fractional semilinear functional differential inclusions. Asian J. Control **17**(5), 1857–1865 (2015)
24. Mur, T., Henriquez, H.R.: Relative controllability of linear systems of fractional order with delay. Math. Control Relat. Fields **5**(4), 845–858 (2015)
25. Nirmala, R.J., Balachandran, K.: The controllability of nonlinear implicit fractional delay dynamical systems. Int. J. Appl. Math. Comput. Sci. **27**(3), 501–513 (2017)
26. Paszke, W., Dabkowski, P., Rogers, E., Gałkowski, K.: New results on strong practical stability and stabilization of discrete linear repetitive processes. Syst. Control Lett. **77**, 22–29 (2015)
27. Schmeidel, E., Zbaszyniak, Z.: An application of Darbo's fixed point theorem in the investigation of periodicity of solutions of difference equations. Comput. Math. Appl. **64**(7), 2185–2191 (2012)
28. Wang, J., Fan, Z., Zhou, Y.: Nonlocal controllability of semilinear dynamic systems with fractional derivative in Banach spaces. J. Optim. Theory Appl. **154**(1), 292–302 (2012)
29. Wang, J., Zhou, Y.: Complete controllability of fractional evolution systems. Commun. Nonlinear Sci. Numer. Simul. **17**(11), 4346–4355 (2012)
30. Xiaoli, D., Nieto, J.J.: Controllability and optimality of linear time-invariant neutral control systems with different fractional orders. Acta Math. Sci. **35**(5), 1003–1013 (2015)

31. Yan, Z.: Existence results for fractional functional integrodifferential equations with nonlocal conditions in Banach spaces. Annales Polonici Mathematici **3**, 285–299 (2010)
32. Yan, Z.: Controllability of fractional-order partial neutral functional integrodifferential inclusions with infinite delay. J. Franklin Inst. **348**(8), 2156–2173 (2011)
33. Yan, Z.: Approximate controllability of partial neutral functional differential systems of fractional order with state-dependent delay. Int. J. Control **85**(8), 1051–1062 (2012)
34. Yang, H., Ibrahim, E.: Approximate controllability of fractional nonlocal evolution equations with multiple delays. Adv. Differ. Equ. **2017**(1), 272 (2017)
35. Zawiski, R.: On controllability and measures of noncompactness. In: AIP Conference Proceedings, vol. 1637, pp. 1241–1246. AIP (2014)
36. Zhang, X., Huang, X., Liu, Z.: The existence and uniqueness of mild solutions for impulsive fractional equations with nonlocal conditions and infinite delay. Nonlinear Anal.: Hybrid Syst. **4**(4), 775–781 (2010)
37. Zhang, X., Zhu, C., Yuan, C.: Approximate controllability of fractional impulsive evolution systems involving nonlocal initial conditions. Adv. Differ. Equ. **2015**(1), 244 (2015)
38. Zhou, X.F., Wei, J., Hu, L.G.: Controllability of a fractional linear time-invariant neutral dynamical system. Appl. Math. Lett. **26**(4), 418–424 (2013)
39. Zhou, Y.: Fractional Evolution Equations and Inclusions: Analysis and Control. Elsevier Science, Amsterdam (2016)
40. Zhou, Y., Jiao, F.: Existence of mild solutions for fractional neutral evolution equations. Comput. Math. Appl. **59**(3), 1063–1077 (2010)
41. Zhou, Y., Vijayakumar, V., Murugesu, R.: Controllability for fractional evolution inclusions without compactness. Evol. Equ. Control Theory **4**(4), 507–524 (2015)

A Data-Driven Approach to Modeling and Solving Academic Teachers' Competences Configuration Problem

Jarosław Wikarek and Paweł Sitek[(✉)]

Department of Information Systems, Kielce University of Technology,
Al. 1000-lecia PP 7, 25-314 Kielce, Poland
{j.wikarek, sitek}@tu.kielce.pl

Abstract. Before starting the subsequent academic year, the management staff from each organizational unit (faculty, department, research unit, etc.) must solve the university course timetabling problem (UCTP). UCTP belongs to the class of NP-difficult computational problems, thus its optimal solution becomes a challenge, even for small organizational units. Additionally, the process of searching for solutions is complicated by the need to take account of additional constraints resulting from local conditions, teachers' and students' preferences, etc. as well as held limited resources. For the management of a given organizational unit at the university, at the beginning of this process, it is important to find an answer to the question: are we able, given the held resources and existing constraints, to find any UCTP problem solution? Only in the next step we can consider finding an optimum solution according to the selected criterion. It seems that every missing resource can be supplemented (acquired, rented, modified etc.). It is a well-known fact from practice that the greatest problem is related to supplementing the missing competences of academic teachers. You can hire new teachers, already employed teachers can acquire new missing competences or you can manage the available set of competences in a different way. However, the solution to this problem (competences configuration) is key even before solving UCTP.

The article presents the data-driven based approach to modeling and solving the academic teachers' competences configuration problem for the university's organizational units of various sizes.

Keywords: Timetabling problem · Data-driven approach ·
Mathematical programming · Optimization

1 Introduction

The task of arranging the student and teacher course timetable in an optimal manner is not a simple or fast problem. Difficulties result, first of all, from using limited resources which are: the number of rooms and laboratories, number of teachers, or number of hours available in the given semester etc. Additionally, there is the need to take account of many competing interests reported by both the students and the teachers. Formally, the above problem is classified as University Course Timetabling Problem (UCTP) and recognized as a problem of operation research (OR). Owing to the computational

© Springer Nature Switzerland AG 2019
N. T. Nguyen et al. (Eds.): ACIIDS 2019, LNAI 11431, pp. 408–419, 2019.
https://doi.org/10.1007/978-3-030-14799-0_35

complexity, UCTP belongs to NP-complete problems i.e. no effective algorithm for its solution is known. For this reason, artificial intelligence methods (AI), heuristics, hybrid methods etc. are very often applied for the solution UCTP. On the very general level, UCTP can be defined as the allocation of student groups and teachers to the courses and lecture halls/laboratories in certain time intervals (time-slots). Additionally, the received allocation needs to take account of the availability of each resource (lecture halls, teachers, equipment etc.) as well as fulfill a lot of various constraints. These are both resource – and time-related constraints, disjunctive constraints and additional constraints resulting from local conditions, university regulations and preferences of both teachers and students.

In many UCTP models presented in the subject literature (Sect. 2), it is assumed that the resources, though limited, are available and their number is known, allowing UCTP to be solved. This usually results from long-term practice of arranging the schedules, using last year's data as well as support for this process from IT tools. It is also assumed that any possible obtained solution will be an acceptable solution and not the optimal one. If, despite that, it is not possible to find even an acceptable solution, and additionally the recruitment has been very successful (more candidates have been recruited than in the previous year), the solution seems obvious. Any missing resources should be obtained (expanded, bought, rented, etc.).

However, this is not always easy and possible to be achieved in such a short time (usually during the summer break); particularly, if lecturers with specific competences (specializations) are not available. An additional difficulty in this area is the fact that in many countries tertiary education has been changing in recent years. More flexible curricula and study forms appear. Many courses and specializations can be selected, which annually results in very variable demand for specific competences of lecturers.

In this context two major questions appear which the managers of the university's organizational unit must answer. First: does our set of teachers' competences guarantee finding an UCTP solution? Second, if our set of competences does not guarantee finding a solution to then: What competences are missing and in which quantity? In order to obtain an answer to this and other questions from this area, the teachers' competences configuration problem and a model proposed for it have been formulated (Sect. 3 Appendix A). Its place and link to UCTP have been presented in Fig. 1.

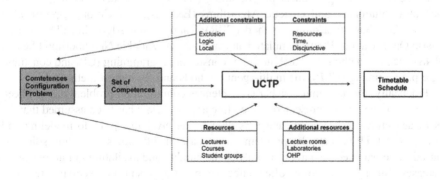

Fig. 1. Place of the competences configuration problem (in gray) in the context of UCTP

The main contribution of the presented research is to propose a new version of the proprietary competence configuration model and the use of a modified hybrid approach (data-driven approach) to solve it. The proposed model is a significant extension of the proprietary model from the first stage of research [1]. A significant innovation is that the model includes logical constraints (Sect. 3) and its implementation is possible using the data-driven approach in the proprietary framework (Sect. 4 Fig. 3). Finding a solution to this problem (obtaining answers to the above questions) may save time and result in appropriate actions being taken even before solving UCTP.

2   Literature Review

Timetabling problem was described in the subject literature for the first time in [2]. It was a relatively simple problem, under which only three data collections were considered: (i) classes (ii) teachers and (iii) timeslots. Since then, the timetabling problem has been the subject of interest of many researchers and practitioners [3]. Many options, classes and sub-classes of this problem have also been developed. One of them is UCTP (University Course Timetabling Problem). A fragment of timetabling problem classification has been presented in Fig. 2.

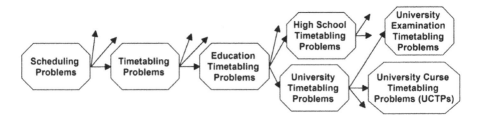

Fig. 2. Part of the classification of timetabling problems.

Over the years, many methods, algorithms and approaches have been developed to modeling and solving variety of UCTPs.

The most important of them include: (a) operational research (OR) methods based on Integer/Mixed Integer Linear programming (IP/MILP), (b) metaheuristic methods, such as Evolutionary and Genetic Algorithms (E&GAs), Ant Colony Optimization (ACO), Case Base Reasoning method (CBR), Tabu Search Algorithm (TS), Partial Swarm Optimization (PSO), Simulated annealing (SA), Variable Neighborhood Search (VNS), etc., (c) methods and techniques of constraint programming (CP) and constraint logic programming (CLP), (d) multi-agent methods and (e) hybrid methods [4–7].

Practically, the academic teacher competences configuration problem has not been considered in any of the presented in literature approaches; it has been assumed that the set of academic teacher competences is definite and available prior to modeling and solving an UCTP or a similar problem. In the case of difficulties in finding solutions caused by constrained resources, in practice the number and availability of all resources increases. Such an action is often effective but this result in excessive resource involvement and increased costs.

3   Academic Teachers' Competences Configuration Problem

After finishing the recruitment and the academic year, organization of the following academic year is planned, this includes division into student groups, specialties, determination of the teaching quotas (loads) for the teachers as well as an initial approach to UCTP. One of key elements of these actions is answer the question: Do we have the appropriate set of teachers with specific competences? This question applies to each organizational unit of the university. The answer is a solution of the specified problem which has been formulated as academic teachers' competences configuration problem (ATCCP).

Formalization of this problem assumes that in the given organizational unit, academic teachers are employed $T = \{a_1, ..., a_i, ..., a_{ZC}\}$ where ZC – number of academic teachers employed in the unit. Each of the academic teacher a has a certain teaching load allocated F_a i.e. the minimum number of hours to be realized in the given period (semester, academic year etc.). For instance, according to the valid law at Polish universities the teaching load is most often: 150, 240, or 360 h per academic year. In practice many academic teacher teach courses in the number of hours exceeding their teaching load. For this reason, W_a coefficient has been introduced. If $W_a = 1$, this means that academic teacher a agrees to teach classes in the number of hours exceeding the teaching load (otherwise $W_a = 0$). WSP coefficient has also been introduced, which determines by which percent an academic teacher's teaching load can be exceeded without the need to obtain his or her consent (currently it is 15%). Certain types of courses are allocated to the given organizational unit (different forms for the given courses: lecturers, projects, laboratory classes etc.) $C = \{b_1, ..., b_j, ..., b_{ZT}\}$ where ZT – number of types of classes in particular courses assigned to the organizational unit. Each type of courses b has a specified number of hours H_b in which it is realized. In addition, for all courses the number of student groups G_b is defined. Academic teacher of a given unit have certain qualifications (competences) to teach certain types of courses (coefficient $Z_{b,a} = 1$ means that academic teacher a, without any further training, courses or postgraduate studies, etc. may offer courses b, otherwise $Z_{b,a} = 0$).

The mathematical model for ATCCP has been formulated in the form of a MILP model (Appendix A). Table A2 presents the description of parameters, indexes and decision variables of the model. Table A1 presents the description of main constraints. The function minimizing the number of missing the academic teachers' competences has been adopted as the objective function (A2).

An important element of the presented model is introducing logical constraints C_Log1, C_Log2 and C_Log3. The constraint C_Log1 formalized as (L1), (L2), (L3) specifies that teacher a should teach all student groups in course b or not teach it at all.

$$X_{b,a} \geq A \cdot F_{b,a} \geq X_{b,a} \tag{L1}$$

$$X_{b,a} = F_{b,a} \cdot h_b \tag{L2}$$

$$F_{b,a} \in \{0, 1\} \tag{L3}$$

The constraint C_Log2 (L4, L5, L6) forces the situation that if teacher a teaches course $b1$, then they must teach course $b2$.

$$X_{b,a} \geq A \cdot F_{b1,a} \geq X_{b1,a}$$
$$X_{b2,a} \geq A \cdot F_{b2,a} \geq X_{b2,a} \tag{L4}$$

$$F_{b1,a} = F_{b2,a} \tag{L5}$$

$$F_{b1,a} \in \{0,1\}; F_{b2,a} \in \{0,1\} \tag{L6}$$

The last of the logical constraints, C_Log3, formalized using (L7, L8, L9) determines the situation when, if the given course b is taught by teacher $a1$, it cannot be, even partially, taught by teacher $a2$. This, of course, does not prevent this course from being taught jointly by teacher $a1$ and another one (apart from $a2$).

$$X_{b,a1} \geq A \cdot F_{b,a1} \geq X_{b,a1}$$
$$X_{b,a2} \geq A \cdot F_{b,a2} \geq X_{b,a2} \tag{L7}$$

$$F_{b,a1} + F_{b,a2} \leq 1 \tag{L8}$$

$$F_{b,a1} \in \{0,1\}; F_{b,a2} \in \{0,1\} \tag{L9}$$

The presented sample logical constraints result from situations existing in practice. They result from local, organizational and personal conditions. Of course, many more logical constraints may be introduced in a similar way, which will affect ATCCP.

4 Data-Driven Approach to Modeling and Solving ATCCP

Due to the great computational complexity of ATCCP (NP-complete) the author's proprietary data-driven approach has been proposed for its implementation, which is a significant extension and generalization of the hybrid approach [8].

The idea of the hybrid approach consisted in the transformation of individual constraints and objective functions and was applied directly to a given model. Each new modeled problem required the development of a new dedicated method of transformation, which involved, among others, the need to modify predicates, facts, etc. The proposed data-driven approach is based on the generalized reduction algorithm and the data instances of the modeled problem. It is universal and can be used for any problem, regardless of its constraints, objective functions, etc.

The general scheme of the proposed data-driven approach has been presented in Fig. 3.

It is based on problem data representation as facts [8, 9]. The structure of the data facts for ATCCP is shown in Fig. 4. The key element in the proposed approach is the reduction algorithm which transforms MILP model into MILP^R model. In simplified terms, this algorithm operates as follows. For each constraint, the model finds such data facts the attributes of which are a sub-set of the constraint attributes (parameters). If a

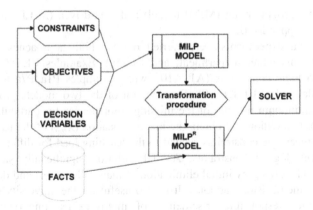

Fig. 3. Proposed data-driven approach concept

row does not exist for the values of these attributes in the given data fact, this constraint is removed from the model. By analogy, the algorithm removes those variables from the constraints for which the parameter/coefficient values at the variables do not exist in the respective data facts. In other words, the algorithm, on the basis of a set of facts and MILP model, generates a new $MILP^R$ model with a reduced number of constraints and decision variables. $MILP^R$ model is solved by any MP-solver [10].

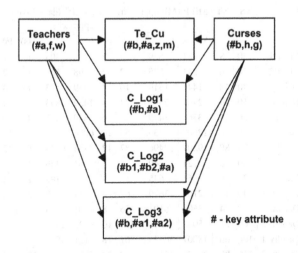

Fig. 4. Structure of the scheme and relations between facts for ATCCP.

5 Computational Experiments and Results

The paper presents the problem concerning academic teacher' competences configuration. A formal model with logical constraints has been proposed for this problem (A1..A10, L1..L9). The proposed model has been implemented using approaches based on mathematical programming (MILP model) and data-driven (MILP^R model) using data facts from Appendix B.

Computational experiments were carried out for both approaches for selected organizational units (chair, department, and faculty). Examples P1..P7 refer to the model with no logic constraints (A1..A10), while examples P1a..P7a to the logical constraint model (A1..A10, L1..L9). The solution of the two models provides information on the minimum number of the missing competences for particular academic teachers ($U_{b,a}$). In addition, we obtain the initial allocation ($X_{b,a}$) of the teachers to the classes which meets the constraints related to the teaching load for different academic teachers as well logical constraints. ATCCP solution significantly simplifies and accelerates UCTP solving by initial elimination of many allocations and determination of values for some decision variables. It is also useful for the university's personnel-accounting services responsible for settlement of employees' teaching loads. Analyzing the problem of computational outlays for solving ATCCP, it is clearly noticeable that the application of data-driven approach (Table 1) has shortened the calculation time by a tier for the model with no logical constraints and by two tiers for the model with logical constraints as compared to the application of MP-solver. For all experiments we used LINGO MP-solver [11] and a PC Intel core (TM2), 2.4 GHz, 4 GB RAM.

Table 1. Computational results for numerical examples

Nr	Unit	NS	NT	MILP (MP-approach)				MILP^R (data-driven approach)			
				V	C	T(s)	Fc	V	C	T(s)	Fc
P1	Chair 1	30	8	481	529	3	1	195	305	1	1
P2	Chair 2	20	6	241	275	2	2	98	163	1	2
P3	Depart. 1	40	18	1441	1519	10	3	359	543	1	3
P4	Depart. 2	50	24	2401	2501	34	4	1459	843	3	4
P5	Depart. 3	60	28	3361	3479	45	6	1680	1367	4	6
P6	Faculty 1	100	60	12007	12236	56	10	6000	4378	4	10
P7	Faculty 2	120	80	19207	19496	87	12	7854	6434	8	12
P1a	Chair 1	30	8	721	1249	18	3	201	346	1	3
P2a	Chair 2	20	6	361	635	17	4	121	187	1	4
P3a	Depart. 1	40	18	2161	3679	67	4	381	578	2	4
P4a	Depart. 2	50	24	3601	6101	256	8	1501	943	4	8
P5a	Depart. 3	60	28	5041	5519	435	10	1723	1479	6	10
P6a	Faculty 1	100	60	18001	30223	867	18	6121	4578	6	18
P7a	Faculty 2	120	80	28801	48283	1489	24	7978	6823	10	24

NS - Number of courses
NT - Number of academic teachers
V - Number of variables
T - Calculation time in seconds
C - Number of constraints
Fc - Number of missing competences

6 Conclusions

The paper presents a modified version of an ATCCP that takes into account logical constraints. The modified (data-driven) solving method has also been proposed, which is the development and generalization of the original hybrid approach [8]. With the proposed data-driven approach, it is possible to solve larger size problems in a shorter (acceptable) time. It is obvious that its application does not decrease computational complexity of the problem but reduces the model dimensions enough to keep it sufficient in practical applications. The ATCCP solution allows for optimal management of teachers' competences and finding preliminary assignments of teachers to courses. What's more, knowledge of these assignments will simplify the UTCP solution.

As part of further research is planned on modeling and solving UCTP integrated with the ATCCP model (Sect. 3, Appendix A). It is also planned to adapt this model to problems related to competences configuration from the production, logistics, vehicle routing problems and supply chain areas [12–16].

Appendix A

Table A1. Description of the constraints of the model for ATCCP

Number	Description
A1	Goal function - minimum number of missing competences
A2	The constraint specifies that only the teachers having appropriate competences can teach the course which requires them
A3	The constraint ensures that each course is provided for the designated number of student groups
A4	The constraint ensures that the teaching load is realized
A5	The constraint does not allow the teaching load to be exceeded by more than wsp, if there is no academic teacher's consent
A6	The constraint ensures that change/extension of the competences may happen only a specified number of times
A7	The constraint allows changing selected competences
A8, 9, 10	Binarity and integrity constraints
A11	No possibility to change competences

Table A2. The decision variables, indices and constraints of the model for ATCCP.

Symbol	Description
Sets	
T	Set of academic teachers
C	Set of courses/classes
Indexes	
a	Index of academic teacher $a \in T$
b	Index of course $b \in C$

(continued)

Table A2. (*continued*)

Symbol	Description
Parameters	
H_b	Number of hours taught under the course b, $b \in C$
G_b	Number of student groups assigned to the course b, $b \in C$
F_a	Number of hours of the teaching load of academic teacher a, $a \in T$
W_a	If the academic teacher a has given their consent to overtime hours $W_a = 1$, otherwise $W_a = 0$
$Z_{b,a}$	If academic teacher a, without any further training, may teach course b $Z_{b,a} = 1$, otherwise $Z_{b,a} = 0$, $b \in C$, $a \in T$
$M_{b,a}$	If academic teacher a, after training, may teach course b $M_{b,a} = 1$, otherwise $M_{b,a} = 0$, $b \in C$, $a \in T$
WSP	Percent exceeded teaching load without academic teacher's necessary consent
O	Percent exceeded teaching load with academic teacher's necessary consent
ST	Arbitrarily large constant (for example $ST = \sum_{b \in C} \sum_{a \in T} (Z_{b,a} + U_{b,a})$)
L	Maximum number of changes in competences for academic teacher a
Decision variables	
$X_{b,a}$	Number of groups for course b which will be taught by academic teacher a $b \in C$, $a \in T$
$F_{b,a}$	If the academic teacher a teach course b $F_{b,a} = 1$ otherwise $F_{b,a} = 0$, $b \in C$, $a \in T$
$U_{b,a}$	If the competences of academic teacher a to teach course b have been changed $U_{b,a} = 1$ otherwise $U_{b,a} = 0$, $b \in C$, $a \in T$
Fc	Number of missing competences

$$Fc = \min \sum_{b \in C} \sum_{a \in T} U_{b,a} \tag{A1}$$

$$X_{b,a} \leq (Z_{b,a} + U_{b,a}) \cdot ST \forall b \in C, a \in T \tag{A2}$$

$$\sum_{b \in C} X_{b,a} = G_b \forall b \in C \tag{A3}$$

$$\sum_{b \in C} H_b \cdot X_{b,a} \geq F_a \forall a \in T \tag{A4}$$

$$\sum_{b \in C} H_b \cdot X_{b,a} \leq WSP \cdot F_a \forall a \in T : W_a = 0$$
$$\sum_{b \in C} H_b \cdot X_{b,a} \leq O \cdot F_a \forall a \in T : W_a = 1 \tag{A5}$$

$$\sum_{b \in C} \sum_{a \in T} U_{b,a} \leq L \tag{A6}$$

$$U_{b,a} \leq M_{b,a} \forall b \in C, a \in T \tag{A7}$$

$$U_{b,a} = 0 \forall b \in C, a \in T : Z_{b,a} = 0 \tag{A8}$$

$$X_{b,a} \in C^+ \forall b \in C, a \in T \quad\quad (A9)$$

$$U_{b,a} \in \{0,1\} \forall b \in C, a \in T \quad\quad (A10)$$

Appendix B Data for Examples P1 and P1a

teachers(#a,F_a,W_a).- fact describing academic teachers
```
teachers(1,150,1). teachers(2,240,1). teachers(3,240,1).
teachers(4,240,1). teachers(5,240,1). teachers(6,360,1).
teachers(2,360,1). teachers(8,360,1).
```
courses(#b,H_b,G_b,) – fact describing courses
```
courses(1,30,1).  courses(2,30,3).  courses(3,30,6).  courses(4,30,1).
courses(5,30,1).  courses(6,30,1).  courses(7,45,8).  courses(8,30,8).
courses(9,15,1).  courses(10,30,4). courses(11,30,1). courses(12,30,2).
courses(13,30,4). courses(14,15,1). courses(15,30,3). courses(16,30,6).
courses(17,30,1). courses(18,15,1). courses(19,30,1). courses(20,30,1).
courses(21,30,1). courses(22,30,2). courses(23,30,1). courses(24,30,4).
courses(25,30,8). courses(26,30,1). courses(27,45,6). courses(28,15,1).
courses(29,30,3). courses(30,30,3).
```
Te_Cu (#b,#a,$Z_{b,a}$,$M_{b,a}$)- fact describing possible teacher assignments for courses
```
Te_Cu(1,1,1,0).   Te_Cu(1,2,0,1).   Te_Cu(1,3,0,1).   Te_Cu(2,2,1,0).
Te_Cu(2,3,1,0).   Te_Cu(2,4,1,0).   Te_Cu(3,2,1,0).   Te_Cu(3,3,1,0).
Te_Cu(3,4,1,0).   Te_Cu(3,6,0,1).   Te_Cu(3,7,0,1).   Te_Cu(3,8,0,1).
Te_Cu(4,1,1,0).   Te_Cu(4,2,0,1).   Te_Cu(4,3,0,1).   Te_Cu(5,1,1,0).
Te_Cu(5,2,0,1).   Te_Cu(5,3,0,1).   Te_Cu(5,6,0,1).   Te_Cu(5,7,0,1).
Te_Cu(5,8,0,1).   Te_Cu(6,1,1,0).   Te_Cu(6,2,0,1).   Te_Cu(6,3,0,1).
Te_Cu(7,1,1,0).   Te_Cu(7,2,1,0).   Te_Cu(7,3,0,1).   Te_Cu(7,6,0,1).
Te_Cu(7,8,0,1).   Te_Cu(8,2,0,1).   Te_Cu(8,3,1,0).   Te_Cu(8,4,1,0).
Te_Cu(8,6,0,1).   Te_Cu(8,7,0,1).   Te_Cu(8,8,0,1).   Te_Cu(9,2,0,1).
Te_Cu(9,3,0,1).   Te_Cu(9,5,1,0).   Te_Cu(9,6,1,0).   Te_Cu(10,2,0,1).
Te_Cu(10,3,0,1).  Te_Cu(10,5,1,0).  Te_Cu(10,6,1,0).  Te_Cu(11,4,0,1).
Te_Cu(11,5,1,0).  Te_Cu(11,6,0,1).  Te_Cu(11,7,0,1).  Te_Cu(11,8,0,1).
Te_Cu(12,4,0,1).  Te_Cu(12,5,1,0).  Te_Cu(12,6,0,1).  Te_Cu(12,7,0,1).
Te_Cu(12,8,0,1).  Te_Cu(13,4,0,1).  Te_Cu(13,5,1,0).  Te_Cu(13,8,1,0).
Te_Cu(14,4,0,1).  Te_Cu(14,5,1,0).  Te_Cu(14,6,0,1).  Te_Cu(14,7,0,1).
Te_Cu(14,8,0,1).  Te_Cu(15,4,0,1).  Te_Cu(15,5,1,0).  Te_Cu(15,6,0,1).
Te_Cu(15,7,0,1).  Te_Cu(15,8,0,1).  Te_Cu(16,4,0,1).  Te_Cu(16,5,1,0).
Te_Cu(17,4,0,1).  Te_Cu(17,5,1,0).  Te_Cu(17,6,1,0).  Te_Cu(17,7,0,1).
Te_Cu(17,8,0,1).  Te_Cu(18,4,0,1).  Te_Cu(18,5,1,0).  Te_Cu(18,6,1,0).
Te_Cu(18,7,0,1).  Te_Cu(18,8,0,1).  Te_Cu(19,4,0,1).  Te_Cu(19,5,1,0).
Te_Cu(19,6,1,0).  Te_Cu(19,7,0,1).  Te_Cu(19,8,0,1).  Te_Cu(20,4,0,1).
Te_Cu(20,5,1,0).  Te_Cu(21,1,1,0).  Te_Cu(21,5,1,0).  Te_Cu(22,2,1,0).
Te_Cu(22,6,1,0).  Te_Cu(22,7,0,1).  Te_Cu(22,8,0,1).  Te_Cu(23,1,1,0).
Te_Cu(23,6,1,0).  Te_Cu(24,2,1,0).  Te_Cu(24,6,0,1).  Te_Cu(24,7,0,1).
Te_Cu(24,8,0,1).  Te_Cu(25,2,1,0).  Te_Cu(25,3,1,0).  Te_Cu(25,6,0,1).
Te_Cu(25,7,0,1).  Te_Cu(25,8,0,1).  Te_Cu(26,1,1,0).  Te_Cu(26,6,1,0).
Te_Cu(27,1,1,0).  Te_Cu(27,2,1,0).  Te_Cu(27,6,1,0).  Te_Cu(27,7,0,1).
Te_Cu(27,8,1,0).  Te_Cu(28,2,1,0).  Te_Cu(28,5,1,0).  Te_Cu(29,2,1,0).
Te_Cu(29,3,1,0).  Te_Cu(29,4,1,0).  Te_Cu(29,5,1,0).  Te_Cu(29,6,1,0).
Te_Cu(29,7,1,0).  Te_Cu(29,8,0,1).  Te_Cu(30,2,1,0).  Te_Cu(30,3,1,0).
Te_Cu(30,4,1,0).  Te_Cu(30,5,1,0).  Te_Cu(30,6,1,0).  Te_Cu(30,7,1,0).
Te_Cu(30,8,0,1).
```
Logic Constraints: `C_log1(27,8). C_log2(1,2,1). C_log3(7,1,2)`

References

1. Wikarek, J.: Lecturers' competences configuration model for the timetabling problem. In: Ganzha, M., Maciaszek, L., Paprzycki, M. (eds.) Proceedings of the 2018 Federated Conference on Computer Science and Information Systems, ACSIS, vol. 15, pp. 441–444 (2018). http://dx.doi.org/10.15439/2018F143
2. Gotlib, C.C.: The construction of class-teacher timetables. Proc. IFIP Cong. **62**, 73–77 (1963)
3. Babaei, H., Karimpour, J., Hadidi, A.: A survey of approaches for university course timetabling problem. Comput. Ind. Eng. **86**, 43–59 (2015)
4. Redl, T.A.: A study of university timetabling that blends graph coloring with the satisfaction of various essential and preferential conditions. Ph.D. Thesis, Rice University, Houston, Texas (2004)
5. Asmui, H., Burke, E.K., Garibaldi, J.M.: Fuzzy multiple heuristic ordering for course timetabling. In: The proceedings of the 5th United Kingdom Workshop on Computational Intelligence (UKCI05), London, UK, pp. 302–309 (2005)
6. Abdullah, S., Burke, E.K., McCollum, B.: Using a randomised iterative improvement algorithm with composite neighbourhood structures for the university course timetabling problem. In: Doerner, K.F., Gendreau, M., Greistorfer, P., Gutjahr, W., Hartl, R.F., Reimann, M. (eds.) Metaheuristics. ORSIS, vol. 39, pp. 153–169. Springer, Boston, MA (2007). https://doi.org/10.1007/978-0-387-71921-4_8
7. Mühlenthaler, M.: Fairness in academic course timetabling. In: Mühlenthaler, M. (ed.) Fairness in academic course timetabling. LNE, vol. 678, pp. 75–105. Springer, Cham (2015). https://doi.org/10.1007/978-3-319-12799-6_3
8. Sitek, P., Wikarek, J.: A hybrid programming framework for modeling and solving constraint satisfaction and optimization problems. Sci. Program. **2016** (2016). Article ID 5102616. https://doi.org/10.1155/2016/5102616
9. Sitek, P., Wikarek, J.: A multi-level approach to ubiquitous modeling and solving constraints in combinatorial optimization problems in production and distribution. Appl. Intell. **48**, 1344–1367 (2018). https://doi.org/10.1007/s10489-017-1107-9
10. Schrijver, A.: Theory of Linear and Integer Programming. Wiley, Hoboken (1998). ISBN 0-471-98232-6
11. Home LINDO. www.lindo.com. Accessed 20 Dec 2018
12. Zwolińska, B., Grzybowska, K.: Shaping production change variability in relation to the utilized technology. In: 24th International Conference on Production Research (ICPR 2017), pp. 51–56 (2017). ISBN 978-1-60595-507-0, ISSN 2475-885X
13. Nielsen, I., Dang, Q.-V., Nielsen, P., Pawlewski, P.: Scheduling of mobile robots with preemptive tasks. In: Omatu, S., Bersini, H., Corchado, J.M., Rodríguez, S., Pawlewski, P., Bucciarelli, E. (eds.) Distributed Computing and Artificial Intelligence, 11th International Conference. AISC, vol. 290, pp. 19–27. Springer, Cham (2014). https://doi.org/10.1007/978-3-319-07593-8_3
14. Bocewicz, G., Nielsen, I., Banaszak, Z.: Production flows scheduling subject to fuzzy processing time constraints. Int. J. Comput. Integr. Manuf. **29**(10), 1105–1127 (2016). https://doi.org/10.1080/0951192X.2016.1145739

15. Sitek, P., Wikarek, J.: Capacitated vehicle routing problem with pick-up and alternative delivery (CVRPPAD) – model and implementation using hybrid approach. Ann. Oper. Res. **273**, 257–277 (2017). https://doi.org/10.1007/s10479-017-2722-x

16. Kłosowski, G., Gola, A., Świć, A.: Application of fuzzy logic in assigning workers to production tasks. Distributed Computing and Artificial Intelligence, 13th International Conference. AISC, vol. 474, pp. 505–513. Springer, Cham (2016). https://doi.org/10.1007/978-3-319-40162-1_54

PID Regulatory Control Design for a Double Tank System Based on Time-Scale Separation

Marian Blachuta$^{(\boxtimes)}$, Robert Bieda, and Rafal Grygiel

Silesian University of Technology, 16 Akademicka St., 44-100 Gliwice, Poland
Marian.Blachuta@polsl.pl

Abstract. An intuitive and simple yet efficient novel approach to the PID control system synthesis for load disturbance rejection in a cascaded tanks system is presented. It is assumed that the controller is fitted with a first order noise filter. The design method consists in choosing appropriate values of filter and controller parameters such that a time-scale separation takes place between control signal and output dynamics while keeping the noise amplification gain at a reasonable value. As a result, low measurement noise results in very high performance of load disturbance rejection. Moreover, the solution shows high degree of robustness against changes of the working point. Simple formulas providing analytical solutions for extrema of time responses are derived that allow to design control system with predefined characteristics.

1 Introduction

Classical PID setpoint control for water tanks systems with a schematic depicted in Fig. 1 is the subject of many control teaching laboratory experiments [1,2,7]. Unfortunately, since the experimental setups usually lack disturbance inputs, the more important regulatory control is not studied there.

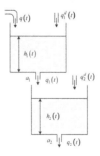

Fig. 1. Schematic diagram of the considered cascaded tanks system. The inlet flow $q(t)$ is the control variable, $q_1^d(t)$ and $q_2^d(t)$ are disturbance inputs, a_1 and a_2 are cross-sectional areas of orifices.

© Springer Nature Switzerland AG 2019
N. T. Nguyen et al. (Eds.): ACIIDS 2019, LNAI 11431, pp. 420–430, 2019.
https://doi.org/10.1007/978-3-030-14799-0_36

Following the contemporary literature, e.g. [9, 10], we assume controller transfer function as a product of the nominal transfer function $C'(s)$ and the noise filter transfer function $F(s)$

$$C(s) = C'(s)F(s), \tag{1}$$

where

$$C'(s) = k_p + \frac{k_i}{s} + k_d s = k_c \frac{(s + c_1)(s + c_2)}{s}, \tag{2}$$

$$F(s) = \frac{1}{\mu s + 1}. \tag{3}$$

Until recently, the traditional approach to the PID control of the cascaded tanks system was the one based on pole placement method, [2,3,7,8]. Unfortunately, this method required the dynamics of the noise filter to be neglected at the design stage. This was admissible since controller gain was small enough to prevent control signal from saturation. A first order filter was then experimentally chosen such that the noise related component of the control signal was sufficiently reduced without affecting the control performance. Unfortunately, the control performance with respect to disturbance rejection was very poor. In [2] the use of relatively small controller gain was partly justified by poor equipment available at that time. Nowadays, due to large technical progress in the area of control components, the noise issue is not even mentioned in [7], since it is simply indiscernible within this tuning method. This is because ADC conversion with 16-bit resolution at sampling frequency of 100 Hz is reported in [9] with the standard deviation less than 3 mV in the measurement range 0–10 V while in [2] a 8-bit ADC's was used with sampling frequency of 10 Hz being itself a source of a rather low frequency quantization noise. This calls for new design methods for high performance control.

In our approach the time constant μ becomes an important fourth PID controller parameter which limits the achievable control performance. Its main idea is that at higher values of controller gain k_c two controller zeros attract two closed-loop system roots making them almost independent of the open loop poles while the remaining roots resulting from $F(s)$ defined in (3) depart to infinity along asymptotes whose crossing point depending on μ^{-1} is far from imaginary axis. Hence, for small μ and k_c big enough, time-scale separation [11] can be induced between the fast varying control signal and the slow varying control error. As a result, parameters of both, control signal and the disturbance output, can be analysed separately. This also guaranties immunity of the control error dynamics against working point changes. Moreover, we are able to relate time responses with poles and zeros not only qualitatively but also quantitatively by calculating parameters of their extrema.

Higher controller gains required by our approach are admissible thanks to contemporary low noise sensors. For example, if the standard deviation of the level sensor noise is less than 3 mV, [10], then even if the noise amplification gain equals to 300 the standard deviation of noise component in the control signal does not exceed a fraction of 1 V. Provided that the pump is propelled by a PWM controlled BLDC motor this also does not make any harm unless the pulse width does not saturate.

The recent methods for simultaneous PID and filter parameters tuning, e.g. [3,9,10], are mainly devoted to higher order and/or delayed systems working in presence of rather severe measurement noise. When applied to the considered system they give much worse results than our method does.

The paper uses the results of [4] where a new approach to various tank systems was outlined, and [5] where the design and a detailed analysis of high performance PID control for a double tank system is presented.

2   Normalized Model of Tanks System

For the sake of brevity equations of tank systems will be written assuming that both tanks are identical, i.e. they have the same cross-sectional areas $A_1 = A_2 = A$ and the same orifices with cross-sectional areas $a_1 = a_2 = a$. In order to make the model independent of particular dimensions, normalized variables

$$u(t) = \frac{q(t)}{q_N}, x_i(t) = \frac{h_i(t)}{h_N}, d_i(t) = \frac{q_i^d(t)}{q_N}, i \in \{1, 2\} \tag{4}$$

are used, where q_N and h_N are certain nominal values that fulfill the equation of equilibrium $q_N = \kappa\sqrt{h_N}$. And assume normalized time $t' = t/2T_N{}^1$. Then the normalized equations of the tank system considered take the form

$$\frac{dx_1(t)}{dt} = 2\left[u(t) - \sqrt{x_1(t)} + d_1(t)\right], \tag{5}$$

$$\frac{dx_2(t)}{dt} = 2\left[\sqrt{x_1(t)} - \sqrt{x_2(t)} + d_2(t)\right]. \tag{6}$$

Assuming $u(t) = u_0 + \Delta u(t), x_1(t) = x_{10} + \Delta x_1(t), x_2(t) = x_{20} + \Delta x_2(t), d_1(t) = \Delta d_1(t)$ and $d_2(t) = \Delta d_2(t)$, the normalized Eqs. (5)–(6) are then linearized

$$u_0\frac{d\Delta x_1(t)}{dt} = 2u_0\left[\Delta u(t) + d_1(t)\right] - \Delta x_1(t), \tag{7}$$

$$u_0\frac{d\Delta x_2(t)}{dt} = \Delta x_1(t) - \Delta x_2(t) + 2u_0 d_2(t) \tag{8}$$

around a working point determined by u_0, x_{10} and x_{20} which fulfills the equilibrium equation $u_0 = \sqrt{x_{10}} = \sqrt{x_{20}}$. The transfer function structure of the linearized system with noise input $n(t)$ are presented in Fig. 3. Both the steady-state gain $2u_0$ and high-frequency gain $2/u_0$ as well as time constant u_0 change with changing u_0. The properties of models linearized at various u_0 of Fig. 3 is characterized by their step responses and Nyquist plots displayed in Fig. 2. For u_0 changing between $1/2$ and 2 the working point water levels $x_{10} = x_{20}$ change between $1/4$ and 4.

---

[1] In order not to increase the number of symbols we will further use variable t instead of t'. Similarly, the Laplace variable $s' = 2T_N s$ corresponding to t' will be denoted by s.

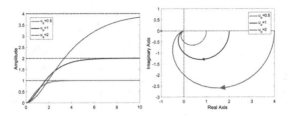

Fig. 2. Step responses and Nyquist plots for system linearized at various u_0.

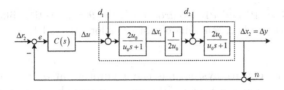

Fig. 3. Control system linearized at u_0; n - measurement noise.

3 Disturbance Attenuation Control

3.1 Ideal Disturbance Rejection

The ideal control for the plant in Fig. 3 which completely rejects the effect of disturbances $d_1(t)$ is $\Delta u(t) = -d_1(t)$ and of $d_2(t)$ is $\Delta u(t) = -u_0 \frac{d}{dt} d_2(t) - d_2(t)$. Stepwise change of $d_1(t)$ requires a compensating stepwise change of $\Delta u(t)$ and does not need any knowledge about the system itself. Rejection of $d_2(t)$ is more complicated. To compensate a stepwise change of $d_2(t)$, a Dirac impuls with the area proportional to u_0 is necessary to instantaneously change the first tank level followed by constant flow compensating $d_2(t)$. In both cases the control signal should change rapidly. Our aim is to design a closed-loop control system in such way that it approaches the ideal solution with control signal changing much faster than the controlled one. This leads to the concept of two time-scale control systems that are robust against working point change.

3.2 Generic Cancellation Controller Design via Root Loci

Root loci for PI and PID controllers are compared in Fig. 4 to show existence of two time scales for PID and only one for PI controller. Just for the sake of simplicity of the root loci and ease of calculations we assume a generic system with controller zeros in (1)–(3) equal to -1 which leads to pole-zero cancellations. This simplifying assumption is not a part of the method, and for PID the setting $c_1 = c_2 = 1$ means that we require two closed-loop system poles to be close to -1 even after change of the working point u_0 according to Fig. 3. This assumption is repealed in [5] and in Sect. 3.5, where the positive effect of changing the position of zeros is studied, see Fig. 10.

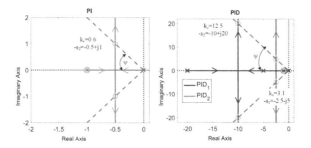

Fig. 4. Root loci for PI and PID control, $\tan(\psi) = \theta = 2$. Notice different scaling of the axes and difference between pairs of fast roots for PID vs PI.

Notice that PI control system has the asymptote at $\sigma_a = -0.5$ while PID_1 at $\sigma_a = -10$. This suggests that PID_1 control systems has two distinct modes: a fast mode determined by the roots on the asymptote and a slow ones determined by pole(s) at -1. In general for the slow roots there is $-s_{3,4} = -\sigma_s \pm j\omega_s = -\sigma_s(1 \pm j\theta_s)$ with $\theta_s = \omega_s/\sigma_s$ and

$$(s + s_3)(s + s_4) = s^2 + 2\sigma_s s + \sigma_s^2(1 + \theta_s^2) \tag{9}$$

The fast roots for PID control denoted by $-s_{1,2} = -\sigma_f \pm j\omega_f = -\sigma_f(1 \pm j\theta_f)$ with $\theta_f = \omega_f/\sigma_f$ form the denominator polynomial

$$(s + s_1)(s + s_2) = s^2 + 2\sigma_f s + \sigma_f^2(1 + \theta_f^2) \tag{10}$$

of the transfer functions G_{ud_1} and G_{ud_2} in (12). The value of θ_f playing the role of oscillability index of the fast mode is used to limit the values of k_c. The value of σ_f is responsible for transients damping rate. The fast mode determines the setpoint response and the slow mode the disturbance one. We assume that the two modes are sufficiently separated if $\sigma_c = (0.1 \div 0.2)\sigma_f$. This is not the case for PID_2 controller with $\mu = 0.2$ and $\sigma_f = 2.5$, where these two modes are not sufficiently separated, and, compared with PID_1, see Fig. 9, the control performance is rather poor. However, compared with PI it is still much better.

3.3   Load Disturbance Responses

Our aim is to study the effect of a stepwise change of disturbances d_1 and d_2 acting in the closed loop system depicted in Fig. 3, where the controller $C(s)$ is defined in (1)–(3). The general form of load disturbance responses is defined by the following equations:

$$\Delta y(s) = G_{yd_1}(s)d_1(s) + G_{yd_2}(s)d_2(s), \tag{11}$$

$$\Delta u(s) = G_{ud_1}(s)d_1(s) + G_{ud_2}(s)d_2(s). \tag{12}$$

Useful Formulas. Due to pole-zero cancellations and possible simplifications, the transfer functions in (11)–(12) are often of second order with a zero. Extrema of step responses for second order systems with complex roots, derived in [6], are summarized for two types of systems. For the unity gain system

$$G_1(s) = \frac{\sigma^2(1+\theta^2)\left(\frac{\alpha}{\sigma}s+1\right)}{s^2 + 2\sigma s + \sigma^2(1+\theta^2)} \tag{13}$$

the maximum $y_m = y(t_m)$ and t_m are expressed as

$$\sigma t_m = \begin{cases} \frac{1}{\theta}\arctan\frac{\alpha\theta}{\alpha-1}, & \alpha \geq 1, \\ \frac{1}{\theta}\left(\pi + \arctan\frac{\alpha\theta}{\alpha-1}\right), & \alpha < 1, \end{cases} \tag{14}$$

$$y_m = 1 + \sqrt{(\alpha-1)^2 + (\alpha\theta)^2}\,e^{-\sigma t_m}. \tag{15}$$

For the derivative system

$$G_2(s) = \frac{s}{s^2 + 2\sigma s + \sigma^2(1+\theta^2)} \tag{16}$$

there is

$$\sigma t_m = \frac{\arctan\theta}{\theta}, \qquad y_m = \frac{1}{\sigma\sqrt{1+\theta^2}}e^{-\sigma t_m}. \tag{17}$$

Therefore the extrema of step responses are easily calculated using the formulas for transfer functions $G_1(s), G_1^\star(s) = G_1(s)|_{\alpha=0}$ and $G_2(s)$ defined above.

3.4 Exemplary Results

We assume $u_0 = 1$ in most considerations, which means that the system is normalized at the working point, and the standard first order noise filter $F(s)$ is employed. Simulation results showing influence of important controller and plant parameters are displayed in Figs. 5, 6 and 7. Data necessary to use the formulas in (13)–(17) are calculated and resulting analytical results for extrema in the format (time, value) are deployed on certain plots.

PI Controller. For the controller

$$C(s) = k_c\frac{s+1}{s} \tag{18}$$

there is

$$G_{yd_1}(s) = \frac{2s}{(s^2+s+k)(s+1)}, \qquad G_{yd_2}(s) = \frac{2s}{s^2+s+k} = 2G_2(s), \tag{19}$$

$$G_{ud_1}(s) = \frac{-k}{s^2+s+k} = -G_1^\star(s), \qquad G_{ud_2}(s) = \frac{-k(s+1)}{s^2+s+k} = -G_1(s). \tag{20}$$

The characteristic polynomial $\chi(s) = (s+s_1)(s+s_2)(s+s_3)$ is of degree 3 with $s_{1,2} = \sigma(1 \pm j\theta)$, and $s_3 = 1$, where $\sigma = \frac{1}{2}$, and $\theta = \sqrt{4k-1}$ for $k \geq \frac{1}{4}$.

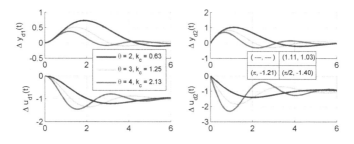

Fig. 5. PI control, $k_c = f(\theta)$. Extrema in the format (time, value) for the blue plots are deployed in the table on the right accordingly to their positions in the array of plots. (Color figure online)

PID Controller. If $c_1 = c_2 = 1$ then

$$C(s) = k_c \frac{(s+1)^2}{s(\mu s + 1)} \tag{21}$$

and

$$G_{yd_1}(s) = \frac{2s(\mu s + 1)}{(\mu s^2 + s + k)(s+1)^2}, \quad G_{yd_2}(s) = \frac{2s(\mu s + 1)}{(\mu s^2 + s + k)(s+1)}, \tag{22}$$

$$G_{ud_1}(s) = \frac{-k}{\mu s^2 + s + k} = -G_1^\star(s), \quad G_{ud_2}(s) = \frac{-k(s+1)}{\mu s^2 + s + k} = -G_1(s). \tag{23}$$

$\chi(s) = (s+s_1)(s+s_2)(s+s_3)(s+s_4)$ is of degree 4, $s_{1,2} = \sigma_f(1 \pm j\theta_f)$ with $\sigma_f = \frac{1}{2\mu}$ and $\theta_f = \sqrt{4k\mu - 1}$ for $k \geq \frac{1}{4\mu}$, and $s_{3,4} = 1$, with $\sigma_s = 1$ and $\theta_s = 0$. Observe that $k_c = (1 + \theta_f^2)/(8\mu)$. For μ small enough and k big enough $G_{yd_1}(s)$, $G_{yd_2}(s)$ and $G_{ud_2}(s)$ can be approximated as follows:

$$G_{yd_1}(s) \simeq \frac{2s}{(s+k)(s+1)^2} \simeq \frac{s}{k_c(s+1)^2} = \frac{1}{k_c} G_2(s), \tag{24}$$

$$G_{yd_2}(s) \simeq \frac{2s}{(s+k)(s+1)}, \quad G_{ud_1}(s) \simeq \frac{-k}{s+k}, \quad G_{ud_2}(s) \simeq -\frac{s+1}{\frac{s}{k}+1} \tag{25}$$

for which

$$\Delta u_{d1}(t) \simeq -(1 - e^{-kt}), \quad \Delta u_{d2}(t) \simeq -[(k-1)e^{-kt} + 1] \tag{26}$$

$$G_{yd_2}(s) \simeq \frac{s}{k_c(s+1)}, \quad \Delta y_{d2}(t) \simeq \frac{1}{k_c} e^{-t}. \tag{27}$$

It is clear that for $d_1(t) = 1(t)$ and $d_2(t) = 1(t)$, μ small enough and k great enough, $\Delta u_{d1}(t) \to -1(t)$ and $\Delta u_{d2}(t) \to -\delta(t) - 1(t)$, i.e. the control signals tend to the ideal ones defined in Sect. 3.1 (Fig. 6).

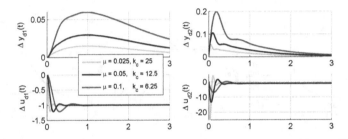

Fig. 6. PID control - influence of k_c as a function of μ; $\theta_f = 2$. Notice performance improvement with decreasing μ and increasing k_c, and a huge performance improvement compared to the PI controller of Fig. 5

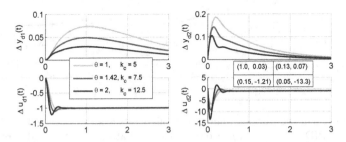

Fig. 7. PID control - influence of k_c as a function of θ_f; $\mu = 0.05$. Extrema in the format (time, value) for the blue plots are deployed in the table on the right accordingly to their positions in the array of plots. (Color figure online)

Frequency Domain Characteristics. Let us define

$$S(s) = \frac{1}{1 + L(s)}, \quad T(s) = \frac{L(s)}{1 + L(s)}, \quad Q(s) = \frac{T(s)}{P(s)} \tag{28}$$

$$L(s) = C(s)P(s), \quad P(s) = \frac{2}{(1 + s)^2}. \tag{29}$$

Function $Q(s)$ is a transfer function between noise input and controller output while $T(s)$ between noise input and system output. Observe, that extrema of $|S(j\omega)|$ and $|T(j\omega)|$ are the same for systems with the same θ_f and different σ_f, and that max $Q(\omega)$ strongly depends on μ (Fig. 8).

Sensitivity to the Working Point. Numerical analysis of the effect of the current working point u_0 while retaining controller settings chosen for the case $u_0 = 1$, is displayed in Fig. 9. In contrast to PI controllers it has almost no influence on PID_1 controlled systems. The later is due to time-scale separation between the fast (control signal) and the slow (output signal) mode. Important observation is that contrary to the popular belief, [10], the value of $\max_\omega |S(j\omega)|$ does not determine robustness against changes of the working point u_0.

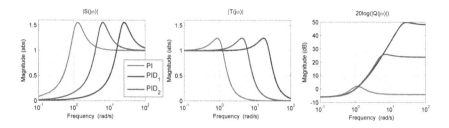

Fig. 8. Plots of $|S(j\omega)|$, $|T(j\omega)|$, and $|Q(j\omega)|$ in the log-log scale.

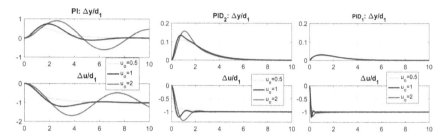

Fig. 9. Influence of the working point. Controllers settings PI ($k_c = 0.63$); PID$_2$ ($k_c = 3.1$, $\mu = 0.2$), PID$_1$ ($k_c = 12.5$, $\mu = 0.05$) for $\theta_f = 2$ at $u_0 = 1$. Notice different scales of Δy and large improvement of performance and sensitivity to the set-point change between PI and PID controllers with increasing degree of time scale separation

Noise Amplification. Assume that the noise is a band-limited stochastic signal with variance var(n), and $k_{un} = \text{sd}(u)/\text{sd}(n) = \sqrt{\text{var}(u)}/\sqrt{\text{var}(n)}$ is the noise amplification gain with

$$\text{var}(u) = \frac{1}{\pi}\int_0^\infty |\frac{Q(j\omega)}{1+j\omega b}|^2 \Phi_0 d\omega, \quad \text{var}(n) = \frac{1}{\pi}\int_0^\infty |\frac{1}{1+j\omega b}|^2 \Phi_0 d\omega = \frac{\Phi_0}{2b}, \quad (30)$$

where $1/b$ defines the band, and Φ_0 is the spectral density of the driving white noise. Then there is $k_{un} < \max|Q(\omega)|$. For PID$_1$ this makes $k_{un} < 300$, which, as discussed in the Introduction, is acceptable for sd(n) = 3 mV and $0 \le u(t) \le 10$ V.

3.5 General Noncancellation Controller

Increasing c and/or making $c_{1,2}$ complex improves the control performance as shown in Fig. 10. For k great enough there is $\sigma_s \to c$ and $\theta_s \to \theta_c$, and the relationships for the fast mode remain approximately unchanged provided that $c \le (0.1 \div 0.2)\sigma_f = (0.05 \div 0.1)\mu^{-1}$. Hence the smaller μ the greater c is allowed, which according to (17) results in smaller maximum error and faster error diminishing, see also [5] for a detailed analysis.

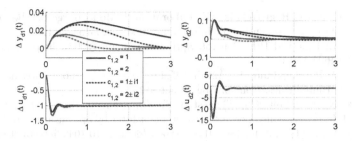

Fig. 10. PID control - notice performance improvement for $c_{1,2} = c(1 \pm j\theta)$, and practical independence of the fast mode from the slow one; $\theta_f = 2, \mu = 0.05, k_c = 12.5$.

4 Conclusion

The paper presents a very simple and intuitive method providing an excellent solution for load disturbance rejection in the double tank system. It illustrates practical use of the text-book notions like system linearization and normalization, root loci, time responses, frequency domain characteristics and stochastic signals to solve a real engineering problem relating the available control performance with the level of sensor noise. Every aspect of the control problem, including the value of the maximum dynamical error, can be easily calculated analytically and compared with experimental results. These features make the presented case study a valuable example for control teaching.

Acknowledgements. The research has been supported by the Department of Automatic Control Grant No. 02/010/BK18/0102.

References

1. Apkarian, J.: Coupled Water Tank Experiments Manual. Quanser Consulting Inc., Markham (1999)
2. Åström, K.J., Österberg, A.-B.: A teaching laboratory for process control. Control Syst. Mag. **6**(5), 37–42 (1986)
3. Åström, K.J., Hägglund, T.: PID Controllers: Theory, Design and Tuning. Instruments Society of America, Pittsburgh (1995)
4. Bieda, R., Blachuta, M., Grygiel, R.: A new look at water tanks systems as control teaching tools. IFAC World Congr. 13480–13485 (2017a). https://doi.org/10.1016/j.ifacol.2017.08.2327. Toulouse, France .
5. Bieda R., Blachuta, M., Grygiel, R.: High performance PID control of a coupled tanks system as an example for control teaching. In: 22nd International Conference on Methods and Models in Automation and Robotics, MMAR 2017, Miedzyzdroje, Poland, pp. 803–808 (2017). https://doi.org/10.1109/MMAR.2017.8046931
6. Blachuta M., Bieda, R., Grygiel, R.: High performance single tank level control as an example for control teaching. In: 25th Mediterranean Conference on Control and Automation, MED 2017, Valletta, Malta, pp. 1053–1058 (2017). https://doi.org/10.1109/MED.2017.7984257

7. Lund University: Reglerteknik AK, Laboration 2. Modellbygge och beräkning av PID-regulatorn, Assistenthandledning, Lund tekniska högskola (2013)
8. Visioli, A.: Practical PID Control. Springer, London (2006). https://doi.org/10.1007/1-84628-586-0
9. Segovia, V.R., Hägglund, T., Åström, K.J.: Measurement noise filtering for common PID tuning rules. Control Eng. Pract. **32**, 43–63 (2014)
10. Soltesz, K., Grimholt, Ch., Skogestad, S.: Simultaneous design of proportional-integral-derivative controller and measurement filter by optimisation. IET Control Theory Appl. **11**(3), 341–348 (2017). https://doi.org/10.1049/iet-cta.2016.0297
11. Yurkevich, V.D.: Design of Nonlinear Control Systems with the Highest Derivative in Feedback. World Scientific, Singapore (2004)

Towards Formal, Graph-Based Spatial Data Processing: The Case of Lighting Segments for Pedestrian Crossings

Sebastian Ernst[(✉)] and Leszek Kotulski

Department of Applied Computer Science,
AGH University of Science and Technology,
Al. Mickiewicza 30, 30-059 Kraków, Poland
{ernst,kotulski}@agh.edu.pl

Abstract. The paper proposes a graph formalism for flexible and efficient manipulation of geospatial data. Its main practical application is preparation of data for lighting optimisation projects in conformance with regulations. The formalism is based on the extended Semantic Environment Graph, already proposed in our previous work. A simple example of a one-way street with a pedestrian crossing is used to illustrate each step of the proposed procedure. The process involves executing a series of graph productions, which introduce the new shapes into the data. Implementation is not the main focus of the paper, but results of conducted studies are provided to present the practical implications of the proposed method, compared to the traditional approach used by lighting designers.

Keywords: Graph transformations · Geospatial modelling ·
Road lighting · Energy efficiency

1 Introduction

As the technology used for street lighting becomes more precise and efficient, with LED fixtures available for a vast spectrum of lighting situations and requirements, the quality of photometric designs becomes ever more important.

The new regulations reflect this trend. Taking the European EN 13201 standard [6–9,11] as example, one can see a wide spectrum of lighting classes along with precise rules when they should be applied. They define the guidelines aimed at maintaining the safety of the road users while making the system as energy efficient as possible. The latter is usually obtained by decreasing the unnecessarily high light intensity, which lowers both the operational (energy) costs and the

This research was supported by the AGH University of Science and Technology grant number 11.11.120.859 and co-financed by the Polish National Centre for Research and Development grant number POIR.01.01.01-00-0037/17 funded from European Funds and granted to GRADIS Ltd.

N. T. Nguyen et al. (Eds.): ACIIDS 2019, LNAI 11431, pp. 431–441, 2019.
https://doi.org/10.1007/978-3-030-14799-0_37

investment expenditure, by allowing the purchase of less powerful luminaires. It also contributes to the reduction of light pollution, which is a very important hazard, both to human health and to other species [19].

LED fixtures can be dimmed with a virtually linear power-to-intensity ratio, and can be matched to fulfil virtually any lighting class on a road of any shape with greater precision than older HPS (high pressure sodium) devices. This gives designers the possibility to vary the illumination of even small parts of streets to reflect fine-grained requirements.

The general requirements specified by the EN 13201 standard are commonly supplemented with those forced by specific local regulations. One notable example is the approach to illuminating pedestrian crossings. Local regulations in this regard have been defined, among otheres, by Belgium [2,21], Czech Republic [14], Germany [4], Italy [5], Norway [10], Poland [20], Sweden [12,13], Switzerland [15] and the U.K. [3].

Such fine-grained regulations increase the complexity of the design process for street lighting, for instance by defining the need to assign different lighting levels for crossings themselves, as well as transition zones located in front of and behind them, in order to avoid sudden changes in lighting for drivers.

The optimisation of photometric calculations is a very intensively researched field [18,22,23]. However, the aforementioned requirements also require the designers to do more 'pre-processing' work, analysing the area of the investment and defining the individual lighting segments[1] in all relevant streets.

The goal of the presented approach is to provide a formal background to automate this process. The benefits are two-fold. First, the automatic procedure will require little or no human interaction, thus reducing the designers' workload. Second, the results will be consistent, which is important especially for large projects, which often get divided among many designers. Practical experience has shown that their interpretation may vary slightly for virtually identical road fragments.

Spatial data processing systems, such as GIS (Geographic Information System) solutions, provide tools for processing of geographic shapes. However, trying to express the complex procedures as code (either in a programming language or a data query language, such as SQL [16]) obscures the intentions, making the process error-prone and verification difficult or impossible. The task is also not trivial, given the multitude of possible spatial layouts of road segments. The goal of this paper is to define a formalism able to express the spatial analysis and transformations needed to generate precise shapes for lighting segments.

2   Problem Statement

To illustrate the mechanism used to define lighting segment transformations in a formal way, we will use the example referred to in Sect. 1, which entails generation of lighting segments for pedestrian crossings.

---

[1] A lighting segment is an area with uniform lighting requirements; this is usually achieved by assigning a single lighting class to the entire area.

While the procedure may vary in different local regulations, the general concept remains unchanged. It involves:

1. significantly increasing the luminance level of the crossing itself,
2. increasing the luminance level of so-called *transition zones*: parts of the road located in front of (run-in) and behind the crossing (run-out).

The length of the transition zones may differ, depending on the location and road parameters. Sometimes, the run-in is omitted, only leaving the run-out to maintain a negative contrast of pedestrians crossing the road [20].

The increase itself is performed by assigning a different lighting class to these segments. The shape of the transition zones is therefore a function of the applied regulations and the structure of the road.

For this paper, we will use an example of a one-way street with three different lighting segments with distinct lighting classes (and, therefore, requirements). This simple example will allow for a clear description of the mechanism.

However, please note that in real life, the input data will rarely be as simple – in fact, it is almost always more complex. Therefore, the rules used to process it will also be more complicated.

3   Formal Model

The proposed approach tries to combine the semantic information about lighting segments and their spatial properties in a graph structure. Spatial relationships obtained using dedicated tools, such as the PostGIS spatial database [1], are transformed into semantic annotations in the graph itself.

The proposed structure must be able to store all data about the relevant area and support the required transformations. It takes the form of a graph, and since it contains a semantic description of the environment (including pre-interpreted spatial relations), it is called the *Semantic Environment Graph* (SEG).

The SEG is generated by a *graph grammar*, denoted as Ω. A detailed definition of a graph grammar is provided e.g. in [17]; therefore, here we will only focus on the actual mechanism for the application of productions.

Productions (denoted as π) are transformation rules, which transform the graph from one coherent state to another. Productions are provided in the form of two graphs, called *lhs* (left-hand side) and *rhs* (right-hand side). Application of π on a graph G involves the following steps:

1. the *lhs* graph is removed from G, creating G';
2. the *lhs* graph is added to G' (however, at this moment these graphs are separated);
3. all edges in G containing one of the nodes belonging to $V_{lhs} \cap V_{rhs}$ and the second to $V_G \setminus V_{lhs}$ are restored in $G' \cup rhs$;
4. all edges in G containing removed nodes ($V_{lhs} \setminus V_{rhs}$) are also removed.

The Semantic Environment Graph was first introduced in [17]; its extended and revised definition is presented below.

Definition 1. SEG_Ω *is defined as an attributed graph over the set of node labels* Σ_Ω *and the set of edge labels* Γ_Ω *such that:*

$$SEG_\Omega = (V_\Omega, E_\Omega, lab_\Omega^V, lab_\Omega^E, \Sigma_\Omega, \Gamma_\Omega, \Delta_\Omega, att_\Omega^V, att_\Omega^E, A_\Omega^V, A_\Omega^E)$$

where:

- V_Ω *is the set of nodes,*
- E_Ω *is the set of edges,*
- $lab_\Omega^V : V_\Omega \to \Sigma_\Omega$ *is the node labelling function,*
- $lab_\Omega^E : E_\Omega \to \Gamma_\Omega$ *is the edge labelling function,*
- $\Sigma_\Omega = \{T, S, F, P, O\}$ *is the set of node label groups, where:*
  - T *represents streets,*
  - S *represents road lighting segments located on streets,*
  - F *represents freeform lighting segments, which are not located on streets (e.g. to represent a parking lot),*
  - P *represents pedestrian crossings, located on road segments,*
  - O *represents other objects, such as buildings, points of interest, etc.*
- $\Gamma_\Omega = \{on, part\_of, spatial\_rel, eq\}$ *is the set of edge labels, where:*
  - *on denotes that a point object (e.g. pedestrian crossing) is located at a given line object (e.g. road segment),*
  - *part_of denotes that a line object is part of another line object,*
  - *spatial_rel denotes that there is a spatial relationship between two objects,*
  - *eq denotes that an object is equivalent to another object.*
- $\Delta_\Omega = \{\alpha, \beta\}$ *is the set of nonterminal nodes,*
- $att_\Omega^V : V_\Omega \times \Sigma_\Omega \to 2^{A_\Omega^V}$ *is the node attributing function, such that for* $x \in V_\Omega, l \in \Sigma_\Omega, a \in A_\Omega^V$ $att_\Omega^V(x, l)(a)$ *is a value of the attribute* a,
- $att_\Omega^E : E_\Omega \times \Gamma_\Omega \to 2^{A_\Omega^E}$ *is the edge attributing function, such that for* $x \in E_\Omega, l \in \Gamma_\Omega, a \in A_\Omega^E$ $att_\Omega^E(x, l)(a)$ *is a value of the attribute* a,
- A_Ω^V *is the set of node attributes, where:*
  - *type denotes the type of an object (e.g., the type of building for O nodes),*
  - *geometry denotes the shape of an object and its geographic location; this can be expressed e.g. as a Well-Known Text (WKT) string,*
  - *name is the name of an object, e.g. the street name or segment label,*
  - *lighting_class is the lighting class assigned to a road or freeform segment,*
- A_Ω^E *is the set of edge attributes, where:*
  - *position denotes the metre within a line object on which a given point is located,*
  - *from and to mark the metres within a line object where another line object begins and ends,*
  - *distance denotes the distance (in metres) between two objects,*
  - *intersects (yes, no) indicates that two objects spatially intersect.*

4   The Segment Generation Procedure

As mentioned in Sect. 2, we will present the mechanism using the example of a one-way street with three lighting segments and a pedestrian crossing.

Let us assume that a street T_1, L_{T1} metres long, is divided into three lighting segments:

- S_1, occupying the initial L_1 metres of the street length, with lighting class M4,
- S_2, occupying the following L_2 metres, with lighting class M3,
- S_3, occupying the final L_3 metres, with lighting class M4.

Additionally, on S_2, there is a pedestrian crossing C_1, with its centre located at the P_{C1}-th metre of the segment. This initial situation has been presented in Fig. 1. For clarity of presentation, attributes denoting the lighting classes are not shown in the graph.

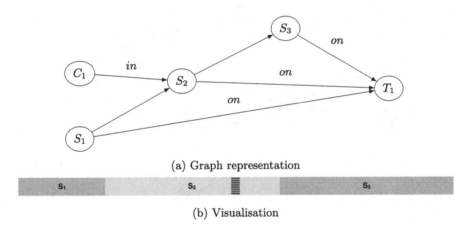

(a) Graph representation

(b) Visualisation

Fig. 1. Initial state of segments

Let us consider the task of defining lighting segments for a street with a pedestrian crossing, along with transition zones recommended by regulations. Then, a series of graph productions is applied, bringing the graph to the desired state. Their description follows.

4.1   Initial Production

Production P_1. The first production consists in generating non-terminal nodes which trigger the generation of the nodes representing the actual lighting segments.

The non-terminal nodes, labelled α and β, are later used to trigger subsequent operations leading to generation of terminal symbols representing new segments (Fig. 2).

The state of the example graph after applying P_1 is presented in Fig. 3.

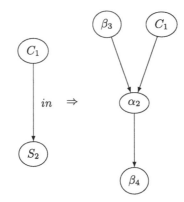

Fig. 2. Production P_1

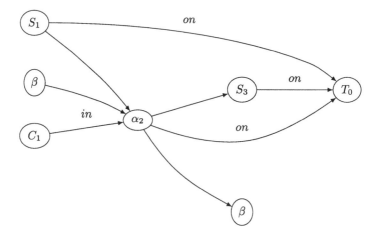

Fig. 3. Example graph after applying P_1

4.2   Generation of the Run-In

This production is triggered by the non-terminal symbol β, generated by P_1. It generates the transition zone *in front of* the pedestrian crossing. The size of the run-in segment is designated by the function $RIL(\ldots)$ (*run-in length*). The parameters for this function are the lighting class associated with the S and α nodes, because the run-in segment's size should be longer with greater differences of illumination levels, due to eye accommodation.

Depending on the geometry of the segments, it can take one of three variants:

1. if the entire transition zone fits in the segment which contains the crossing and a part of the segment is left (i.e. it is shorter than part of the segment which lies before the crossing), production P_{2a} is used;
2. if the transition zone does not fit in the segment with the crossing, but the segment is preceded by other segments in the same street, it will cover it and

needs to be further propagated to preceding segments – production P_{2b} is used;
3. if the transition zone does not fit in the segment with the crossing, the preceding segment must be shortened – production P_{2c} is used;

We assume priority of application of the mentioned productions, i.e. we will apply production P_{2a} before P_{2b} and P_{2c}, and P_{2b} before P_{2c}. A detailed description of the productions follows.

Production P_{2a}. In this case (described by item 1 above), the non-terminal symbol β is replaced with two segments:

- S_1, which represents the run-in segment in front of the crossing,
- S_2, which represents the remaining part of the segment which originally contained the crossing (Fig. 4).

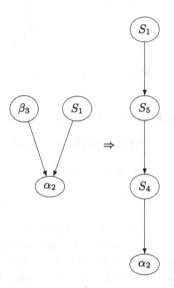

Fig. 4. Production P_{2a}

The applicability predicate is defined as:

$$\Pi_{2a} : | \ S_1.to - \alpha_2.from \ | > RIL(...)$$

The attributes of the right hand side of the production are defined as follows:

$$S_5.from = S_1.to$$
$$S_4.from = S_5.to = \alpha_2.from + RIL(...)$$
$$S_4.to = \alpha_2.from$$

Production P_{2b}. This production is similar to P_{2a}, but differs in that no part of the original segment is left before it (see item 2 above), hence only one new segment S_1 is created, representing the run-in segment. The applicability predicate is defined as (Fig. 5):

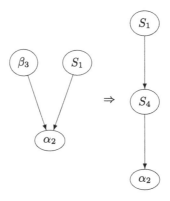

Fig. 5. Production P_{2b}

$$\Pi_{2b} :\mid S_1.to - \alpha_2.from \mid =< RIL(...) <\mid S_1.from - \alpha_2.from \mid$$

The attributes of the right hand side of the production are defined as follows:

$$S_1.to = S_4.from = \alpha_2.from + RIL(...)$$
$$S_4.to = \alpha_2.from$$

Production P_{2c}. In this case (item 3), the run-in covers its part of the original segment and must be further propagated to the preceding segment.

The preceding segment (S) is 'consumed' by α, and the non-terminal symbol β remains in the graph to force its further processing by one of the other productions. The attributes of β representing the current length of the run-in and the remaining length to be assigned must be updated accordingly. The predicate of applicability is as follows:

$$RIL(...) >=\mid S_1.from - \alpha_2.from \mid$$

The attributes of the right hand side of the production are defined as follows:

$$S_2.to = S_1.to$$

4.3   Generation of the Run-Out

The procedure for run-out generation is analogous to that described in Sect. 4.2, but takes the segment located *behind* the crossing into consideration (Fig. 6).

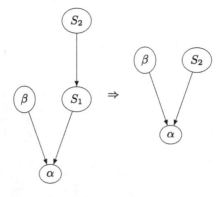

Fig. 6. Production P_{2c}

5   Results and Practical Implications

The presented approach provides means for flexible, formal definition of spatial data transformation rules using a graph formalism. Obviously, the performance of the transformations depends on the implementation, which is not relevant since the paper focuses on the formal aspects of the method.

The *lhs* and *rhs* graphs in production definitions can be arbitrarily complex. This means that the expressive power of the proposed formalism is limited only by the assumed graph model (SEG), which can be freely extended. Therefore, the main contribution of this approach lies in the ability to define virtually any procedure to modify the map data and apply it accordingly, in a consistent manner.

However, it is important to stress the practical implications of using the describe method for execution of lighting modernisation projects. To demonstrate the outcome, let us refer to a lighting modernisation project carried out by AGH University of Science and Technology in cooperation with the City of Kraków. The project involved replacing almost 4,000 old fixtures with LED-based ones, along with introduction of an innovative, real-time control system. Simulations using a prototype system showed that a simpler version of the proposed method yields over 20 times more lighting segments that the traditional approach executed by a professional lighting designer [17]. The obtained results are presented in Fig. 7.

Finally, to provide a view on real-life applicability of the proposed approach, a few remarks will be provided. It must be stressed that the *transformation rules* need to be defined manually at the moment. This means that processing of other objects, such as intersections, will require precise definition of appropriate productions, and may also require extension of the SEG with regard to labels and attributes. However, the *contents* of the graph can be automatically generated from map data, e.g. using OpenStreetMap[2] data. If the transformation rules are detailed enough, further processing of the graphs is completely automatic.

---

[2] http://www.openstreetmap.org.

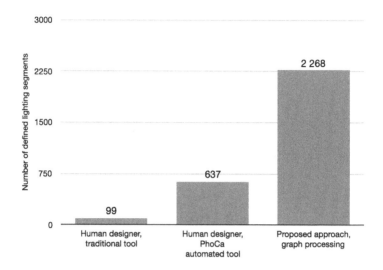

Fig. 7. Number for segments defined in the area of the pilot project [17]

6   Conclusions and Future Work

The paper extends the formal methods first proposed in [17] by providing a detailed study of a spatial transformation procedure applied in real-world lighting modernisation projects.

It provides a formal definition of all steps necessary to describe a transformation procedure, using an example of a one-way street with one pedestrian crossing for clarity. The implementation is not the main focus of the paper, but practical implications of the proposed method have been outlined.

The method has been verified using a software prototype built for this purpose. Future work involves migrating other transformations currently used in the prototype system to the graph formalism and fully integrating the graph processing engine with the database used for the production system.

References

1. PostGIS 2.4 Documentation. https://postgis.net/docs/manual-2.4/. Accessed 4 Oct 2018
2. NBN L 18–002: Recommendations for special cases of public lighting. Technical report, Bureau voor Normalisatie, January 1988
3. Technical Report 12: Lighting of Pedestrian Crossings. Technical report, Institution of Lighting Engineers (2007)
4. DIN 67523–2:2010–06: Beleuchtung von Fußgängerüberwegen (Zeichen 293 StVO) mit Zusatzbeleuchtung - Teil 2: Berechnung und Messung. Technical report, Deutsches Institut für Normung, June 2010
5. Linee guida per la progettazione degli attraversamenti pedonali. Technical report, Automobile Club d'Italia (2011)

6. CEN/TR 13201-1: Road lighting – Part 1: Guidelines on selection of lighting classes. Technical report, European Committee for Standarization, December 2014
7. EN 13201-2: Road lighting – Part 2: Performance requirements. Technical report, European Committee for Standarization, December 2014
8. EN 13201-3: Road lighting – Part 3: Calculation of performance. Technical report, European Committee for Standarization, December 2014
9. EN 13201-4: Road lighting – Part 4: methods of measuring lighting performance. Technical report, European Committee for Standarization, December 2014
10. Håndbok N100: Veg-og gateutforming. Technical report, Statens vegvesen (2014)
11. prEN 13201-5: Road lighting – Part 5: Energy performance indicators. Technical report, European Committee for Standarization, December 2014
12. Vägbelysningshandboken. Technical report, Trafikverket (2014)
13. Krav för vägars och gators utformning. Technical report Trafikverkets publikation 2015:086, Trafikverket, Sveriges Kommuner och Landsting (2015)
14. Osvětlení pozemních komunikací. In: Technické Kvalitativní Podmínky Staveb. Ministerstvo Dopravy (2015)
15. Beleuchtung von Fussgänger-Überwegen. In: SLG Richtlinie 202:2016. Schweizer Licht Gesellschaft (2016)
16. Date, C.J.: A Guide to the SQL Standard: A User's Guide to the Standard Relational Language SQL. Addison-Wesley Longman Publishing Co., Inc., Boston (1987)
17. Ernst, S., Łabuz, M., Środa, K., Kotulski, L.: Graph-based spatial data processing and analysis for more efficient road lighting design. Sustainability 10(11), 3850 (2018). https://doi.org/10.3390/su10113850
18. Gómez-Lorente, D., Rabaza, O., Espín Estrella, A., Peña-García, A.: A new methodology for calculating roadway lighting design based on a multi-objective evolutionary algorithm. Expert Syst. Appl. 40(6), 2156–2164 (2013). https://doi.org/10.1016/j.eswa.2012.10.026
19. Hölker, F., Wolter, C., Perkin, E.K., Tockner, K.: Light pollution as a biodiversity threat. Trends Ecol. Evol. 25(12), 681–682 (2010). https://doi.org/10.1016/j.tree.2010.09.007
20. Jamroz, K., Tomczuk, P., Mackun, T., Chrzanowicz, M.: Wytyczne prawidłowego oświetlenia przejść dla pieszych. Technical report, Ministerstwo Infrastruktury (2018)
21. Ministerie van de Vlaamse Gemeenschap Ministerie van de Vlaamse Gemeenschap: Ontwerprichtlijnen voor Voetgangersvoorzieningen. In: Vademecum Voetgangersvoorzieningen (2003)
22. Rabaza, O., Peña-García, A., Pérez-Ocón, F., Gómez-Lorente, D.: A simple method for designing efficient public lighting, based on new parameter relationships. Expert Syst. Appl. 40(18), 7305–7315 (2013). https://doi.org/10.1016/j.eswa.2013.07.037
23. Sędziwy, A.: Sustainable street lighting design supported by hypergraph-based computational model. Sustainability 8(1), 13 (2015). https://doi.org/10.3390/su8010013
24. Wojnicki, I., Kotulski, L.: Empirical study of how traffic intensity detector parameters influence dynamic street lighting energy consumption: a case study in Krakow, Poland. Sustainability 10(4), 1221 (2018). https://doi.org/10.3390/su10041221

Multi-agent Support for Street Lighting Modernization Planning

Adam Sędziwy$^{(\boxtimes)}$, Leszek Kotulski, and Artur Basiura

AGH University of Science and Technology, Kraków, Poland
{sedziwy,kotulski,basiura}@agh.edu.pl

Abstract. The inherent problem related to the maintenance of public lighting infrastructures is replacing (retrofitting) aging fixtures with the new ones. Sometimes municipalities decide to retrofit installations which did not reach their lifespan limit. The example is replacing the high-intensity discharge lamps with much more energy-efficient LEDs (*Light Emitting Diodes*) or induction bulbs. Any retrofit approach, however, generates costs which are to be reduced. Thus, the retrofit strategy is aimed at finding a trade-off between financial outlays and potential benefits expressed in terms of power savings, payback period etc. Financial constraints imposed on an investment usually delimit a retrofit scope to restricted parts of entire public lighting installation. A key problem which arises then is: which luminaires should be selected for retrofitting, to make it the most profitable? In this work we propose using an agent-based approach to response this question by preparing a high-level strategy for optimal retrofits of large-scale lighting systems (e.g., covering entire city area).

Keywords: Multi-agent system · Public lighting · LED lighting · Agent-based optimization

1 Introduction

One of the most important services ensured by municipalities is a maintenance of public lighting. Its importance origins from the fact that a proper illumination of public spaces improves citizen's comfort, safety, wellbeing and sustainability of local environments. Additional considerable factor is reducing both light pollution and power consumption. The LED (light emitting diodes) technology strongly supports those objectives: a lot of authorities decide to retrofit public lighting installations based on high-intensity discharge (HID) lamps with LED-based ones as those latter yield a significant reduction of the power usage. Frequently such retrofits are made even if existing installations are relatively new. It is because the expected savings prevail over the modernization costs.

In 2016 AGH University completed R&D project [10,12] aimed at implementing the intelligent, dynamically controlled street lighting system in the city

© Springer Nature Switzerland AG 2019
N. T. Nguyen et al. (Eds.): ACIIDS 2019, LNAI 11431, pp. 442–452, 2019.
https://doi.org/10.1007/978-3-030-14799-0_38

of Cracow, Poland, co-funded by the European Union. The pilot project, carried out with the municipality of Cracow, relied on replacing approximately 3,700 HID (precisely, high-pressure sodium) lamps with LEDs and, in the next phase, introducing an adaptive control system for those luminaires. After completing the works we started analyzing the first results of modernization which showed the significant power usage reduction. The HID to LED replacement only, reduced the power consumption by 56%. A dynamic lighting system, controlling lamps installed along an arterial road, brought 35% of additional savings (calculated with reference to a non-controlled LED installation). Besides such straightforward observations, it was also revealed that the important challenge is selecting particular installations to retrofit. It is because financial limitations do not allow for a full-scale retrofit. Actually AGH is involved in another R&D project co-financed from resources of the Polish National Center of Research and Development. The important role in this project plays a multi-agent system which carries out a complex, multivariate optimization of lighting installations, which is necessary for a correct control system performance. As the "side effect" of these works the concepts presented in this paper arose. They are intended to be a conceptual base for development of the software solution supporting municipalities in the maintenance of lighting infrastructure.

The scale of modernizations of public lighting infrastructures is demonstrated for cities like New York where the number of installed LEDs exceeded the stunning number of 240,000 [3], for Los Angeles and Houston it is over 160,000 LED-based street lights. As the demand for such modernizations is large but financial resources of municipalities are usually limited, the authorities have to make a decision what retrofit strategy should be implemented and which installations have to be renewed with the highest priority. Making such decisions is not a simple task as multiple factors have to be taken into account: installation's age, investment costs, payback period etc. Additionally, municipalities can use various sophisticated business models which enable covering those outlays with future savings, for example the ESCO (Energy Service Company) model [1,13]. Hence, the crucial elements for decision making are quantitative data, describing all potential variants (strategies), and the support for effective processing of those data. Due to the potentially large size of a search space some method of an automated assessment of retrofit strategies is necessary. To handle this requirement we propose using a multi-agent system which analyzes a search space and selects top retrofit strategies.

The article's structure is following. In the next section the brief overview of the state of art is presented. The scheme of calculations is discussed in Sect. 3. In Sect. 4 we present a graph model used in agent-based calculations for representing an area being illuminated. The architecture and performance of the proposed multi-agent system are presented in Sect. 5. The final conclusions and the plans of further works can be found in Sect. 6.

2   Related Works

Retrofits of public lighting installations are carried out for the financial and environmental profits. The white paper [6] issued by Clinton Climate Initiative presents a comprehensive overview of this subject. It discusses different aspects of retrofits such as barriers of their implementation, financing options, typical retrofit roadmaps, technical details related to light sources. Yet another white paper, published by the South-central Partnership for Energy Efficiency as a Resource [2], is a guide highlighting the major steps and key problems concerning public lighting modernization. The particularly interesting paper [3] prepared by World Bank Group represents a business outlook on public lighting for the case of Brazil. It covers such issues as, for example, the structure of responsibility for public lighting maintenance, the changes in electric energy prices, statistical characteristics of cities and their resources, business models for public lighting financing, risk factors and so on. This detailed, in-depth analysis, although made for the particular country, gives a view on non-technological aspects which usually play a crucial role in decision making.

The recommendations and good practices [1] are complemented by case studies. The interesting one was published by the Caribbean Development Bank (CDB) which financed the retrofit (21,000 HID street lamps) made in Saint Lucia. The CDB confirms that power savings reached 58% thanks to migrating to LEDs. The main quantitative data characterizing this project are following. The approximated investment cost is USD 12 million, annual energy savings are 6,500,000 kWh what yields cost savings at the level of USD 1.9 million per annum and the payback period about $6^1/2$ years. The expected annual reduction of green house gases emission is 3,900 tonnes. Further details can be found in [5].

Similar results were announced by the Worcester Energy initiative [15] which expects to achieve annual savings yielded by retrofitting 13,300 fixtures at the level of USD 910,000 and over 6,000 MWh in electricity what is a 60% reduction. It's worthy to remark that these financial profits include the longer lifespan and lower maintenance costs. In this case the payback period is estimated as 10.9 years.

In the following sections we propose the computational model, based on a multi-agent system, which supports making decision by taking into account both technological and financial demands of municipalities.

3   Materials and Methods

The goal of a retrofit is modernization of an existing lighting installation by performing one or more actions such as replacing poles, arms, fixtures, changing luminaire locations (rarely applied due to the costs and spatial constraints), adjusting fixture settings (e.g., dimming), introducing a range of system control capabilities. Retrofitting, besides being a renewal of an infrastructure, yields reduction of the power usage and light pollution. It also increases public safety and well being. The most important requirement which must be satisfied is compliance with relevant lighting standards (e.g., EN 13201-2 in Europe, ANSI/IES

RP-8-14 in the United States or AS/NZS 1158.1.1 in Australia and New Zealand) [4,7,8] defining what are the expected luminosity levels for particular *lighting situations*, such as road junctions, residential areas, bike lanes, parkings and so on [7]. The example of such requirements, defined for traffic routes, is shown in Table 1.

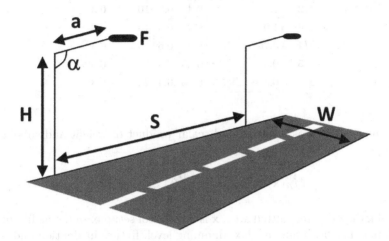

Fig. 1. The sample lighting situations and some parameters being taken into account during computations: H – pole height, S – luminaire spacing, a – arm length, α – arm inclination, F – fixture model, W – carriageway width

An "atomic" portion of computations, underlying retrofit works, is made for a single lighting situation (Fig. 1) representing given street or square. Those computations are aimed at ensuring that an installation's setup, including such parameters as pole height, arm length, fixture model and others (see Fig. 1), will yield correct photometric conditions for a given area and its lighting class, as defined in the standard (Table 1). If those requirements are fulfilled, then a setup is tested against compliance with problem specific assumptions like expected investment outlays or an annual power usage. Otherwise, the next possible setup has to be verified. That "atomic" portion of computations is made for each street, walkway, square etc.

The important element to be defined, before launching a multi-agent-aided optimization, are metrics used by agents to evaluate if a given standard-compliant setup obtained in computations is satisfactory with respect to final costs, power consumption or another criteria. Note that such metrics can be quite complex, as multiple aspects have to be convoluted: economic, technological, environmental or those related to local regulations. For example, the special illumination (e.g., in terms of the color temperature) is required for areas at risk of violence or being under surveillance cameras monitoring. As particular agents will operate in different areas, their metrics will also differ. In a general case, one

Table 1. M-lighting classes (for traffic routes) according to the EN 13201-2:2015 standard. L_{avg}–average luminance, U_o – overall uniformity, U_l – longitudinal uniformity, f_{TI}– disability glare, R_{EI}– lighting of surroundings

Class	L_{avg} [cd/m^2]	U_o	U_l	f_{TI} [%]	R_{EI}
M1	2.0	0.4	0.7	10	0.35
M2	1.5	0.4	0.7	10	0.35
M3	1.0	0.4	0.6	15	0.3
M4	0.75	0.4	0.6	15	0.3
M5	0.5	0.35	0.4	15	0.3
M6	0.3	0.35	0.4	20	0.3

can define a metrics as a normalized scalar product of weight and sub-metrics vectors:

$$M_A(\mathbf{x}) = [w_e, w_f] \cdot \begin{bmatrix} M_A^e(\mathbf{x}) \\ M_A^f(\mathbf{x}) \end{bmatrix} \in [0, 1],$$

where A indicates a retrofitted area, $\mathbf{x}$ is a vector of setup parameters like fixture model, pole height, luminous flux, dimming level, fixture inclination and so on; w_e and w_f ($w_e + w_f = 1$) are non-negative weights reflecting importance of particular metrics; e and f refer to energy and financial components of a metrics, for instance, an expected power usage reduction or a return of investment value (ROI). It can be assumed, without loss of generality, that $M_A(\mathbf{x}) \to 1$ for top solutions (and a corresponding setup $\mathbf{x}$) and, conversely, $M_A(\mathbf{x}) \to 0$ for the worst ones. The above formula is limited to two-component vectors only but it can be easily extended by adding other criteria being important for decision making.

Remark. Note that separation of M_A into M_A^f and M_A^e is rather arbitrary. For example, the energy usage reduction which apparently influences the latter component impacts also financial results by decreasing energy related costs. We will not discuss, however, this problem which is beyond the scope of this work.

Example. Let us consider an area containing a road junction. For simplicity, we define two objectives only for this problem, i.e., minimizing power usage and minimizing payback period of an investment. Additionally, for the decision making purposes it is assumed that the first objective is taken with the weight $w_e = 0.4$ and the second one with $w_f = 0.6$.

For the above objectives we define two metrics

$$M_A^e(\mathbf{x}) = \frac{P_{\text{old}} - P_{\text{new}}(\mathbf{x})}{P_{\text{old}}}$$

and

$$M_A^f(\mathbf{x}) = 1 - \exp[-\Delta_e(\mathbf{x}) - \Delta_m(\mathbf{x})],$$

where $P_{old}(\mathbf{x})$, $P_{new}(\mathbf{x})$ denote a total installation power, before and after retrofit respectively, and $\mathbf{x}$ is a vector of installation settings defined for this retrofit project; $\Delta_e(\mathbf{x})$, $\Delta_m(\mathbf{x})$ denote annual savings in energy and maintenance costs respectively. Note that those metrics are highly simplified for demonstration purposes. The above form of $M_A^f(\mathbf{x})$ was taken to satisfy the assumption that $M_A(\mathbf{x}) \to 1$ (or $\to 0$) for the best (or the worst) solutions. A payback period can be calculated as the ratio:

$$T_{Payback}(\mathbf{x}) = \frac{\text{Total investment cost}(\mathbf{x})}{\Delta_e(\mathbf{x}) + \Delta_m(\mathbf{x})}.$$

Finally, the complete metrics function is

$$M_A(\mathbf{x}) = 0.4 \cdot \frac{P_{old} - P_{new}(\mathbf{x})}{P_{old}} + 0.6 \cdot \Big(1 - \exp[-\Delta_e(\mathbf{x}) - \Delta_m(\mathbf{x})]\Big).$$

The list of solutions obtained during computations is descendingly ordered by $M_A(\mathbf{x})$ value and reported to a dispatching agent (further details will be discussed in Sect. 5).

4 Graph Representation of a Multi-agent System Environment

The natural representation for modeling both the roadway and lighting installation networks are graphs. Besides reflecting physical and logical structures of entities, graphs allow assigning additional properties to particular objects. It is possible thanks to using so called *labeled* and *attributed* graphs [9].

Having in mind that considered physical systems are structurally static we decided to chose the replicated complementary graphs (RCG)[1] representation which supports graph partitioning and replication of the shared areas [9]. In this approach an urban space will be defined as a graph $G = (V, E, a_V, a_E)$, where V is a finite nonempty set of vertices representing road junctions and other areas such as squares or roundabouts, which cannot be classified as streets, walkways, bike lanes etc., which, in turn, are represented by graph edges $e \in E$ of the form $e = \{u, v\}$, where $u, v \in V$; $a_V : V \longrightarrow A_V$ and $a_E : E \longrightarrow A_E$ are node and edge attributing functions respectively which assign certain attributes from A_V and A_E to each node and edge. Those attributes can describe such properties as traffic capacity, reflective properties of a road surface, maximum vehicle speed etc. We introduce yet another, lightweight edge type, representing neighborhood relations. Edges of this type have the informative purpose rather than the computational one, in the sense that they carry auxiliary information, necessary for calculations made on their endpoints, u and v. This situation is shown in Fig. 2b where, among others, edges $\{v_1, v_3\}$, $\{v_2, v_4\}$ are examples (dotted lines).

As mentioned previously, the example of an attribute of a vertex $v \in V$ representing a junction point is a lighting class ascribed to it or its daily traffic flow profile; for an edge $e \in E$ modeling a carriageway, it can be a number of carriageway's lanes or its width.

---

[1] Otherwise the problem of synchronization time overhead arises.

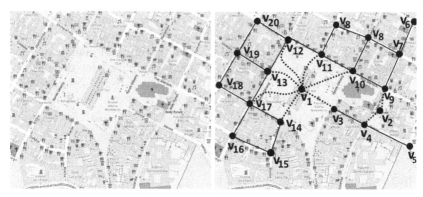

(a) The sample portion of a city map   (b) The graph modeling the map shown
in Fig. 2a

Fig. 2. The sample map and its graph representation

Graph partitioning is made according to the following scheme. A line delimiting two subgraphs is spanned on the graph nodes which finally become a shared (and replicated) vertices for two or more subgraphs (black dots in Fig. 3). If two neighboring nodes, say u, v are replicated between two graphs, say G_1 and G_2, then an edge $e = \{u, v\}$ will be also replicated to G_1 and G_2 (see Fig. 3).

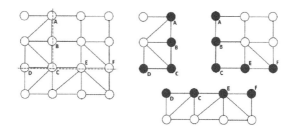

Fig. 3. The sample centralized graph (at left) and its distributed form. Dashed line indicates the partitioning scheme. For better readability the labels of shared vertices (black dots) only are left.

5   Multi-agent System Architecture, Performance and Computation Model

The architecture of a multi-agent system optimizing a retrofit's setup relies on two types of agents: *dispatcher agent* (shortly: DA) and *retrofitting agent* (RA). A role of the former agent type is threefold: (i) partitioning a retrofit problem into subproblems which are defined for all subareas to be illuminated (e.g., city districts or sets of neighboring streets), (ii) gathering results from retrofitting agents in order to find a global optimum, and (iii) terminating an optimization process.

Retrofitting agents, in turn, are aimed at finding optimal setups of lighting installations in their subareas. It is accomplished by:

1. Making photometric computations for subsequent streets, squares, walkways etc.
2. Finding an optimal result with respect to some criterion (discussed in Sect. 3).
3. Reporting it to DA.

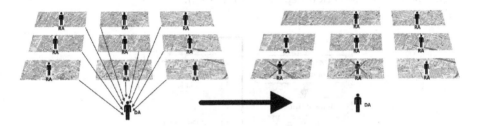

Fig. 4. Possible actions undertaken by a DA on the basis results obtained from RAs, preceding transition to a next iteration: merging areas; moving an agent into the passive state; passing an agent to a next iteration

An inherent property of an RA is maintaining a knowledge related to context specific constraints and requirements. For example, the maximum allowed luminous fluxes in the historical areas, the highest permitted glare index for roadways or preferred correlated color temperatures in pedestrian traffic zones. Moreover, this knowledge has to contain such elements as actual power prices, infrastructure-related costs (fixtures, poles etc.) which are crucial for decision making.

The objective for the considered retrofit problem is to find a modernization scenario which (i) covers a continuous area, (ii) maximizes a return on investment (ROI), and (iii) minimizes an investment cost, C_I, ensuring that it is not greater than the cost limit, C_L.

The above objectives are accomplished in the following steps (see Fig. 5). Initially, a multi-agent system (MAS) consists of a single DA which transforms a map into its centralized graph representation, then a DA distributes it using the RCG model. The partitioning strategy is context specific. In the simplest case it is a geographically-based partition. After this step, a DA creates and deploys RAs to distributed graphs. Each retrofitting agent, RA_i, gets the financial resources from a DA, equal to $R_i = (n_i/N) \times S$, where n_i is a number of luminaires in its area, N denotes a total number of lamps and S stands for a value of total available financial resources. In this moment, a MAS is ready to start computing. Each an RA is equipped with a knowledge consisting of relevant infrastructure data (control cabinets, luminaires etc.) and metrics necessary to rank particular configurations, obtained in a design process, in the order imposed by a given objective function (see Sect. 3). Retrofitting agents make lighting projects for all available configurations of parameters (pole heights, fixture models, lamp dimming levels and so on) and order them using given metrics. After completing computations, an RA sends the top solution to a dispatcher agent. Special attention should be paid to the projects covering shared areas, i.e., areas represented by replicated nodes and edges. Note that agents (two or more) processing such areas can obtain different results due to the different metrics. Such cases can be handled in two ways. The first strategy is a negotiation, between conflicting RAs, leading to some trade-off. The other method is delegating conflict resolution to a DA. The method selection depends on a system.

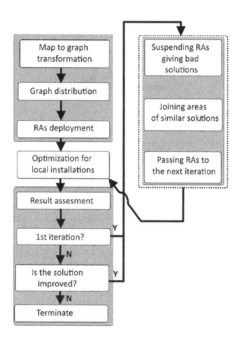

Fig. 5. Retrofit planning process flow. The scenarios shown in dotted box are illustrated in Fig. 4

After receiving the results from RAs, a DA takes one of the following actions:

- Accepts a given solution and passes an RA to the next iteration.
- If two agents processing adjacent areas produced solutions of the same quality then their areas are merged and only one agent is left for further processing.
- If a given solution is not acceptable then a relevant agent, RA_i, gets the status *passive*, what means that it cannot continue processing anymore, but it is obligated to accept any change made to elements on its area, which are

shared with other (non-passive) agents. The resources R_i, assigned previously to RA_i, are redistributed over other agents.

After collecting all data, a DA calculates an overall solution quality. If it is improved, then it proceeds to the next iteration. Otherwise, the process is terminated.

6 Discussion, Conclusions and Further Works

In this work we defined the multi-agent-based framework for determining an optimal public lighting retrofit strategy. This problem arose during analysis of the results of the real-life installation modernization, made within R&D project, carried out in the city of Cracow, Poland.

The algorithmic approach presented here is intended to be a base for the future software implementation of a computer system, supporting lighting design, control and maintenance, which is developed at AGH University. Among other properties of the proposed framework, two ones are particularly important: scalability and well defined flow in terms of the process convergence. Two agents operating on shared areas (graph edges or vertices) can potentially conflict while providing different setups to shared luminaires. Usually, such a behavior results in a flickering effect which violates process convergence. A dispatcher agent, however, is able to arbitrary break such an unstable process.

The multi-agent systems are a handy tool for solving problems related to large-scale lighting installations. In the future works we intend to involve MAS's in performing large-scale customized designs and control tasks [11,14]. These works have already been initiated in the R&D project, supported by the Polish National Center of Research and Development, mentioned in the Introduction section.

Acknowledgments. This work was supported from resources of the project POIR.01.01.01-00-0037/17-00.

References

1. Asian development Bank: Led Street Lighting Best Practices (2017). http://bit.ly/2Ox4Wh6. Accessed 2 Oct 2018
2. South-central Partnership for Energy Efficiency as a Resource: Street Lighting Retrofit Implementation Guide (2017). http://bit.ly/2OxFEzt. Accessed 2 Oct 2018
3. World Bank Group: Lighting Brazilian Cities: Business Models For Energy Efficient Public Street Lighting (2016). http://bit.ly/2OtHHoc. Accessed 2 Oct 2018
4. Australian/New Zealand Standard: AS/NZS 1158.1.1:2005 Lighting for roads and public spaces Vehicular traffic (Category V) lighting - Performance and design requirements. SAI Global Limited (2005)
5. Caribbean Development Bank: Making The Switch: Energy-Efficient Street Lighting (2016). http://bit.ly/2OtU2Zr. Accessed 2 Oct 2018

6. Clinton Climate Initiative: Street Lighting Retrofit Projects: Improving Performance, while Reducing Costs and Greenhouse Gas Emissions (2016). https://goo.gl/iV32FL. Accessed 2 Oct 2018
7. European Committee for Standarization: Road Lighting. Performance requirements, EN 13201–2:2015 (2015)
8. Illuminating Engineering Society of North America (IESNA): American National Standard Practice For Roadway Lighting, RP-8-14. IESNA, New York (2014)
9. Kotulski, L., Sędziwy, A.: Parallel graph transformations supported by replicated complementary graphs. In: Dobnikar, A., Lotrič, U., Šter, B. (eds.) ICANNGA 2011. LNCS, vol. 6594, pp. 254–264. Springer, Heidelberg (2011). https://doi.org/10.1007/978-3-642-20267-4_27
10. Polish National Fund for Environmental Protection and Water Management: ISE Project (Polish) (2014). http://bit.ly/2Ou7QDu. Accessed 2 Oct 2018
11. Sędziwy, A.: A new approach to street lighting design. LEUKOS 12(3), 151–162 (2016)
12. Sędziwy, A., Basiura, A.: Energy reduction in roadway lighting achieved with novel design approach and LEDs. LEUKOS 14(1), 45–51 (2018)
13. The European Parliament and The Council: Directive 2006/32/EC on energy end-use efficiency and energy services and repealing Council Directive 93/76/EEC (2006). http://bit.ly/2Ox5dRa
14. Wojnicki, I., Kotulski, L.: Street lighting control, energy consumption optimization. In: Rutkowski, L., Korytkowski, M., Scherer, R., Tadeusiewicz, R., Zadeh, L.A., Zurada, J.M. (eds.) ICAISC 2017. LNCS (LNAI), vol. 10246, pp. 357–364. Springer, Cham (2017). https://doi.org/10.1007/978-3-319-59060-8_32. http://bit.ly/2QnbhZV
15. Worcester Energy: Street Lighting Retrofit Project. Progress Report (2017). http://bit.ly/2Ov5Cnj. Accessed 2 Oct 2018

An Application a Two-Level Determination Consensus Method in a Multi-agent Financial Decisions Support System

Adrianna Kozierkiewicz[1] , Marcin Hernes[2][(⊠)] ,
and Thanh Tung Nguyen[3]

[1] Faculty of Computer Science and Management,
Wroclaw University of Science and Technology, Wrocław, Poland
adrianna.kozierkiewicz@pwr.edu.pl
[2] Wrocław University of Economics, Wrocław, Poland
marcin.hernes@ue.wroc.pl
[3] Nguyen Tat Thanh University, Ho Chi Minh City, Vietnam
nttung@ntt.edu.vn

Abstract. Supporting financial decisions, including Foreign Exchange Market (Forex) is recently realized more often by using multi-agent systems. On this market currencies are traded against one another in pairs. The quotations are in High Frequency Trading. The supporting decisions on Forex rely on provide, by a multi-agent system, as soon as possible advice on what position should be taken: long, short or none. This advice is given by different investment strategies based on agents running based on statistics, economics, mathematics or artificial intelligence methods. However, these strategies are not optimized in terms of HFT (the computational complexity of these strategies is too high and the signals for open/close positions are too late in many cases). Due this fact, the main problem which appear concerns decreasing a computational complexity of the strategy applied in a multi-agent financial decisions support system. In our research we have assumed that for building HFT investment strategy in a-Trader system, the consensus methods are applied. In many cases determination a consensus is impossible to achieve in the required time using only a one-level determination method in which all of the incoming agents' knowledge is processed at the same time. A decomposition of a task of determining a consensus into smaller subtasks and their parallelization can solve the mentioned problem. The aim of this paper is to develop a method for a multi-level consensus determining, in order to build a HFT investment strategy in a-Trader multi-agent system.

Keywords: Consensus · Multi-agent system · Forex

1 Introduction

A process of financial decision taking is more often supported by using multi-agent systems. Supporting decisions on Foreign Exchange Market (Forex) can serve as an example. On this market currencies are traded against one another in pairs, for instance

© Springer Nature Switzerland AG 2019
N. T. Nguyen et al. (Eds.): ACIIDS 2019, LNAI 11431, pp. 453–463, 2019.
https://doi.org/10.1007/978-3-030-14799-0_39

GBP/USD, EUR/USD. The quotations are in High Frequency Trading (HFT – online, real time trading) and aggregated in different periods, such as M1 (one minute), M30 (thirty minutes), H1 (one hour), W1 (one week). In comparison with traditional trading, High Frequency Trading puts strong attention on the short-term positions, high rate of quotation changes, and sophisticated algorithms based on efficient and robust indicators and modern IT. On FOREX long/short positions are open/close in order to trading. Where we buy a currency pair at a certain price and we hope to sell it later at a higher price – the long position is open. On the other hand, the short position is open where we want to *"buy high, sell low"*. Therefore, supporting decisions on Forex rely on rapid providing, by a multi-agent system, advices on what position should be taken: long, short or none. Such advice is given by different investment strategies, that are created on top of statistics, economics, mathematics or artificial intelligence methods. One of the example of such system is a-Trader [1]. There are different investment strategies in a-Trader, such as: Consensus, Candle genetic algorithm Kohonen network, Growing neural gas, Fundamental back propagation network, Evolutionary algorithm and Deep Learning. However, these strategies are not optimized in terms of HFT (the computational complexity of these strategies is too high and the signals for open/close positions are too late in many cases)

Due this fact, the main problem which appears concern on how to decrease the computational complexity of the strategy applied in a multi-agent financial decisions support system. In our research, we have assumed that for building the HFT investment strategy in a-Trader system, consensus methods are applied. In many cases, a determination of a consensus is impossible to achieve in the required time, using only a one-level determination method, in which all the incoming agents' knowledge is processed at the same time. A division of a task of determining a consensus into smaller subtasks and their parallelization can solve the mentioned problem. Thus, the aim of this paper is developing a method for a multi-level consensus determining in order to building the HFT investment strategy in a-Trader multi-agent system.

The remaining part of this paper is organized as follows. The first part of paper presents introduction to the problem and state-of-the-art in the considered field. Next the introduction to a multi-level determination method and a-Trader system is presented. Last section recalls the results of the conducted experiment and focuses on our upcoming research plans and concludes the paper.

2   Related Works

Taking into consideration multi-agent financial decision support systems, the work [2] presents a solution, where agents incorporate a set of classification and regression models, a case-based reasoning system and an expert system. This system supports trading on six selected pairs of currencies. Specialized financial models are used to build agents in work [3]. In a multi-agent system, presented in [4], fuzzy agents are used for supporting trading decisions. In a system presented by [4, 5] agents are divided into two groups: the ones based on a fundamental analysis and based on a technical analysis. Agents making decisions in a multi-agent system presented in [6] use neural networks and neuro-fuzzy computing. In work [7], in turn, agents use multiple behavioral techniques. An evaluation of the shares portfolio optimization strategy by a rational agent, an interference agent and a technical analysis agent is presented in [8]. The works [9, 10] present using the consensus methods for a financial decision supporting in multi-agents' systems. The consensus is used as a trading strategy and it is calculated based on decisions provided by several agents, which are built on top of the fundamental analysis, the technical analysis and behavioral methods. However, these solutions use a one-level consensus determining algorithm. There are two main problems related to this approach. The first one is long time of a consensus determining in HFT conditions (this time should be close to a real time). The second is related to a small number of "open/close" "short/long" positions, because large number of agents lead to harder consensus determining in the same tick (for example, "buy" decisions of selected agent may be shifted about one tick and in consequence a consensus should be "open" long position, but it is "do nothing".

To resolve these problems, a multi-level consensus determining is proposed in this paper. The multi-level consensus determination is a quite new idea and so far, it has not been widely investigated in the literature. There are some theoretical papers like [11–13] where the general idea of one- and multi-level consensuses is defined. The experimental verification of multi-level integration methods is proved that the division of integration task into smaller sub-problems gives similar results as the one-level approach, but improves a time performance. The multi-level idea is applied in the ontology integration task [14]. The analytical analysis pointed out that for the presented algorithm the one- and multi-level integration processes give the same final ontology. However, the multi-level integration allows to decrease the time of data processing even by 20% in comparison to the one-level approach. The preliminary success of a multi-level approach has encouraged us to apply it in the practical solution for supporting financial decisions making.

3   Basic Notions

3.1   A-Trader Multi-agent System

A-Trader is a simulation environment implemented as a-multi-agent system for experiments and verification purpose. The a-Trader system has been detailed described in [1, 9]. Due to page limit, in this paper we provide only a general information about this system. The a-Trader consist of about 1500 agents, divided into following groups:

- Notification Agent - receives the quotations, distributes messages and data to various agents, and controls the system operation,
- Market Communication Agents - deliver news from financial markets and quotations of the available securities and transmit open and close position orders,
- Basic Agents – calculate basic technical and fundamental analysis indicators,
- Intelligent Agents – generate a buy/sell decisions on the basis of different methods,
- Supervisor Agents - generate trading strategies, coordinate the computing Basic and Intelligent Agents, and provide the final advice to the trader; they are charged to resolve conflicts and to assess the effectiveness in investing and risk.

The main task of a-Trader system is to provide users a decision about buy or sell a financial instrument. Thus, in our system the agents' decision is defined as follows [1]:

Definition 1. *Decision* D *about* *a finite* *set* *of financial instruments* $E = \{e_1, e_2, \ldots, e_N\}$ *is defined as a set:*

$$D = \langle EW^+, EW^\pm, EW^-, Z, SP, DT \rangle \tag{1}$$

where:

(1) $EW^+ = \{\langle e_o, pe_o \rangle, \langle e_q, pe_q \rangle, \ldots, \langle e_p, pe_p \rangle\}$
A couple $\langle e_x, pe_x \rangle$, where: $e_x \in E$ and $pe_x \in [0, 1]$ denote a financial instrument and this instrument's participation in set EW^+, $x \in \{o, q, \ldots, p\}$. The financial instrument $e_x \in EW^+$ is denoted by e_x^+. The set EW^+ is called a positive set; in other words, it is a set of financial instruments about which the agent knows the decisions to buy, and the volume of this buying.

(2) $EW^\pm = \{\langle e_r, pe_r \rangle, \langle e_s, pe_s \rangle, \ldots, \langle e_t, pe_t \rangle\}$
A couple $\langle e_x, pe_x \rangle$, where: $e_x \in E$ and $pe_x \in [0, 1]$ denote a financial instrument and this instrument's participation in set $EW^\pm$, $x \in \{r, s, \ldots, t\}$. The financial instrument, $e_x \in EW^\pm$ will be denoted by $e_x^\pm$. The set $EW^\pm$ is called a neutral set, in other words, it is a set of financial instruments, about which the agent does not know that buy or sell. If these instruments are held by an investor, that they should not be sold, or if they are not in possession of the investor, should not be bought by them.

(3) $EW^- = \{\langle e_u, pe_u \rangle, \langle e_v, pe_v \rangle, \cdots, \langle e_w, pe_w \rangle\}$
A couple $\langle e_x, pe_x \rangle$, where: $e_x \in E$ and $pe_x \in [0, 1]$, denote a financial instrument and this instrument's participation in set $EW^-, \in \{u, v, \ldots, w\}$. The financial instrument $e_x \in EW^-$ will be denoted by e_x^-. The set EW^- is called a negative set; in other words it is a set of financial instruments of which the agent knows that these elements should sell.

(4) $Z \in [0, 1]$ - predicted rate of return.
(5) $SP \in [0, 1]$ - degree of certainty of rate Z. It can be calculated on the basis of the level of risk related with the decision.
(6) DT - date of decision.

A situation in which the structures of a decision in the system differ, or the values of their attributes are different, is called a knowledge conflict of these agents. This conflict results in the taking by agents of various, often contradictory decisions concerning buying and selling a financial instrument.

3.2   Consensus Determination Methods

In a multi-agent system for supporting the financial decision many investment strategies could be applied, where a consensus determination method is one of them. Let U be a set of objects representing the potential elements of knowledge referring to a concrete subject in the real world. Let 2^U be the powerset of set U that is the set of all subsets of U. Then the set of all k-element subsets (with repetitions) of set U is $\prod_k(U)$ (for $k \in N$), and let $\prod(U) = \bigcup_{k=1}^{\infty} \prod_k(U)$ be the set of all non-empty finite subsets with repetitions of set U. A set $A \in \Pi(U)$ is called a knowledge profile involving the knowledge states given by members on the same subject in the real world. Elements of U have two structures (i.e. macrostructure and microstructure). The microstructure is considered as the representations of elements in the set U such as: linear orders, n-tree, tuples, etc. In this paper the macrostructure of the set U is defined in the previous Section (Definition 1).

The macrostructure is understood as a relationship between elements and often defined as a distance function with a signature $\delta : U \times U \to [0, 1]$. In this paper the function δ is assumed according to [15].

Definition 2. *For an assumed distance space (U, δ), the consensus choice problem requires establishing the consensus choice function. By a consensus choice function in a space (U, δ) we mean a function:*

$$C : \Pi(U) \to 2^U$$

By $C(X)$ we denote the representation of $A \in \Pi(U)$ and for each $c \in C(X)$ we denote a consensus of a profile A.

A consensus is determined based on a decision generated by different agents running in a system. A consensus is interpreted as the final decision about buying or selling some financial instruments. A final decision is designated based on all agents' knowledge. The profile is a set of decisions generated by Base Agents and Intelligent Agents are defined as follows [6]:

Definition 3. *A profile $A = \{A^{(1)}, A^{(2)}, ..., A^{(M)}\}$ is called a set of M decisions of a finite set of a financial instrument E, such that:*

$$
\begin{aligned}
A^{(1)} &= \left\langle EW^{+(1)}, EW^{\pm(1)}, EW^{-(1)}, Z^{(1)}, SP^{(1)}, DT^{(1)} \right\rangle \\
A^{(2)} &= \left\langle EW^{+(2)}, EW^{\pm(2)}, EW^{-(2)}, Z^{(2)}, SP^{(2)}, DT^{(2)} \right\rangle \\
&\cdots\cdots\cdots\cdots \\
A^{(M)} &= \left\langle EW^{+(M)}, EW^{\pm(M)}, EW^{-(M)}, Z^{(M)}, SP^{(M)}, DT^{(M)} \right\rangle
\end{aligned}
\tag{2}
$$

where: $E = \{e_1, e_2, \ldots, e_N\}$. In the case of a-Trader it is a set of pairs of currencies, e.g. GBP/PLN, USD/GBP.

For the assumed distance space, the consensus determined in the one level (called the one-level consensus) is appointed in the following steps [9]:

Algorithm 1. The one-level consensus determination method
Data: The profile $A = \{A^{(1)}, A^{(2)}, ... A^{(M)}\}$ consists of M agents' decisions.
Result: Consensus decision generated by the Supervisor Agent
$CON = \langle CON_+, CON_\pm, CON_-, CON_Z, CON_{SP}, CON_{DT} \rangle$ according to A.

Begin
1: $CON_+ = CON_\pm = CON_- = \varnothing, CON_Z = CON_{SP} = CON_{DT} = 0$
2: $j := 1$.
3: $i := +$.
4: If $t_i(j) > M/2$ then $CON_i := CON_i \cup \{e_j\}$
    Go to:6.
5: If $i = +$ then $i := \pm$
    If $i = \pm$ then $i := -$
    If $i = -$, then Go to:6
    Go to:4.
6: If $j < N$ then $j := j+1$ Go to:3
    If $j \geq N$ then Go to:7.
7: $i := Z$.
8: Determine $pr(i)$.
9: $k_i^1 = (M + 1)/2$, $k_i^2 = (M + 2)/2$.
10: $k_i^1 \leq CON_i \leq k_i^2$.
11: If $i = Z$ then $i := SP$
    If $i := SP$ then $i := DT$
    If $i := DT$ then End
    Go to: 8.
End

The computational complexity of this algorithm is $O(3NM)$.

The general idea of the two-level consensus determination method is based on an initial division of the sequence of n agents into k classes. For each class, the one-level algorithm is applied. The final consensus (final decision) is designated as a consensus of these partial consensuses. The procedure can be repeated many times where outputs obtained in the previous stage serve as inputs of subsequent steps then we can say about a multi-level approach. The general idea of the two-level consensus determination method is presented in Fig. 1.

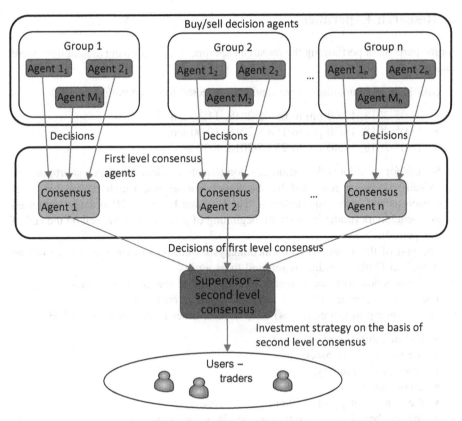

Fig. 1. The two-level consensus determining in a-Trader.

The idea presented in Fig. 1 is realized by the Algorithm 2.

Algorithm 2. The two-level consensus determination method
Data: The profile $A = \{A^{(1)}, A^{(2)}, ... A^{(M)}\}$, k
Result: $CON = \langle CON_+, CON_\pm, CON_-, CON_Z, CON_{SP}, CON_{DT} \rangle$ according to A.

Begin
1: Divide a profile A into k class in a random way
2: For each class apply Algorithm 1 and determine CON_1,
   $CON_2, \ldots, CON_k$
3: For a profile $\{CON_1, CON_2, \ldots, CON_k\}$ apply the Algorithm 1
End

4   Research Experiment

For the purpose of performing the research experiment, the following assumptions were made:

1. GBP/PLN M1 quotations were randomly selected for three periods:

   - 07-05-2018, 17:00 pm to 08-05-2018, 11:30 am,
   - 14-05-2018, 18:30 pm to 15-05-2018, 10:00 am,
   - 22-05-2018, 14:40 pm to 23-05-2018, 8:30 am,

2. Signals for open long/close short position (equals 1), close long/open short position (equals −1) were generated by the Supervisor agents, which realize two-level consensus and consensus strategies. The Buy and Hold (B&H) strategy were used as a benchmark (trader buys on the beginning of given period and sell on the end of this period).
3. The unit of the agents' performance analysis were pips (a change in price of one "point" in FOREX trading is referred to as a pip).
4. The transaction costs were directly proportional to the number of transactions.
5. The investor engages 100% of the capital held in each transaction.
6. The following measures (ratios) were used in order to performance analysis:

   - rate of return (ratio x_1),
   - the number of transactions,
   - gross profit (ratio x_2),
   - gross loss (ratio x_3),
   - the number of profitable transactions (ratio x_4),
   - the number of profitable transactions in a row (ratio x_5),
   - the number of unprofitable transactions in a row (ratio x_6),
   - Sharpe ratio (ratio x_7),
   - the average coefficient of volatility (ratio x_8),
   - the average rate of return per transaction (ratio x_9), counted as the quotient of the rate of return and the number of transactions.

7. To compare the agents' performance, the following evaluation function was used [9]:

$$y = (a_1x_1 + a_2x_2 + a_3(1 - x_3) + a_4x_4 + a_5x_5 + a_6(1 - x_6) + a_7x_7 + a_8(1 - x_8) + a_9x_9) \tag{3}$$

where: x_i denotes the normalized values of ratios mentioned in item 6 from x_1 to x_9. It was adopted in the test that coefficients a_1 to a_9 = 1/9. It should be mentioned that these coefficients may be modified with the use of, for instance, an evolution-based method, or determined by the trader in the accordance with his or her preferences.
8. The times for the consensus determination were calculated as the average time of each signals generated in particular periods. These included following processes: receiving signals from source agents, calculating consensus, sending signals to the Notification Agent and storing them in the database.

Table 1 presents results of performance analysis.

Table 1. The trading performance

Ratio	One-level consensus			Two-level consensus			B & H		
	Period 1	Period 2	Period 3	Period 1	Period 2	Period 3	Period 1	Period 2	Period 3
Rate of return [Pips]	-173	267	287	-92	244	397	-577	79	198
Number of transactions	17	14	19	50	46	67	1	1	1
Gross profit [Pips]	163	112	98	116	91	117	0	79	198
Gross loss [Pips]	154	61	127	135	54	90	-119	0	0
Number of profitable transactions	11	9	13	34	38	54	0	1	1
Number of profitable consecutive transactions	5	5	7	17	15	10	0	1	1
Number of unprofitable consecutive transactions	1	2	1	2	3	3	1	0	0
Sharpe ratio	1.52	0.75	1.19	1.90	0.80	1.12	0	0	0
Average coefficient of volatility	0.11	0.22	0.48	0.47	0.30	0.59	0	0	0
Average rate of return per transaction	-10.18	19.07	15.11	-1.84	5.30	5.93	-577	79	198
Value of evaluation function (y)	0.42	0.47	0.39	0.49	0.39	0.52	0.02	0.21	0.30
Time of determining [ms]	317	286	335	97	112	117	-	-	-

Taking into consideration values of the evaluation function, in the first and the third period, the two-level consensus was the best one. In the second period, the one-level consensus was the best. The two-level consensus strategy was ranked highest most often (2 out of 3 periods), although the rate of return of this strategy was not always the highest. The B&H benchmark has been evaluated lowest in each periods. Values of particular ratios are differ in case of analyzed strategies. For example, the two-level consensus in more cases was characterized by the highest rate of return, however the one-level consensus was characterized by lower values of risk measures in more cases. It may result from lower number of transactions in case the one-level consensus.

Taking into consideration the time required for the consensus determination, the two-level consensus calculation time was about 30% shorter than the one-level consensus approach. These decrease of time is very important in case of HFT. A-Trader receives data from the data provider about ten signals (ticks) per second, therefore, the one-level consensus calculation time (about 300 ms) is too long. The two-level consensus determination time was about 100 ms and it is adequate for HFT.

5   Conclusions

Some financial decisions have deadlines and may become useless if these boundaries are exceeded. Thus, in financial decision support systems not only the accuracy of the provided decisions, but also the time of their designations play the important role. This paper has been devoted to the reduction of the time of the final decision determination. It was achieved by applying the two-level consensus determination method in the multi-agent financial decision support system. The conducted experiment demonstrates that the two-level approach allows us to decrease almost one third of the time of the decision determination. It is caused by the parallelization of calculations. Moreover, the evaluation of the one- and two-level consensus decisions has proved us that the two-level strategy was ranked the highest most often and allows users to take more profitable financial decisions.

In our upcoming publications, we plan to conduct experiments for more levels of the consensus determination. We would like to find some dependencies between the number of levels and time of determining and the accuracy of designated decisions. Additionally, in our previous work we have proved that the initial division of the profile in the first level has significant influence on the final decision. Random division of agents profile does not guaranty optimal solution. It is quite good for regular situations on market but it can be not good for some irregular situations. Thus, in our future work we would like to examine this aspect as well.

Acknowledgement. This research was performed in the frame of the project "Business Data Mining A-Trader" realized in cooperation with 4-TUNE IT s.c. (http://4tune.pl) and project no. 4005/0011/17, "Smart University", Nguyen Tat Thanh University, Vietnam.

References

1. Korczak, J., Hernes, M., Bac, M.: Collective intelligence supporting trading decisions on FOREX market. In: Nguyen, N.T., Papadopoulos, George A., Jędrzejowicz, P., Trawiński, B., Vossen, G. (eds.) ICCCI 2017. LNCS (LNAI), vol. 10448, pp. 113–122. Springer, Cham (2017). https://doi.org/10.1007/978-3-319-67074-4_12
2. Barbosa, R.P., Belo, O.: Multi-agent forex trading system. In: Håkansson, A., Hartung, R., Nguyen, N.T. (eds.) Agent and Multi-agent Technology for Internet and Enterprise Systems. SCI, vol. 289, pp. 91–118. Springer, Heidelberg (2010). https://doi.org/10.1007/978-3-642-13526-2_5
3. Sycara, K.P., Decker, K., Zeng, D.: Intelligent agents in portfolio management. In: Jennings, N., Wooldridge, M. (eds.) Agent Technology, pp. 267–282. Springer, Heidelberg (2002). https://doi.org/10.1007/978-3-662-03678-5_14
4. Tatikunta, R., Rahimi, S., Shrestha, P., Bjursel, J.: TrAgent: a multi-agent system for stock exchange. In: Proceedings of the 2006 IEEE/WIC/ACM International Conference on Web Intelligence and Intelligent Agent Technology (WI-IATW 2006), pp. 505–509. IEEE Computer Society, Washington, DC, USA (2006)
5. Ivanović, M., Vidaković, M., Budimac, Z., Mitrović, D.: A scalable distributed architecture for client and server-side software agents. Vietnam J. Comput. Sci. 4(2), 127–137 (2017)

6. Sher, G.I.: Forex trading using geometry sensitive neural networks. In: Soule, T. (ed.) Proceedings of the 14th Annual Conference Companion on Genetic and Evolutionary Computation (GECCO 2012), pp. 1533–1534. ACM, New York (2012)
7. Khosravi, H., Shiri, Mohammad E., Khosravi, H., Iranmanesh, E., Davoodi, A.: TACtic- a multi behavioral agent for trading agent competition. In: Sarbazi-Azad, H., Parhami, B., Miremadi, S.-G., Hessabi, S. (eds.) CSICC 2008. CCIS, vol. 6, pp. 811–815. Springer, Heidelberg (2008). https://doi.org/10.1007/978-3-540-89985-3_109
8. Bohm, V., Wenzelburger, J.: On the performance of efficient portfolios. J. Econ. Dyn. Control 29(4), 721–740 (2005)
9. Korczak, J, Hernes, M., Bac M.: Risk avoiding strategy in multi-agent trading system. In: Proceedings of Federated Conference Computer Science and Information Systems (FedCSIS), Kraków, pp. 1131–1138 (2013)
10. Hernes, M., Sobieska-Karpińska, J.: Application of the consensus method in a multiagent financial decision support system. IseB 14(1), 167–185 (2016)
11. Kozierkiewicz-Hetmańska A., Nguyen N.T.: A comparison analysis of consensus determining using one and two-level methods. In: Advances in Knowledge-Based and Intelligent Information and Engineering Systems, vol. 243, pp. 159–168 (2012)
12. Nguyen, V.D., Nguyen, N.T.: A two-stage consensus-based approach for determining collective knowledge. In: Le Thi, H., Nguyen, N., Do, T. (eds.) Advanced Computational Methods for Knowledge Engineering, pp. 301–310. Springer, Heidelberg (2015). https://doi.org/10.1007/978-3-319-17996-4_27
13. Kozierkiewicz-Hetmańska, A., Sitarczyk, M.: The efficiency analysis of the multi-level consensus determination method. In: Nguyen, N.T., Papadopoulos, George A., Jędrzejowicz, P., Trawiński, B., Vossen, G. (eds.) ICCCI 2017. LNCS (LNAI), vol. 10448, pp. 103–112. Springer, Cham (2017). https://doi.org/10.1007/978-3-319-67074-4_11
14. Kozierkiewicz-Hetmanska, A., Pietranik, M.: The knowledge increase estimation framework for ontology integration on the concept level. J. Intell. Fuzzy Syst. 32(2), 1–12 (2016)
15. Hernes, M., Sobieska-Karpińska, J., Kozierkiewicz, A., Pietranik, M.: A new distance function for consensus determination in decision support systems. In: Nguyen, N.T., Pimenidis, E., Khan, Z., Trawiński, B. (eds.) ICCCI 2018. LNCS (LNAI), vol. 11056, pp. 155–165. Springer, Cham (2018). https://doi.org/10.1007/978-3-319-98446-9_15

Implementing Smart Virtual Product Development (SVPD) to Support Product Manufacturing

Muhammad Bilal Ahmed[1(✉)], Cesar Sanin[1],
and Edward Szczerbicki[2]

[1] The University of Newcastle, Callaghan, NSW, Australia
muhammadbilal.ahmed@uon.edu.au,
cesar.sanin@newcastle.edu.au
[2] Gdansk University of Technology, Gdansk, Poland
edward.szcerbicki@newcastle.edu.au

Abstract. This paper illustrates the concept of providing the manufacturing knowledge during early stages of product life cycle to experts working on product development. The aim of this research is to enable a more collaborative product development environment by using Smart Virtual Product Development (SVPD) system, which is powered by Set of Experience Knowledge Structure (SOEKS) and Decisional DNA (DDNA). It enhances the industrial product development process by storing, using and sharing previous manufacturing experience and knowledge. This knowledge is stored in form of formal decisional events after being collected from the set of similar products having some common functions and features. The proposed system uses a collective, team-like knowledge developed by product designers, manufactures, and metrologists. Implementing this system in the process of product development enables the small and medium enterprises (SMEs) to take proper decisions at appropriate time by reducing mistakes at an early stages of product development.

Keywords: Smart virtual product development · Product development ·
Manufacturing capability analysis and process planning ·
Set of experience knowledge structure · Decisional DNA

1 Introduction

Product development is a multidisciplinary work which requires vigorous information and knowledge [1] including customers' demands during product design and manufacturing. In order to develop highly successful new products, it is necessary to integrate both the engineering design and manufacturing knowledge. Superior product development capabilities of a firm are highly dependent on its ability to create, distribute and utilize knowledge throughout the product development process. There is a substantial form of literature on work integration in product development, but much less attention has been focused on knowledge integration or knowledge sharing [2]. Product development does not only rely on the knowledge of new technological advancements but also the complete knowledge of all the past and present of similar

© Springer Nature Switzerland AG 2019
N. T. Nguyen et al. (Eds.): ACIIDS 2019, LNAI 11431, pp. 464–475, 2019.
https://doi.org/10.1007/978-3-030-14799-0_40

products. This knowledge is possessed by a group of experts in differing fields, i.e. marketing, design, manufacturing, and metrology. In practice, there is no clear dividing line between design, manufacturing, and other life-cycle stages as they are connected with each other by certain means. For example, a designer may need to work out even on the selection of suitable manufacturing process plan and logistics matters in order to make cost-effective design decisions. Design of products and systems requires involvement from people with wide ranging areas of expertise. They often need to collaborate over long distances through virtual technologies [3].

In order to design and develop high quality of products, experts working on product development have to utilize a wide range of sources of knowledge i.e. knowledge related to product design and manufacturing, selection of measuring equipment, and knowledge related to quality procedures. Companies involved in new product development acquire at least two types of design knowledge; first one is regarding the product itself, and the second one is concerned with how the product will be manufactured effectively to meet cost, quality and short product development cycle time [4]. Manufacturing knowledge is an expression with vast meanings, which may include knowledge on the effects of material properties decisions, machine and process capabilities or understanding the unintended consequences of design decisions on manufacturing [5]. Manufacturing planning at the conceptual or early design stage is the key for designers to evaluate manufacturability in terms of criteria and metrics such as cost and time. However, there are not many techniques and tools for conceptual manufacturing planning. Therefore, the use of inappropriate manufacturing knowledge can lead to mistakes during product development and harm the environment. These mistakes are likely to be caused by designers relying on poor or inadequate knowledge during the design process. Time spent searching for "reliable" knowledge can also contribute to delays in the design process and may affect overall quality and product lead times. Evidences suggest that designers spend up to forty percent of their time searching for the right information. This has obvious effects on the productivity of the company investing in new product development processes [6].

Electronics and IT have played a key role in the 3rd industrial revolution by increasing automation of production. Now, the world is moving towards Industry 4.0, which is the 4th revolution of the industry. Conventional manufacturing processes will be replaced by smart manufacturing, which consists of new concepts, i.e. Internet of Things (IoT), Internet of Services (IoS), Cyber-Physical Systems (CPS), Mass collaboration, High-speed internet and affordable 3-D printing; thus, creating great potentials for the development of new smart knowledge-based product development frameworks [7]. The concept of smart manufacturing is also closely related to knowledge-driven decision making to meet customers' demands for new products. In order to make decisions at various stages of product development, it is very important to have complete knowledge of each manufacturing process and its possible outcomes. Furthermore, engineering knowledge is embedded in various stages in the product lifecycle in forms of rules, logical expressions, ontologies, predictive models, statistics, and information extracted from previous experiences and sensors in real-world situations, such as production, inspection, product use, supplier networks, and maintenance. Currently, knowledge is not completely captured and stored in a digital form during all phases of product life cycle. Therefore, organizations are aiming at achieving streamlined knowledge capture and curation through knowledge management. This problem can be solved by using an applicable form of knowledge management [8].

We try to overcome this problem by proposing a system that uses a collective, team-like knowledge created from relevant past experiences; we call this system as Smart Virtual Product Development (SVPD). This approach uses a smart knowledge management technique called Set of Experience Knowledge Structure (SOEKS or SOE in short) and Decisional DNA [9]. It captures, stores, and shares the experiential knowledge in the form of set of experiences. Whenever a similar query is presented during the problem solving process, this stored knowledge is recalled to overcome the problem. It provides a list of proposed optimal solutions according to the priorities set by the user. By the passage of time, system achieves more expertise in specific domains as it stores relevant knowledge and experience related to formal decision events.

The structure of the paper includes the background in Sect. 2, which presents the basic concepts of product and product families, product development, and set of experience knowledge structure and decisional DNA. Introduction to SVPD describing architecture of proposed system and design of test case study is presented in Sect. 3, and finally the concluding remarks are presented in Sect. 4.

2   Background

2.1   Concept of Product and Product Families

A product is something sold by an enterprise to its customers in the form of a good, service, place, organization or an idea. In this research, products are objects which are manufactured for the end users. Our research is based on developing products from evolving range of products. We use an approach to develop a new product from existing product families and part hierarchies [10]. The composition of the new product derived from product families can be formulated in terms of equations as follows [11].

$$[Product\ Family] \Leftarrow [Prod]_1 + [Prod]_2 + [Prod]_3 + .. + [Prod]_n \tag{1}$$

Based on [12], a product can be further described as being made up of the structured assembly of part objects, which can be expressed as:

$$[Product] \Leftarrow [Part]_1 + [Part]_2 + [Part]_3 + .. + [Part]_n \tag{2}$$

Equation (2) illustrates the idea that a product is made up of a structured assembly of n-number of different parts.

The next step is to define part objects. Part object [Part] is considered to consist of a set of part properties [12]. These properties include six basic properties as shown in the expression (3):

$$[Part] \Leftarrow \{M \wedge F \wedge D \wedge T \wedge SF \wedge Q\} \tag{3}$$

As illustrated by Eq. (3), each [Part] has a set of properties, these include Material (M), Form (F), Dimensions (D), Tolerances (T), Surface Finish (SF), and Quantity (Q).

Manufacturing processes require to produce a range of products are defined by process requirements and are basically modelled as a number of steps within a process plan model:

$$[Process\ Plan\ Model] \Leftarrow [Process\ Step]_1 + [Process\ Step]_2 + .. + [Process\ Step]_n \quad (4)$$

whereas the process steps can be further defined in terms of their properties as follows:

$$[Process\ Step]_1 \Leftarrow [(Tp) \wedge (Ts) \wedge (SN) \wedge (P)] \quad (5)$$

where $[Process\ Step]_1$ has its unique type (Tp), time (Ts), step number (SN) and properties (P).

2.2 Product Development

Product development processes are basic approaches and procedures that companies use to design and manufacture newly introduced products and bring them to market. Companies are forced to develop new products due to certain factors, e.g. competition, technological advancement and market changes, etc. [13]. It is basically a series of interconnected processes and sub-processes, which covers product introduction, product design, production system design, and the start of its production [14]. The main aim of product development process is to integrate engineering and industrial design requirements through a structural process that allows the achievement of lower production cost, higher quality and shorter development time with quick access to market, so that it can contribute to customers' satisfaction and companies' financial benefits [15]. It is very important for companies with short product life cycles, to quickly and safely develop new products and new product platforms that fulfill reasonable demands on quality, performance, and cost.

Classical methods such as Stage-Gate model process, Product development process by Ulrich and Eppinger, Development funnel product model process, Simultaneous engineering (SE), Concurrent engineering (CE), Integrated product development (IPD), and lean product development have played key roles in product development process in past [16]. Now the world is moving towards fourth industrial revolution, which encompasses a set of advancements in both products and manufacturing processes. Therefore, the adaptation of this new industrial pattern and the production of smart and connected products means deep changes in the whole organizations value chain, especially in product development process. Henceforth, organizations that produce smart products need to adopt the most suitable product development approaches. Resource optimization and waste elimination are one of the important factors to consider during development of the smart products in order to increase company's competitiveness. This can be achieved by introducing and developing new technological tools which can eliminate mistakes during early stages of product development [17].

Lean product and process development have achieved a great success in recent past as it integrates engineering knowledge into product development process. Knowledge based engineering, mistake proofing (Poka-Yoke), and continuous improvement (Kaizen) culture are the core lean enablers for lean product development process [18].

Knowledge is an important factor to get competitive advantage over competitors, therefore organizations are concentrating on their competencies of knowledge generating, saving and sharing in product development process. It is very important to identify the required knowledge and the ability to utilize it in an effective way. Furthermore, the majority of commercially available software systems used to support project collaboration and reuse of past experiences are mainly restricted to document management and sharing. Very little literature is available on what engineering knowledge is required to support product development [19].

2.3   Set of Experience Knowledge Structure and Decisional DNA

Set of experience knowledge structure (SOEKS) is a smart knowledge management technique. It collects and analyses formal decision events and uses them to represent experiential knowledge. A formal decision is defined as a choice (decision) made or a commitment to act that was the result (consequence) of a series of repeatable actions performed in a structured manner. A set of experience (SOE, a shortened form of SOEKS) has four components: Variables (V), functions (F), Constraints (C) and Rules (R). Each formal decision is represented and stored in a unique way based on these components. Variables are the basis of the other SOEKS components, whereas functions are based upon the relationships and links among the variables. The third SOEKS component is constraints, which, like functions, are connected to variables. They specify limits and boundaries and provide feasible solutions. Rules are the fourth components and are conditional relationships that operate on variables. Rules are relationships between a condition and a consequence connected by the statements 'if/then/else' [20]. The four components of a SOE and its structural body can be defined by comparing it with some important features of human DNA. First, just as the combination of its four nucleotides (Adenine, Thymine, Guanine, and Cytosine) makes DNA unique, the combination of its four components (Variables, Function, Constraints, and Rules) makes an SOE unique.

Each formal decision event is deposited in a structure that combines these four SOE components. Several interconnected elements are visible in the structure, resembling part of a long strand of DNA, or a gene. Thus, a SOE can be associated to a gene and, just as a gene produces a phenotype, a SOE creates a value for a decision in terms of its objective function. Hence, a group of SOEs in the same category form a kind of chromosome, as DNA does with genes. Decisional DNA contains experienced decisional knowledge and it can be categorized according to areas of decisions. Further, just as assembled genes create chromosomes and human DNA, groups of categorized SOEs create decisional chromosomes and DDNA. In short, a SOEKS represents explicit experiential knowledge which is gathered from the previous decisional events [21]. SOE and DDNA have been successfully applied in various fields such as industrial maintenance, semantic enhancement of virtual engineering applications, state-of-the-art digital control system of geothermal and renewable energy, storing information and making periodic decisions in banking activities and supervision, e-decisional community, virtual organization, interactive TV, and decision-support medical systems, etc. [22].

3 Introduction to Smart Virtual Product Development

Smart Virtual Product Development (SVPD) is a decision support system for industrial product development process. It stores, uses and shares the experiential knowledge of past decisional events in the form of set of experiences (SOEs) as mentioned above. It is developed to overcome the need for capturing knowledge in the digital form in product design, production planning, and inspection planning in smart manufacturing [8]. It will help in enhancing the product quality and development time in Industry 4.0 perspective.

3.1 Architecture of Smart Virtual Product Development

Our proposed Smart Virtual Product Development system consists of three main modules, i.e. design knowledge management (DKM), manufacturing capability analysis and process planning (MCAPP), and product inspection planning (PIP). These modules interact with the decisional DNA which hold all the relevant knowledge of similar products. This knowledge repository is filled with past formal decisional events involved in manufacturing of these similar products in existing facility. The proposed system stores decisional DNA knowledge in the form of SOEs. The architecture of the SVPD system is shown in Fig. 1.

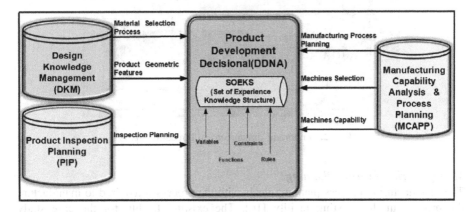

Fig. 1. The architecture of a Smart Virtual Product Development (SVPD).

These modules have further different steps to perform particular decisional activity. Design knowledge management deals with material selection process and product geometric features generation; manufacturing capability analysis and process planning provides solution regarding manufacturing process planning, machines' selection, and machines' capability to perform particular manufacturing operations; and inspection planning involves selection of different measuring equipment for product inspection during manufacturing and at final stage. Once all of these modules are successful, it provides validation that a product can be easily manufactured in an existing facility. Different variables involved in the design knowledge management module have been explained in our previous work [23]. The aim of this study is to deal with how to provide the manufacturing knowledge to product development experts during early stages of product development; therefore, it covers only those steps which are involved in manufacturing capability analysis and process planning module.

3.2   Design of a Test Case Study for the MCAPP-Module

Design and development of a threading tap (a tool to create screw threads which is called threading) is our case study, as it was also used in our previous work [23]. We are considering a machine use threading tap, as shown in Fig. 2 with few important dimensions. As part of this case study, we will take into consideration some important variables involved in the manufacturing capability analysis and process planning module, and in particular, how it deals with manufacturing knowledge management in three steps as follows: (i) manufacturing process planning, (ii) machines' selection, and (iii) machines' capability.

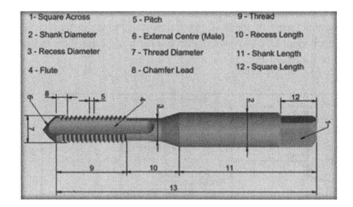

Fig. 2.   Important dimensions in threading tap.

Manufacturing Process Planning:
The manufacturing process planning deals with the processes required to manufacture the product in the existing facility [16]. The existing facility for the case study (Threading Tap) is a small tool manufacturing factory which comprises of a design office, a well-equipped machine shop (including conventional and non-conventional machines), metrology (inspection unit), and heat treatment section. In this step, we have to decide which machining operations are required to manufacture the product under consideration. Every process will start by inputting the query into the decisional DNA of the system based on initial objectives.

For the present case study of threading tap, let us consider that selected material is high speed steel and the initial objective is to define manufacturing processes for this product. These manufacturing operations can be simply recalled from an existing virtual engineering process (VEP) of a family of similar products. Whereas, a VEP is a knowledge representation of manufacturing process-planning involving the required operations, their sequence, and resources [24]. Final solution is selected from the list of proposed solutions by the user based on priorities to complete the process.

If no relevant VEP is found, a new SOEKS is generated and used to form a new VEP which is saved for future reference. The new manufacturing process planning query information is in the form of SOE (variables, functions, constraints, and rules) and is shown in Table 1.

Table 1. SOE for manufacturing process planning as a combination of variables, functions, constraints and rules.

Manufacturing process planning			
Variables	**Functions**	**Constraints**	**Rules**
TypeOfThreadingTap MachineUse SelectedMaterial HighSpeedSteel TungstenCarbideFacility ManufacturingProcesses MaterialCutting Turning Milling HeatTreatmentProcesses CylindricalGrinding ThreadGrinding Marking		TungstenCarbideFacility = NotAvailable	IF SelectedMaterial=HighSpeedSteel& TypeOfThreadingTap=MachineUse THEN ManufacturingProcessesAre MaterialCutting Turning CMilling Heat treatment Processes Cylindrical Grinding ThreadGrinding Marking

Machines' Selection:

In this step, all machines required for manufacturing processes are checked for their availability. If any of the machine is not available in terms of physical availability or due to a busy schedule, it suggests to be outsourced. An existing set of VEPs is searched based on the manufacturing resources query. If it is not found, the new machine selection information is generated in the form of SOE as it is structured in Table 2.

Table 2. SOE for machines' selection as a combination of variables, functions, constraints and rules.

Machines' selection			
Variables	**Functions**	**Constraints**	**Rules**
TypeOfThreadingTap MachineUse SelectedProcess MillingProcess SelectedMachine CNCMilling SkilledLevel HighSkilled MachineEfficiency			IF TypeOfThreadingTap=MachineUse& SelectedProcess=MillingProcess THEN SelectedMachine=CNCMilling& SkillLevel=HighSkilled

Machines' Capability:

It is not only machine's availability which needs to be considered by designers and manufacturers. They should also be well aware of each machine's capability. It is measured in terms of its maximum and minimum processing limits. Therefore, the capability of each machine will be checked in this step. A virtual engineering object (VEO) can be recalled from the product development DDNA, which will provide the required upper and lower limits of the selected machine. Whereas, a VEO is the knowledge representation of an engineering object that embodies its associated knowledge and experience [25].

If it is not found, the new machine capability information is generated in the form of SOE as it is structured in Table 3.

Table 3. SOE for machines' capability as a combination of variables, functions, constraints and rules.

Machines' capability			
Variables	**Functions**	**Constraints**	**Rules**
SelectedProcess MillingProcess MachineType MachineEfficiency MachineServiceDue MaximumCapacityXmm MaximumCapacityYmm MaximumCapacityZmm MinimumCapacityXmm MinimumCapacityYmm MinimumCapacityZmm	MachineEfficiency= WorkDone/WorkInput	MachineServiceDue=Nill MachineEfficiency<=90%	IF SelectedProcess=MillingProcess MachineType=CNCMilling THEN MaximumCapacityXmm=500 MaximumCapacityYmm=250 MaximumCapacityZmm=100 MinimumCapacityXmm=10 MinimumCapacityYmm=10 MinimumCapacityZmm=10

For evolving products, knowledge representation structures are never completed as they also keep evolving. They are updated with new decisions that are captured by the SOE and added to the system's decisional DNA. In this way, the decisional DNA continues to gain new and updated experiential knowledge, which helps it to support and enhance future decisions related to any of the step mentioned above. Similarly, SOEs are generated for each individual step of MCAPP module having specific weights for the variables. A combination of all the individual SOEs are under the MCAPP module. The other two modules of system, i.e. design knowledge management (DKM) and product inspection planning (PIP) work in the same manner. Whereas, the whole necessary knowledge and experience for supporting product development process is accommodated in the SVPD system.

As the decisional DNA is constructed in JAVA [9] and has been applied successfully in various other fields of application as discussed earlier, the SVPD system is also implemented using JAVA as its programing language. For illustration purposes, a SOE variable of selecting manufacturing process is shown below (it is generated in XML):

```
<set_of_variables>
<! -- Variables included in the model -->
<Variable>
    <var_name>ppf_ManufacturingProcesses</var_name>
    <var_type>CATEGORICAL</var_type>
    <var_cvalue>2</var_cvalue>
    <var_evalue>2</var_evalue>
    <unit></unit>
    <internal>true</internal>
    <weight>0.5</weight>
    <l_range>0.0</l_range>
    <u_range>0.0</u_range>
    <categories>
        <category>CATEGORY UNDEFINED</category>
        <category>2</category>
    </categories>
    <priority>0.0</priority>
</variable>
```

Describing the variables includes the fields of the name of the variable (<var_name>); the type of the variable (<var_type>) that can be numerical or categorical; cause value of the variable (<var_cvalue>) which is the starting value of the variable before being optimized; effect value of the variable (<var_evalue>) which is the final value of the variable after being optimized; the measurement unit of the values (<unit>); and the boolean value internal variables (<internal>). Moreover, variables includes fields that allow them to participate in the processes of similarity, uncertainty, impreciseness, or incompleteness measures, they are: weight (<weight>), priority (<priority>), lower range (<l_range>), upper range (<u_range>), and categories (<categories>), which contains the different values which a categorical value can have [26].

4 Conclusion

In this research, we presented a concept of enhancement of product development process by providing manufacturing knowledge during early phases of product life cycle. This is achieved by introducing the concept of Smart Virtual Product Development (SVPD) system. Presented research has addressed different steps involved in one of the key modules of the proposed system, i.e. manufacturing capability analysis and process planning. Provision of manufacturing knowledge to experts working on product development is illustrated with the example of a case study, which also helps to understand the architecture and working of the proposed system.

The proposed system is dynamic in nature as it updates itself every time a new decision is taken. It will benefit the entrepreneurs and manufacturing organizations involved in new product development process by reducing the extent of dependability on experts and the fast computational capabilities of the system. In this way, Smart Virtual Product Development system is one step forward in the direction of automation and Industrie 4.0. The next step will be the refinement of the algorithm in more detail and its translation into JAVA platform.

References

1. Clark, K.B., Fujimoto, T.: Product Development Performance: Strategy, Organization, and Management in the World Auto Industry. Harvard Business School, Boston (1991)
2. Hong, P., Doll, W.J., Nahm, A.Y., Li, X.: Knowledge sharing in integrated product development. Eur. J. Innov. Manag. **7**, 102–112 (2004)
3. Hayes, C.C., Goel, A.K., Tumer, I.Y., Agogino, A.M., Regli, W.C.: Intelligent support for product design: looking backward, looking forward. J. Comput. Inf. Sci. Eng. **11**, 021007 (2011)
4. Nonaka, I., Takeuchi, H.: The Knowledge-Creating Company: How Japanese Companies Create the Dynamics of Innovation. Oxford University Press, Oxford (1995)
5. Hedberg Jr., T.D., Hartman, N.W., Rosche, P., Fischer, K.: Identified research directions for using manufacturing knowledge earlier in the product life cycle. Int. J. Prod. Res. **55**, 819–827 (2017)
6. Rodgers, P.A., Clarkson, P.J.: Knowledge usage in new product development (NPD). In: IDATER 1998 Conference. Loughbourogh University, Loughborough (1998)
7. Forbes, H., Schaefer, D.: Social product development: the democratization of design, manufacture and innovation. Prosedia CIRP **60**, 404–409 (2017)
8. Feng, S.C., Bernstein, W.Z., Hedberg, T., Feeney, A.B.: Toward knowledge management for smart manufacturing. J. Comput. Inf. Sci. Eng. **17**, 031016 (2017)
9. Sanin, C., Szczerbicki, E.: Towards the construction of decisional DNA: a set of experience knowledge structure Java class within an ontology system. Cybern. Syst. Int. J. **38**, 859–878 (2007)
10. Simpson, T.W., Maier, J.R., Mistree, F.: Product platform design: method and application. Res. Eng. Des. **13**, 2–22 (2001)
11. Francalanza, E., Borg, J., Constantinescu, C.: A knowledge-based tool for designing cyber physical production systems. Comput. Ind. **84**, 39–58 (2017)
12. Tjalve, E.: A Short Course in Industrial Design. Elsevier, Amsterdam (2015)
13. Unger, D., Eppinger, S.: Improving product development process design: a method for managing information flows, risks, and iterations. J. Eng. Des. **22**, 689–699 (2011)
14. Johansen, K.: Collaborative product introduction within extended enterprises. Doctoral dissertation. Institutionen för konstruktions-och produktionsteknik (2005)
15. Cagan, J., Vogel, C.M.: Creating breakthrough products: innovation from product planning to program approval. Financial Times Prentice Hall Press, Upper Saddle River (2002)
16. Wasim, A., Shehab, E., Abdalla, H., Al-Ashaab, A., Sulowski, R., Alam, R.: An innovative cost modelling system to support lean product and process development. Int. J. Adv. Manuf. Technol. **65**(1–4), 165–181 (2013)
17. Nunes, M.L., Pereira, A., Alves, A.: Smart products development approaches for Industry 4.0. Procedia Manuf. **13**, 1215–1222 (2017)
18. Khan, M.S., et al.: Towards lean product and process development. Int. J. Comput. Integr. Manuf. **26**, 1105–1116 (2013)

19. Brown, J.S., Duguid, P.: Balancing act: how to capture knowledge without killing it. Harvard Bus. Rev. **78**, 73–80 (2000)
20. Sanin, C., Szczerbicki, E.: Set of experience: a knowledge structure for formal decision events. Found. Control Manag. Sci. **3**, 95–113 (2005)
21. Sanin, C., Szczerbicki, E.: Experience-based knowledge representation: SOEKS. Cybern. Syst. Int. J. **40**, 99–122 (2009)
22. Shafiq, S.I., Sanín, C., Szczerbicki, E.: Set of experience knowledge structure (SOEKS) and decisional DNA (DDNA): past, present and future. Cybern. Syst. **45**, 200–215 (2014)
23. Ahmed, M.B., Sanin, C., Szczerbicki, E.: Experience based decisional DNA (DDNA) to support sustainable product design. In: Dao, D., Howlett, R.J., Setchi, R., Vlacic, L. (eds.) KES-SDM 2018. SIST, vol. 130, pp. 174–183. Springer, Cham (2019). https://doi.org/10.1007/978-3-030-04290-5_18
24. Shafiq, S.I., Sanin, C., Toro, C., Szczerbicki, E.: Virtual engineering process (VEP): a knowledge representation approach for building bio-inspired distributed manufacturing DNA. Int. J. Prod. Res. **54**, 7129–7142 (2016)
25. Shafiq, S.I., Sanin, C., Toro, C., Szczerbicki, E.: Virtual engineering object (VEO): toward experience-based design and manufacturing for Industry 4.0. Cybern. Syst. **46**, 35–50 (2015)
26. Sanin, C., Szczerbicki, E.: Using XML for implementing set of experience knowledge structure. In: Khosla, R., Howlett, R.J., Jain, L.C. (eds.) KES 2005. LNCS (LNAI), vol. 3681, pp. 946–952. Springer, Heidelberg (2005). https://doi.org/10.1007/11552413_135

PID Tuning with Neural Networks

Antonio Marino and Filippo Neri[(✉)]

Department of Electrical Engineering and Information Technologies,
University of Naples, Naples, Italy
marinoantonio.96@gmail.com, filippo.neri.email@gmail.com

Abstract. In this work we will report our initial investigation of how a neural network architecture could become an efficient tool to model Proportional-Integral-Derivative controller (PID controller). It is well known that neural networks are excellent function approximators, we will then be investigating if a recursive neural networks could be suitable to model and tune PID controllers thus could assist in determining the controller's proportional, integral, and the derivative gains. A preliminary evaluation is reported.

Keywords: PID tuning and approximation · Machine learning · Neural networks

1 Introduction

Industrial plants are characterized by a variety of processes [1] that require the design of a controller able to complete the control task and respecting the required performances. In this research activity, we will study if some neural network architectures might be an efficient tool to model Proportional- Integral-Derivative controller (PID controller) [2]. It is well known that neural networks are excellent function approximators we are then interested in studying if some of their topologies could be suitable to model PID controllers. A variety of approaches to model or tune PID controllers exist: some works explore architectures for neural networks, other uses fuzzy logic, other evolutionary computation [3–10]. This paper reports our initial investigation in this area. To begin with we will empirically investigate recursive multi layer perceptrons in order to synthesize a PID controller and its gains by determining the proportional, integral, and derivative actions. In the paper we will report the performances of a recursive neural network for tuning PID controller on testcase system developed in Symulink and we will compare the architecture's performances with the methods by Zigler and Nichols (ZN) [2] and Extremum Seeking (ES) [11]. The paper reports a preliminary empirical investigation of the novel methodology. Given the great variety of Artificial Intelligence techniques that have been applied in several context, including financial time series [12,13], one has to wonder if such techniques could be leveraged to also model PID controllers.

© Springer Nature Switzerland AG 2019
N. T. Nguyen et al. (Eds.): ACIIDS 2019, LNAI 11431, pp. 476–487, 2019.
https://doi.org/10.1007/978-3-030-14799-0_41

2   The PID Controller

The PID controller is usually the best choice when designing a control system because of its simple implementation and design. Indeed its standard architecture is usually the best choice because it consists of a linear combination of three components: the proportional, the integral and the derivative actions as in

$$u = K_p e_p + K_i e_i + K_d e_d \tag{1}$$

where u is the control output and e_{index} are the error components (actions) which are defined as follows:

$$e_p = e, e_i = \int_0^t e \, dt, e_d = \frac{de}{dt} \tag{2}$$

where e is the input error.

The great diffusion of the PID controller is due to its low cost (with respect to the cost of a microprocessor) and also is due to the possibility of utilization without the exact knowledge of the industrial process to be controlled. This last characteristic is not to be undervalued considering the great variety of industrial processes. Possibly controllers designed ad hoc for a specific industrial process, maybe even including a model for it or even its simulation, could result in a better controller than a PID but would require a significant increase in the design and realization costs.

The components of Eq. (1) are the essential actions for a linear time invariant controller which allows for the proportional action, an appropriate gain of reference signal, for the integral action, an overtime error converging to zero, and for the derivative action, a correction speed which adjust over time. It is not necessary for all the actions to occur at the same time. In fact by setting to zero a K coefficient a specific action gain can be canceled. In reality, it is in fact common to use just P and PI controllers.

The techniques by Ziegler e Nichols (ZN) [2] are some of the classic techniques for tuning PID controllers. Of heuristic nature, they are based on a preliminary evaluation of the plant for determining the initial values of the controller. The first technique uses an approximation of the plant to a system of the first order with an apparent delay to calculate the controller parameters. The second technique is concerned with different plant typologies of higher order on which a closed loop test with a purely proportional gain is executed. This leads to estabilish the proportional gain that brings the system on the bound of its stability. Then this initial value will help to obtain the optimized target gains.

The ZN's techniques are usually suitable for many kind of systems, producing however a relatively ready response with overshooting and sustained oscillations since they have been designed to obtain relatively low dampings.

Another classic technique for PID tuning is Extremum Seeking (ES) [14] which is a widely used technique used for real time tuning. It does not need any knowledge about the system or about the error function. However it assumes that the error function respects the preconditions of the gradient theorem since ES uses the minimization of the gradient to compute the three PID gains. The ES can be realized in hardware with a sequence of filter, modulator and integrator pointing out the trend of the error and an estimation of the parameters. Both approaches, ZN and ES, are largely overtaken by more sophisticated techniques based on system identification or by optimization techniques, still they remain important methods to benchmark the performance of new PID tuning methods.

3   Neural Networks

Under the supervised learning paradigm, an artificial agent received a set of labeled instances formed by couples <inputs, output> where output is the classification label to be predicted. The output values are produced by an unknown control function f. The agent has as objective the learning of the function f which, according to the machine learning terminology, is called hypothesis by searching an hypothesis space for a function f' where at least a good approximation of f is supposed to exist. The quality or performance of the discovered f' is then evalutated on a separate set of instances called test set not occurring in the set of instances used for finding f'. As stated above, the selection of the hypothesis space determines the possibility to find a solution f' which is consistent (explains) the training set. An acceptable solution f' may not be perfect as f but will explain the majority of the learning instances and perform well on the test set.

A Neural Network (NN) [15] is a non linear mathematichal structure made of a set of interconnected neurons or nodes. The properties of the NN are determined by its topology and by the features of the nodes. Feed forward NNs can be represented as a set of node levels where we can identify the input level, the subset of nodes receiving the inputs, the output level, the set of nodes producing the NN's outputs, and the intermediary levels, all the remaining nodes. In a feed forward network the NN is traversed only in one direction from the input layer to the output layer, it contains no cycles and thus no internal memory.

Other types of NNs exist: for instance in a recursive NN the outputs are fed again to the input nodes thus determining a memory effect where previous inputs/outputs can also affect the present outputs together with the current inputs. Because, in the scope of this research, we are interested in modeling a PID controller, which is a dynamic model, we will consider recursive NNs.

4   PID Neural Network

In this section, we describe the PID Neural Network (PIDNN) used in our experimentation. The PIDNN uses three special types of neurons: P-neuron, I-neuron and D-neuron which realize the three fundamental controlling actions. The network is built with three level. The input level is made up by two neurons which takes as input the output of the plant and the reference input. The second level is made up by the P-,I-,D-neurons. And the third level combines the three controlling action components in the controlling action to be applied to the plant (Fig. 1).

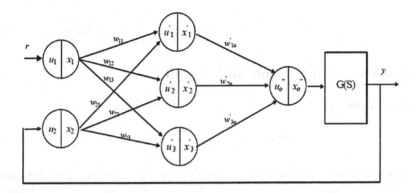

Fig. 1. PIDNN where G(s) is the system to be controlled.

The functions of the neurons are defined as follows:

$$P - neuron\ output = \sum_i input_i \tag{3}$$

$$I - neuron\ output = \int_0^t \sum_i input_i \tag{4}$$

$$D - neuron\ output = \frac{d}{dt} \sum_i input_i \tag{5}$$

The integration and derivative functions can be implemented in different way: we have chosen a numerical derivation (adoptable even in the simplest embedded system) whereas the integration function consists in the integration done till the current time.

These choices has been guided by the results obtained during preliminary experimentation. We notice that the implementation choices impact on the efficiency of the PIDNN because they require different sampling methods of the input signals.

The weights are defined as follows:

$$w_{i,j} = +1, w_{2,j} = -1, w_{1o} = K_P, w_{2o} = K_I, w_{3o} = K_D \tag{6}$$

Going across the network from the input layer to the output layer:

$$x_1' = u_1' = w_{11}x_1 + w_{21}x_2 = r - y = e \tag{7}$$

$$x_2' = \int_0^t u_2 d\tau = \int_0^t e d\tau; with\ e = w_{12}x_1 + w_{22}x_2 = r - y \tag{8}$$

$$x_3' = \frac{du_3}{dt} = \frac{de}{dt}; with \ e = w_{12}x_1 + w_{22}x_2 = r - y \qquad (9)$$

$$x_o'' = u_o'' = \sum_{j=1}^{3} w_{jo}'x_j = w_{1o}'x_1' + w_{2o}'x_2' + w_{3o}'x_3' \qquad (10)$$

$$= K_p e + K_i \int_0^t e d\tau + K_d \frac{de}{dt} \qquad (11)$$

5 Experimental Setting and Results

The plant model chosen for the experimental evaluation is:

$$P(s) = \frac{(1 - 5s)}{(1 + 10s)(1 + 20s)}$$

This system is a non-minimum phase one with the zero with a positive real part. Moreover it has two small poles, although asymptotically stable, it is really slow. We choose a Linear Time Invariant system to begin with our study although our neural network based methodology should also allow to synthesize controller for non-linear system.

An example of the development workspace in Simulink used to run the experiments is shown in Fig. 2.

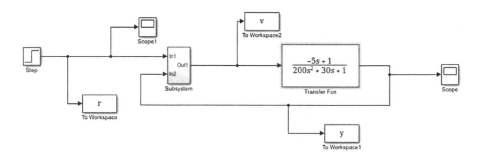

Fig. 2. Simulation workspace

This is a method for off-line auto-tuning the controller based on repeated trials to improve the controller's performances. During the trials the convergence of the controller is not guaranteed hence it would be impossible to perform the trials on a real plant. In Fig. 3, the architecture of the NN based PID used in our experiments is reported. It will be included in the architecture of Fig. 2 in the subsystem box.

In the first set of experiments, the error back-propagation algorithm, written in Matlab, changes the weights of the links initialized with fixed values. The initial weight values come out directly from the second techniques of Ziegler-Nichols which has provides as initial values: $Kp = 3.53$, $Ki = 0.2101$, $Kd =$

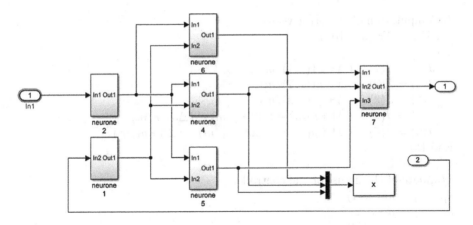

Fig. 3. PIDNN implemented in Simulink. PIDNN is implemented in the subsystem box of Fig. 2.

14.8260 for the weights on the input edges of the output neurons (5, 6, 7) while leaving the initial values of 1 and -1 for the others. The algorithm follows.

// Global variables
// m is the current sample of the signals
// n_{i1} learning rate for the first layer
// n_i learning rate for the second layer

//Definitions of the input parameters
//r, the reference input of the plant
//y, the real output of the closed loop system
//v, the control input of the plant
//x, the output vector of the PID neurons

$BackPropagation(r, y, v, x, K)$
    $J = 0$ // vector of upgrades set initially to zero
    $y_1 = y(m + 1)$ // the following output
    $y_2 = y(m)$ // the current output
    $v_1 = v(m - 1)$ // the previous control input
    $v_2 = v(m)$ // the current control input
    $x1 = x(m)$ // the vector of the output signal at current time of the
                  three neurons
    $x2 = x(m + 1)$ // the vector of the output signal at next time of the three
                  neurons
    $\Delta = 2 * (r - y_1) * (y_2 - y_1)/(v_1 - v_2)$
    $\Delta 1 = \Delta * k_p * (x2(1) - x(1))/((r - y_2) - (r - y_1))$
    $\Delta 2 = \Delta * k_i * (x2(2) - x(2))/((r - y_2) - (r - y_1))$
    $\Delta 3 = \Delta * k_d * (x2(3) - x(3))/((r - y_2) - (r - y_1))$

```
//Computation of the error vector
For Each i From 0 To m
Do
```

$$J(1) = J(1) - (\Delta * x1(1,1)/m)$$
$$J(2) = J(2) - (\Delta * x1(1,2)/m)$$
$$J(3) = J(3) - (\Delta * x1(1,3)/m)$$
$$J(4) = J(4) - (\Delta 1 * r/m) - (\Delta 2 * r/m) - (\Delta 3 * r/m)$$
$$J(5) = J(5) - (\Delta 1 * y_1/m) - (\Delta 2 * y_1/m) - (\Delta 3 * y_1/m)$$

```
End Do
```

```
//update of the gains for the controller
```
$$k_p = k_p - ni * J(1)$$
$$k_i = k_i - ni * J(2)$$
$$k_d = k_d - ni * J(3)$$
$$k_j 1(1) = k_j 1(1) - ni1 * J(4)$$
$$k_j 2(1) = k_j 2(1) - ni1 * J(5)$$

$$K = [k_p, k_i, k_d, k_j 1(1), k_j 2(1)]$$
$$m = m + 1$$
$$return(K)$$

The proposed algorithm monitors the output of the system, the output of the controller, and of the individual neurons comparing the result with the reference input. Then the algorithm learns by sampling these signals whose number depends on the sampling rate. The learning rate ni and ni1 are at the core of this procedure because they establish the learning rate of the PIDNN separately for the input and output neurons.

The effect of choosing the values for the learning rates will be explored in a future work, yet we already have observed that a suitable choice can produce a robust controller, or a performant one, and also that out of some ranges the PIDNN would not converge thus producing a controller causing instability in the system.

During the preliminary experimentation, we also learned that one factor affecting the instability in the system/controller is due to an high sampling rate. It appears in fact that an high sampling rate would produce a signal sampling whose beginning portion bears no revelance with the end of it.

For the experimentation performed, we selected a sampling rate of 0.1 seconds and we utilized the above algorithm in parallel on different portions of the signal curve and we averaged the gains obtained to produce the final ones. The following code show how we proceeded.

initialization of the PID gain
$K = [K_p, K_i, K_d, 1, -1]$

acquiring signal from the simulation

m = 0;
for each j in range of 1 and $NUMBER\_OF\_SCAN$
do

for each i in range of 1 and $NUMBER\_OF\_ITERATIONS/5$
do

$m = i$
$K1 = BackPropagation(r, y, v, x, K(1), K(2), K(3), K(4), K(5))$

$m = NUMBER\_OF\_ITERATIONS/5 + i$
$K2 = BackPropagation(r, y, v, x, K(1), K(2), K(3), K(4), K(5))$

$m = 2 * NUMBER\_OF\_ITERATIONS/5 + i$
$K3 = BackPropagation(r, y, v, x, K(1), K(2), K(3), K(4), K(5))$

$m = 3 * NUMBER\_OF\_ITERATIONS/5 + i$
$K4 = BackPropagation(r, y, v, x, K(1), K(2), K(3), K(4), K(5))$

$m = 4 * NUMBER\_OF\_ITERATIONS/5 + i$
$K5 = BackPropagation(r, y, v, x, K(1), K(2), K(3), K(4), K(5))$

$K(1) = (K1(1) + K2(1) + K3(1) + K4(1) + K5(1))/5$
$K(2) = (K1(2) + K2(2) + K3(2) + K4(2) + K5(2))/5$
$K(3) = (K1(3) + K2(3) + K3(3) + K4(3) + K5(3))/5$

$K(4) = (K1(4) + K2(4) + K3(4) + K4(4) + K5(4))/5$
$K(5) = (K1(5) + K2(5) + K3(5) + K4(5) + K5(5))/5$

$k_1(i) = K(1)$
$k_2(i) = K(2)$
$k_3(i) = K(3)$
$k_4(i) = K(4)$
$k_5(i) = K(5)$

// starting simulation and acquisition of new singals
end
$f = f + 1$
end

The repetition of the algorithm on a singular sampling at time proved ineffi-cient, therefore it has been thought to parallelize the action of the algorithm to five samplings of the signal. In that way each piece of the signal gives a greater contribution to the definition of the final gains.

The final result shows significant improvements with respect to the perfor-mances observed using the ZN and ES methods with a settling time reduced by 60 and 20 s respectively, Fig. 4.

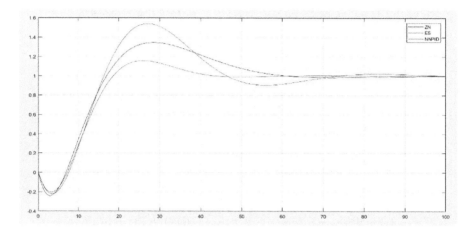

Fig. 4. Compared response for the PIDNN, ZN, and ES techniques

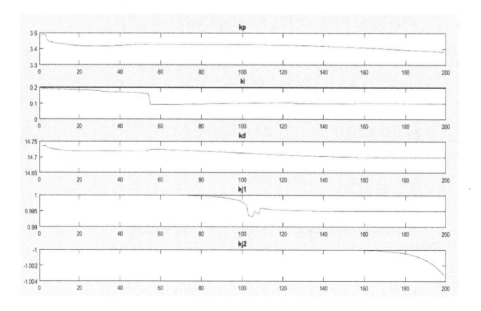

Fig. 5. Learning rate of our PIDNN network

As it's can be seen form the outputs reported, the solution found also improves the undershoot of a few percentage points and, above all, the overshoot that records 30% and 10% less than the first two methods. In order to train the network, 1000 as NUMBER_OF_ITERATIONS was used, after which a widespread overfitting phenomenon was appearing, so it was useless to continue. The learning curves used to evaluate the velocity of convergence for the learning rates are reported in Fig. 5.

Where kp, ki, and kd are the gains for the controller while kj1 and kj2 are the weights of the forward and feeback edges respectively.

6 Conclusions

The goal of this work is to explore a new method PIDNN for the tuning of PID controllers based on neural networks. We investigate both an architectural choice for the NN-based controller and we study its performances in learning the controller's gains in a given setting. The empirical results shows that the PIDNN method is very promising in term of a faster learning of the controller gains with respect to more traditional methods such as ZN and ES. We report in details the algorithms we used as well as we show the behavior of the controller's gains during the learning of PIDNN. As future works we will consider to expand the set of cases where PIDNN could be applied such as the investigation of the modeling of non-linear and Multiple Input - Multiple Output systems. Also we will consider how different levels of noise and disturbances will affect the stability and the convergence of PIDNN in a given amount of time. Another research direction that we will pursue is to understand how alternative artificial intelligence techniques may be used, like agent based modeling, in modeling a plant controller. Agent based modeling has in fact been successfully used to model time series [13, 16-22] and then might be used to model the behavior of a PI controller over time. Also in terms of optimization of the PI controller, one may explore how meta-learning and hyper-heuristics [23, 24] could be used to improve the neural network learning in PIDNN.

References

1. Huailin Shu, Y.P.: Decoupled temperature control system based on PID neural network. In: ACSE 05 Conference, CICC, Cairo, Egypt, 19–21 December 2005 (2005)
2. Ziegler, J.G., Nichols, N.B.: Optimum settings for automatic controllers. Trans. ASME **64**, 759–768 (1942)
3. Boubertakh, H., Tadjine, M., Glorennec, P.Y., Labiod, S.: Tuning fuzzy PD and PI controllers using reinforcement learning. ISA Trans. **49**, 543–551 (2010)
4. Carlucho, I., Paula, M.D., Villar, S.A., Acosta, G.G.: Incremental Q-learning strategy for adaptive pid control of mobile robots. Expert Syst. Appl. **80**, 183–199 (2017)

5. Zhang, J., Wang, N., Wang, S.: A developed method of tuning PID controllers with fuzzy rules for integrating processes. In: Proceeding of the 2004 American Control Conference Boston, Massachusetts, 30 June -2 July 2 (2004)
6. Kim, J.S., Kim, J.H., Park, J.M., Park, S.M., Choe, W.Y., Heo, H.: Auto tuning PID controller based on improved genetic algorithm for reverse osmosis plant. Eng. Technol. Int. J. Comput. Electr. Autom. Control Inf. Eng. **211** (2008)
7. Salem, A., Hassan, M.A.M., Ammar, M.E.: Tuning PID controllers using artificial intelligence techniques applied to DC-motor and AVR system. Asian J. Eng. Technol. **22** (2014). ISSN 2321–2462
8. Muderrisoglu, K., Arisoy, D.O., Ahan, A.O., Akdogan, E.: PID parameters prediction using neural network for a linear quarter car suspension control. Int. J. Intell. Syst. Appl. Eng. (2014)
9. Scott, G.M., Shavlik, J.W., Ray, W.H.: Refining PID controllers using neural networks. National Science Foundation Graduate Fellowship, pp. 555–562 (1994)
10. Shen, J.C.: Fuzzy neural networks for tuning PID controller for plants with underdamped responses. IEEE Trans. Fuzzy Syst. **9**(2), 333–342 (2001)
11. Killingsworth, N., Krstic, M.: PID tuning using extremum seeking: online, model-free performance optimization. IEEE Control Syst. **26**, 70–79 (2006)
12. Papoutsidakis, M., Piromalis, D., Neri, F., Camilleri, M.: Intelligent algorithms based on data processing for modular robotic vehicles control. WSEAS Trans. Syst. **13**, 242–251 (2014)
13. Neri, F.: PIRR: a methodology for distributed network management in mobile networks. WSEAS Trans. Inf. Sci. Appl. **5**, 306–311 (2008)
14. Draper, C., Li, Y.: Principles of optimalizing control systems and an application to the internal combustion engine. Optimal and Selfoptimizing Control (1951)
15. Rumelhart, D.E., Widrow, B., Lehr, M.A.: The basic ideas in neural networks. Commun. ACM **37**, 87–92 (1994)
16. Neri, F.: Learning and predicting financial time series by combining natural computation and agent simulation. In: Di Chio, C., et al. (eds.) EvoApplications 2011. LNCS, vol. 6625, pp. 111–119. Springer, Heidelberg (2011). https://doi.org/10.1007/978-3-642-20520-0_12
17. Neri, F.: A comparative study of a financial agent based simulator across learning scenarios. In: Cao, L., Bazzan, A.L.C., Symeonidis, A.L., Gorodetsky, V.I., Weiss, G., Yu, P.S. (eds.) ADMI 2011. LNCS (LNAI), vol. 7103, pp. 86–97. Springer, Heidelberg (2012). https://doi.org/10.1007/978-3-642-27609-5_7
18. Staines, A., Neri, F.: A matrix transition oriented net for modeling distributed complex computer and communication systems. WSEAS Trans. Syst. **13**, 12–22 (2014)
19. Neri, F.: Agent-based modeling under partial and full knowledge learning settings to simulate financial markets. AI Commun. **25**, 295–304 (2012)
20. Neri, F.: Case study on modeling the silver and nasdaq financial time series with simulated annealing. In: Rocha, Á., Adeli, H., Reis, L.P., Costanzo, S. (eds.) WorldCIST 2018. AISC, vol. 746, pp. 755–763. Springer, Cham (2018). https://doi.org/10.1007/978-3-319-77712-2_71
21. Neri, F.: Combining machine learning and agent based modeling for gold price prediction. In Cagnoni, S. (ed.) WIVACE 2018, Workshop on Artificial Life and Evolutionary Computation, vol. tbd. Springer (2018, in press)
22. Neri, F.: Can agent based models capture the complexity of financial market behavior. In: 42nd Annual Meeting of the AMASES Association for Mathematics Applied to Social and Economic Sciences, Napoli. University of Naples and Parthenope University Press (2018, in press)

23. Burke, E.K., Hyde, M., Kendall, G., Ochoa, G., Ozcan, E., Woodward, J.: A classification of hyper-heuristics approaches. In: Gendreau, M., Potvin, J.Y. (eds.) Handbook of Metaheuristics. International Series in Operations Research and Management Science, vol. 146, pp. 449–468. Springer, Heidelberg (2009). https://doi.org/10.1007/978-1-4419-1665-5_15. In press

24. Camilleri, M., Neri, F., Papoutsidakis, M.: An algorithmic approach to parameter selection in machine learning using meta-optimization techniques. WSEAS Trans. Syst. **13**, 203–212 (2014)

Repeatable Trust Game – Preliminary Experimental Results

Anna Motylska-Kuzma[1]([✉]) [iD], Jacek Mercik[1] [iD],
and Aleksandra Sus[2] [iD]

[1] WSB University in Wroclaw, Wroclaw, Poland
{anna.motylska-kuzma, jacek.mercik}@wsb.wroclaw.pl
[2] Wroclaw University of Economics, Wroclaw, Poland
aleksandra.sus@ue.wroc.pl

Abstract. This paper discusses the so-called trust game between two players. Player 1 is given some amount of money, which he can transfer in some part or whole to Player 2. The value of the money he transfers will be multiplied by multiplier. Player 2 then decides how much money to transfer back to Player 1. Using computer aided system in our schematic in contrast to the previous research devoted to such games, it is possible (and intentionally it was done) to repeat the matches of the same player pairs with both the same and changed values of general game parameters, i.e. basic amount of money, as well as multiplier. Note also that it is advisable to change roles in individual pairs, i.e. to be the first or the second player, so that it is possible to examine players' behaviour in the awareness of the role played in a given game not only at a given moment, but also taking into account possible roles in the future. In the classic approach to the trust game, there was no such possibility

Keywords: Trust game · Computer experiments

1 Introduction

How to develop the new entrepreneurial opportunities is the crucial knowledge in today's economy. The understanding of how they get brought forward is limited. One concept that is starting to gain some attraction involves alertness, the assumption being that entrepreneurs tend to be more alert to possibilities for new entrepreneurial ventures [22]. Alertness is the concept that has the potential to bring us closer to understanding of how new ideas get initiated and pursued.

Kirzner [11] defined alertness as an individual's ability to identify opportunities which are overlooked by others. Hence, the important component of alertness is a judgment which focuses on evaluating the new changes, shifts and information and deciding if they would reflect a business opportunity with profit potentials. Tang et al. [22] suggest that alertness is consisting on three elements: scanning and searching for information, connecting the previous-disparate information, and making evaluation of existence of profitable business opportunities. From the other point of view, social cognition theory suggests the decision and inference process can be improved with training and the appropriate inferential techniques [6]. Thus, alertness represents a

© Springer Nature Switzerland AG 2019
N. T. Nguyen et al. (Eds.): ACIIDS 2019, LNAI 11431, pp. 488–498, 2019.
https://doi.org/10.1007/978-3-030-14799-0_42

capability that can be learned and improved, and may offer guidance to aspiring entrepreneurs in how to mindfully discover opportunities with business potential.

2 Opportunities and Alertness

Opportunity is a conceptual construct close to the chance category. These concepts should not be treated as synonyms, as there are differences between them. A chance can be attributed to a certain probability of success (big or small), while the opportunity is accentuated by the occurrence of a certain event, which is associated with favorable conditions - rather, without grading the potential benefits (see Krupski [12], p. 6). The first is the opportunity, i.e. the situation that may turn out to be a chance or a threat if its potential is noticed in time (or not). In the theory of statistics, chance is a measure of probability, commonly used in games of chance and gambling. Chance is the ratio of the probability of opportunities favorable to the player, the possibility of disadvantage (the chance that we throw out the number of points - 3 in the cube is like 1:5). Opportunities, on the other hand, are certain combinations of circumstances, time and place that can be used to bring benefits, which should also include no loss [13].

Defining the initial assumptions of the game, the authors of this paper assumed that the opportunity is not only earnings, but also no loss, which in the light of considerations is not a threat. Lack of losses is a chance for decision-makers to maintain the existing status quo, which in the future gives him the opportunity to benefit (earn). In general, the chance accentuates the probability of success, and the opportunity does not graduate the potential benefit, which is why the authors decided to adopt such predetermined conditions.

The emergence of new ideas and how they can lead commercializable opportunities are central to the fields of entrepreneurship [3, 17, 20]. Explanations of how the new opportunities emerge include prior experiences, personal disposition, changes in the broader environment, gaining specific information, and being the frustrated user [7, 18, 19, 23]. Furthermore, discover the new opportunities had been linked to personal awareness, skills and insights [9, 10]. From this point of view, alertness is a process and perspective that helps some individuals to be more aware of changes, shifts, opportunities, and overlooked possibilities [22].

The research on the entrepreneurship alertness was initiated by Kirzner [11], who recognized individuals with higher level of awareness of opportunities as group of people who has a special kind of skills that permits recognition of gaps with limited clues. Furthermore, according to McMullen and Shepherd [15], alertness is not entrepreneurial unless it involves judgement and a movement toward action. "To act on the possibility that one is identified an opportunity worth pursuing" is the heard of being entrepreneur.

Some of research argues that the opportunities are discovered or created (see [20]). Another approach parcels it into the three areas of opportunity recognition, opportunity discovery and opportunity creation [16]. In our work we will concern on the recognition and discovery of the opportunity.

3   The Trust Game

The game is based on Berg et al. [4]. Player 1 is given some amount of money (a), which he can transfer in some part or whole to Player 2 (x^1). The value of the money he transfers will be multiplied by multiplier (m). Player 2 then decides how much money to transfer back to Player 1 (x^2). Thus the payoff resulting from the decisions are given by

$$v^1(x^1, x^2) = a - x^1 + x^2 \tag{1}$$

$$v^2(x^1, x^2) = m x^1 - x^2 \tag{2}$$

where $x^1 \epsilon \{0, 1, 2, ..., a\}$, $x^2 \epsilon \{0, 1, 2, ..., m \cdot a\}$, $a > 0$, $m > 0$.

In one-shot game under the classical assumptions of "economic rationality", i.e. each player maximizes his/her expected reward, Player 2 should not return any money to Player 1 and hence Player 1 should not transfer any money.

However, results from experiments indicate that humans do not behave in this way. Their actual behavior in one-shot games results from the level of trust and level of aversion to inequality. Studies have shown also that the amount of transferred money by Player 1 is associated with age and knowledge regarding the responder [1, 8, 21]. According to Markowska – Przybyła and Ramsey [14]:

- when Player 1 is confident that Player 2 will reciprocate, then Player 1 will transfer all his money;
- when Player 1 is sufficiently egalitarian but is not confident enough that Player 2 will reciprocate, then he will transfer some of his funds,
- when Player 1 is neither sufficiently egalitarian, nor confident enough that Player 2 will reciprocate, then he will not transfer any of his funds.

When interactions are repeated, if cooperation brings benefits, then this can give Player 1 an incentive transfer money to Player 2 (even under the assumptions of economic rationality), so that Player 2 will be positively inclined to Player 1 in the future. Information regarding the likelihood of future interactions is important here, e.g. if Player 2 knows that the present interaction will be the last interaction, then he/she has no incentive to return money to Player 1 (under the assumptions of economic rationality).

In our research Player 1 does not know who is the Player 2 and additionally, he does not know if he plays with the same person or not. Thus, Player 1 is not able to determine the probability of reciprocation of Player 2. Hence, when Player 1 is relatively altruistic, he will transfer some of his funds but if the main goal of Player 1 is to achieve as high profit as possible and he is alert to opportunities, then he will not transfer any of his funds.

It has to be clearly emphasized that in contrast to the previous research devoted to such games in our schematic, it is possible (and intentionally it was done) to repeat the matches of the same player pairs with both the same and changed values of general game parameters, i.e. basic amount of money (a), as well as multiplier (m). Note also that it is advisable to change roles in individual pairs, i.e. to be the first or the second player, so that it is possible to examine players' behavior in the awareness of the role played in a given game not only at a given moment, but also taking into account possible roles in the future. In the classic approach to the trust game, there was no such possibility.

Thus, we have two players appearing in pair: (p_i^1, p_i^2) where $i = 1, \ldots, \frac{n}{2}$, and n is the total number of players. The upper index 1 and 2 mean the player starting or ending respectively. j is the index of the round of game, $j = 1, \ldots, k$. In our schemes $k \leq 4$. We set the basic amount a_j and multiplier m_j, which are stable for j-th game.

In j-th game player p_i^1 chooses the value x_j^1. We know about it, that:

1. $0 \leq x_j^1 \leq a_j$
2. x_j^1 is the realisation of a random variable X_j^1 describing the behavior of the first players in j-th game. For a given game, the realization of variable X_j^1 can be determined from observation.

In j-th game, player p_i^2 chooses the value x_j^2. We know about it, that:

1. $0 \leq x_j^2 \leq m_j a_j$
2. x_j^2 is the realization of a random variable X_j^2 describing the behavior of the second players in j-th game. For a given game, the realizations of variable X_j^2 can be determined from observation.

Thus, in the j-th game, the opportunity for the first player in the pair means that

$$x_j^1 \leq x_j^2 \tag{3}$$

The optimal value x_j^{1opt} for the first player from the i-th pair in j-th game is such amount, for which $(x_j^2 - x_j^1)$ achieves maximum. It is worth notice, that Player 1 does not know the value x_j^2. He only knows that it is some value $x_j^2 \in [0, m_j a_j]$.

It is obvious, that the optimal value for the Player 2 in every game is $x_j^{2opt} = 0$. It results in $x_j^{1opt} = 0$.

The only thing, which could stop the Player 2 from using the optimal strategy is the information that the roles can be reversed and the Player 2 will turn in Player 1. Hence he should know the information about the number of games k. In the last play in the given game he uses of course the optimal strategy $x_j^{2opt} = 0$. In other games $(j < k)$ he should assume some reciprocity. We can reproduce these assumptions by examining the distribution of the random variable X_j^2. Correspondingly, by examining the distribution of a random variable X_j^1 we can find the Player 1's image about the Player 2.

The lack of information about the number of games means that $\left(x_j^{1opt}, x_j^{2opt}\right) = (0, 0)$.

The difference between X_j^1 and X_j^2 is a measure of the divergence of mutual ideas about themselves, hence we should study the distribution $(X_j^1 - X_j^2)$, which is possible on the basis of empirical observations. This divergence can be called the "unbelief", "lack of trust", etc.

It could be assumed that both players do not behave strategically, it means they behave independently. In such situation, we count separately X_j^1 and X_j^2 and we wonder why they did not use optimal strategies $\left(x_j^{1opt}, x_j^{2opt}\right) = (0, 0)$.

4 Experimental Design

For the implementation of experiments we have created the appropriate software to support and automate the course of the game. At the beginning, the participants of the game made registration with the provision of personal data: name, surname, education, age, working status (e.g. employee, student, etc.), the employer's data (e.g. sector, size) and the position (i.e. manager or not). Each participant had a unique login and password to enable bidding and viewing the results of previous games. Then, a random selection of player pairs and setting of game parameters was made, i.e. basic amount (a_j) and multiplier (m_j). The course of the game was as follows:

- Player 1 was informed by the system about the conditions of the game. He/she knew the basic amount (a_j), multiplier (m_j), the maximum number of rounds (k). He/she had also the information that the roles will be changed and if the player is first, in the next round he/she could be the second. According the data, he/she could decide to transfer the amount x_j^1 to the Player 2. For a decision the Player 1 had one week, which means that he could seriously think over his/her response. Not all randomly drawn participants of the pairs cooperated with us and performed the first moves appropriate for the Player 1. Such players were gradually eliminated from the system, which meant that the number of players taking part in the competition finally decreased. As part of the whole experiment, new players were not added after the experiment began,
- Player 2 was informed by the system about the decision of Player 1. He/she knew the basic amount (a_j), multiplier (m_j), the maximum number of rounds (k). He/she had also the information that the roles will be changed and if the player is second, in the next round he/she could be the first. Player 2 had also one week to make a decision and transfer of the selected amount x_j^2 to Player 1 from the same pair. It happened that the Player 2 did not cooperate. Then only the value x_j^1 was used in statistical analyzes. An inactive second player was not eliminated from the set of players and could in the future take part in games as both the first and the second player,
- after the end of a given game individual results were saved on the account of each participant of the game. Each of them could at any time view their results and analyze their previous behavior.

Technically speaking, we recorded the following values in the game:

multiplier	value of multiplier (m_j) in j-th game. In this preliminary experiments we chose the tree different values of multiplier: 3, 2.75 and 0.8. The first two are greater than one. First is simply the basic multiplier in the classic trust game. The second one has value between the classic multiplier and the one. Besides that, it has unobvious value. The third one is below the one, what results decrease of the payoff
basic amount	basic amount (a_j) in j-th game, which was chosen randomly but should be not a round number

decision_p1	decision of Player 1 in *j*-th game – the amount transferred to the player 2 $\left(x_j^1\right)$
step1_%	percentage of the initial value transferred to the Player 2
decision_p2	decision of Player 2 in *j*-th game – the amount given back to the player 1 $\left(x_j^2\right)$
step2_%	percentage of the obtained value transferred to the Player 1 (the obtained value is x_j^1 multiplied by the multiplier - m_j)
result_p1	Player 1's score calculated by the system (without multiplier in the amount transferred by the Player 2 to Player 1)
result_p1_ multi	Player 1's score including the multiplier
result_p2	Player 2's score including the amount given back to the Player 1
get%_nor	the difference between what the Player 1 got in the given game and what he gave in the previous game (with the same Player 2). The components of this difference are expressed as a percentage of the points (capital) received by the given player
multiplier_ dyn_value	dynamics of the multiplier in the given game versus the previous game, which was played by the same pair of players (the difference of value)
amount_dyn_ value	dynamics of the basic amount in the given game versus the previous game, which was played by the same pair of players (the difference of value)
effect%_nor	the difference between the decision of Player 1 in the given game and in the previous game (with the same Player 2). The components of this difference are expressed as a percentage of the points (capital) available by Player 1 in the given game
efficiency_p1	Player 1's efficiency, that is the ratio of the player's score for the Player 1 to the funds received
efficiency_p2	Player 2's efficiency, that is the ratio of the player's score for the Player 2 to the funds received (including multiplier)
result_p1_ dyn	a change in the score for the Player 1 in the given game against the result of the previous game (with the same Player 2). The ratio of the result of the given game to the result of the previous game: 1 – no change, >1 – increase in value, <1 – decrease in value

5   Experimental Results

During the experiment, we received the following general (Table 1) and detailed (Table 2) results.

The average decision of Player 1 is about 134 and this is 46% of the basic amount. Such behavior is similar to that observed by Ashraf et al. [2], where there were high concentration of transfers from Player 1 at the level 0%, 50% and 100%. Camerer [5] notes also, that in the range of studies, on average Player 1 transfers around 50% of his money and Player 2 returns 37% of the money received. In our experiment this value is a little bit less, but still it is about 32% for Player 2.

Table 1. Descriptive values of the results of all games

	N	Minimum	Maximum	Mean	Std. Deviation	Skewness	Kurtosis
decision_p1	173	0	328	134.39	102.25	0.4571	−0.8355
step1_%	173	0	1	0.4628	0.3306	0.2041	−1.1265
decision_p2	45	0	630	103.62	148.84	2.0138	3.6660
step2_%	45	0	1	0.3164	0.2963	0.7981	0.0020
result_p1	173	81	630	288.33	112.59	1.0052	1.9122
result_p1_multi	173	0	1890	230.93	257.87	3.6333	17.2660
result_p2	43	0	847	187.87	201.62	1.4615	2.1139
get%_nor	51	−0.7967	0.9177	−0.0447	0.3409	0.1384	0.7483
multiplier_dyn_value	171	0.2667	1	0.8259	0.2896	−1.3066	−0.2341
amount_dyn_value	171	0.6585	1.5185	1.0430	0.2855	0.5146	−0.5899
effect%_nor	173	−1	0.7735	−0.0246	0.2863	−1.2343	3.4461
efficiency_p1	173	0	5.7622	0.7866	0.8315	3.2339	13.3985
efficiency_p2	39	0	1	0.6349	0.2769	−0.7152	0.0066
result_p1_dyn	91	0	29.1696	1.4624	3.1430	8.0234	69.9016

Table 2. Decisions and results for players

	Basic amount	Basic statistics	Multiplier		
			3	2.75	0.8
Analyses of Player 1's decisions (absolute values)	328	Mean	157.39		140.55
		Std.Dev.	115.76		112.44
		Count	51		67
	216	Mean		105.56	
		Std.Dev.		63.65	
		Count		55	
Analyses of Player 1's decisions (relative values in %)	328	Mean	0.48		0.43
		Std.Dev.	0.35		0.34
		Count	51		67
	216	Mean		0.49	
		Std.Dev.		0.29	
		Count		55	
Analyses of Player 2's decisions (absolute values)	328	Mean	154.30		38.55
		Std.Dev.	194.12		64.38
		Count	20		13
	216	Mean		89.63	
		Std.Dev.		95.50	
		Count		12	
Analyses of Player 2's decisions (relative values in %)	328	Mean	0.30		0.27
		Std.Dev.	0.29		0.36
		Count	20		13
	216	Mean		0.39	
		Std.Dev.		0.19	
		Count		12	

(continued)

Table 2. (*continued*)

	Basic amount	Basic statistics	Multiplier		
			3	2.75	0.8
Player 1's results	328	*Mean*	352.14		193.43
		Std.Dev.	405.83		155.73
		Count	51		67
	216	*Mean*		164.21	
		Std.Dev.		155.73	
		Count		55	
Player 2's results	328	*Mean*	285.80		35.78
		Std.Dev.	241.99		53.79
		Count	20		11
	216	*Mean*		164.06	
		Std.Dev.		101.19	
		Count		12	

The average efficiency of Player 1's decision, counted as a ratio of final result of Player 1 and the basic amount is about 80%. Thus, the final result is below the basic amount and can be recognize as a loss. From the point of view of Player 2 the situation is not better or even worse. The Player 2's efficiency is lower that Player 1's and is 63.5%, although he returns less to Player 1 than he obtained (as a percentage), what we can observe in value of variable *get%_nor*.

6 Conclusions and Future Research

The obtained results allow on the following conclusions:

- the average value (in %) of the Player 1's decision does not depend on the multiplier value ($\alpha = 0.05$, one-way ANOVA, p-value = 0.5531)
- the average value (in %) of the Player 1's decision does not depend on the basic amount ($\alpha = 0.05$, one-way ANOVA, p-value = 0.4828)
- the average value (in %) of the Player 2's decision does not depend on the multiplier value ($\alpha = 0.05$, one-way ANOVA, p-value = 0.5743)
- the average value (in %) of the Player 2's decision does not depend on the basic amount ($\alpha = 0.05$, one-way ANOVA, p-value = 0.3190)
- the correlation between the decision of the Player 2 and the decision of the Player 1 (counted in amount) is 0.7061. Observed variation in amount of the Player 2 is explained by the variation of the Player 1 only in 48.7% (p-value = 5.9779E−08). Such low determination suggests existence of another factors influencing Player 2's decisions.

- at the level of α = 0.02, we can conclude that Player 2 is not guided by the values received from Player 1, although he knows how the amount transferred to him is generated
- in both cases (amount and percentage) at the level of α = 0.02, it cannot be excluded that Player 1 is not guided by efficiency in his decisions

From Eq. (3) arises that opportunity for the Player 1 appears when the value of Player 1's decision is less or equal to the value of Player 2's decision. In our case it is not true. The mean value of Player 1's decision is 134.39 whereas the average value of Players 2's decision is 103.62. Hence, we could assume that Player 1 was not noticed the opportunity. Furthermore, if we take into account the optimal value of the Player 1's decision $\left(x_j^{1opt} = 0 \right)$, our experiment shows that Player 1 is far away from this value. Neither Player 1 nor Player 2 does not strive to achieve optimum, although as we noticed in conclusions above, we cannot exclude that Player 1 is not guided by efficiency in his decisions. He certainly does not aspire to the optimum, thus he does not see the opportunity, as well as the Player 2. However, Player 2 is not guided by the values received from Player 1, thereby to the certain extent he is sensitive to opportunity. His alertness is higher than the Player 1's.

It should be noted that our research is only preliminary experiment and do not analyse every possibilities and results. First of all, we checked only one attitude to the "optimal behavior" and the "opportunity". From this point of view, we obtained the results which could be confusing. Maybe, through changing the above definitions and turn more to the "social optimality", could give us better knowledge about the seizing the opportunities and the alertness of our players. Considering the Nash equilibrium behavior, i.e. the behavior that maximizes the sum of the players' payoffs, the results could be more understandable.

Next limitation is the attitude or understanding the basic amount. Our players were informed about number of points, which are given them on the beginning. "Points" are not money, thus the players could behave in another way. Besides that, they did not obtained any material rewards what could significantly influence on their motivations and decisions.

The last but not least was the problem of willingness to consistent participation in the games. We observed that some of our players did not answer if they were a Player 1 although they take part in the games where they were a Player 2 and vice versa. Thus we had to delete the data from such games and could not analyse them. Finally, from the 1025 initial observations we obtained only 173. It is a huge number of wasted data.

For the future research it will be interesting to compare the results of the game within the special target group of players, e.g. managers vs. students, employed vs. unemployed, etc. Especially, it will be interesting to know how behave the entrepreneurs and the students after the special courses dedicated to seizing the opportunities. We plan also compare the results across the countries, which could shed light on cultural elements.

References

1. Akai, K., Netzer, R.J.: Trust and reciprocity among international groups. experimental evidence from Austria and Japan. J. Soc.-Econ. **41**, 266–276 (2012)
2. Ashraf, N., Bohnet, I., Piankov, N.: The composing the trust and trustworthiness. Exp. Econ. **9**, 193–208 (2006)
3. Baron, R.A.: Opportunity recognition as pattern recognition: how entrepreneurs "connect the dots" to identify new business opportunities. Acad. Manag. Perspect. **20**(1), 104–119 (2006)
4. Berg, J., Dickhaut, J., McCabe, K.: Trust, reciprocity and social history. Games Econ. Behav. **10**, 122–142 (1995)
5. Camerer, C.F.: Behavioral Game Theory. Experiments in Strategic Interaction. Princeton University Press, Princeton (2003)
6. Fiske, S.T., Taylor, S.E.: Social Cognition: From Brains to Culture. Sage, Thousand Oaks (2013)
7. Gaglio, C.M., Katz, J.A.: The psychological basis of opportunity identification: entrepreneurial alertness. Small Bus. Econ. **16**(2), 95–111 (2001)
8. Guillen, P., Ji, D.: Trust, discrimination and acculturation. Experimental evidence on Asian international and Australian domestic university students. J. Soc.-Econ. **40**, 594–608 (2001)
9. Kaish, S., Gilad, B.: Characteristics of opportunities search of entrepreneurs versus executives: sources, interests, general alertness. J. Bus. Ventur. **6**(1), 45–61 (1991)
10. Kirzner, I.M.: Creativity and/or alertness: a reconsideration of the Schumpeterian entrepreneur. Rev. Austrian Econ. **11**(1–2), 5–17 (1999)
11. Kirzner, I.M.: Perception, Opportunity, and Profit. University of Chicago Press, Chicago (1979)
12. Krupski, R.: Rodzaje okazji w teorii i praktyce zarządzania (Types of opportunities in management theory and practice). Prace Naukowe Wałbrzyskiej Wyższej Szkoły Zarządzania i Przedsiębiorczości, t.21, Wałbrzych, Polska (2013)
13. Krupski, R.: Elementy koncepcji zarządzania okazją w organizacji (Elements of the opportunity management concept in the organization), [in:] Dynamika zarządzania organizacjami. Paradygmaty – metody – zastosowania. Prace Naukowe Akademii Ekonomicznej w Katowicach, Katowice, Polska (2007)
14. Markowska-Przybyła, U., Ramsey, D.: A game theoretical study of generalised trust and reciprocation in Poland. I. Theory and experimental design. Oper. Res. Dec. **24**(3), 59–76 (2014)
15. McMullen, J.S., Shepherd, D.A.: Entrepreneurial action and the role of uncertainty in the theory of the entrepreneur. Acad. Manag. Rev. **31**(1), 132–152 (2006)
16. Sundaramurthy, C., Lewis, M.: Control and collaboration: paradoxes of governance. Acad. Manag. Rev. **28**, 397–415 (2003)
17. Shane, S., Venkataraman, S.: The promise of entrepreneurship as a field of research. Acad. Manag. Rev. **25**(1), 217–226 (2000)
18. Shane, S.: Prior knowledge and the discovery of entrepreneurial opportunities. Organ. Sci. **11**(4), 448–469 (2000)
19. Shepherd, D.A., McMullen, J.S., Jennings, P.D.: The formation of opportunity beliefs: overcoming ignorance and reducing doubt. Strat. Entrepreneurship J. **1**(1–2), 75–95 (2007)

20. Short, J.C., Ketchen Jr., D.J., Shook, C.L., Ireland, R.D.: The concept of "opportunity" in entrepreneurship research: past accomplishments and future challenges. J. Manag. 36(1), 40–65 (2010)
21. Slonim, R., Garbarino, E.: Increases in trust and altruism from partner selection. Experimental evidence. Exp. Econ. 11, 134–153 (2008)
22. Tang, J., Kacpar, K.M., Busenitz, L.: Entrepreneurial alertness in the pursuit of new opportunities. J. Bus. Ventur. 27(1), 77–94 (2012)
23. Tripsas, M.: Customer preference discontinuities: a trigger for radical technological change. Manag. Decis. Econ. 29(2–3), 79–97 (2008)

Modeling of Uncertainty with Petri Nets

Michal Kuchárik and Zoltán Balogh[✉]

Department of Informatics, Faculty of Natural Sciences,
Constantine the Philosopher University in Nitra,
Tr. A. Hlinku 1, 949 74 Nitra, Slovakia
{mkucharik, zbalogh}@ukf.sk

Abstract. This paper deals with the idea of calculating probabilities and percentages with Petri nets. Uncertainty can be expressed with Petri nets as one place with multiple output transitions. In that case, the transition that fires are selected randomly while each transition has the same chance to fire. This paper presents the idea of assigning a weight to a transition that will be used to modify the chance at which a concurrent transition can fire. Higher weight increases the chance of firing when an uncertain situation occurs. We want to later use this to simulate university students chance to successfully complete a university course.

Keywords: Petri nets · Fuzzy logic · Simulation · Petri net tool · Transitions · Probability · Concurrency

1 Introduction

Petri Nets are a graphical tool for the formal description of the flow of activities in complex systems. With respect to other more popular techniques of graphical system representation, Petri Nets are particularly suited to represent in natural way logical interactions among parts or activities in a system. Typical situations that can be modeled by Petri Nets are synchronization, sequentially, concurrency and conflict. Petri net was first proposed by Carl Adam Petri in his PhD thesis "Kommunikation mit Automaten" (Communication with Automata) in 1962. Petri nets have become an important computational paradigm to represent and analyze a broad class of systems. As a computational paradigm for intelligent systems, they provide a graphical language to visualize, communicate and interpret engineering problems [1, 2]. One of the major advantages of using Petri net models is that the same model is used for the analysis of behavioural properties and performance evaluation, as well as for the systematic construction of discrete-event simulators and controllers [3, 4].

Petri net is a bipartite graph composed of places and transitions. Places and transitions are connected by weighted arcs. Every place has a capacity and can hold a non-negative number of tokens.

Petri nets are defined as PN = {P, T, A, W, M0} where P = {p1, p2, ..., pm} is finite set of places. T = {t1, t2, ... tn} is a finite set of transitions. $F \subseteq (P \times T) \cup (T \times P)$ is a finite set of arcs. W is a weight function and M0 is initial marking [5]. Marking assigns a non-negative number of tokens to a place P. Each simulation step active

© Springer Nature Switzerland AG 2019
N. T. Nguyen et al. (Eds.): ACIIDS 2019, LNAI 11431, pp. 499–509, 2019.
https://doi.org/10.1007/978-3-030-14799-0_43

transitions may or may not fire. If all conditions for the firing of a transition are fulfilled transition becomes active.

Depending on arc type, arc weight and arc direction, an active transition that fires will remove or add a number of tokens to its input and output places. When one transition fires, some other transition may become in-active due to their common conditions. In the case where multiple transitions are active and firing of one will disable other transitions, the transition that fires first is selected randomly. Each active transition has an equally high chance to fire first. This is something we want to change by introducing weighted transitions. When multiple transitions are active, the ones with higher weight will have a higher chance to fire first. This should be useful when simulating uncertain concurrent situations where the chance for the individual condition is known [6, 7].

In this case, however, we still need to be able to compute the probability of the final state.

The application of Petri nets to degradation models is a recent research field, but this modeling technique has shown several advantages relative to the more traditional Markov chains. The graphical representation can be used to describe the problem in an intuitive way; PN are more flexible than the Markov chains, allowing the incorporation of a multitude of rules in the model to accurately simulate complex situations and keeping the model size within manageable limits. Moreover, with this modeling technique, transition times are not required to be exponential distributed [8].

Article [9] deals with Probabilistic Time Petri nets, but does not deal with the calculation of state probability. Paper [10] deals with concurrency, and probability but does not present any way of calculating state probability.

Paper [11] states hat Petri networks have been very useful for modelling processes for studying synchronization behaviours and properties such as deadlock and starvation. Time is implicitly represented and there is no notion of the duration of a task that is being modeled, which hinders its utility for broad classes of problems, such as planning and scheduling, in which duration of tasks and time play a major role [12]. This paper dealt with timed Petri nets, but again does not present a way of calculating probabilities.

Paper [13] dealt with similar issue however their approach to computing state probabilities was different.

As stated in [14] Petri nets can be used to simulate educational processes. Book [6] also shows how Petri nets with fuzzy logic can be used to calculate the final grade of students based on their individual nature. In an example in this book a Petri net model is presented where the student has different characteristics, learning time and methods. The result of this model and simulation is the student's final grade. However, this model is only an example as in our previous work we tried to simulate this model using the application TransPlaceSim and the outcome that we got indicated that not all character trait combinations yield a proper result. We assume hat a proper model like this could be used to calculate the probability of a student's final grade based on student's personality traits and more important, based on the university course itself.

2 Materials and Methods

In this part of the article, we deal with weighted transitions. If we at this moment want to simulate higher chance of an event to occur, with Petri nets we can do by adding more transitions between places with a higher chance. This is demonstrated by Fig. 1.

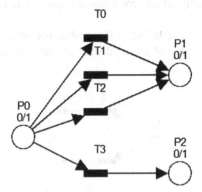

Fig. 1. Petri net model with random choices

When we add weights to transitions we can exchange the first 3 transitions with one that has the weight of 3. With this, there is a chance 3:1 that the token will be transferred to P1 [15–17]. This should prove useful when we try to compute the final probability of long chains of events.

These weights are used in case of conflicts in stochastic Petri nets, in a case when it is possible to activate multiple transitions at once but the activation of one transition will disable the activation of other transitions. With weighted transitions, a model from Fig. 1 can be changed into a model shown in Fig. 2.

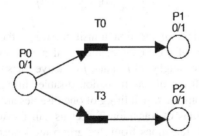

Fig. 2. Petri net model with weighted transitions

In Fig. 2 transition T0 has the weight of 3 and transition T3 has the weight of 1. The probability of token going from P0 to P1 is the same for both models. Ability to assign weights to transitions allows us to create models where a situation can occur with a different probability [18]. In the case of a simple decision, the probability of the result can be calculated by dividing the weight of the firing transition by the sum of all concurrent transitions:

$$p = \frac{w0}{\sum_{i=0}^{n} wi} \tag{1}$$

Where $w0$ is the weight of current transition and wi is the weight if i-th active transition.

In case of concurrent models Fig. 3 the probability of final state depends on the combination of the probability of the final state of each concurrent model.

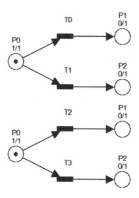

Fig. 3. Petri net model with multiple concurrent parts

In case of concurrent models, Fig. 3 is the activation probability of two indepen-dent transitions is calculated as follows:

$$p = \frac{w0 * w3}{\sum_{i=0}^{n} w0i * \sum_{j=0}^{m} w3j} \tag{2}$$

Where $w0$ is the weight of the first transition, $w3$ is the weight of the second transition such that firing of the first transition will not disable firing of the second transition, $w0i$ and $w3j$ are weights of i-th and j-th active transition from the first group of transitions where the firing of one transition disables firing of all other transitions from that group. Transitions for which firing of one disables all other transitions will be called a group of transitions of transition T. Let us call the weight of this transition w. Let us call weights of transitions from this group ws. Equations from the previous example can be generalized to the form:

$$p = \prod_{i=0}^{n} \frac{wi}{\sum_{j=0}^{m} wij} \tag{3}$$

Where p is the probability of state to which the model can get from the current state through activation of specific transition, wi is the weight of a transition from one group, that fired in this step and wij is the weight of transitions from a group of transition Ti.

This equation works only in the case when models are concurrent and independent. If these models are connected it is necessary to use different equations. The problem in this case is the increasing number of possible combinations of active transitions. Example from Fig. 4 presents 2 different equations for calculating probabilities of final states.

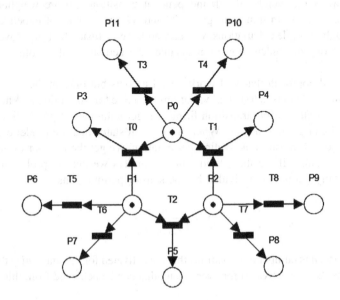

Fig. 4. Petri net model with mutually exclusive choices

For the case from Fig. 4, it is necessary to consider 2 different outcomes: only independent transitions are activated (T3, T4, T5, T6, T7, T8), one dependent transition (T0, T1, T2) and 2 independent transitions are activated. The dependent transition is a transition that has more than 1 input place. Independent transition is transitions that have only 1 input place.

Let us define s as the sum of weights of all active transitions from the current step.

$$s = \sum_{i=0}^{n} wi \tag{4}$$

Then the activation probability of 1 dependent and 2 independent transitions, for example transitions T0 and T7 or T8 is calculated as:

$$p = \frac{wn}{s} * \frac{wi}{\sum_{k=0}^{n} wik} \tag{5}$$

Where wn is the weight of a dependent transition, wi is the weight of independent transition and wik are weights from the group of transition Ti.
Activation probability of all independent transitions can be calculated as:

$$p = \prod_{i=0}^{n} \frac{wi}{\sum_{j=0}^{m} wij} * \frac{s - \sum_{k=0}^{o} wk}{s} \qquad (6)$$

Where wi is the weight of i-th independent transition, wij are weights of independent transitions from transition group Ti and wk are weights of dependent transitions. Through multiple calculations, we reached the conclusion that Eq. 7 works when activation of one dependent transition disables the activation of all other dependent transitions.

The second approach that will work for all models but only in the case when we ignore transition weights is to get all variations of active transition firing. When we take permutations of all possible transition firing, we get a number of possibilities through which model can get into a state. When we take all states that the model can get into and the sum of all transition activation permutation, we get the number of all possible transition activation. If we then divide these numbers we get the probability of that state. Or simply permutation divided by the sum of permutations:

$$p = \frac{n!}{\left(\sum_{i=0}^{m} ni!\right)} \qquad (7)$$

In this case, n is number of transitions that are activated for this state, ni is the number of transitions that are activated for every state, that can be achieved from this state.

3   Model of E-Course in Learning Management System

We decided to use weighted transitions in a model of the new e-course for learning management system (LMS). We want to model student choices in this e-course and then, based on his decisions, use Fuzzy logic to determine his final grade. Figure 5 shows the entire model in Petri nets.

Student choices for a single part of the e-course (single lecture) are shown in Fig. 6. A student can only access a part of the e-course if he accessed the previous part. This happens to force students to work during seminars. Once student visits the lecture, he can visit it repeatedly in order to complete all assignments for the lecture. Once a student finishes these assignments, he can attempt the lecture test that he otherwise could not attempt. Completing assignments and lecture test will add to student's knowledge. Students are not going to be forced to complete all assignments and tests in order to attempt the final exam. This is because the course, on which we apply this model is theoretical, and the final exam consists of theoretical questions only. Therefore it shouldn't really matter whether the student finishes assignments and tests from lectures or not. What matters is whether the student learns for the final exam. On the other hand, we believe that the work on assignments will help the student to better understand the theory of this course and therefore increases his chance to successfully complete the final exam.

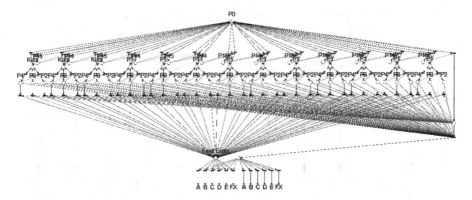

Fig. 5. Petri net model of the entire e-course

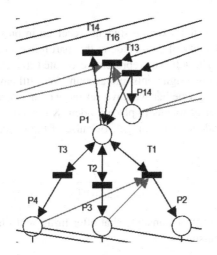

Fig. 6. Single lecture from e-course

In this part of our model, P3 and P4 represent study materials in LMS and P2 represents course test. Only when students use course learning materials they are allowed to take that courses test. P1 represents students visit this specific course. In the last part of our model we have an evaluation of students' choices and their impact on their final grade. Similar to the paper [13], where authors used Fuzzy logic to predict student final grade based on its attributes, we will be using Fuzzy logic interpreted through weighted transitions to predict students final grade based on knowledge he earned from our e-course. Figure 7 shows the simplified version of the final part from a model from Fig. 5.

A place called Student Knowledge can contain from 0–40 tokens, depending on the assignments student completed during one semester, 40 in total. This part of our model contains two different grading methods. The one on the right is based on the current system that is independent of student knowledge. Transitions T6–T11 have weight

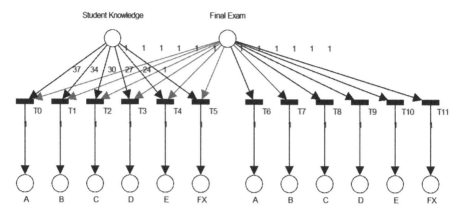

Fig. 7. Impact of student's choices on his final grade

modified based on grades that existing students received in previous years. Thanks to these weights, the probability of final grade in this part reflects the probability students had in previous years. The left part represents estimated grades students will receive when taking their work on assignments into account. We still have to test this model to see how accurate it is and whether we have set the weights correctly.

Based on the number of tokens in place Student Knowledge we can determine the probability of every grade student can get. Because of the structure of this model, we will be using Eq. 1:

$$p = \frac{wj}{\sum_{i=0}^{n} wi} \qquad (8)$$

Where wj is the weight of transition that after firing will fill the place and wi are weights of all active transitions.

3.1   Experimental Values

We decided to make an experiment for this part of our model. Based on the construction of our course from previous years, based on the construction of the final exam for previous years and based on discussion with students from previous years we can assume that the activity done during courses had no impact on students final grade, in other words, grades that students received from the final exam were based solely on their ability to memorize definitions, not on our e-course. That is why we decided to use a number of students as weights for transitions T6–T11 from Fig. 7.

We decided to make an experiment for specific knowledge ranges. These ranges were selected based on the standard scoring system for a final exam where the grade is given for percentages and for student knowledge these values will be (Table 1):

- A – 40 – 37
- B – 36 – 34
- C – 33 – 30

- D – 29 – 27
- E – 26 – 24
- FX – 23 – 0

Table 1. Percentages and student knowledge values

A	40	100%
	37	93%
B	36	90%
	34	85%
C	33	83%
	30	75%
D	29	73%
	27	68%
E	26	65%
	24	60%
FX	23	58%
	0	0%

We have to correctly set weights for transitions T0–T11. For this experiment, we want to assume that completing assignments really increase student's knowledge and their final grade will correspond to their knowledge. However, we still want to leave the possibility that even if they completed a certain number of assignments their final grade can still be lower than what we could expect due to the number of factors.

For this experiment, we will set the weights of transitions T6–T11 to 1 and the weights of transitions T0–T5 will go down from 10 to 5. In case of the model from Fig. 7, we will need to sum Eq. 8 for all grade pairs, Eq. 9.

$$p = \frac{wj}{\sum_{i=0}^{n} wi} + \frac{wk}{\sum_{i=0}^{n} wi} \qquad (9)$$

The following table shows probabilities for grades based on a number of tokens in the place Student Knowledge (Table 2):

Table 2. Probabilities for final exam grades based on student knowledge

	A	B	C	D	E	FX	Verification
40-37	0,215686	0,196078	0,176471	0,156863	0,137255	0,117647	1
36-34	0,02439	0,243902	0,219512	0,195122	0,170732	0,146341	1
33-30	0,03125	0,03125	0,28125	0,25	0,21875	0,1875	1
29-27	0,041667	0,041667	0,041667	0,333333	0,291667	0,25	1
26-24	0,058824	0,058824	0,058824	0,058824	0,411765	0,352941	1
23-0	0,090909	0,090909	0,090909	0,090909	0,090909	0,545455	1
	0,077121	0,110438	0,144772	0,180842	0,22018	0,266647	1

In this table, we can see the probability for each grade based on the student's knowledge. Verification column is used to make sure the sum of probabilities for every row equals 1.

4   Discussion

Adding weights to Petri net transitions can be used to modify their firing probability in concurrent situations. The example used in this paper demonstrates how to use these weights in order to make a model that will help us to compute the final probability of a different concurrent event. In the model from Fig. 7 we had to assign correct weights to individual transitions. Different weights will mean different results in the table from Table 2. The correctness of proposed equations can be verified simply by the fact that no matter what weights we assign to these transitions, the verification column will always be 1. Weights used in our example are based on our assumptions based on student's results from previous years. However, the results that we obtain this way do not reflect real student grades. In the future, we plan to develop a method to set exact weights to these transitions, so that we can make a model that would represent real students. Weighted transitions in Petri nets can be used in situations where it is necessary to model probability, concurrency and mutually exclusive events.

5   Conclusion

In this article, we talked about the possibility to add weights to Petri net transitions to modify their firing probability in concurrent situations. We demonstrated how this could be useful when modeling student behavior in an e-course. This example proves that equation we propose work for this type of model and is useful for this and similar situations. In the future, we plan to use this method to simulate student's probability to successfully complete a Moodle e-course.

Acknowledgements. This research has been supported by University Grant Agency under the contract No. VII/12/2018 and KEGA 036UKF-4/2019.

References

1. Suraj, Z., Grochowalski, P., Bandyopadhyay, S.: Flexible generalized fuzzy petri nets for rule-based systems. In: Martín-Vide, C., Mizuki, T., Vega-Rodríguez, M.A. (eds.) TPNC 2016. LNCS, vol. 10071, pp. 196–207. Springer, Cham (2016). https://doi.org/10.1007/978-3-319-49001-4_16
2. Bandyopadhyay, S., Suraj, Z., Grochowalski, P.: Modified generalized weighted fuzzy petri net in intuitionistic fuzzy environment. In: Flores, V., et al. (eds.) IJCRS 2016. LNCS (LNAI), vol. 9920, pp. 342–351. Springer, Cham (2016). https://doi.org/10.1007/978-3-319-47160-0_31
3. Zurawski, R., Zhou, M.C.: Petri nets and industrial applications: a tutorial. IEEE Trans. Industr. Electron. **41**(6), 567–583 (1994). https://doi.org/10.1109/41.334574

4. Suraj, Z., Bandyopadhyay, S.: Generalized weighted fuzzy petri net in intuitionistic fuzzy environment. In: 2016 IEEE International Conference on Fuzzy Systems, FUZZ-IEEE 2016, pp. 2385–2392 (2016)

5. Balogh, Z., Turčáni, M.: Possibilities of modelling web-based education using IF-THEN rules and fuzzy petri nets in LMS. In: Abd Manaf, A., Zeki, A., Zamani, M., Chuprat, S., El-Qawasme, E. (eds.) ICIEIS 2011. CCIS, vol. 251, pp. 93–106. Springer, Heidelberg (2011). https://doi.org/10.1007/978-3-642-25327-0_9

6. Klimeš, C., Balogh, Z.: Modelovanie procesov pomocou Petriho sietí. Univerzita Konštantína Filozofa v Nitre, Nitra (2012)

7. Zhu, D., Tan, H., Yao, S.: Petri nets-based method to elicit component-interaction related safety requirements in safety-critical systems. Comput. Electr. Eng. **71**, 162–172 (2018). https://doi.org/10.1016/j.compeleceng.2018.07.019

8. Ferreira, C., Canhoto Neves, L., Silva, A., de Brito, J.: Stochastic Petri net-based modelling of the durability of renderings. Autom. Constr. **87**, 96–105 (2018). https://doi.org/10.1016/j.autcon.2017.12.007

9. Emzivat, Y., Delahaye, B., Lime, D., Roux, O.H.: Probabilistic time petri nets. In: Kordon, F., Moldt, D. (eds.) PETRI NETS 2016. LNCS, vol. 9698, pp. 261–280. Springer, Cham (2016). https://doi.org/10.1007/978-3-319-39086-4_16

10. Katoen, J.-P., Peled, D.: Taming confusion for modeling and implementing probabilistic concurrent systems. In: Felleisen, M., Gardner, P. (eds.) ESOP 2013. LNCS, vol. 7792, pp. 411–430. Springer, Heidelberg (2013). https://doi.org/10.1007/978-3-642-37036-6_23

11. Kurkovsky, S., Loganantharaj, R.: Extension of petri nets for representing and reasoning with tasks with imprecise durations. Appl. Intell. **23**(2), 97–108 (2005). https://doi.org/10.1007/s10489-005-3415-8

12. Stencl, M., Stastny, J.: Neural network learning algorithms comparison on numerical prediction of real data. In: 16th International Conference on Soft Computing Mendel 2010, Brno, pp. 280–285 (2010)

13. Liu, Y., Miao, H.-K., Zeng, H.-W., Ma, Y., Liu, P.: Nondeterministic probabilistic petri net — a new method to study qualitative and quantitative behaviors of system. J. Comput. Sci. Technol. **28**(1), 203–216 (2013). https://doi.org/10.1007/s11390-013-1323-7

14. Balogh, Z., Magdin, M., Turcani, M., Burianova, M.: Interactivity elements implementation analysis in e-courses of professional informatics subjects. In: 2011 8th International Conference on Efficiency and Responsibility in Education, Efficiency and Responsibility in Education, pp. 5–14 (2011)

15. Tian, Y., Wang, X., Jiang, Y., You, G.: A distributed probabilistic coverage sets configuration method for high density WSN. In: Proceedings - 2017 Chinese Automation Congress, CAC 2017, pp. 2312–2316 (2017)

16. Kodamana, H., Raveendran, R., Huang, B.: Mixtures of probabilistic PCA with common structure latent bases for process monitoring. IEEE Trans. Control Syst. Technol. (2017) https://doi.org/10.1109/tcst.2017.2778691

17. Rossi, M., Vigano, G., Moneta, D., Clerici, D.: Stochastic evaluation of distribution network hosting capacity: evaluation of the benefits introduced by smart grid technology. In: 2017 AEIT International Annual Conference: Infrastructures for Energy and ICT: Opportunities for Fostering Innovation, AEIT 2017, pp. 1–6 (2017)

18. Dehban, A., Jamone, L., Kampff, A.R., Santos-Victor, J.: A deep probabilistic framework for heterogeneous self-supervised learning of affordances. In: 2017 IEEE-RAS International Conference on Humanoid Robots, pp. 476–483 (2017)

A Collaborative Approach Based on DCA and VNS for Solving Mixed Binary Linear Programs

Sara Samir[✉] and Hoai An Le Thi

Computer Science and Applications Department, LGIPM, University of Lorraine, Metz, France
{sara.samir,hoai-an.le-thi}@univ-lorraine.fr

Abstract. This paper addresses the Mixed Binary Linear Programming problems (MBLPs) by a collaborative approach using two component algorithms. The first is DCA (Difference of Convex functions Algorithm), an efficient deterministic algorithm in nonconvex programming framework, and the second is VNS (Variable Neighborhood Search), a well known metaheuristic method. The DCA and VNS are executed in parallel. At the end of each cycle, the best-found solution is exchanged between these algorithms via MPI (Message Passing Interface) library. The next cycle starts with the previous best-found solution as an initial solution. The performance of the proposed approach is tested on a set of benchmarks of the Capacitated Facility Location Problem. Numerical experiments show the efficiency of our approach.

Keywords: Mixed Binary Linear Programming problems ·
DC programming and DCA · Metaheuristics ·
Parallel and distributed programming

1 Introduction

Mixed Binary Linear Programs (MBLPs) are NP-hard optimization problems which take the form:

$$(MBLP) \begin{cases} \min Z = c^T x + d^T y, \\ Ax + By \leq b, \\ x \in \{0,1\}^n, \quad y \in \mathbb{R}_+^p, \end{cases} \tag{1}$$

where $c \in \mathbb{R}^n$, $d \in \mathbb{R}^p$, $b \in \mathbb{R}^m$, $A \in \mathbb{R}^{m*n}$ et $B \in \mathbb{R}^{m*p}$.

Let C be the set of feasible solutions of MBLP and let Ω be the corresponding linear relaxation set, say

$$\Omega = \left\{ (x,y) \in \mathbb{R}^n * \mathbb{R}^p | Ax + By \leq b, x \in [0,1]^n, y \in \mathbb{R}_+^p \right\}. \tag{2}$$

MBLP plays a central role in combinatorial optimization and is a common model of several applications including scheduling, transportation, routing, assignment, etc.

© Springer Nature Switzerland AG 2019
N. T. Nguyen et al. (Eds.): ACIIDS 2019, LNAI 11431, pp. 510–519, 2019.
https://doi.org/10.1007/978-3-030-14799-0_44

This problem has been studied by many searchers for a long time. Numerical approaches for MBLP can be divided into three categories: exact methods such as Brand and Bound, Cutting plan, Branch and Cut aiming to find an optimal solution, heuristic/metaheuristic approaches search an *approximate* (a good feasible) solution, and local deterministic approaches such as DCA achieve a local solution (which could be a global solution if the algorithm starts from a good initial point).

The main contribution of our work relies on a collaborative approach based on the deterministic algorithm DCA and variable neighborhood search (VNS) metaheuristic, using paradigms of parallel and distributed programming, for solving MBLP. Our work is inspired by the idea of MetaStorming approach introduced in [12] (a collaboration between metaheuristics). The particularity of our approach is the collaboration between a metaheuristic and a deterministic algorithm, which has never been proposed before. In the present approach, the cooperation is expressed by exchanging information between collaborating methods using MPI (Message Passing Interface) Library. The choice of component algorithms is motivated by the successful application of VNS in combinatorial optimization and the power of DCA in nonconvex programming. Being a continuous approach, DCA works on continuous domains but it achieves a very good integer solution if an integer solution is found during the algorithm. Therefore, a collaboration between VNS, an efficient method to find an integer solution, and DCA could be effective for MBLP.

DC programming and DCA were introduced by Pham Dinh Tao in 1985 and extensively developed since 1994 by Le Thi Hoai An and Pham Dinh Tao. DCA has been successfully applied to large-scale non-convex programs in various areas. An excellent review on thirty years of developments of DC programming and DCA can be found in Le Thi and Pham Dinh [5]. To treat MBLP as a DC program, we will use exact penalty techniques (see [6,7]).

VNS was presented first by Mladenovic in 1995 in a conference and then in 1997, by Mladenovic and Hansen (see [2,8,10,11]). Unlike many other metaheuristics, VNS exploits the idea of the systematic change of neighborhood and local search.

As an application, we consider the capacitated facility location problem (CFLP), to evaluate the performance of our approach.

This paper is organized as follows. Section 2 gives an overall description of the component algorithms including VNS, DCA, and their application to solve MBLP. Section 3 is devoted to the collaborative approach VNS-DCA for MBLP. Numerical results are given and discussed in Sect. 4 and Sect. 5 concludes the paper.

2   DCA and VNS for Solving MBLP

To facilitate the reader, let us introduce briefly DC programming and DCA, VNS and their solution methods to MBLP, before discussing the collaborative VNS-DCA.

2.1   DC Programming and DCA

DC programming and DCA constitute the backbone of smooth/non-smooth non-convex programming and global optimization. They address the problem of minimizing a so called DC function f (which is a difference of convex functions) on the whole space $\mathbb{R}^n$ or on a convex set $C \in \mathbb{R}^n$. A standard DC program takes the form

$$\min \{f(x) = g(x) - h(x) : x \in \mathbb{R}^n\}, \tag{3}$$

where $g, h \in \Gamma_o(\mathbb{R}^n)$, the set contains all lower semi-continuous proper convex functions on $\mathbb{R}^n$. g and h are called DC components, while $g - h$ is a DC decomposition of f. In DC programming, the following usual convention [9] is used:

$$(+\infty) - (+\infty) = +\infty, \tag{4}$$

A constrained DC program is defined by

$$\min \{f(x) = g(x) - h(x) : x \in C\}. \tag{5}$$

When C is a nonempty closed convex set, a constrained DC program can be transformed into an unconstrained DC program using $\chi_C(x)$, the indicator function of the set C, i.e. $\chi_C(x) = 0$ if $x \in C$, $+\infty$ otherwise:

$$(5) \Leftrightarrow \min \{f(x) = \chi_C(x) + g(x) - h(x) : x \in \mathbb{R}^n\}. \tag{6}$$

DCA is an iterative algorithm based on local optimality. At each iteration k, DCA replaces h with its affine minorization by taking $y^k \in \partial h(x^k)$ and then solving the resulting convex program

$$\min\{g(x) - \left[h(x^k) + \langle x - x^k, y^k \rangle\right] : x \in \mathbb{R}^n\}. \tag{7}$$

Standard DCA Scheme [9]
  Initialization: Set an initial solution x^0, $k = 0$, $\epsilon_1 > 0$, $\epsilon_2 > 0$.
  1. Compute $y^k \in \partial h(x^k)$.
  2. Compute $x^{k+1} \in argmin\{g(x) - \langle x, y^k \rangle : x \in \mathbb{R}^n\}$
  3. If($\|x^{k+1} - x^k\| \leq \epsilon_1(\|x^k\| + 1)$ or $|f(x^{k+1}) - f(x^k)| \leq \epsilon_2(|f(x^k)| + 1))$, then **stop.**
  4. Otherwise, $k \leftarrow k + 1$; go to 1.

2.2   DCA for Solving MBLP

DCA was first investigated for MBLP in [3] and then for several applications of MBLP (e.g. [4,5,13]). We present below one version of DCA for MBLP [3].

DC Reformulation and DCA for Solving MBLP. To treat (1) as a DC program, we will reformulate it as a concave quadratic program with continuous variables by using exact penalty function. We consider p the penalty function defined by

$$p(x) = \Sigma_{i=1}^{m} x_i(1 - x_i). \tag{8}$$

The problem (1) is equivalent to

$$\min \left\{ c^T x + d^T y : (x, y) \in \Omega; p(x) \le 0 \right\}. \tag{9}$$

Thus, we can define MBLP$_t$ the penalized program of (1) by

$$\min \left\{ F(x, y) = c^T x + d^T y + tp(x) : (x, y) \in \Omega \right\}. \tag{10}$$

Where $t \in \mathbb{R}$ and $t > 0$.

Theorem 1 [6]. *Let $\mathbb{K}$ be a non-empty bounded polyhedral convex set on $\mathbb{R}^n$. Let f be a finite concave function on $\mathbb{K}$ and p a finite non negative concave function on K. Then there exists $t_0 \ge 0$ such that for $t \ge t_0$, the following problems have the same optimal value and the same solution set:*

(i) $\alpha(t) = inf\{f(x) + tp(x) : x \in \mathbb{K}\}$.
(ii) $\alpha = inf\{f(x) : x \in \mathbb{K}, p(x) \le 0\}$.

Furthermore, if the vertex set of $\mathbb{K}$, denoted with $V(\mathbb{K})$ is contained in $\{x \in \mathbb{K}, p(x) \le 0\}$, then $t_0 = 0$, otherwise $t_0 = \min\{\frac{f(x)-\alpha(0)}{\xi} : x \in \mathbb{K}, p(x) \le 0\}$, where $\xi := \min\{p(x) : x \in V(\mathbb{K}), p(x) > 0\} > 0$.

DC Decomposition for MBLP$_t$
According to Theorem 1, there exists $t_0 \ge 0$ such that (1) and (10) have the same optimal solution for $t \ge t_0$.

We can propose the following DC decomposition for MBLP$_t$:

$$\begin{cases} g(x, y) = \chi_K(x, y), \\ h(x, y) = -c^T x - d^T y - tp(x). \end{cases} \tag{11}$$

The first sequence $(w_x^k, w_y^k) \in \partial(h(x^k, y^k))$ is calculated as

$$\begin{cases} w_{x_i}^k = -c_i - t(1 - 2x_i^k), \forall i \in \{1, \dots, n\}. \\ w_{y_j}^k = -d_j, \qquad \forall j \in \{1, \dots, p\}. \end{cases} \tag{12}$$

The second sequence is

$$(x^{k+1}, y^{k+1}) \in argmin \left\{ -\langle (x, y), (w_x^k, w_y^k) \rangle : (x, y) \in K \right\}. \tag{13}$$

The DCA scheme for solving MBLP can be described as follows

Convergence properties of DCA for MBLP can be found in [3]. For instance, it is worthwhile to note that, if at iteration r of DCA, x^r is integer, then x^k is integer for all $k \ge r$. This property is very interesting and useful for the collaborative VNS-DCA scheme.

Algorithm 2.1. DCA for solving MBLP

Initialization: (x^0, y^0), $k = 0$, $\epsilon > 0$, $\epsilon_2 > 0$, $t > 0$, $\theta > 0$.
1. Compute $(w_x^k, w_y^k) \in \partial(h(x^k, y^k))$ via (12).
2. Compute (x^{k+1}, y^{k+1}) by solving (13).
3. If $|f(x^{k+1}, y^{k+1}) - f(x^k, y^k)| \leq \epsilon(|f(x^k, y^k)| + 1)$, then return (x^k, y^k).
4. Otherwise,
4.1. If$(p(x^{k+1}) \geq \epsilon_2)$, then $t \leftarrow t + \theta$.
4.2. $k \leftarrow k + 1$.
4.3. go to 1.

2.3   Variable Neighborhood Search

VNS exploits the idea of the systematic change of neighborhood within a local search. The intensification in VNS is reflected in the application of a local search method whose idea is to perform some changes to the current solution until finding a local minimum and no improvement is possible. As for the diversification, it is expressed by using several neighborhood structures, changing systematically this neighborhood and adopting a randomized functions. We can summarize all the steps of VNS in the next algorithm.

Algorithm 2.2. VNS

Initialization: Initial solution x, $N_{k(k \leftarrow 1, \ldots, k_{max})}$, k← 1.
1. $x \leftarrow$ Local-search(x).
repeat
3. $x^{'}$ is a randomly generated solution in the kth neighborhood of x ($N_k(x)$).
4. $x^{''} \leftarrow$ Local-search($x^{'}$).
5. If ($x^{''}$ is better than x) then $x \leftarrow x^{''}$.
6. Change the neighborhood.
until a stopping condition.

2.4   VNS for Solving MBLP

VNS is a standard metaheuristic method for MBLP. Two main points to be determined in VNS are local search methods and neighborhood structures. Several local search methods can be taken on for MBLP such as k-opt, another metaheuristic, complementing a value of a binary variable, etc. As for neighborhood structures, suppose that we have a solution (x, y), we can use the idea of generating a new solution $(x^{'}, y^{'})$, such that:

$$\Sigma_i |x_i - x_i^{'}| + \Sigma_j |y_j - y_j^{'}| \geq MinVal. \text{ Or, } \Sigma_i |x_i - x_i^{'}| \geq MinVal.$$

3   The Collaborative Approach VNS-DCA for Solving MBLP

VNS-DCA is the brainstorming of VNS and DCA. The collaborative scheme is inspired by the Metastorming approach [12] in which only metaheuristics have been used. In the literature, there exist many paradigms of parallel programs. For our approach, we opted for the master-slave model. As for exchanging and distributing information between processors, we use the MPI library. MPI is defined in [1] as a library containing a set of standardized functions, available on distributed memory architectures with many languages such as C++. We are now in the position to describe VNS-DCA (an illustrative scheme is found in Fig. 1).

- **Step 0:** The master reads data and generates an initial solution. Then it distributes the solution (the value of the objective function and the variables), to the slaves, using MPI.
- **Step 1:** The slaves receive the solution and start running, in parallel.
  * **Slave 1:** Performs some iterations of DCA until it gets a binary solution. Then, it sends the objective function to the master.
  * **Slave 2:** Performs one iteration of VNS. Then, it sends the objective function to the master.
- **Step 2:** The master receives the informations. It determines the best slave and the worst one and distributes this information. If stopping condition is satisfied, it gives the stop order to the two slaves.
- **Step 3:** The slaves receive information (best slave and worst one). The slave which gets the best result will distribute the solution to the master and the slave which gets the worst solution.
- **Step 4:** The master and the worst slave receive the solution. The master after receiving the solution, makes an evaluation and saves the best solution.
- **Step 5:** The slaves restart running using the best current solution, from the second instruction of Step 1.

To evaluate the efficiency of our approach, VNS-DCA is tested on a set of benchmarks the capacitated facility location problem (CFLP) which is composed of two sets of m facilities and n customers. Each facility has a location and a limited capacity. However, a customer has only one demand. We distinguish two types of cost. The opening cost of the facility and the connection cost between facilities and customers. The objective of the CFLP is to assign customers to facilities, with respect to capacity constraints, in such a way that the total cost is the minimum.

We propose a heuristic to get an initial solution for CFLP. Let $I = \{1, 2, \ldots, m\}$ be the set of facilities and $J = \{1, 2, \ldots, n\}$ be the set of customers.

4 Numerical Results

The proposed approach VNS-DCA was implemented using C++ and compiled within Microsoft Visual Studio 2017. Experiments were carried out on a Dell desktop computer with Intel Core(TM) i5-6600 CPU 3.30 GHz and 8 GB RAM under Windows 10. Linear programs were solved using the software CPLEX version 12.6. The parameters of CPLEX solver were set at their default. The

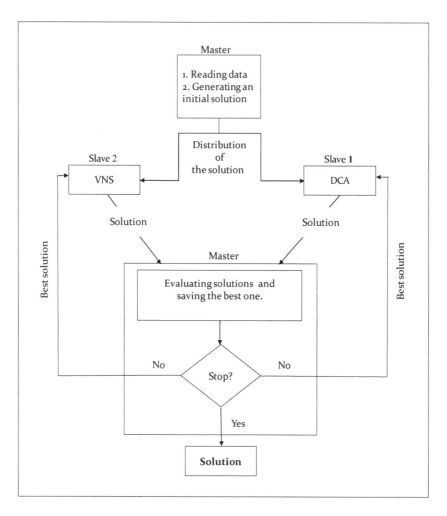

Fig. 1. VNS-DCA: cooperative/collaborative scheme based on VNS and DCA.

Algorithm 3.1. Constructive heuristic for solving the CFLP

---

1. While($J \neq \phi$), do
2. Choose $(i, j) \in I \times J$ having the minimum connection cost.
3. Assign j to i, update J and the capacity of the facility i.
repeat
4. Assign customers in J to facility i, starting with the customers having the smallest connection cost and satisfying the capacity constraint.
5. Update J and the capacity of the facility i.
until Facility i is saturated.

Table 1. Comparison between VNS-DCA, DCA, VNS and CPLEX.

Code	Objective							CPU(s)						
	VNS-DCA	Gap%	DCA	Gap%	VNS	Gap%	CPLEX	VNS-DCA	Ratio	DCA	Ratio	VNS	Ratio	CPLEX
1	1880.6	0.04	1882	0.11	2555	26.42	1879.9	40	1.5	4	14.5	102	0.57	58
2	1950.5	0.00	1991.3	2.05	2485	21.51	1950.5	21	1.3	4	7	102	0.27	28
3	1997.6	0.39	2700.4	26.31	2457	19.01	1989.9	32	2.2	0.61	113.11	107	0.64	69
4	1987.6	0.68	2024.6	2.50	2685	26.48	1974	17	1.2	4	5	41	0.49	20
5	1868	0.42	1868	0.42	2687	30.77	1860.2	26	0.8	9	2.22	76	0.26	20
6	2054.8	0.83	2165.1	5.88	2608	21.86	2037.8	26	3.8	2	49	129	0.76	98
7	1979.1	0.18	2027	2.54	2693	26.64	1975.6	28	122.8	7	491.14	67	51.31	3438
8	1923.4	1.34	2458.9	22.82	2653	28.47	1897.7	37	2.3	1	86	67	1.28	86
9	1945.8	1.55	1969.6	2.74	2485	22.91	1915.7	60	0.7	5	7.8	92	0.42	39
10	1990	0.01	1991.4	0.08	2638	24.57	1989.9	27	2.5	2	34	70	0.97	68
11	1963.5	1.27	2088.1	7.16	2393	18.99	1938.5	51	0.5	2	14	160	0.18	28
12	2019.7	2.69	2484.7	20.90	2499	21.36	1965.3	36	1.9	1	67	102	0.66	67
13	1920.7	0.11	2454.7	21.84	2518	23.81	1918.5	76	1	0.86	90.70	70	1.11	78
14	1995.9	0.01	2104.4	5.16	2669	25.22	1995.8	22	3.1	2	34.5	52	1.33	69
15	1887.6	0.42	1892.5	0.68	2678	29.81	1879.7	24	0.8	5	3.8	74	0.26	19
16	1960.5	0.35	2000.4	2.33	2591	24.60	1953.7	31	1.3	1	39	111	0.35	39
17	2029.7	0.78	2040.6	1.31	2803	28.15	2013.9	44	3.5	3	50.67	63	2.41	152
18	1993.8	0.98	2057.3	4.04	2647	25.42	1974.2	28	5.8	1	161	69	2.33	161
19	1915.3	0.00	1918.5	0.17	2626	27.06	1915.3	19	0.9	3	6	63	0.29	18
20	1939.3	0.01	2580.3	24.85	2761	29.76	1939.2	22	2.5	0.61	90.16	80	0.69	55
Average:		0.60		7.69		25.14		33.35	8.01		68.33		3.33	230.5

CFLP instances used to perform VNS-DCA are available at Benchmarks library. In Table 1, dataset consists of 20 instances having the same size ($n = 100$, $m = 100$).

For each instance we run the algorithms 5 times and take the best solution within 5 runs. We compare VNS-DCA with the results of CPLEX, DCA and VNS as indicated in Table 1. More specifically, the columns from two to eight show the objective value and the gap obtained by VNS-DCA, DCA, VNS. Here the "Gap" of Algorithm (A) is defined by

$$Gap = \frac{|Objective\ value\ given\ by\ (A) - Optimal\ value\ (given\ by\ CPLEX)|}{Objective\ value\ given\ by\ (A)} * 100\%.$$

The other columns provide the corresponding running time in seconds taken by each algorithms to get these solutions, and "Ratio" of algorithm (A) is calculated as

$$Ratio(A) = \frac{CPU(CPLEX)}{CPU(A)}.$$

From the numerical experiments, we observe that the solution given by the collaborative approach VNS-DCA is very close to the optimal solution. In all 20 instances, the gap is less than 2.7% and the average gap is 0.6%. In particular, the gap is less than 1% in 14/20 instances and 0% in two instances. As for the component algorithms, DCA is much better than VNS: the average gap is 7.69% versus 25.14%.

Regarding the running time, DCA is the fastest and VNS is the slowest. It can be seen that VNS-DCA is faster than CPLEX in 15/20 instances. In particular, the seventh dataset, VNS-DCA runs approximately 123 times faster than CPLEX and averagely, the CPU time is 33.35 s for VNS-DCA and 230.50 s for CPLEX.

5   Conclusion

We have presented in this paper a new distributed cooperative and collaborative approach (VNS-DCA) based on DCA and VNS for solving MBLP. To evaluate the efficiency of VNS-DCA, the CFLP was considered for its application. The numerical results confirm the efficiency of the proposed approach. Through an extensive numerical analysis, it can be seen that the results obtained by the collaboration VNS-DCA are satisfactory. The technique introduced in this paper can be extended to collaborate more methods for solving larger classes of problems.

References

1. Centre de Calcul de l'université de Bourgogne: Bases de la parallélisation par passage de message: MPI (Message Passing Interface) (2016)
2. Hansen, P., Mladenović, N.: Variable neighborhood search: principles and applications. Eur. J. Oper. Res. **130**(3), 449–467 (2001)
3. Le Thi, H.A., Pham Dinh, T.: A continuous approach for large-scale constrained quadratic zero-one programming. Optimization **45**(3), 1–28 (2001). In honor of Professor ELSTER, Founder of the Journal Optimization
4. Le Thi, H.A., Pham Dinh, T.: The DC (Difference of convex functions) programming and DCA revisited with DC models of real world nonconvex optimization problems. Ann. Oper. Res. **133**, 23–46 (2005)
5. Le Thi, H.A., Pham Dinh, T.: DC programming and DCA: thirty years of developments. Math. Program. **169**(1), 5–68 (2018)
6. Le Thi, H.A., Pham Dinh, T., Muu, L.: Exact penalty in DC programming. Vietnam J. Math. **27**(2), 169–178 (1999)
7. Le Thi, H.A., Pham Dinh, T., Van Ngai, H.: Exact penalty and error bounds in DC programming. J. Glob. Optim. **52**(3), 509–535 (2011)
8. Mladenović, N., Hansen, P.: Variable neighborhood search. Comput. Oper. Res. **24**(11), 1097–1100 (1997)
9. Pham Dinh, T., Le Thi, H.A.: Convex analysis approach to D.C. programming: theory, algorithm and applications. Acta Mathematica Vietnamica **22**(1), 289–355 (1997)
10. Polacek, M., Hartl, R.F., Doerner, K., Reimann, M.: A variable neighborhood search for the multi depot vehicle routing problem with time windows. J. Heuristics **10**(6), 613–627 (2004)
11. Vidović, M., Popović, D., Ratković, B., Radivojević, G.: Generalized mixed integer and VNS heuristic approach to solving the multisize containers drayage problem. Int. Trans. Oper. Res. **24**(3), 583–614 (2016)

12. Yagouni, M., Le Thi, H.A.: A collaborative metaheuristic optimization scheme: methodological issues. In: van Do, T., Le Thi, H.A., Nguyen, N.T. (eds.) Advanced Computational Methods for Knowledge Engineering. AISC, vol. 282, pp. 3–14. Springer, Cham (2014). https://doi.org/10.1007/978-3-319-06569-4_1
13. Ta, A.S., Le Thi, H.A., Khadraoui, D., Pham Dinh, T.: Solving partitioning-hub location-routing problem using DCA. J. Ind. Manag. Optim. 8(1), 87–102 (2012)

Adoption of Cloud Business Intelligence in Indonesia's Financial Services Sector

Elisa Indriasari[✉], Suparta Wayan, Ford Lumban Gaol,
Agung Trisetyarso, Bahtiar Saleh Abbas, and Chul Ho Kang

Computer Science Depatment, BINUS Graduate Program – Doctor of Computer
Science, Bina Nusantara University, Jakarta, Indonesia
elisa.indriasari@yahoo.com, wayan.suparta@upj.ac.id,
{fgaol,atrisetyarso,bahtiars}@binus.edu,
chkang5136@kw.ac.ar

Abstract. Business Intelligent (BI) tools adopted in many companies as an effort to develop new strategies and survive in the rapid changes and agility situation. New horizons emerged with the implementation of the BI concept using "Cloud Computing". The leading IT companies compete to develop BI cloud-based services (cloud BI). This study aimed to determine the factors influencing manager decision to adopt cloud BI in Indonesia's financial service sector, using the diffusion of innovation (DOI) model and technology organization environment (TOE) framework. The research also obtains prediction of the cloud BI marketing trend in Indonesia. The survey conducted with 30 participants at the senior management level in Indonesia's financial services sector. The findings reveal, the adoption rate of cloud BI in Indonesia financial service sector still very low. Only 3.4% of firms have implemented cloud BI, while 10% not considering the cloud BI adoption. The results of this research can be used by vendors to develop cloud BI tools that fit into the customer's need. It is also very useful for marketers to know the characteristics and demographics of cloud BI users. For regulators, this research illustrates how and what drives regulators to make policies that can encourage companies to use the latest technology.

Keywords: Cloud business intelligence · Cloud BI · DOI framework · TOE framework

1 Introduction

In the current world situation, the decision-making process has become more complicated in the last few decades, caused by the unpredictable situation of world change related to markets, competition, and technology. It is stated that a critical component of an organization's success is the capability to capitalize on all available information and knowledge [1]. Each organization has tended to become more scalable, flexible and intelligent, using new BI solutions [2]. The superior solutions accomplished by integrating different big data platforms to achieve the compatibility with any existing BI solutions [3]. Business intelligence (BI) has grown as main of the solution to provide companies with crucial information. The BI tools help the decision-making process in all

© Springer Nature Switzerland AG 2019
N. T. Nguyen et al. (Eds.): ACIIDS 2019, LNAI 11431, pp. 520–529, 2019.
https://doi.org/10.1007/978-3-030-14799-0_45

level of management to ensure sustainability and generate value for shareholders [4]. In many cases, the implementation of traditional BI is considered complex, expensive, and inflexible. On the other side of the coin, cloud computing nowadays is the new buzzword in the ICT industry [5]. Cloud computing has made BI tools more accessible [6]. The integrating cloud computing and BI technologies generated the cloud BI [7]. The main benefits driven by this model are business agility with lower costs, which allows organizations to respond quickly and effectively to the constantly changing business environment [8]. However, currently, there is still a serious lack of conceptualization in the cloud BI area. The cloud BI implementation strategy still needs to be investigated [9]. Since then, there are many obstacles for companies adopting this new technology.

This research will investigate the adoption of cloud BI according to an empirical study in Indonesia's financial services sector. The objectives of the research are investigating the factors that influence the decision of managers when adopting the cloud BI and predicting its marketing trend in Indonesia's financial services sectors.

2 Literature Review

2.1 Definition and Benefits of Cloud Business Intelligence

Cloud BI represents a way for reporting and analysis solutions to be developed, installed, and consumed more easily due to its lower cost and easier deployment [10]. Many leading IT providers developed cloud-based BI tools as a solution to create a more accessible and accessible solution in BI (see Fig. 1).

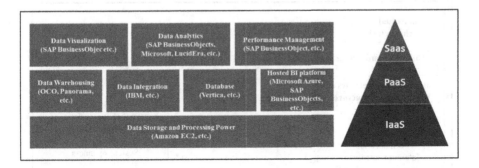

Fig. 1. Cloud BI, Source: Infosys Research.

Nowadays, the popularity of cloud BI solutions are gradually increase among businesses, as many businesses are realizing the benefits of data analytics [11].

2.2 Empirical Cloud BI Research in Indonesia Financial Sector

Financial Services Authority (*Otoritas Jasa Keuangan* - OJK) stated, there are 115 commercial banks, 137 insurance companies, 172 Multi finance, and 63 Fintech operated in Indonesia (OJK Report, June 2018). Prior studies have performed to explore the adoption of Business Intelligence and cloud BI in many sectors, including financial institutions around the world. However, no empirical study research has been conducted on the adoption of BI and the cloud BI in Indonesia's financial sector. Limited access to and regulation of privacy and data security are the barriers for the academy to conduct an empirical study on the financial service in Indonesia. With respect to data privacy and customer safety of commercial banks, POJK 38 issued by OJK contains the Regulation number 38/POJK.03/2016 of the financial services authority on the implementation of the use of the risk management of the information technology of the commercial bank. Indonesia Government through the Ministry of Communication and Informatics issued PP82 on the government regulation of the Republic of Indonesia number 82/2012 on "Management of electronic systems and transactions".

2.3   DOI Framework

The DOI framework built principally on the characteristics of the technology and the perceptions of the users on the implementation of innovation strategy. According to Rogers [13], the diffusion of Innovation (DOI) theory and the adoption rate of innovations were impacted by five factors: (1) compatibility, (2) the relative advantage, (3) observability, (4) complexity, and (5) trialability [12]. Rogers further explains that complexity has a negative impact related to the rate of adoption, trialability, relative advantage, observability, and compatibility are usually positively related with the rate of adoption. The DOI Framework proposed by Rogers is illustrated in Fig. 2.

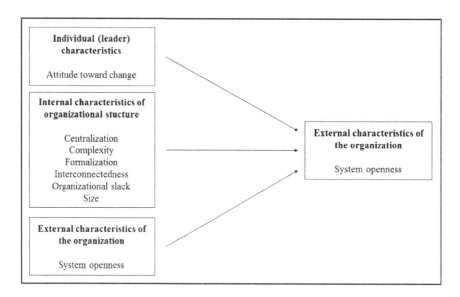

Fig. 2. Diffusion of innovations [13].

2.4 TOE Framework

Oliveira and Martins [13] advised the TOE framework to define the process of innovation in the context of a company. Figure 3 illustrated the TOE framework [13]. Based on this framework, the process of adopting technological innovation is influenced by three aspects of the context of a company:

- Technological context describes the internal and external technologies related to the organization; both the technologies that are already in use in the company and those that are available in the market but that is not currently in use. These technologies can include equipment or practice.
- The organizational context defines as the resources and the characteristics of the company, for example, size and management structure.
- Environmental context, which refers to the area in which a company conducts its business; It may be related to surrounding elements, such as industry, competitors and the presence of technology service providers.

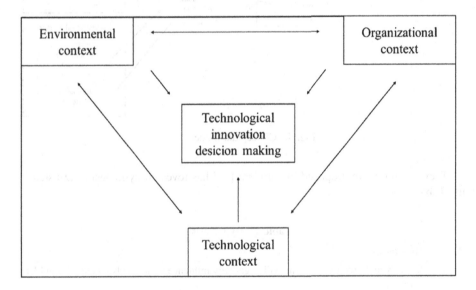

Fig. 3. TOE framework [13].

The TOE framework already adopted in a variety of Information System (IS) adoption settings including open systems, ERP, E-Commerce, and Cloud Computing [14].

2.5 Conceptual Model and Hypothesis

In this era, the complexity of the new technology adoption required people to combined more than one theoretical model to achieve a better understanding of IT/IS adoption phenomenon.

In this research, we assume that the DOI-TOE framework developed by Oliveira [15] is suitable to be adapted to describe the variable relationship of the implementation of cloud BI to the financial service sector in Indonesia. We use the framework and conduct an empirical study to find out the factors that influence BI's cloud adoption. The integrative research model is shown in Fig. 4.

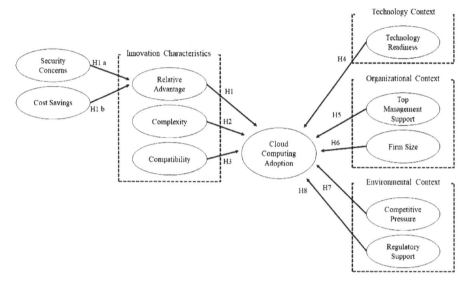

Fig. 4. Conceptual model [15]

Previous literature proposed by Oliviera [15] has revealed hypothesis and listed in the Table 1.

Table 1. Hypothesis

H#	Hypothesis
H1a	Security and privacy concerns will negatively influence relative advantage of cloud BI adoption
H1b	Cost saving will positively influence relative advantage of cloud BI adoption
H1	Relative advantage will positively influence cloud BI adoption
H2	Complexity will negatively influence cloud BI adoption
H3	Compatibility will positively influence cloud BI adoption
H4	Technology readiness will positively influence cloud BI adoption
H5	Top management support will positively influence cloud BI adoption
H6	Firm size will positively influence cloud BI adoption
H7	Competitive pressure will positively influence cloud BI adoption
H8	Regulatory support will positively support cloud BI adoption

3   Research Methods

3.1   Research Design

The research conducted is employed two research methods: interviews and the questionnaires. The purpose of research, methods, and participants of the research is illustrated in Table 2.

Table 2. Research design

Purpose of research	Research methods	Research group
Previous theories considering cloud BI. Gaining new insights of the cloud BI market trend	Interviews	Study case. Participants: 2 CIO
Ranking the importance of key adoption factors	Questionnaire	CIO or Head of IT of 30 financial firms (banking, insurance Multi finance, Fin-tech)

3.2   Data Collection and Analysis

To identify the variables in Cloud BI adoption, interviews are performed and used as the main method in the data collection. The first draft of the questionnaire was an experiment with two participants from different companies. The personal interview with the Director of Information Technology (CIO) in the two firms were involved in the adoption decision-making process of cloud BI.

The interviews are performed for one hour. Interview questions included (1) the background of the company, (2) BI, cloud computing, and the cloud BI implementation, and (3) the impact of the factors in the cloud BI adoption.

After conducting a face-to-face interview, the next step is conducted a survey. We distributed questionnaires to CIOs of 30 financial service firms in Indonesia. Adoption is generally measured on the Likert scale. In the questionnaire, we used a 5-point Likert scale to measure the degree of adoption. 26 questions were asked in the survey. The questions measure 10 variables represented in the research model.

3.3   Validity and Reliability

The operational measures applied in this investigation were taken from prior work. Statistic software is exercised to run multiple tests. The validity and reliability tests are used to ensure that the measurements were accurate. The results of the Cronbach's scores were used to evaluate the reliability of the items constructed. We eliminate questions that have low reliability. After eliminating CP1 and FS2, all the constructs have a high-reliability value >0.6 as shown in the Table 3.

Table 3. Cronbach's α score for each factor DOI-TOE framework

Factors	Factor code	Item/question no. (questioner)	Cronbach's α score
Security concerns	SC	SC1, SC2, SC3	0.927
Cost savings	CS	CS1, C2, CS3	0.703
Relative advantage	RA	RA1, RA2, RA3	0.778
Complexity	CX	CX1, CX2	0.795
Compatibility	CM	CM1, CM2, CM3	0.691
Technology readiness	TR	TR1, TR2, TR3	0.774
Top management support	TS	TS1,TS2,TS3	0.751
Firm size	FS	FS2	1.000
Competitive pressure	CP	CP1	1.000
Regulatory support	RS	RS1, RS2	1.000
Cloud BI adoption	CBI		1.000

4   Finding and Dicussion

4.1   Survey Result Discussion

A prototype questionnaire had been floated in November 2018. The survey conducted in 10 days, with 30 participants at the senior management level (Chief or Head of Division) in Indonesia's financial services sector, including (1) Banking, (2) Insurance, (3) Multi finance, (4) Fintech, and (5) Securitas. Most participants are from banking firms that have annual revenues more than IDR 50 Billion. Currently, most Indonesian financial firms have implemented Business Intelligence. 76.7% of firms have implemented BI, and 23% of firms not yet implemented BI. Most of the firms also have adopted cloud computing for a particular type of services. Total 56.7% of firms have implemented cloud computing, and 43.3% of firms have not yet implemented cloud computing. Figure 5 illustrated cloud BI features that appropriate to financial service firms in Indonesia.

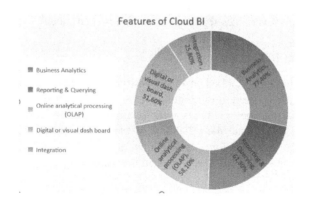

Fig. 5. Features of Cloud BI.

The functional areas of cloud BI are illustrated in Fig. 6. The result of the survey represents the adoption rate of cloud BI in Indonesia's financial service sector still very low. Only 3.4% of firms have implemented Cloud BI, while 10% not considering Cloud BI adoption. The future of cloud BI is still promising, most the firms have a positive perspective of cloud BI integration soon. 60% of firms will adopt BI between 1–2 years, while 13.3% of firms will adopt more than 1 year, and 13.3% of firms will adopt less than 1 year.

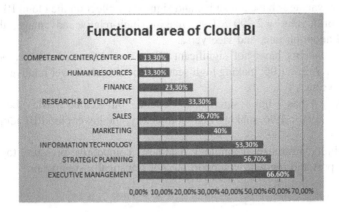

Fig. 6. Functional areas of Cloud BI.

We perform identification and validation of ten factors through data analysis, and the result is as follow. Table 4 shows the estimated coefficients and statistics calculation for each variable path.

Table 4. The estimated coefficients and statistics

	Original Sample (O)	Sample Mean (M)	Standard Deviation (STDEV)	T Statistics	P Values	Sign Level
RA → CBI	−0.556	−0.530	0.186	2.990	**0.003**	S
SC → RA	0.337	0.305	0.194	1.733	**0.084**	NS
CS → RA	0.767	0.751	0.085	8.975	**0.000**	S
CM → CBI	0.117	0.077	0.361	0.323	**0.747**	NS
CX → CBI	0.275	0.257	0.230	1.194	**0.233**	NS
TR → CBI	0.335	0.334	0.407	0.823	**0.411**	NS
TS → CBI	0.114	0.169	0.250	0.456	**0.649**	NS
FS → CBI	−0.194	−0.172	0.205	0.948	**0.344**	NS
CP → CBI	0.307	0.298	0.178	1.726	**0.085**	NS
RS → CBI	0.097	0.099	0.209	0.464	**0.643**	NS

NS=Not significant; S=Significant;

The t test deliberated is to test whether the independent variables partially have a significant effect on the dependent variable. The variable is significant if p-values < 0.05. We found two paths: RA → CBI and CS → RA have p-values < 0.05. The summary result from the Table 4 can be clarified as follows:

- Cost savings have a positive and significant effect on the relative advantage, because t value = 8.975 > 1.96, making high-cost savings, the higher relative advantage, and vice versa.
- Relative advantages have a positive and significant effect on the cloud BI adoption, because of t value = 2.990 > 1.96, making high-relative advantage, the higher cloud BI adoption rate, and vice versa.
- Security concerns have not significant effect on relative advantage, because of t value = 1.733 < 1.96, making high-security concern, the lower relative advantage, and vice versa.
- Complexity has not significant effect on cloud BI, because of t value = 1.194 < 1.96, making high-complexity, the lower cloud BI adoption, and vice versa.
- Compatibility, firm size, technology readiness, competitive pressure, top management support and regulatory support are not statistically significant.

5 Conclusion and Recommendation

Recommendations based on the findings of research provide an overview of cloud BI's acceptance in Indonesia's financial service sector. The results of this research can be used by application designers to create applications that focus on user needs. IT engineers can produce an application with better functions and designs. Vendors can use this as a marketing trend data. So that they can see the priority of problems encountered on the cloud BI adoption. Thus, vendors can focus on existing problems, developing applications, and roll out new versions. This research is very useful for marketers to know the characteristics and demographics and types of BI cloud user companies in the financial services sector.

For regulators, this research illustrates how and what drives regulators to make policies that can encourage companies to use the latest technology, but still focus on high levels of privacy and security. The implementation of Information Technology or Information Technology able to encourage companies to achieve high performance and ultimately be able to improve Indonesia's economy on a macro basis. The government can find out what problems the company faces in adopting a technology, especially the cloud BI. So that companies, especially the financial services sector in Indonesia can increase efficiency by using cloud BI. Development of a policy plan can use this research to identify gap areas and implement policies that are in line with technological developments.

In the end, the company can analyze the problems and issues faced. Cloud BI implementation is part of the implementation strategy plan and system integration with cloud computing. As part of the IT infrastructure strategy that is more integrated and efficient.

Further research is needed with a greater number of participants, besides requires empirical research that focuses on cloud BI integration strategy with traditional BI and infrastructure in the company.

References

1. Olszak, C.M.: Toward better understanding and use of business intelligence in organizations. Inf. Syst. Manag. **33**(2), 105–123 (2016)
2. Kasem, M., Hassanein, E.E.: Cloud business intelligence survey. Lect. Notes Bus. Inf. Process. **183**, 307–317 (2014)
3. Chang, B.R., Wang, Y.A., Lee, Y.D., Huang, C.F.: Development of multiple big data analysis platforms for business intelligence. In: Proceedings of 2017 IEEE International Conference on Applied System Innovation, ICASI 2017, vol. 1, pp. 1930–1933 (2017)
4. Dawson, L., Van Belle, J.-P.: Critical success factors for business intelligence in the South African financial services sector. SA J. Inf. Manag. **15**(1), 1–12 (2013)
5. Lindner, M.A., Vaquero, L.M., Merino, L.R., Caceres, J.: Cloud economics: dynamic business models for business on demand. Int. J. Bus. Inf. Syst. **5**(4), 373 (2010)
6. Olszak, C.M.: Business intelligence in cloud. Polish J. Manag. Stud. **10**(2), 115–125 (2014)
7. Ricardo, J., Bernardino, J., Almeida, A.: Cloud business intelligence for virtual organizations. In: Proceedings of 2014 International C* Conference on Computer Science and Software Engineering - C3S2E 2014, pp. 1–7 (2008)
8. Gurjar, Y.S., Rathore, V.S.: Cloud business intelligence - is what business need today. Int. J. Recent Technol. Eng. **1**(6), 81–86 (2013)
9. Baars, H., Kemper, H.-G.: Business intelligence in the cloud? In: PACIS 2010 Proceedings, vol. 1, pp. 1528–1539 (2010)
10. Ouf, S., Nasr, M.: Business intelligence in the cloud. In: 2011 IEEE 3rd International Conference on Communication Software Networks, ICCSN 2011, vol. 12, no. 1, pp. 650–655 (2011)
11. Al-Aqrabi, H., Liu, L., Hill, R., Antonopoulos, N.: Cloud BI: future of business intelligence in the cloud. J. Comput. Syst. Sci. **81**(1), 85–96 (2015)
12. Acheampong, O., Moyaid, S.A.: An integrated model for determining business intelligence systems adoption and post-adoption benefits in banking sector. J. Adm. Bus. Stud. **2**(2), 84–100 (2016)
13. Oliveira, T., Martins, M.F.: Literature review of information technology adoption models at firm level. Electron. J. Inf. Syst. Eval. **14**(1), 110–121 (2011)
14. Mckinnie, M.: Cloud computing: TOE adoption factors by service model in manufacturing, p. 131 (2016)
15. Oliveira, T., Thomas, M., Espadanal, M.: Assessing the determinants of cloud computing adoption: an analysis of the manufacturing and services sectors. Inf. Manag. **51**(5), 497–510 (2014)

Stakeholder Impact on the Success and Risk of Failure of ICT Projects in Poland

Kazimierz Frączkowski⬤ and Barbara Gładysz$^{(\boxtimes)}$⬤

Faculty of Information Technology and Management, Wrocław University of
Science and Technology, Wybrzeże Wyspiańskiego 27, 50-370 Wrocław, Poland
{kazimierz.fraczkowski,barbara.gladysz}@pwr.edu.pl

Abstract. The analysis of collective cooperation of various stakeholder groups
in terms of its effectiveness in Information and Communication Technologies
projects in Poland was the subject of surveys. In addition to unclear requirements, inefficiencies in communication between project stakeholders were
identified as the main causes of problems in project implementation. The impact
of cooperation of different stakeholder groups: project team, management of the
project implementation unit, suppliers and end users of the final product on the
success/failure of the project was examined. The impact of stakeholder identification before the start of the project on the chances of success or failure of the
project was also analysed. The results of the research are recommendations for
improving the communication management model in which the process of
stakeholder identification, cooperation with contractors, suppliers, end users of
innovative services or products is an important factor in gaining knowledge
necessary for the success of project in Poland.

Keywords: ICT project management · Project success · Communication ·
Stakeholder

1 Introduction

Communication is a necessary condition for the transfer of knowledge between
stakeholders, the basis for their involvement, which significantly influences the success
of projects implemented using Information and Communication Technologies (ICT).
The term Information and Communication Technologies was first used in 1997 in the
UK (Stevenson 1997). This term refers to the use of the Internet, wireless networks,
Bluetooth, cellular networks, computer media, personal computers, workstations, ser-
vers, IT applications and complex IT systems. The term ICT is broader and includes
also IT projects (Information Technology Project, IT project), although in the case of
information functions, services, supporting business processes and others, it seems
more accurate to call them information systems in which IT systems are a separate part
(Kuraś 2009). Since the projects examined meet the features that are covered by the
term ICT, we will use this term in this paper.

Poor communication has been identified as a major obstacle to the success of the
project. Lack of required cooperation or effective communication and its consequences
are discussed at work (Reed and Knight 2009; Ober 2013; Soczówka 2017). The

© Springer Nature Switzerland AG 2019
N. T. Nguyen et al. (Eds.): ACIIDS 2019, LNAI 11431, pp. 530–540, 2019.
https://doi.org/10.1007/978-3-030-14799-0_46

importance of the role of communication in the project team and the generated problems of knowledge transfer management are taken up in the works (Kisielnicki 2011; Frączkowski et al. 2019). Deficiencies in popular project management methodologies are indicated (Mckay et al. 2014). It proves that project management methodologies do not show how to manage communication depending on the size of the project team, the number of stakeholders, the complexity of the project and the scale of impact of the results. In large projects where there is a conflict of interests of stakeholder groups, communication is not only an element of project knowledge management, but also an element of influence on overcoming resistance. Knowledge management through effective communication channels is an element of overcoming the resistance of stakeholder groups (Vann 2004, pp. 49–50). It points to the problem of the failure of the ICT project in the public sector, where there are a number of cultural, social and linguistic "clashes" between civil servants who work according to a particular style and ICT specialists (both internal and external). Particular problems are visible outside the organization implementing the project at the stage of trying to implement large integrated software systems. The author argues that a significant part of the perceived resistance of government officials was "inadvertent and unconscious", because employees permeated with one work model and language of concepts tried to adapt to "an incomprehensible range of new terminology, abstract concepts and specialised disciplines", to the introduction of "new and radically different methods in the working environment of public organisations". An example is the project **"Electronic Platform for Collection, Analysis and Sharing of Digital Medical Records". (P1)**, in which national coordinators for cooperation with the main stakeholder groups were appointed, a broad information campaign was undertaken using various channels of communication with stakeholders and providing a dedicated portal www.csioz.gov.pl/projekty/realizowane/projekt-p1/ and sending out a newsletter to thousands of health care entities in electronic form (Frączkowski et al. 2011). The analysis of the reasons for the mediocre use of the systems by Polish hospitals (Bartczak and Barańska 2015) indicates mainly the state of equipping hospitals with IT solutions, barriers to access to technology and a shortage of financial resources. On the other hand Głuszyński and Kowalewska (2012, pp. 8–9) write, among other things, that despite a long period of preparation and considerable information effort (during 6 months of 2012 nearly 1,400 different types of articles were published, of which over 500 in the press and nearly 900 on the Internet), service providers for which the P1 project is being prepared, usually do not know anything about it yet. The research showed that the knowledge about the largest project in the health care sector, with a budget of over PLN 790 million, was low. The knowledge about the key service to be, among others, access to Electronic Medical Documentation (EMD) was as follows: "I've heard" indicated every fifth person surveyed over 15 years of age. Only 4% of respondents declared that they have "heard a lot about it". Knowledge about e-Prescriptions was developing at a similar level. Creating project initiation documents and involving the sponsor, contractors and other stakeholders in the process is one of the approaches to forcing communication through a formal channel. However, documentation is a tedious task and is usually underestimated or performed in a superficial manner. This is one of the reasons why the creative approach suggests focusing on direct communication. The complexity of communication channels, used technologies and problems in agreeing on uniform communication within IT companies and between teams, subcontractors is indicated by work (Muszyńska 2011). Knowledge transfer is much more complex than indicated by the definition, hidden knowledge and its acquisition and the transfer of classified knowledge and informal knowledge causes that not all available channels of communication can be effectively used, as is indicated by Reed and Knight (2009). Special attention and support is required from beneficiaries of social projects with EU funds in health care where motivation of medical personnel and support of the project manager is a necessary factor. Such an example is the CareWell project carried out in the Lower Silesian Voivodship (Frączkowski et al. 2015). Recent research by the authors indicates that the most effective method of communication are face-to-face meetings (F2F) followed by videoconferencing, telephone calls, text chat or e-mail. However, the most effective channel is F2F communication (Frączkowski et al. 2019). This is not possible between dispersed - virtual teams in a global software project or projects carried out in cooperation with numerous subcontractors, as well as in one company with numerous territorially distant branches.

The aim of this paper is to verify the impact of the environment of various stakeholder groups on the success and failure of ICT projects. To this end, questionnaire surveys were carried out. The responses were received from 125 respondents who represented main stakeholders of the project. The respondents assessed the success of the project, as well as the factors influencing this success using the five-stage Likert scale. The following stakeholder groups were distinguished: management of the entity in which the project is implemented, external contractors, end users of the project product, suppliers. The success/failure of the ICT project was investigated by factors such as: identification of stakeholders before the start of the project and the nature of their cooperation during the implementation. Appropriate research hypotheses were formulated. In order to verify them, statistical tests were used: non-parametric Kruskal-Wallis test and Spearman's rank correlation coefficients test. Statistical analysis was carried out with the use of SPSS software.

2 Test Methods and Test Results

The identification of factors that affect the chances of success and constitute a risk of project failure is based on our own research conducted in between April to June 2010 and repeated on a wider scale in the years 2015–2017. For the purpose of the research into the success factors of IT projects carried out in 2010, a synthetic measure of project success was created, called the success rate. This measure was constructed on the basis of five criteria:

- compliance of the actual budget of the project with the planned one,
- compliance of the actual duration with the planned duration,
- compliance of the implemented scope with the planned one,
- project appraisal by the project manager,
- customer satisfaction assessment by the project manager.

On the basis of the success rate, each project was included in one of the following groups: 'successful', 'partially successful' or 'failure'.

The research was carried out within the framework of the grant "Study of critical success factors and failures of Polish IT projects 2015–2017". The aim of the research was to improve project management on the basis of experience in conducting IT projects in Poland. The analysis involved 125 respondents, each of whom described one ICT project as a whole, answering 60 questions, in closed or open categories, with one or more answers, mandatory or optional, providing 496 unit information about the project attributes. For the purposes of the research, paper questionnaires were used, which were made available during direct meetings with project managers and in electronic form through the portal interankiety.pl. In the evaluation of the project's success there were questions that allowed to evaluate the success or failure of the project according to quantitative criteria as in the previous 2010 survey and there was a question on the basis of which the respondent assessed the degree of success on the Likert scale: 'definitely no', 'rather not', 'I have no opinion', 'rather yes', 'definitely yes' (see Table 1).

Table 1. Achieving project success and objectives

Evaluation of success	Definitely NO	Rather NOT	I have no opinion	Rather YES	Definitely YES
Was the project a success?	10%	3%	16%	43%	28%
Has the objective of the project been achieved?	6%	4%	11%	34%	46%
Have partial objectives been achieved?	2%	8%	14%	49%	28%
Was the final result consistent with assumptions at the planning stage?	7%	13%	18%	41%	22%

Source: own elaboration

Respondents indicated that 60% of the projects are implemented in cooperation. In this group of 21.6% of the projects, subcontractors perform more than 50% of the scope of work. A significant part of the surveyed projects, 57.4% in total were carried out for organizations from 251 to 2,000 people, and over 2,000 people were carried out 36.6% of the projects. In addition, it was identified that 41% of the projects involved foreigners. Taking into account the rate of cooperation with subcontractors and the percentage of foreigners, it was necessary to assess the reason why as many as 48% of the respondents indicated that the reason for the problems during the project implementation was communication. Communication and unclear requirements had the highest percentage share in indicating the reason for problems with project implementation (see Table 2).

56% of the respondents indicated that the support of the sponsoring client has an impact (assessment: Rather YES/Definitely YES) on the success of the project. Customer involvement as important for the success of the project (Rather YES/Definitely YES) was indicated by 66% of customers, see Table 3.

Out of 125 projects surveyed, 28% of respondents assessed the described project as successful, 43% as projects with certain problems (rather successful), 10% as unsuccessful, 3% as rather unsuccessful, and 16% had no opinion on the subject. In the group

Table 2. Factors indicated as a cause of problems in project implementation

No.	Reason	Percentage share
1	Changing composition of the team (large rotation)	26%
2	Lack of experience in project implementation	30%
3	Location	14%
4	Working style	30%
5	**Unclear requirements**	**57%**
6	**Communication**	**48%**
7	Other	23%

Source: own elaboration

Table 3. Assessment of the impact of customer/user involvement in the project on the success of the project

Impact assessment for the success of the project	Definitely NO	Rather NOT	No opinion	Rather YES	Definitely YES
Support and involvement of the sponsoring client	11%	12%	21%	31%	25%
User involvement	7%	15%	12%	35%	31%
Appropriate cooperation	6%	10%	19%	38%	26%

Source: own elaboration

of 35 projects completed successfully, 60% of the projects were carried out in teams of up to 10 persons (17% - in teams of up to 5 persons, 43% in teams of 6–10 persons), and 23% - in teams of 11–25 persons. On the other hand, out of 54 projects completed, according to the respondents, 50% of the projects were carried out in teams of up to 10 people (22% - in teams of up to 5 people, 28% in teams of 6–10 people), 31% - in teams of 11–25 people. This has become the basis for the identification of a stakeholder group that has a significant impact on the end result of the project in the form of success or failure.

3   Research Hypotheses and Their Evaluation

At the project initiation stage, it is extremely important to identify project stakeholders because their number and characteristics of their role in the project allow for the identification of risks related to communication, knowledge transfer and allow for the preparation of an effective stakeholder management plan. It seems reasonable to make a hypothesis:

H1: Identification of stakeholders before the start of a project increases the chances of success.

The hypothesis was verified with the Kruskal-Wallis test. There are no grounds to reject the hypothesis of H1 (sign. level ≈ 0).

The relevance of the correlation coefficient between the identification of stakeholders before the start of the project and its success was also examined. The Spearman's correlation coefficient is r = 0.421 - is positive and significantly different from zero (sign. level ≈ 0).

Identifying stakeholders before the project starts increases the chances of success. Among the projects where stakeholders were correctly identified (5 on the Likert 1-5 scale), 88% of the projects were successful or almost successful.

Surveys allowed us to distinguish that the feature of the surveyed ICT projects is that they are carried out in dispersed teams, a significant part of the projects is carried out with a large share of subcontractors, who are service providers (outsourcing) or products, and with the participation of foreigners. Respondents assessed the impact of the "right cooperation" factor on the success of the project respectively "Rather YES" - 38%, "Definitely YES" - 26%. Since the questions and answers were related to these key stakeholders, it was justified to make hypotheses:

H2: The success of an IT project depends on proper cooperation with external contractors.

H3: The success of an IT project depends on the correct cooperation with suppliers.

The hypotheses were verified with the Kruskal-Wallis test. There are no grounds for rejection of hypotheses H2, H2 (sign. level ≈ 0) In addition, the relevance of correlation coefficients between proper cooperation with external contractors and the success of the project and between proper cooperation with suppliers and the success of the project was examined. Spearman's rank correlation coefficients are positive and equal respectively r = 0.327 and 0.261. Both coefficients are positive and significantly different from zero (unilateral significance (sign. level ≈ 0). Therefore, proper cooperation with external contractors increases the chances for the success of the project. Among the projects where cooperation with external contractors was correct (5 on the Likert 1-5 scale), 70% of the projects were successful or almost successful. Similarly, proper cooperation with suppliers increases the chances of project success. Among the projects where cooperation with suppliers was correct (5 on the Likert 1-5 scale), 71% of the projects were successful or almost successful.

Projects with a budget ranging from PLN 0.5 million to PLN 10 million accounted for 32.2%, and organisations which were beneficiaries of more than 2,000 users accounted for 36.6% of the total number of projects examined. In large projects, due to the number of end users, groups of end users and sometimes conflicting interests, there are numerous problems related to the acquisition of knowledge, communication, commitment to cooperation as well as difficulties in motivation. It cannot also be ruled out that the degree of involvement depends on trust and the level of conflict of interest. The resultant of the above conditions is the involvement of end users and the quality of cooperation with them. It is therefore reasonable to examine the hypotheses:

H4: The success of an IT project depends on the involvement of the end users of the project.

H5: The success of an IT project depends on the correct cooperation with end users.

The hypotheses were verified using the Kruskal-Wallis test. There are no grounds for rejection of hypotheses H4, H5 (sign. level $\approx$ 0).

The importance of correlation coefficients between the involvement of end users and the success of the project and between the correct cooperation with end users and the success of the project was also examined. Spearman's rank correlation coefficients are positive and equal respectively r = 0.363 and 0.344. Both coefficients are positive and significantly different from zero (sign. level $\approx$ 0).

Proper cooperation with end users also increases the chances of success of the project. Among the projects in which the recipients were involved in the implementation of the project (5 on the Likert 1-5 scale), 67% of the projects were successful or almost successful.

Similarly, proper cooperation with end product customers increases the chances of project success. Among the projects where the cooperation with end product customers was correct (5 on the Likert 1-5 scale), 70% of the projects were successful or almost successful.

Another extremely important shareholder of the projects is the sponsor, who is the beneficiary of the project results. Opinions on the impact of the support of the management of the entity for which the project was implemented on the success of the project were examined. "Rather YES" was answered by 31% of respondents, "Definitely YES" was answered by 25%. 21% of respondents had no opinion. The above results were the basis for the hypothesis:

H6: The support of project contractors during project implementation by the management of the unit where the project is implemented increases the chances of success of the project.

The hypothesis was verified with the Kruskal-Wallis test. There are no grounds to reject the hypothesis of H6 (sign. level $\approx$ 0). The relevance of the correlation coefficient between management support and project success was also examined. The Spearman's rank correlation coefficient is r = 0.475 - is positive and significantly different from zero (sign. level $\approx$ 0).

The support of the project contractors during the project implementation by the management of the unit where the project is implemented increases the chances for the success of the project. Among the projects where management supported contractors (5 on the Likert 1-5 scale), 76% of the projects were successful or almost successful.

4 Discussion of Results

The results of the study indicate that over the last few years there has been a slight improvement in the success rate of projects, which now stands at 28%. This confirms a trend that coincides with the results of the reports (Chaos Report 2015). Research conducted in 2010 under the name PMresearch.pl (Frączkowski et al. 2011, Gładysz and Frączkowski 2015) showed that 21% of the projects were successful. This confirms a trend that coincides with the results of the reports (Chaos Report 2015), see Fig. 1. As progress is unsatisfactory, it is appropriate to examine the reasons for this low percentage of successful projects. Over the last few years, the nature of the projects has changed, their scale, scope, complexity, size of project teams, number of stakeholders. ICT technologies and communication tools and means (Reed and Knight 2009) have developed and despite this, the unsuccessful project indicator is unsatisfactory, which leads to an intensive search for causes and means and methods of its improvement.

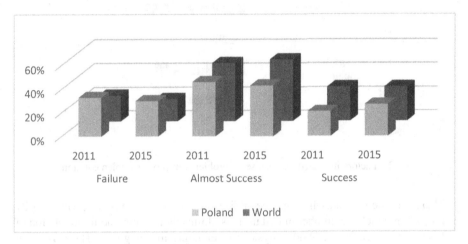

Fig. 1. Success and failure of ICT projects in the World and in Poland

1. The number of contractors and sub-contractors, users as one of the characteristics of the surveyed projects: 60% of the projects are carried out in cooperation and in this group 21.6% of the projects, sub-contractors perform more than 50% of the scope of work. This justifies the need for organizations and project managers to pay attention to the need to build effective communication channels that guarantee proper cooperation. Different suppliers are usually IT companies with their characteristic work model and the use of different tools to document the results of work or communication, therefore respondents indicated that the reason for problems in the project is 14% location and 30% working style (see Fig. 2 and Table 2). It is only in the course of joint work on the project that standards are developed which facilitate cooperation. This element should be taken into account already at the project initiation stage, where communication management should be extended to include a communication project that takes into account tools, standards and the way of building trust (Frączkowski et al. 2019). The conclusions arrived at by Reed and Knight (2009) should be accepted that planning communication between stakeholders such as external contractors and suppliers is a more complex problem than it would seem.

2. The proper cooperation with end users has an important influence on success of ICT project. Due to the development of technology and social and economic needs, ICT projects are growing in terms of the number of end users. A significant part of the surveyed projects, 57.4% in total were carried out for organizations from 251 to 2000 people, and over 2000 people were carried out 36.6% of the projects.

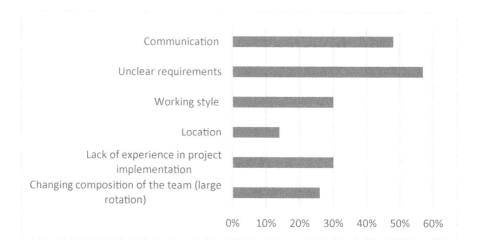

Fig. 2. Factors indicated as a cause of problems in project implementation

Therefore, the project, which has been discussed a lot for almost 10 years, is a P1 project, in which the number of end users is ultimately the entire adult population of Poles with more than 179,000 physicians belonging to this group. The reason for the delay of ICT projects (Frączkowski et al. 2015) lies in insufficient involvement and improper cooperation with end users of the results of these projects, i.e. services using ICT technology, see Fig. 3.

Fig. 3. Factors of ITC project success

3. Studies showing the importance of management's support and involvement in the Polish environment are consistent with the continuing trend of this factor, which is

confirmed in subsequent studies (Chaos Report 2015), it is a leading factor of one of three (management support, maturity of the organization, user involvement of 15% each). In our research, the support of project contractors during project implementation by the management of the unit in which the project is carried out increases the chances of project success. Among the projects where management supported contractors, 76% of the projects were successful or almost successful, see Fig. 3.

5 Conclusions

The conducted research and verified hypotheses confirm that communication and its conditions significantly influence the success and failure of projects. These conditions, which should be taken into account in the individual phases of the project implementation, are the following theses:

T1: Stakeholder identification prior to the project increases the chances of success.
T2: The success of an IT project depends on proper cooperation with external contractors.
T3: Success of an IT project depends on proper cooperation with suppliers.
T4: The success of an IT project depends on the involvement of the end users of the project.
T5: Success of an IT project depends on the correct cooperation with end product customers.
T6: The support of project contractors during project implementation by the management of the unit where the project is implemented increases the chances of project success.

In further work, the results obtained entitle us to recommend modifications and additions to the process of project management communication in project management methodologies. The main area of change is the processes related to stakeholder identification, the extension of the communication function to ensure cooperation and involvement with end users and the flow of knowledge between all stakeholder groups. The importance of the role and the need to involve end users of ICT products and services, subcontractors and service providers throughout the life cycle of the project should be postulated more strongly than hitherto.

References

Bartczak, K., Barańska, A.: Analysis of the use of IT systems by Polish hospitals. Acta Universitatis Nicolai Copernici, EKONOMIA 46(2), 186–193 (2015)
Frączkowski, K., Zwiefka, A., Zaremba, M., Sikora, K.: Patient with complex needs: experience in implementation of LSV - Carewell platform. In: Proceedings of the Fourth International Conference on Telecommunications and Remote Sensing, ICTRS 2015, Rhodes, Greece, pp. 122–128 (2015)

Frączkowski, K., Gładysz, B., Mazur, H., Prałat, E.: Behavioural aspects of communication in the management process of IT projects. In: Wilimowska, Z., Borzemski, L., Świątek, J. (eds.) ISAT 2018. AISC, vol. 854, pp. 335–347. Springer, Cham (2019). https://doi.org/10.1007/978-3-319-99993-7_30

Frączkowski, K., Dabiński, A., Grzesiek, M.: Report on the Polish survey of IT projects 2010, Wrocław (2011). http://pmresearch.pl/wp-content/downloads/report_pmresearchpl.pdf. Accessed 20 Nov 2018

Głuszyński, J., Kowalewska, A.: The public opinion is waiting for the implementation of project P1. Information Bulletin, pp. 8–9. Center of Information Systems for Health Care, Ministry of Health, Warsaw (2012). https://csioz.gov.pl/o-nas/informacja-i-promocja/biuletyninfor macyjny/biuletyny-informacyjny-csioz-wydanie-siodme/. Accessed 18 Nov 2012

Gładysz, B., Frączkowski, K.: Multidimensional analysis of success factors of IT projects. Economic Studies. Scientific Papers of the University of Economics in Katowice 248, pp. 80–89 (2015)

Kisielnicki, J.: The communication system in project teams: problem of transfer of knowledge and information for the management of IT project. Issues Informing Sci. Inf. Technol. 8, 351–361 (2011)

Kuraś, M.: The information system and the information system - apart from the names, what is different between these two objects? Sci. Pap. Cracow Univ. Econ. 770, 259–275 (2009)

Muszyńska, K.: Communication management in Polish IT Companies. Ann. UMCS Informatica AI 11(4), 89–101 (2011)

Mckay, J., Marshall, P., Grainger, N.: Computing and Processing. In: 47th Hawaii International Conference on System Sciences, Waikoloa, HI, USA, pp. 4315–4324 (2014)

Ober, J.: Function and role of effective communication in management, Scientific Papers of the Silesian University of Technology. Organisation and Management 65, pp. 257–266 (2013)

Reed, A.H., Knight, L.V.: Effect of virtual project team environment on communication-related project risk. Int. J. Project Manag. 28, 422–427 (2009)

Stevenson, D.: The Independent ICT in Schools Commission (1997) Information and Communications Technology in UK Schools, an independent inquiry, pp. 36–38 (1997). http://web.archive.org/web/20070104225121/, http://rubble.ultralab.anglia.ac.uk/stevenson/ICT.pdf. Accessed 25 Nov 2018

Soczówka, A.: Effectiveness of the communication process in the scrum team. Acta Universitatis Nicolai Copernici. Management 44(1), 131–141 (2017)

Vann, J.: Resistance to change and the language of public organizations: a look at clashing grammars in large-scale information technology projects. Public Org. Rev.: Global J. 4(1), 47–73 (2004)

The Standish Group Homepage. https://www.standishgroup.com/sample_research_files/CHAOS Report2015-Final.pdf. Accessed 25 Nov 2018

Analysis of the Influence of ICT and Public Recognition on University Credibility

Irvan Santoso[✉], Wayan Suparta[✉], Agung Trisetyarso,
Bahtiar Saleh Abbas, and Chul Ho Kang

Computer Science Department, BINUS Graduate Program –
Doctor of Computer Science, Bina Nusantara University,
Jakarta 11480, Indonesia
{isantoso,atrisetyarso,bahtiars}@binus.edu,
wayan.suparta@upj.ac.id, chkang5136@kw.ac.ar

Abstract. University Credibility is one of the important factors in maintaining the sustainability of the university. However, to maintain credibility, trust from the public on service and learning capabilities as well as recognition extensively are needed. To measure capability in service and learning can be displayed by the technology used during all activities at the university. In this study, an analysis was conducted on the influence of ICT and Public Recognition on University Credibility by using Multiple Linear Regression test. The data used are questionnaire with 13 items representing three variables, namely Information and Communication Technology (ICT), Public Recognition, and University Credibility. Each instrument that is distributed and obtained was used to test the validity and reliability of the items. The results show that 61% of ICT and Public Recognition are proved to influence the credibility of the university. This shows that ICT and Public Regulations play a significant role in showing University Credibility where this can be used as a benchmark in improving good perceptions of university selection.

Keywords: Credibility · ICT · Public · Recognition · Regression

1 Introduction

Inevitably, education is an important factor that supports a person's needs [1]. Education can be obtained both formally and informally [2]. Formal education can be obtained by participating in programs that have been planned and arranged neatly by schools, universities, departments or ministries, and other educational institutions [3]. Whereas non-formal education can be obtained from daily life and various things experienced or learned from others [4].

Formal education is a type of education that is tiring and has certain conditions in registering. This is applied in various private and public schools and universities. Universities as a provider of education for students have different services according to the capacity of their respective universities [5]. Those services are one of the public benchmarks needed to build trust in the quality of the university that involve the ability to obtain awards, achievements of alumni/students, the curriculum used, and others. In addition, technology is also one of the main factors that support the university in

© Springer Nature Switzerland AG 2019
N. T. Nguyen et al. (Eds.): ACIIDS 2019, LNAI 11431, pp. 541–551, 2019.
https://doi.org/10.1007/978-3-030-14799-0_47

demonstrating its reliability and strength in providing maximum educational services [6]. It is also used to facilitate the distribution of important information to students and lecturers known as Information and Communication Technology (ICT) [7]. ICT both directly and indirectly provide tremendous benefits in improving the way people think, learn and work. ICT has become the core that must be considered in all fields including in looking at the capability of the university.

Research that analyzes the capability and credibility of universities is still very minimal. However, there is one study that is quite relevant, namely "An Empirical Study on Credibility of China's University Rankings" [8]. The study discusses a similar topic which is universities compete to provide their best capability for increasing the credibility of their respective universities. In this study, there are three variables used, namely Authority, Trustworthiness, and Level of Social Concern that are tested to obtain the mean of three influential universities in China.

Therefore, this study aims to analyze the influence of ICT and Public Recognition on the University Credibility by using different variables. The first step performed is to determine the dependent variable and the independent variable which are estimated to correlate with each other. Then, number of questions were made to represent each variable that has been determined and distributed publicly. The questions were tested using Pearson Product Moment Correlation test to check the validity and Cronbach Alpha test to check the reliability of items. Furthermore, the correlation between both variables was obtained by using Multiple Linear Regression test. Multiple Linear Regression also produces an equation to see whether the instrument is spread normally/evenly. Thus, we will find out whether ICT and Public Recognition variables have a significant influence on the University Credibility or vice versa.

2   Information and Communication Technology (ICT)

ICT is a term describing a set of techniques to process and convey all forms of information widely [9]. ICT consists of two aspects that cannot be separated from each other, namely information technology and communication technology. Information technology helps in processing information while communication technology helps in the process of transferring information.

In education, ICT has a significant in increasing the efficiency and effectiveness of learning and as a storehouse of knowledge that continues to grow as information increases [10]. This provides the ability to find data in the research process and the dissemination of information quickly and accurately.

3   Research Concerns and Hypothesis

This study uses two independent variables (ICT and Public Recognition) and one dependent variable (University Credibility) which is represented into 13 items that have their particular concerns in accordance with those listed in Table 1. The concern of this research is to prove whether the variables that have been determined have influence with each other or vice versa.

Table 1. Research concerns

Variable	Items	Concerns
ICT	A1	Computer facilities
	A2	Learning media
	A3	Internet media
	A4	Application for learning
Public Recognition	A5	Accreditation of university
	A6	The name of university
	A7	Ranking of university
	A8	Track record of alumni/student achievement
	A9	Track record of awards
University Credibility	A10	Teaching and learning curriculum
	A11	Services provided to students
	A12	Level of security in university
	A13	Level of comfort and ease in getting information

ICT was chosen as one of the variables because most of universities have used technology as a base for learning and information exchange for all their students and lecturers. Then, Public Recognition is also chosen as one of the variables because in determining credibility it requires trust from other parties. This public trust becomes an acknowledgment and recognition of the credibility of the university. However, ICT and Public Recognition are factors that have different standards and assessments, so that these two variables become independent variables which influence the University Credibility variable. University Credibility variable shows the quality, capability, and strength possessed by the university to create public trust. Therefore, this variable is specified as a dependent variable.

Furthermore, the following are hypothesis based on items acquired from the respondents:

- H_1: there is significant influence between ICT and University Credibility.
- H_2: there is significant influence between Public Recognition and University Credibility.
- H_3: there is significant influence between ICT, Public Recognition, and University Credibility.

4   Methodology

4.1   Data Collection

The framework of this research is presented in Fig. 1. Based on the figure, a hypothesis has been formulated after the variable is determined. Data are collected among the students and lectures. The validity and reliability of the data are tested. From the tested data, the analysis of the correlation between the variables will be performed.

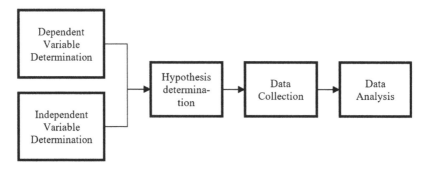

Fig. 1. Research framework

Data was collected using an online questionnaire that was distributed publicly. This questionnaire was conducted to determine the public's views on the factors that affect the university's credibility. Respondents who participated were 204 people with three backgrounds, scilicet Senior High School Students, University Students, and Lecturers. The reason for choosing Senior High School Students as respondents is because their views need to be considered in the selection of higher education in a university, whether the items have a significant value on their decision to choose a university or not. Furthermore, the reason for choosing University Students is because the views of those who have studied also need to be considered. This also provides input to the university on the needs or concerns asked. In addition, Lecturers were also selected as respondents to give their views on the need to maintain the credibility of the university. Table 2 shows the demographics of the respondents.

Table 2. Respondents demography

	Total respondents	Percentage [%]
Age		
17–24	134	66
25–34	31	15
35–44	29	14
45–54	10	5
Gender		
Men	98	48
Women	106	52
Job		
Senior high school student	81	40
University student	60	29
Lecturer	63	31

In data collection, several items are made to represent each variable. Those items are distributed to several parties who have a background related to the dependent variable. Then, after the authors get feedback from the respondents, an analysis of data will be performed by using validity and reliability test to check which items are invalid and unreliable. If there is an invalid or unreliable item, the authors need to replace it or fix it. The last step that must be performed is to test the correlation between the variables that have been determined, whether the independent variables have significant influence and value on the dependent variable or vice versa. The results obtained are used to prove the hypothesis that was made at the beginning.

Table 2 shows that Senior High School Students are the largest respondents who participated at 40%, then Lecturers by 31% and the least is University Students by 29%. In addition, the number of women and men who participated in this questionnaire was almost balanced with a presentation that was not much different, 48% for men and 52% for women of quite diverse ages. In addition, the answers provided in this questionnaire apply the 4 Points Likert Scale method [11] which has four values as follows, Strongly Disagree, Disagree, Agree, and Strongly Agree. Table 3 shows the demography of respondents' answer.

Table 3. Respondents' answers demography

Items	Strongly disagree (SD)	Disagree (D)	Agree (A)	Strongly agree (AS)	Percentage of SD & D [%]	Percentage of A & AS [%]
A1	–	3	75	126	2	98
A2	–	1	111	92	1	99
A3	–	3	88	113	2	98
A4	–	5	95	104	3	97
A5	–	5	115	84	3	97
A6	–	10	111	83	5	95
A7	–	11	95	98	6	94
A8	–	9	100	95	4	96
A9	–	12	100	92	6	94
A10	–	4	95	105	2	98
A11	–	–	88	116	–	100
A12	–	15	81	108	7	93
A13	–	–	90	114	–	100

Based on Table 3, the results obtained from the respondent's answers for each item with Agree and Strongly Agree is more than 90%. Therefore, it can be concluded that each item in question has a concern in accordance with the views of the respondents. However, the data still needs to be tested to obtain the correlation between the variables.

4.2 Measurement Model

In determining the feasibility and reliability of the instruments used, a validity test is conducted using the Pearson Product Moment Correlation (Eqs. 1 and 2) [12] and reliability test by using Cronbach Alpha (Eqs. 3, 4 and 5) [13].

$$r_{XY} = \frac{N \sum XY - \sum X \sum Y}{\sqrt{(N \sum X^2 - (\sum X)^2)(N \sum Y^2 - (\sum Y)^2)}} \tag{1}$$

where:

r_{XY} = correlation coefficient between variables X and Y
N = number of respondents
$\sum X$ = number of question/item
$\sum Y$ = total number of questions/items
$\sum X^2$ = square of number of question/item
$\sum Y^2$ = square of total number of questions/items

$$t_{hit} = \frac{r_{XY}\sqrt{n-2}}{\sqrt{(1 - r_{XY}^2)}} \tag{2}$$

From Eq. 2, if $t_{hit} > t_{table}$ [14], then the item tested is valid.

$$\alpha = \frac{K}{K-1}\left[1 - \frac{\sum S_i^2}{S_t^2}\right] \tag{3}$$

where:

α = instrument reliability coefficient
K = number of question/item
$\sum S_i^2$ = total variance of questions/items
S_t^2 = total variance

Calculates the total variance of questions/items $(\sum S_i^2)$

$$\sum S_i^2 = \frac{\sum X_i^2 - \frac{(\sum X_i)^2}{N}}{N} \tag{4}$$

Calculates the total variance (S_t^2)

$$S_t^2 = \frac{\sum X_t^2 - \frac{(\sum X_t)^2}{N}}{N} \tag{5}$$

Then from each resulting value are processed using the Multiple Linear Regression test [15] which produces the regression equation (Eq. 6) to see whether the distribution of data obtained is normal or vice versa. Multiple Linear Regression is used because there are two independent variables calculated. The authors also calculate F distribution by using ANOVA test [16] where if $F_{hit} > F_{table}$ [17], then there is a significant value between independent variables simultaneously on the dependent variable. In addition, the purpose of this test is to prove the hypothesis that has been determined at the beginning.

$$Y = a + b1.X1 + b2.X2 \tag{6}$$

where:
Y = dependent variable
a = coefficient of dependent variable
b = coefficient of independent variable
$X1$ = first independent variable
$X2$ = second independent variable.

5   Result and Discussion

5.1   Validity and Reliability Test

Before obtaining the correlation between ICT, Public Recognition, and University Credibility, an analysis is carried out using the Pearson Product Moment Correlation and the Cronbach Alpha test. Table 4 shows the results of the Pearson Product Moment Correlation test.

Table 4. Validity of items statistic

Items	t_{hit}	Items	t_{hit}
A1	7.005241	A8	15.5973
A2	12.9047	A9	15.5579
A3	17.47388	A10	16.11135
A4	14.98926	A11	17.39792
A5	12.18284	A12	22.17813
A6	11.64134	A13	10.09163
A7	14.84123		

Because the value of t_{hit} obtained for each question is greater than t_{table} ($t_{hit} > t_{table}$), then the items tested are valid. Furthermore, the Cronbach Alpha value obtained from 13 items is 0.911. Based on [18] then all items tested are acceptable.

5.2   Multiple Linear Regression Test

In this study, Multiple Linear Regression test is calculated using SPSS software [19] which is used to determine the relationship of influence between one variable and another variable. In this study, the variables that influence dependent variable are ICT and Public Recognition, while for the variable that is affected is University Credibility. Table 5 describes the summary model obtained.

Table 5. Model summary of multiple linear regression test

R	R square	Adjusted R square	Std. error of the estimate
0.784	0.615	0.611	1.111

Based on Table 5, the value of R Square generated is 0.615 from the correlation coefficient of 0.784. Therefore, in this case, ICT and Public Recognition affect the University Credibility by 61% while the rest can be explained by other variables.

To find out whether ICT and Public Recognition variables have significant value simultaneously to University Credibility or not, it can also be done by calculating ANOVA and comparing F_{hit} value with F_{table}. Table 6 below shows the result of ANOVA test.

Table 6. ANOVA test

Model	Sum of squares	Mean square	F	Sig
Regression	396.441	198.221	160.458	0.000
Residual	248.304	1.235		

Based on Table 6, the significant value is 0.000, which is smaller than 0.05 ($0.000 < 0.05$) and F_{hit} is greater than F_{table} ($F_{hit} > F_{table}$), it can be concluded that the University Credibility variable is influenced by ICT and Public Recognition variable simultaneously. Therefore, the results obtained from Multiple Linear Regression Test meet H_3, i.e. there is significant influence between ICT, Public Recognition, and University Credibility.

Next, a comparison is made by checking the significant value of each variable to obtain the regression equation. Table 7 shows the coefficients used for the regression equation.

Table 7. Coefficients for regression equation

Variable	Unstandardized coefficients		Sig.
	B	Std. error	
University Credibility	1.552	0.716	0.031
ICT	0.419	0.062	0.000
Public Recognition	0.390	0.045	0.000

Because the significant value obtained is 0.000 for each independent variable and 0.0031 for dependent variable, which is smaller than 0.05 (0.000 < 0.05), it can be concluded that the value in column B is significant. So, the most appropriate equation for those variables are $Y = 1.552 + 0.419X_1 + 0.390X_2$, where $Y =$ University Credibility, $X_1 =$ ICT and $X_2 =$ Public Recognition.

From the equation that has been obtained, Fig. 2 shows Normal Q-Q Plots for each independent variable (ICT in Fig. 2(a) and Public Recognition in Fig. 2(b)) towards dependent variable (University Credibility). The graph is separated with the intention to show the status of its distribution, whether the distribution is normal or vice versa. The spreads can be concluded as normal because the graphs formed are approximately in a straight line. Then, the two graphs are combined to see the overall picture as in Fig. 3.

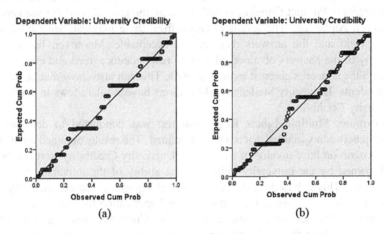

(a)   (b)

Fig. 2. Normal Q-Q plots for each independent variable

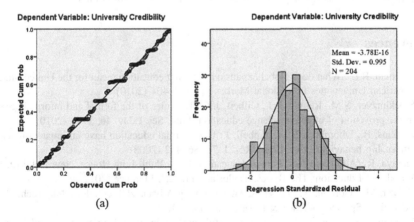

(a)   (b)

Fig. 3. Normal Q-Q plots and curve histogram

Figure 3(a) shows Normal Q-Q Plots and produces normal values because the combined independent variables value points on the dependent variable are approximately in a straight line. Then, Fig. 3(b) is another form of interpretation using a curve, and it can be explained that the data is normal because the curve formed is symmetrical like a bell. This shows that the distribution of data is evenly distributed.

6 Conclusion

This research was conducted to find out whether the ICT and Public Recognition have a significant influence on a university's credibility or not. This research was conducted by distributing questionnaires by determining three variables that will be tested by using the Pearson Product Correlation and checking the reliability of the answers with the Cronbach Alpha test. The results obtained from these two tests are the questions asked are valid and the answers received are acceptable. Moreover, based on the demography of the answers obtained, almost all respondents agreed and even strongly agreed with the concerns given at more than 90%. This can also show that Senior High School Students, University Students, and Lecturers have similar views in determining the University Credibility.

Furthermore, Multiple Linear Regression test was conducted to determine the relationship between variables that have been defined. The results obtained are ICT and Public Recognition have a value of 61% of the University Credibility so it can be said that ICT owned by the university influences the ability of the university to provide better services. This is based on the concerns of the questions raised, including computer facilities, learning media, internet media, and application for learning. Moreover, the influence of Public Recognition on University Credibility also has a significant effect because without the Public Recognition a university cannot maintain its continuity. In addition, because the public is now more thorough and wiser in choosing, these factors can become a public reference in choosing a university.

References

1. Freeman, R.B.: What does global expansion of higher education mean for the United States? American Universities in a Global Market, pp. 373–404 (2010)
2. Stocklmayer, S.M., Rennie, L.J., Gilbert, J.K.: The roles of the formal and informal sectors in the provision of effective science education. Stud. Sci. Educ. **46**, 1–44 (2010)
3. Hoskins, B., Dhombres, B., Campbell, J.: Does formal education have an impact on active citizenship behaviour? Eur. Educ. Res. J. **7**, 386–402 (2008)
4. Pandya, R.: Adult and Non Formal Education. Gyan Publishing House, New Delhi (2010)
5. Wood, K.: Education: The Basics. Taylor and Francis, Florence (2012)
6. Hacker, M., Gordon, A., de Vries, M: Integrating Advanced Technology Into Technology Education. Springer Science & Business Media (2012)
7. Bingimlas, K.A.: Barriers to the successful integration of ICT in teaching and learning environments: a review of the literature. Eurasia J. Math. Sci. Technol. Educ. **5**, 235–245 (2009)

8. Ying, Y., Jingao, Z.: An empirical study on credibility of Chinas university rankings. Chin. Educ. Soc. **42**, 70–80 (2009)
9. Linawati, Mahendra, M.S., Neuhold, E.J., Tjoa, A.M., You, I.: Information and Communication Technology. Springer, Berlin (2014). https://doi.org/10.1007/978-3-642-55032-4
10. Huang, R., Kinshuk, P., Jon, K.: ICT in Education in Global Context: Comparative Reports of Innovations in K-12 Education. Springer, Heidelberg (2016). https://doi.org/10.1007/978-3-662-47956-8
11. Leung, S.-O.: A comparison of psychometric properties and normality in 4-, 5-, 6-, and 11-point Likert scales. J. Soc. Serv. Res. **37**, 412–421 (2011)
12. Pearson Product-Moment Correlation Coefficient: Encyclopedia of Research Design (2010)
13. Bonett, D.G., Wright, T.A.: Cronbachs alpha reliability: interval estimation, hypothesis testing, and sample size planning. J. Organ. Behav. **36**, 3–15 (2014)
14. StatPrimer (Version 7.0). http://www.sjsu.edu/faculty/gerstman/StatPrimer/t-table.pdf. Accessed 07 Dec 2018
15. Making sense of Cronbach's alpha – IJME. https://www.ijme.net/archive/2/cronbachs-alpha.pdf. Accessed 07 Dec 2018
16. Gamst, G., Meyers, L.S., Guarino, A.J.: Analysis of Variance Designs: A Conceptual and Computational Approach with SPSS and SAS. Cambridge University Press, Cambridge (2008)
17. StatPrimer (Version 7.0). http://www.sjsu.edu/faculty/gerstman/StatPrimer/F-table.pdf. Accessed 07 Dec 2018
18. Yan, X., Su, X.: Linear Regression Analysis: Theory and Computing. World Scientific, Hackensack (2009)
19. Field, A.: Discovering Statistics Using IBM SPSS Statistics. Sage Publications, Los Angeles (2018)

Quantum Game-Based Recommender Systems for Disruptive Innovations

Agung Trisetyarso[1]([⊠])[iD] and Fithra Faisal Hastiadi[2][iD]

[1] Computer Science Department, BINUS Graduate Program,
Doctor of Computer Science, Universitas Bina Nusantara,
Daerah Khusus Ibukota Jakarta, 11480 Jakarta, Indonesia
atrisetyarso@binus.edu
[2] Faculty of Economic and Business, Universitas Indonesia,
Kampus Universitas Indonesia, Depok, Indonesia
fithra@ui.ac.id

Abstract. A recommender system based on quantum game for disruptive innovations is proposed. Permutation operators are used to determine the dominance of players on the game, while several possibilities of measurement scheme of quantum game are utilized for harnessing disruptive innovations. It is shown in this manuscript that mathematical properties in quantum mechanics can be implemented for the extensions of recommender system capabilities.

Keywords: Recommender systems · Quantum game ·
Disruptive innovations

1 Introduction

The clash between innovation-led start ups and incumbent monopoly power has been happening for ages. Jolly *et al.* argue that the aggressivity of these start ups was the major impetus that built up persistent challenge to the incumbents since the third industrial revolution in 1969 [14]. The period of 1980's was the time when these startups grooming into global competitiveness. They also argue that these start ups overcame two drawbacks simultaneously, one is becoming the start ups itself the latter was to be able compete with the already established global players. They said start ups were able not only to survive but also to do the leapfrogging some of the intermediate stages of internationalization through technological advancement. Moreover, what makes it count are complex strategies that involved the whole market ecosystem. Smink *et al.* shed the light of institutional strategies of incumbent firms to cope with the persistent disruptions by the new entrants [19]. Nevertheless, the continuous scuffle between incumbents and new entrants has been leading to market transformation just as

Supported by Bina Nusantara University.

N. T. Nguyen et al. (Eds.): ACIIDS 2019, LNAI 11431, pp. 552–561, 2019.
https://doi.org/10.1007/978-3-030-14799-0_48

Hockerts and Wustenhagen famous interplay of 'Greening Goliath' (incumbents) with 'Emerging Davids' that is well structuring the transformation [12].

The emerging of innovations is not uniform and a non-linear problem: for example, the rise of e-commerce in developing countries such as Indonesia [17]. Another example is peer-to-peer markets which are internet-based. Including in these emerging peer-to-peer markets are ride sharing services, goods and services marketplaces, information technology freelancer, peer-to-peer loans, local crafts, new entrants financing, accommodation, automotive service, and currency exchange [7]. Therefore, the emerging of innovations depends on complex parameters.

2   Related Works

One of the problem in disruptive innovations is how to catch the next wave [6]. The problems rely on the Christensen effect "trap": the incumbents are always trying to keep their customer within the high end market regardless the ongoing disruptions from the new entrants. Let us assume that there is a black box which belongs to an incumbent: the parameters of black box inside the incumbents include administration, egotism, tired managerial blood, lack of planning, and short-term investment horizons [4]. The dynamics of market is non-linear which open the possibility for new entrants and technology to catch the new wave of innovations. The disruptive innovations ride new entrants to grow exponentially and overtaking the market of incumbents.

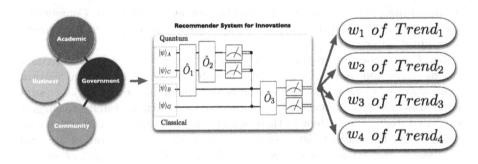

Fig. 1. A framework of a recommender system based on quantum game of innovations ecosystem.

In previous works [23], we showed the quantum game of quadruple innovations ecosystem and its quantum algorithm variations. We already show in ISESD 2018 [23] that non-credible threat-inspired quantum game is possible to model the ecosystem. However, it is still unclear to obtain a recommender system for disruptive innovations, especially when the gamification between the members of quadruple innovations ecosystem is taken into account.

3   Description of the Contribution

In this manuscript, we develop a recommender system which is based on *Eisert-Wilkens-Lewenstein* quantum strategies [8]; the users in this context are the members of quadruple helix innovations ecosystem and the items are innovation trends. The recommendation system of disruptive innovations are achieved through measurement strategy over the player: on the *half-round* of the process or on every knot of helix during gamification. During the gamification process of quadruple innovations ecosystem, collaborative ratings on innovation trends are occurred. Information of recommender system arrives from communication overlay each time players corresponding via social media, workshops, conferences or other meeting. The outcomes of this gamification are estimated ratings of innovation trends. This approach proposes less intrusive and flexible recommendation system.

4   Theoretical Verification

Let $\mathcal{C}$ be the set of all members of quadruple helix innovation ecosystem and let $\mathcal{S}$ be the set of all possible innovation trends that can be suggested, such as Deep Learning, Connected Home, or Blockchain. Let ν be a utility function that measures the disruptness of innovation trend κ to member ζ, i.e., $\nu: \mathcal{C} \times \mathcal{S} \to \mathcal{R}$, where $\mathcal{R}$ is a recommender system matrix. For each member $\zeta \in \mathcal{C}$, there is such trend $\kappa' \in \mathcal{S}$ that maximizes the member's utility. One can write,

$$\forall \zeta \in \mathcal{C}, \quad \kappa'_\zeta = \arg \max(\zeta, \kappa) \tag{1}$$

In this manuscript, $|\psi\rangle_\alpha \in \mathcal{C}$, where α can be *Academics, Business, Community, and Government*. Equation (12) is related to *Christensen effect* in the context of disruptive innovations [6].

Let $|Content(\kappa)\rangle$ be an innovation trend profile, i.e., a set of features characterizing innovation trend κ. It is mostly calculated by extracting a set of features from innovation trend κ (its content) and is used to obtain the appropriateness of the item for recommendation goals. Due to the construction systems to suggest text-based items, the content in these systems is usually stated with keywords. For instance, the characteristics of technologies for Open Innovation are *idea submission, problem submission, problem solving and analysis, evaluation, collaboration, and marketplace* [13]. The expression is as follows,

$$|Content(\kappa)\rangle = \sum_i^6 w_i |\kappa_i\rangle \tag{2}$$

The 'significance' (or 'meaningfulness') of word ω_j in document π_j is determined with several weighting measurement ξ_{ij} that can be explained in many ways.

In this proposition, quantum strategies correspond to $|ContentBased Profile(\nu)\rangle$ as follows,

$$|\psi\rangle_\alpha = \sum_n |ContentBasedProfile(\nu)_n\rangle \tag{3}$$

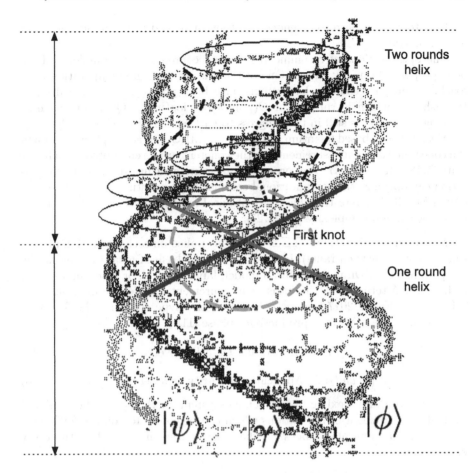

Fig. 2. The measurement and dominance factor of players in data topology [9] of triple helix innovations ecosystem depicting layer of communications is explored using equipments in quantum theory; in this approach, the dominance of first player over other players are explained by permutations operators. Measurement is conducted on the helix to forecast the evolution of innovations. $|\psi\rangle, |\gamma\rangle$, and $|\phi\rangle$ are quantum and classical state of innovations ecosystem members.

The variables of $|ContentBasedProfile(\nu)_n\rangle$ are administration, emotions, managerial blood, planning, and investment horizons [4] (Fig. 2).

Collaborative recommender system in this context attempts to forecast the utility of innovation trend for a particular quadruple helix innovation ecosystem member based on the items previously used by other members. For example, midsize enterprises trends in United States small and medium enterprises, such as virtual machine backup and recovery, disaster recovery as a service, application performance monitoring suites, and information technology event correlation and analysis tools, can be recommended to other small and medium enterprises in other countries [11].

The optimisations of recommender system computational complexity have been proposed by many efforts. One of them is by quantum computation approach. Quantum recommender systems have been proposed by Kerenedis and Prakash [15] which is running in $O(\text{poly}(k)\text{polylog}(mn))$. Quantum machine learning in general is still in the progress to cutting the edge the field [3].

Measurement-based quantum computation (MBQC) is accomplished through the concatenation of measurements on the coherent qubits encrypted in a cluster state [5, 18]. Raussendorf *et al.* prove that measurement-based quantum computation can optimize the computational complexity of quantum circuit. The novelty of MBQC is interesting due to its physical implementations and advantage in computational complexity [21, 24, 25].

Measurement operation, ⊣⊠⊢, here is applied on each innovation players (incumbents or new entrants). In this context, the basis of measurement can be the parameters in Dirac-Solow-Swan model, labor, capital, or labor-augmenting technology. Alternatively, the basis of measurement can be one of technological trends [10]. For 2017 midsize enterprises, the hype cycle has the following innovations state of plateau productivity phase [11]:

$$|Trend_1\rangle := \text{Visual Data Discovery}. \tag{4a}$$

$$|Trend_2\rangle := \text{Application Security as a Service}. \tag{4b}$$

Less intrusive and flexible recommender system methods can be achieved through measurement of innovations which can be conducted on the *half-round* of helix of communications between the ecosystem members and on the every knot of helix.

5   Results Analysis

We improve our quantum game model, so that the vector effect can be taken into account,

$$\varepsilon_{\alpha\beta\gamma\delta}\mathcal{U}|\psi\rangle_{A,\{\alpha\}}|\psi\rangle_{B,\{\beta\}}|\psi\rangle_{C,\{\gamma\}}|\psi\rangle_{G,\{\delta\}}. \tag{5}$$

Odd permutation, $\{-1\}$, corresponds to $\otimes$ or out of the surface, while even permutation, $\{+1\}$, corresponds to $\odot$, or into the surface. We argue that this permutation operator can be implemented into the type of knot in helix form of ecosystem: if the knot of the helix out of plane, it is interpreted that the player is in dominance, vice versa.

The initial composition from previous work [23] is described as follows,

$$\varepsilon_{3124}\mathcal{U}|\psi\rangle_{A,\{3\}}|\psi\rangle_{B,\{1\}}|\psi\rangle_{C,\{2\}}|\psi\rangle_{G,\{4\}} \tag{6a}$$

$$\to \varepsilon_{1234}\mathcal{U}|\psi\rangle_{A,\{1\}}|\psi\rangle_{B,\{2\}}|\psi\rangle_{C,\{3\}}|\psi\rangle_{G,\{4\}} \tag{6b}$$

$$\to \varepsilon_{3142}\mathcal{U}|\psi\rangle_{A,\{3\}}|\psi\rangle_{B,\{1\}}|\psi\rangle_{C,\{4\}}|\psi\rangle_{G,\{2\}} \tag{6c}$$

Quantum

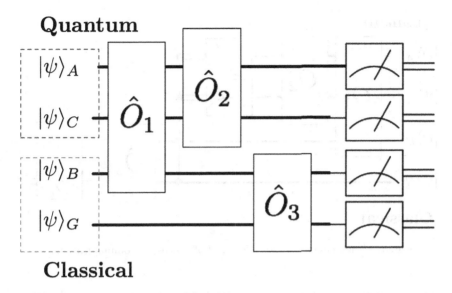

Classical

Fig. 3. A circuit of quantum games for a measurement of *full-round* helix dynamics.

This chain is performing one round of knot. Another possibility, if the knowledge creation is initiated by Academics and Community, followed by Business and Government, we would have the following scheme:

$$\varepsilon_{3124}\mathcal{U}|\psi\rangle_{A,\{1\}}|\psi\rangle_{B,\{3\}}|\psi\rangle_{C,\{2\}}|\psi\rangle_{G,\{4\}} \tag{7a}$$

$$\rightarrow \varepsilon_{1234}\mathcal{U}|\psi\rangle_{A,\{1\}}|\psi\rangle_{B,\{2\}}|\psi\rangle_{C,\{3\}}|\psi\rangle_{G,\{4\}} \tag{7b}$$

$$\rightarrow \varepsilon_{3142}\mathcal{U}|\psi\rangle_{A,\{3\}}|\psi\rangle_{B,\{4\}}|\psi\rangle_{C,\{1\}}|\psi\rangle_{G,\{2\}} \tag{7c}$$

Therefore, the architecture would be depicted in Fig. 3.

As shown in Fig. 3, a measurement-based strategy on quantum games of innovations ecosystem to determine the knowledge creation which leads into disruptive innovations. This kind of circuit performs *"global"* disruptive innovations in which has the mathematical expression of recommender system as follows,

$$\forall \zeta \in \{|\psi\rangle_A, \psi\rangle_C, |\psi\rangle_B, \psi\rangle_G\}, \quad (Trend)'_\zeta = \arg\max(\zeta, Trend). \tag{8}$$

In this scheme, innovation ecosystem is represented by quantum state $|\psi\rangle = e^{i\omega_k t}|\phi\rangle_k$. Quantum game operations on the quantum state leads $|\psi'\rangle = e^{i\omega'_k t}|\phi\rangle_k$. Measurement on the outcomes leads $\prod_{n=1}^{4}\langle\hat{e}_n|\psi'\rangle$. The measurement outcomes on a player is $\{w_k, s_k\}$, where w_k is the weight and s_k is the player strategy. In this sense, a quantum strategy state of a player can be decomposed into a superposition of content based profiles as follows,

$$|\psi\rangle_{i,\{j\}} = \sum_n w'_n |ContentBasedProfile(\nu)_n\rangle \tag{9}$$

Quantum

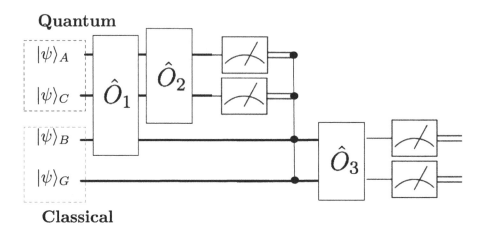

Classical

Fig. 4. Measurement on the *half-round* of process of communications.

In this context, Eq. (9) can be implemented to improve recommender system capabilities in order to understand innovation trends [1]. One can perform a measurement of a quantum strategy state of a player by an innovation trend state, $|Trend\rangle_m$, to obtain the weight of the state, w_m

$$_m\langle Trend|\psi\rangle_{i,\{j\}} = w_m. \tag{10}$$

Therefore, for n number of trends in innovations, one can measure as following to obtain the weight of every trend,

$$\prod_m^n \langle Trend|\psi\rangle_{i,\{j\}} = \sum_m^n w_m. \tag{11}$$

From Eq. (11), one can find the recommendation list of trend in disruptive innovations as shown in Fig. 1.

In Fig. 4, it is shown an instance for *"local"* disruptive innovations which is occurred on *Business* and *Government*,

$$\forall \zeta \in \{|\psi\rangle_B, \psi\rangle_G\}, \quad (Trend)'_\zeta = \arg \max(\{|\psi\rangle_B, \psi\rangle_G\}, Trend). \tag{12}$$

As shown in Eqs. (8) and (12), the way we harness innovation in this context can be seen through the measurement schemes that determine the type of disruptive innovations. They said innovations can take two unique forms as *global* or *local* innovations.

The key take-away from the above wide range of equations lies on the interplay between the incumbents and the new entrants within the ecosystem. Thus said, this triangulation creates efficiency in the market that serves toward sustainable innovation. Here we define incumbents take form as government and business while new entrants are attributed to University and Community. The continuous innovation from new entrants may lead the incumbents to be part of the innovation ecosystem. The incumbents-although lagging in the process of innovation-may occasionally utilize the innovation that is driven by the new entrants. The incumbents can accentuate the innovation through their already established system thanks to the economies of scale. This is serving as a very fundamental basis towards sustainability development of each innovations. Hockerts and Wustenhagen [12] put the interplay as co-evolution, where the incumbents and the new entrants moves independently to support market transformation.

This overall participation for the ecosystem can be best coined through the perspective of Uber in the US. Since their successful milestone, many have copied their business model in what we call now as the "Uberification" phenomenon. Governments have been generally not able to monitor Uber's operations in their jurisdictions due to its operations are conducted mainly over the Internet. Certainly we have found the notion of disruptive innovation characteristics in Uber: cheaper and better quality of services. The key message, as economists we need time to really understand the phenomenon let alone to justify the good and better recommendation on any events surrounding the technological determinism.

6 Conclusions

The architecture of a quantum game based-recommender system, utilizing measurement scheme and permutation operators, are presented. The proposed system may be included into a type of recommender systems for investors and useful to explain the complexity of innovations ecosystem in developing countries [1]. This approach may explain the emerging of local innovations in developing countries, such as innovations in peer-to-peer food delivery, ridesharing, automotive service, and transportation services [16]. The nonlinearity of the local innovations dynamics and its effect on society is still open problem to be observed. The proposed framework will be integrated into Dirac-Solow-Swan model which is suitable for describing of incumbents and new entrants dynamics [22]. Moreover, it is interesting to observe the correlation between Nash equilibrium, capital accumulation, and estimated ratings in recommender systems.

For future work, we would like to implement Benjamin-Hayden quantum strategies [2], in order to optimize the computational complexity. This approach may provide novelties for more recommendation system capabilities extensions. It would be interesting if this quantum strategies is reconciled with a quantum-inspired classical algorithm for recommendation systems to meet recent state-of-the-art technology [20].

Acknowledgements. This research was funded by Directorate Research and Community Services (Universitas Indonesia) under PITTA grant.

References

1. Adomavicius, G., Tuzhilin, A.: Toward the next generation of recommender systems: a survey of the state-of-the-art and possible extensions. IEEE Trans. Knowl. Data Eng. **6**, 734–749 (2005)
2. Benjamin, S.C., Hayden, P.M.: Multiplayer quantum games. Phy. Rev. A **64**(3), 030301 (2001)
3. Biamonte, J., Wittek, P., Pancotti, N., Rebentrost, P., Wiebe, N., Lloyd, S.: Quantum machine learning. Nature **549**(7671), 195 (2017)
4. Bower, J.L., Christensen, C.M.: Disruptive Technologies: Catching the Wave. Harvard Business Review Video, Boston (1995)
5. Briegel, H.J., Browne, D.E., Dur, W., Raussendorf, R., Van den Nest, M.: Measurement-based quantum computation. Nat. Phys. 19–26 (2009). https://doi.org/10.1038/nphys1157
6. Christensen, C.M.: The ongoing process of building a theory of disruption. J. Prod. Innov. Manag. **23**(1), 39–55 (2006)
7. Einav, L., Farronato, C., Levin, J.: Peer-to-peer markets. Ann. Rev. Econ. **8**, 615–635 (2016)
8. Eisert, J., Wilkens, M., Lewenstein, M.: Quantum games and quantum strategies. Phys. Rev. Lett. **83**(15), 3077 (1999)
9. Etzkowitz, H., Leydesdorff, L.: The dynamics of innovation: from national systems and "mode 2" to a triple Helix of university-industry-government relations. Research Policy **29**(2), 109–123 (2000)
10. Gartner: Hype cycle. Hype Cycle Special Report **1** August 2014
11. Gartner: Hype cycle. Hype Cycle Midsize Enterprises **19**, July 2017
12. Hockerts, K., Wüstenhagen, R.: Greening goliaths versus emerging Davids–theorizing about the role of incumbents and new entrants in sustainable entrepreneurship. J. Bus. Ventur. **25**(5), 481–492 (2010)
13. Hrastinski, S., Kviselius, N.Z., Ozan, H., Edenius, M.: A review of technologies for open innovation: characteristics and future trends. In: 2010 43rd Hawaii International Conference on System Sciences (HICSS), pp. 1–10. IEEE (2010)
14. Jolly, V.K., Alahuhta, M., Jeannet, J.P.: Challenging the incumbents: how high technology start-ups compete globally. Strat. Change **1**(2), 71–82 (1992)
15. Kerenidis, I., Prakash, A.: Quantum recommendation systems. arXiv preprint arXiv:1603.08675 (2016)
16. Martin, C.J.: The sharing economy: a pathway to sustainability or a nightmarish form of neoliberal capitalism? Ecol. Econ. **121**, 149–159 (2016)
17. Ndou, V.: E-government for developing countries: opportunities and challenges. Electron. J. Inf. Syst. Dev. Countries **18**(1), 1–24 (2004)
18. Raussendorf, R., Briegel, H.J.: A one-way quantum computer. Phys. Rev. Lett. **86**(22), 5188–5191 (2001)
19. Smink, M.M., Hekkert, M.P., Negro, S.O.: Keeping sustainable innovation on a leash? Exploring incumbents' institutional strategies. Bus. Strat. Environ. **24**(2), 86–101 (2015)
20. Tang, E.: A quantum-inspired classical algorithm for recommendation systems. arXiv preprint arXiv:1807.04271 (2018)
21. Trisetyarso, A.: Theoretical study towards realization of measurement-based quantum computers. Ph.D. thesis, Keio University (2011)
22. Trisetyarso, A., Hastiadi, F.F.: Disruptive innovations dynamics. In: 2016 11th International Conference on Knowledge, Information and Creativity Support Systems (KICSS), pp. 1–4. IEEE (2016)

23. Trisetyarso, A., Hastiadi, F.F.: Quantum games of quadruple helix ecosystem. In: 2018 International Symposium on Electronics and Smart Devices (ISESD). IEEE (2018)
24. Walther, P., et al.: Experimental one-way quantum computing. Nature **434**(7030), 169–176 (2005). https://doi.org/10.1038/nature03347
25. You, J.Q., Wang, X., Tanamoto, T., Nori, F.: Efficient one-step generation of large cluster states with solid-state circuits. Phys. Rev. A 75(5) (2007). https://doi.org/10.1103/PhysRevA.75.052319

Computer Vision Techniques

Matching Topological Structures
for Handwritten Character Recognition

Daw-Ran Liou, Yang-En Chen, and Cheng-Yuan Liou$^{(\boxtimes)}$

Department of Computer Science and Information Engineering,
National Taiwan University, Taipei, Taiwan
cyliou@csie.ntu.edu.tw

Abstract. This work presents a locking and deforming process to accomplish the recognition. The best matched features of the template are locked to their target features of the unknown pattern. The whole template is then deformed and calibrated according to these features. Improved similarity score can be obtained from the deformed template. This work illustrates this process and its operations. This process indirectly overcomes difficult distortion problems.

Keywords: Pattern recognition · Image recognition · Image classification · Image restoration · Robot vision · Handwritten character recognition

1 Introduction

The spatial topology distance method (STD) [9, 10] used the self-organizing algorithm (SOM) [4] to evolve the template toward its target, unknown handwritten character. This evolution can preserve the topological structure of the template and resolve serious distortions in the character. The computation cost of the evolution is high. To reduce the number of iterations in the evolution, geometrical constraints among features are manually constructed to reduce the searching operation in STD. These constraints can be converted into a set of rules [6, 7], Fig. 1, and saved in a Hopfield network [7, 11]. The number of such rules is countless. It is tedious to list the enormous number of rules manually. The compatibility of the connection rules for a whole template can be loaded in the Hopfield network [7, 11]. It can be obtained from the converged network [12, 13]. So far, there is no efficient procedure for the construction of such rules. This work devises an alternative way to skip those rules in the recognition. The same features used to represent the character and the template [11] will be used in this work. We briefly review the feature and related works in this section.

The unknown character and all its templates are prepared in advance by their feature collections that resemble the receptive fields [1–3] of visual system. All collections are loaded with flexible and variable features along their skeletons. The global operations, shift, rotation, and scaling, are applied to deform the template according to certain highly matched, locked, features. After these operations, the locked features in the deformed template will align with their corresponding features in the unknown character perfectly. The improved whole similarity score can be obtained for all features of the deformed template. This kind deforming and locking process can be

N. T. Nguyen et al. (Eds.): ACIIDS 2019, LNAI 11431, pp. 565–575, 2019.
https://doi.org/10.1007/978-3-030-14799-0_49

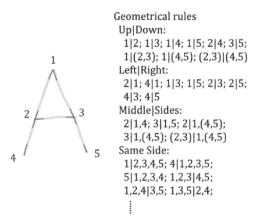

Fig. 1. Geometrical rules for the five features of character "A".

operated iteratively. The recognition is accomplished for the template which received the highest whole score.

The templates and unknown character are both represented by collections of bended-ellipse features [10, 11], Fig. 2. Each bended-ellipse feature is represented by a vector which includes its coordinates, direction, angle, and arm length size. To save space, we skip the length portion in this work. This kind feature representation is much stable than the rigid skeleton representation. Each template outline is used to normalize the unknown by properly shifting, rotating, and scaling the unknown. The outline can be accomplished by thickening the template strokes.

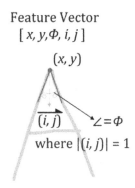

Fig. 2. The feature vector for the top portion of A.

Note the CNN [5] can extract local features in its front end layer as elements for constructions in its successive layers. High level correlations of such elements are picked in its deep layers. Recognition of unknown character can be accomplished from such correlations. Various distortions will increase its correlation complexity.

Representation by Bended-Ellipse Features

Suppose there are J templates $\{\mathbf{T}^1, \mathbf{T}^2, \ldots, \mathbf{T}^J\}$. Each template is represented by a collection of its all features. For each template $\mathbf{T}^j$ we will generate L^j bended-ellipse features. Write $\mathbf{T}^j = \{\mathbf{t}_l^j | 1 \leq l \leq L^j\}$ where $\mathbf{t}_l^j$ is the lth feature vector of the jth template. Let $\mathbf{t}_l^j = [x_l^j, y_l^j, \phi_l^j, i_l^j, j_l^j]$ be a five-dimensional row vector. The elements x_l^j and y_l^j in $\mathbf{t}_l^j$ are the x-y coordinates of the seed (corner) of the angle. Let (i_l^j, j_l^j) be the unit vector that extends from the seed and equally divides the angle ϕ_l^j, Fig. 1.

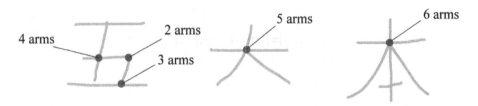

Fig. 3. Branch and cross.

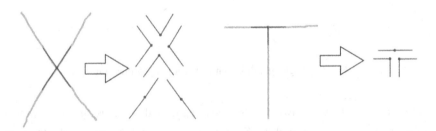

Fig. 4. Cross and branch points are represented by multiple features with a shared seed position.

The branch point and cross point will have three or more arms, Fig. 3. They will be represented by collections of bended-ellipse features with shared seed positions, Fig. 4. For example, a branch with 3 arms will be represented by 3 bended-ellipses and a cross with 4 arms will be represented by 6 bended-ellipses. A fork with more branches is important in recognition.

The feature collection that represents the target character, unknown handwritten character, is $\mathbf{U} = \{\mathbf{u}_m | 1 \leq m \leq M\}$, where $\mathbf{u}_m = [x_m, y_m, \phi_m, i_m, j_m]$ is the mth feature vector of the unknown character.

2   Locking and Deforming Process

All templates and their various histograms must be well prepared in advance for candidate template screening. The unknown character must have more number of features, two to three times, than that of the template. To save computation, each template is carefully selected from the character corpus by the screening procedure.

It is expected that each selected template is additive to the recognition rate [10]. The template outline is used in the screening. The distribution of the template features in the i-j-ϕ diagram, Fig. 5, is also included as one related histogram.

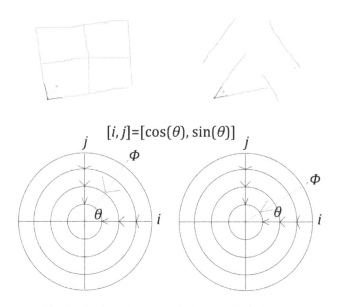

Fig. 5. Angle and direction histogram, i-j-ϕ diagram.

The similarity between two features, $s\left(\mathbf{t}_l^j, \mathbf{u}_m\right)$, is defined as the inner product between the two features, $\mathbf{t}_l^j$ and $\mathbf{u}_m$. Set parameters that will be used in the following context, $\mathbf{t}_l^j = [x^t, y^t, \phi^t, i^t, j^t]$; $\mathbf{u}_m = [x^u, y^u, \phi^u, i^u, j^u]$;

$$
\begin{aligned}
\theta &= \cos^{-1}\left(\frac{\|(i^u, j^u)\|^2 + \|(i^t, j^t)\|^2 - \|(i^t, j^t) - (i^u, j^u)\|^2}{2\|(i^t, j^t)\|\|(i^u, j^u)\|}\right) \\
&= \cos^{-1}\left(\frac{2 - \|(i^t, j^t) - (i^u, j^u)\|^2}{2}\right);
\end{aligned}
\tag{1}
$$

$$
s_1 = \max\left\{0, \left(1 - \frac{\sqrt{(x^t - x^u)^2 + (y^t - y^u)^2}}{d}\right)\right\};
\tag{2}
$$

$$
s_2 = \max\left\{0, \left(1 - \frac{|\phi^t - \phi^u|}{\frac{\pi}{2}}\right)\right\}; \quad s_3 = \max\left\{0, \left(1 - \frac{|\theta|}{\frac{\pi}{2}}\right)\right\}; \quad \text{and}
\tag{3}
$$

$$
s_4 = \max\{0, (1 - |i^t i^u + j^t j^u|)\}.
\tag{4}
$$

$$\text{Then write } s\left(t_l^j, \mathbf{u}_m\right) = \begin{cases} \frac{1}{2}(s_1 + s_4), & \phi^t \geq \phi_{\text{line}} \\ \frac{1}{3}(s_1 + s_2 + s_3), & \text{otherwise} \end{cases}.$$

$$\text{Note that } 0 \leq s\left(t_l^j, \mathbf{u}_m\right) \leq 1. \tag{5}$$

ϕ_{line} is a threshold to discriminate a straight line feature from a bended feature. We also use an alternate function in this work,

$$s\left(t_l^j, \mathbf{u}_m\right) = \min\left\{s_p\left(t_l^j, \mathbf{u}_m\right), s_d\left(t_l^j, \mathbf{u}_m\right)\right\}, \tag{6}$$

where s_p is the similarity between positions, s_d is the similarity between both angles and directions. Let

$$p(\mathbf{v}) = p([x, y, \phi, i, j]) = [x, y]. \tag{7}$$

Then the similarity of positions between two feature vectors is

$$s_p\left(t_l^j, \mathbf{u}_m\right) = f_p\left(\frac{1}{w}\left\|p\left(t_l^j\right) - p(\mathbf{u}_m)\right\|\right),$$

where

$$f_p(x) = 2^{-8x^3} \text{ and}$$
$$w = d_1 \text{ or } d_2. \tag{8}$$

The similarity between both angles and directions is

$$s_d(\mathbf{t}, \mathbf{u}) = \frac{1}{2} + \frac{1}{4}\max\{a_1(\mathbf{t}) \cdot a_1(\mathbf{u}) + a_2(\mathbf{t}) \cdot a_2(\mathbf{u}), a_1(\mathbf{t}) \cdot a_2(\mathbf{u}) + a_2(\mathbf{t}) \cdot a_1(\mathbf{u})\},$$

where

$$a_1(\mathbf{v}) = a_1([x, y, \phi, i, j]) = [i, j]\begin{bmatrix} \cos\frac{\phi}{2} & \sin\frac{\phi}{2} \\ -\sin\frac{\phi}{2} & \cos\frac{\phi}{2} \end{bmatrix} \text{ and}$$
$$a_2(\mathbf{v}) = a_2([x, y, \phi, i, j]) = [i, j]\begin{bmatrix} \cos\frac{-\phi}{2} & \sin\frac{-\phi}{2} \\ -\sin\frac{-\phi}{2} & \cos\frac{-\phi}{2} \end{bmatrix}. \tag{9}$$

a_1 and a_2 are functions to reconstruct the direction vector of one arm of the feature. The whole similarity score between a template and the unknown is the sum of all similarities between every corresponding pairs, $S\left(\mathbf{T}^j, \mathbf{U}\right) = \sum_{l=1}^{L^j} s\left(t_l^j, \mathbf{u}_l^j\right)$. Note that we set

$$s\left(t_l^j, \mathbf{u}_l^j\right) = \begin{cases} s\left(t_l^j, \mathbf{u}_l^j\right) & \text{when there exists a matched } \mathbf{u}_l^j; \\ 0 & \text{otherwise} \end{cases} \tag{10}$$

to include the case that t_l^j does not have a matched feature.

The process between the jth template, and the unknown character, $\mathbf{U}$, is divided into three phases. In the first phase, each template feature, $\mathbf{t}_l^j \in \mathbf{T}^j$, finds its corresponding similar feature $\mathbf{u}_l^j$ in $\mathbf{U}$ within a small radius, d_1, where d_1 is a designed parameter, Fig. 6. In other words,

$$\mathbf{u}_l^j = \arg \max_{\mathbf{u}_m} \left\{ s\left(\mathbf{t}_l^j, \mathbf{u}_m\right) \right\} \text{ where } \left\| p\left(\mathbf{t}_l^j\right) - p(\mathbf{u}_m) \right\| < d_1. \tag{11}$$

In the second phases, pick the pair of the locked features, $\left(\mathbf{t}_{l_a}^j, \mathbf{u}_{l_a}^j\right)$, which is defined as

$$l_a = \arg \max_{l=1,..,L^j} \left\{ s\left(\mathbf{t}_l^j, \mathbf{u}_l^j\right) \right\}. \tag{12}$$

This locked pair receives the highest similarity value, $s\left(\mathbf{t}_{l_a}^j, \mathbf{u}_{l_a}^j\right)$, among all pairs $\left\{\left(\mathbf{t}_l^j, \mathbf{u}_l^j\right), l = 1, .., L^j\right\}$. Then shift the whole template to a new position and overlap the seeds of these two features, $\mathbf{t}_{l_a}^j$ and $\mathbf{u}_{l_a}^j$, Fig. 7. Let $[x_a, y_a] = p\left(\mathbf{u}_{l_a}^j\right)$, which is the seed position of the locked feature, $\mathbf{u}_{l_a}^j$. The shifted template, $\left(\mathbf{T}^j\right)^s$, has new features $\left\{\left(\mathbf{t}_l^j\right)^s = [(x_l^j)^s, (y_l^j)^s, \phi_l^j, i_l^j, j_l^j], l = 1, \ldots, L^j\right\}$, where $(x_l^j)^s = x_l^j + \left(x_a - x_{l_a}^j\right)$ and $(y_l^j)^s = y_l^j + \left(y_a - y_{l_a}^j\right)$. Then we repeat the first phase and find the corresponding features $\left(\mathbf{u}_l^j\right)^s$ in $\mathbf{U}$ for every feature in the shifted template, $\left(\mathbf{t}_l^j\right)^s \in \left(\mathbf{T}^j\right)^s$. That is

$$\left(\mathbf{u}_l^j\right)^s = \arg \max_{\mathbf{u}_m \in \mathbf{U}} s\left(\left(\mathbf{t}_l^j\right)^s, \mathbf{u}_m\right) \text{ where } \left\| p\left(\left(\mathbf{t}_l^j\right)^s\right) - p(\mathbf{u}_m) \right\| < d_2. \tag{13}$$

Let $\left\{\left(\left(\mathbf{t}_l^j\right)^s, \left(\mathbf{u}_l^j\right)^s\right), l = 1, .., L^j\right\}$ be the corresponding pairs between the shifted template $\left(\mathbf{T}^j\right)^s$ and unknown $\mathbf{U}$. The new whole similarity score is $S\left(\left(\mathbf{T}^j\right)^s, \mathbf{U}\right) = \sum_{l=1}^{L^j} s\left(\left(\mathbf{t}_l^j\right)^s, \left(\mathbf{u}_l^j\right)^s\right)$. Note that the locked pairs, $\left(\mathbf{t}_{l_a}^j, \mathbf{u}_{l_a}^j\right)$, is in some sense similar to the resonance portion in ART [12].

In the third phase, pick the second locked feature pair that received the highest similarity value, $\left(\left(\mathbf{t}_{l_b}^j\right)^s, \left(\mathbf{u}_{l_b}^j\right)^s\right)$, where l_b is not equal to l_a,

$$l_b = \arg \max_{l=1,..,L^j} \left\{ s\left(\left(\mathbf{t}_l^j\right)^s, \left(\mathbf{u}_l^j\right)^s\right) \right\}, \text{ and } \left(\mathbf{u}_{l_b}^j\right)^s \neq \mathbf{u}_{l_a}^j \tag{14}$$

Let $[x_b, y_b] = p\left(\left(\mathbf{u}_{l_b}^j\right)^s\right)$ be the seed position of the feature $\left(\mathbf{u}_{l_b}^j\right)^s$. Using the position $[x_a, y_a]$ as the rotation center and the scaling center, rotate and scale the whole template $\left(\mathbf{T}^j\right)^s$ so that the two seeds $p\left(\left(\mathbf{t}_{l_a}^j\right)^s\right) = [x_a, y_a]$ and $p\left(\left(\mathbf{t}_{l_b}^j\right)^s\right) = [x_b, y_b]$ will

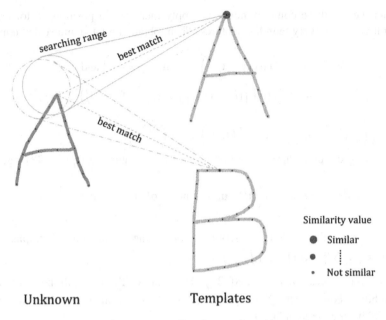

Fig. 6. Searching corresponding features in unknown character.

coincide with their corresponding seeds in the unknown character. After rotation and scaling, we obtain a refined template collection, $(\mathbf{T}^j)^r$, where

$$(\mathbf{T}^j)^r = \left\{ (\mathbf{t}_l^j)^r = \left[(x_l^j)^r, (y_l^j)^r, \phi_l^j, (i_l^j)^r, (j_l^j)^r \right], l = 1, \dots, L^j \right\} \tag{15}$$

In order to facilitate the operation of the rotation and scaling procedure for the whole template, set complex numbers,

$$\begin{aligned} \mathrm{cpx}(\mathbf{v}) &= \mathrm{cpx}([x, y]) = [x, y] \cdot [1, i] = x + yi, \\ \mathrm{vec}(x + yi) &= [x, y] = \mathbf{v}. \end{aligned} \tag{16}$$

Then, operate the rotation and scaling according to the seed positions of the two locked feature pairs. Set

$$\mathbf{p}_t = \mathrm{p}\left(\left(\mathbf{t}_{l_b}^j \right)^s \right) - \mathrm{p}\left(\mathbf{u}_{l_a}^j \right) \text{ and } \mathbf{p}_u = \mathrm{p}\left(\left(\mathbf{u}_{l_b}^j \right)^s \right) - \mathrm{p}\left(\mathbf{u}_{l_a}^j \right).$$

Also set

$$c_{r,s} = \frac{\mathrm{cpx}(\mathbf{p}_u)}{\mathrm{cpx}(\mathbf{p}_t)} \quad \text{and} \quad c_r = \frac{\mathrm{cpx}\left(\frac{\mathbf{p}_u}{|\mathbf{p}_u|} \right)}{\mathrm{cpx}\left(\frac{\mathbf{p}_t}{|\mathbf{p}_t|} \right)}.$$

After we get these complex numbers, apply them to the position vector and the direction vector of every template features. Then, obtain a rotated and scaled template,

$$
\begin{aligned}
\left[(x_l^j)^{\mathrm{r}}, (y_l^j)^{\mathrm{r}}\right] &= \mathrm{vec}\Big(\mathrm{cpx}\Big(\mathrm{p}\big((\mathbf{t}_l^j)^{\mathrm{s}}\big) - \mathrm{p}\big(\mathbf{u}_{l_a}^j\big)\Big) \times c_{\mathrm{r,s}}\Big) \text{ and} \\
\left[(i_l^j)^{\mathrm{r}}, (j_l^j)^{\mathrm{r}}\right] &= \mathrm{vec}\Big(\mathrm{cpx}\big([(i_l^j)^{\mathrm{s}}, (j_l^j)^{\mathrm{s}}]\big) \times c_{\mathrm{r}}\Big), \quad \text{for } \forall(\mathbf{t}_l^j)^{\mathrm{s}} \in (\mathbf{T}^j)^{\mathrm{s}},
\end{aligned}
\tag{17}
$$

where $\mathrm{p}\Big(\big(\mathbf{t}_{l_a}^j\big)^{\mathrm{r}}\Big) = \mathrm{p}\big(\mathbf{u}_{l_a}^j\big)$ and $\mathrm{p}\Big(\big(\mathbf{t}_{l_b}^j\big)^{\mathrm{r}}\Big) = \mathrm{p}\Big(\big(\mathbf{u}_{l_b}^j\big)^{\mathrm{s}}\Big)$.

Repeating the procedure in the first phase for all features, $\forall(\mathbf{t}_l^j)^{\mathrm{r}} \in (\mathbf{T}^j)^{\mathrm{r}}$, get

$$
(\mathbf{u}_l^j)^{\mathrm{r}} = \arg \max_{\mathbf{u}_m \in \mathbf{U}} \mathrm{s}\Big((\mathbf{t}_l^j)^{\mathrm{r}}, \mathbf{u}_m\Big), \text{ where } \Big\|\mathrm{p}\big((\mathbf{t}_l^j)^{\mathrm{r}}\big) - \mathrm{p}(\mathbf{u}_m)\Big\| < d_3,
\tag{18}
$$

The refined whole score for the refined template is $\mathrm{S}\big((\mathbf{T}^j)^{\mathrm{r}}, \mathbf{U}\big) = \sum_{l=1}^{L^j} \mathrm{s}\Big((\mathbf{t}_l^j)^{\mathrm{r}}, (\mathbf{u}_l^j)^{\mathrm{r}}\Big)$.

The three scores, $\mathrm{S}(\mathbf{T}^j, \mathbf{U})$, $\mathrm{S}\big((\mathbf{T}^j)^{\mathrm{s}}, \mathbf{U}\big)$, and $\mathrm{S}((\mathbf{T})^{\mathrm{r}}, \mathbf{U})$, will be used as the compatibility scores for the jth template. Pick the largest value of the three as the final compatibility for the template $\mathbf{T}^j$,

$$
\mathrm{S}^*(\mathbf{T}^j, \mathbf{U}) = \max\big\{\mathrm{S}(\mathbf{T}^j, \mathbf{U}), \mathrm{S}\big((\mathbf{T}^j)^{\mathrm{s}}, \mathbf{U}\big), \mathrm{S}\big((\mathbf{T}^j)^{\mathrm{r}}, \mathbf{U}\big)\big\}.
\tag{19}
$$

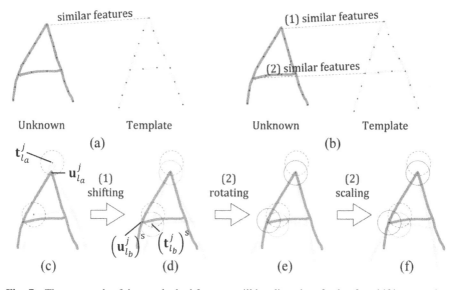

Fig. 7. The two seeds of the two locked features will be aligned perfectly after shifting, rotation, and scaling procedures.

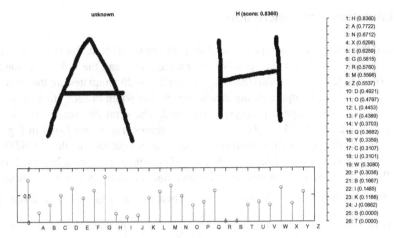

unknown

H (score: 0.8360)

1: H (0.8360)
2: A (0.7722)
3: N (0.6712)
4: X (0.6296)
5: E (0.6259)
6: G (0.5815)
7: R (0.5780)
8: M (0.5596)
9: Z (0.5537)
10: D (0.4921)
11: O (0.4797)
12: L (0.4453)
13: F (0.4389)
14: V (0.3703)
15: Q (0.3682)
16: Y (0.3359)
17: C (0.3107)
18: U (0.3101)
19: W (0.3080)
20: P (0.3036)
21: B (0.1667)
22: I (0.1485)
23: K (0.1166)
24: J (0.0862)
25: S (0.0000)
26: T (0.0000)

Fig. 8. The unknown character 'A', 256 by 256 pixels.

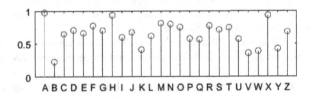

Fig. 9. Images of the 26 templates.

Fig. 10. Similarity scores (Eq. 7).

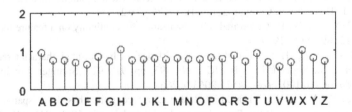

Fig. 11. Similarity scores (Eq. 6).

3 Simulation and Discussion

We tested the process to examine its capability. In our simulation, the size of the image of each character is 256-by-256 pixels for both templates and the unknown character, Fig. 8. The whole scores are also listed in the figure. All 26 templates are the 26 images of handwritten English uppercase alphabets, Fig. 9. The width of each stroke is roughly 5 to 15 pixels. Each image is properly normalized. The similarity scores, Eq. 7, of all 26 templates are listed in Fig. 10. The similarity scores, Eq. 6, are listed in Fig. 11.

The recognition rate of the proposed process is comparable to that of STD. The computation cost is much less than that of STD. Since each template is designed manually, several fixed key features can be specified manually in advance. These keys have priority in the process. The highest locked pair $\left(\mathbf{t}^j_{l_a}, \mathbf{u}^j_{l_a}\right)$ may be picked only from those keys. The second highest pair l_b, $\left(\left(\mathbf{t}^j_{l_b}\right)^s, \left(\mathbf{u}^j_{l_b}\right)^s\right)$, may also be picked from them. The whole scores S may be accumulated only for those key features.

The two locked pairs can be picked simultaneously in the first phase. When one picked three different locked pairs, one can apply the conformal mapping technique [8] to deform the whole template and accomplish the score.

References

1. Blasdel, G.G., Salama, G.: Voltage-sensitive dyes reveal a modular organization in monkey striate cortex. Nature **321**, 579–585 (1986)
2. Carpenter, G.A., Grossberg, S.: ART 2: self-organization of stable category recognition codes for analog input patterns. Appl. Opt. **26**(23), 4919–4930 (1987)
3. Hubel, D.H., Wiesel, T.N.: Brain mechanisms of vision. In: Rock, I. (ed) The Perceptual World: Readings from Scientific American Magazine, pp. 3–24 (1990)
4. Kohonen, T.: Self-Organization and Associative Memory, 3rd edn. Springer, Berlin (1989). https://doi.org/10.1007/978-3-642-88163-3
5. Krizhevsky, A., Sutskever, I., Hinton, G.E.: ImageNet classification with deep convolutional neural networks. In: Advances in Neural Information Processing Systems, pp. 1097–1105 (2012)
6. Liou, D.-R., Lin, C.-C., Liou, C.-Y.: Setting shape rules for handprinted character recognition. In: Pan, J.-S., Chen, S.-M., Nguyen, N.T. (eds.) ACIIDS 2012. LNCS (LNAI), vol. 7197, pp. 245–252. Springer, Heidelberg (2012). https://doi.org/10.1007/978-3-642-28490-8_26
7. Liou, C.-Y., Shih, H.-Y., Liou, D.-R.: Finite geometrical relations loading in Hopfield model. J. Theor. Appl. Comput. Sci. **6**(4), 59–76 (2012)
8. Liou, C.-Y., Tai, W.-P.: Conformal self-organization for continuity on a feature map. Neural Netw. **12**, 893–905 (1999)
9. Liou, C.-Y., Yang, H.-C.: Spatial topology distance for handprinted character recognition. In: Gielen, S., Kappen, B. (eds.) ICANN 1993, pp. 918–921. Springer, London (1993). https://doi.org/10.1007/978-1-4471-2063-6_266
10. Liou, C.-Y., Yang, H.C.: Handprinted character recognition based on spatial topology distance measurement. IEEE Trans. Pattern Anal. Mach. Intell. **18**(9), 941–945 (1996)

11. Liou, C.Y., Yang, H.C.: Selective feature-to-feature adhesion for recognition of cursive handprinted characters. IEEE Trans. Pattern Anal. Mach. Intell. **21**(2), 184–191 (1999)
12. Nasrabadi, N.M., Li, W., Choo, C.Y.: Object recognition by a Hopfield neural network. In: ICCV, pp. 325–328 (1990)
13. Suganthan, P.N., Teoh, E.K., Mital, D.P.: Pattern recognition by homomorphic graph matching using Hopfield neural networks. Image Vis. Comput. **13**(1), 45–60 (1995)

Enhancing the Resolution of Satellite Images Using the Best Matching Image Fragment

Daniel Kostrzewa(✉) [ID], Pawel Benecki [ID], and Lukasz Jenczmyk

Institute of Informatics, Silesian University of Technology,
Akademicka 16, 44-100 Gliwice, Poland
{daniel.kostrzewa,pawel.benecki}@polsl.pl

Abstract. Due to very high costs and a long revisit time, it is challenging to obtain good quality satellite images of the area of interest. As a result, super resolution reconstruction (SRR) methods which allow for creating a high-resolution (HR) image based on single or multiple low-resolution (LR) observations are being extensively developed. In this paper, we propose a few improvements to well-known single-image SRR technique based on a dictionary of pairs of matched LR and HR image fragments. The modifications concern both increasing the number of pairs of images fragments and the reconstruction algorithm itself in order to achieve visually pleasing results. This allows us to increase the quality of newly produced HR satellite images what is supported by conducted experiments.

Keywords: Dictionary of matched fragments · Image processing ·
Satellite image · Single-image super-resolution reconstruction

1 Introduction

It is often very challenging to obtain satellite images of high quality of the area of interest. This is due to very high costs or a long revisit time. The price of taking one kilogram of equipment into a low Earth orbit starts at approximately $2700 (using the Falcon 9 rocket) [21]. The main reason of such high prices is that most of the rockets that fly today are used only once. After their mission ends, some of them fall into the ocean, stay in orbit or burn down as a result of the deorbitation when entering the atmosphere with high speed. This situation is slowly changing. After unsuccessful attempts of reducing the cost of space travel by space shuttles, the greatest progress at this area can be seen in SpaceX company, which regularly re-used the first stage of the Falcon 9 shuttle in 2017. SpaceX is currently developing a new BFR rocket, which is designed to be fully re-usable.

Despite falling prices in the space sector, the construction and launching of new satellites are still very costly. It is one of the main answers to the question: Why is it worthwhile to deal with algorithms that improve the quality

© Springer Nature Switzerland AG 2019
N. T. Nguyen et al. (Eds.): ACIIDS 2019, LNAI 11431, pp. 576–586, 2019.
https://doi.org/10.1007/978-3-030-14799-0_50

of satellite images rather than allocate resources to better equipment? Luckily, there are many existing satellites, whose images can be improved by appropriate algorithms.

1.1 Related Work

Single-image super resolution reconstruction (SRR) is a branch of image processing that aims to create a high-resolution image (HR) based on a single low-resolution (LR) image [17,27]. It is important that the produced image should be visually pleasing and of high quality.

Unlike image deblurring, SRR techniques assume that the input image is sharp in the original resolution. SRR methods are concerned with increasing image resolution without losing sharpness.

There are many different approaches to resolve the task of single-image SRR. One of them is prediction modeling. SRR algorithms in this category produce HR images from LR inputs through a predefined mathematical formula without training data (e.g., bilinear, bicubic, and Lanczos interpolation methods).

Edge based methods are important because edges itself are key primitive image structures. Fattal in his work [5] proposed to learn from the depth and width of edge features for reconstructing HR images. On the other hand, Sun et al. [22] and Tai et al. [23] exploited the parameter of a gradient profile.

Other techniques include but are not limited to usage of: a heavy-tailed gradient distribution [19], sparsity property of large gradients in generic images [13], discrete and stationary wavelet decomposition [2,3], and a universal hidden Markov tree model [15].

One of the biggest branch of single-image SRR methods is exploiting a dictionary of pairs of matched LR and HR image fragments. Some of them use an external data [1,6], while the other ones use the input data itself [7,10]. A number of different learning methods of the mapping functions have been proposed, e.g., weighted average [1], kernel regresion [13], or sparse dictionary representation [8,28].

Moreover, many recent works employ deep learning to model the relation between the LR and HR image fragments [4,9,12,16,24,25].

1.2 Contribution

In this paper, we propose a few improvements to well-known single-image SRR technique based on a dictionary of pairs of matched LR and HR image fragments. The modifications are focused on increasing the number of pairs of images fragments as well as the reconstruction algorithm itself. The main goal is to achieve visually pleasing results. This allows us to increase the quality of newly produced HR satellite images what is supported by the conducted experiments. A part of the presented research is included in Lukasz Jenczmyk's bachelor thesis.

1.3   Paper Structure

The paper is structured as follows. Section 2 describes the original well-known algorithm of SRR by example method altogether with introduced modifications. Section 3 presents results of conducted experiments, while Sect. 4 concludes the paper and outlines the main goals of our ongoing research.

2   Overview of the Enhancing Strategy

The developed algorithm is composed of two phases. The first, which is a preparation phase, consist in the creation of a dictionary (i.e., set of matching pairs of LR and HR fragments of images). The second phase consists in the proper reconstruction of the HR image by using the LR image and the dictionary prepared in the first phase.

2.1   Creation of the Dictionary

In order to create a set of matching LR and HR fragments of images it is needed to obtain pairs of satellite images, which have to depict the same area of the surface. Those images should be taken with cameras of different resolution. As an example, Fig. 1 shows two images of Temecula, CA, USA, taken by the Landsat-8 OLI & TIRS (LR, Fig. 1a) and Sentinel-2 (HR, Fig. 1b). Besides satellite images of the Earth's surface, we also used images of the Moon [11,18][1] where LR images have been obtained by degradation of HR images which is a common test scenario in SRR methods (LR – Fig. 1c, HR – Fig. 1d).

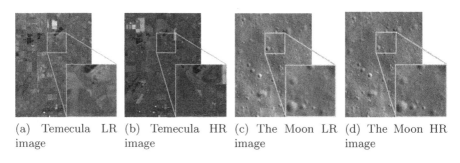

(a) Temecula LR (b) Temecula HR (c) The Moon LR (d) The Moon HR
image   image   image   image

Fig. 1. Example of satellite images

The next step is to divide the lower resolution images into squares of size of n pixels, while the higher resolution images are divided into larger ones representing the same area as the square obtained from LR images. The last part is

---

[1] Images taken by Lunar Reconnaissance Orbiter Camera, NASA/GSFC/Arizona State University.

to cut out and save the corresponding squares (fragments of LR and HR images) as a set of pairs.

It is difficult to get sufficient number of good satellite images to create a suitable dictionary. This is caused by the changeability of the Earth's surface (e.g., some seas are characterized by large inflows and outflows, vegetation is constantly changing, new buildings and roads are being built). Therefore, images from different satellites should be taken in a small time frame. Additionally, the quality of the images is also affected by the weather conditions, especially cloud coverage, which eliminates certain fragments of the images from further analysis. In some other cases acquisition of new images can be extremely expensive (e.g., the Moon images).

In order to solve this problem it is possible to obtain more pairs of fragments from the same images. Instead of splitting the image according to the grid (what is well-known from the original method, Fig. 2a), fragments can be extracted by moving the window of size of the fragment along the image. In a LR image, the window is moved by one pixel, while in a HR image it moves by the number of pixels equal to the value of the increase in resolution factor. Modified version of dictionary creation is depicted on Fig. 2b.

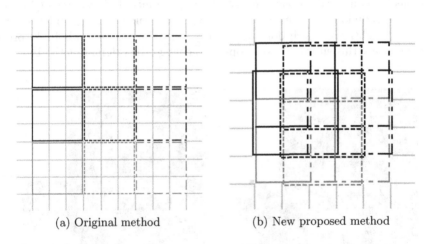

(a) Original method   (b) New proposed method

Fig. 2. Dictionary creation method.

2.2   Enhancing Image Resolution

When the dictionary (a collection of matching parts) is created, and an image, whose resolution is to be increased, then the actual SRR method can be performed. The original well-known method can be described as follows. At the beginning, it is necessary to divide the input image into squares of size of n pixels, as it was done when creating the dictionary. Then, for each fragment of the image, the method is searching for the most similar square of the LR in the collection. Different functions can be used as a measure of similarity. The simplest and most popular are a mean square error (MSE)

$$\text{MSE} = \frac{1}{n} \sum_{i=1}^{n} (a_i - b_i)^2, \tag{1}$$

and a mean absolute error (MAE)

$$\text{MAE} = \frac{1}{n} \sum_{i=1}^{n} |a_i - b_i|, \tag{2}$$

where a_i – brightness of the i-th pixel of the fragment a, b_i – brightness of the i-th pixel of the fragment b, n – number of pixels composing fragments a and b. When the most similar fragment is found, a HR counterpart is used to create a part of the new image.

In Fig. 1a and b, it can be noted that the brightness of the corresponding images may vary considerably (e.g. due to time of the day of the image acquisition or inequalities of the camera characteristics). It would not be a problem if the images were always different in brightness in the same way. In practice, this effect has a great impact on the obtained results. It is visible in Fig. 3a, where the individual fragments of the new reconstructed image differ significantly in brightness, which does not create a visually coherent whole. In order to compensate this situation, we propose to change the original method.

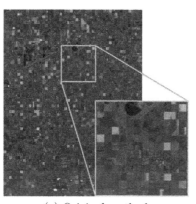 

(a) Original method   (b) The method with introduced
                                   modifications

Fig. 3. Image of Temecula, CA, USA, reconstructed by the original and modified method.

Brightness Adjustment

Brightness of the inserted fragment is modified in such a way that the MSE between it and the scaled original fragment is as small as possible. For this purpose, the adjusted MSE was introduced (Eq. 3).

$$\text{MSE}_{adj} = \frac{1}{n} \sum_{i=1}^{n} (a_i - s \cdot b_i)^2, \tag{3}$$

where a_i – brightness of the ith pixel of the fragment a, b_i – brightness of the ith pixel of the fragment b, n – number of pixels composing fragments a and b, s – brightness change coefficient.

Determining the MSE_{adj} derivative at s coefficient and comparing its value to zero, the minimum value can be defined as it is shown in Eq. 4.

$$s = \frac{\sum_{i=1}^{n}(a_i \cdot b_i)}{\sum_{i=1}^{n}(b_i^2)}. \tag{4}$$

It is possible to scale the brightness of the part which have to be inserted using this factor (Eq. 5).

$$b_i' = s \cdot b_i, \tag{5}$$

where s – brightness change coefficient for which the MSE_{adj} has the minimum value, b_i – brightness of the ith pixel of the fragment, b_i' – brightness of the ith pixel with scaled brightness value.

Pixel values can range from 0 to 255, and this condition can be broken using the formula 5. Therefore, that equation should be modified (Eq. 6).

$$b_i' = \min(255, s \cdot b_i). \tag{6}$$

If the brightness of the fragments is irrelevant, because they can be easily made brighter or darker, it should not have any influence when the most similar fragments are found. As a result, the similarity function used so far (MAE) can be replaced by the MSE_{adj}. An example of the result of using MSE_{adj} is shown in Fig. 3b.

Bicubic Scaling

The value of the similarity function indicates the quality of the most suitable HR fragment. In most cases it is possible to find very good fragment in the dictionary. On the other hand, there is a chance that the best matching fragment is not as similar as it is supposed.

It can be assumed that if the best fragment found exceed a certain value of MSE_{adj}, another (simpler) method of increasing the resolution (e.g., scaling with bicubic interpolation) would be better. Applying this improvement with a suitable similarity threshold should increase the overall quality of the reconstructed image. As a result, a modification has been introduced which, in the case of not finding a sufficiently good fragment in the dictionary, uses a scaled original fragment with bicubic interpolation as a HR fragment.

Figure 4 shows the original algorithm (continuous lines) altogether with introduced modifications (dashed lines).

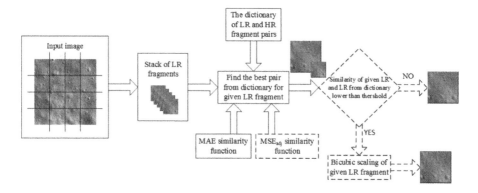

Fig. 4. Image enhancement method with introduced modification.

3   Results of Experiments

For the validation process we used satellite images of a few areas (e.g., Sydney and fragments of the Moon). In the case of images of Earth's surface LR images were taken by Sentinel-2, while HR images are downscaled from images taken by Spot. For the Moon images LR fragments were created by HR image downscaling (we do not have images obtained by different cameras). As a dictionary we exploited a database of 500000 image fragments made from LR and HR images of another areas of the Earth and the Moon. The input images were partially taken from the B4MultiSR dataset [14][2]. Each LR image of the Earth is of Ground Sampling Distance (GSD) equal to 10 m (i.e., one pixel represents area of 10 m) and HR has GSD of 5 m. Thus input images were to be enlarged by a factor of 2 (GSD should be improved from 10 m to 5 m).

Tests were performed on a machine equipped with a 4-core Intel i7 processor with 32 GB RAM but the characteristics of the machine far exceeded the requirements of the method. Input images were processed using two search-fragment-by-similarity functions (i.e., MAE and MSE_{adj}) defined in this paper. We conducted experiments using following values for maximal allowed errors:

- MAE: 2, 3, 4, 5, 6, 7, 9 – which has given us coverage of replaced patches of 2%–95%,

---

[2] B4MultiSR dataset is available at https://research.future-processing.com/sispare/dataset.

– MSE_{adj}: 2, 5, 10, 20, 30, 40, 50 – which gave a similar coverage.

The quality was assessed using the structural similarity index measure (SSIM) [26] and the information fidelity criterion (IFC) [20] which are known for taking into account the same things as perceived by human eyes. We analyzed quality of the reconstruction as a function of amount of replaced image fragments which is presented in Figs. 5 and 6 (the graphs show average values for all reconstructed satellite images). The reasoning behind that approach is the better similarity function used in the method, the better the image quality is produced while replacing the same amount of image fragments. On the Figs. 5 and 6 it can be seen that the obtained quality of reconstructed images is higher for our new MSE_{adj} similarity function than for well-known MAE.

The improvement cannot be clearly seen with the naked eye (Fig. 7), however results show that images were enhanced better using method introduced in this paper than a standard similarity function. Advantageous results could be observed while using a larger database of image fragments but in this research we were limited by performance of the prototype software.

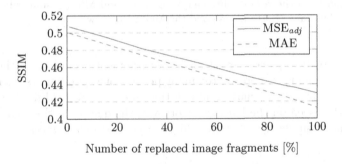

Number of replaced image fragments [%]

Fig. 5. Reconstruction quality (SSIM) in function of amount of patches replaced in images.

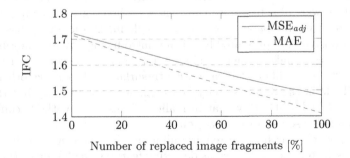

Number of replaced image fragments [%]

Fig. 6. Results for IFC.

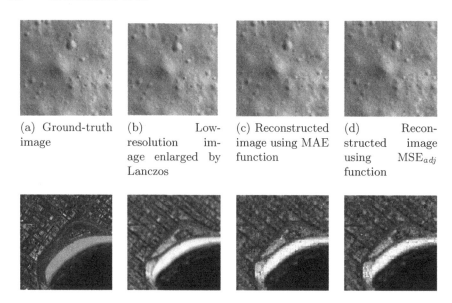

(a) Ground-truth image

(b) Low-resolution image enlarged by Lanczos

(c) Reconstructed image using MAE function

(d) Reconstructed image using MSE_{adj} function

Fig. 7. Example of the results obtained using MAE and MSE_{adj} similarity functions for Sydney, Australia (bottom row) and the Moon (top row).

While a plain replacement of all image fragments could drive the process towards poor quality if there are too many poorly corresponding fragments in the dictionary (right side of charts), adding the similarity treshold limits this effect which can bee seen on the left side. Moreover, the quality of the reconstruction using modified similarity function is higher than a standard method.

4 Conclusion and Future Work

The method presented in this paper is an efficient single-image SRR strategy. Two improvements added to the standard method: threshold limiting replacement of image fragments and better similarity function greatly improve the simple algorithm at a very low cost. Building much larger database of image fragments and software able to use it in efficient way should give much better results than presented in this paper. As in other single-image dictionary-based methods, the reconstruction presented here is not introducing any real information to images but rather makes image looking better than simple zoom. For example, in satellite imaging, different areas on Earth surface will look like other similar areas available in the database (city areas will still look like city but the exact locations will look differently that in reality). Thus such methods could never be used a reliable source of information.

Our ongoing research is centred around improving the results by introducing more sophisticated methods such as detecting features on the image (i.e., edges, corners, lines, etc.) and using the extracted knowledge to enhance the quality of the reconstructed image. These strategies often depend on many parameters. We intend to determine their values using evolutionary heuristics.

Acknowledgements. This work was supported by research funds of Institute of Informatics, Silesian University of Technology, Gliwice, Poland (grant no. BKM-556/RAU2/2018).

References

1. Chang, H., Yeung, D.Y., Xiong, Y.: Super-resolution through neighbor embedding. In: Proceedings of the 2004 IEEE Computer Society Conference on Computer Vision and Pattern Recognition, CVPR 2004, vol. 1, p. I. IEEE (2004)
2. Chavez-Roman, H., Ponomaryov, V.: Super resolution image generation using wavelet domain interpolation with edge extraction via a sparse representation. IEEE Geosci. Remote Sens. Lett. **11**(10), 1777–1781 (2014)
3. Demirel, H., Anbarjafari, G.: Image resolution enhancement by using discrete and stationary wavelet decomposition. IEEE Trans. Image Process. **20**(5), 1458–1460 (2011)
4. Dong, C., Loy, C.C., He, K., Tang, X.: Image super-resolution using deep convolutional networks. IEEE Trans. Pattern Anal. Mach. Intell. **38**(2), 295–307 (2016)
5. Fattal, R.: Image upsampling via imposed edge statistics. ACM Trans. Graph. (TOG) **26**, 95-1–95-8 (2007). Article No. 95
6. Freeman, W.T., Jones, T.R., Pasztor, E.C.: Example-based super-resolution. IEEE Comput. Graph. Appl. **22**(2), 56–65 (2002)
7. Glasner, D., Bagon, S., Irani, M.: Super-resolution from a single image. In: 2009 IEEE 12th International Conference on Computer Vision, pp. 349–356. IEEE (2009)
8. He, L., Qi, H., Zaretzki, R.: Beta process joint dictionary learning for coupled feature spaces with application to single image super-resolution. In: 2013 IEEE Conference on Computer Vision and Pattern Recognition (CVPR), pp. 345–352. IEEE (2013)
9. Heinrich, L., Bogovic, J.A., Saalfeld, S.: Deep learning for isotropic super-resolution from non-isotropic 3D electron microscopy. In: Descoteaux, M., Maier-Hein, L., Franz, A., Jannin, P., Collins, D.L., Duchesne, S. (eds.) MICCAI 2017. LNCS, vol. 10434, pp. 135–143. Springer, Cham (2017). https://doi.org/10.1007/978-3-319-66185-8_16
10. Huang, J.B., Singh, A., Ahuja, N.: Single image super-resolution from transformed self-exemplars. In: Proceedings of the IEEE Conference on Computer Vision and Pattern Recognition, pp. 5197–5206 (2015)
11. Humm, D., et al.: Flight calibration of the LROC narrow angle camera. Space Sci. Rev. **200**(1–4), 431–473 (2016)
12. Kim, J., Kwon Lee, J., Mu Lee, K.: Accurate image super-resolution using very deep convolutional networks. In: Proceedings of the IEEE Conference on Computer Vision and Pattern Recognition, pp. 1646–1654 (2016)
13. Kim, K.I., Kwon, Y.: Single-image super-resolution using sparse regression and natural image prior. IEEE Trans. Pattern Anal. Mach. Intell. **32**(6), 1127–1133 (2010)

14. Kostrzewa, D., Skonieczny, Ł., Benecki, P., Kawulok, M.: B4MultiSR: a benchmark for multiple-image super-resolution reconstruction. In: Kozielski, S., Mrozek, D., Kasprowski, P., Małysiak-Mrozek, B., Kostrzewa, D. (eds.) BDAS 2018. CCIS, vol. 928, pp. 361–375. Springer, Cham (2018). https://doi.org/10.1007/978-3-319-99987-6_28

15. Li, F., Jia, X., Fraser, D.: Universal HMT based super resolution for remote sensing images. In: 2008 15th IEEE International Conference on Image Processing, ICIP 2008, pp. 333–336. IEEE (2008)

16. Liu, D., Wang, Z., Wen, B., Yang, J., Han, W., Huang, T.S.: Robust single image super-resolution via deep networks with sparse prior. IEEE Trans. Image Process. **25**(7), 3194–3207 (2016)

17. Nasrollahi, K., Moeslund, T.B.: Super-resolution: a comprehensive survey. Mach. Vis. Appl. **25**(6), 1423–1468 (2014)

18. Robinson, M., et al.: Lunar reconnaissance orbiter camera (LROC) instrument overview. Space Sci. Rev. **150**(1–4), 81–124 (2010)

19. Shan, Q., Li, Z., Jia, J., Tang, C.K.: Fast image/video upsampling. ACM Trans. Graph. (TOG) **27**(5), 153 (2008)

20. Sheikh, H.R., Bovik, A.C., De Veciana, G.: An information fidelity criterion for image quality assessment using natural scene statistics. IEEE Trans. Image Process. **14**(12), 2117–2128 (2005)

21. SpaceX website. http://www.spacex.com. Accessed 30 Dec 2017

22. Sun, J., Xu, Z., Shum, H.Y.: Image super-resolution using gradient profile prior. In: 2008 IEEE Conference on Computer Vision and Pattern Recognition, CVPR 2008, pp. 1–8. IEEE (2008)

23. Tai, Y.W., Liu, S., Brown, M.S., Lin, S.: Super resolution using edge prior and single image detail synthesis. In: 2010 IEEE Conference on Computer Vision and Pattern Recognition (CVPR), pp. 2400–2407. IEEE (2010)

24. Wang, L., Huang, Z., Gong, Y., Pan, C.: Ensemble based deep networks for image super-resolution. Pattern Recogn. **68**, 191–198 (2017)

25. Wang, Z., Liu, D., Yang, J., Han, W., Huang, T.: Deep networks for image super-resolution with sparse prior. In: Proceedings of the IEEE International Conference on Computer Vision, pp. 370–378 (2015)

26. Wang, Z., Bovik, A.C., Sheikh, H.R., Simoncelli, E.P.: Image quality assessment: from error visibility to structural similarity. IEEE Trans. Image Process. **13**(4), 600–612 (2004)

27. Yue, L., Shen, H., Li, J., Yuan, Q., Zhang, H., Zhang, L.: Image super-resolution: the techniques, applications, and future. Sig. Process. **128**, 389–408 (2016)

28. Zeyde, R., Elad, M., Protter, M.: On single image scale-up using sparse-representations. In: Boissonnat, J.-D., et al. (eds.) Curves and Surfaces 2010. LNCS, vol. 6920, pp. 711–730. Springer, Heidelberg (2012). https://doi.org/10.1007/978-3-642-27413-8_47

Multi-scale Autoencoders in Autoencoder for Semantic Image Segmentation

John Paul T. Yusiong[1,2(✉)] and Prospero C. Naval Jr.[1]

[1] Computer Vision and Machine Intelligence Group, Department of Computer Science, College of Engineering, University of the Philippines, Diliman, Quezon City, Philippines
jtyusiong@up.edu.ph, pcnaval@dcs.upd.edu.ph
[2] Division of Natural Sciences and Mathematics, University of the Philippines Visayas Tacloban College, Tacloban City, Leyte, Philippines

Abstract. Semantic image segmentation is essential for scene understanding. Several state-of-the-art deep learning-based approaches achieved remarkable results by increasing the network depth to improve performance. Using this principle, we introduce a novel encoder-decoder network architecture for semantic image segmentation of outdoor scenes called SAsiANet. SAsiANet utilizes multi-scale cascaded autoencoders at the decoder section of an autoencoder to achieve high accuracy pixelwise prediction and involves exploiting features across multiple scales when upsampling the output of the encoder to obtain better spatial and contextual information effectively. The proposed network architecture is trained using the cross-entropy loss function but without incorporating any class balancing technique to the loss function. Our experimental results on two challenging outdoor scenes: the CamVid urban scenes dataset and the Freiburg forest dataset demonstrate that SAsiANet provides an effective way of producing accurate segmentation maps since it achieved state-of-the-art results on the test set of both datasets, 72.40% mIoU, and 89.90% mIoU, respectively.

Keywords: Autoencoders in autoencoder ·
Semantic image segmentation · Stacked autoencoders

1 Introduction

Semantic image segmentation also referred to as scene parsing, is essential for scene understanding because it can depict the complex relationships of the various semantic entities. This task is also crucial in modern autonomous driving systems, specifically for navigation and action planning, since it can provide a means to correctly understand the surrounding environment [27,29,37]. Semantic image segmentation is a challenging task because it involves three sub-tasks, object detection, image segmentation, and multi-class recognition, to obtain spatial and contextual representations from an image [20]. Essentially, semantic

© Springer Nature Switzerland AG 2019
N. T. Nguyen et al. (Eds.): ACIIDS 2019, LNAI 11431, pp. 587–599, 2019.
https://doi.org/10.1007/978-3-030-14799-0_51

image segmentation involves pixel-wise prediction such that each pixel in an image is assigned to one of the many possible class labels [27,29,37].

Pixel-wise semantic image segmentation methods must be able to model appearance and shape, understand the spatial relationships among the different classes, retain boundary information and produce smooth segmentations [2]. As a result, several techniques have been developed, and these techniques can be divided into two broad categories: classical approaches and deep learning-based approaches. Classical semantic image segmentation methods heavily rely on hand-crafted features to learn the representation while deep learning-based approaches usually involve end-to-end trainable frameworks that employ convolutional neural networks to extract rich features hierarchically and directly learn to map pixels to class labels [4,33]. The network architectures of most deep learning techniques for semantic image segmentation are derived from the fully convolutional network (FCN) architecture [19] and its variants such as the convolutional autoencoder [2]. The typical approach is to reconfigure existing network architectures that have been proven to work well in image classification for semantic image segmentation task in such a way that the reconfigured network can accept an input image and generate a spatial map for each class label instead of classification scores. To obtain a better and more consistent output some researchers used recurrent neural networks or probabilistic graphical models [2,16,27,32].

Recent approaches in semantic image segmentation also focused on increasing the network depth and very deep network architectures performed very well on standard benchmarks datasets [15]. One way to increase network depth is to stack multiple autoencoders together end-to-end consecutively. Stacked autoencoders such as SDNs [12] and SUNets [28] capture more contextual information when processing and combining features across scales because of its ability to refine the predictions stage-by-stage.

Unlike the existing approaches [12,28] that stack autoencoders with the same network structure end-to-end, our key contribution is the introduction of the autoencoders in autoencoder architecture called SAsiANet, which stack autoencoders with different network structure depending on the scale, as shown in Fig. 1(d). This design enables the network to exploit features at multiple scales when upsampling the output of the encoder section and generate better segmentation outputs. As a result, incorporating class balancing techniques [10,25] to the cross-entropy loss function becomes unnecessary when training the network. We demonstrate the effectiveness of the new approach on two challenging outdoor scenes, the CamVid urban scenes dataset [3] and the Freiburg forest dataset [31].

2   Related Works

This section focuses on the recent advances in fully-supervised semantic segmentation using deep learning approaches. The most common approach in semantic image segmentation is to employ existing CNN classifiers and convert them to

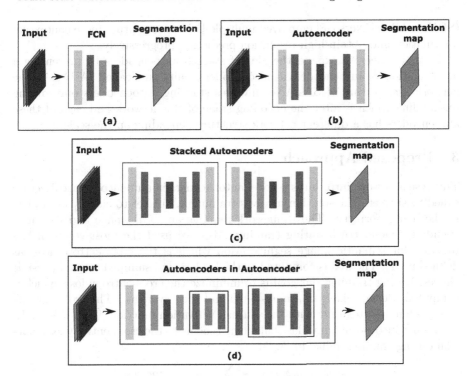

Fig. 1. Semantic image segmentation architectures. (a) Fully Convolutional Network, (b) Autoencoder, (c) Stacked Autoencoders, (d) Autoencoders in Autoencoder (our proposed architecture).

fully convolutional networks (FCNs) to obtain a spatial map for each class label. Moreover, the introduction of FCNs by Long *et al.* [19] inspired many researchers to develop segmentation techniques with better accuracy by training deeper FCN models [27,33,37]. Another approach is the use of convolutional autoencoders, also known as encoder-decoder networks which have an encoder module that systematically reduces the spatial maps to obtain higher contextual representations and a decoder module that gradually recovers the spatial information [2,8,13,21]. On the other hand, some researchers focused on the task of real-time semantic segmentation where computational efficiency is of utmost importance but without severely degrading the performance of the model [5,25]. Other researchers introduced recurrent neural networks (RNNs) in their framework to improve the performance of the model further [32,35]. Another approach is to combine FCNs and CRFs to obtain better feature and contextual representations thereby significantly improving the accuracy of the model [20].

Recent approaches involve stacking several encoder-decoder structures end-to-end [12,28] to generate high-resolution segmentation maps with detailed boundaries. Fu *et al.* [12] used multiple stacks of deconvolutional networks

(SDNs) while SUNets [28] employed multiple U-Nets to capture more contextual information and obtain high-resolution prediction progressively.

In this paper, we expand the idea of stacked autoencoders by presenting a novel multi-scale autoencoders in autoencoder architecture called SAsiANet. The network architecture of SAsiANet involves stacking autoencoders only during upsampling which is after the encoding stage of the network and each of these autoencoders has a different network structure depending on the scale.

3   Proposed Approach

This research work introduces a novel autoencoders in autoencoder architecture called SAsiANet. The solution involves training SAsiANet to classify each pixel in the image correctly. The framework requires an RGB image and its corresponding ground truth during training. Also, we used the most common loss function for semantic image segmentation which is the cross-entropy loss, as defined in (1). The loss is computed as the log loss and summed over all possible classes. During training, our goal is to minimize the cross-entropy loss, which is computed over all pixels in a mini-batch and then averaged. The cross-entropy loss is often used together with the softmax activation function [9]. In this study, we did not incorporate any class balancing scheme to the cross-entropy loss function during training unlike in [5, 25].

$$loss(y_{true}, y_{pred}) = - \sum_{classes} y_{true} \times log(y_{pred}) \qquad (1)$$

3.1   Network Architecture

Figure 2 depicts the overall network architecture of SAsiANet while Fig. 3 shows its building blocks. As shown in the previous studies [12, 28], stacking autoencoders enables the model to obtain better feature and contextual representations.

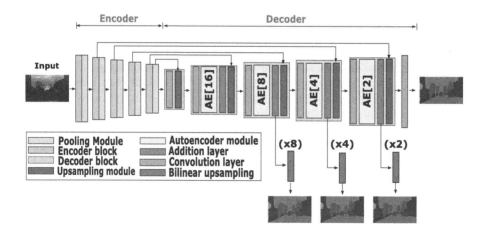

Fig. 2. SAsiANet for semantic image segmentation of outdoor scenes.

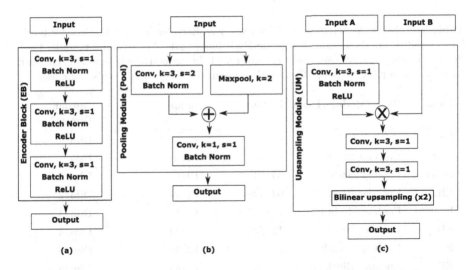

Fig. 3. Building blocks of SAsiANet. (a) Encoder block, (b) Pooling Module (+ means add layers), (c) Upsampling Module (× means concatenate layers)

Based on this principle, the main idea in our network design is to employ multi-scale cascaded autoencoders at the decoder section of an autoencoder with long skip connection between the encoder and decoder to exploit features at multiple scales when upsampling the output of the encoder. The Encoder section consists of several encoder blocks. The decoder section of SAsiANet does not merely perform upsampling; it also performs "autoencoding" before each upsampling step to exploit features at multiple scales and obtain better segmentation outputs arising from improved spatial and contextual information. A significant amount of effort was taken to highlight the details of the network for reproducibility. Table 1 defines the SAsiANet architecture while Table 2 details the structure of the autoencoder (AE) module at different scales. SAsiANet generates a segmentation map at four different scales. Each of this segmentation maps is upsampled to the original image resolution using bilinear interpolation when computing the cross-entropy loss, and then we calculate the average loss. However, at testing time, only the final segmentation map (*segmap*1) is relevant.

4 Experiments

In this section, we present an extensive evaluation of our proposed method on two challenging outdoor scenes, the CamVid urban scenes dataset [3] and the Freiburg forest dataset [31]. We briefly describe the standard datasets and metric used in evaluating the performance of our proposed method.

The Cambridge-driving Labeled Video (CamVid) [3] dataset is an urban driving scenes dataset consisting of 701 high-resolution video frames. This dataset is based on a ten-minute video footage comprising of five video sequences. The video footage was recorded by driving around a challenging urban environment

Table 1. Network Architecture of SAsiANet. EB: Encoder block; DB: Decoder block; Dim: Input image dimension; Image: an RGB image; AE: Autoencoder module; Pool: Pooling module; Conv: Convolutional layer with linear activation function; Bilinear: Bilinear upsampling; segmap: Predicted segmentation map; n_c: Number of classes.

Name	Block/Module/Layer	In/Out	Input	Output Dimension
Image	-	-/3	-	Dim
EB1	Encoder Block	3/64	Image	Dim
Pool1	Pooling Module	64/64	EB1	Dim/2
EB2	Encoder Block	64/128	Pool1	Dim/2
Pool2	Pooling Module	128/128	EB2	Dim/4
EB3	Encoder Block	128/256	Pool2	Dim/4
Pool3	Pooling Module	256/256	EB3	Dim/8
EB4	Encoder Block	256/512	Pool3	Dim/8
Pool4	Pooling Module	512/512	EB3	Dim/16
EB5	Encoder Block	512/1024	Pool4	Dim/16
Pool5	Pooling Module	1024/1024	EB5	Dim/32
DB5	Conv, k = 1, s = 1	1024/n_c	Pool5	Dim/32
	Upsampling Module	[1024, n_c]/n_c	Pool5, DB5_Conv	Dim/16
DB4	Conv, k = 1, s = 1	n_c/n_c	DB5	Dim/16
	AE[16]	n_c/n_c	DB4_Conv	Dim/16
	Add	[n_c, n_c]/n_c	DB4_Conv, AE[16]	Dim/16
	Upsampling Module	[1024, n_c]/n_c	EB5, DB4_Add	Dim/8
DB3	Conv, k = 1, s = 1	n_c/n_c	DB4	Dim/8
	AE[8]	n_c/n_c	DB3_Conv	Dim/8
	Add	[n_c, n_c]/n_c	DB3_Conv, AE[8]	Dim/8
	Upsampling Module	[512, n_c]/n_c	EB4, DB3_Add	Dim/4
segmap4	Bilinear (x8)	n_c/n_c	DB3_Add	Dim
DB2	Conv, k = 1, s = 1	n_c/n_c	DB3	Dim/4
	AE[4]	n_c/n_c	DB2_Conv	Dim/4
	Add	[n_c, n_c]/n_c	DB2_Conv, AE[4]	Dim/4
	Upsampling Module	[256, n_c]/n_c	EB3, DB2_Add	Dim/2
segmap3	Bilinear (x4)	n_c/n_c	DB2_Add	Dim
DB1	Conv, k = 1, s = 1	n_c/n_c	DB2	Dim/2
	AE[2]	n_c/n_c	DB1_Conv	Dim/2
	Add	[n_c, n_c]/n_c	DB1_Conv, AE[2]	Dim/2
	Upsampling Module	[128, n_c]/n_c	EB2, DB1_Add	Dim
segmap2	Bilinear (x2)	n_c/n_c	DB1_Add	Dim
segmap1	Conv, k = 1, s = 1	n_c/n_c	DB1	Dim

Table 2. Autoencoder (AE) Modules. EB: Encoder block; Pool: Pooling Module; Dim: Input image dimension; AE: Autoencoder module; UM: Upsampling Module; Conv: Convolutional layer with linear activation function; n_c: Number of classes.

AE module	Composition	Output channels	Output dimension
AE[16]	EB-Pool]- [Conv-UM]	1024-1024]- [n_c-n_c]	Dim/ [16-32- 32-16]
AE[8]	EB-Pool-EB-Pool]- [Conv-UM-Conv-UM]	512-512-1024-1024]- [n_c-n_c-n_c-n_c]	Dim/ [8-16-16-32- 32-16-16-8]
AE[4]	EB-Pool-EB-Pool-EB-Pool]- [Conv-UM-Conv-UM-Conv-UM]	256-256-512-512-1024-1024]- [n_c-n_c-n_c-n_c-n_c-n_c]	Dim/ [4-8-8-16-16-32- 32-16-16-8-8-4]
AE[2]	EB-Pool-EB-Pool-EB-Pool-EB-Pool]- [Conv-UM-Conv-UM-Conv-UM-Conv-UM]	128-128-256-256-512-512-1024-1024]- [n_c-n_c-n_c-n_c-n_c-n_c-n_c-n_c]	Dim/ [2-4-4-8-8-16-16-32- 32-16-16-8-8-4-4-2]

with variable light settings. The dataset provides ground-truth labels that associate each pixel to one of the 32 semantic classes. Following [2], we use a subset of 11 semantic classes: building, tree, sky, car, sign-symbol, road, pedestrian, fence, column-pole, sidewalk, and bicyclist. A small number of pixels labeled as *void* are ignored during training and testing because these pixels do not belong to one of the classes. The dataset is split into three sets following [30], which consists of 367 training images, 100 validation images, and 233 test images. During training, image patches of dimension 360×480 are randomly cropped from the original image dimension of 720×960, but at testing time, the trained model is evaluated on the original image resolution of the test set.

On the other hand, the Freiburg Multispectral segmentation benchmark (Freiburg forest) [31] is the first-of-a-kind dataset of unstructured forested environments where objects are very diverse and the objects' appearance change due to seasonal variations. The dataset contains over 15,000 images, equivalent to traversing 4.7 km daily. However, only 366 images have pixel-wise ground truth segmentation masks for 6 classes: sky, trail, grass, vegetation, obstacles, and void. Similar to the CamVid dataset, pixels labeled as *void* are ignored during training and testing. Also, following [24] the class *obstacles* is ignored during training and testing since this class has a frequency of 0.85% and is considered an outlier. The dataset is split into two sets which consist of 230 training images, and 136 test images. Moreover, the images need to be resized to a fixed dimension before using the dataset. Instead of resizing the image to a fixed dimension during training, we randomly crop the original image to obtain image patches of dimension 192×384, but at testing time, we resized the test images to 384×768.

To evaluate the performance of our model we used the mean intersection over union (mIoU), which is the standard metric for segmentation tasks. The IoU measures the similarity between the intersection and union of the ground truth and the predicted segmentation map, and is calculated on a per-class basis and then averaged as defined in (2). The IoU can also be computed as the ratio between the number of true positives (intersection) over the sum of true positives (TP), false positives (FP) and false negatives (FN) (union).

$$mIoU = \frac{1}{n_{cl}} \sum_i \frac{n_{ii}}{\sum_j n_{ij} + \sum_j n_{ji} - n_{ii}} = \frac{TP}{TP + FP + FN} \qquad (2)$$

where n_{cl} is the total number of classes, n_{ii} represents the true positives, n_{ij} represents the false positives, and n_{ji} represents the false negatives.

4.1   Implementation Details

The semantic segmentation network is implemented in Tensorflow [1] using a single GTX 1080 Ti (11 GB) GPU. We used the Adam optimizer for loss minimization with $B_1 = 0.9$, $B_2 = 0.999$, $\epsilon = 1e^{-8}$, and a mini-batch size of 4. We set the initial learning rate of the encoder layers to $\lambda = 0.0001$, and the decoder layers to $\lambda = 0.001$ for the weight parameters and $\lambda = 0.0002$ for the biases. We employed the "poly" learning rate policy instead of the usual "step" policy (i.e., the learning rate is multiplied by $(1.0 - \frac{iter}{max\_iter})^{power}$) to update the learning rate because the "poly" policy is more effective for semantic image segmentation [6]. The *power* variable is set to 0.9, and the maximum number of epochs (*max_iter*) is set to 300. The weight parameters are initialized randomly according to the Xavier initialization. To avoid overfitting, we applied L_2 regularization on all the weight parameters by adding a small constant, 0.00001. Furthermore, we applied mean image subtraction to the images in the dataset and performed a series of data augmentation techniques on the fly, such as random cropping, horizontal flipping, and random scaling (0.6 to 1.4). The SAsiANet architecture is trained using the augmented training set.

4.2   Results and Discussion

We present several experimental results demonstrating the effectiveness of our proposed method in semantic image segmentation for outdoor scenes. To compare the performance of our proposed method with the existing methods on the CamVid dataset, we present the quantitative results of SAsiANet in Table 3. This table shows that our method outperforms the previous methods in terms of the mean intersection over union (mIoU) metric because our method achieved an mIoU of 72.4%. Our method also obtained the highest IoUs on 7 out of 11 object classes. Furthermore, the quantitative results reveal that SAsiANet was able to improve on the results of the challenging classes: sign-symbol and column-pole. The experimental evaluation demonstrates that SAsiANet is a useful model for semantic segmentation since it was able to deal with a highly imbalanced dataset such as the CamVid dataset, even without using any class balancing scheme during training. Figure 4 shows visual samples of the segmentation results on the CamVid test set. The qualitative results show remarkable details and the ability of the SAsiANet architecture to segment smaller classes such as the column-pole and sign-symbol. Furthermore, the network can easily distinguish between larger classes, such as the building, sky, and road.

Table 3. Semantic segmentation results on the CamVid test set. The results of the other models are taken directly from the published works. The red and **bold** values indicate the best results.

Method	IoU											mIoU
	Building	Tree	Sky	Car	Sign	Road	Pedestrian	Fence	Pole	Sidewalk	Bicyclist	
SegNet [2]	68.7	52.0	87.0	58.5	13.4	86.2	25.3	17.9	16.0	60.5	24.8	46.4
STFCN-8 [11]	73.5	56.4	90.7	63.3	17.9	90.1	31.4	21.7	18.2	64.9	29.3	50.6
DPDB-Net [24]	72.4	62.9	88.6	61.9	30.0	88.8	44.8	26.1	23.6	69.4	33.1	54.7
DeepLab-LFOV [7]	81.5	74.6	89.0	82.2	42.3	92.2	48.4	27.2	14.3	75.4	50.1	61.6
LRN [13]	78.6	73.6	76.4	75.2	40.1	91.7	43.5	41.0	30.4	80.1	46.5	61.7
Dilation8 [36]	82.6	76.2	89.0	84.0	46.9	92.2	56.3	35.8	23.4	75.3	55.5	65.3
STDilation8 [11]	83.4	76.5	90.4	84.6	50.4	92.4	56.7	36.3	22.9	75.7	56.1	65.9
Dilation8+FSO [17]	84.0	77.2	91.3	85.6	49.9	92.5	59.1	37.6	16.9	76.0	57.2	66.1
FC-DenseNet103 [15]	83.0	77.3	93.0	77.3	43.9	94.5	59.6	37.1	37.8	82.2	50.5	66.9
FCCN [34]	79.7	77.2	85.7	86.1	45.3	94.9	45.8	69.0	25.2	86.2	52.9	68.0
G-FRNet [14]	82.5	76.8	92.1	81.8	43.0	94.5	54.6	47.1	33.4	82.3	59.4	68.0
SDN [12]	84.5	76.9	92.2	88.2	51.6	93.4	62.4	37.0	37.5	77.8	64.4	69.6
Ours (SAsiANet)	85.6	79.1	93.3	82.6	54.3	95.6	66.5	51.0	44.7	85.2	57.9	72.4

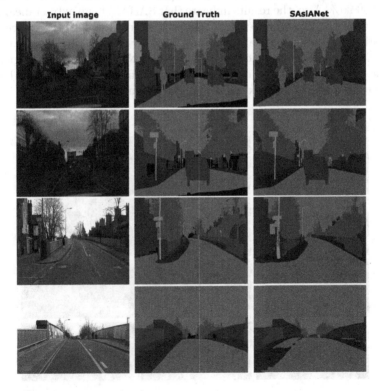

Fig. 4. Qualitative results of SAsiANet on the CamVid test set.

Table 4. Semantic segmentation results on the Freiburg forest test set. The results of the other models are taken directly from [24]. The red and **bold** values indicate the best results.

Method	IoU				mIoU
	Sky	Trail	Grass	Vegetation	
ParseNet [18]	87.78	81.82	85.20	85.20	85.00
M-Net [22]	89.26	82.41	84.93	88.70	86.30
Fast-Net [23]	90.46	84.51	86.72	90.66	88.00
GCN [26]	91.94	86.29	86.44	88.73	88.30
DPDB-Net [24]	92.30	87.28	87.80	90.14	89.40
Ours (SAsiANet)	92.50	88.00	88.20	90.80	89.90

We also performed a quantitative evaluation on SAsiANet using the Freiburg forest test set, and the results in Table 4 show the effectiveness of SAsiANet in semantic image segmentation. Our proposed method achieved the highest mIoU of 89.90%. Also, the results indicate that SAsiANet can extract meaningful features from the input image and classify these features into the correct class labels. Figure 5 shows sample predictions made by SAsiANet. The qualitative results of the proposed approach reveal that SAsiANet can clearly distinguish the different classes. These experiment results show that SAsiANet works because it benefits from the greater expressive ability of stacked autoencoders.

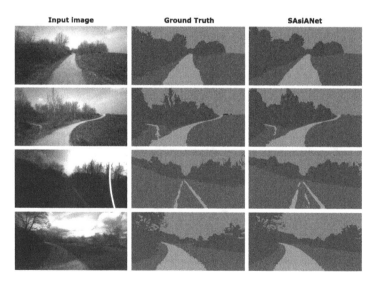

Fig. 5. Qualitative results of SAsiANet on the Freiburg forest test set.

5   Conclusions

In this work, we proposed to address the problem of semantic image segmentation using a novel multi-scale autoencoders in autoencoder architecture called SAsiANet. By incorporating the multi-scale cascaded autoencoders at the decoder section enables the network to effectively obtain better spatial and contextual information by exploiting features at multiple scales when upsampling the output of the encoder. As a result, SAsiANet achieved excellent performance even without incorporating any class balancing techniques to the cross-entropy loss function during training. The experiment results show that SAsiANet improves on the state-of-the-art performance on two challenging outdoor scenes, the CamVid urban scenes dataset as well as the Freiburg forest dataset.

References

1. Abadi, M., et al., M.D.: TensorFlow: large-scale machine learning on heterogeneous distributed systems. arXiv:1603.04467 (2016)
2. Badrinarayanan, V., Kendall, A., Cipolla, R.: SegNet: a deep convolutional encoder-decoder architecture for image segmentation. IEEE Trans. Pattern Anal. Mach. Intell. **39**(12), 2481–2495 (2017)
3. Brostow, G.J., Fauqueur, J., Cipolla, R.: Semantic object classes in video: a high-definition ground truth database. Pattern Recogn. Lett. **30**(2), 88–97–272 (2009)
4. Caesar, H., Uijlings, J., Ferrari, V.: Region-based semantic segmentation with end-to-end training. In: Leibe, B., Matas, J., Sebe, N., Welling, M. (eds.) ECCV 2016. LNCS, vol. 9905, pp. 381–397. Springer, Cham (2016). https://doi.org/10.1007/978-3-319-46448-0_23
5. Chaurasia, A., Culurciello, E.: Linknet: exploiting encoder representations for efficient semantic segmentation. In: IEEE Visual Communications and Image Processing (VCIP), pp. 1–4 (2017)
6. Chen, L., Papandreou, G., Kokkinos, I., Murphy, K., Yuille, A.L.: Deeplab: semantic image segmentation with deep convolutional nets, atrous convolution, and fully connected CRFs. IEEE Trans. Pattern Anal. Mach. Intell. **40**(4), 834–848 (2018)
7. Chen, L.C., Papandreou, G., Kokkinos, I., Murphy, K., Yuille, A.L.: Semantic image segmentation with deep convolutional nets and fully connected CRFs. In: ICLR (2015)
8. Chen, L.C., Zhu, Y., Papandreou, G., Schroff, F., Adam, H.: Encoder-decoder with atrous separable convolution for semantic image segmentation. arXiv:1802.02611 (2018)
9. Dunne, R.A., Campbell, N.A.: On the pairing of the softmax activation and cross-entropy penalty functions and the derivation of the softmax activation function. In: Proceedings of the 8th Australian Conference on the Neural Networks (1997)
10. Eigen, D., Fergus, R.: Predicting depth, surface normals and semantic labels with a common multi-scale convolutional architecture. In: ICCV, pp. 2650–2658 (2015)
11. Fayyaz, M., Saffar, M.H., Sabokrou, M., Fathy, M., Huang, F., Klette, R.: STFCN: spatio-temporal fully convolutional neural network for semantic segmentation of street scenes. In: Chen, C.-S., Lu, J., Ma, K.-K. (eds.) ACCV 2016. LNCS, vol. 10116, pp. 493–509. Springer, Cham (2017). https://doi.org/10.1007/978-3-319-54407-6_33

12. Fu, J., Liu, J., Wang, Y., Lu, H.: Stacked deconvolutional network for semantic segmentation. arXiv:1708.04943 (2017)
13. Islam, M.A., Naha, S., Rochan, M., Bruce, N.D.B., Wang, Y.: Label refinement network for coarse-to-fine semantic segmentation. arXiv:1606.07415 (2017)
14. Islam, M.A., Rochan, M., Bruce, N.D.B., Wang, Y.: Gated feedback refinement network for dense image labeling. In: CVPR (2017)
15. Jégou, S., Drozdzal, M., Vazquez, D., Romero, A., Bengio, Y.: The one hundred layers tiramisu: fully convolutional densenets for semantic segmentation. In: Proceedings of the IEEE Conference on Computer Vision and Pattern Recognition Workshops, pp. 1175–1183 (2017)
16. Krähenbühl, P., Koltun, V.: Efficient inference in fully connected CRFs with Gaussian edge potentials. In: Advances in Neural Information Processing Systems 24, pp. 109–117 (2017)
17. Kundu, A., Vineet, V., Koltun, V.: Feature space optimization for semantic video segmentation. In: CVPR, pp. 3168–3175 (2016)
18. Liu, W., Rabinovich, A., Berg, A.C.: ParseNet: looking wider to see better. arXiv:1506.04579 (2015)
19. Long, J., Shelhamer, E., Darrell, T.: Fully convolutional networks for semantic segmentation. In: CVPR, pp. 3431–3440 (2015)
20. Nguyen, K., Fookes, C., Sridharan, S.: Deep context modeling for semantic segmentation. In: IEEE Winter Conference on Applications of Computer Vision (WACV), pp. 56–63 (2017)
21. Noh, H., Hong, S., Han, B.: Learning deconvolution network for semantic segmentation. In: ICCV, pp. 1520–1528 (2015)
22. Oliveira, G.L., Bollen, C., Burgard, W., Brox, T.: Efficient and robust deep networks for semantic segmentation. Int. J. Robot. Res. **37**, 472–491 (2017)
23. Oliveira, G.L., Burgard, W., Brox, T.: Efficient deep models for monocular road segmentation. In: IEEE/RSJ International Conference on Intelligent Robots and Systems (IROS) (2016)
24. Oliveira, G.L., Burgard, W., Brox, T.: DPDB-Net: exploiting dense connections for convolutional encoders. In: IEEE International Conference on Robotics and Automation (ICRA) (2018)
25. Paszke, A., Chaurasia, A., Kim, S., Culurciello, E.: ENet: a deep neural network architecture for real-time semantic segmentation. arXiv:1606.02147 (2016)
26. Peng, C., Zhang, X., Yu, G., Luo, G., Sun, J.: Large kernel matters-improve semantic segmentation by global convolutional network. In: CVPR (2017)
27. Pohlen, T., Hermans, A., Mathias, M., Leibe, B.: Full-resolution residual networks for semantic segmentation in street scenes. In: CVPR, pp. 3309–3318 (2017)
28. Shah, S.A., Ghosh, P., Davis, L.S., Goldstein, T.: Stacked U-Nets: a no-frills approach to natural image segmentation. arXiv:1804.10343 (2018)
29. Shelhamer, E., Long, J., Darrell, T.: Fully convolutional networks for semantic segmentation. IEEE Trans. Pattern Anal. Mach. Intell. **39**(4), 640–651 (2017)
30. Sturgess, P., Alahari, K., Ladicky, L., Torr, P.H.S.: Combining appearance and structure from motion features for road scene understanding. In: BMCV (2009)
31. Valada, A., Oliveira, G.L., Brox, T., Burgard, W.: Deep multispectral semantic scene understanding of forested environments using multimodal fusion. In: Kulić, D., Nakamura, Y., Khatib, O., Venture, G. (eds.) ISER 2016. SPAR, vol. 1, pp. 465–477. Springer, Cham (2017). https://doi.org/10.1007/978-3-319-50115-4_41
32. Visin, F., et al.: ReSeg: a recurrent neural network-based model for semantic segmentation. In: Proceedings of the IEEE Conference on Computer Vision and Pattern Recognition Workshops, pp. 426–433 (2016)

33. Wang, P., Chen, P., Yuan, Y.: Understanding convolution for semantic segmentation. In: IEEE Winter Conference on Applications of Computer Vision (WACV), pp. 6230–6239 (2017)
34. Wu, Y., Yang, T., Zhao, J., Guan, L., Li, J.: Fully combined convolutional network with soft cost function for traffic scene parsing. In: Huang, D.-S., Bevilacqua, V., Premaratne, P., Gupta, P. (eds.) ICIC 2017. LNCS, vol. 10361, pp. 725–731. Springer, Cham (2017). https://doi.org/10.1007/978-3-319-63309-1_64
35. Yan, Z., Zhang, H., Jia, Y., Breuel, T., Yu, Y.: Combining the best of convolutional layers and recurrent layers: a hybrid network for semantic segmentation. arXiv:1603.04871 (2016)
36. Yu, F., Koltun, V.: Multi-scale context aggregation by dilated convolutions. In: ICLR (2016)
37. Zhao, H., Shi, J., Qi, X., Wang, X., Jia, J.: Pyramid scene parsing network. In: CVPR, pp. 6230–6239 (2017)

Simultaneous Localization and Segmentation of Fish Objects Using Multi-task CNN and Dense CRF

Alfonso B. Labao and Prospero C. Naval Jr.$^{(\boxtimes)}$

Computer Vision and Machine Intelligence Group, Department of Computer Science,
College of Engineering, University of the Philippines, Diliman,
Quezon City, Philippines
pcnaval@dcs.upd.edu.ph

Abstract. We propose a deep learning tool to localize fish objects in benthic underwater videos on a frame by frame basis. The deep network predicts fish object spatial coordinates and simultaneously segments the corresponding pixels of each fish object. The network follows a state of the art inception resnet v2 architecture that automatically generates informative features for object localization and mask segmentation tasks. Predicted masks are passed to dense Conditional Random Field (CRF) post-processing for contour and shape refinement. Unlike prior methods that rely on motion information to segment fish objects, our proposed method only requires RGB video frames to predict both box coordinates and object pixel masks. Independence from motion information makes our proposed model more robust to camera movements or jitters, and makes it more applicable to process underwater videos taken from unmanned water vehicles. We test the model in actual benthic underwater video frames taken from ten different sites. The proposed tool can segment fish objects despite wide camera movements, blurred underwater resolutions, and is robust to a wide variety of environments and fish species shapes.

Keyword: Fish object localization

1 Introduction

1.1 Overview and Motivation

The research of this paper is motivated by the relatively old problem of fish-counting. The problem of how to accurately count estimate fish populations is important for scientists and fishermen as it provides the basis for local and national government decisions as to how much fish can be harvested from the ocean. [23]. From [5], measuring the distribution and abudance of fish organisms in marine environments is of fundamental importance in assessing an ecosystem's health, since it provides criteria for adjusting fishing quotas. If fish abundance

© Springer Nature Switzerland AG 2019
N. T. Nguyen et al. (Eds.): ACIIDS 2019, LNAI 11431, pp. 600–612, 2019.
https://doi.org/10.1007/978-3-030-14799-0_52

counts are left unchecked, it may lead to overfishing which disrupts the natural dynamics of fish populations.

However, the fish-counting problem involves the sub-routine of Fish object localization, and localization of fish objects in underwater video frames is non-trivial due to several challenges unique to underwater videos. [17] mentions that underwater mediums are characterized by light absorption and scattering caused by dissolved constituents and suspended particulates that produce marine snow artifacts. This causes several false positives on fish objects. Other problems include strong similarities between fish-object colors and background colors - which are both dominant in the blue channel. In some videos - particularly those that are taken near the surface - strong illumination changes are prevalent, which could distort methods that rely on brightness and motion for fish localization.

1.2 Related Methods

Earlier methods for fish species localization [21], rely on background-subtraction methods. This approach usually gathers background information per pixel in the form of average values in the RGB channels. Pixels that show significant deviations from the average RGB values are classified as moving objects - which are usually fish objects. A limitation of this approach is its dependence on fixed-camera set-ups and static backgrounds. But for videos with dynamically changing backgrounds - as in cases of sudden illumination caused by turbulent surface waves, or by moving cameras, background subtraction is no longer useful. Other methods rely on manually-defined features and thresholding to detect fish objects such as [11], which relied on CLAHE-thresholded features. These methods however are limited by the information set by the manually-crafted features and may not generalize to a wide variety of fish appearances and shapes. In addition, such methods do not perform in a unified manner since the feature extraction, object localization and classification methods are separate, and a single frame has to pass through different methods to predict the desired outputs.

Recent methods based on deep-learning show some promise to address the limitations of the above approaches. Particularly, deep learning combines both the feature generation process and the localization process into a unified network that could be elegantly trained via backpropagation. Since features are learned from the data, it overrides the limitations caused by manually-defined features. In addition, several features of deep learning are robust to changes in brightness or motion since they are edge-based. This overcomes the need to rely on motion-sensitive background subtractors - which would be applicable to underwater video applications.

Some prior works that uss deep learning methods for fish detection is by [14] and [13]. However, both methods perform species classification with bounding-box estimation, while our method performs pixel-wise instance segmentation with bounding-box estimation. In addition, the previous works rely on manually-labeled ground-truth from the FishCLEF database [8]. The work of [12] uses a deep learning network (Residual Network) to compute for the semantic segmentation of fish object pixels based on weak labels provided by Gaussian background-subtraction. However, the deep network model in [12] does not perform a bounding-box localization task but limits itself to semantic segmentation. A box localization sub-routine can be performed on the output of [12]'s network but it is an external sub-routine that is not embedded on the deep network and may lead to additional processing delays.

For this paper, the deep learning methods that are proposed are networks that could perform both localization, classification and segmentation tasks, [18], [3] - since the problem is to both detect and segment a large number of fishes in each video frame, and differentiate each fish instance. Hence, the proposed method can perform multiple localization and segmentation tasks all at the same time - without need to call methods external to the core architecture. This simplifies the process, and given modern GPU hardware, prediction time is lessened - enabling fast video processing of thousands of frames. To further refine prediced masks, we pass the network's predicted mask to a dense Conditional Random Field (CRF) [9] for post-processing as is done in several other methods. Relating our model to the motivation provided in the previous sub-section, the localization sub-task provides fish-count estimates. The segmentation sub-task on the other hand provides an easier visual assessment of the fish contours which could be used for fish size estimation in future works. For our experiment, we use actual underwater video data that is subject to several underwater illumination distortions and with large numbers of fish objects per frame. We also rely on weak-labels to train our algorithm since manufacturing manual ground truth for our large data is impractical. Models trained with only weak-labels have to be more powerful to generalize well despite inherent errors in training.

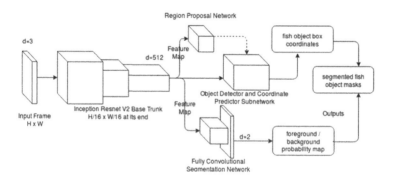

Fig. 1. Network architecture

2   Methodology

The base model of the proposed tool is a convolutional neural network (CNN). Convolutional neural networks first gained popularity in [10], which won the

ImageNet classification task [19] by accurately classifying 1000 classes with 90% accuracy. Recent models extended the classification CNN to localization tasks, where several objects can be detected per input frame, and the CNN localizer outputs cartesian box coordinates (x1, y1, x2, y2) per object. CNNs are trained through the backpropagation algorithm and several iterations are required before achieving meaningful results. For this paper, the CNN follows an architecture [3] that can simultaneously perform several tasks. The only required input during is an RGB video-frame during test-time (of arbitrary size), and the output is a list of fish-objects, along with their corresponding box coordinates, and instance masks (where an instance mask pinpoints the corresponding pixels of the fish object). Our proposed CNN model is subdivided into three parts: (1) the base trunk, (2) the region-proposal network, (3) the pixel segmentation network, and (4) the objectness detector and box-coordinate predictor network.

2.1 Base Trunk Subnetwork

In most CNN architectures, the base trunk consists of a very deep network responsible for generating data-driven features to be fed to object detector and segmentation networks. As per [18], a deep network can generate meaningful features starting from basic RGB values through several convolution operators that generate features from neighboring spatial information. Unlike traditional neural networks, each layer in a CNN has three dimensions (height, width and depth), corresponding to a feature-map. Experimental results show that the first layers of deep networks generate patterns on edges of small regions of the input frame. Later layers combine several small patterns to form larger and more meaningful shapes (i.e. fish eyes, fins, mouth).

For this model, we design the base trunk according to the inception-resnet-v2 architecture [22] that consists of more than a hundred layers. It has achieved state-of-the-art results in the ImageNet challenge. Beginning with the input layer, the base-trunk passes the input frame through several inception blocks. As the layers progress, the height, and width of the feature map decrease, while the depth increases. Following [18], features generated at the very end of the base-trunk are general enough to be useful for a wide variety of tasks, and can be combined for both localization and segmentation. To save memory and for computational efficiency, the base trunk shrinks the original input image to size $H/16 \times W/16 \times D$, where $D = 512$ as shown in Fig. 1.

Similar to [18], the base trunk network does not have any sub-loss function on its own. Weights in the base trunk merely receive gradient corrections from the proposal, segmentation and box coordinate predictor sub-networks.

2.2 Region Proposal Subnetwork

The region proposal network (RPN) is responsible for generating hundreds of region proposals per input frame. A region proposal is a box-coordinate set $[x1, y1, x2, y2]$ corresponding to a sub-region in the input frame, that is likely to contain a fish-object. Some region proposal methods rely on external superpixel proposal algorithms [6], [20]. But the Faster R-CNN work of [18], incorporates the proposal stage as a subnetwork task in the overall CNN architecture. The RPN in Faster R-CNN relies on "anchors", which are spaced evenly every 16 pixels. Each anchor location is assigned with 9 boxes of different sizes and aspect ratios. The RPN operates by finding those anchor-boxes that have the highest intersection over union (or IoU) over ground-truth bounding boxes, as well as those that have an IoU above 70%. Each anchor-box outputs a regressed box coordinate set $[x1, y1, x2, y2]$ that roughly approximates the actual coordinates of the object it overlaps. But these coordinates are usually not accurate, and [18] resorts to a secondary coordinate refinement procedure through the R-CNN sub-network that is connected with the RPN.

RPNs are trained using a loss function formed from a linear combination of objectness classification losses and smooth L1 loss functions [6]. Formally, let $F(\Theta)$ denote the shareable feature map received from the base trunk subnetwork of size of $H/16 \times W/16$, and where Θ denotes all network parameters. Similar to the Faster R-CNN RPN architecture, the RPN passes a 3×3 convolution over $F(\Theta)$, followed by two 1×1 convolutions to produce $p_i^a(\Theta)$ and $t_i^a(\Theta)$, where i denotes an anchor location. The RPN's compound loss function L_i^{rpn} for a certain anchor location i is as follows:

$$L_i^{rpn} = l_{cls}(p_i^a(\Theta)) + l_{reg}(t_i^a(\Theta)) \tag{1}$$

where $p_i^a(\Theta)$ is an 18-d vector of object probabilities for anchor i's 9 scale and aspect ratios. $t_i^{anchor}(\Theta)$ is a 36-d vector of bounding box parameter adjustments. l_{cls} is softmax, while l_{reg} is the smoothL1 loss function. Both $p_i^a(\Theta)$ and $t_i^{anchor}(\Theta)$ involve two inputs relating to ground-truth inputs and network prediction outputs. The details over $t_i^{anchor}(\Theta)$ are a bit involved but are all written in [6] and [18]. The loss function in Eq. 1 is summed up over all i anchor boxes, which reaches an average of 12,000 boxes, followed by averaging. We let A define the set of all anchor box locations such that for each anchor $i \in A$. The RPN loss function is then:

$$L^{rpn} = \frac{1}{|A|} \sum_i L_i^{rpn} \tag{2}$$

Bounding box parameter adjustments $t_i^a(\Theta)$ can be transformed to bounding box coordinates $[x1, y1, x2, y2]$ using a box-inversion function shown in [6]. For each anchor i, the corresponding regressed bounding box coordinates $[x1, y1, x2, y2]$ are used by the R-CNN in [18] to extract sub-feature maps from $F(\Theta)$. This procedure is described in more detail in the following subsections.

2.3   Segmentation Subnetwork

The segmentation task consists of labeling each pixel in the input frame as either belonging to a fish-object or not. The expected output is a binary mask where pixels of fish objects are correspondingly labeled as 1 (or foreground), and all other pixels are labeled as 0 (or background). For this model, segmentation is first performed in the entire frame through a fully-convolutional network (or FCN) [16]. Segmentation could be performed on a local per-region basis as has been done in [3] or [7]. But we prefer to perform global segmentation through an FCN subnetwork since it is able to benefit from a combination of both global and local information. From [16] FCN takes features generated at the very last feature map of the base trunk (with $H/16 \times W/16$ dimensions) and estimates a foreground-background per-pixel probability map with the original $H \times W$ input size. Estimation of the foreground-background probability map is performed through a series of bilinear-interpolation and mini inception blocks. Bilinear-interpolation is responsible for restoring the small $H/16 \times W/16$ base trunk end-feature map to the original $H \times W$ input size.

The segmentation sub-network has its own set of loss functions. We let $F^{fcn}(\Theta)$ describe the FCN output tensor of size $H \times W \times 2$ that is normalized to form a probability defined over foreground and background classes (2 channels). To make the notation simple, we let $|F^{fcn}(\Theta)|$ denote the number of pixels corresponding to $F^{fcn}(\Theta)$. The loss function is then:

Fig. 2. Large frame fish detection and instance segmentation

$$L^{seg} = \frac{1}{|F^{fcn}(\Theta)|} \sum_{|F^{fcn}(\Theta)|} log(F^{fcn}(\Theta))\mathrm{I}[\mathrm{fg}] \qquad (3)$$

From Eq. 3, L^{seg} is the cross entropy of the $F^{fcn}(\Theta)$ foreground/background probability predictions per pixel with the true label I[fg] which takes the value of 1 if foreground and 0 o.w. The resulting foreground-background per-pixel probability map provides the probability that a certain pixel in the input frame belongs to a foreground fish object (or not). However, the FCN output map does not yet differentiate whether a certain pixel belongs to a certain fish object or to another. In other words, the predictions are instance-insensitive. To assign foreground pixels among the various fish objects detected is the role of the objectness detector and box-coordinate predictor network.

2.4   Dense Conditional Random Field Post-processing and Fish Object Instance Differentiation

Before combining the mask and box-coordinate outputs from the segmentation and box-coordinate subnetworks respectively, we first perform an additional mask refinement scheme. Specifically, after the FCN subnetwork produces a 2-channel foreground/background probability mask for all pixels in the input frame, the mask is passed through a dense conditional random field (CRF) for post-processing [9]. This has been performed as a mask segmenter refinement procedure in [2]. The dense CRF associates the foreground/background per-pixel probabilities to the RGB color-space of the input frame to come up with more refined contours and shapes of fish objects. Unary potentials in the dense CRF are the log outputs of the FCN segmentation sub-network $log(F^{fcn}(\Theta))$. Pairwise potentials consist of euclidian distances between pixels in the RGB space following the sample from [9]. By using RGB values, learning dense CRF pairwise potentials is greatly simplified, and inference can be done linearly using convolutional filters following from [9] in a transformed permutahedronal space. After dense CRF post-processing, each pixel is assigned to its corresponding fish object using the box coordinates produced by the box coordinate predictor sub-network (BCPS). This creates an instance-sensitive classification of the pixel mask.

2.5   Objectness Detector and Box Coordinate Predictor Subnetwork

The box coordinate predictor subnetwork (BCPS) follows the refinement process in Faster R-CNN [18], but with additional cascades that perform stepwise corrections on box predictions. The BCPS subnetwork begins with extraction of sub-feature maps from the shareable feature map $F(\Theta)$. For each predicted anchor location $p_i^a(\Theta)$ with $\geq 70\%$ objectness probability, the BCPS subnetwork uses RoI-Cropping [3] to extract a set of sub feature maps $f(\Theta)$ from $F(\Theta)$. The initial dimensions of $f(\Theta)$ are dependent on the regressed anchor box coordinates $[x1, y1, x2, y2]$ from the RPN, but RoI-cropping bilinearly interpolates all $f(\Theta)$ to a uniform $h \times w$ - following the R-CNN structure in [18].

We model the BCPS subnetwork to follow a 4-cascade process [3] for better accuracy. The first cascade predicts preliminary box-coordinates of fish objects along with their objectness information, where high objectness scores represent a region-of-interest with a high probability of containing a fish species. The next three cascades serve to refine the box-coordinates predicted in the previous step. For increased accuracy, each subnetwork incorporates local mask information as well following the architecture described in [3]. In mathematical notation, we let $t_c(f_c(\Theta))$ denote the bounding box parameter adjustments for cascade c. Using the inversion function shown in [6], these bounding box adjustments are transformed to bounding box coordinates $[x1, y1, x2, y2]$. Using these coordinates, the next cascade $c + 1$ re-extracts sub-feature maps $f_{c+1}(\Theta)$ and passes them on to the next cascade sub-network for another regression that outputs the next bounding box adjustments $t_{c+1}(f_{c+1}(\Theta))$. Following the cascade algorithm in [3], bounding box adjustments in $t_{c+1}(f_{c+1}(\Theta))$ should contain a tighter box around the fish object.

For the loss function of BCPS, we let f denote the set of sub-feature maps extracted by the BCPS sub-network where $f_c(\Theta) \in f$ for all cascades $c = 1...4$. The sub loss function of the BCPS sub-network is a summation of losses contributed by each subfeature map, with a total of $4|f|$ sub-feature maps (given 4 cascades). Similar to the RPN, the losses are composed of a smooth L1 regression loss $l_{reg}^c(t_c(f_c(\Theta)))$ and a softmax cross-entropy loss $l_{cls}^c(p_c(f_c(\Theta)))$. The sub-loss function of the BCPS sub-network is as follows:

$$L^{bcps} = \frac{1}{4|f|} \sum_{c=1}^{4} \sum_{f} \left[l_{reg}^c(t_c(f_c(\Theta))) + l_{cls}^c(p_c(f_c(\Theta))) \right] \tag{4}$$

Where c defines the position of the cascade, which in this case begins from 1 to 4. As per [3], incorporating more cascades in the box coordinate regression process should provide more accuracy gains.

2.6   Total Loss Function

During training of the network, the overall loss function supervises gradient corrections for all network parameters Θ, using losses formed in a linear combination of the sub-loss functions (Eq. 5). We train the network using stochastic gradient descent (SGD) with learning rate 0.001 and Nesterov Momentum with 0.9 momentum parameter.

$$L^{all} = L^{rpn} + L^{seg} + L^{bcps} \tag{5}$$

2.7   Prediction Outputs

After training, the network outputs a set of results during prediction (test) time as listed below. The 1st and 2nd outputs are from the BCPS subnetwork (as the RPN subnetwork outputs are merely intermediate outputs for generating proposals). The 3rd output uses the FCN global mask and takes only those pixels that are identified to be within a box coordinate set [x1, y1, x2, y2].

1. Coordinates [x1, y1, x2, y2] of pixels with high probabilities of containing fish objects ($\geq 70\%$)
2. Probability of fish objectness given each coordinate set [x1, y1, x2, y2]
3. Per pixel foreground-background probability for each coordinate set [x1, y1, x2, y2], following FCN and dense CRF post-processing.

2.8   Data Sources and Training Details

The network is trained with benthic underwater video frames taken from ten (10) different sites in central Philippines. Each video frame has different background, underwater illumination, and hues. Some video frames have static camera set-ups while some video frames experience rapid camera movement. In some sites, only a few fish species can be visually detected, but other sites have more than 20 fish objects per frame. To train the proposed model, the needed information are the box coordinates of fish objects per frame (to train the object and coordinate predictor subnetwork) along with the corresponding ground truth binary mask per frame (to train the fully convolutional segmentation sub-network).

The usual method to train a deep learning model is through manually-labeled ground-truth. However, given the large number of frames that have to be processed, it is impractical and time-consuming to manufacture manually-labeled ground truth. We settled for training the model with weakly-labeled data derived from motion-based background subtraction tools. We first extract motion masks from an adaptive Gaussian background subtractor implemented in OpenCV [1], and then refine it using a 152-layer fully convolutional network [15]. Afterwards, we pass the refined motion masks to a contour detector to extract fish blobs. The detected fish blobs are then passed to a deep network classifier for prediction of the true fish blobs.

We select 1,600 weakly-labeled training frames for the training pool, where each site has at least 10 frames included. Training of the network is done with a 12 GB Titan X GPU, with tensorflow deep learning computer code. Training time took around four (4) days, since the network went over the entire training set 80 times.

3   Results and Discussion

Figure 2 depicts a sample frame taken from one of the sites, where fish objects are simultaneously enclosed in bounding boxes and segmented pixel-wise. For testing the model, we collected 30 frames randomly sampled from the different sites and calculated their precision and recall values in terms of bounding box-localization. We consider a detected fish object as a true positive (TP) if the bounding box visually covers more than 50% of the fish's pixel area. If two bounding boxes cover the same fish, only one is counted as a true positive, while the duplicate box is penalized as a false positive. These rules are similar to the standard PASCAL VOC localization challenge [4]. We manually as many fish objects as we could find in the 30 test-frames, but some fish objects that are very small and blurry are no longer counted as they already lie very far from the camera center (Table 1).

Upon test-time, we gathered a precision of 93.77% and a recall of 48.51%, where the average number of fish objects detected per frame number up to 12 fishes per frame. We would like to note however, that the 48.51% recall measure includes fishes of all sizes. However, this measure includes fish objects that are already located far from the camera, and have pixel sizes that are already $\leq 30 \times 30$. If the experiment's scope is limited to fishes that are within a certain perimeter radius from the camera, i.e. a 5-meter box perimeter radius, then the recall of fish objects should be much higher, and we have a conservative estimate

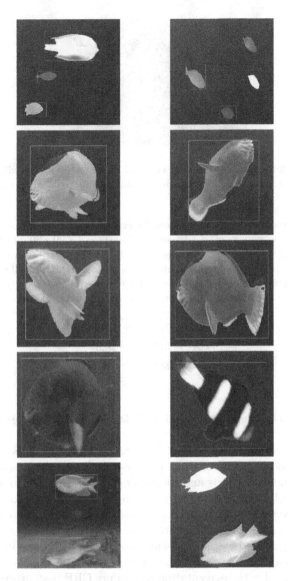

Fig. 3. Sample fish detections with simultaneous bounding box and pixel instance segmentation. The figures above depict sample detections produced by the model. The model simultaneously estimates both the fish object's bounding box (green box outline) and segments the pixels belonging to the fish object within the bounding box. The results above include Dense CRF post-processing which facilitate more accurate segmentation. Segmentation is robust whether the fish object is located in the water column or along the seabed with coral background (i.e. the bottom left picture) (Color figure online)

Fig. 4. Sample detection with moving camera. Since our model predicts on a frame-by-frame basis, it does not need temporal information to localize and segment fish objects

Table 1. Set A includes fishes of all sizes discernable by a human observer. Several of the fishes in set A however are very small with pixel sizes ≤30 × 30. Set B only includes fishes that have pixel sizes that are roughly ≥100 × 100 and are seen within a 5-meter volume radius from the camera viewpoint. Given this constraint, recall increases to 60% estimate. We consider an object to be correctly detected if the predicted bounding box is 50% the area of the ground-truth box, following PASCAL VOC.

Test set	Precision	Recall
Set A	93.77	48.51
Set B	93.77	60 (est.)

of more than 60% recall. This recall includes most of the larger fishes of pixel sizes ≥100 × 100.

From our experiments on test-set frames, several of the detected fish objects lie along the middle portion of the frame's y-axis. This section is where the contrast between the fish objects and the water column are easier to detect. Several of the missed fish objects lie along the frame's base, where most of the corals are located. Several of these missed fish objects are hard to detect even from a human observer. Most of the large and noticeable fishes however are adequately detected and segmented by the model as shown in Fig. 3. The results here include dense CRF post-processing, which greatly increases segmentation precision. If we compare our model with segmentation results pre CRF post-processing, the segmentation with CRF post-processing has more noticeable edges and very finely outlines the fish's contour relative to either a water column or seabed background. Our model is generally high in precision and 93.77% of the bounding boxes are able to cover the fish object by more than 50% - analogously matching the PASCAL VOC requirement for a true-positive in terms of box area coverage.

We tested the model in another set of test frames with rapid camera movement, and in a completely different site. Our model is able to localize and segment the fishes. A sample is in Fig. 4, where individual fishes were segmented. Recall and precision are lower however than the original test set since the frames are from completely different environments. But it is notable that the model is able to operate despite camera movements which shows its independence from motion-based information. Independence from camera movements, shows that our network model can have applications to videos taken from unmanned water vehicles, where jittery camera movements are expected.

4 Conclusion

In this paper, we proposed a deep network model that is able to simultaneously localize fish objects, estimate their bounding boxes and segment their corresponding pixels using a unified multi-task convolutional neural network. Estimation is independent of motion-based information and processing can be performed on a frame-by-frame basis. The model can detect a wide variety of fish object shapes and sizes against different backgrounds. Our proposed model provides a solution for the problem of fish counting in marine environments through its localization feature, which allows automated counting of fish organisms represented by their bounding boxes. Precision is high at 93.77%, with a 60% estimated recall for fish within a 5-meter volume radius. In addition, the added segmentation feature of our model provides a visual description of fish object contours - which would be useful in aiding the visual verification of a human expert or for future work related to fish size estimation.

References

1. Bradski, G., Kaehler, A.: Learning OpenCV: Computer Vision with the OpenCV Library. O'Reilly Media Inc., Sebastopol (2008)
2. Chen, L.C., Papandreou, G., Kokkinos, I., Murphy, K., Yuille, A.L.: Deeplab: Semantic image segmentation with deep convolutional nets, atrous convolution, and fully connected CRFs. arXiv preprint arXiv:1606.00915 (2016)
3. Dai, J., He, K., Sun, J.: Instance-aware semantic segmentation via multi-task network cascades. arXiv preprint arXiv:1512.04412 (2015)
4. Everingham, M., Van Gool, L., Williams, C.K., Winn, J., Zisserman, A.: The Pascal visual object classes (VOC) challenge. Int. J. Comput. Vis. **88**(2), 303–338 (2010)
5. Fier, R., Albu, A.B., Hoeberechts, M.: Automatic fish counting system for noisy deep-sea videos. In: Oceans-St. John's 2014, pp. 1–6. IEEE (2014)
6. Girshick, R.: Fast R-CNN. In: Proceedings of the IEEE International Conference on Computer Vision, pp. 1440–1448 (2015)
7. He, K., Gkioxari, G., Dollár, P., Girshick, R.: Mask R-CNN. In: 2017 IEEE International Conference on Computer Vision (ICCV), pp. 2980–2988. IEEE (2017)
8. Joly, A., et al.: Lifeclef: multimedia life species identification. In: EMR@ ICMR, pp. 7–13 (2014)
9. Koltun, V.: Efficient inference in fully connected CRFs with Gaussian edge potentials. Adv. Neural Inf. Process. Syst. **2**(3), 4 (2011)
10. Krizhevsky, A., Sutskever, I., Hinton, G.E.: Imagenet classification with deep convolutional neural networks. In: Advances in Neural Information Processing Systems, pp. 1097–1105 (2012)
11. Kumar Rai, R., Gour, P., Singh, B.: Underwater image segmentation using clahe enhancement and thresholding. Int. J. Emerg. Technol. Adv. Eng. **2**(1), 118–123 (2012)

12. Labao, A.B., Naval, P.C.: Weakly-labelled semantic segmentation of fish objects in underwater videos using a deep residual network. In: Nguyen, N.T., Tojo, S., Nguyen, L.M., Trawiński, B. (eds.) ACIIDS 2017. LNCS (LNAI), vol. 10192, pp. 255–265. Springer, Cham (2017). https://doi.org/10.1007/978-3-319-54430-4_25

13. Li, X., Shang, M., Hao, J., Yang, Z.: Accelerating fish detection and recognition by sharing CNNs with objectness learning. In: OCEANS 2016-Shanghai, pp. 1–5. IEEE (2016)

14. Li, X., Shang, M., Qin, H., Chen, L.: Fast accurate fish detection and recognition of underwater images with fast R-CNN. In: OCEANS 2015-MTS/IEEE Washington, pp. 1–5. IEEE (2015)

15. Lin, G., Shen, C., van den Hengel, A., Reid, I.: Efficient piecewise training of deep structured models for semantic segmentation. In: Proceedings of the IEEE Conference on Computer Vision and Pattern Recognition, pp. 3194–3203 (2016)

16. Long, J., Shelhamer, E., Darrell, T.: Fully convolutional networks for semantic segmentation. In: Proceedings of the IEEE Conference on Computer Vision and Pattern Recognition, pp. 3431–3440 (2015)

17. Negahdaripour, S., Yu, C.H.: On shape and range recovery from image shading for underwater applications. Underwater Robot. Veh.: Des. Control 221–250 (1995)

18. Ren, S., He, K., Girshick, R., Sun, J.: Faster R-CNN: towards real-time object detection with region proposal networks. In: Advances in Neural Information Processing Systems, pp. 91–99 (2015)

19. Russakovsky, O., et al.: Imagenet large scale visual recognition challenge. Int. J. Comput. Vis. 115(3), 211–252 (2015)

20. Sermanet, P., Eigen, D., Zhang, X., Mathieu, M., Fergus, R., LeCun, Y.: Overfeat: Integrated recognition, localization and detection using convolutional networks. arXiv preprint arXiv:1312.6229 (2013)

21. Spampinato, C., Chen-Burger, Y.H., Nadarajan, G., Fisher, R.B.: Detecting, tracking and counting fish in low quality unconstrained underwater videos. VISAPP 2(2008), 514–519 (2008)

22. Szegedy, C., Ioffe, S., Vanhoucke, V., Alemi, A.: Inception-v4, inception-resnet and the impact of residual connections on learning. arXiv preprint arXiv:1602.07261 (2016)

23. Twilley, N., Graber, C.: Gastropod: How many fish are in the sea? counting fish is a daunting but essential task in protecting aquatic ecosystems-and now artificial intelligence, autonomous submarines, and drones can help. https://www.theatlantic.com/science/archive/2016/10/how-many-fish-are-in-the-sea/502937/

Violent Crowd Flow Detection
Using Deep Learning

Shakil Ahmed Sumon, MD. Tanzil Shahria, MD. Raihan Goni,
Nazmul Hasan, A. M. Almarufuzzaman,
and Rashedur M. Rahman[✉]

Department of Electrical and Computer Engineering, North South University,
Plot-15, Block-B, Bashundhara Residential Area, Dhaka, Bangladesh
{shakil.sumon,tanzil.shahria,raihan.goni,
a.marufuzzaman,rashedur.rahman}@northsouth.edu,
Edufornazmul@gmail.com

Abstract. This research aims in detecting violent crowd flows in the context of Bangladesh. For this purpose, we have collected a dataset which includes both violent and non-violent crowd flows. Different deep learning algorithms and approaches have been applied on this dataset to detect scenarios which contain violence. Convolutional neural networks (CNN) and long short-term memory network (LSTM) based architectures have been experimented separately on this dataset and in combination as well. Moreover, a model that was already pretrained on violent movie scenes has been used to leverage transfer learning which outperformed all other experimented approaches with an accuracy of 95.67%. Surprisingly, the sequence model alone or in combination with CNN has not performed well on this particular dataset. The proposed model is lightweight hence it can be deployed easily in any security systems consisting of CCTV cameras or unmanned aerial vehicles (UAVs).

Keywords: Deep learning · Violence · CNN · LSTM · Transfer learning

1 Introduction

The complex socio-economical structure has made security a major concern for the societies all over the world. The terrorist attacks on crowded places have seen a dramatic rise over the recent years. However, the weak security measures and differences in the classes of societies have made the developing countries like Bangladesh more vulnerable to crowded violence. The lack of resources of the security forces has made it extremely difficult to detect violent crowd flows. Government and private organizations have installed CCTV cameras in some part of the cities. Moreover, in time of political and religious assemblies in open fields, security forces tend to deploy unmanned aerial vehicles (UAVs) to detect unrest in the crowd. However, these security installments need constant human supervision which is costly and prone to human error.

This paper proposes a deep learning model which detects violent crowd flows. A dataset has been collected from YouTube containing violent and non-violent videos

© Springer Nature Switzerland AG 2019
N. T. Nguyen et al. (Eds.): ACIIDS 2019, LNAI 11431, pp. 613–625, 2019.
https://doi.org/10.1007/978-3-030-14799-0_53

in the context of Bangladesh. The dataset has an equal distribution of violent and non-violent videos. However, the model which has been trained on this dataset is light-weight and hence can be deployed in security systems which consist of CCTV cameras as well as in UAVs. As the model is automated, it does not need any human intervention to detect violent crowd flows.

This study contributes in the following fields: (i) It proposes a dataset containing violent and non-violent videos on the context of Bangladesh. (ii) It builds a deep learning model to detect violence on Bangladeshi crowd. (iii) It finds out that, in this dataset, 2D convolutional neural networks perform better than sequence models like long short-term memory (LSTM) networks.

The paper is organized as follows: Sect. 2 explores similar experiments concerning violent crowd flow detection, Sect. 3 gives an overview of the dataset we have collected, Sect. 4 describes the methodology of our experiments, Sect. 5 has the results of our experiments and the analysis of the results and we conclude our paper in Sect. 6.

2   Related Works

In this paper [5], the authors introduced a 3D convolution neural network instead of a 2D convolution neural network. 2D CNN only provides spatial information of the video, where temporal information is important as well to classify violent video. They used hockey dataset [1] to train their 3D CNN model having 9 layers. Output node has only one node which gives only true or false. The sigmoid function is used for activation function while stochastic gradient descent is used for cost minimization. Using this architecture with a batch size of 20, accuracy of 91% was achieved.

Another approach was made to predict violence [3] using a different procedure. Four types of audio-visual features were taken to introduce this model. Trajectory-based motion features, spatial-temporal interest points (STIP), attribute features and audio features. STIP algorithm searches were used to find interest points in both spatial and temporal dimensions. Audios from the violence videos are also taken as a feature. For classification purpose, they used Support Vector Machine (SVM). They ran classification on various pairs of features and got accuracy ranging from 31.2% to 68.2%.

Authors of this paper [11] presented a model using Convolutional Long Short-Term Memory (convLSTM). Series of convolutional layers and pooling layers are used to extract features and convLSTM is provided with the extracted features to classify violence. The main purpose of using convLSTM is to learn both spatial and temporal features from the videos. 256 filters were used in convLSTM, and Rectified linear unit (ReLU) were used as the activation function. As it was a binary classification problem, binary cross entropy was used as a loss function. In the input layer, the difference of adjoining frames was given which made sure to identify the changes that took place in videos. Three open data sets were used to train and test. Hockey fight dataset [1], movie dataset [2], and violence-flow data set. Before training, various pre-processing techniques like resizing, mean zero, and variance unity are used. The authors compared LSTM and convLSTM using the same data set and found out 94.6% and 97.1% accuracy respectively.

To solve the problem, an approach was made in [4]. The research introduced an algorithm named Motion Weber Local Descriptor (MoWLD). This algorithm can find spatial and temporal information from the interesting point in videos. By collecting the WLD histograms of adjacent regions the authors in [4] reconstructed the WLD histogram and oriented the WLD histograms to their governing orientation. Multi-scale optical flow was also included to adopt WLD features. Kernel density estimation (KDE) was used to reduce the dimension of the data and smooth pdf curve was produced. Later they used a sparse coding scheme to process the MoWLD features to obtain more dominant features for violence detection. Then max pooling approach was used to obtain a more compact representation of the features. The authors used three datasets: hockey fight dataset, the BEHAVE dataset, and crowd violence dataset. With their proposed model they gained 91.9%, 94.9% and 89.78% accuracy respectively for three dataset mentioned before.

Some researchers took the problem of real-time violent detection and got some amazing results from crowded places. In this study [6], the authors made the widely used surveillance video camera system as their research field and got real-time unique data. By using the Violent Flows (ViF) descriptor they made a statistic which was based on changing of magnitudes of flow-vector over time. The authors used the statistics for short frame sequences which was collected from the dataset. Their accuracy rate was around 82.9%, which was very impressive.

Some other researchers also worked on the videos of the surveillance system but in a different way. The authors in [7] applied CENTRIST-based features on video scenes which were used to identify many important contexts, like violence. They divided the whole process into several sub-processes: preprocessing, feature extraction, data normalization, feature reduction, and classification. They used 2 kinds of dataset. They used 246 videos of crowded scenes as violent flow dataset and 1000 clips from hockey game as Hockey Fights Dataset. Each of the datasets had exactly 2 categories. They got very good accuracy. The best accuracy they got from their first dataset was 91.46% and from the second dataset was 92.79% respectively.

There are good number of interesting works on violent flow, especially in crowded scenes. A group of researchers solved the problem in a different way. Instead of following the standard approaches, in this experiment [8], the authors followed a particular concept of fluid mechanics, which was a substantial derivative. They analyzed the fluid property of a video and captured the rate of change of that property. To form the final descriptor, two histograms were concatenated, those produced from two different optic flows. The authors used 4 different kinds of datasets: Violence in Movies, Crowds, Riot in Prison and Panic. They got an average accuracy of 95% from their datasets.

This paper [9] introduced a new algorithm for violent flow detection. The algorithm was based on optical flow and local spatiotemporal features. This was not like other works in this field because of the representation. The authors in [9] combined the optical flow method and Harris 3D spatiotemporal interest point detector with physical contact detection algorithm to make their own algorithm. Unfortunately, they got a very conflicting result. Their result fluctuated rapidly. Their accuracy for violence started at 21% and their best accuracy was around 69.43%

Some researchers of this field defined their objective differently though the aim was the same. The authors in [10] tried to detect robbery and improve security system of self-service banks. First, they identified the motion region and drew a rectangle around that region. Then, from the selected region, they evaluate the energy flow and optical flow which should be very high in a motion region. The authors used a very small amount of data for this research. They only used 13 videos. 8 of them were used as training videos which produced 2163 samples. The accuracy they got was 76.9% while using the whole dataset and 70% while using only test data.

3   Dataset

The data set has been collected from YouTube. The collected videos are based on Bangladesh context. We have collected videos of different types of violence such as street fighting, chaos on political precisions, fighting with police etc. and labelled them as violent videos. Moreover, we have collected videos of peaceful precision, cricket match gallery, cricket match winning celebration etc. and labeled them as non-violent videos. We have collected 80 videos of each class which gives us 160 videos in total. Each of the videos is approximately 10 s long.

The dataset has been split into two parts; the training set and testing set. The training part consists of 70 videos per class and 10 videos per class are dedicatedly reserved for the testing purpose.

This is the first of such dataset which is dedicatedly proposed for violence detection in Bangladeshi crowd. We encourage further research on similar fields and hence open sourced the data set.

Moreover, we have used this [12] dataset to pretrain the model. This dataset has been collected from different Hollywood movies. It consists of crowded scenes of violence and non-violence from various movies.

4   Methodology

The video dataset which we have used in this study, as it has been collected from YouTube, has varying resolutions ranging from 1024 pixels to 72 pixels. At first, we have extracted 30 frames per second from the videos and downsampled them to 28 * 28 pixels as larger resolution slows down the training process. After that, these frames are extracted from the violent and nonviolent videos are used as inputs to our proposed network.

However, a convolutional neural network (CNN) architecture has been proposed to detect violent flows. CNN has a different architectural infrastructure than other artificial neural networks (ANN) and deep neural networks (DNN). However, a typical CNN has a stack of layers such as an input layer, a combination of several convolutional and pooling layers and at the end, output layer. Moreover, the CNN layer architecture is followed by several fully connected layers which play a vital role as the CNN works as a feature extractor and ANN acts as a classifier. However, the model which is proposed for violent flow detection has 2 convolutional layers with 32 filters on each of them.

The filters used in the model have the size of 3 * 3. Moreover, a batch normalization layer has been used after each convolutional and fully connected layer. The Rectified Linear Unit (ReLU) is used as the activation function of this model.

Table 1 demonstrates the layered architecture and the parameters those are to be trained in each layer. The architecture incorporates different regularization techniques like kernel regularizes l2 and dropout of .5 to prevent the model from overfitting. Figure 1 shows the higher-level architecture of the convolutional neural network we have used.

Table 1. CNN layer architecture

Layer type	Parameters
Convolutional	320
Dropout	0
Batch normalization	128
Convolutional	9248
Dropout	0
Batch normalization	128
Flatten	0
Fully connected	184330
Dropout	0
Fully connected (softmax layer)	22
Total parameters	194176

A pretrained model on the movie violence dataset has been used to leverage transfer learning as the data set has been used in this study are not fairly large enough. A model with the similar architecture as of Table 1 has been retrained on our collected dataset. In both of the models discussed above, we have used Adam as the optimizer and binary cross-entropy as the loss function.

Additionally, the data set has experimented with Long Short-Term Memory (LSTM) network which is a variant of the Recurrent Neural Network (RNN). LSTM is considered as a recurrent neural network architecture which has memory cells in it. These memory cells are the central idea behind the LSTM architecture as they can maintain their state over a certain span of time. LSTMs have regulators unit called gates which control what will get in and out of the memory cell. Table 2 shows the LSTM network architecture and the parameters per layer of the network that has been used in this experiment. The model has two LSTM layers with the first layer having 10 and the second having 50 LSTM units in them.

Additionally, it has two fully connected layers with 10 neurons each and a softmax layer on top of them. However, the activation function for the LSTMs in this model is hyperbolic tangent function and ReLU for the fully connected layer. Figure 2 shows the higher-level architecture of the LSTM network that has been used in this experiment.

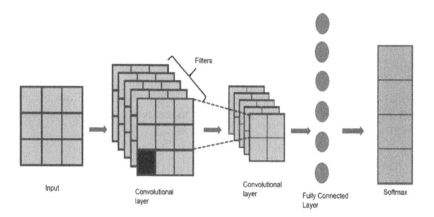

Fig. 1. Convolutional neural network architectures

Table 2. LSTM layer architecture

Layer	Parameters
LSTM	1600
Dropout	0
LSTM	12400
Dropout	0
Batch normalization	200
Fully connected	5100
Dropout	0
Fully connected	1010
Dropout	0
Fully connected (softmax)	22
Total parameters	20332

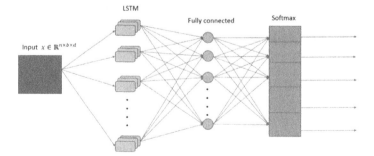

Fig. 2. Long short-term memory architecture

Moreover, we have experimented by combining the convolutional neural networks and LSTMs in one model where the CNN extracts spatial features from the video frames and passed the extracted features to the LSTMs. Additionally, the model has two fully connected layers after the LSTM layers which act as the classifiers. Finally, a softmax layer has been added at the end of the architecture which gives the probability distribution of the classes.

However, in LSTM and CNN-LSTM model, like two other models, regularizes techniques such as kernel regularizes and dropout have been applied. The optimizer and loss function for these models are Adam and binary cross entropy respectively.

5   Results and Analysis

The models have been trained on the training portion of the data set and after that they have been tested against the videos which were reserved for the testing purpose. Table 3 shows the performance of the CNN model after 10, 20, 50, 100, 200 and 500 epochs in terms of percentage accuracy.

We see that after 50 epochs the accuracy is not converging instead decreasing to some margin. So, we have taken the accuracy after 50 epochs as the benchmark for the CNN model. However, Fig. 3. demonstrates the epochs vs accuracy graph of the CNN model. The graph shows that the model is guilty of overfitting to some scale and the accuracy is fluctuating in the range of 76% to 96%. Figure 4 is the epoch vs loss graph of the CNN model which shows us the fluctuating loss of test set.

However, we have used the pretrained model to leverage transfer learning in two different approaches. Firstly, we have loaded the pretrained model in our own architecture and have frozen all the layers of the network. This prevents the model to retrain on our collected data and enables the network to detect violent flow in our test data with what it had learned from the movie dataset.

Table 3. Performance of CNN model in terms of accuracy

Epochs	Training accuracy	Testing accuracy
10	94.25	92.90
20	94.23	93.93
50	95.57	94.86
100	95.72	94.47
200	94.23	91.53
500	94.59	92.27

Secondly, we have retrained the model with our dataset without freezing any layer. Table 4 shows the accuracy of the models. Figures 5 and 6 represent the epochs vs accuracy and epoch vs loss graphs of the transfer learning model with freezing all the layers respectively. The graphs show that the accuracy and the loss are static which are supposed to happen as the model has not learned anything new rather has predicted with what it had learned previously from the movie dataset.

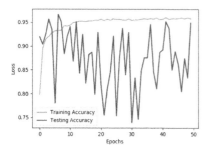 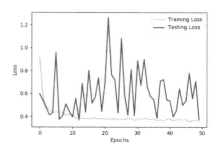

Fig. 3. Epochs vs accuracy graph of CNN model

Fig. 4. Epochs vs loss graph of the CNN model

On the other hand, Figs. 7 and 8 show the epochs vs accuracy and epochs vs loss graph of the model that has been retrained without freezing any layer. This model, like the CNN model, is guilty of overfitting too and the accuracy and loss are not consistent as well.

Table 4. Accuracy of the transfer learning models

Approach	Training	Testing
Freezing the layers	96.85	92.27
Without freezing layers	95.67	95.70

However, we have taken two different approaches while dealing with LSTMs as well. In one model we have used LSTMs with fully connected layers whereas in another model we have added CNN layers before the LSTMs. Table 5 shows the performance of the models in terms of percentage accuracy.

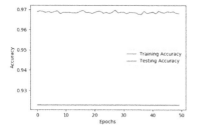 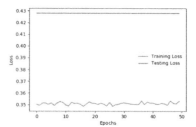

Fig. 5. Epochs vs accuracy graph of the transfer learning model with all layers frozen

Fig. 6. Epochs vs loss graph of the transfer learning model with all layers frozen

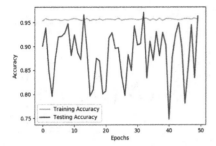

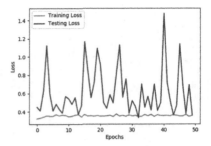

Fig. 7. Epochs vs accuracy of the transfer learning model without freezing layers

Fig. 8. Epochs vs loss of the transfer learning model without freezing layers

Figures 9 and 10 show the epochs vs accuracy and the epochs vs loss graph of the LSTM model whereas Figs. 11 and 12 show the epochs vs accuracy and epochs vs loss graph of the CNN + LSTM model respectively. We see that the models are not performing well in terms of accuracy but their metrics are fluctuating relatively less than the other CNN models. We also observe that adding convolutional neural network layer in front of the LSTM layers improves the accuracy by a large margin.

Table 5. Performance of the LSTM and CNN + LSTM model

Approach	Training	Testing
LSTM	87.23	67.41
CNN + LSTM	89.79	75.10

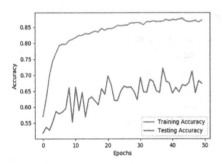

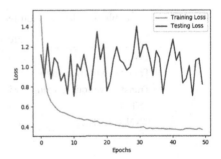

Fig. 9. Epochs vs accuracy of the LSTM model

Fig. 10. Epochs vs loss of the LSTM model

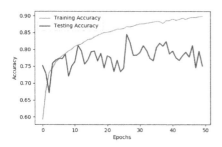

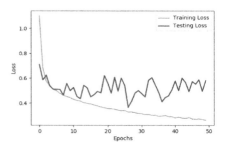

Fig. 11. Epochs vs accuracy of the CNN + LSTM model

Fig. 12. Epochs vs loss of the CNN + LSTM model

However, the transfer learning model which has been retrained without freezing any layer seems to perform better in terms of accuracy among all the models. Additionally, we have calculated the precision, recall and f1-support of the models we have experimented with. Precision gives us the idea about the classifier's ability of not labeling a positive sample as a negative one. Recall is the metrics by which we know how good the model is in finding all the positive samples of the dataset.

F1-support is the weighted average of the above two. Tables 6 and 7 shows the metrics per class and Table 8 shows the overall classification report of the models.

The classification report shows that the CNN and transfer model without freezing layers are consistent in detecting violent and non-violent videos whereas the transfer model with all layers frozen has difficulties in detecting violent videos. The LSTM models have done fairly good in predicting nonviolent videos but have performed miserably while detecting violent videos.

Table 6. Classification report of nonviolent class

Type	Precision	Recall	F1 support
CNN	.96	.94	.95
Transfer with freezing	.89	.98	.93
Transfer without freezing	.95	.99	.97
LSTM	.65	.84	.73
CNN + LSTM	.69	.96	.80

Table 7. Classification report of violent class

Type	Precision	Recall	F1 support
CNN	.93	.95	.94
Transfer with freezing	.97	.86	.91
Transfer without freezing	.97	.92	.94
LSTM	.73	.49	.58
CNN + LSTM	.92	.51	.66

Figures 13 and 14 shows some examples of the predictions of the transfer learning model without freezing layers on video frames. Figure 13 shows 9 non-violent video frames of a particular video from the dataset. The model has been predicted all of the video frames correctly as non-violent.

Table 8. Overall classification report of the models

Type	Precision	Recall	F1 support
CNN	.95	.95	.95
Transfer with freezing	.93	.92	.92
Transfer without freezing	.96	.95	.95
LSTM	.69	.67	.66
CNN + LSTM	.80	.75	.74

On the other hand, Fig. 14 shows 9 video frames of a violent video from the collected dataset. The model has been predicted 7 of the video frames correctly as violent but it failed in predicting 2 violent video frames and labelled them as non-violent.

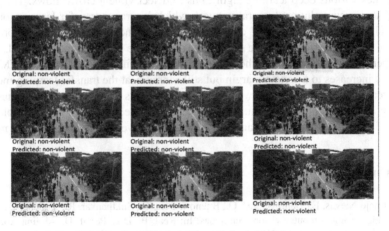

Fig. 13. Predictions on non-violent video frames

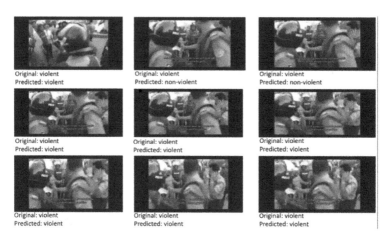

Fig. 14. Predictions on violent video frame

6 Conclusion and Future Work

The paper explores a relatively small dataset containing violent and non-violent videos and applies various deep learning algorithms to detect violent crowd flows. The paper reveals that a convolutional neural network which leverages transfer learning outperforms all the other variance of convolutional neural networks and long short-term memory networks. Moreover, the study finds out that the sequence models like LSTMs have performed worse than other models. By combining CNN with LSTM, the accuracy increases to a certain margin but still cannot beat the transfer learning models. However, in our future study, we will be making this model more lightweight by pruning and will deploy it in an unmanned aerial vehicle. Moreover, we have plans to host our model in a web server and make an API to give people access to our model.

References

1. Bermejo Nievas, E., Deniz Suarez, O., Bueno García, G., Sukthankar, R.: Violence detection in video using computer vision techniques. In: Real, P., Diaz-Pernil, D., Molina-Abril, H., Berciano, A., Kropatsch, W. (eds.) CAIP 2011. LNCS, vol. 6855, pp. 332–339. Springer, Heidelberg (2011). https://doi.org/10.1007/978-3-642-23678-5_39
2. Claire-Heilene, D.: VSD, a public dataset for the detection of violent scenes in movies: design, annotation, analysis and evaluation. In: The Handbook of Brain Theory and Neural Networks, vol. 3361 (1995)
3. Dai, Q., Tu, J., Shi, Z., Jiang, Y.G., Xue, X.: Fudan at MediaEval 2013: violent scenes detection using motion features and part-level attributes. In: MediaEval, October 2013
4. Zhang, T., Jia, W., Yang, B., Yang, J., He, X., Zheng, Z.: Mowld: a robust motion image descriptor for violence detection. Multimed. Tools Appl. **76**(1), 1419–1438 (2017)
5. Ding, C., Fan, S., Zhu, M., Feng, W., Jia, B.: Violence detection in video by using 3D convolutional neural networks. In: Bebis, G., et al. (eds.) ISVC 2014. LNCS, vol. 8888, pp. 551–558. Springer, Cham (2014). https://doi.org/10.1007/978-3-319-14364-4_53

6. Hassner, T., Itcher, Y., Kliper-Gross, O.: Violent flows: real-time detection of violent crowd behavior. In: 2012 IEEE Computer Society Conference on Computer Vision and Pattern Recognition Workshops (CVPRW), pp. 1–6. IEEE, June 2012

7. De Souza, F., Pedrini, H.: Detection of violent events in video sequences based on census transform histogram. In: 2017 30th SIBGRAPI Conference on Graphics, Patterns and Images (SIBGRAPI), pp. 323–329. IEEE, October 2017

8. Mohammadi, S., Kiani, H., Perina, A., Murino, V.: Violence detection in crowded scenes using substantial derivative. In: 2015 12th IEEE International Conference on Advanced Video and Signal Based Surveillance (AVSS), pp. 1–6. IEEE, August 2015

9. Lyu, Y., Yang, Y.: Violence detection algorithm based on local spatio-temporal features and optical flow. In: 2015 International Conference on Industrial Informatics-Computing Technology, Intelligent Technology, Industrial Information Integration (ICIICII), pp. 307–311. IEEE, December 2015

10. Xu, Y., Wen, J.: Detecting robbery and violent scenarios. In: 2013 Second International Conference on Robot, Vision and Signal Processing (RVSP), pp. 25–30. IEEE, December 2013

11. Sudhakaran, S., Lanz, O.: Learning to detect violent videos using convolutional long short-term memory. In: 2017 14th IEEE International Conference on Advanced Video and Signal Based Surveillance (AVSS), pp. 1–6. IEEE, August 2017

12. Violent Scenes Dataset: Technicolor. https://www.technicolor.com/dream/research-inno vation/violent-scenes-dataset

Drowsiness Detection in Drivers Through Real-Time Image Processing of the Human Eye

Erick P. Herrera-Granda[1], Jorge A. Caraguay-Procel[1],
Pedro D. Granda-Gudiño[1], Israel D. Herrera-Granda[1],
Leandro L. Lorente-Leyva[1(✉)], Diego H. Peluffo-Ordóñez[2],
and Javier Revelo-Fuelagán[3]

[1] Facultad de Ingeniería en Ciencias Aplicadas, Universidad Técnica del Norte,
Av. 17 de Julio, 5-21 y Gral. José María Cordova, Ibarra, Ecuador
{epherrera,jacaraguay,pdgranda,idherrera,
lllorente}@utn.edu.ec
[2] Escuela de Ciencias Matemáticas y Tecnología Informática, Yachay Tech,
Hacienda San José s/n, San Miguel de Urcuquí, Ecuador
dpeluffo@yachaytech.edu.ec
[3] Universidad de Nariño, Pasto, Colombia
javierrevelof@udenar.edu.co

Abstract. At a global level, drowsiness is one of the main causes of road accidents causing frequent deaths and economic losses. To solve this problem an application developed in Matlab environment was made, which processes real time acquired images in order to determine if the driver is awake or drowsy. Using AdaBoost training Algorithm for Viola-Jones eyes detection, a cascade classifier finds the location and the area of the driver eyes in each frame of the video. Once the driver eyes are detected, they are analyzed whether are open or closed by color segmentation and thresholding based on the sclera binarized area. Finally, it was implemented as a drowsiness detection system which aims to prevent driver fall asleep while driving a vehicle by activating an audible alert, reaching speeds up to 14.5 fps.

Keywords: Drowsiness detection · Image processing · Artificial intelligence · Human eye · Alarm

1 Introduction

Globally, drowsiness in drivers is one of the main causes of road accidents which result in heavy human and economic losses [1, 2]; therefore, to contribute to the solution of this problem, many devices have been developed in the way of alarms for the driver, whose purpose is to prevent a driver from falling asleep while driving a vehicle. Many of these devices focus on the analysis of the driver's eyes and their movement patterns in order to prevent drowsiness in the driver [3–6].

Drowsiness is considered an intermediate state between wakefulness and deep sleep, and for this reason, it can be easily interrupted by visual, auditory or sensory stimuli, so that the person can return to an apparent waking state by a certain time that depends on fatigue and the state of health of the person [7]. Recent studies reveal that many people

© Springer Nature Switzerland AG 2019
N. T. Nguyen et al. (Eds.): ACIIDS 2019, LNAI 11431, pp. 626–637, 2019.
https://doi.org/10.1007/978-3-030-14799-0_54

who maintain a state of forced wakefulness can enter a state of drowsiness and stay there even though their behavior proves otherwise [8]. Drowsiness is defined as the feeling of heaviness and clumsiness of the senses motivated by sleep. This feeling is determined as the intermediate state between waking state and deep sleep [9, 10]. The beginning of this stage is not a unique event, but is caused by changes in various neurological functions, sensory changes occur in memory, consciousness, and loss of logical thinking, latency in response to stimuli and alterations in the brain potentials.

2 Related Works

In this chapter, we summarize previous approaches on drowsy driver's detection. Dajeong et al. research [11] proposes a method of drowsiness detection with eyes open using EEG-based power spectrum analysis. Sahayadhas et al. [3] developed a drowsiness detector based on multiple sensors, which measure facial movement, driving direction and hearth signals. Also, Vural et al. [4] developed a study based on facial expression focused on assumptions of behavior that might predict drowsiness. Borghini et al. [6] developed a study detecting drowsiness using measurements of brain activity during drive for aircraft pilots and car drivers. Also embedded devices, like Reddy et al. [12], have been developed running on Jetson TK1 using deep learning Multi-Task Cascaded Convolutional Networks, and identifying eyes and mouth, determining eye condition and yawning to evaluate drowsiness in the driver. Recent studies have been developed for portable devices on android environment like El-den et al. [5], which application is based on monitoring eye movement testing eye blinking using VIDMIT database at speed of 2 fps, and Montanini et al. [13], which system is based on detecting eyes in the image and evaluating the distribution of dark pixels in the eye, when the eyes are open the dark pixels are in the center of the image and when the eye is closed dark pixels have an elongated form.

This study presents a simple novel approach to detect driver's drowsiness by applying two distinct methods in computer vision and image processing. With the objective to combine both methods under one single profile instead of relied solely on one a detection method [14], with similar precision but reaching sclera detection speed of 13.4 to 14.5 fps, which makes this solution relevant compared whit [12, 13] that reached the speeds of 14.9 fps on a NVIDIA Jetson TK1 and 2.5 fps on a smartphone respectively. In addition the study of a solution based on sclera segmentation, corresponding to the white pixels in the binarized image, has not been deepened and is proposed in this paper.

3 Face and Eyes Detection

Viola Jones algorithm has been one of the most used detection methodology which is based on the Adaboost machine learning algorithm constraining each classifier on a single feature, which achieved 77.8% accuracy with 5 false positives and 15 fps [15]. With the objective of solving boosting over fitting problem the impact of gradient ascent in the Adaboost machine learning is studied in [16] and conclude that Boosting algorithm outperforms Random Forests with lesser Mean squared Test Errors. In [17] the author proposed a robust multi-class AdaBoost with a noise-detection based multi-class loss function and a new weight updating scheme. Other authors [18] present an Adaboost and Contour Circle algorithm for the problem of traffic accidents, based on a traditional Adaboost method for recognizing whether eyes are in open or closed state. Also in [19] Viola Jones object boosted cascade classifier accuracy was improved by using a set of 15 features for the classifier, including 45° rotated, line and center-surround features, improving false alarm rate from 10% to 12.5%. In [20] proposed a new weight adjustment factor as a weak learner of the AdaBoost algorithm that allows achieving a highly accurate, robust and fast classification, so nowadays many studies have been developed in order to improve Viola Jones algorithm accuracy.

Although it has been mentioned that certain visual, auditory and/or sensory stimuli may seem to remove a person from their drowsy state, there are others that could worsen this situation and deepen this state until finally the person falls into deep sleep.

The Viola-Jones algorithm is based on three main ideas: the representation of the image in a new format of processing known as integral image, the extraction of simple but effective characteristics of Haar-type and the use of a combined application method of the cascade classifiers. It allows to quickly eliminate the regions of the background image to preserve only those corresponding to a face [21, 22].

The obtaining of the integral image competes with a methodology of the image computation to facilitate a fast processing of the characteristics. The integral image in a location (x, y), of $h \times w$ rectangle dimensions, contains the sum of the pixels of it and those that are on the left and below the defined area. This is due to the Eq. (1) and the recurrences that are detailed below:

$$ii(x, y) = \sum_{x' \le x, y' \le y} i(x', y') \tag{1}$$

$$s(x, y) = s(x, y - 1) + i(x, y) \tag{2}$$

$$ii(x, y) = ii(x - 1, y) + s(x, y) \tag{3}$$

Where (x, y) are the coordinates of the integral image $ii(.)$ in the integral image, $i(.)$ is the original grayscale image and $s(x, y)$ is the cumulative sum of the rows [15]. Thus, the integral image can be processed using the recurrences (2) and (3), when passing over the original image, as shown in Fig. 1.

For example, to obtain the new sum for the element D of the integral image, the values obtained are: 1 which is the sum of the pixels in the rectangle A, 2 which is the

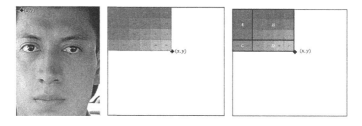

Fig. 1.   (a) Obtaining the integral image in a location (x, y), (b) integral image.

sum of the pixels of the rectangles A + B, 3 which is the sum of the regions A + C, 4 which is the sum of the pixels A + B + C + D and thus the sum for D can be obtained as 4 + 1−(2 + 3) [15].

The characteristics that this algorithm uses for learning are basic Haar-type functions, which use the operations of type: characteristic of two rectangles Fig. 2a, b, difference between the sum of the values of the pixels of two rectangular regions; a feature of three rectangles Fig. 2c, sum between two external rectangles subtracted from the sum of the central rectangle; characteristic of four rectangles Fig. 2d, difference between pairs of rectangles in diagonal.

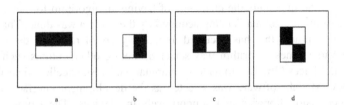

a   b   c   d

Fig. 2. Haar-type functions used in the Viola-Jones algorithm

The obtaining of characteristics by means of this method, allows to get a high number of them. For example, for an image of only 384 × 288 pixels as proposed in [15], considering that the resolution of this detector is in regions of 24 × 24 pixels, it has a total of 160000 characteristics. In this way, an efficient method was required to process and classify them, which in this case is AdaBoost [23–25]. These characteristics are combined to form the classifiers, but very few of these characteristics can be combined to form an effective classifier [15]. This is how we proceed to identify and organize these characteristics. For example, the first two characteristics selected by the AdaBoost algorithm for facial detection are obtained as shown in Fig. 3.

The first characteristic obtained by means of two rectangles shows that the region of the eyes is darker than the region of the cheeks and nose. While the second characteristic obtained from the Haar function of three rectangles, it shows that the eyes are darker than the bridge of the nose. Next, the Viola-Jones algorithm applies the classifiers in cascade, so that the simplest classifiers are the ones that are used first and are

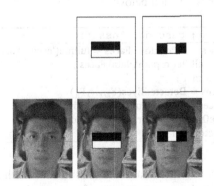

Fig. 3. Obtaining the first characteristics for the AdaBoost classifier

responsible for eliminating most sub-regions of the image, for example, the first classifier could be responsible for eliminating the 50% of the elements that are not faces. While the more complex classifiers will be used later in order to reduce false positives. It is so that this algorithm achieves great efficiency, since it is armed by layers and does not necessarily need to enter the complex layers to obtain good results.

4 Eyes Detection Algorithm

For the development of the detection and analysis algorithm for the eyes present in a human face, in the first stage the challenge of having an algorithm for detecting these components of the image that is being acquired by the camera was done. Thanks to the extensive development that has occurred in recent years of real-time detection algorithms, the Viola Jones algorithm was selected, among other options such as Eigen Faces or Fisher Faces [26], due to its low computational cost, excellent detection speed and accuracy. Matlab software was selected as the development interface, because it has computer vision libraries and support webcam devices. The Computer Vision System toolbox includes the Viola Jones algorithm implemented in the "Vision Cascade Object Detector" function. In this way, its application and acquisition of results, through a script type code, is simple.

The development of the code starts with the acquisition of the image, it can be started selecting the video device source, where the element 'winvideo', 1, 'RGB24_384 × 288' represents the entry of the PC webcam. In addition, a detector configured for the recognition of the eyes is applied in each image that will be acquired, for this purpose the.xml file 'EyesDetector.xml' is used. This detector was trained using, the following elements: a database of the images that contain the faces and eyes of the person to be detected, a stored ".mat" variable file that contains the bounding boxes, where the person's eyes are located in each image. A database of images that do not contain the face to be detected so that the classifier can learn from reducing the false positives rates. Finally, the result obtained from the training using the Adaboost methodology was the file in ".xml" format that contains the method as cascade classifiers will be combined, so that the simplest ones are organized in the first place, and the more complex ones later as it was mentioned previously. The code used for the training of the classifier is detailed below:

Algorithm 1. Training cascade classifier
---
1. **Input:**   Faces   folder:Faces = fullfile('c:path\data\Faces'),without
    faces: NoFaces = fullfile('c:path\data\Faces')
2. **Load** ('DetectedEyes.mat')
3. **Set** positiveInstances = DetectedEyes($F_1,...,F_k$)
4. **Set** negativeInstances = imageDataStore(negativeFolder)
5. **Train** trainCascadeObjectDetector('EyesDetector.xml')
6. **Output:** EyesDetector.xml

The file "EyesDetector.xml" contains the classification method, which classifiers of Haar-type will be used and the configuration of the cascade sequence, in other words which features will be classified in each layer in order to eliminate the regions where the eyes are not located, so the final image contains only the eyes. Once trained, the eyes detection classifier it can be called using the function "vision.CascadeObjectDetector" as follows:

ConfiguredDetector = vision.CascadeObjectDetector('DetectorOjos.xml')

It should be mentioned that the training process can take several minutes depending on the computer performance, the alarm range that indicates that learning continues until a maximum admitted percentage of false positives and the number of learning stages is reached, which indicates how many times the learning process will be repeated. In the tests developed with the configuration shown above, learning took times of less than five minutes. Finally, the results obtained for the detection of faces by the trained cascade detector were satisfactory, as shown in Fig. 4.

Fig. 4. Example of detection results obtained through the trained detection algorithm.

It should be noted that, at this stage, excellent results were found, because as shown in [15]; the detector can be used as a preventive warning element that will allow the driver to remain alert and with his sight to the front visualizing the road, based on the failure modes of Viola Jones algorithm the detector supports rotation head angles up to $\pm 15°$ moving up or down, and rotations from $\pm 45°$ side to side. For example, an alarm will be executed, if the driver does not have his gaze to the front. The code that was used to detect the driver's eyes and extract the area of the image where they are located, is detailed below:

Algorithm 2. Face detection

```
1.Input: Video imput device, N desired execution time
2.while i < N do
3.Capture each frame of the camera Im = getdata(vid, 1)
4.Transform pixel values on uint8 format: Detectorinput = unit8(im)
5.Get the position of the center of the image matrix: centerx =
size(im, 2)/2
6. centery = size(im, 1)/2
7.Generate the bounding box where the detected eyes are
      located: Boundig box = step(DetectorConfigured, inputDetect)
8. if Bounding box = empty
9.        Show image: imshow(Detectorinput)
10.       k = k + 1
11.       if k > 15
12.            Generate alert
13.       end if

14. else
|
```

Algorithm 3. Image Eye detection and Show
15. **for** 1: number of detected faces
16. subplot (obtained image, image of eye obtained)
17. Show image: imshow(inputDetector)
18. Draw box: box = Bounding box of each pair of eyes
19. Horizontal position of detected eyes: $\text{centroid}_x = \text{box}(1) + \text{box}(3)/2$
20. Vertical position of detected eyes: $\text{centroid}_y = \text{box}(3) + \text{box}(4)/2$
21. Generate vector: $\text{line}_x = [\text{center}_x, \text{centroid}_x]$
22. Generate vector: $\text{line}_y = [\text{center}_y, \text{centroid}_y]$
23. Draw line between of each centroid line: $\text{line}(\text{line}_x, \text{line}_y)$
24. Calculate distance from the center of the image:
25. $\text{centroid distance}_x = \text{centroid}_x - \text{center}_x$
26. $\text{centroid distance}_y = \text{centroid}_y - \text{center}_y$
27. Concatenate vector position
28. Draw rectangle: $('Position', BB(i,:))$
29. Plot: (point red)
30. Crop eyes area from image: $Ic = \text{imcrop}(\text{inputDetector}, \text{ojos})$
31. Show detected eyes: imshow(Ic)
32. **end for**
33. **end if else**
34. **end while**
35. Stop camera: stop(vid)
36. Reset of memory: flushdata(vid)
37. **Output:** Ic

5 Sclera Segmentation

Once the eyes of the driver are detected in the image, the extracted image is segmented, to identify only the clearest area of color, corresponding to the sclera of the eye that is only visible when eyes are open. Finally, a sound action can be triggered to alert the driver from falling asleep.

For the white color segmentation in the eyes extracted image, a segmentation algorithm was used, which allows, by means of the pixels tool in the Computer Vision package of Matlab, to extract an array with the values of the colors in RGB of each pixel $p(i,j)$ that is selected in an image. This tool greatly facilitates the segmentation of the image obtaining the matrix $S = [(R_1, G_1, B_1); (R_2, G_2, B_2); \ldots; (R_n, G_n, B_n)]$, which contains the color values of the selected pixels on RGB, for the $n = 1, 2, \ldots$ times that the user clicked the desired sclera area, which vary for each user and depend on the lighting conditions. Finally, selecting several of the corresponding pixels of the area of interest to be segmented, a range R is defined by the maximum and minimum values of the extracted pixels.

$$R = \begin{pmatrix} \max(S_{i,1}) - \min(S_{i,1}) \\ \max(S_{i,1}) - \min(S_{i,2}) \\ \max(S_{i,1}) - \min(S_{i,3}) \end{pmatrix}$$

By means of two loops executed simultaneously, the value of each pixel is replaced by 255, which is the maximum value for images in uint8 format, if the pixel values are within the established range R, and zero if they aren't within.

$$\{\forall p(i,j) \in I_c : [\min(S_{i,1}) \le p_R(i,j) \le \max(S_{i,1}) \wedge \min(S_{i,2}) \le p_G(i,j)$$
$$\le \max(S_{i,2}) \wedge \min(S_{i,3}) \le p_B(i,j) \le \max(S_{i,3})] \rightarrow p(i,j)$$
$$= (255, 255, 255)\}$$

The algorithm of segmentation and binarization, of the sclera area in the image of the eyes, is presented below:

Algorithm 4. Sclera color segmentation

---

1. **Input:** Cropped image containing the identified eyes area: Im
2. Obtain the maximum value vector of each matrix array: Max = max(max(Im))
3. Obtain the maximum value vector of each matrix array: Max = min(min(Im))
4. Run the Matlab pixels tool in order to get the pixel values of the sclera in the image: pixels = impixel(im)
5. Define the minimum value of the Red, Green and Blue matrixes in the pixels array as the minimum for the segmentation loop:
6. minR = min(pixels(:,1))
7. minG = min(pixels(:,2))
8. minB = min(pixels(:,3))
9. Define the maximum value of the Red, Green and Blue matrixes in the pixels array as the maximum for the segmentation loop:
10. maxR = max(pixels(:,1))
11. maxG = max(pixels(:,2))
12. maxB = max(pixels(:,3))

13. Get the size of the eyes image: s = size(im)
14. Start the segmentation loop using the pixel values of the sclera area:
15. **for** x = 1:1:s(1) do
16. **for** y = 1:1:s(2) do
17.   **if**
  im(x,y,1) >= minR && im(x,y,1) <= maxR && im(x,y,2) >= minG && im(x,y,2) < = maxG && im(x,y,3) >= minB && im(x,y,3) <= maxB do
18. Replace the pixel value with the 255 value in order to get the sclera area:
19. im(x,y,1) = 255
20. im(x,y,2) = 255
21. im(x,y,3) = 255
22. Binarize the matrix where white represents the sclera area: imbw = imbinarize(im(:,:,1))
23. Show the results: imshow(im); imshow(imbw)
24. Sum all the values of the binarized sclera matrix: m = sum(sum(imbw))
25. **Output:** m, s(1), s(2)

The binarized image obtained is a color segmentation of the eyes where the sclera is represented in white and all the rest of the pixels in the image that not correspond to the sclera are eliminated by the zero value, an example of the results that can be obtained by this algorithm is shown in Fig. 5.

Finally, the closed eyes are determined by the sum of all the pixels p_{bw} of the $h \times w$ size binarized eyes image, so if this value is less than 2.5% an audible alert will trigger, this value is defined in 2.5% as an example empirically, but it can be adjusted by its coefficient in the code.

$$\sum_{i=1,j=1}^{w,h} p_{bw}(i,j) \cdot \frac{100}{w \times h} \leq 2.5 \rightarrow Alert$$

This way as proposed in [5], if the detected white area is less than 2.5% of the image of the identified eyes during an interval of 0.5 s, the audible alert will be executed. This value can be adjusted by the variable n. The code for the audible alert is presented below.

Algorithm 5. Audible alert

1. **Input:** Total amount of values in the binarized sclera matrix
   m; eyes matrix size for rows s(1) and columns s(2).
2. If m is less or equal to 2,5% of the eyes image area, then
   the driver could be asleep, so a counter is started: n = 1
3. **if** m ≤ 0,025(s(1) × s(2)) do
4.   imshow(inputDetector)
5.   n = n + 1
6. **if** the counter n takes values higher than **15**, equivalent to
   0,5 seconds, then the driver is asleep, so a sound alert is
   started:
       if m > 15 do
7.   beep
8. **end if**
9. **end if**
10. **Output:** Audible alert

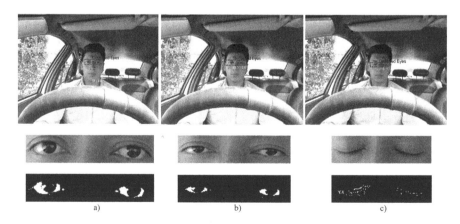

Fig. 5. Segmentation and binarization of the sclera applied to the image of the detected eyes: (a) corresponds to an awaken driver (sclera pixels 6.3%), (b) corresponds to possible drowsiness started (sclera pixels 3.03%); (c) corresponds to driver closed eyes (sclera pixels 1.37%)

6 Results and Discussion

To test the proposed algorithm, we used a Toshiba Satellite S55-C5161 computer with Intel Core i7 6700HQ 2.6 GHz processor, 8 GB DDR3 RAM, and a score of 276577 points on Antutu V6 benchmark. On Matlab environment, the proposed code was executed, and correct functioning of the algorithms implemented could be verified. Tests performed in this device shown that as expected for lower size of the image faster detection can be reached. The accuracy improves significantly as the acquired image size is reduced, so as proposed in [15] the selected size for the image was 384 × 288 pixels which for original Viola Jones algorithm reached speeds up to 15 fps with an accuracy of 77.8%. In drowsiness applications the system Baseline-2 model proposed on [12] reached 12.5 fps and 93.84% of accuracy and the compressed-2 model 14.9 fps and 89.5% of accuracy. Also, the android detector of [13] reached 2.5 fps. For our device the drowsiness detection speed varied from 134 to 145 processed and binarized sclera images in an interval of 10 s, which means detection speeds of around 13.4 to 14.5 fps with pc camera running at 30 fps. From the processed images a maximum of 14 to 18 faces, drowsy faces or distracted faces weren't detected bringing an accuracy of 87.58% up to 90.14% for the processed images. This drowsiness detector also as mentioned above can detect distraction on the driver based on head rotation angle.

During executed tests a threshold for the audible alert was set based on experimental results where: for an attentive awaken driver the sum of white pixels present in the binarized image represented values in the range of 5.27 to 7.63% Fig. 5a. For a started drowsiness process values of 2.94 to 3.89% were found, and closed eyes gave values of 0.74 to 1.56%. In this way the thresholding selected values were: for less than 4% of white pixels a preventive audible alert must be executed, for a drowsy driver; and for less than 2% a strong audible warning alert must be executed for closed eyes.

7 Conclusions

The proposed system uses computer vision for eyes and distraction detection, and image processing for drowsiness detection by sclera segmentation. The development of the algorithm was successfully implemented and compared to previous related works proves that has similar or even better accuracy and speed, and represents a simple and effective alternative to determine drowsiness and distraction on drivers.

Nowadays a device can be considered a real time detector depending on the application that is made for. As mentioned in previous related works like [5, 12, 13], drowsiness detection speeds of 0.5 fps on [5], 2.5 fps on [13], 12.5 fps on [12] Baseline-2 model, and 14.9 fps on [12] Compressed-2 model, are considered real-time detectors. So, in conclusion our system can be considered a real time drowsiness detector.

For future related work, it is recommended to use thermal cameras for drowsiness detection, so that the detection will not be affected by the level of luminosity, which is one of the failure modes of Viola Jones algorithm. In this way this devices will be able to work in dimly illuminated environments for example, at night.

Acknowledgment. The authors acknowledge to the research project "Desarrollo de una metodología de visualización interactiva y eficaz de información en Big Data" supported by Agreement No. 180 November 1st, 2016 by VIPRI from Universidad de Nariño. Also, authors thank the valuable support given by the SDAS Research Group (www.sdas-group.com) and Facultad de Ingeniería en Ciencias Aplicadas from Universidad Técnica del Norte, Ibarra, Ecuador.

References

1. Horne, J., Reyner, L.: Driver sleepiness. J. Sleep Res. **4**, 23–29 (1995)
2. Garcés, M., Salgado, J., Cruz, J., Cafión, W.: Sistemas de detección de somnolencia en conductores: inicio, desarrollo y futuro. Ingeniería y Región **13**, 159–168 (2015)
3. Sahayadhas, A., Sundaraj, K., Murugappan, M.: Detecting driver drowsiness based on sensors: a review. Sensors (Switzerland) **12**(12), 16937–16953 (2012)
4. Vural, E., Cetin, M., Ercil, A., Littlewort, G., Bartlett, M., Movellan, J.: Drowsy driver detection through facial movement analysis. In: Lew, M., Sebe, N., Huang, T.S., Bakker, E. M. (eds.) HCI 2007. LNCS, vol. 4796, pp. 6–18. Springer, Heidelberg (2007). https://doi.org/10.1007/978-3-540-75773-3_2
5. El-Den, B.M., Mohamed, M.A., AbdelFattah, A.I.: Safe vehicle driving using android based smartphones. In: Abraham, A., Jiang, X.H., Snášel, V., Pan, J.-S. (eds.) Intelligent Data Analysis and Applications. AISC, vol. 370, pp. 291–303. Springer, Cham (2015). https://doi.org/10.1007/978-3-319-21206-7_25
6. Borghini, G., Astolfi, L., Vecchiato, G., Mattia, D., Babiloni, F.: Measuring neurophysiological signals in aircraft pilots and car drivers for the assessment of mental workload, fatigue and drowsiness. Neurosci. Biobehav. Rev. **44**, 58–75 (2014)
7. Liu, C., Hosking, G., Lenné, M.: Predicting driver drowsiness using vehicle measures: recent insights and future challenges. J. Safety Res. **40**(4), 239–245 (2009)
8. Hyun, J., Mao, Z.H., Tijerina, L., Pilutti, T., Coughlin, J., Feron, E.: Detection of driver fatigue caused by sleep deprivation. IEEE Trans. Syst. Man Cybernetics Part A: Syst. Humans **39**(4), 694–705 (2009)
9. Powell, B., Chau, K.: Sleepy driving. Med. Clin. North Am. **94**(3), 531–540 (2010)
10. Correa, A.G., Orosco, L., Laciar, E.: Automatic detection of drowsiness in EEG records based on multimodal analysis. Med. Eng. Phys. **36**(2), 244–249 (2014)
11. Dajeong, K., Hyungseob, H., Sangjin, C., Uipil, C.: Detection of drowsiness with eyes open using EEG-based power spectrum analysis. In: 7th International Forum on Strategic Technology (IFOST), pp. 1–4 (2012)
12. Reddy, B., Kim, Y.H., Yun, S., Seo, C., Jang, J.: Drowsiness detection for embedded system using model compresssion of deep neuronal networks. In: Computer Vision Foundation CVF, pp. 121–128 (2017)
13. Montanini, L., Gambi, E., Spinsante, S.: An OpenCV based android application for drowsiness detection on mobile devices. In: Conti, M., Martínez Madrid, N., Seepold, R., Orcioni, S. (eds.) Mobile Networks for Biometric Data Analysis. LNEE, vol. 392, pp. 145–158. Springer, Cham (2016). https://doi.org/10.1007/978-3-319-39700-9_12
14. Lee, B., Jung, S., Chung, W.: Real-time physiological and vision monitoring of vehicle driver for non-intrusive drowsiness detection. IET Commun. **5**(17), 2461–2469 (2011)
15. Viola, P., Jones, M.: Rapid object detection using a boosted cascade of simple features. In: IEEE CVPR (2001)

16. Muzamil, S., Singh, D., Vandhan, V.: Impact of gradient ascent and boosting algorithm in classification. Int. J. Intell. Eng. Syst. **11**(1), 41–49 (2018)
17. Wang, M., Guo, L., Chen, W.Y.: Blink detection using Adaboost and contour circle for fatigue recognition. Comput. Electr. Eng. **58**(1), 502–512 (2017)
18. Bo, S., Chen, S., Wang, J., Chen, H.: A robust multi-class AdaBoost algorithm for mislabeled noisy data. Knowl.-Based Syst. **102**(1), 87–102 (2016)
19. Lienhart, R., Maydt, J.: An extended set of haar-like features for rapid object detection. In: IEEE ICIP, pp. 900–903 (2002)
20. Wonji, L., Jun, C.H., Lee, J.S.: Instance categorization by support vector machines to adjust weights in AdaBoost for imbalanced data classification. Inf. Sci. **381**(1), 92–103 (2017)
21. Egorov, A.: Algorithm for optimization of Viola–Jones object detection framework parameters. J. Phy.: Conf. Ser. 945(1) (2018)
22. Lescano, G., Santana, P., Costaguta, R.: Analysis of a GPU implementation of Viola-Jones' algorithm for features selection. J. Comput. Sci. Technol. 17(1), (2017)
23. Wu, B., Ai, H., Huang, C., Lao, S.: Fast rotation invariant multi-view face detection based on real adaboost. In: Proceedings of Sixth IEEE International Conference on Automatic Face and Gesture Recognition, pp. 79–84 (2004)
24. Freund, Y., Schapire, R.: A short introduction to boosting. J. Japanese Soci. Artif. Intell. **14** (5), 771–780 (1999)
25. Tsai, P., Hsu, Y., Chiu, C., Chu, T.: Accelerating AdaBoost algorithm using GPU for multi-object recognition. In: 2015 IEEE International Symposium on Circuits and Systems (ISCAS), pp. 738–741 (2015)
26. Lorente-Leyva, L.L., et al.: Developments on solutions of the normalized-cut-clustering problem without eigenvectors. In: Huang, T., Lv, J., Sun, C., Tuzikov, A. (eds.) ISNN 2018. LNCS, vol. 10878, pp. 318–328. Springer, Cham (2018). https://doi.org/10.1007/978-3-319-92537-0_37

Software Techniques to Reduce Cybersickness Among Users of Immersive Virtual Reality Environments

Kazimierz Choroś[(⊠)] and Piotr Nippe

Faculty of Computer Science and Management,
Wrocław University of Science and Technology, Wybrzeże Wyspiańskiego 27,
50-370 Wrocław, Poland
kazimierz.choros@pwr.edu.pl

Abstract. The virtual worlds are photorealistic, the users have difficulty distinguishing the reality and virtuality. The high degree of realism of artificial virtual worlds results in deep immersion in virtual environments. The users immerse themselves in the virtual world. The immersion may, however, provoke different undesirable side-effects. The symptoms of motion sickness occurring mainly in transport means (buses, planes, ships, etc.) are also observed among users of virtual worlds where the users usually view moving scenes while they remain physically stationary. In the case of virtual reality it is called cybersickness. The paper presents software techniques to reduce cybersickness among the users of immersive virtual reality environments and analyses the results of experiments showing their efficiency and their impact on the users of virtual worlds.

Keywords: Virtual reality · Virtual worlds · Video games ·
Immersion side-effects · Cybersickness · Cybersickness reduction ·
Motion sickness · Simulator sickness · Virtual environment interfaces

1 Introduction

Virtual reality is more and more popular in education, architecture, video games, simulators, robotics, commerce, therapeutics, health, army, fashion, tourism, cinema, as well as in many other different areas. Generated by computers they are very attractive. Virtual worlds are usually interactive and immersive. Users of virtual worlds can move or manipulate in computer-generated 3D scenes and explore their surroundings. This manipulation can be assured through different multimedia tools, such as handheld controllers, virtual goggles, head-mounted displays, sense gloves. The virtual worlds became very photorealistic, so, their users have difficulty distinguishing the reality and virtuality. We say that the users immerse themselves in the virtual worlds. The high degree of realism of artificial virtual worlds results in deep immersion in virtual environments. The level of visual immersion reflects how the visual output is close to real-world. Visual immersion is only one part of the overall level of immersion. Its components are [1]: field of view, i.e. the size of the visual field (in degrees of visual angle) that can be viewed instantaneously, field of regard, i.e. the total size of the visual

© Springer Nature Switzerland AG 2019
N. T. Nguyen et al. (Eds.): ACIIDS 2019, LNAI 11431, pp. 638–648, 2019.
https://doi.org/10.1007/978-3-030-14799-0_55

field (in degrees of visual angle) surrounding the user, display size, display resolution, stereoscopy, head-based rendering, realism of lighting, frame rate, and refresh rate.

High levels of immersion can make applications more efficient and more attractive. The immersion may, however, provoke different undesirable side-effects. We can observe among users such phenomena as fatigue, nausea, fainting, weariness, difficulty with concentration, breath awareness, somnolence, deterioration of the mental state, headache, "heavy thinking", loss of appetite, increased appetite, overwork, misted eyes, feeling of waves at the opening eyes, the feeling of waving at closed eyes, gastric problems, the need to empty the intestines, excessive salivation, reduced salivation, confusion, dizziness, bounce, sweating, visual disturbances, vomiting. These are well-known symptoms of motion sickness occurring mainly in transport means (buses, planes, ships, etc.). In virtual worlds, the users usually view moving scenes while they remain physically stationary. In the case of virtual reality it is called simulator sickness or cybersickness. Although the symptoms are very similar, their origins seem to be different [2]. There is no one exact cause of cybersickness. Cybersickness is simultaneously a result of the visual aspect as well as motion reason and not due to merely the motion alone. The users of VR systems consider ascending and descending stairs inside buildings as the most cybersickness-inducing part of the virtual worlds [3].

The most frequent advise in such a situation is to limit the time of using computer virtual systems. Nevertheless, using some software techniques we can reduce these undesirable side-effects. The paper discusses these software techniques and presents the results of a series of experiments which show their impact on the user comfort.

The paper is structured as follows. The next section describes recent related work on cybersickness reduction. The tested software techniques leading to the reduction of cybersickness are described in the third section. The fourth section presents the results of experiments verifying their impact on the user comfort during the immersion in virtual worlds. The final conclusions and suggestions for developers of virtual reality systems are discussed in the last fifth section.

2   Related Work

When the manipulators for virtual operating environment have been used the cyber-sickness was observed. To investigate this phenomenon [4] the ratings of a number of symptoms were reported before and after an head-mounted display (HMD) was used for one hour. The results were compared with ratings when using a visual display unit, and where the tasks were the same. The head-mounted displays produced a significantly greater number of symptoms, such as: general discomfort, fatigue, headache, nausea, dizziness, stomach awareness, hot or burning eyes, double vision, and general visual discomfort. The results of these experiments lead to the suggestion that specific symptoms of cybersickness are associated with specific causal factors. Certain symptoms occur more frequently when particular problems are observed. For example sensory conflict stimulates the genesis of nausea and stomach awareness. Whereas motion, including head movement, is a cause of disorientation feelings, and inappropriate optical design may result in eyestrain and associated ocular symptoms.

The experiments described in [5] were performed with 23 healthy young male volunteers using head mounted displays. The subjects were previously informed about the virtual system and the details of the experiment. The authors compared the results of the tests performed without HMD, with transparent HMD but not immersed in VR, and then with HMD and immersed in VR.

Other experiments [6] were based on the observation that visual discomfort when viewing stereoscopic content on monitors can be decreased by dynamic depth blur of field. The change in discomfort over 30 min exposure was examined by aggregating the results for such symptoms as: general discomfort, fatigue, headache, eyestrain, difficulty focusing, blurred vision, salivation increase, sweating, nausea, difficulty concentration, fullness of head, dizziness, vertigo, and stomach awareness. 20 participants took part, with an age range of 18 to 50 years old. Six of them were female, and 14 were male. All subjects have no experience with head mounted displays. When dynamic depth of field was enabled mean nausea discomfort, mean oculomotor discomfort, and mean disorientation discomfort decreased.

The paper [7] presents a comparative analysis between two in-place navigation techniques, a navigation technique that reduces postural instability with a technique that reduces sensory conflicts enhancing motion. In-place navigation techniques have allow users to freely move in virtual worlds without physical translation. However, these techniques generate in users the illusory perception of self-motion, resulting in cybersickness symptoms. It was then suggested that the reduction self-motion changes could be a more effective method to reduce cybersickness symptoms than a interaction technique based on postural instability reduction.

The authors of [8] analyzed the accuracy of different extrapolation and filtering techniques to predict head movements, reducing the impact of latency. Latency is one of the main causes of cybersickness. In the tests 10 participants played a virtual reality game that required quick and subsequent head rotations. A total of 150,000 head positions were captured in the pitch and yaw rotation axes. These rotational movements were then extrapolated and filtered. The final conclusion was that significant reduction of latency by extrapolating head movements provides a low-cost solution with an acceptable prediction error.

In [9] it was shown that in a 3D virtual environment, when motion sickness appears, the area of the body's center of gravity is dilated and the shape of the area tends to change from an ellipse to a circle. Moreover, the difference between low frequency and high frequency components of the center of gravity's signals increases. Also a speed incremental variation has a significant effect on sickness.

It has been observed that decreasing field of view tends to decrease cybersickness. In [10] it was suggested that by automatically manipulating field of view during a virtual session, the degree of cybersickness perceived by VR users can be reduced and help them adapt to virtual reality environment.

Many other solutions have been proposed to reduce the cybersickness symptoms, for example new VR navigation techniques were proposed [11], introducing static and dynamic rest frames [12], or using peripheral visual effects [13].

In the experiments described in [14] the level of cybersickness among younger and older adults, men and females was compared during simulated driving under different multisensory conditions. Also time needed to fully recover from the driving session was compared for these groups of VR users.

Whereas in [15] the great similarity has been shown of cybersickness noticed in virtual reality environments and standard well-known motion sickness observed in transport means. The main conclusion of this research was that symptoms and physiological changes occurring during cybersickness and standard motion sickness are quite similar, at least during advanced stages.

Latency is may be the most important but it is not the only one cause of cybersickness. The others are [16]: eyewear with image separation between eyes, incorrect calibration or poor focus simulation, and also convergence accommodation conflict. In health applications the discomfort and sickness symptoms become critical. In [16] a design guideline is proposed to help developers to minimize sickness problems in virtual systems when head-mounted displays are used.

Many other research investigations, tests, experiments have been undertaken to find the efficient solutions to reduce cybersickness among the user of virtual reality systems. The papers presenting recent achievements in this field are reported in the review papers [17] and [18].

3 Techniques to Reduce Cybersickness

Seven techniques were implemented as well as analyzed for further investigations and evaluations of the reduction of cybersickness symptoms. These techniques are:

- adequate navigation technique

Usually three techniques of movements in virtual worlds are distinguished: artificial, physical, and teleportation. Teleportation seems to be the most secure for user comfort, however, it decreases the level of immersion because it is not natural. Artificial techniques realized by using special keys or manipulators ensure smooth movements but provoke strong user discomfort. Physical techniques offer the highest interaction of the user with virtual worlds.

- simplified textures of virtual environments

The discomfort of users of virtual worlds can be caused by the great richness of shapes and colors and in consequence the high level of photorealism. The simplified textures of virtual environment can result in decreasing cybersickness symptoms (Fig. 1).

Fig. 1. Examples of realistic virtual city and simplified virtual environment.

- limitation of the time of the user immersion in virtuality

Too long immersion in virtual worlds may be danger for the user and then it requires much time for disappearing of cybersickness symptoms after the end of virtual session.

- dynamic reduction of field of view

It was observed that usually the greater field of view in virtual worlds provokes the stronger symptoms of cybersickness. On the other hand the limited field of view significantly diminishes the immersion. The best solution is to reduce the field of view only in the moments of a great risk of cybersickness appearance (Fig. 2).

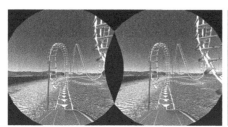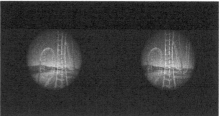

Fig. 2. Example of the reduction of field of view.

- insertion of reference frame

The insertion of a static frame, such as vehicle cockpit, helmet, or nose both reduces the cybersickness symptoms and increases the immersion. The user receives a static point ensuring a sense of stability.

- predetermined path of movement

It was observed that motion sickness is less frequent among drivers than among passengers. It is due to the fact that the drivers follow landmarks and may easily predict the future situations (Fig. 3).

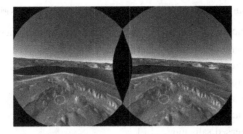

Fig. 3. Example of the image with reference point on the predetermined path.

- image quality and frame rate

The great number of details in virtual environments and also the high frame rate can result in the delay and loss of synchronization between the user manipulation and the virtual image generated by a computer. One of the solution is to reduce the details of virtual world to ensure the adequate frame rate during the virtual session.

4 Experimental Evaluation of Techniques of Cybersickness Symptoms Reduction

In the experiments the virtual reality headset HTC Vive (2016) was used. In this package there were headset, two controllers, two base station, control box, motion tracker. This headset was chosen for the experiments because it is fully immersive. Then three virtual worlds were implemented using Unity – a cross-platform game engine. The first virtual world was a relatively small scene with seven virtual panels describing briefly seven software techniques of sickness symptoms reduction. The second world was a virtual city with moving cars on the streets in two versions: very detailed realistic virtual world and smooth world with simple shapes and simple textures. Finally, the third world was a huge green nature valley with lakes.

To evaluate the software techniques presented in Sect. 3 the questionnaire SSQ (Simulator Sickness Questionnaire) was used. The cybersickness questionnaire proposed in [19] is still relevant today [20]. 16 symptoms of cybersickness are included in the questionnaire: general discomfort, fatigue, headache, eye strain, difficulty focusing, increased salivation, sweating, nausea, difficulty concentrating, fullness of head, blurred vision, dizzy (eyes open), dizzy (eyes closed), vertigo, stomach awareness, burping. The symptoms were evaluated by eight subjects from 26 to 30 years old on a four point Liekert scale: none, slight, moderate, and severe. The subjects declared that they are not sensitive to motion sickness in standard transport means. They did not observed motion sickness symptoms during travels.

These 16 symptoms can be placed into only three general categories: nausea, oculomotor, and disorientation (Table 1). Weights can be assigned to each of these categories and summed together to obtain a single score. This score becomes a indicator of overall cybersickness scores for a given virtual environment [19].

Table 1. Assigning symptoms to three general categories.

	Symptom	Nausea	Oculomotor	Disorientation
1.	General discomfort	1	1	
2.	Fatigue		1	
3.	Headache		1	
4.	Eye strain		1	
5.	Difficulty focusing		1	1
6.	Increased salivation	1		
7.	Sweating	1		
8.	Nausea	1		1
9.	Difficulty concentrating	1	1	
10.	Fullness of head			1
11.	Blurred vision		1	1
12.	Dizzy (eyes open)			1
13.	Dizzy (eyes closed)			1
14.	Vertigo			1
15.	Stomach awareness	1		
16.	Burping	1		

CSI – cybersickness indicator

CSI = [(sum of values of observed symptoms included in nausea category) × 9.54 + (sum of values of observed symptoms included in oculomotor category) × 7.58 + (sum of values of observed symptoms included in disorientation category) × 13.92] × 3.74

The lower value od cybersickness indicator the better comfort of the user in virtual worlds and the lower probability of the occurring of cybersickness.

Tables 2, 3, 4, 5, 6, 7 and 8 presents aggregated scores for cybersickness indicator received for seven tested software techniques to reduce cybersickness among users of immersive virtual reality environments.

Table 2. Cybersickness indicator received for three different navigation techniques.

Symptom category	Average score		
	Using touchpad	Using manipulator	Running in place
Nausea	118.06	82.28	107.33
Oculomotor	90.01	75.80	71.06
Disorientation	170.52	100.92	106.14
Global CSI	136.51	96.77	105.66

Manipulators seem to be the least affecting the cybersickness. Whereas using touchpads more intensively provokes the symptoms of cybersickness (Table 2).

Table 3. Cybersickness indicator received for simple and realistic textures.

Symptom category	Average score	
	Simple textures	Realistic textures
Nausea	79.89	87.05
Oculomotor	76.75	78.64
Disorientation	114.84	151.38
Global CSI	100.05	113.60

Table 4. Cybersickness indicator received when the time of the user immersion in virtuality was limited.

Symptom category	Average score	
	Long virtual session	Sequence of short sessions
Nausea	157.41	89.44
Oculomotor	122.23	72.96
Disorientation	208.80	114.84
Global CSI	178.12	101.92

Cybersickness indicators received for simple and realistic textures are rather a surprising result. Realistic textures (Table 3) have not significant influence on the cybersickness among the users of virtual worlds.

Whereas it is not surprising that the length of virtual session has great impact on the occurring of cybersickness symptoms (Table 4).

Table 5. Cybersickness indicator received when the field of view is reduced.

Symptom category	Average score	
	Full field of view	Reduced field of view
Nausea	150.26	96.59
Oculomotor	131.70	66.33
Disorientation	212.28	120.06
Global CSI	180.92	102.85

Table 6. Cybersickness indicator received for flights without and with visible nose.

Symptom category	Average score	
	Virtual flight without visible nose	Virtual flight with visible nose
Nausea	115.67	87.05
Oculomotor	85.28	72.01
Disorientation	128.76	125.28
Global CSI	122.02	103.32

Table 7. Cybersickness indicator received for the flight without and with predetermined path of movement.

Symptom category	Average score	
	Without predetermined path	With predetermined path
Nausea	90.63	87.05
Oculomotor	72.96	78.64
Disorientation	116.58	151.38
Global CSI	102.85	113.60

Also the field of view (Table 5) has a great impact on the occurring of cyber-sickness symptoms. Wide field of view may generate discomfort among the users of virtual reality systems.

Table 8. Cybersickness indicator received with and without controlling the image quality and frame rate.

Symptom category	Average score	
	Without controlling the image quality and frame rate	With controlling the image quality and frame rate
Nausea	174.11	107.33
Oculomotor	131.70	82.43
Disorientation	217.50	123.54
Global CSI	191.68	115.94

Usually the nose visible during the virtual flight is thought to be a very efficient solution to weaken the cybersickness symptoms. The results of the experiments show that this solution is not so useful as it could be expected. Although, the global score is lower, so, the inserting of static elements visible on the display such as nose diminishes the probability of occurring the symptoms of cybersickness, its impact is relatively less significant in comparison to other software techniques tested in the reported experiments (Table 6). Similarly inserting landmarks in predetermined path of movement has medium impact on the cybersickness (Table 7).

The results presented in Table 8 prove that controlling the image quality and frame rate has a particularly great impact on the occurring of cybersickness symptoms.

5   Conclusion

The immersion of users in virtual worlds is a very significant property of virtuality. The immersion may, however, provoke different undesirable side-effects. The symptoms of cybersickness are observed among users of immersive virtual reality environments. Software techniques described in this paper and tested in the reported experiments are useful for the reduction of these undesirable cybersickness symptoms. The developers of the applications and systems of virtual reality should introduce these software techniques to ensure better comfort for the users.

References

1. Bowman, D.A., McMahan, R.P.: Virtual reality: how much immersion is enough? Computer **40**(7), 36–43 (2007)
2. Kim, Y.Y., Kim, H.J., Kim, E.N., Ko, H.D., Kim, H.T.: Characteristic changes in the physiological components of cybersickness. Psychophysiology **42**(5), 616–625 (2005)
3. Pouke, M., Tiiro, A., LaValle, S.M., Ojala, T.: Effects of visual realism and moving detail on cybersickness. In: Proceedings of the IEEE Conference on Virtual Reality and 3D User Interfaces (VR), pp. 665–666. IEEE (2018)
4. Howarth, P.A., Costello, P.J.: The occurrence of virtual simulation sickness symptoms when an HMD was used as a personal viewing system. Displays **18**(2), 107–116 (1997)
5. Akizuki, H., et al.: Effects of immersion in virtual reality on postural control. Neurosci. Lett. **379**(1), 23–26 (2005)
6. Carnegie, K., Rhee, T.: Reducing visual discomfort with HMDs using dynamic depth of field. IEEE Comput. Graphics Appl. **35**(5), 34–41 (2015)
7. Dorado, J.L., Figueroa, P.A.: Methods to reduce cybersickness and enhance presence for in-place navigation techniques. In: Proceedings of the IEEE Symposium on 3D User Interfaces (3DUI), pp. 145–146. IEEE (2015)
8. Garcia-Agundez, A., Westmeier, A., Caserman, P., Konrad, R., Göbel, S.: An evaluation of extrapolation and filtering techniques in head tracking for virtual environments to reduce cybersickness. In: Alcañiz, M., Göbel, S., Ma, M., Fradinho Oliveira, M., Baalsrud Hauge, J., Marsh, T. (eds.) JCSG 2017. LNCS, vol. 10622, pp. 203–211. Springer, Cham (2017). https://doi.org/10.1007/978-3-319-70111-0_19
9. Chardonnet, J.R., Mirzaei, M.A., Merienne, F.: Visually induced motion sickness estimation and prediction in virtual reality using frequency components analysis of postural sway signal. In: Proceedings of the International Conference on Artificial Reality and Telexistence Eurographics Symposium on Virtual Environments, pp. 9–16 (2015)
10. Fernandes A.S., Feiner S.K.: Combating VR sickness through subtle dynamic field-of-view modification. In: Proceedings of the IEEE Symposium on 3D User Interfaces (3DUI), pp. 201–210. IEEE (2016)
11. Kemeny, A., George, P., Mérienne, F., Colombet, F.: New VR navigation techniques to reduce cybersickness. Electron. Imaging **3**, 48–53 (2017)
12. Cao Z., Jerald J., Kopper R.: Visually-induced motion sickness reduction via static and dynamic rest frames. In: Proceedings of the 25th IEEE Conference on Virtual Reality and 3D User Interfaces (VR 18), pp. 105–112. IEEE (2018)
13. Buhler, H., Misztal, S., Schild, J.: Reducing VR sickness through peripheral visual effects. In: Proceedings of the IEEE Conference on Virtual Reality and 3D User Interfaces (VR), pp. 517–519. IEEE (2018)
14. Keshavarz, B., Ramkhalawansingh, R., Haycock, B., Shahab, S., Campos, J.L.: Comparing simulator sickness in younger and older adults during simulated driving under different multisensory conditions. Transp. Res. Part F: Traffic Psychol. Behav. **54**, 47–62 (2018)
15. Gavgani, M.A., Walker, F.R., Hodgson, D.M., Nalivaiko, E.: A comparative study of cybersickness during exposure to virtual reality and "classic" motion sickness: are they different? J. Appl. Physiol. (2018). https://doi.org/10.1152/japplphysiol.00338.2018

16. Porcino, T.M., Clua, E., Trevisan, D., Vasconcelos, C.N., Valente, L.: Minimizing cyber sickness in head mounted display systems: design guidelines and applications. In: Proceedings of 5th International Conference on Serious Games and Applications for Health (SeGAH), pp. 1–6. IEEE (2017)
17. Rebenitsch, L., Owen, C.: Review on cybersickness in applications and visual displays. Virtual Reality **20**(2), 101–125 (2016)
18. Martirosov, S., Kopecek, P.: Cyber sickness in virtual reality – literature review. In: Proceedings of the 28th DAAAM International Symposium, DAAAM International, Vienna, Austria, pp. 0718–0726 (2017)
19. Kennedy, R.S., Lane, N.E., Berbaum, K.S., Lilienthal, M.G.: Simulator sickness questionnaire: an enhanced method for quantifying simulator sickness. Int. J. Aviat. Psychol. **3**(3), 203–220 (1993)
20. Balk, S.A., Bertola, M.A., Inman, V.W.: Simulator sickness questionnaire: twenty years later. In: Proceedings of the Seventh International Driving Symposium on Human Factors in Driver Assessment, Training, and Vehicle Design, Public Policy Center, University of Iowa, pp. 257–263 (2013)

Semi-supervised Learning
with Bidirectional GANs

Maciej Zamorski[1,2]($\boxtimes$) and Maciej Zięba[1,2]

[1] Department of Computer Science, Faculty of Computer Science and Management,
Wrocław University of Science and Technology, Wrocław, Poland
{maciej.zamorski,maciej.zieba}@pwr.edu.pl
[2] Tooploox Ltd., Wrocław, Poland

Abstract. In this work we introduce a novel approach to train Bidirectional Generative Adversarial Model (BiGAN) in a semi-supervised manner. The presented method utilizes triplet loss function as an additional component of the objective function used to train discriminative data representation in the latent space of the BiGAN model. This representation can be further used as a seed for generating artificial images, but also as a good feature embedding for classification and image retrieval tasks. We evaluate the quality of the proposed method in the two mentioned challenging tasks using two benchmark datasets: CIFAR10 and SVHN.

Keywords: Generative models · Triplet learning ·
Generative adversarial networks · Image retrieval

1 Introduction

One of the most common and important tasks of machine learning is building generative models, that can capture and learn a wide variety of data distributions. Recent developments in generative modeling concentrate around two major areas of research: variational autoencoders (VAE) [5] that aim at capturing latent representations of data, while simultaneously keeping it restricted to known distribution (e.g., normal distribution) and generative adversarial networks (GANs) [3,8] with grounds in game theory, having strong emphasis on creating realistic samples from underlying distributions.

These kind of models are not only known from generating data from the distribution represented by data examples but are also used to train informative and discriminative feature embeddings. It can be obtained either only with unsupervised data by using good discriminative properties of the GAN's discriminator achieved during adversarial training [8,12] or using some subset of labeled data and incorporating semi-supervised mechanisms during training the generative model [9,13].

In this work we concentrate on obtaining better feature representation for image data using semi-supervised learning with a model based on Bidirectional Generative Adversarial Networks (BiGANs) [1]/Adversarially Learned Inference

© Springer Nature Switzerland AG 2019
N. T. Nguyen et al. (Eds.): ACIIDS 2019, LNAI 11431, pp. 649–660, 2019.
https://doi.org/10.1007/978-3-030-14799-0_56

(ALI) [2]. In order to incorporate semi-supervised data into training procedure, we propose to enrich the primary training objective with an additional triplet loss term [4] that operates on the labeled examples.

Our approach is inspired by the work [13] where the triplet loss was used to increase the quality of features representation in the discriminator. Contrary to this approach, we make use of an additional model in BiGAN architecture - encoder and aim at increasing the quality of feature representation in the coding space, that is further used by a generator to create artificial samples. Practically, it means that the feature representation can be used not only for classification and retrieval purposes but also for generating artificial images similar to existing.

The contribution of the paper is twofold. We introduce a new GAN training procedure for learning latent representations that extends the models presented in [1,2] and inspired by [13] for semi-supervised learning. We show that Triplet BiGAN will result in superior scores in classification and image retrieval tasks.

This work is organized as follows. In Sect. 2 we present the basic concepts related to GAN models and triplet learning. In Sect. 3 we describe our approach - Triplet BiGAN model. In Sect. 4 we provide the results obtained by Triplet BiGAN on two challenging tasks: image classification and image retrieval. This work is summarized in Sect. 5.

2   Related Works

2.1   Generative Adversarial Networks

Since their inception, Generative Adversarial Networks (GANs) [3] have become one of the most popular models in a field of generative computer vision. Their main advantages come from their straightforward architecture and ability to produce state-of-the-art results. Studies performed in recent years propose many performance, stability and usage improvements to the original version, with Deep Convolutional GAN (DCGAN) [8] and Improved GAN [9] being used most often as architectural baselines in pure image generation learning tasks.

The main idea of GAN is based on game theory and assumes training of two competing networks: generator $G(\mathbf{z})$ and discriminator $D(\mathbf{x})$. The goal of GANs is to train generator G to sample from the data distribution $p_{data}(\mathbf{x})$ by transforming the vector of noise $\mathbf{z}$ to the data space. The discriminator D is trained to distinguish the samples generated by G from the samples from $p_{data}(\mathbf{x})$. The training problem formulation is as follows:

$$\min_G \max_D V(G, D) = \mathbb{E}_{x \sim p_{data}}[\log D(x)] + \mathbb{E}_{z \sim p_z}[\log(1 - D(G(z)))] \quad (1)$$

where: p_{data} - true data distribution, p_z - prior to the data space.

The model is usually trained with the gradient-based approaches by taking minibatch of fake images generated by transforming random vectors sampled from $p_{\mathbf{z}}(\mathbf{z})$ via the generator and minibatch of data samples from $p_{data}(\mathbf{x})$. They are used to maximize $V(D, G)$ with respect to parameters of D by assuming a constant G, and then minimizing $V(D, G)$ with respect to parameters of G by assuming a constant D.

2.2 Bidirectional Generative Adversarial Networks

BiGAN model, presented in [1,2] extends the original GAN model by an additional encoder module $E(\mathbf{x})$, that maps the examples from data space $\mathbf{x}$ to the latent space $\mathbf{z}$. By incorporating the encoder into the GAN architecture, we can code examples in the same space that is used as a seed for generating artificial samples.

The objective function that is used train the BiGAN model can be defined in the following manner:

$$V(G, D, E) = \mathbb{E}_{x \sim p_x}[\log D(x, E(x))] + \mathbb{E}_{z \sim p_z}[\log(1 - D(G(z), z))]. \quad (2)$$

The adversarial paradigm applied to train the model BiGAN is analogous as for GAN model. The goal of training is to solve the min-max problem stated below:

$$\min_{G,E} \max_{D} V(G, D, E). \quad (3)$$

Practically, the model is trained in alternating procedure, where the parameters of discriminator D are updated by optimizing the following loss function:

$$L_D = \mathbb{E}_{x \sim p_x}[\log D(x, E(x))] + \mathbb{E}_{z \sim p_z}[\log(1 - D(G(z), z))], \quad (4)$$

and the parameters of generator G and encoder E are jointly trained by optimizing the following loss:

$$L_{EG} = \mathbb{E}_{z \sim p_z}[\log D(G(z), z)] + \mathbb{E}_{x \sim p_x}[\log(1 - D(x, E(x)))]. \quad (5)$$

Since (as authors showed in [1]) the encoder, in order to be an optimal one, learns to invert the examples from true data distribution, the same loss can be applied to the encoder and the generator parameters.

Experiments show that the encoder despite learning in a pure unsupervised way was able to embed meaningful features, which later show during reconstruction. The inclusion of additional module raises a question about the quality of this feature representation for classification and image retrieval tasks. The approach of combining objectives seems promising, as the encoder module is explicitly trained for feature embedding, as opposed to the discriminator, which main task is to categorize samples into real and fake.

2.3 Triplet Networks

Triplet networks [4] are among the most used methods for deep metric learning [6,10,11]. Triplet networks consist of three instances of the same neural network, that share parameters among themselves. During training, the triplet model T receives three examples from the training data: the reference sample x^q, the positive sample (the sample that is in some way similar to the reference sample, f.e. it belongs to the same class) x^+ and the negative sample (that is dissimilar to the reference sample) x^-. The goal is to train the triplet network T in such a way, that the distance d^- between encoded query example $T(x^q)$ to the encoded negative example $T(x^-)$ is greater than the distance d^+ form the encoded query example $T(x^q)$ to the encoded positive example $T(x^+)$. In general case, this distances are computed as a L2-norm between feature vectors, i.e.: $d^- = \|T(x) - T(x^-)\|_2$ and $d^+ = \|T(x) - T(x^+)\|_2$.

During the training the triplet model makes use of the probability p_T that the distance of the query example to the negative example is greater than its distance to the positive one which can be defined in the following way:

$$p_T = \frac{\exp(d^-)}{\exp(d^+) + \exp(d^-)} \qquad (6)$$

We formulate the objective function for a single triplet (x, x^+, x^-) in a following manner [13]:

$$L_T = -\log(p_{T(x,x^+,x^-)}) \qquad (7)$$

The parameters of the model T are updated according to the gradient-based approach that is used to optimize the objective function L_T by utilizing the minibatches of triplets (x, x^+, x^-) selected from data. Usually, the procedure of triplet selection is performed randomly (assuming that x, x^+ are closer than x, x^-) but there are some other approaches that speed-up the training process. The most popular is to construct the triplets for training taking under consideration the hardest negative samples x^-, which are the closest to the currently selected reference sample x.

3   Triplet BiGANs

In this work we introduce Triplet BiGAN model that combines the benefits of using BiGAN in terms of learning interesting representations in latent space and the superior power of the triplet model that trains well using supervised data. The core idea of our approach is to incorporate the encoder model of BiGAN to act as triplet network on the labeled part of training data (see Fig. 1).

In terms of training the Triplet BiGAN we simply modify the L_{EG} (see Eq. (5)) criterion by incorporating an additional triplet term:

$$L_{TEG} = L_{EG} + \lambda \cdot \mathbb{E}_{(x,x^+,x^-) \sim p_{triplet}}[-\log(p_{T(x,x^+,x^-)})], \qquad (8)$$

where $p_{T(x,x^+,x^-)}$ is triplet loss defined by Eq. (7), λ is a hyperparameter that represents the impact of the triplet loss on the global criterion and $p_{triplet}$ is the distribution that generates triplets, where x, x^+ are from the same class and x, x^- are from different classes.

Triplet BiGAN model is dedicated to solving semi-supervised problems, where only some portion of labeled data is available. Practically, we do not have access to $p_{triplet}$, therefore we are sampling the triplets (x, x^+, x^-) from some portion of an available labeled dataset, X_q.

The training procedure for the model is described in Algorithm 1. We assume that G, E, D are neural networks, that are described by the parameters θ_G, θ_E,

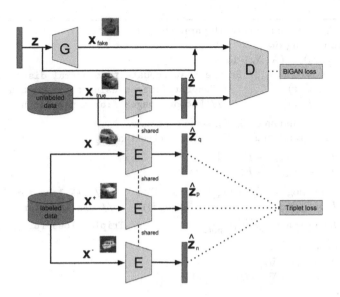

Fig. 1. The scheme presents the architecture of Triplet BiGAN model. We modify the BiGAN architecture by incorporating the encoding model in additional triplet task. We use the labeled part of the training data to construct triplets (x, x^+, x^-), where we assume that x and x^+ are from the same class and x and x^- are from different classes. Each of component of the triplet (x, x^+, x^-) is passed thru the encoding module to obtain the corresponding coding vectors (z, z^+, z^-) that are further used to construct the triplet loss. The remaining part of the model operates on the unsupervised data as in basic BiGAN model.

θ_D respectively. For the training procedure we assume, that we have access to unsupervised data X and some portion of supervised data X_q. For each training iteration, we randomly sample noise vector z from the normal distribution, pass it thru generator G to obtain the fake sample $\hat{x}$. We select x from unlabeled data X and triplet (x, x_+, x_-) from labeled data X_q. Using encoder E we receive the coding vector z corresponding to the sample x. Next, we update the parameters θ_D of discriminator D by optimizing the criterion L_D. During the same iteration we update the parameters of generator G and encoder E by optimizing $L_{TEG} = L_{EG} + L_T$. The procedure is repeated until convergence.

In practical implementation, we make use stochastic gradient optimization techniques and perform gradient updates using ADAM method. We also initialize parameters of the Triplet BiGAN by training simple BiGAN without triplet term for given number of epochs without triplet term ($\lambda = 0$).

The motivation behind this approach is to increase the discriminative capabilities of the codes obtained from latent space for BiGAN model using some portion of labeled examples involved in triplet training. As a result, we obtain the encoding model E, that is not only capable of coding the data examples for further reconstruction but also can be used as good quality feature embedding for the tasks like image classification or retrieval.

---

1 θ_G, θ_E, θ_D ← initialize network parameters
2 **while** converging **do**
3   **for** (x, x_q) *in* (X, X_q) **do**
4        $z \sim \mathcal{N}(0, I)$ // Sample feature data from normal distribution
5        $\hat{x} \leftarrow G(z)$ // Generate images using sampled data
6        $\hat{z} \leftarrow E(x)$ // Encode images from a minibatch
7        $x_+ \leftarrow$ random $x'_q \in X_q$, same class as x_q
8        $x_- \leftarrow$ random $x'_q \in X_q$, different class than x_q
9        $d_+ \leftarrow \|E(x_q) - E(x_+)\|_2$
10      $d_- \leftarrow \|E(x_q) - E(x_-)\|_2$
11      $L_D \leftarrow \log D(x, \hat{z}) + \log(1 - D(\hat{x}, z))$ // GAN loss for D
12      $L_{EG} \leftarrow \log D(\hat{x}, z) + \log(1 - D(x, \hat{z}))$ // GAN loss for G, E
13      $L_T \leftarrow -\log \frac{\exp(d_-)}{\exp(d_+) + \exp(d_-)}$ // Triplet loss for E
14      $\theta_D \leftarrow \theta_D - \nabla_{\theta_D} L_D$
15      $\theta_G \leftarrow \theta_G - \nabla_{\theta_G} L_{EG}$
16      $\theta_E \leftarrow \theta_E - \nabla_{\theta_E}(L_{EG} + L_T)$
17   **end**
18 **end**

Algorithm 1. Training procedure for Triplet BiGAN.

4 Experiments

The goal of the experiments is to evaluate the discriminative properties of the encoder in two challenging tasks: image retrieval and classification. We compare the results with the two reference approaches: triplet network trained only with supervised data and simple BiGAN model, where the latent representation of encoder is used for evaluation.

Datasets. The model was trained on two datasets: Street View House Numbers (SVHN) and CIFAR10. In each dataset, 50 last examples of each class were used as a validation set and were not used for training the models. During training only selected portion of training set have assigned labels. The next subsection presents results obtained when using only 100, 200, 300, 400 or 500 labeled examples per class. For testing purposes, we trained classifier only on the images from the training split, that were given a label for triplet training.

Metrics. Retrieval evaluation was done with accuracy and mean average precision (mAP). For classification, 9-nearest neighbors classifier was used with weighted by the distance-based importance of neighbors. Mean average precision was calculated at length of encoded data. Cluster visualization was performed by applying t-SNE [7] with Euclidean metric, perplexity 30 and Barnes-Hut approximation for 1000 iterations.

Architecture. The architectures of discriminator, encoder, and generator were as presented in [2].

The encoder network E is a 7-layer convolutional neural network, that learns the mapping from image space X to feature space z. After each convolutional layer (excluding the last), a batch normalization is performed, and the output is passed through a leaky relu activation function. After penultimate convolutional block (meaning convolutional layer with normalization and activation function) a reparametrization trick [5] is performed.

The generator network G is a neural network with seven convolution transposition layers. After each layer (except the last) a batch normalization is performed, and the output is passed through a leaky relu activation function. After the last convolution-transposition layer, we squash the features to $(0, 1)$ range with the sigmoid function.

The discriminator part D consists of three neural networks – D_x discriminates in the image space, D_z discriminates in the encoding space. Both of them map their inputs into a discriminative latent space and each of them returns a same-size vector. Third network D_{xz} takes concatenation of said vectors as an input and returns a decision, whether an input tuple (image, encoding) comes from encoding or generative part of the Triplet BiGAN network. Image discriminator D_x is made of five convolution layers with Leaky Relu nonlinearity after each of them. Encoding discriminator is represented as two convolution layers with Leaky Relu nonlinearity after each of them and Joint discriminator D_{xz} is another three convolutional layers with Leaky Relu between them and the sigmoid nonlinearity at the end.

4.1   Results

Classification. For assessing classification accuracy quality, the experiments were done to test the influence of feature vector size and images per class taken for semi-supervised learning. For each of the model, the experiments were done, when the feature vector consisted of either 16, 32, 64 or 128 (256 for SVHN) variables and 500 labeled images per class were taken. On the other hand, using feature vector size of 64, the experiments measured the impact of a number of labeled examples available during training, with possible values being 100, 200, 400 and 500 (only for Cifar10). The experiments were conducted on Cifar10 and SVHN datasets.

Table 1. Classification results on CIFAR10 dataset for different sizes of encoder feature vector using 500 labeled examples per class in the training set. m - vector size. Only labeled samples from training set used.

Model	m = 16	m = 32	m = 64	m = 128
Triplet	44.42	45.56	52.32	46.15
BiGAN	41.30	43.81	48.50	49.13
Triplet BiGAN	61.08	62.40	63.14	53.92

Table 2. Classification results on SVHN dataset for different sizes of encoder feature vector using 200 labeled examples per class in the training set. m - vector size. Only labeled samples from training set used.

Model	m = 16	m = 32	m = 64	m = 256
Triplet	66.12	69.96	71.11	71.54
BiGAN	44.65	53.80	62.35	17.48
Triplet BiGAN	71.43	75.65	79.12	78.86

Table 3. Classification results on CIFAR10 dataset for different portions of labeled samples per class and feature vector size $m = 64$. n - number of labeled samples per class. Only labeled samples from training set used.

Model	n = 100	n = 200	n = 400	n = 500
Triplet	28.80	34.81	40.12	52.32
BiGAN	42.73	45.06	47.67	41.30
Triplet BiGAN	53.15	52.02	57.45	63.14

Table 4. Classification results on SVHN dataset for different portions of labeled samples per class and feature vector size $m = 64$. m - vector size. Only labeled samples from training set used.

Model	n = 100	n = 200	n = 400
Triplet	66.48	71.11	73.76
BiGAN	61.16	62.35	58.94
Triplet BiGAN	72.40	79.12	82.21

Retrieval. For assessing image retrieval quality, the experiments were made to test an influence of feature vector size and images per class taken for semi-supervised learning. For each sample in the testing data, an algorithm sorts the images from the training dataset from closest to most further. Distances are calculated basing on Euclidean distances between images' feature vectors to check if samples that are close to each other in data space (images belong to the same class) are close to each other in feature space (their representation vectors are similar). In ideal type situation (mAP = 1), all of the relevant training images would be put first and only then training images that belong to the same class. With 10 classes in each dataset mAP = 0.1 may be considered random ordering, as it roughly means that on average, only every tenth image was of the same class as the test image (Tables 2 and 4).

Table 5. Image retrieval results on CIFAR10 dataset for different sizes of encoder feature vector using 500 labeled examples per class in the training set. m - vector size. Only labeled samples from training set used.

Model	m = 16	m = 32	m = 64	m = 128
Triplet	0.4993	0.5134	0.5235	0.5197
BiGAN	0.1458	0.1433	0.1634	0.1620
Triplet BiGAN	0.6292	0.6457	0.6528	0.3748

Table 6. Image retrieval results on SVHN dataset for different sizes of encoder feature vector using 200 labeled examples per class in the training set. m - vector size. Only labeled samples from training set used.

Model	m = 16	m = 32	m = 64	m = 256
Triplet	0.6855	0.7224	0.7474	0.6989
BiGAN	0.1492	0.1582	0.1748	0.1633
Triplet BiGAN	0.7201	0.7633	0.8002	0.7931

Table 7. Image retrieval results on CIFAR10 dataset for different portions of labeled samples per class and feature vector size $m = 64$. n - number of labeled samples per class. Only labeled samples from training set used.

Model	n = 100	n = 200	n = 400	n = 500
Triplet	0.3732	0.4027	0.4778	0.5235
BiGAN	0.1681	0.1666	0.1645	0.1634
Triplet BiGAN	0.5628	0.5487	0.5966	0.6528

Table 8. Image retrieval results on SVHN dataset for different portions of labeled samples per class and feature vector size $m = 64$. n - number of labeled samples per class. Only labeled samples from training set used.

Model	n = 100	n = 200	n = 400
Triplet	0.6323	0.7474	0.7490
BiGAN	0.1737	0.1748	0.1655
Triplet BiGAN	0.7300	0.8002	0.8198

Results presented in tables below indicate the increased average precision of image retrieval when using Triplet BiGAN method as opposed to using only labeled examples by 0.05–0.15 in all, but one experiment (Tables 6 and 8).

Figure 2 presents a visualization of the closest images from the restricted training set (used only 500 examples per class from the original training set, the same examples that were used in semi-supervised learning). The closeness of the image was decided between each image from the randomly chosen sample of 5 images from the test set and each image from the restricted training set.

Fig. 2. Visualization presenting image retrieval results. Left-most column of each section contains five randomly chosen images from the Cifar10 test set. Columns from second-left to right-most present images from the restricted (500 examples per class) training set, from the closest to 5th-closest image.

The distance between images was calculated by encoding each image to feature vector form and calculating Euclidean distance between selected test and training images.

As seen in the visualization, BiGAN model, despite the fact of being trained in a pure unsupervised way, can still embed similar images to similar vectors. However, the closest pictures tend to contain occasional errors, which is not the case with retrieval using triplet models, that tend to contain errors sparingly.

The notable example, showing better results of Triplet BiGAN in comparison to regular Triplet model is the 4th image from the selected test pictures (grey frog). Using 32 and 64 size features vectors Triplet BiGAN was able to retrieve other frog and toad images correctly. The same image caused problems for the original Triplet model, not to mention BiGAN. This shows that additional unsupervised learning of underlying data architecture is indeed beneficial to finding subtle differences in images and can improve the quality of feature embedding.

Clustering. Figure 3 shows visualizations of embedding quality of tested models. Each sub-figure present embedding mapped to 2 dimensions using the t-SNE algorithm with one of the three models: Triplet, BiGAN and Triplet BiGAN on the training set and the test set. For Triplet and Triplet BiGAN models, 500 labeled samples per class were used. Two experiments were performed: one with the feature vector of size 32, and one with the feature vector of size 64, as mentioned in figure captions. T-SNE algorithm ran for 1000 epochs using perplexity of 30 and Euclidean metric for distance calculation. In the visualization each class was marked with own color, that was preserved through all sub-figures.

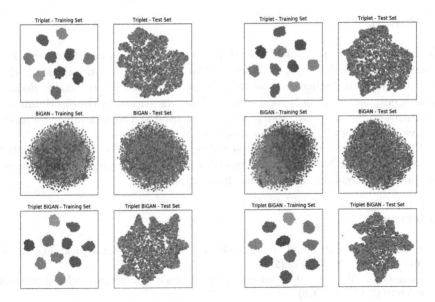

Fig. 3. (a) Clusterization results for models with feature vector size $m = 32$ and trained (Triplet and Triplet BiGAN) using 500 labeled examples per class. (b) Clusterization results for models with feature vector size $m = 64$ and trained (Triplet and Triplet BiGAN) using 500 labeled examples per class. (Color figure online)

Discussion. In classification and retrieval experiments Triplet BiGAN achieved worse results (Tables 1, 3, 5 and 7) than a Triplet GAN presented in [13]. However, we believe than our proposed model has still several advantages in comparison to the reference method. Since in Triplet BiGAN, we perform metric learning on the Encoder (unlike in [13] where metric learning is done on the Discriminator features) we are able to regularize our features to any distribution (e.g. normal distribution) and make them available to use in other methods.

As visualizations in Fig. 3 suggest, both Triplet and Triplet BiGAN models did not have any problems with learning clusterization on training sets. The output from the t-SNE clearly shows separate group for each class of the samples for this models. This is not the case in BiGAN model. However, while BiGAN was trained without distance-based objective, one can still spot concentration of particular colors. This aligns with observations [2] that the encoder learns to embed meaningful features into the feature vector, including those, that are somewhat characteristic for specific classes.

Regarding test sets, Triplet and Triplet BiGAN did not generalize to create perfect separations of classes. The models learn to rather bind particular classes into small, homogeneous groups, which are not clearly visible on visualizations but are enough to perform classification using the nearest neighbor algorithm. In the case of BiGAN model the embedding features from the training set do not translate well to the test set, creating a somewhat chaotic collection of points, that is able to generate image retrieval results that are close to random.

5 Summary

This work presents the Triplet BiGAN model that uses joint optimizing criteria: to learn to generate and encode images and to be able to recognize the similarity of given objects. Experiments show that features extracted by an encoder, despite learning only on true data (in opposition to features learned by the discriminator, that learns on real and generated data), may be used as a basis of image classifier, retrieval, grouping or autoencoder model.

Also included in this work are descriptions of the models that were essential milestones in the field of generative models and distance learning models and an inspiration for creating the presented framework.

References

1. Donahue, J., Krähenbühl, P., Darrell, T.: Adversarial feature learning. arXiv preprint arXiv:1605.09782 (2016)
2. Dumoulin, V., et al.: Adversarially learned inference. arXiv preprint arXiv:1606.00704 (2016)
3. Goodfellow, I., et al.: Generative adversarial nets. In: Advances in Neural Information Processing Systems, pp. 2672–2680 (2014)
4. Hoffer, E., Ailon, N.: Deep metric learning using triplet network. In: Feragen, A., Pelillo, M., Loog, M. (eds.) SIMBAD 2015. LNCS, vol. 9370, pp. 84–92. Springer, Cham (2015). https://doi.org/10.1007/978-3-319-24261-3_7
5. Kingma, D.P., Welling, M.: Auto-encoding variational Bayes. arXiv preprint arXiv:1312.6114 (2013)
6. Kumar, B., Carneiro, G., Reid, I., et al.: Learning local image descriptors with deep siamese and triplet convolutional networks by minimising global loss functions. In: Proceedings of the IEEE Conference on Computer Vision and Pattern Recognition, pp. 5385–5394 (2016)
7. van der Maaten, L., Hinton, G.: Visualizing data using t-SNE. J. Mach. Learn. Res. 9(Nov), 2579–2605 (2008)
8. Radford, A., Metz, L., Chintala, S.: Unsupervised representation learning with deep convolutional generative adversarial networks. arXiv preprint arXiv:1511.06434 (2015)
9. Salimans, T., Goodfellow, I., Zaremba, W., Cheung, V., Radford, A., Chen, X.: Improved techniques for training GANs. In: Advances in Neural Information Processing Systems, pp. 2234–2242 (2016)
10. Yao, T., Long, F., Mei, T., Rui, Y.: Deep semantic-preserving and ranking-based hashing for image retrieval. In: IJCAI, pp. 3931–3937 (2016)
11. Zhuang, B., Lin, G., Shen, C., Reid, I.: Fast training of triplet-based deep binary embedding networks. In: Proceedings of the IEEE Conference on Computer Vision and Pattern Recognition, pp. 5955–5964 (2016)
12. Zieba, M., Semberecki, P., El-Gaaly, T., Trzcinski, T.: BinGAN: learning compact binary descriptors with a regularized GAN. arXiv preprint arXiv:1806.06778 (2018)
13. Zieba, M., Wang, L.: Training triplet networks with GAN. arXiv preprint arXiv:1704.02227 (2017)

Sharp Images Detection for Microscope Pollen Slides Observation

Aysha Kadaikar[1,2(✉)], Maria Trocan[2], Frédéric Amiel[2],
Patricia Conde-Cespedes[2], Benjamin Guinot[1], Roland Sarda Estève[3],
Dominique Baisnée[3], and Gilles Oliver[4]

[1] Laboratoire d'Aérologie, Toulouse, France
aysha.kadaikar@gmail.com, benjamin.guinot@gmail.com
[2] Institut Supérieur d'Electronique de Paris, Paris, France
{maria.trocan,frederic.amiel,patricia.conde-cespedes}@isep.fr
[3] Laboratoire des Sciences du Climat et de l'Environnement, Gif-sur-Yvette, France
{sarda,dominique.baisnee}@lsce.ipsl.fr
[4] Réseau National de Surveillance Aérobiologique, Brussieu, France
gilles.oliver@rnsa.fr

Abstract. In this paper, a new preprocessing algorithm to qualify images of different pollen grains for further processing is proposed. This algorithm provides a score related to the sharpness of the image and will be used to automatically adjust the focal length of a microscope that magnifies the image. The obtained score has been compared to four quality metrics generally used to estimate the clarity of an image and to a reference made by a human. The results of the simulations show that the proposed algorithm combines better performance with low complexity on the set of images.

Keywords: Microscope slide image acquisition ·
Sharp image detection · Fourier transform

1 Introduction

Allergic rhinitis concerns between 8 and 25% of the world population [1,2] and it is caused in particular by pollen. Pollen detection is thus an important issue to prevent this allergic rhinitis. Usually a human operator uses a microscope to observe slides containing particles present in the ambient air. Then, he can determine the presence or not of pollen particles, and also count them. One of the difficulties of this observation is that the viewer always has to set the focus knob of the microscope to get a clear image of the view to make this analysis. An example is shown in Fig. 1: each image from (a) to (e) represents the same view acquired by increasing the focus knob from $4\,\mu m$ each time. The sharpest view corresponds in this example to the view (c). One can see that setting the focus knob on a lower or a higher value leads to a blurred image (see images (a) and (e), respectively, as example). Moreover, finding a good setting for one view

© Springer Nature Switzerland AG 2019
N. T. Nguyen et al. (Eds.): ACIIDS 2019, LNAI 11431, pp. 661–671, 2019.
https://doi.org/10.1007/978-3-030-14799-0_57

is still not enough: as the viewer changes the view to observe different parts of the slide, one has to set the focus knob again.

In this paper, we propose a new algorithm to automatically set the focus knob in order to get the sharpest image. It is based on a metric which gives the highest score to the image containing the highest frequencies, and has thus the least blur. Other no-reference quality metrics can be found in the literature: some of them are especially dedicated to the evaluation of the sharpness of an image [3]. Some others give a score for the global quality of an image (including sharpness) using for example a perception based quality evaluator [4], or a comparison to a default model computed from natural images [5–7]. These latter have the advantage of taking into account the natural perception. However, they could fail in evaluating the images acquired from the microscope as they take into account many others criteria in addition to the sharpness.

The rest of the paper is organized as follows: Sect. 2 presents the developed Sharp Image Detection algorithm (denoted SID-algorithm). Section 3 discusses the simulation performance by comparing the use of the SID-algorithm with others state-of-the-art no references quality metrics. Finally, Sect. 4 concludes this work.

2   Selection of the Sharpest Image

The first subsection describes the metric which is used to evaluate the image sharpness. Next, the sharp image detection algorithm is presented in the second subsection. We define the following notations: z corresponds to a given setting of the focus knob. I_z represents the image obtained from the microscope with the setting z. I_z is an image of $M \times N$ pixels resolution.

2.1   Image Sharpness Evaluation Metric

This section describes the Fourier-based Image Sharpness Evaluation Metric (denoted as FISEM in the rest of the paper) that is used by the SID-algorithm to determine the clearest image by varying the focus knob in the microscope. The FISEM relies on the fact that sharp images contain less blur and thus higher frequencies. The computation of the metric for an given image I_z is composed of three steps:

– First, a Laplacian high-pass filter (denoted H) is applied to the image I_z. The obtained filtered image I_z' contains only the higher frequencies of the image and is given by:

$$I_z'(i,j) = \sum_{m=-1}^{1} \sum_{n=-1}^{1} I_z(m-i, n-j)H(m,n) \tag{1}$$

$$\text{where} \quad H(i,j) = \frac{1}{8} \begin{bmatrix} 0 & 1 & 0 \\ 1 & -4 & 1 \\ 0 & 1 & 0 \end{bmatrix}$$

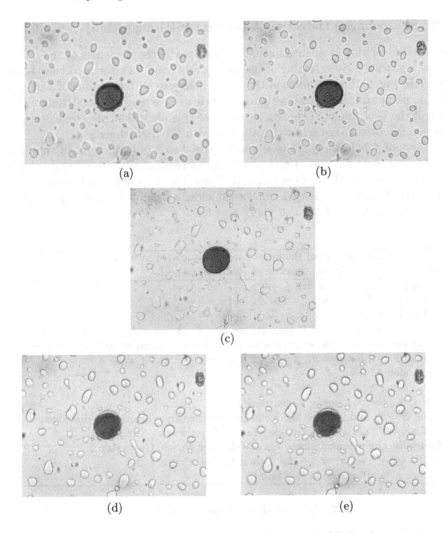

Fig. 1. Microscope views by increasing the focus knob with (c) the sharpest view.

- Then, a 2D Fourier-transform is applied to the filtered image. The transformed image is given by:

$$I_z''(u, v) = \sum_{m=0}^{M-1} \sum_{n=0}^{N-1} I_z'(m, n) e^{-j\left(\frac{2\pi}{M}\right)um} e^{-j\left(\frac{2\pi}{N}\right)vn} \tag{2}$$

- Finally, the score S_z of the image I_z is computed as the sum of the absolute values of the Fourier coefficients:

$$S_z = \sum_{m=0}^{M-1} \sum_{n=0}^{N-1} |I_z''(n, m)| \tag{3}$$

2.2 Sharp Image Detection Algorithm (SID-algorithm)

The SID-algorithm is at first initialized with the current settings of the micro-scope, e.g. with a given setting of the focus knob z and its associated view I_z. As shown in Fig. 1, the sharpness of the image increases with the value of z up to a certain point and then decreases again. This feature is used by the SID-algorithm to find the best image. The detailed steps of this algorithm are given in Fig. 2. Finally, the best image selected is the one with the highest score.

The flag f is initially set to 0 and is turned to 1 only once, when the focus knob is changing direction (example: firstly increasing, then decreasing it). The variable o contains the last operation performed on z : + if it has been increased and − if it has been decreased. Δz corresponds to a defined value (in μm) which is added or deducted from z at every new test.

Input: Current setting of the focus knob z, associated view I_z
Output: Best setting $\mathbf{z}^*$, associated sharpest image $I_{\mathbf{z}^*}$

1. Initialization: set the flag $f = 0$.
2. Compute $S = S_z$. Set $z = z + \Delta z$ and $o = $ ”+”.
3. Compute S_z.
If $S < S_z$, the new image is sharper. Set $S = S_z$, $z = z + \Delta z$, $o = $ ”+” and $f = 1$.
Otherwise the previous image was sharper: the focus knob should be turned in the other way. Set $S = S_z$, $z = z - 2 \times \Delta z$, $o = $ ”-” and $f = 1$.
Go to step 4.
4. Compute S_z.
If $S < S_z$, the new image is sharper.
    Set $S = S_z$ and increase the focus knob in the same direction as previously:
    if $o = $ ”+”, set $z = z + \Delta z$, else if $o = $ ”-”, set $z = z - \Delta z$.
Otherwise the previous image was the sharpest image. Go to step 5.
5. If $o = $ ”+”, set $\mathbf{z}^* = z - \Delta z$. Otherwise, set $\mathbf{z}^* = z + \Delta z$.
Return $\mathbf{z}^*$ and associated $I_{\mathbf{z}^*}$. End.

Fig. 2. Pseudo-code of the proposed SID-algorithm

3 Discussion of the Experimental Results

This section discusses the simulation results of the SID-algorithm for finding the best images in microscope observations. This algorithm was tested on 20 sets of 11 images each of 1280×960 pixels resolution. For each set, the 11 images were acquired by setting manually the focus knob such as to have the globally sharpest image denoted "0". Then, increasing the focus knob of 2 μm at each time, 5 more images containing an increasing amount of blur were obtained. These images are denoted from "+1" to "+5". In a same way, from the position of the focus knob leading to the best image, 5 more images with increasing blur are obtained by decreasing the focus knob of 2 μm at each time. These latter are denoted "−1" to "−5".

The efficiency of SID-algorithm has been compared to four state-of-the-art no reference quality metrics. The first metric denoted PIQE relies on a perception based quality evaluator [4], while the second and the third one denoted NIQE and BRISQUE use a comparison to a default model computed from natural images to evaluate the quality of the images [5–7]. These three metrics can be found in the Matlab processing toolbox [8]. Finally the fourth metric denoted as CPBD deals with to the images sharpness assessment and it is based on the cumulative probability of blur detection [3].

Table 1. Best image selected by the SID-algorithm using the FISEM, PIQE, NIQE, BRISQUE and CPBD no-references quality metrics.

Images set	Image selected by the SID-algorithm using				
	FISEM	PIQE	NIQE	BRISQUE	CPDB
1	0	−5	−4	+5	−5
2	+2	−5	−1	−5	−5
3	−2	+5	+3	+5	+5
4	−1	+5	−4	+5	+5
5	−2	+5	+2	+5	+5
6	0	−5	+4	+3	+5
7	−2	+4	−1	+4	+5
8	−2	+5	+4	−5	+5
9	−2	+4	+3	−5	+5
10	−1	+5	+2	+5	+5
11	+2	−5	−2	+3	−5
12	0	−3	+4	−1	−1
13	+1	−5	−3	−5	−1
14	0	−5	−3	−5	−5
15	+1	−5	−3	−5	−1
16	0	+5	+3	+5	+5
17	−1	−5	−3	−5	+5
18	+1	−5	+4	−5	−1
19	0	−5	−3	−5	+5
20	0	+5	+2	+5	+5

The best images selected by the SID-algorithm using the different metrics including FISEM are given in Table 1. One can clearly see that the images selected by the SID-algorithm using FISEM are very close to the sharpest image selected by the viewer. The four other metrics and mainly the PIQE and BRISQUE metrics often choose the most blurred image as the best on of the set. Figures 3, 4, 5, 6 and 7 (respectively Figs. 8, 9, 10, 11 and 12) give an example of

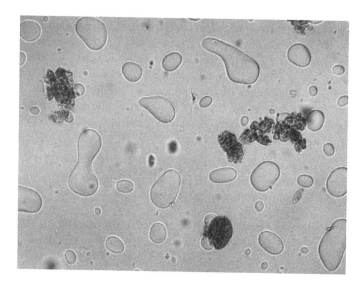

Fig. 3. Best image using FISEM in the set of images 12

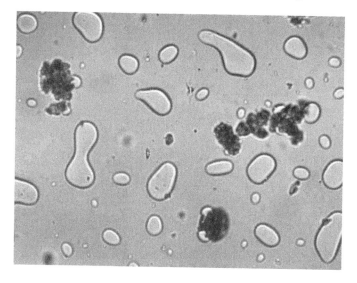

Fig. 4. Best image using PIQE in the set of images 12

the images selected by the SID algorithm using the five metrics corresponding to the set number 12 (respectively the set number 6) in the table. It seems that the three global quality metrics often consider the huge amount of particles in the image as degradations. Thus they prefer blurred images where these particles can not be seen properly. Of course the FISEM metric is not appropriated if the degradations in the images have very high frequencies like salt and pepper noise. But in the case of the acquisition of sharp images for microscope obser-

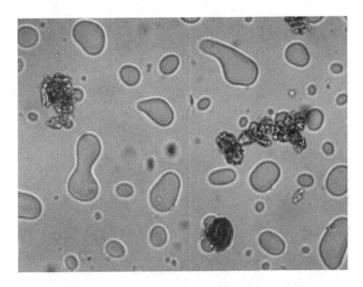

Fig. 5. Best image using NIQE in the set of images 12

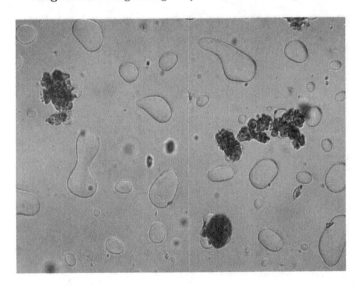

Fig. 6. Best image using BRISQUE in the set of images 12

vation, this metric leads to good results. Concerning the CPBD-metric, it does not allow to find the sharpest image in most cases also. It seems that this metric is more adapted to assess the sharpness of images having very different levels of blurriness between the objects present in the images and the surrounding areas, for example between the foreground and the background.

The SID-algorithm has also the advantage of being of low complexity. Indeed for each image, the complexity is equal to $O(Mlog(M)Nlog(N))$ depending on

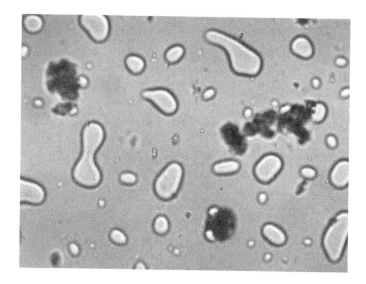

Fig. 7. Best image using CPBD in the set of images 12

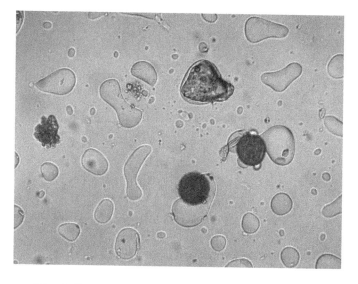

Fig. 8. Best image using FISEM in the set of images 6

the size ($M \times N$ pixels) of the image. Using a computer with an Intel Core i5 processor of 2.4 GHz and 4 GB of RAM, the FISEM code running under the version R2014a of Matlab takes an average of 0.04 s to find the sharpest image for a given set of images.

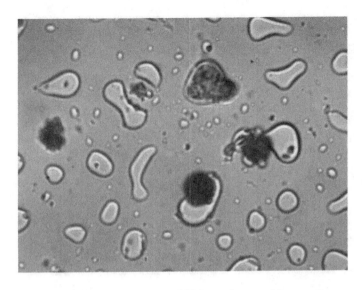

Fig. 9. Best image using PIQE in the set of images 6

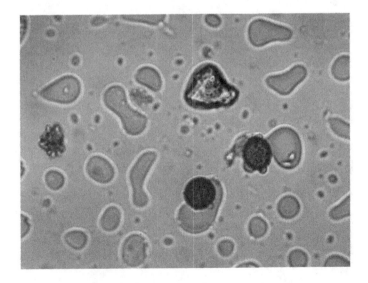

Fig. 10. Best image using NIQE in the set of images 6

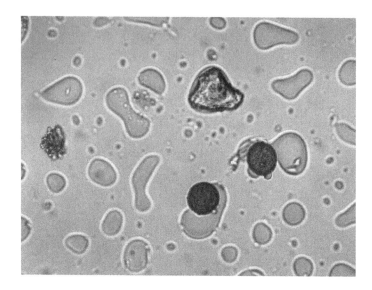

Fig. 11. Best image using BRISQUE in the set of images 6

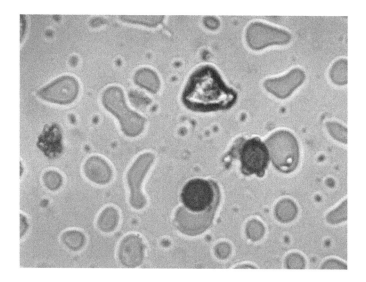

Fig. 12. Best image using CPBD in the set of images 6

4 Conclusion

This paper presents a new method to automatize the acquisition of sharp images using a microscope. One of the major constraints using a microscope is that the viewer always has to manually set the focus knob in order to get the sharpest view. The developed method allows to select the best settings leading to very sharp images by using a metric which gives higher scores to images containing higher frequencies and thus less blur. This metric performs better than state-of-the-art no-references quality metrics for the images acquired using a microscope. Moreover, the proposed algorithm has also the advantage of low time complexity.

Acknowledgment. The authors would like to gratefully acknowledge the support of the National Research Agency and the STAE foundation under the auspices of the Saint-Exupery Technological Research Institute without which the present study could not have been completed.

References

1. Pawankar, R., et al.: State of world allergy report 2008: allergy and chronic respiratory diseases. World Allergy Organ. J. **1**(1), S4 (2008)
2. Dykewicz, M.S., Hamilos, D.L.: Rhinitis and sinusitis. J. Allergy Clin. Immunol. **125**(2), S103–S115 (2010)
3. Narvekar, N.D., Karam, L.J.: A no-reference perceptual image sharpness metric based on a cumulative probability of blur detection. In: 2009 International Workshop on Quality of Multimedia Experience. IEEE (2009)
4. Venkatanath, N., Praneeth, D., Chandrasekhar, B.M., Channappayya, S.S., Medasani, S.S.: Blind image quality evaluation using perception based features. In: Proceedings of the 21st National Conference on Communications (NCC). IEEE, Piscataway (2015)
5. Mittal, A., Moorthy, A.K., Bovik, A.C.: Referenceless image spatial quality evaluation engine. In: Presentation at the 45th Asilomar Conference on Signals, Systems and Computers, Pacific Grove, CA (2011)
6. Mittal, A., Moorthy, A.K., Bovik, A.C.: No-reference image quality assessment in the spatial domain. IEEE Trans. Image Process. **21**(12), 4695–4708 (2012)
7. Mittal, A., Soundararajan, R., Bovik, A.C.: Making a completely blind image quality analyzer. IEEE Signal Process. Lett. **22**(3), 209–212 (2013)
8. Image processing toolbox description page (2018). https://www.mathworks.com/help/images/index.html Accessed 20 Dec 2018

Databases and Intelligent Information Systems

Slowly Changing Dimension Handling in Data Warehouses Using Temporal Database Features

Thanapol Phungtua-Eng[1] and Suphamit Chittayasothorn[2(✉)]

[1] Faculty of Business Administration and Information Technology,
Rajamangala University of Technology Tawan-Ok,
Chakrabongse Bhuvanarth Campus, Bangkok 10400, Thailand
thanapol.phu@cpc.ac.th
[2] Faculty of Engineering, King Mongkut's Institute of Technology Ladkrabang,
Bangkok 10520, Thailand
suphamit.ch@kmitl.ac.th

Abstract. This paper presents the use of temporal database features to solve the Slowly Changing Dimension (SCD) problem of data warehouses. The SCD problem is presented and existing solutions, together with their limitations are shown. Temporal database features of SQL are described. Temporal data retrieval and temporal data manipulations, together with illustrated examples are demonstrated. The solution to the SCD problem is shown with illustrated examples. The data warehouse whose dimension tables are validtime state tables, but the fact table is a conventional fact table without any timestamp or validtime period, is proposed. The identifier integrity of dimension instances is preserved. The sample fact table, dimension tables, and the SQL codes which perform temporal operations to solve the problem are presented. The proposed solution gives correct results regardless of the number of changes made to the attribute of the dimension table, thus completely solves the Slowly Changing Dimension problem.

Keywords: Slowly Changing Dimension · Temporal data warehouse · Temporal SQL

1 Introduction

A data warehouse is a centralized data source for data analytics and management information making supports. Typical data warehouses comprises facts and dimensions.

Data warehouses help ease the data integration issues in organizations that employ individual special purpose information systems, which are separately developed. Many organizations deploy such information systems from many vendors or software developing teams without any prior agreement on data formats and value standards.

For many other organizations, which may not afford the costly centralized database, or have to deploy ready-to-use information systems from several vendors due to the deployment time constraint, building data warehouses is the alternative solution. Data

© Springer Nature Switzerland AG 2019
N. T. Nguyen et al. (Eds.): ACIIDS 2019, LNAI 11431, pp. 675–687, 2019.
https://doi.org/10.1007/978-3-030-14799-0_58

warehouses employ star schemas or snowflake schemas, which comprise fact and dimension table schemas. Dimension table schemas are schemas that describe object types, which are of particular interest to the organization. Each row of the dimension table represents an object instance, which is uniquely identified by a surrogate key. The use of surrogate keys ease the join operations between the dimension tables and the fact tables. A fact table schema comprises surrogate keys from the associated dimension tables and at least one attribute called the measure. Typically, there is a time dimension. The smallest time granularity is normally at the day level since a data warehouse mostly keep summarized data as opposed to detailed transactions. A data warehouse fact table can also be represented as a cube. Each axis represent a dimension and the value at a coordinate is a measure value.

2   The Slowly Changing Dimensions Problem

Naturally, attribute values of existing rows of dimension tables may be updated. This could affect the analytic results. The changes made to an attribute value of a dimension, in principle, should not affect the past information of the data warehouse. However, if not handled properly, the changes could yield incorrect analytic results from the data warehouse. This dimensional attribute's value change problem is widely known in the data warehouse area as the Slowly Changing Dimensions (SCD) problem [1, 2]. The term "slowly" reflects that the values are updated occasionally and infrequently.

A classic example is the case where a sale person changes his team. The person used to belong to a sale team but later moved to a new team. The team name of this person is therefore changed at a particular point of time. An ideal data warehouse should be able to show that the person's sale amount before the change belongs to the former team, and the sale amount after the change belongs to the new team. This problem sounds easy to handle but it is not. Current solutions have limitations.

Solutions have been proposed and referred to as Type 1, Type 2 and Type 3 solutions. They are approaches based on the conventional relational database. All of them has some limitations. In this paper, we propose an approach to solve the SCD problem using the temporal relational database, which solves the SCD problem and all the limitations of the existing solutions.

Kimball [1] a pioneer of data warehousing, and Jensen et al. [2], classify three typical approaches for the SCD problem handling. They are summarized as follow:

2.1   Type 1 (Overwrite)

The type one approach simply overwrite the value of the changed attribute. The new value replaces the old value. This is a simple approach to the problem. It does not solve the problem but it is simple and easy to implement. All related facts, which are referred to by the dimension instance now refer to the new value. Figure 1 illustrates this approach. The sale person S1 moves from IT department to Retail department. This approach simply replaces IT by Retail. All the fact instances of S1 are now refer to the new department Retail even though they are old facts.

Fig. 1. The overwrite approach to the SCD problem

2.2   Type 2 (Add New Dimension Instance)

The type two approach is an attempt to preserve the history. It creates a new row in the dimension table for the dimension instance with the newly updated value. The newly created row with the new column value also has a new surrogate key. The dimension instance therefor has a new row and a new surrogate key for each change. This simple approach works but lacks the identification integrity. Each object instance should have a unique identifier; otherwise, the two instances are interpreted as different object instances. Figure 2 demonstrates the type two approach.

Dimension Table

Surrogate Key	Sales_ID	Depart	Start	End	Current
12345	S1	IT	1/1/2017	31/12/9999	Current

Dimension Table

Surrogate Key	Sales_ID	Depart	Start	End	Current
12345	S1	IT	1/1/2017	1/1/2018	Expired
23456	S1	Retail	1/1/2018	31/12/9999	Current

New record

Fig. 2. The 'add new dimension instance' approach to the SCD problem

2.3   Type 3 (Add a New Attribute for Each Version)

The Type 3 approach add a new attribute to the dimension instance row. Each instance has only one row and one unique identifier. The identification integrity is preserved. However, this approach can handle only limited number of changes. Figure 3 demonstrates the Type 3 approach. Here only two versions of the department value can be accommodated.

Dimension Table

Surrogate Key	Sales_ID	Depart	Prior_dep
12345	S1	IT	None

Dimension Table

Surrogate Key	Sales_ID	Depart	Prior_dep
12345	S1	Retail	IT

Fig. 3. The 'add new attribute' approach to the SCD problem

There are other attempts to solve the SCD problems. Snapshot tables [3] are used to represent a dimension instance row at a point of time. The separation of a historical table from the current table is also suggested [4]. However, the instances from the two tables use different surrogate keys.

Surrogate key-based temporal data warehouse [5], another interesting approach, is an attempt to use the validtime concept of temporal database to show the validity of the changed value. However, different surrogate keys are still used for different version of the changed dimension instance. This simple approach works but still lacks the identification integrity.

Versioning tables [6, 7] are extensions to the surrogate key-based temporal data warehouse approach. A current flag is introduced to ease programming tasks and can be viewed according to applications' requirements. Different surrogate keys are still used for different versions of the changed dimension instance.

In this paper, we present an approach to the Slowly Changing Dimension problem. This approach completely solves the problem without any previous limitations. We do not replace the old value by the new value and lost track of the old one likes the Type 1 approach. We do not use different surrogate keys to refer to the same object instance likes the Type 2 approach. This identification integrity problem of the Type 2 approach obviously lead to the inability to track information of the same object instance over different periods since different identifiers are used to identify the same dimensional object instance. We do not have the limitation on number of versions of the changes that the Type 3 solution has.

Our approach does not use different surrogate key value for different changed versions. The unique identifier of each dimension instance remains unchanged. This is feasible due to the deployment of temporal database technology in modern relational database systems and the temporal features of the SQL language as described in the following section.

3   Temporal Database Operations

The handling of the SCD problem without the well-known limitations, and still maintain the identification integrity so that tuples which refer to the same object instance, but have different attribute values due to the changes, and still use the same identifier is a big challenge in the data warehouse area. In this paper, we propose the use of temporal database technology to handle the problems. In order to understand our proposed solution, temporal database operations must be well understood first. This section gives a brief summary of temporal database operations in SQL.

In conventional database systems, only facts that are currently true are stored. In reality, facts in the databases change from time to time. Previously true facts are also considered important and might be referred to by applications. The database should contain not only the current information of an employee but also previous versions of the information as well. Validtime denotes the time when a fact is true in reality. Relational database tables that support validtime is called validtime state tables. Nowadays several commercially available relational database systems support validtime state tables and their SQL languages support temporal aspects of validtime state tables.

Validtime state tables keep time varying facts and the time that the fact is valid. For each time varying fact type, there are validtime start and validtime end, in the closed-open format. Each validtime interval is closed at its lower bound, and open at its upper bound to facilitate the continuity checking. Facts which are currently valid, have the upper bound valid, to-date, set to either null or 'infinity', which is the maximum date 31/12/9999. Removed facts, are not actually deleted but instead are marked as invalid facts. Figure 4 shows the validtime state table used to present the sale person dimension of a data warehouse. The validtime is in the close-open format. The two rows of each sale person with sales_id S1 and S3 are the rows before and after the changes of their department. The previous department value was RETAIL and the current one is IT. A new row is inserted for each modification. The row with the old value is marked as valid up to the day the update took place and a new row with the new value is inserted. The end date of the new row is set to 31/12/9999 to show that the row is the current one.

	SAL...	SALES_NAME	DEPARTMENT	START_DATE	END_DATE
1	S01	John	IT	01/05/2018	31/12/9999
2	S01	John	RETAIL	01/01/2018	01/05/2018
3	S02	Smith	IT	01/01/2015	01/10/2018
4	S03	TIM	RETAIL	01/01/2018	01/06/2018
5	S03	TIM	IT	01/06/2018	31/12/9999
6	S04	Tom	Engineer	01/06/2017	31/12/9999
7	S05	Jane	IT	01/06/2017	31/12/9999

Fig. 4. A validtime state table SALES_DIMENSION shows sale persons change departments

We use the Oracle Workspace Manager [8] to demonstrate relevant temporal database operations. The temporal features in this section are used to implement our approach to solve the SCD problem.

3.1   Validtime State Table Creation

The concept of validtime state table enables temporal queries; as well as temporal insert, delete and update operations. These temporal database operations are tedious to implement using conventional relational database technology [9]. The validtime period, if handled by conventional relational systems, needs long SQL codes to manipulate and queries. Instead, if the validtime period is handled by the temporal feature of the supported DBMS, the SQL codes are much more concise. The following commands as shown in Fig. 11 create the validtime state table sales_dimension which is to be used as our temporal dimension table for the SCD problem handling. The EXECUTE DBMS_WM.EnableVersioning command enables validtime versioning.

The attribute WM_VALID is automatically created and handled by the DBMS. It comprises the validtime attributes (WM_VALID.VALIDFROM) and (WM_VALID. VALIDTILL).

For each temporal operation, a validtime period, the period of applicability, needs to be specified. The EXECUTE DBMS_WM.SetValidTime (START_DATE, END_ DATE) command therefore needs to be issued before each temporal operation. In the Fig. 5, sales_id is the apparent primary key of the validtime state table. The actual primary key, according to the conventional relational database model, is the combination of the apparent primary key sales_id and the validtime period WM_VALID. This actual primary key is handled by the temporal feature of the DBMS (the Oracle Workspace Manager).

```
CREATE TABLE sales_dimension (
    sales_id VARCHAR2(20) PRIMARY KEY,
    sales_name varchar2(20),
    department varchar2(20)
);
EXECUTE DBMS_WM.EnableVersioning ('sales_dimension', 'VIEW_WO_OVERWRITE', FALSE, TRUE);
```

	SALES_ID	SALES_NAME	DEPARTMENT	WM_VALID
1	S01	John	IT	[WMSYS.WM_PERIOD]
2	S01	John	RETAIL	[WMSYS.WM_PERIOD]

Fig. 5. The SQL command for validtime state table creation and the table with some rows

3.2   Temporal Query Features

Since our proposed solution to the SCD problem needs the temporal query features to handle the problem, dimension tables are created as validtime state tables. Temporal queries on them are now possible. Temporal operators comparable to the Allen's Interval Algebra [10] are available. Oracle's WM_OVERLAPS is Alllen's P1 overlaps P2. WM_CONTAINS is P1 contains P2. WM_MEETS is P1 Meets P2. WM_EQUALS is P1 equals P2.

Figure 6 demonstrates the WM_OVERLAPS operation. The SQL select statement retrieves rows of the validtime state table SALES_DIMENSION from Fig. 5 where the valid period overlaps 01/02/2018 and 31/05/2018 (the close-open format excludes 01/06/2018). The result shows two rows which belong to the sale person S01. The first row is his current row where he belongs to the IT department since 01/05/2018 until now (31/12/9999). The second row shows his previous appointment at the RETAIL department between 01/01/2018 and 30/04/2018 (01/05/2018).

Fig. 6. An SQL command that demonstrates the WM_OVERLAPS operation

Figure 7 demonstrates the join of the SALES_DIMENSION validtime state table to itself with the WM_CONTAINS condition. The query gives a list of sale persons who work during the same period of time for the same department. Temporal join operations are required to join the dimension tables with valid time and the fact tables of our proposed temporal data warehouse.

```
SELECT p1.sales_id p1_sales_id,
    p1.sales_name p1_sales_name,
    p1.department p1_department,
    TO_CHAR(p1.wm_valid.validFrom,'DD/MM/yyyy') p1_start_date,
    TO_CHAR(p1.wm_valid.validTill,'DD/MM/yyyy') p1_end_date,
    p2.sales_id p2_sales_id,
    p2.sales_name p2_sales_name,
    p2.department p2_department,
    TO_CHAR(p2.wm_valid.validFrom,'DD/MM/yyyy') p2_start_date,
    TO_CHAR(p2.wm_valid.validTill,'DD/MM/yyyy') p2_end_date
FROM sales_dimension p1
INNER JOIN sales_dimension p2
ON (WM_CONTAINS (p1.wm_valid,p2.wm_valid) = 1
OR WM_CONTAINS (p2.wm_valid,p1.wm_valid) = 1)
AND p1.sales_id                            != p2.sales_id
AND p1.department                          = p2.department
order by P1_Sales_id, p1.wm_valid.validFrom, P1_department;
```

Query Result ×

SQL | All Rows Fetched: 24 in 0.352 seconds

	P1_SALES_ID	P1_SALES_NAME	P1_DEPARTMENT	P1_START_DATE	P1_END_DATE	P2_SALES_ID	P2_SALES_NAME	P2_DEPARTMENT	P2_START_DATE	P2_END_DATE
1	S01	John	RETAIL	01/01/2018	01/05/2018	S03	TIM	RETAIL	01/01/2017	01/06/2018
2	S01	John	RETAIL	01/01/2018	01/05/2018	S09	Anna	RETAIL	01/01/2017	01/11/2018
3	S01	John	RETAIL	01/01/2018	01/05/2018	S08	Jo	RETAIL	01/02/2018	01/03/2018
4	S01	John	IT	01/05/2018	31/12/9999	S03	TIM	IT	01/06/2018	31/12/9999
5	S01	John	IT	01/05/2018	31/12/9999	S05	Jane	IT	01/06/2017	31/12/9999
6	S02	Smith	IT	01/01/2015	01/10/2018	S11	Roy	IT	01/01/2015	01/10/2018
7	S03	TIM	RETAIL	01/01/2018	01/06/2018	S01	John	RETAIL	01/01/2018	01/05/2018
8	S03	TIM	RETAIL	01/01/2018	01/06/2018	S09	Anna	RETAIL	01/01/2017	01/11/2018

Fig. 7. An SQL query which demonstrates the join of the SALES_DIMENSION validtime state table to itself with the WM_CONTAINS condition

3.3 Temporal Data Manipulations

Insertion of new rows to the validtime state table is a straight forward operation. A period of validity needs to be declared before the insert operation. Update and delete operations on validtime state tables, however, are complicated and require careful considerations [9]. Period of applicability needs to be declared and considered when update or delete operations are issued. It is then checked against the period of validity of existing data. There could be several overlap patterns between the two periods. The temporal database feature of the DBMS handles them automatically. In this section, the update of attribute value is demonstrated. It will be used to modify attributes of dimension tables which are validtime state tables, thus leads to the solving of the SCD problem.

The commands in Fig. 8 demonstrate the department change of the sale person S05 to the MECHANIC department from 01/01/2018 to 30/12/2018. Note the close-open format, 31/12/2018 is not included in the period. The left table shows the sale

```
EXECUTE DBMS_WM.SetValidTime(TO_DATE('01/01/2018', 'DD/MM/YYYY'),TO_DATE('31/12/2018', 'DD/MM/YYYY'));
    update SALES_DIMENSION set DEPARTMENT = 'MECHANIC'
    where SALES_ID ='S05';
```

	SALES_ID	DEPARTMENT	START_DATE	END_DATE
1	S01	RETAIL	01/01/2018	01/05/2018
2	S01	IT	01/05/2018	31/12/9999
3	S02	IT	01/01/2015	01/10/2018
4	S03	RETAIL	01/01/2018	01/06/2018
5	S03	IT	01/06/2018	31/12/9999
6	S04	Engineer	01/06/2017	31/12/9999
7	S05	IT	01/06/2017	31/12/9999

	SAL...	DEPARTMENT	START_DATE	END_DATE
1	S01	RETAIL	01/01/2018	01/05/2018
2	S01	IT	01/05/2018	31/12/9999
3	S02	IT	01/01/2015	01/10/2018
4	S03	RETAIL	01/01/2018	01/06/2018
5	S03	IT	01/06/2018	31/12/9999
6	S04	Engineer	01/06/2017	31/12/9999
7	S05	IT	01/06/2017	01/01/2018
8	S05	MECHANIC	01/01/2018	31/12/2018
9	S05	IT	31/12/2018	31/12/9999

Fig. 8. The temporal update operation that moves S05 to the MECHANIC department for a year

dimension table before the update. S05 was in the IT department since 01/01/2017. The update result on the right table shows three rows of S05. The END_DATE of row number 8 is updated to 01/01/2018 and two new rows are inserted.

4   Data Warehouses with Temporal Dimensions

Based on the temporal database technology, we propose the data warehouse with temporal dimensions. The dimension tables whose attributes' values can be modified and referred to when queried, are validtime state tables. SQL queries on them only consider the apparent primary keys. Temporal operators can be employed. The fact table is a conventional fact table without the validtime.

Figure 9 shows the star schema whose dimension SALES_DIMENSION is a validtime state table to accommodate the department changes. SALES_ID is the sale person key. It is the apparent primary key, which is unique for each dimension instance. Unlike other related works in the literatures, the primary key value will not be changed after updates on the temporal attribute. The data warehouse that employs such validtime dimension tables, can now be referred to as a temporal data warehouse. Note that there are no such validtime periods in dimension tables of other conventional data warehouses.

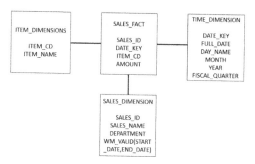

Fig. 9. A star schema whose dimension SALES_DIMENSION is a validtime state table

Figure 10 shows the time dimension table TIME_DIM. Note that the surrogate key DATE_KEY is used instead of the actual date. This is a common practice in data warehousing. The attribute FULL_DATE is the actual date. When the SCD problem is detected, this date will be used to check validity of the corresponding fact instances. Figure 11 shows the sale persons dimension SALES_DIM, which is a validtime state table. The item dimension table ITEM_DIM is a conventional dimension table. Figure 12 is the fact table SALES_FACT together with its fact instances of S01. The fact table does not have a validtime but it is possible to include one here. Note also the use of surrogate keys for easy references.

DATE_KEY	FULL_DATE	D..	DAYS_NAME	DAYS_OF_WEEK	FISCAL_QUARTER	YEAR_MONTH	MONTH_SHORT
5	05/01/2018	05	FRIDAY	6	2018/1	2018/01	Jan
21	21/01/2018	21	SUNDAY	1	2018/1	2018/01	Jan
81	22/03/2018	22	THURSDAY	5	2018/1	2018/03	Mar
110	20/04/2018	20	FRIDAY	6	2018/2	2018/04	Apr
121	01/05/2018	01	TUESDAY	3	2018/2	2018/05	May
192	11/07/2018	11	WEDNESDAY	4	2018/3	2018/07	Jul
229	17/08/2018	17	FRIDAY	6	2018/3	2018/08	Aug
265	22/09/2018	22	SATURDAY	7	2018/3	2018/09	Sep
282	09/10/2018	09	TUESDAY	3	2018/4	2018/10	Oct
352	18/12/2018	18	TUESDAY	3	2018/4	2018/12	Dec

Fig. 10. The Time dimension table TIME_DIM

SALES_ID	SALES_NAME	DEPARTMENT	START_DATE	END_DATE
1 S01	John	RETAIL	01/01/2018	01/05/2018
2 S01	John	IT	01/05/2018	31/12/9999

SALES_DIM

ITEM_CD	ITEM_NAME
1 P1	COMPUTER
2 P2	HAMMER

ITEM_DIM

Fig. 11. Rows of S01 from the dimension tables SALES_DIM, and rows of ITEM_DIM

SALES_ID	DATE_KEY	ITEM_CD	AMOUNT
1 S01	5	P1	200
2 S01	21	P2	1000
3 S01	81	P2	120
4 S01	110	P1	1000
5 S01	121	P1	40
6 S01	192	P2	200
7 S01	229	P2	300

Fig. 12. Rows of the fact table SALES_FACT which show fact instances of S01

5 Slowly Changing Dimension Handling

In this section, we present the use of temporal database features to solve the Slowly Changing Dimension (SCD) problem. We propose that the dimension tables, whose attributes are significant enough so that their changes are to be tracked and shown as output attributes of the data warehouse, be validtime state tables. The surrogate keys of such dimension tables are the apparent primary keys. Each entity instance has the surrogate primary key, which does not changed with time. Attribute value updates and the validtime, are handled by the temporal feature of the DBMS as described in the previous section.

The key idea is to relate each row of the fact table to the correct row of the dimension table whose attribute values have been changed. In our approach, since the apparent primary key remains the same regardless of the number of changes made to an attribute value, and the number of attribute value versions, simple joining of common surrogate keys of the fact table and the corresponding dimension table is not enough. The join must be a temporal join with time overlap or contain checking.

The next issue is the time validity of the fact instance rows of the fact table. In principle, a row of a fact table refers to the time key of the time dimension. This time key represents a point of time, not a period of time, and is a surrogate key. The SCD dimension table, on the other hand, has a validtime period. A possible representation is to attach a time period for the day to each fact instance row, thus complicate the Extract

Transfer Load (ETL) task of the data warehouse. This validtime on the fact table would also lead to the wrong interpretation that each fact instance is only valid for the day, which is not true. We therefore keep the fact table as a conventional relational database table.

To check that the date on the fact instance row corresponds with the validtime period of a corresponding dimension row, we extract the full date of the fact instance row by using a join with the time dimension table via the time surrogate key. Once the full date of the fact instance is obtained, it can be checked if the date and the date plus one (due to the close-open format) contains in a validtime period of a dimension instance row with the same dimension surrogate key.

An illustrated example in Fig. 11 shows that the sale person S01 John was with the RETAIL department from 01/01/2018 to 30/04/2018. He then moved to the IT department from 01/01/2018 until now. Figure 10 shows some rows of the time dimension table TIME_DIM. The surrogate key, DATE_KEY, is used instead of the actual date timestamp. This is a typical data warehouse identification. Date keys 5, 21, 81, 110 represent the days when the sale person S01 was with the RETAIL department. The rest of the date keys, 121 to 352 are the days when S01 belongs to the IT department.

The fact table SALE_FACT in Fig. 12 shows sale fact instances some of which belong to S01 when he was with RETAIL and some when he was with IT. The first four rows of the fact table shows the sale records on the days 5, 21, 81, and 110, which are the days when S01 was with RETAIL. The other rows show the sale records on the days 121, 192, and 229 when he was already moved to IT. Notice that all these sales fact instances refer to the person as S01. The facts that he belongs to different departments during different periods of time are in the dimension table SALES_DIM which is a validtime state table.

The Slowly Changing Dimension (SCD) problem, according to the above sample data, is the case when the supplier S01 changed his department and the system cannot keep track of the changes, thus gives incorrect total values of his sales amount for each department. Figure 13 shows the correct result obtained from our data warehouse with temporal dimensions. The total sales amount of S01 when he was with the RETAIL department is 2,320 units. The first fiscal quarter of 2018 is 1,320 units and the second quarter of 2018 is 1,000 units. The total sales amount of S01 when he was with the IT department is 540 units. The first fiscal quarter of 2018 is 40 units and the second quarter of 2018 is 500 units. The fiscal quarter information is obtained from the time dimension table TIME_DIM. The SQL query, which employs the WM_CONTAINS to check fact instance and dimension instance periods of validity, is shown in Fig. 14.

DEPARTMENT	FISCAL_QUARTER	SUM(AMOUNT)
1 RETAIL	2018/1	1320
2 RETAIL	2018/2	1000
3 IT	2018/2	40
4 IT	2018/3	500

Fig. 13. The query result demonstrates that the SCD problem is solved.

```
SELECT department,
    d.fiscal_quarter,
    SUM(amount)
FROM sum_sales s
INNER JOIN TIME_DIMENSION d
ON s.date_key = d.date_key
INNER JOIN ITEM_DIMENSION i
ON i.item_cd = s.item_cd
INNER JOIN sales_dimension sd
ON s.SALES_ID                                      = sd.sales_id
AND WM_CONTAINS (sd.wm_valid,wm_period(d.full_date,d.full_date+1)) = 1
where s.sales_id = 'S01'
GROUP BY department,
    d.fiscal_quarter
order by fiscal_quarter,department;
```

Fig. 14. The SQL query which produces correct result despite the change in the dimension.

To illustrate the consequence of a further update to the attribute, suppose the sale person S01 changes his department once again. This time he moved from IT to MECHANIC department since 01/09/2018. The temporal SQL update statement marks the END_DATE of the IT row to be 01/09/2018. The sale persons dimension table SALES_DIM now has a new row as shown in Fig. 15.

	SALES_ID	SALES_NAME	DEPARTMENT	START_DATE	END_DATE
1	S01	John	RETAIL	01/01/2018	01/05/2018
2	S01	John	IT	01/05/2018	01/09/2018
3	S01	John	MECHANIC	01/09/2018	31/12/9999

Fig. 15. Rows of S01 from the dimension table SALES_DIM after S01 moved to MECHANIC

The sale person makes more sales and corresponding new fact instance rows are shown in the fact table SALES_FACT as shown in Fig. 16. The date 265, 282, 352 are actually 22/09/2018, 09/10/2018, and 18/12/2018 respectively. By applying the SQL query with temporal feature from Fig. 14, the query result which shows sales by department and fiscal quarter of S01 is correctly shown in Fig. 17.

	SALES_ID	DATE_KEY	ITEM_CD	AMOUNT
1	S01	265	P2	790
2	S01	282	P1	1000
3	S01	352	P2	40

Fig. 16. New fact rows of S01 after moving to MECHANIC department

◊ DEPARTMENT	◊ FISCAL_QUARTER	◊ SUM(AMOUNT)
1 RETAIL	2018/1	1320
2 IT	2018/2	40
3 RETAIL	2018/2	1000
4 IT	2018/3	500
5 MECHANIC	2018/3	790
6 MECHANIC	2018/4	1040

Fig. 17. The query result, which shows correct sales by department and fiscal quarter of S01

6   Conclusions

This paper presents the use of temporal database technology to solve the Slowly Changing Dimension problem. A temporal data warehouse, which has validtime dimension tables is proposed. The identifier integrity of dimension instances is preserved. The same surrogate key identifier is used to identify the same object instance even after a dimension attribute value is changed. The dimension attribute value can be changed several times without any limitations. The query results are always correct regardless of the number of changes made, thus completely solves the Slowly Changing Dimension problem. All limitations of the previously well-known solutions of the problem are lifted. Our implementation is based on the Oracle Workspace Manager. However, relational DBMSs that adhere to the SQL standards since 2011 have the temporal SQL features as presented in the paper and therefore can be used to implement the proposed solution.

References

1. Kimbal, R., Ross, M.: The Data Warehouse Toolkit: The Complete Guide to Dimensional Modeling, 3rd edn. Wiley, Indianapolis (2013)
2. Jensen, C., Pederson, T.B., Thomsen, C.: Multidimensional Databases and Data Warehousing. Lexington, Morgan Claypool (2010)
3. Nguyen, T.M., Tjoa, A.M., Nemec, J., Windisch, M.: An approach towards an event-fed solution for slowly changing dimensions in data warehouses with a detailed case study. Data Knowl. Eng. 63(1), 26–43 (2007)
4. Santos, V., Belo, O.: No need to type slowly changing dimensions. In: Proceedings of IADIS International Conference Information Systems 2011, Avila, Spain, pp. 11–13 (2011)
5. Faisal, S., Sarwar, M.: Handling slowly changing dimensions in data warehouses. J. Syst. Softw. 94, 151–160 (2014)
6. Ravat, F., Teste, O., Zurfluh, G.: A multiversion-based multidimensional model. In: Tjoa, A. M., Trujillo, J. (eds.) DaWaK 2006. LNCS, vol. 4081, pp. 65–74. Springer, Heidelberg (2006). https://doi.org/10.1007/11823728_7
7. Golfarelli, M., Lechtenbörger, J., Rizzi, S., Vossen, G.: Schema versioning in data warehouses. In: Wang, S., et al. (eds.) ER 2004. LNCS, vol. 3289, pp. 415–428. Springer, Heidelberg (2004). https://doi.org/10.1007/978-3-540-30466-1_38

 8. Workspace Manager Valid Time Support. https://docs.oracle.com/database/121/ADWSM/long_vt.htm
 9. Snodgrass, R.T.: Managing temporal data – a five part series, Database programming and design, TimeCenter technical report (1998)
10. Allen, J.F.: Maintaining knowledge about temporal intervals. Commun. ACM **26**, 832–843 (1983)

Tackling Complex Queries
to Relational Databases

Octavian Popescu$^{(\boxtimes)}$, Ngoc Phuoc An Vo, Vadim Sheinin,
Elahe Khorashani, and Hangu Yeo

IBM Research, Yorktown Heights, USA
{octavian.popescu,ngocphuoc.anvo,vadim.sheinin,
elahe.khorashani,hangu.yeo}@us.ibm.com

Abstract. Most people who want to get an answer from a structured
repository, such as a database, are agnostic of both the formal language
requested by database Structured Query Language (SQL), and of the particular structure of specific databases. On the other hand, processing arbitrary queries in natural language to automatically get the SQL is very challenging, especially due to the fact that most of the most frequent queries
lead to Nested Logic Queries (NLQs). While most of the Natural Language
Interface to Databases systems (NLIDB) may put severe restrictions on
the form of the acceptable input queries, QUEST can deal with large variability in input. QUEST is a semi-supervised system which can encode the
information about any database and process complex queries via an unsupervised learning methodology which addresses the problem of NLQs. We
report a significant improvement in accuracy over other approaches.

1 Introduction

Transforming natural language queries (NTQs) into SQL queries is a difficult
process, due to the fact that understanding the logical form of natural language
questions leaves no room for errors in the lexical, syntactic, and semantic analysis. While in some cases, where a more or less simple approach is effective, in
most of the cases, especially those in which the corresponding SQL query is of
the nested logic query (NLQ), only a powerful mechanism of language understanding may have a chance of success. From the SQL perspective, an NLQ
needs more than one **select** command, and very often multiple join commands
over different tables with different keys. From the natural language perspective, a natural language question requiring an NLQ has a very complex form
since in natural language we employ a series of complex linguistics phenomena
(CLP), like anaphora, ellipsis, relative clause and preposition attachment, constituent boundaries, logical vs. syntactic head identification, garden path, etc.
This complexity is currently beyond the available tools for sentence processing,
like (semantic) parsers, coreference resolution, word disambiguation – which have
a low rate of accuracy for sentences involving CLP.

In fact, not all the NTQs are equally complex. There are simple NTQs, mostly
containing one relationship expressed via one unique predicate in the query,

© Springer Nature Switzerland AG 2019
N. T. Nguyen et al. (Eds.): ACIIDS 2019, LNAI 11431, pp. 688–701, 2019.
https://doi.org/10.1007/978-3-030-14799-0_59

which can be directly linked to information stored in the databases, like *What is John's salary*. These are simple logic queries, SLQs. However, the majority of NTQs are NLQs, that is, the associated SQL formula is nested and to get it right, a deep and complex treatment of query is required. For example, *In which shop the quantity of stocked Iphones is higher than in Bestbuy* is an example of NTQ that requires multiple `select` commands, and a fairy complex processing to determine the relationship between the nominal phrases. We will show that there is way to manage the complexity of NTQs leading to NLQs by building a model of language understanding able to decompose the original complex NTQ into a a series of simple NTQs.

We present an NLIDB system, named QUEST, which (1) builds in semi-automatic way a representation of any databases which allows (2) an effective way to generate an intermediate form for simple, one predicate, queries, (3) an unsupervised model which manages the NTQs leading to NLQs by decomposing them into a series of simple, one predicate, queries, and (4) an SQL query generator. The core mechanism in QUEST is a model of linguistics knowledge derived from analyzing a large corpus coupled with a technology able to determine relations and sentence classification. In a nutshell, the system constructs first a model that links a reduced set of simple English sentences to a particular database. A rule base module identifies the `data items` which make the connection between the relationships between nominal phrases occurring in a NTQ and the rules describing a particular database (1) At the heart of processing simple sentences resides a mechanism of transforming the query into a sequence of `data items`, i.e. terms that have a direct and specific connection to the databases and SQL query (2) A deep natural language understanding module is employed to decompose the complex sentences into an ordered sequence of simple queries (3) The SQL generator component is based on the Cognos server[1], an IBM product designed to offer a convenient and efficient access to structured data via automatically generated reports.

In Sect. 2 we present the general system architecture. In Sect. 3 we cover the details for processing the simple queries, while in Sect. 4 we focus on the complex queries. In Sect. 5 we carry out evaluation experiments and error analysis. The paper concludes with the Conclusion and Further research section.

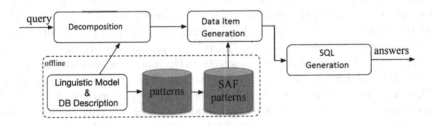

Fig. 1. QUEST framework.

[1] http://www.ibm.com/analytics/us/en/technology/cognos-software/.

2 Framework

QUEST is a modular system that combines four modules: (1) A module that represents the relationships between table and columns in a database as one predicate sentences, like *shop sell product; shop in TableX-ColumnY, product TableZ-ColumnT* (2) a module that extracts from simple queries the `data items`, which are nominal phrases that have a direct representation in the database, and link them to the database representation based module (3) a Decomposition module which is in charge of decomposing a CLP question into a set of ordered simple sub-questions, (4) an SQL Query Engine Cognos, which is able to produce full fledged SQL queries and return an answer (see Fig. 1).

All the modules have an `off-line` component, where the deep natural language understanding models are created. The `on-line` component applies the learned model to a specific NTQ.

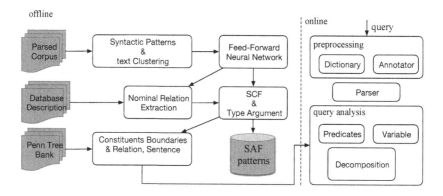

Fig. 2. Deep understanding model for CLP questions.

3 Semantic Parsing of One Predicate Queries

With progress in the structured databases such as Freebase [1] and DBpedia [2], semantic parsing against these databases has attracted more and more research interests. Almost all of the existing semantic parsers need to be trained either with the direct supervision such as logical forms [3] or with the weak supervision such as question-answer pairs [4–6]. However, in real application scenarios, collecting sufficient question-answer pairs to train the model remains a challenge. To address this, QUEST implements a rule-based semantic parser for one predicate NTQs. The fundamental building block is a template pattern file (TPF). TPF is a database-independent rule file and contains a set of rule patterns for generic sentence structures. There is no lexical information in the TPF, only grammatical information that is applicable to any relationships in any database. Quest `data item` is a key-value item that contains information for creating one or more SQL clauses (e.g. SELECT, WHERE etc.). The TPFs are general patterns that link data items to SQL commands. However, for a particular database we need to have databases dependent lexicalized rules, that include reference to particular tables and columns.

off-line

- **DataBase Description.** For each database, the system requires a database description. In its basic form, this description contains an English nominal phrase associated with columns in the database tables, like id of employee, date of sale, etc.
- **Schema Annotation File.** The English sentences generated by the syntactic patterns according to the database description are stored in a file, called SAF. It represents the main repository of the information that makes the connection between questions expressed in natural languages and the structure of a particular database. TPF patterns are applied to the SAF entries, and the database-dependent lexicalized rules are created.

on-line

- **NTQ Annotation.** After detecting entities, i.e nominal phrases that refers to database entries, the question is processed through a sequence of annotators, such as numerical annotator, date annotator, comparator annotator and predicate argument structure (PAS) annotator.
- **Sub Tree Pattern Matching.** The last step of our semantic parser is to search previous annotation results to find all the patterns in the lexicalized rules that match the input question. These patterns determine the types of the SQL formula that should be produced and the values of the filter items if there are any.
- **SQL Generation and Execution.** The extracted data items via PAS and lexicalized rules are sent to the Cognos server, which generates the final SQL query and executes it. We implemented Cognos Interface Library which is comprised of three modules, i.e., the connector module, builder module and execute module. The connector module connects to the Cognos services and submits report requests, and the builder module creates the report. The execute module uses the other two modules, connects to the server, builds and submits the report, and receives the query results.

4   CLP Question Decomposition

In a simple question, the entities and the relevant relation among them are explicit and the SQL query can be derived from the dependency parses via extraction of data items. However, in a CLP question, relevant information is not overtly expressed, and must be inferred via a non trivial processing module. see Table 1.

Table 1. Simple vs. CLP questions.

	Simple questions	CLP questions
1	What is John's salary?	What is the average salary of the employees working in the same department as John?
2	What shops sell Iphone?	In how many shops the price of Iphone is greater than Galaxy's?
3	How many Iphone were sold in Bestbuy in 2017?	What product was sold in shops in NY more in 2017 than in 2018?
4	What product has a price greater than 600?	In what shop in which John bought Galaxy, at least 10 types of phones are there with prices greater than Iphone X?
5	What vendor stocked Iphones in 2017?	What vendor sold more products in 2016 than it stocked in same year?
6	What is the number of sales that Bestbuy made?	What shop made more sales in 2017 than in 2016?
7	In what department does John work?	What department has an average salary greater than the one of the department in which John works?

In order to get the right arguments of the predicates, such as *sell*, or of operators, such as *more than*, we have to deal with anaphora (1, 5, 7 from Table 1), ellipsis (2–6), relative clause and preposition attachment (1–7), constituent boundaries (1–7), logical vs. syntactic head identification, etc.

Our solution for NLQs rests on decomposing the initial query into a set of simple queries via linguistics knowledge under the shape of regular patterns [7–11]. For an input NTQ, which contains one or many CFPs, a new query is obtained by augmenting the original query with the information from the patterns matched against it. The new augmented query contains all the explicit information required for decomposition into simple queries. The generation of patterns and their matching against an NTQ couples two procedures respectively, each one implementing statistical and deep learning methods: (i) **generation** – The word embedings are used to generate patterns via statistical inference; (ii) **matching** – we first augmented the model proposed in [12] to predict the pattern involved at each level of the parse tree. When the pattern is incomplete or span across different levels it means that the sentence contains a CLP. By determining the relationship between nominal phrases as dictated by the pattern, CLP may be resolved and also possible parser errors are corrected. Our system contains an off-line phase when the patterns and the data base description is generated and an on-line phase when the system matches the input query is decomposed (see Fig. 2). Below we present schematically each component and at the end of this section we present in details the two approaches to generation and match of the patterns.

```
Pattern: X pay Y for Z from U
X buy Y for Z           X write check for Y      X make payment of Y for Z
X obtain Y for Z        U offer Z for Y to X     X authorize payment of Y for Z
U have Z for Y          X spend Y for Z          U receive Y for Z from X
X acquire Z for Y       X receive Z for Y        U accept payment from X for Z
X get Z for Y           U sell Z for Y to X      X cover expenses for Z
U charge X with Y for Z                          X give amount of Y for Z to U

Y is [MONEY] & X is [PER] & Z is [ARTIFACT] & U is [ORG]
X in {client, customer} & Y in {commodity, merchandise, artifact} & U in {shop, store}
```

Fig. 3. Syntactic patterns.

off-line

- **Syntactic Patterns.** From large parsed corpora, we extract parts of dependency chains and cluster them considering the similarity of the syntactic slots and the semantic distance of the lexical items that occur there. These are fed into a feed forward neural network which creates embeddings over these special constructs. The output is a large resource of pattern paraphrases presented in Fig. 3.
- **Linguistic Model.** A seq2Tree LSTM is trained on a syntactic annotated corpus Penn Tree Bank [13]. However, the resulting model needs to be improved with linguistic knowledge in order to deal with complex phenomena. The consistency of the model is checked against the English sentences generated by syntactic patterns and the result is back propagated into the network. The final model is used to correct the parser's decisions in the cases of structural ambiguities generated by complex linguistic phenomena.

on-line

- **Pre Processing.** The input query is preprocessed. At this step, the type of entities is recognized by using a dictionary and an annotator. A dictionary created from the database is used to replace the proper nouns with their column description, like replacing *John* with employee, *iPhone* with product, etc.
- **Question Analysis.** The question is parsed. If it is a simple question, the default processing is invoked. The dependency tree is matched against the dependency trees of the sentences from SAF and the data items are identified. From data items, a report is generated which is passed to the SQL engine in order to get the answer.

If it is a CLP question, the linguistics model constructed off-line is invoked in order to find the relations and their arguments, and the question is decomposed into simple sub-questions. The assignment of variables is used to mark the relevant arguments of the relations in the subsequent questions as needed.

4.1 Pattern Generation

RNN. We implemented a recursive neural network (RNN) that maximizes the probability of a certain value for each dependency slot given a verbal phrase.

For a given sequence of vectors representing parts of a dependency tree, $r_{n=1..N}$ computes the following functions:

$$r_n = f(WX_{n-1} + Vr_{n-1} + b) \tag{1}$$
$$y_n = softmax(Ur_n + c) \tag{2}$$

where f is a pairwise linear function, W, V, U are weight matrices, U is the output one, b and c are the bias. In a nutshell , we compute a dependency embedding for each verbal phrase and output the vectors that are most similar according to the Euclidean metrics.

Show me in which day before Paul bought Iphone, Bestbuy sold a number of these which higher than one of radioshack.

Operator *more than*
PATTERN [NUMERIC] MORE THAN [NUMERIC]
Syntactic arguments: *number* and *Radioshack*
Logical head analysis: *these* and *Radioshack*

PATTERN [SHOP] SELL [PRODUCT]
Coreference *these → [product] ← Iphone*

Pattern more than has no numerical arguments
Ellipsis
Iphone → number of Iphone sold by bestbuy
Radioshack → number of iphone sold by radioshack

Operator *before*
PATTERN [DATE] BEFORE [DATE]
Syntactic arguments: day and bought
Boundary detection and SCF: [Customer]] buy [Product]
Parser correction: *in which day* attached to *buy* not to *show*

Decomposition
queryOT: what is the day when paul bought iphone;
assigment day as Q0XA60

query1: what is the number of product that radioshack sold per day;
assignment radioshack as Q1XA1; assignment product as Q1XA2; assignment per day as Q1XA40
query2: what is the number of product that bestbuy sold per day;
assignment bestbuy as Q2XA1; assignment product as Q2XA2; assignment per day as Q2XA40
query3: connection Q1 and Q2; computation Q1XA2 and Q2XA2 with operator MoreThan;
selection Q2XA40; same(Q2XA40, Q1XA40); Before(Q2XA40,Q0XA60)

Fig. 4. Decomposition by pattern matching.

The output of the RNN is a huge list of candidates for each verbal phrase. The level of noise can be really high, that is , it is hard to make a decision on the quality of the proposed paraphrases just on the distance of the vectors. A filtering procedure must be implemented that does two things in parallel:

FILTER1 the noise depends on two parameters that we can control: frequency and number of senses. In a nutshell the higher the number of occurrences of a candidate and the less ambiguous it is, the higher the chances that it is a valid candidate.

FILTER2 each candidate is checked for semantic compatibility, that is, the probability of the semantic role correspondence is checked between the candidate and the target verbal phrase. For example, the candidates *pay* and *sell* are considered valid if the subject of *sell* is likely to be the prepositional object of *from* of the *pay*, that is *shop*. We computed the conditional probabilities from each slot and value and we applied the chain formula (see Eq. 3).

$$\begin{aligned}
Argmax\ p(X_c, Y_c, Z_c \mid V_c, X_t, Y_t, Z_t) &= \\
p(V_c, X_c, Y_c, Z_c, V_t, X_t, Y_t, Z_t) * const &= \\
p(V_c, X_c, Y_c, Z_c, V_t, X_t, Y_t, Z_t) * p(V_c \mid V_t, X_t, Y_t, Z_t) &* \\
p(X_c \mid V_t, V_c, X_t, Y_t, Z_t) &* \\
p(Y_c \mid V_t, V_c, X_c, X_t, Y_t, Z_t) &* \\
p(Z_c \mid V_t, V_c, X_c, Y_c, X_t, Y_t, Z_t) &\quad (3)
\end{aligned}$$

4.2   Decomposition by Pattern Matching

In Fig. 4 we present schematically an example of the on-line decomposition. The NTQ *Show me in which day before Paul bought Iphone, Bestbuy sold a number of these that is higher than the one of Radioshack* has a series of CLPs.

The analysis starts by identifying the operators at lexical level. There are two *before* and *higher than*. The patterns associate with these operators are $[DATE]before[DATE]$ and $[NUMBER]higherthan[NUMBER]$. However, the syntactic fillers of these slots do not match. This fact shows that we have a NTQ with CLP. Th procedure is the same for both of them. For **higher than** the syntactic fillers are *number* and *one* respectively. The fact that *number* is a semantically void element triggers a procedure that identifies the real head, that is *these*. As *these* is the object of the verb *sold*, according to the pattern of sold in **SAF**, $[SHOP]sell[PRODUCT]$, this pronoun must be indexed to a nominal phrase that has its type, namely *Iphone*. So the filler of the **sell** pattern are *Bestbuy* and *iphone* respectively, and the left argument of **more than** is Iphone, but it must be a number. For the right argument of the operator **more than**, the ellipsis procedure is triggered, and *one* is indexed with *number of iphone* and *Radioshack* with $[SHOP]$. These leads to two simple subqueries, *What is the number of Iphones that Bestbuy sold* and *What is the number of Iphones that Radioshack sold*. But this the partial form, because the WH-head and the temporal operator have not been analyzed yet. The same analysis is applied to the operator **before**, its left argument is a $[DATE]$, but the right argument is **Paul** so the whole clause *Paul bought Iphone* must be converted to $[DATE]$ in decomposition. So the pattern for **before** creates a simple query, *In which day did Paul bought Iphone*. Because the WH-head is a $[DATE]$, the two partial subqueries must be considered for *day*. The connection between subqueries is maintained via variables. The fact that the ellipsis was triggered for operator **more than** imposes that the variable representing the dates must have the same value, same(Q2XA40, Q1xA40) and the **before** imposes the comparison between.

5   Experiments and Evaluation

In this section we introduce the experiment setup, the main results and detailed analysis of our Quest system. Quests system variants have been evaluated on complex datasets The baseline we use is a state of the art system on language to logical form conversion.

5.1   Baseline and Evaluation Corpora

We implemented a sequence to tree Long Short Term Memory network (seq2tree) based on [14,15] as the baseline. We evaluated it on the GeoQuery dataset [16]. This corpus is used only to evaluate the effectiveness of the baseline system as there are no SQL query associated with these questions, but only their logical form. It yielded 92% accuracy which proves it a strong baseline.

Then we manually created two datasets GolfQuery on golf tournaments (consists of GF_NLQ and GF_SLQ) and SalesQuery on sales/warehouse schema for testing QUEST. Table 2 shows the details of our datasets.

5.2   Results and Discussion

Our first experiment focused on the baseline system on the GeoQuery dataset. The accuracy is computed by ratio between the number of correct resolved queries vs. the total number of queries. It was 92%, which differs insignificantly from the current state of the art performances for this corpus, which proves that the baseline is indeed a competitive system.

Table 3 presents the accuracy of the baseline system (Seq2Tree), and three settings of our Quest system: Quest_v0 (without decomposition module), Quest_v1 (decomposition without relative clause and constitute boundaries), and Quest_v2 (decomposition with relative clause and constitute boundaries) for the GolfQuery and SalesQuery datasets. Overall, Quest systems outperformed the Seq2Tree baseline on all datasets. Especially, the Quest_v2 significantly improved the accuracy on CLP queries for both GolfQuery and SalesQuery datasets over the LSTM baseline.

As Quest_v1 and Quest_v2 are empowered by the decomposition module to handle CLP questions, we do not test them for GE_SLQ. Looking at the results on GF_SLQ vs. GF_CLPQ we can see that there is a large difference between the accuracy for SLQs vs. CLPQs. We wanted to understand how this gap is related to the number of training examples, and if there is a scalability problem for the baseline system.

Scalability. To evaluate the scalability of the baseline approach, we investigated the relation between the number of training examples vs. SLQ and NLQ accuracy. As shown in Fig. 5, the accuracy of the baseline system grows quickly as the number of training examples increases for SLQs, both for GeoQuery and GF_SLQ datasets. However, for GF_NLQ, the learning curve looks different both for the direct test on the chosen 249 testing queries (GF_TST) of GF_NLQ, and for the 10-fold cross validation (GF_CRV) for the whole GF_NLQ dataset. This result suggests that the number of training examples needed for a higher accuracy may be very high. This is a serious bottleneck, as in all real scenarios we are aware of, training data is unavailable.

Nested Question Decomposition. The Quest's accuracy gain is due to the linguistic processing of the question, which consists of constituents attachment

Table 2. Datasets.

	Total	Train	Test
GeoQuery	880	600	280
GF_CLPQ	1,200	951	249
GF_SLQ	1,150	800	350
SalesQuery	1,000	800	200

Table 3. Baseline vs. quest in performance.

	GF_SLQ	GF_CLPQ	SalesQuery
Quest_v2	N/A	61%	75%
Quest_v1	N/A	38%	68%
Quest_v0	98%	30%	50%
Seq2Tree	92%	21%	22%

disambiguation, ellipsis, anaphora resolution, etc. The interaction of these linguistic operations within a single question may lead to a very complex analysis of the query. In fact, linguistic complexity plays a determinant role as the following experiment shows. We selected a set of 64 sentences from the GF_CLPQ test set that are particularly complex. Then we run Quest on a 10-fold cross validation for the GF_CLPQ corpus, and we impose that a third of the test questions at each fold is randomly chosen from the 64 complex questions. The performance on each fold varied considerably from 30% to 63%. The most unfavorable case for now seems to be when the operator scope disambiguation, coupled with right most ellipsis, fail to produce the right decomposition. For example the question *"Which is the player having an average of birdies higher than Tigers?"* was incorrectly decomposed by computing *"the birdies that were longer than Tiger's birdies"*, while the intended reading was to compare two averages.

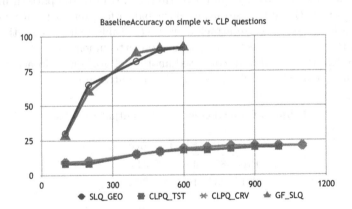

Fig. 5. Seq2Tree learning curve.

5.3   Error Analysis

We carefully analyze the 155 questions that are not answered correctly by our system.

- The query data is not available in DB (43 queries, 27.7%): In this case, the query data is missing in the DB, and QUEST cannot return the query results.
- Current limitation of QUEST (40 queries, 25.8%): QUEST does not support negation (NOT) or 'OR' in the query. The performance can be improved by adding more functionalities or by creating additional views in the Cognos model.
- Template rules do not support the query (26 queries, 16.8%): The QUEST could answer *Who won the most US Open?*, but couldn't answer *Who won the US Open the most?* because the incorrect 'most' scope assignment. This is a difficult problem, we hope to find a general way to address it via linguistic processing.
- The query is not clearly stated (24 queries, 15.5%): The QUEST cannot generate data items when a query or its context is unclear, like in *Where was the event held?*.
- Annotations do not support the query (21 queries, 13.5%): Although the QUEST was able to answer *Who was the runner up at the 1950 US Open?*, it couldn't answer *Who was the second place in the 2015 US Open?* because the semantics of *second* was not yet defined in schema annotation file.

Most of the errors in pattern matching come form incorrectly determining the phrase boundaries. The reason for this type or error is the fact the nominal phrases may be complex by using relative clauses, or by interleaving operators. A percentage of these errors, under 20%, is due to the fact that the relationship between patterns - like in resolving ellipsis or connecting the WH-head to the left most pattern - is wrongly obtained. However, overall the pattern matching approach corrects the imprecise output of the parser for prepositional attachment (*Prep*) and for relative clause attachment (*Rel*). Table 4 shows that the parser, PR column, resolved correctly 48 preposition attachment out of 87, and the pattern matching, PT column, corrected another 29 for a gain in accuracy, Acc Gain column, of 31% and 24% for relative clause attachment.

Table 4. Correction for parser's output on sales.

	Occ	PR	PT	Acc gain
Prep	87	48	27	31%
Rel	33	16	8	24%

These are all issues that we plan to prioritize and work on. Some of them do not need to implement in the present framework, but for a context clarification, we plan to develop a whole dialog component.

6   Related Work

There is a huge literature on NLIDB. We are just pointing here some general trends, being aware that a recognition of merits would require a large space.

Initial systems developed on the basis of hand written rules, were very accurate and able to resolve complicated sentences, but lack of coverage. Basically they could respond to a very limited (type) of natural language questions. The problem relating to the complexity of logical quantifiers and their scope was very well studied and interesting solutions, not generalizable though, were found. The literature [17] provides an excellent overview of the state of the art before the 1990's.

The focus of research in the nineties was on less NLIDB type of problems. But important steps were made [18] for a relevant survey. It was recognized that the current system back then "may use some form of English" to interact with users.

The paper [19] adopts a pragmatic thinking, by pointing out that the existence of very complex natural language queries does not preclude building automatic systems that are able to process simple queries. The recent developments in deep learning algorithms seem to bring back into the foreground the NLIDBs. However, there are a lot of achievements in between. Working on related topics, such as learning inferences over constrained outputs, global inference in natural languages [20,21], open domain relation extraction [22], or more recently, resolving algebraic problems expressed in natural language [23], semantic parser [14] provide a strong ground for approaching the NLIDB by taking advantage of the power of semi-supervised and unsupervised techniques. Some papers, among which [24,25] have directly approached the NLIDB problem. However, there is no explicit reference to CLP questions, because the evaluations were carried out in an undifferentiated manner and the examples described were not nested SQL questions.

7 Conclusion and Further Research

In this paper we present a complete system QUEST to an old open problem in NLP, namely natural language interface to databases. QUEST was designed and indeed managed to get a significant improvement in handling of complex linguistic phenomena questions. As such, QUEST is an effective solution that overcomes difficulties that have constituted a major bottleneck in the development of these systems.

We want to improve further the linguistics knowledge model and to improve further the accuracy of the system. We plan to improve the process of generating syntactic patterns and to increase the power of the system in dealing with combinations of CLPs. In particular, we focus on the structure of the sub-queries resulted from the decomposition of a CLP questions. We have identified a dozen of combinations of CLP, to which a particular structure of sub-questions corresponds and we will implement these structures in the future.

References

1. Bollacker, K.D., Evans, C., Paritosh, P., Sturge, T., Taylor, J.: Freebase: a collaboratively created graph database for structuring human knowledge. In: SIGMOD (2008)
2. Auer, S., Bizer, C., Kobilarov, G., Lehmann, J., Cyganiak, R., Ives, Z.: DBpedia: a nucleus for a web of open data. In: Aberer, K., et al. (eds.) ASWC/ISWC -2007. LNCS, vol. 4825, pp. 722–735. Springer, Heidelberg (2007). https://doi.org/10.1007/978-3-540-76298-0_52
3. Kwiatkowski, T., Zettlemoyer, L.S., Goldwater, S., Steedman, M.: Lexical generalization in CCG grammar induction for semantic parsing. In: ACL, pp. 1512–1523 (2011)
4. Berant, J., Chou, A., Frostig, R., Liang, P.: Semantic parsing on freebase from question-answer pairs. In: EMNLP (2013)
5. Berant, J., Liang, P.: Semantic parsing via paraphrasing. In: ACL (2014)
6. Reddy, S., Lapata, M., Steedman, M.: Large-scale semantic parsing without question-answer pairs. Trans. Assoc. Comput. Linguist. 2, 377–392 (2014)
7. Pustejovsky, J., Hanks, P., Rumshisky, A.: Automated induction of sense in context. In: Proceedings of the 20th International Conference on Computational Linguistics, Association for Computational Linguistics, p. 924 (2004)
8. Popescu, O., Tonelli, S., Pianta, E.: IRST-BP: preposition disambiguation based on chain clarifying relationships contexts. In: Proceedings of the 4th International Workshop on Semantic Evaluations, Association for Computational Linguistics, pp. 191–194 (2007)
9. Popescu, O.: Regular patterns-probably approximately correct language model. In: Proceedings of the Joint Symposium on Semantic Processing. Textual Inference and Structures in Corpora, p. 12 (2013)
10. Kawahara, D., Peterson, D., Popescu, O., Palmer, M.: Inducing example-based semantic frames from a massive amount of verb uses. In: Proceedings of the 14th Conference of the European Chapter of the Association for Computational Linguistics, pp. 58–67 (2014)
11. Popescu, O., Hanks, P., Jezek, E., Kawahara, D.: Corpus patterns for semantic processing. In: Tutorials, pp. 12–15 (2015)
12. Collobert, R., Weston, J., Bottou, L., Karlen, M., Kavukcuoglu, K., Kuksa, P.: Natural language processing (almost) from scratch. J. Mach. Learn. Res. 12(Aug), 2493–2537 (2011)
13. Marcus, M., et al.: The penn treebank: annotating predicate argument structure. In: Proceedings of the workshop on Human Language Technology, Association for Computational Linguistics, pp. 114–119 (1994)
14. Wang, Y., Berant, J., Liang, P.: Building a semantic parser overnight. In: ACL (2015)
15. Dong, L., Lapata, M.: Language to logical form with neural attention. In: Proceedings of the 54th Annual Meeting of the Association for Computational Linguistics, ACL 2016, 7–12 August 2016, Berlin, Germany, vol. 1: Long Papers (2016)
16. Zelle, J.M., Mooney, R.J.: Learning to parse database queries using inductive logic programming. In: AAAI, pp. 1050–1055 (1996)
17. Copestake, A., Spärck Jones, K.: Inference in a natural language front end for databases. Technical report, University of Cambridge, Computer Laboratory (1989)

18. Androutsopoulos, I., Ritchie, G.D., Thanisch, P.: Natural language interfaces to databases - an introduction. Nat. Lang. Eng. **1**(1), 29–81 (1995)

19. Popescu, O.: Learning corpus patterns using finite state automata. In: Proceedings of the 10th International Conference on Computational Semantics, pp. 191–203 (2013)

20. Roth, D., Yih, W.: A linear programming formulation for global inference in natural language tasks. In: HLT-NAACL, pp. 1–8 (2004)

21. Punyakanok, V., Roth, D., Yih, W., Zimak, D.: Learning and inference over constrained output. In: IJCAI, vol. 5, pp. 1124–1129 (2005)

22. Fader, A., Soderland, S., Etzioni, O.: Identifying relations for open information extraction. In: EMNLP (2011)

23. Koncel-Kedziorski, R., Hajishirzi, H., Sabharwal, A., Etzioni, O., Ang, S.D.: Parsing algebraic word problems into equations. TACL **3**, 585–597 (2015)

24. Condoravdi, C., Richardson, K., Sikka, V., Suenbuel, A., Waldinger, R.: Natural language access to data: it takes common sense! In: 2015 AAAI Spring Symposium Series (2015)

25. Li, F., Jagadish, H.: Constructing an interactive natural language interface for relational databases. Proc. VLDB Endow. **8**(1), 73–84 (2014)

Approximate Outputs of Accelerated Turing Machines Closest to Their Halting Point

Sebastien Mambou[1], Ondrej Krejcar[1(✉)], and Ali Selamat[1,2]

[1] Center for Basic and Applied Research, Faculty of Informatics
and Management, University of Hradec Kralove, Rokitanskeho 62,
500 03 Hradec Kralove, Czech Republic
{jean.mambou, ondrej.krejcar}@uhk.cz
[2] Malaysia Japan International Institute of Technology (MJIIT),
Universiti Teknologi Malaysia, Jalan Sultan Yahya Petra,
Kuala Lumpur, Malaysia
aselamat@utm.my

Abstract. The accelerated Turing machine (ATM) which can compute super-tasks are devices with the same computational structure as Turing machines (TM) and they are also defined as the work-horse of hypercomputation. Is the final output of the ATM can be produced at the halting state? We supported our analysis by reasoning on Thomson's paradox and by looking closely the result of the Twin Prime conjecture. We make sure to avoid unnecessary discussion on the infinite amount of space used by the machine or considering Thomson's lamp machine, on the difficulty of specifying a machine's outcome. Further-more, it's important for us that a clear definition counterpart for ATMs of the non-halting/halting dichotomy for classical Turing must be introduced. Con-sidering a machine which has run for a countably infinite number of steps, this paper addresses the issue of defining the output of a machine close or at the halting point.

Keywords: Thomson's paradox · Super-task · Halting problem ·
Ultrafilter accepting computations · Accelerating turing machines ·
Non-standard output concepts

1 Introduction

A possible primary setting of a Turing machine (TM) can be to inscribe on its tape, some sequence of binary digits. This is the input, the data upon which a TM is to operate, encoded in binary form. Call this number the machine's data number. For the most part, the data number will alter from run to run of the device. One may speak of the machine being set in motion bearing such-and-such a data number. Undoubtedly, an apparatus capable of "solving the Halting problem" can inform us, concerning a given Turing machine, whether this machine would halt or not when it was set in motion with a given data number. The halting theorem indicates that no Turing machine can compute the value of the halting function for each pair of integers x, y. Despite the theorem, an Accelerated Turing Machine (ATM) can produce the values of the halting function.

© Springer Nature Switzerland AG 2019
N. T. Nguyen et al. (Eds.): ACIIDS 2019, LNAI 11431, pp. 702–713, 2019.
https://doi.org/10.1007/978-3-030-14799-0_60

Can the final output of the ATM be produced at or close to the halting state? The resolution of this interrogation is the motivation of this paper. Indeed, the Accelerated Turing Machine can be defined as a Turing machine that performs the second primitive operation requested by the program twice as fast as the first time, the third time, and three times faster [1]. For the goal of simplicity and clarity, we will refer to ordinary Turing machines with single working/Input tape and output tape which will therefore maps each input, to a sequence of configuration of the machine, where (q, z) is a description at any point of time (q is a description of the position of the heads, machine state and content of the working/input tape; and z is a description of the content of the output tape). An ordinary description of a specific machine M accordingly gives rise to a mapping:

$$M : x \rightarrow \left(q_n^x, z_n^x\right)_n^\infty$$

Subject to the usual rules for operation of a Turing Machine, including that q_1^x has the machine in the initial state and x on the input/working tape etc., and the values $x, z, q \in \mathbb{N}$. We exempt, without loss of generality, with the designation of a specific set of accepting states of the machine, since a single accepting state can always be introduced artificially without changing the input/output (I/O) behavior of the machine for conventionally halting computations. It is good to mention that every succession of prompt description

$$\left(q_n^x, z_n^x\right)_n^\infty$$

After computation on the input x, lead to Output (run of the machine):

$$\lim_{n \to \infty} z_n^x$$

Our research does not specify any new feature of the machine, but rather proposes a definition of the output at very close halting computations. In the same way, it is akin to the infinite time Turing machines [2], despite very close similarities our approach differs. Based on ultra-filters Comfort and filters [3], we attempt to respectively put the defined output of a non-halting machine (a kind of machine in infinite time) and the halting of classical Turing machines within similar frameworks. We call it "hyper-computation" in the same sense that any super-task machine (e.g. the infinite-time machines of Seabold and Hamkins [4] is. Obviously to propose a (pseudo-)physical model of our machines, any of the speculative models for realizing super-task machines (e.g. based on Relativity Theory) could be used.

2 Related Work

In these days of ultrafast computers whose speed seems to increase without limit, the most philosophical of us are perhaps pushed to ask: what could we calculate with a computer infinitely fast? By proposing a natural model for super-tasks, calculations involving infinite stages, the authors in [2] provided a theoretical basis on which to answer this question. Also, they gave an in light on one common problem face in classical computation theory which is the "halting problem". This last, is defined as the sets

$$H = \{(p,x)|p \ halts \ on \ input \ x\}$$

or in another words

$$h = \{p|p \ halts \ on \ input \ 0\}.$$

Moreover, the authors in [5] explicitly characterize the decidable predicates on the integers of infinite time Turing machines, in terms of eligibility theory and constructible hierarchy. They do this by defining delta, the least ordinal and not the length of any eventual output of a Turing machine in infinite time (Halt or otherwise); using this, the infinite duration degrees were considered, and it was shown how the jump operator coincides with the generation of master codes for the constructible hierarchy.

In addition, the author [6] shows that there is a dependency between the computability of algorithmic complexity and the decidability of different algorithmic problems. He mentioned that the computability of the algorithmic complexity C(x) is equivalent to the decidability of the Halting problem for Turing machines. He extended this result to the domain of super-recursive algorithms, considering the algorithmic complexity of inductive Turing machines. He studied two types of algorithmic complexity: recursive (classical) and inductive algorithmic complexities. The relationships between these types of algorithmic complexity and the decidability of algorithmic problems for Turing machines and inductive Turing machines were examined. Another author [7] has elaborated on Turing assemblers which are Turing machines that operate on n-dimensional bands with restrictions that characterize an assembly procedure rather than computation and that are designed as an abstraction of some algorithmic processes of molecular biology. It has been demonstrated previously that Turing assemblers with n-dimensional bands can simulate arbitrary Turing machines for any n > 1. They showed in their paper that for n = 1, even non-deterministic Turing assemblers have a computation capacity distinctly reduced, unable to assemble only regular sets. The Halting problem for linear Turing assemblers can, therefore, be solved algorithmically, and characterization of the set of feasible final assemblies was presented in the form of a subclass of context less languages.

In the paper [8], the authors present a detailed and enlightening discussion on "accelerated Turing machines" (ATMs), machines that take exponentially decreasing intervals for each of their discrete computationally consecutive steps, and can, therefore, complete an infinite series of such steps in a finite period. In particular, the authors consider whether it can be said that Accelerated machines like ATMs calculate non-calculable functions, such as the halt function. Furthermore, they mentioned ATM paradox: a Turing machine derives functions that have proven to be beyond the computational capabilities of any Turing machine. Similarly, the author of [9] considers some models of "hypercomputation" capable to address the halting problem, for example Zeno machines which are Turing machine with accelerated clock. Also, in

[10] the authors point the fact that Alan Turing introduced a specific imaginary machine which can take input representing various mathematical objects and process it with some precise steps until a value is obtained as final output. In addition, the authors mentioned that Turing through his research found a machine of type "hypercomputation" corresponding to many of the traditional mathematical algorithms and provided convincing reasons to believe that any mathematical procedure that was precise enough to be recognized as an algorithm or "effective procedure" would have a corresponding Turing machine. They also referred to Hypercomputation as hyper machines which are, like the Turing machine, theoretical machines using abstract resources to manipulate abstract objects such as symbols or numbers.

3 Hypercomputation Illustrate by Accelerated Turing Machine

Let assume that in one way or another, a Turing machine (M) was able to achieve an infinite number of steps (considerate in our point of view as a finite amount of time) and then emit a signal including the answer to a yes/no question. This kind of Accelerated Turing Machine (ATM) would be able to provide a solution to innumerable questions and can be used as support for several models of relativistic space-time [11]. An ATM could easily solve the latter conjecture as an assertion where every even number larger than two is the sum of two primes. The accelerating machine would operate through this algorithm (Simplifying matters, we could assume the functions "sumOfTwoPrimes", "Exit", "time" and "overallTime" which return respectively True if there are two prime number equal to $2(n + 1)$ when they are added, or it returns False; Exit from the loop when an exception occurs; the duration of the iterations and the total time spend for those iterations; and Minutes (min) uses here as unity of time):

```
Algorithm:
Set n:=1;
while time =< 1 min do:
    write "1" on the designated output square the symbol 1
(TRUE);
End while;
While time =< 1/2^n and overallTime =< 1 min do:
    If sumOfTwoPrimes(2(n+1)) <> True then:
    write "0" in the output tap at the actual position of
the head;
    Exit;
    End if;
n = n+1;
End While;
```

One way or another, the machine would complete its task within 2 min, providing the truth-value of Goldbach's conjecture [12, 13]. However, without a boundary of time in our algorithm, a conceptual difficulty could rise for confirmation of Goldbach's conjecture, as the absence of a signal from this machine, requires the machine to use an infinite amount of space on the working tape. It is important to recall that any Turing Machine (TM) will eventually:

1. halt,
2. go into a loop after enumerable computation, or
3. use an infinite amount of space on the tap.

It has been proved by Turing [14, 15] that an ordinary TM cannot compute a halting function H (x, y), a reason why we pay particular attention to ATM which characterizes the halting states of the set of TM. Two values 0 and 1 are expected respectively, 0 if the machine never halts and 1 if providing the input y to the machine on index $\{x|$ x belongs to set of TMs$\}$, it halts. Considering M_1 as that ATM, it will simulate the operation by applying the abstract algorithm bellow (Simplifying matters, we could assume the function "time", "index" which respectively return the duration of the computation, the current index of the machine M_1 and "haltingState" which checks if the machine halts after countable infinite iteration):

```
Algorithm:
set x and y;
write "0" in design output square;
while time>=2min do:
if index(M₁) == x and input == y then:
  if haltingState == True then:
  write "1" in design output square;
  exit;
  end if;
end if;
end while;
```

The algorithm can be explained as follow: initially, our machine M_1 will print 0 in the designated output scare. Since M_1 operates as a universal TM, it will process similar to the actions of the xth TM operating on input y. In the same idea, the actions of M_1 will be executed in speed up manner, in such a way that the halting state of M_1 will be verified (if it reached to halting state or not) after each operation. The machine M_1 will halt and 0 will be replaced by 1 at the output square as soon as M_1 enters in his halting state otherwise it will, run infinitely with 0 at the output square. Furthermore, as the time constraint is put in the algorithm, the device which operates on input y at index "x" will tell after 2 min whether the xth TM halts or not. Next section will emphasize on the paradox which resides on the determination of the current state of the machine at the end of the computation of a super-task thus, without boundary of time.

4 Thomson's Paradox

In his paper, Thomson maintained that there are "reasons for supporting that super-tasks are not possible of performance". The essence of his point is that we cannot design a comprehensible method which will determine the state of the machine after the completion of the super-task. He refers in his main example to a reading lamp but by analogy, we can easily extend it to an ATM. Considering an accelerated Turing Machine (M_2) able to compute the partial sum of an infinite series: $1, -1, 1, -1, \ldots$. An output 1 is print out at the end of the first period, which lasts 1 min. another change of 1 to 0 occurred at the second period which lasts ½ min. Instead of 1 at the third period which lasts ¼ min, 0 is printed out, and so ahead, infinitely. As summary, M_2 oscillates between two values 0 and 1. After completing its super-task, what would be insert in the tape after 2 min? 0 cannot be that output as M_2 always prints 1 immediately after. Neither 1 can be printout as it always prints 0 after. Even though, we might think that 0 or 1 will be printout after 2 min. 0 or 1 cannot be printed. This is the paradoxical situation, so what is the way out? Considering a diverging series, it is naïve to believe that there is a non-halting machine able to compute a partial sum. This let us see the necessity of a clear explanation about the concept of super-task.

Considering the logical setup of a machine introduced later by Thompson. It is now possible for those 2 min of super-task, to tell what the actions of the machine M_2 would be. However, those actions refer to the period of the temporal segment [0, 2[but nothing is specified at exactly 2 min, thus the function might lose its continuity at 2 or conserves it as well. Hence, all value at 2 would be coherent with the previous specifications. Similarly, after 2 min, the machine M_2 might print out on the tap value 0, 1, 17 or nothing at all. In addition, there is a consistency between each printout and the specification of the machine.

5 Use of Frechet Filter's Ordinary Accepting Computations (OAC)

Given the input x, we have described a conventional halting computation on a Turing machine M as the sequence

$$\left(q_n^x, z_n^x\right)_n^{\infty}$$

Which is probably constant. We will further emphasize on this concept of **eventually (probably) constant**. It is good to precise that the sequence is completely characterized by the rules that determined the machine M and the input x.

It is also important to recall the definition of a filter, which is in mathematics a non-void family of subsets of X (where X is the variable on which we apply the filter), not containing the empty set, and closed with respect to the taking of finite intersections and the forming of supersets. An example can be, the Frechet filter

$$F_r = \{L \sqsubseteq \mathbb{N} | \mathbb{N} \backslash L \text{ is finite}\}$$

Is composed of, all co-finite subsets of the natural numbers N, where $0 \in \mathbb{N}$, as customary in Computer science.

Definition 1: Considering some elements of Frechet filter, a sequence

$$\left(q_n^x, z_n^x\right)_n^\infty$$

Which is eventually constant, can be considerate as an accepting computation.

By avoiding the loss of generality, we have relinquished the notion of explicitly defined accepting state. Considering a Turing Machine (t) with an accepting computation as defined in Definition 1 and with an explicitly defined accepting state that certainly satisfied the Definition 1, we can reformulate it so that it contends an explicitly defined accepting state. Furthermore, by looking at filters larger than F and by considering Definition 1, we can extend the notion of an accepting computation as shown in the next section.

6   Application of an Ultrafilter $F_r \subseteq Z$ for the Accepting Computations

6.1   Avoid Thomson's Lamp in the Approach

It is good to remember our goal which is to determine an output close or at the halting point, for that our approach has been since to find first the set of accepting computation that a given Accelerate Turing Machine can support.

As mentioned in the previous section, F_r is a Frechet filter. Further, we assume that

$$\{\exists G | F_r \in G \text{ and } \emptyset \notin G\}$$

Where G is infinite subsets of N which included all co-finite sets. To increase the number of runs of a Turing machine (M) which eventually computes something using elements of G instead of F, we redesign a new accepting computation.

Definition 2: Considering some element of G represented by E(G), a sequence $\left(q_n^x, z_n^x\right)_n^\infty$ which is constant on E(G), is qualified as a G-accepting computation.

This definition is very interesting when we consider Z-accepting computations, where Z is an ultrafilter. It is good to remember that an Ultra-filter (in this case Frechet filter F_r) on Y is a filter with the property that for each

$A \subseteq$ Y, either A or A^c (*complement of A*) $\in F_r$. The general truth of choice insures that $\{\exists Z \text{ } applied \text{ } on \text{ } \mathbb{N} |$ $F_r \in \mathbb{N}\}$ where Z is an ultrafilter. Further, it is good to mention that the subsets belonging to a precise filter are most of the time seen as the large sets. In the same idea, we can say that the F_r-large sets represent the co-finite subsets of $\mathbb{N}$ and Z-large sets can be defined as the natural generalization of the co-finite sets.

To keep the order of the idea expressed, let's resume few keys words:

- *Z-accepting computation* is when we have a Z-large set of points at one point in time when the machine state, as well as the output tape content, remain constant.
- *Accepting computation* is when there exists a F_r-large set of points at one point of time when the machine state as well as the output tape content remain constant.

Case study: Avoid a Thomson's lamp by Turing machine

Consider a machine M_{TL} with alphabet $\{0, 1\}$ which has a minimal number of states and at time t writes

$$\frac{1 - (-1)^t}{2}$$

To the first position of the output tab. considering the properties of ultrafilter where either the set of even or point the set of odd points in time, belongs to Z, we can clearly see that the output machine is a Thomson's lamb but any computation done by this machine is a Z-accepting computation. Furthermore, M_{TL} is with respect to Z-accepting computation, equivalent to a machine outputting a constant bit on the tape (where 1 or 0 is that bit, depending on the filter Z). So, by applying the principle of an ultrafilter, we can assign with a lot of reserves an output to a machine M_{TL} when it performs numerous infinite computation. By assuming that the Goldbach conjecture machine is considered to have a valid output, we are deeply convinced that it should be the same for the Thomson lamp machine.

6.2 Contrast of Our Approach with the Infinite Machine Framework of Hamkins and Lewis

We can resume on the Table 1 below the keys deference among the two methods.

Table 1. Keys deference among two frameworks

	Our infinite machine framework	The infinite machine framework of Hamkins and Lewis (2000)
State of the machine	Z-accepting computation	Define at the first, limit ordinal ω
Primary output in the tap	1 or 0 based on the choice of Z	+1 *if* lim sup *is chosen as limit operator for each cell*

Considering a slightly Accelerated Turing Lamp machine (M_{ATL}) with alphabet {0, 1} which has a minimal number of states and at time t writes

$$\frac{1 - (-1)^{t+1}}{2}$$

To the first position of the output tab. This machine is the 'Mirror' of the Turing Lamp machine (M_{TL}), always have the exact opposite on the tape if the machines are run concurrently. In other words, depending on the choice of Z, the 'output' of M_{ATL}, is either 0 or 1. So in infinite time, both machine M_{ATL} and M_{TL} have opposite output when the run concurrently and reach to the same state at ordinary ω close to halting state.

6.3   An Ultrafilter-Based Acceptance Notion Using the Output Tape

We are introducing here the notion of "limit computable".

Definition 3: A function g is limit computable allowed there exists a Turing machine M such that for each x, the run

$$\left(q_n^x, z_n^x\right)_n^\infty$$

of M on x has $\left(z_n^x\right)_n^\infty$ constant on a set of $G \in F_r$. Considering that M is run on input x, $g(x)$ is the limit of the content of the output tab with no constant constraint on $\left(q_n^x\right)_n^\infty$.

Definition 4: Considering each sequence of $\{A_t \in \mathbb{N}\}$, we set $(A_t)_Z = \{(B_t) | \{w | A_w = B_w\} \in Z\}$ which is the equivalence class of all sequences that agree with A_t on a Z-large set of indices. Furthermore, we denote H^* as the equivalence class of the unary sequence having H as the unique element. Based on the previous discussion, we can derive this result: if $\forall H \in \mathbb{N}, (A_t)_Z = H^* \Rightarrow \{(A_t)_Z$ is finite}.

Definition 5: Given a run $\left(q_n^x, z_n^x\right)_n^\infty$ of some Accelerate Turing machine (ATM), we define:

- The non-standard output of the machine, corresponding to input x to be the equivalence class $\left(z_n^x\right)_n^\infty$ and
- The non-standard terminal internal configuration of the machine, corresponding to input $x = \left(q_n^x\right)_n^\infty$.

Definition 4 favorites the following straight-forward observation.

7 Experiment

Considering Thomson's lamp paradox [1] and the Thompson's convergence, we try to determine the state of the light at 2 min (on or off). We tried as many researchers to solve it only in mathematics way. We first focus our attention on finding the number of time the state of the lamp has been changing during the two minutes of the experiments but with soon release the complexity to determine (n) in the Thomson's lamp equation:

$$\sum_{i=0}^{n}\left(\frac{1}{2^i}\right) = 2 \tag{1}$$

$$\frac{1 - \left(\frac{1}{2}\right)^{n+1}}{1 - \frac{1}{2}} = 2 \tag{2}$$

$$\left(\frac{1}{2}\right)^{n+1} = 0 \tag{3}$$

As the total time of the experiment is 2 min. Equation (3) let us understand that (n) is close to ∞ value. This makes the determination of the state of the lamp nearly impossible in this way.

We propose another point of View for our experiment. We would like to base our analyst by adding another parameter the time spend by the electrons to go through the Tungsten filament (Fig. 1).

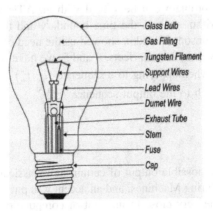

Fig. 1. Part of a bulb

Base on the argumentations [16, 17] and the assumption that the distance between the electric source and the lamp is less than 1 miles. We know that the speed S miles/s of the electrons is:

$$S > \frac{1}{180000} \tag{4}$$

Discussion 1: Considering the visible rate of the human eye which is 1000 frames/s and the part of the Thomson's lamp statement "0 cannot be that output as M_2 always prints 1 immediately after. Neither 1 can be printout as it always prints 0 after" due to the small interval close to 0 s. The Eq. (4), let us understand that the lamp will appear to the observer as it is at the state "ON".

It is also good to mention that the electricity in a wire acts as a vast chain of electrons close to each other and the close/open of the circuit at a rate extremely small can be seen as "no interruption of the flow of electron".

It seems clear that following this idea, the state of the lamp will be "ON" at *two min*.

Proposition: Any classically accepting run of a Turing machine has a finite terminal state and has finite output.

Discussion 2: Let's recall that the Accelerated Turing machine (ATM) which can compute super-tasks are devices with the same computational structure as Turing Machines (TM) reason why through the Definitions 3 to 5, we have explored several Turing Machine outputs obtained when a sequence is processed through countable infinite steps. In Sect. 2, Hypercomputation has been introduced and we saw that Goldbach's conjecture problem can be solved with an ATM by introducing further constraints in the algorithm such as the time boundary and the exit value 0 or 1. In addition to that, the Thomson's paradox showed us the need of a hypercomputation to solve complex tax. Never the last, in Sects. 4 and 5, we have introduced the notion of halting function which when applying to a sequence $\left(q_n^x, z_n^x\right)_n^\infty$ as a configuration of a Machine M, we can reach to the output sequence $\left(z_n^x\right)_n^\infty$.

8   Conclusion

The paper has explored possible output of certain non-classical computation schemes based on Accelerated Turing Machines, and an accent was put on the output, as well as the state of the machine very close to its limit of computation. Furthermore, several computations problems as the Goldbach's conjecture and the Thomson's paradox have been used to support the argumentation. The problematic exposed at the introduction was addressed and base on the two discussions elaborated in this paper, the output sequence $\left(z_n^x\right)_n^\infty$ can be computed. Consequently, the Accelerated Turing Machine can solve a complex problem by giving output (decision) even close to its limit of computation assimilate here as the halting point. Besides, we have seen in recent publication, an interest to the field of machine learning. One invent derivation of Turing machine was presented in [18] and similar to that, our future work will take advantage of the power of Accelerated Turing Machines, and we will see how to combine it to the state-of-the-art implementation in machine learning such as the models described in [19] and [20].

Acknowledgement. The work and the contribution were supported by the SPEV project "Smart Solutions in Ubiquitous Computing Environments", 2019, University of Hradec Kralove, Faculty of Informatics and Management, Czech Republic.

References

1. Copeland, B.: Accelerating turing machines. Mind. Mach. **12**(2), 281–300 (2002)
2. Hamkins, J.: Infinite time turing machines. Mind. Mach. **12**(4), 521–539 (2002)
3. Comfort, W.: Ultrafilters: some old and some new results. Bull. Am. Math. Logic Q. **83**, 417–456 (1977)
4. Hamkins, J.D., Seabold, D.E.: Infinite time turing machines with only one tape. Math. Log. Q. **47**, 271–287 (2001). https://doi.org/10.1002/1521-3870(200105)47:2<271::AID-MAL Q271>3.0.CO;2-6
5. Welch, P.: Eventually infinite time turing machine degrees: infinite time decidable reals. J. Symbolic Logic **65**(03), 1193–1203 (2000)
6. Burgin, M.: Algorithmic complexity as a criterion of unsolvability. Theor. Comput. Sci. **383** (2), 244–259 (2007)
7. Baer, R., Leeuwen, J.: The halting problem for linear turing assemblers. J. Comput. Syst. Sci. **13**(2), 119–135 (1976)
8. Copeland, B., Shagrir, O.: Do accelerating turing machines compute the uncomputable. Mind. Mach. **21**(2), 221–239 (2011)
9. Potgieter, P.: Zeno machines and hypercomputation. Theor. Comput. Sci. **358**(1), 23–33 (2006)
10. Ord, T.: The many forms of hypercomputation. Appl. Math. Comput. **178**(1), 143–153 (2006)
11. Hogarth, M.: Does general relativity allow an observer to view an eternity in a finite time? Found. Phys. Lett. **5**, 173–181 (1992)
12. Shagrir, O.: Super-tasks, accelerating turing machines and uncomputability. Theor. Comput. Sci. **317**(1), 105–114 (2004)
13. Pitowsky, I.: The physical church thesis and physical computational complexity. Iyyun: Jerusalem Philos. Q. **39**, 81–99 (1990)
14. Turing, A.: On Computable Numbers, with an Application to. http://www.cs.virginia.edu/~robins/Turing_Paper_1936.pdf
15. Turing, A.: On computable numbers, with an application to the "Entscheidungsproblem". Proc. Lond. Math. Soc. **42**, 230–265 (1937)
16. How Light Bulbs Work. In: HowStuffWorks. https://home.howstuffworks.com/light-bulb.htm
17. Circuits and the Speed of Light. In: allaboutcircuits. https://www.allaboutcircuits.com/textbook/alternating-current/chpt-14/circuits-and-the-speed-of-light/
18. Graves, A., Wayne, G., Danihelka, I.: Neural turing machines. arXiv: Neural and Evolutionary Computing (2014)
19. Mambou, S., Krejcar, O., Kuca, K., Selamat, A.: Novel cross-view human action model recognition based on the powerful view-invariant features technique. Future Internet **10**(9), 89 (2018)
20. Mambou, S., Maresova, P., Krejcar, O., Selamat, A., Kuca, K.: Breast cancer detection using infrared thermal imaging and a deep learning model. Sensors **18**, 2799 (2018)

Development of Seawater Temperature Announcement System for Improving Productivity of Fishery Industry

Yu Agusa[✉], Takuya Fujihashi, Keiichi Endo, Hisayasu Kuroda, and Shinya Kobayashi

Graduate School of Science and Engineering, Ehime University, Matsuyama, Japan
agusa@koblab.cs.ehime-u.ac.jp

Abstract. Real-time visualization of seawater temperature information is strongly desired by aquaculture fishermen because water temperature fluctuation caused by tide inflow greatly affects fish and shellfish farming. In this research, we developed an announcement system for seawater temperature based on the demands from fishery researches and fishery workers. This system visualizes seawater temperature in Uwa Sea and conveys seawater temperature information to fishermen. The developed system consists of two sub-systems: seawater temperature accumulation and seawater temperature visualization. In the first system, the seawater temperature data measured by marine buoys deployed on the sea is stored on the server. In the second system, the latest and past seawater temperature can be visualized by using tables, graphs, and the three-dimensional (3D) map to provide the tendency of seawater temperature variation over time to users, i.e., fishery researches and fishery workers. In addition, the past seawater temperature information at arbitrary buoys and dates can be downloaded by the users. From experimental evaluations, we discuss the effectiveness of our developed system for users.

Keywords: IoT · Big data · Visualization · Information system

1 Introduction

Seawater temperature is very important in the sea area's aquaculture fishery. There are two reasons. First reason is there is a water temperature zone where fish diseases that cause great damage to aquaculture fishery are likely to occur. For fish weakened by the onset of fish diseases, it is necessary to control the feeding amount. If the fish eat the feed more than necessary, it leads to worsening of the health condition of the fish and disease death. However, if it is possible to know the water temperature at the site, it will be an indicator to know the health condition of the fish and it will be possible to suppress damage caused by fish diseases. Second reason is it is possible to grasp the current of the tide by

© Springer Nature Switzerland AG 2019
N. T. Nguyen et al. (Eds.): ACIIDS 2019, LNAI 11431, pp. 714–725, 2019.
https://doi.org/10.1007/978-3-030-14799-0_61

knowing the temperature of multiple depths. The degree of the influence of the tidal current on the aquaculture fishery depends on the time, the thickness and the range of the inflow, and it is necessary to adjust the feeding amount according to the tidal current situation. If it is possible to know the water temperature at the site, it is possible to respond flexibly.

The seawater temperature information is important for the fishery research because the red tide occurrence can be predicted by the tidal current obtained from the water temperature variation over time. The red tide causes a significant damage on the fishery.

We deal with the issue of water temperature information announcement in the aquaculture fishery in Ehime prefecture in this paper.

We describe the situation before this research started, from the viewpoint of fisherman and fishery researcher. Prior to the start of this research, fishermen had no means of knowing the water temperature of their own fishing ground, predicting the water temperature from the temperature at that point. However, in Ehime prefecture, especially Uwa Sea, the actual water temperature was greatly different from the prediction of the fishermen because warm tides and cold tides flow in Uwa Sea [1].

In response to this problem, fishery researchers installed five water temperature continuous observation devices called marine buoys on the sea, and measured and accumulated the water temperature of 5 m depth every hour. However, there were problems that it was impossible to cover the entire Uwa Sea with only 5 units, and only the water temperature information of 5 m depth was collected, so it was not possible to grasp the lower tide current. In addition, method for notifying the measured water temperature information to fishermen immediate were not established.

Therefore, we built a multi-depth sensor network (15 marine buoys installed in September 2018) in Uwa Sea in this research. In addition, we developed Seawater Temperature Announcement System visualizes the collected data that collected and accumulated on this network system as current status information and variation over time.

We describe the overview of Seawater Temperature Announcement System at first in this paper. Next, we report the overview and results of the evaluation experiments that we need to do in this system construction. Finally, we report on the overview and results of the quantitative evaluation of the effect of this system intervention.

2 Our System

We develop an announcement system of seawater temperature to notify the current and past temperature to the users. Figure 1 shows the overview of our developed system. Each marine buoy measures the temperature and sends it to a mail server using e-mail. A server accumulates the temperature from multiple marine buoys and visualizes the accumulated temperatures on a web page. Each user can see the current and past temperatures by accessing the web page. The

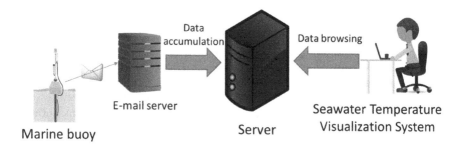

Fig. 1. Overview of our developed system

developed system consists of two sub-systems: seawater temperature accumulation and seawater temperature visualization systems. In the first sub-system, when measuring the water temperature, the data is immediately decrypted and accumulated on the server. The second sub-system is the web application that reads the data accumulated on the server by Seawater Temperature Accumulation System and displays it on the web page in the form such as tables or graphs.

3 Seawater Temperature Accumulation System

The marine buoy measures every 30 min, or hourly. After the measurement by the buoy, the e-mail written measurement data transmitted from the different send-only e-mail address for each buoy to pre-specified e-mail address.

Seawater Temperature Accumulation System decrypts the contents of the e-mail sent from the marine buoy at the timing of the reception, and extracts the data portion. When the extracted data is the data measured the first time after the buoy was installed, this system make a text file named the same as the sender's e-mail address on the server and write the data in the file. Otherwise, this system add data to the text file named the same as the sender's e-mail address on the server. With the above method, the data is accumulated on the server.

4 Seawater Temperature Visualization System

4.1 Overview

Fisheries researchers and fishermen can browse the data accumulated on the server using both sub-systems. To satisfy all the above-mentioned requirements, we realize the announcement system as an Web application because it can provide seawater temperature information regardless of platforms. When this system is started (accessed), the data stored on the server is read. Thereafter, this system create dynamic tables, graphs and the like from the read data and display it on the Web page. We implemented read processing by using PHP and the processing to create dynamic tables and graphs by using JavaScript.

4.2   Requirements

There are two requirements in our system based on hearing from fishermen and fishery researchers. Requirement from both fishermen and fishery researchers:

- seawater temperature should be displayed under the consideration of spatial spread of marine area.
- The visualized temperatures should be easy to understand with simple operation and displayed with short computation time.
- The current information and past information should be prepared in different pages and the both pages can be easily switched each other.

Requirement from fishery researchers:

- The past seawater temperature should be stored in the server.
- In contrast to the fishermen, the detail temperature information should be provided to the researchers.

4.3   Functions

To satisfy the above-mentioned requirements, we implement the following functions in the developed system. Specifically, we consider the requirements from both fishermen and fishery researchers for implementing the following:

Function (a) Displaying the location of the observation point on the map
Function (b) Displaying measurement data in tabular form
Function (c) Displaying current situation of seawater temperature as a graph
Function (d) Displaying seawater temperature variation over time as a graph
Function (e) Displaying seawater temperature variation over time in three dimensions by using distribution chart

We also consider the requirements from both fishermen and fishery researchers for implementing the following:

Function (f) Saving measured data as a file in csv format to the user's terminal.

4.4   Function Details

4.4.1   Displaying the Location of the Observation Point on the Map
This system can display the position of each measurement point as a point on a map in Function (a). Before this function was implemented, fishermen were unable to grasp the location of the installed marine buoys. However, by providing this function, it became possible to know the measurement points located near the fishing grounds themselves and the installation situation of the marine buoy in real time.

Location	Fukuura	Shiokojima	Shimonada	Kitanada	Hiburijima	Shimonami	Miura	Komobuchi	Yusu	Uwajima	Yoshida	Akehama	Mikame	Yawatahama	Saijo
Update date	2018/02/16 09:00:00	2018/02/16 09:00:00	2018/02/16 09:00:00	2018/02/16 09:32:51	2018/02/16 09:00:00	2018/02/16 09:00:00	2018/02/16 09:03:25	2018/02/16 09:00:00	2018/02/16 09:00:00	2018/02/16 09:15:53	2018/02/16 09:24:38	2018/02/16 09:00:00	2018/02/16 09:00:00	2018/02/16 09:00:00	2018/02/16 09:00:33
1m	15.9°C	15.3°C	NA	NA	12.9°C	NA	NA	NA	13.7°C	12.3°C	NA	12.1°C	11.5°C	NA	8.5°C
3.5m	NA	NA	NA	NA	NA	NA	NA	NA	NA	NA	NA	NA	NA	NA	9.9°C
5m	16.0°C	15.4°C	14.6°C	13.1°C	13.0°C	13.7°C	12.1°C	13.6°C	12.3°C	11.0°C	11.9°C	12.2°C	11.5°C	11.1°C	NA
10m	15.9°C	15.4°C	14.7°C	13.1°C	12.9°C	NA	12.1°C	13.6°C	12.3°C	11.2°C	12.2°C	12.1°C	11.5°C	NA	NA
20m	15.9°C	15.3°C	14.7°C	13.2°C	12.9°C	NA	12.1°C	13.6°C	12.3°C	11.9°C	11.9°C	12.2°C	11.4°C	NA	NA
30m	15.9°C	15.2°C	NA	NA	13.0°C	NA	NA	NA	13.5°C	12.4°C	NA	12.2°C	11.3°C	NA	NA
40m	15.9°C	15.1°C	NA	NA	12.8°C	NA	NA	NA	13.3°C	12.2°C	NA	12.2°C	11.2°C	NA	NA
50m	15.8°C	15.0°C	NA	NA	12.8°C	NA	NA	NA	NA	NA	NA	NA	NA	NA	NA
60m	15.7°C	15.0°C	NA	NA	12.9°C	NA	NA	NA	NA	NA	NA	NA	NA	NA	NA

(a) "current status"

Location	Fukuura	Shiokojima	Shimonada	Kitanada	Hiburijima	Shimonami	Miura	Komobuchi	Yusu	Uwajima	Yoshida	Akehama	Mikame	Yawatahama	Saijo
1m	15.9°C	15.3°C	NA	NA	12.9°C	NA	NA	NA	13.6°C	12.3°C	NA	12.1°C	11.5°C	NA	8.5°C
3.5m	NA	NA	NA	NA	NA	NA	NA	NA	NA	NA	NA	NA	NA	NA	9.9°C
5m	16.0°C	15.3°C	14.6°C	12.9°C	13.0°C	13.7°C	12.1°C	13.5°C	12.3°C	11.0°C	12.0°C	12.2°C	11.5°C	11.0°C	NA
10m	15.9°C	15.4°C	14.7°C	13.5°C	13.0°C	NA	12.1°C	13.6°C	12.3°C	11.2°C	12.2°C	12.1°C	11.5°C	NA	NA
20m	15.9°C	15.3°C	14.7°C	13.2°C	12.9°C	NA	12.1°C	13.5°C	12.3°C	11.9°C	11.9°C	12.2°C	11.5°C	NA	NA
30m	16.0°C	15.3°C	NA	NA	12.9°C	NA	NA	NA	13.5°C	12.4°C	NA	12.2°C	11.2°C	NA	NA
40m	15.9°C	15.2°C	NA	NA	12.8°C	NA	NA	NA	13.4°C	12.2°C	NA	12.2°C	11.2°C	NA	NA
50m	15.8°C	15.0°C	NA	NA	12.8°C	NA	NA	NA	NA	NA	NA	NA	NA	NA	NA
60m	15.7°C	15.0°C	NA	NA	12.9°C	NA	NA	NA	NA	NA	NA	NA	NA	NA	NA

Measuring date : 2018/02/06 10:00:00

(b) "past status"

Fig. 2. displaying measurement data in a tabular

4.4.2　Displaying Measurement Data in Tabular Form

In Function (b), this system can display the measurement data of each measurement point in a table format as shown Fig. 2(a) in the "current status" or as shown Fig. 2(b) in the "past status". In each table, the horizontal items is the location (measurement point) and the vertical items the water depth in order to be able to grasp seawater temperature variation due to differences in latitude of the measurement point. In "past status", it is possible to measurement data of arbitrarily specified date and time by a simple operation such as sliding a slider or pressing a button. It becomes easy to grasp the current situation and past situation of seawater temperature by using this function.

4.4.3　Displaying Current Situation of Seawater Temperature as a Graph

In Function (c), as shown Fig. 3, the latest measurement data can also be displayed as a line graph in which lines are color-coded according to the water depth, with the seawater temperature as the vertical axis and the measurement point as the horizontal axis. The users can change the upper and lower limits of the seawater temperature to be displayed by sliding the slider, and the water depth to be displayed can be selected by unchecking the check box. The users can simultaneously grasp variation in seawater temperature between measurement points and variation in seawater temperature due to difference in water depth with this function.

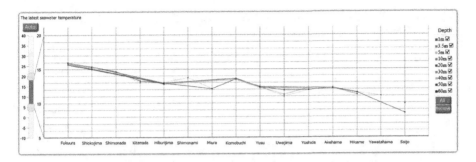

Fig. 3. Displaying the latest measurement data as a graph

4.4.4 Displaying Seawater Temperature Variation over Time as a Graph

Function (d) is the function to display the variation of the past seawater temperature graphically for each measurement point, as shown in Fig. 4. Each graph is a line graph with the seawater temperature on the vertical axis and the time on the horizontal axis because it becomes possible to grasp the variation of seawater temperature due to the passage of time and the difference of water depth at the same time. In addition, like the function (c), the color of the line on the graph is divided, and the user can select the depth of water to be displayed and change display setting of the upper and lower limits of the seawater temperature arbitrarily. Furthermore, the users can select the display period from 24 h, 48 h, 7 days, 30 days, 60 days because fishery researchers desire to grasp the seawater temperature variation over time in short-term or long-term. Selection of the display period can be switched by a simple operation by using the slider method. It is possible to visualize the time change of the seawater temperature at each depth at each measurement point with this function.

Before this function was implemented, the seawater temperature information was only known by fishery researchers involved in installation of marine buoys. However, by providing the function to display the current condition and variation over time in a table or graphs form, fishermen enable to browse the information easily. In addition, by providing a function to display dynamic tables and graphs, it becomes unnecessary to perform the work of sequentially graphing and the like from measurement data.

4.4.5 Displaying Seawater Temperature Variation over Time in Three Dimensions by Using Distribution Chart

In Function (e), the seawater temperature variation over time can be visualized in three dimensions by displaying the temperature distribution map on the plane of the specified water depth and the cross-sectional temperature distribution map of the two points specified on the map, as shown in Fig. 5. The users can change the display water depth and the display time arbitrarily. Changing the display water depth and the display time can be changed by a simple operation by using

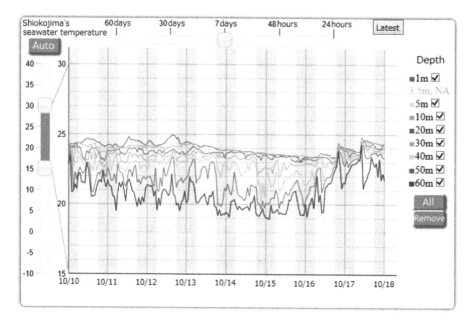

Fig. 4. Displaying seawater temperature variation over time

the slider system. In addition, we implemented a function to switch the displayed temperature distribution chart in constant time intervals (animation function) as a device to allow users to easily grasp the seawater temperature variation over time. The users can animate the temperature distribution chart with the same feeling as the video playback software and the video playback interface often seen in the Web site. Incidentally, we selected Inverse Distance Weighted (IDW) as the interpolation method on the plane and linear interpolation as the interpolation method for the depth direction, as the interpolation method of data. In the IDW, the estimated value at the point s is calculated as follows:

$$\mu(s) = \frac{\sum_{i=1}^{n} w_i(s)\mu_i}{\sum_{j=1}^{n} w_j(s)}, w_i(s) = \frac{1}{d(s, s_i)^2}$$

The symbol n means the number of measurement points, the symbol μ_i means the real value of the i-th measurement point, the function $d(s, s_i)$ means the distance between the points and the measurement point s_i, and the function $w_i(s)$ means the weight of the measurement point s_i at the point s, in the equation.

Before the function to visualize the seawater temperature variation over time in three dimensions is provided (this function is the newest function among the functions implemented in this system), the user could browse the current situation and variation over time at each measurement point. However, due to the provision of this function, fishermen have become able to grasp the seawater temperature of their fishing ground with an accuracy of about 0.5 °C error, although it is the estimation.

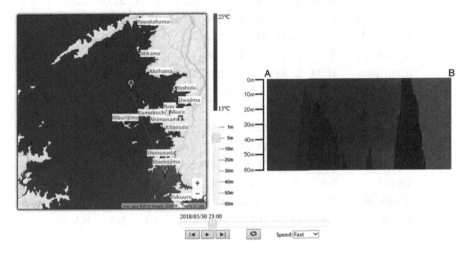

Fig. 5. Displaying seawater temperature variation over time in three dimensions

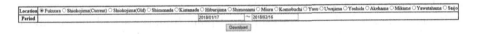

Fig. 6. Saving measured data as a file in csv format

4.4.6 Storing Past Data

As shown in Fig. 6, specifies the measurement place and the measurement period and presses the button written as download so that the measurement data within the measurement place and the measurement period is downloaded as a csv format file in Function (f). Specifically, we made it possible to designate the location by radio button method, designate the period by text box, and display the calendar when entering the period, so that the user can download by simple operation. Past measurement data list can be stored arbitrarily by the user in the terminal with this function.

5 Evaluation

In this section, we describe the overview and the results of the evaluation experiment with Seawater Temperature Announcement System.

Table 1. Actual measurement data used for evaluation of interpolation accuracy

Depth	1 m	5 m	10 m	20 m	30 m	40 m	50 m	60 m
Fukuura	23.26	23.26	23.21	23.24	23.18	23.11	23.00	22.01
Shiokojima	22.49	22.50	22.52	22.46	22.45	22.41	21.68	21.11
Shimonada	NA	21.61	21.65	21.63	NA	NA	NA	NA
Kitanada	NA	21.71	21.69	21.58	NA	NA	NA	NA
Hiburijima	21.39	21.43	21.41	21.40	21.44	21.38	21.41	21.42
Shimonami	NA	21.60	NA	NA	NA	NA	NA	NA
Miura	NA	21.57	21.55	21.53	NA	NA	NA	NA
Komobuchi	21.66	21.59	21.62	21.57	21.56	21.49	NA	NA
Yusu	21.70	21.65	21.65	21.63	21.66	21.64	NA	NA
Uwajima	NA	21.71	21.66	21.47	NA	NA	NA	NA
Yoshida	NA	21.72	21.66	21.62	NA	NA	NA	NA
Akehama	21.65	21.70	21.54	21.51	21.47	21.38	NA	NA
Yawatahama	NA	21.01	NA	NA	NA	NA	NA	NA

5.1   Evaluation of Interpolation Accuracy

To prove the validity of the function to visualize seawater temperature three-dimensionally, we evaluated interpolation accuracy by the procedures follows:

1. Select one from all measured points
2. Remove the selected actual measurement point from the actual measurement point used for interpolation (assume that the measured value is unknown)
3. Calculate the estimated value at the coordinates of the selected actual measurement point and compare it with the actual measurement value (calculate the error)
4. Execute procedures 1 to 3 for all measured points

We used the data actually measured at 6 p.m. on November 17, 2017 (JST) in this experiment. Table 1 shows the data details. "NA" in Table 1 means that there is no sensor for measuring the water temperature of the corresponding water depth.

Table 2 shows the difference between measured value and interpolated value of each station/each depth. Among the areas surrounding each measurement point, Fukuura, which is the outer part, has an error of more than 1 °C, and in Komobuchi, the central part, the accuracy is less than 0.2 °C at the maximum. In Akehama, which is the middle part of them, the accuracy is less than 0.3 °C at maximum. Lower accuracy of the outer part is the future task.

5.2   Quantitative Evaluation of System Effectiveness

We analyzing the access log for Seawater Temperature Announcement System in order to investigate the influence after intervention of the system. Specifically,

Table 2. Interpolation accuracy

Depth	1 m	5 m	10 m	20 m	30 m	40 m	50 m	60 m
Fukuura	1.143	1.390	1.305	1.388	1.091	1.067	1.357	0.858
Shiokojima	0.310	0.805	0.797	0.764	0.321	0.342	0.841	0.722
Shimonada		0.619	0.629	0.591				
Kitanada		0.039	0.004	0.063				
Hiburijima	0.473	0.284	0.334	0.293	0.347	0.350	0.625	0.068
Shimonami		0.014						
Miura		0.097	0.116	0.076				
Komobuchi	0.064	0.029	0.033	0.041	0.124	0.166		
Yusu	0.010	0.032	0.040	0.070	0.065	0.115		
Uwajima		0.042	0.016	0.141				
Yoshida		0.042	0.010	0.084				
Akehama	0.074	0.070	0.128	0.103	0.224	0.268		
Yawatahama		0.700						

we analysis about the total number of accesses that counted as multiple accesses from the same IP address in the same time zone as once (the unique access per a hour) every day and the total number of accesses that counted as multiple accesses from the same IP address in the same day as once (the unique access per a day) similarly. The access from IP addresses corresponding to Google crawlers etc. was excluded and totalized in order to investigate the number of users as accurate as possible.

Figure 7 shows the graph about the number of access variation over time from March 6, 2017 to October 1, 2018 in JST. The blue line represents the number

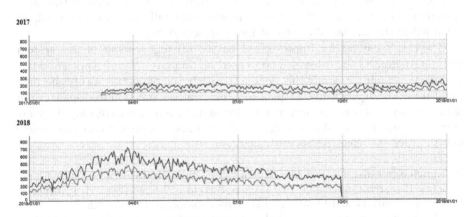

Fig. 7. Unique access variation over time (Color figure online)

of unique access per a hour, and the red line represents the number of unique access per a day.

Figure 7 shows the number of access peaked around April 2018, and thereafter it tends to decrease. The frequency of occurrence of red tide in Ehime prefecture is high in the summer (June to August), so fishery workers and fishery researchers become more interested in water temperature in around March to April. Therefore, it is considered that the demand for this system is increasing when fishermen in particular need water temperature information. In the future, we intend to demonstrate the effectiveness of this system by investigating this trend in more detail in comparison with seawater sample survey frequency variation over time by fishery researchers.

6   Conclusion

We developed Seawater Temperature Announcement System aiming to improving productivity of fisher industry in this research. The accurate water temperature information can be announced to fishermen immediately after measurement by intervening this system.

This system can be used regardless of the type of terminal such as PC, smartphone, tablet etc. by being developed as web application. We realized the visualization of the present condition of seawater temperature and seawater temperature variation over time considering spatial spread of marine area by incorporating requests from researchers and fishermen and implementing the function to display the table, the graph, and the distribution map. In developing this system, for the display setting of the graph, we devised so that visualization of seawater temperature information can be made with simple operation (e.g. the selection of the display water depth is made the check box type, the setting of the temperature display range and the selection of the display period are made the slider type). Furthermore, We also devised ingenuity on the layout on the web page (e.g. the background color of the "Update date" column in the table is made red in order to know the buoy in which the failure or the measurement is paused in the display of the latest measurement data).

There is a request for improvement on the briefing sessions of this system and questionnaires for fishermen concerned on January 12 and April 26, 2018. Therefore, we must upgrade this system to satisfy the requirements of fishermen as future works. Specifically, we plan to implement the function to display graph of the average water temperature every day and the function to display the year-over-year ratio in the table and graph. We will also continue to analyze access logs to prove the effectiveness of this system.

Acknowledgments. I would like to thank Mr. Hidetaka Takeoka of South Ehime Fisheries Research Center and Mr. Akihiko Takechi of Ehime Research Institute of Agriculture, Forestry and Fisheries for their cooperation in this research.

This work was partially supported by Strategic Information and Communications R&D Promotion Programme (SCOPE) (152309003), the Ministry of Internal Affairs and Communications, Japan.

References

1. Takeoka, H.: Fisheries support by advanced altitude information (Kodo kaikyo joho ni yoru suisangyo shien). Ehime J. **31**(5), 84–87 (2017)
2. Agusa, Y., Fujihashi, T., Endo, K., Kuroda, H., Kobayashi, S.: Development of seawater temperature announcement system for quick and accurate red tide estimation. In: Sieminski, A., Kozierkiewicz, A., Nunez, M., Ha, Q.T. (eds.) Modern Approaches for Intelligent Information and Database Systems. SCI, vol. 769, pp. 245–253. Springer, Cham (2018). https://doi.org/10.1007/978-3-319-76081-0_21

Big5 Tool for Tracking Personality Traits

Binh Thanh Nguyen[(⊠)] and Dang Ngoc Dung

Duy Tan University, Danang, Vietnam
ttb_2001@gmail.com

Abstract. In the big data era, understanding consumers through digital data is as important as the approach and exploitation of customers through their behavioral and personality traits in the digital world. First, a data warehouse has been studied and developed to extract, transform and load big mobile log data. Afterwards, the data warehouse's data cubes are aggregated and used to calculate a set of Big5 indicators. Hereafter, Big5 traits can be predicted based on those just-specified indicators. To proof of our concepts, implementation results will be presented in the context of the Big5 tool, which has been designed and developed to predict Big5 personalities in a representative manner.

Keywords: Big5 traits · Personality · Indicators · Data management ·
Mobile logs · Machine learning · Naive Bayes classification

1 Introduction

As the number of mobile phone users in the world is expected to pass the five billion mark by 2019 [25] and carriers have increasingly made available phone logs to researchers [3], data patterns retrieved from mobile phone user logs can open the door to exciting avenues for future research in social sciences [6]. On another hand, five-factor model (Big5) of personality, which has been introduced as *extraversion, agreeableness, conscientiousness, ceuroticism,* and *openness to experience* [11, 12, 20, 23], can be used to explore the relationship between personality and various behaviors [23]. In [14], we indicated that understanding clients through these Big5 factors is a commonly used method of business people, e.g. service providers and marketing agent, who need information about their customers in investigating, planning for policies, development campaigns.

In this context, determining the personality of a mobile phone user simply through standard carrier's logs has became a topic of tremendous interest [3]. Based on the predicted results, mobile phones datasets could thus provide a valuable unobtrusive and cost-effective alternative to survey-based measures of personality [3]. These data permit fine-grained, continuous collection of people's social interactions (e.g., speaking rates in conversation, size of social groups, calls, and text messages), daily activities (e.g., physical activity and sleep), and mobility patterns (e.g., frequency and duration of time spent at various locations) [7].

In this paper, first, our methodology for processing, storing and retrieving the Call Detail Records (CDR) datasets provided by Orange Senegal [4] has been described. The big data sets are the basis for our analysis into calling and texting behaviors, spatial

© Springer Nature Switzerland AG 2019
N. T. Nguyen et al. (Eds.): ACIIDS 2019, LNAI 11431, pp. 726–736, 2019.
https://doi.org/10.1007/978-3-030-14799-0_62

interactions to predict Big5 traits. As a result, a data warehouse, namely Big5DW has been studied and developed to store big mobile log data in multi-dimensional data structure [7, 8, 15]. Hereafter, data in data cubes can be used to calculate a set of Big5 indicators. Furthermore, the data sets of Bandicoot tool [3] also provide many meaningful indicators for our study. In this context, machine learning algorithms, especially, Naive Bayes classification [22, 24] are studied and applied to predict if a phone user has feature of low, average, or high in the framework of Big5 traits based on the just-specified indicators. To proof of our concepts, the Big5 tool has been developed to present implementation results of our research in a representative manner.

The rest of this paper is organized as follows: Sect. 2 introduces some approaches and projects related to our work; after an introduction of the Big5 conceptual model, i.e. Big5 data management, Big5DW, indicators, and how to predict Big5 traits based on Naive Bayes classification in Sect. 3, Sect. 4 will present our implementation results with their main use cases. And lastly, Sect. 5 gives a summary of what have been achieved and future works.

2   Related Work

This work has been proposed by the requirements of analyzing user's mobile log data, which are basic resources with the more precise and uptodate information provided by Orange Senegal. According to [27], determining the personality of mobile phone users, besides being important solely from the psychological point of view, can also provide an interesting application framework for wearable computing. Furthermore, partterns retrieved from mobile phone logs contain many features, which are very useful for researchers, marketing agents as well as phone companies [2].

In this context, in [9], various features extracted from Facebook data can be explored to predict personality. In [10], the authors classified author's personality from weblog texts by using the n-grams as the features and the Naïve Bayes algorithm as the classification algorithm. They performed experiments on the authors with the highest and lowest scores and reported how to automatically select features that yield the best performance. In [22], personality is believed to be an important factor in determining individual variation in thoughts, emotions and behavior patterns.

Also, [22] predicted conscientiousness by exploiting the nuances on the usages of the verbs, in other words, measuring the specificity and objectivity of the verbs taken from WordNet and Senti-WordNet. In [1], personality is predicted by different classification methods such as Support Vector Machine, Bayesian Logistic Regression. However, most research used social texts as their input sources. On another hand, the ability to draw connections between behavioral aspects derived through contextual data collected by mobile phones, as well as personality, could lead to designing and applying machine learning methods to classify users into personality types [2, 5].

In our current research [2, 3, 15], we are focusing on main key features: developing a data warehouse [9, 16–19] to manage big mobile log data, specifying new Big5 indicators, which could be retrieved from the data warehouse' data cubes; then implementing a tool to predict personalities.

3 Big5 Conceptual Model

The objective of our research is to build a framework application and provide a quantitative assessment methodology for the use of mobile phone log data to retrieve Big5 traits distributed in developing countries in general, and Senegal in particular. Based on those data sets, a data warehouse, namely Big5DW has been designed. Afterwards, multi-dimensional data cubes can be defined and used to store aggregated data. In this context, Big5 indicators can be calculated from those data cubes. Futhermore, there are many useful indicators generated by Bandicoot tool [3] and inherited by our study. Based on those just-specified indicators, Naive Bayes classification [22, 24] are studied and applied to predict whether a phone user has feature of low, average, or high in the farmework of Big5 traits. Afterwards, the implementation results will be presented in Sect. 4 to proof of our concepts.

3.1 Big 5 Data Management

As mentioned above, mobile phone use logs could provide key insights into personality traits of Senegal with their interesting data patterns retrieved by data science approaches. In this section, the Orange mobile data will be introduced briefly. Afterwards, our data warehouse approach will be used to manage such big data files. Furthermore, the data cube concepts [7, 8, 15] will be applied to aggregate data according dimensional hierarchies specified in the next section.

3.1.1 Orange Mobil Data

The anonymized mobile Call Detail Records (CDR) are collected in Senegal between January 1, 2013 and December 31, 2013 and provided by Orange in the context of the Data for Climate Action challenge [3]. In this study we analyzed the mobile data acquired from the 1666 towers distributed across Senegal. According to [3], those datasets are based on phone calls and SMS exchanged between 9 million users during the year 2013. In the context of this challenge, those data have been properly anonymized before being handled to researchers [4]. The Fig. 1 shows a subset of *SMS* antenna to antenna mobile log data. According to [3], the data are organized into three sets:

- Set 1 contains the hourly voice and text traffic between Outgoing_site and Incoming_site. The information includes total call duration, number of calls and total number of text messages.
- Set 2 contains the site-based fine-grained mobility data of about 300,000 randomly sampled and anonymized users for each two week period. This data set also includes bandicoot user-based behavioral indicators. Figure 1 shows a subset of indicators provided in this data set from Badicoot tool.
- Set 3 contains one year (2013) coarse-grained trajectories of 123 arrondissements. There are also bandicoot behavioral indicators at individual level for about 150,000 randomly sampled users as shown in Fig. 2.

sms_timestamp	outgoing_site_id	incoming_site_id	number_of_sms
2013-01-02 13	24	393	2
2013-01-02 13	24	394	1
2013-01-02 13	24	396	1
2013-01-02 13	24	408	1
2013-01-02 13	24	415	1
2013-01-02 13	24	420	3
2013-01-02 13	24	427	1
2013-01-02 13	24	438	1
2013-01-02 13	24	440	1
2013-01-02 13	24	441	1
2013-01-02 13	24	447	1

Fig. 1. A subset of SMS antenna to antenna log data

userid	activedays_all	durationofcalls_call_mean	durationofcalls_call_median	durationofcalls_call_std
240	10	131.97	73	166.21
480	5	27.77	20.5	24.27
720	14	28.7	20.5	29.59
960	12	30.13	21	41.58
1200	14	44.23	27.5	56.32
1440	13	53.58	32	60.04
1680	13	25.15	18	21.44
1920	8	71.37	31.5	104.17
2160	12	34.44	23	37.22
2400	12	45.99	22.5	59.94
2640	0	83.55	26.5	127.02
2880	8	19.87	5	30.04
3120	13	55.59	23	132.08
3360	6	26.64	18	24.72

Fig. 2. An example list of indicators calculated by Bandicoot tool.

3.1.2 Data Preprocessing

First, the Orange data CSV files are imported into fact tables of our Big5DW data warehouse, the concepts of which will be introduced in the next section. Afterwards, the data in those fact tables are aggregated into multi-dimensional data cubes as shown in the following Fig. 3.

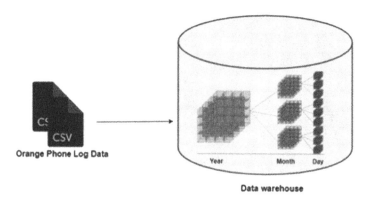

Fig. 3. Pre-processing data

3.2 Big5DW Data Warehouse

First, the Big5 data warehouse (*Big5DW*) is formalized based on the basis of mathematical model. The purpose of this conceptual model is to provide an extension of the standard data warehouse architecture used in our previous studies [7, 8, 13, 14]. As a result, the *Big5DW* can be defined as follows:

$$Big5DW = <Big5Dims, Big5Facts, Big5FTs, Big5Gbys>$$

where:

- *Big5Dims* = {*Site, Time*} is a set of dimensions.
- *Big5Facts* = {*number_of_calls, Total_call_duration, number_of_sms*} is a set of log variables.
- *Big5FTs* = {*FTCalls, FTSMSs*}.
- *Big5Gbys* is a set of data cubes grouped by the hierarchical levels of *Time* and *Site* dimensions, e.g. *CallbyNighttime, CallbyDay, CallbyMonth, CallbyYear, SMSby-NightTime, SMSbyDay, SMSbyMonth, SMSbyYear*, etc.

In the next sections, we will introduce main components of the Big5 multidimensional data model, i.e. dimension and facts (variables) and their related elements.

3.2.1 Big5Dims

In *Big5DW* data warehouse, *Site* and *Time* are main dimensions. Furthermore, the *Site* dimension is specified into *Outgoing_site, Incomming_site* dimensions where:

- *Outgoing_site_id*: id of site the call/text originated from.
- *Incoming_site_id*: id of site receiving the call/text.

According the *SITE_ARR_LON_LAT* and *SENEGAL_ARR_V2* tables [3], the hierachical levels of the *Site* dimension can be defined as follows:
{*Senegal->Regions->arrondissement_id ->Site_ID*}

Based on the *timestamp* format [3], the *Time* dimension can be specified as:

$$\{Year \rightarrow Month \rightarrow Week \rightarrow Day \rightarrow$$
$$DayNight \rightarrow TimeStamp\}$$

3.2.2 Big5 Fact Tables and Related Elements

Based on Antenna-to-antenna traffic data provided in Set 1 [3], two main fact tables, namely *FTCalls, FTSMSs* are defined by having *Outgoing_site_id, Incomming_site_id, timestamp* as dimensions and *number_of_calls, Total_call_duration, number_of_sms* as facts. Figure 4 shows an example of *FTCalls* fact table. So, the *Big5Gby* set can be specified from *FTCalls, FTSMSs* as aggregations of hierarchical levels of those dimensions.

call_timestamp	outgoing_site_id	incoming_site_id	number_of_calls	total_call_duration
2013-01-01 00	1	1	1	54
2013-01-01 00	1	2	1	39
2013-01-01 00	1	24	1	2957
2013-01-01 00	1	186	1	56
2013-01-01 00	2	2	22	418
2013-01-01 00	2	3	2	53
2013-01-01 00	2	4	4	455
2013-01-01 00	2	5	8	386
2013-01-01 00	2	6	7	204
2013-01-01 00	2	8	13	314
2013-01-01 00	2	9	2	60
2013-01-01 00	2	10	6	229
2013-01-01 00	2	11	2	27

Fig. 4. An example of fact table *FTCalls*

3.3 Big5 Indicators

In [14], we denote a set of Big5 personality traits, which can be classified into agreeableness (A), conscientiousness (C), extraversion (E), neuroticism (N) and openness to experience (O) as follows:

$$Cl = \{A, C, E, N, O\} \qquad (1)$$

The Big5 indicators have been calculated based on our multidimensional data cubes *Big5Gbys* as shown in this Fig. 5:

user_id	place_entropy	daily_distance_traveled	ar_phi_1	ar_phi_2
1	-5.005598118718...	71.92328767123287	0.25180698321524836	0.113083191634
2	-3.896590857599...	23.21095890410959	0.2673719172618684	0.024961369067
3	-3.206328155547898	30.78082191780822	0.31101570008379353	0.123927304107
4	-5.08267736873818	4.087671232876712	0.2062824161731841	0.152268452901
5	-4.712199345240127	29.484931506849314	0.32871850552170434	-0.03745448665
6	-5.44808243664946	5.36712328767123	0.2218102873762684	0.078320791618
7	-5.378101888323574	10.413698630136986	0.23877061164227445	0.131103132375
8	-4.997456605596454	5.134246575342465	0.23962476592563403	0.044161007177
9	-4.914782647282899	24.057534246575344	0.24744959054205756	0.093481135335
10	-5.373774757874341	262.32876712328766	0.23274960187241073	0.041518603257
11	-4.19942508380227	13.564383561643835	0.1617105006071068	0.042860187285
12	-5.072567383826007	12.841095890410958	0.1341174494720068	0.057471098000

Fig. 5. Exanple list of indicators calculated by user from Orange data sets.

3.4 Predict Big5

We have presented how Big5 personality can be predicted by applying Naive Bayes classification [21, 23] in [14]. This approach can be summarised as follows:

Given a class variable $cl \in CL$ defined in formula (1) and a dependent feature vector x $(x_1, .., x_n)$, the following relationship is defined:

$$P(cl|x_i, \ldots, x_n) = \frac{P(cl)P(x_i, \ldots, x_n|cl)}{P(x_1, \ldots, x_n)} \quad (2)$$

Then, following *Uniform distribution* [27], the probability are equal each other's traits as:

$$P(cl|x_i, \ldots, x_n) = \frac{\frac{1}{5}\prod_{i=1}^{n} P(x_i|cl)}{P(x_1, \ldots, x_n)} \quad (3)$$

In this context, indicators x_i are mapped into low (l) and high (h) degrees [26]. Then, *Multinomial Naive Bayes method* [23] has been applied to calculate $P(x_i|cl)$ by mean of low or high in each personality's dimension.

4   Big5 Tool

The Big5 tool has been developed to enables user(s) to explore personality traits. First, mobile phone log data provided by Orange Senegal [3] are extracted, transformed, and then loaded into the PostgreSQL data warehouse as illustrated in Fig. 6. Afterwards, Big5 indicators are calculated and stored in MongoDB. Those indicators are used to predict Big5 traits as presented in [14] and summarised in the previous section. The below paragraphs will present Big5 actors and their main use cases and some typical examples of the Big5 tool.

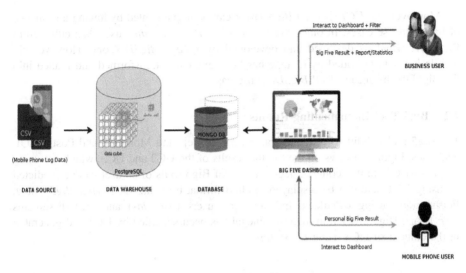

Fig. 6. Big5 system architecture

4.1 Big5 Dashboard Use Cases

Big5 dashboard use cases have been designed for two main actors, i.e. *Mobile Phone Provider*, and *Marketing Agents* as shown in Fig. 7.

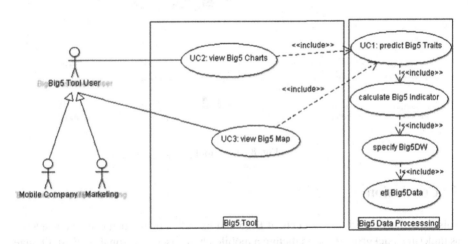

Fig. 7. Big5 tool use case diagram

First, main actors can be aggregated into an abstract actor, namely *Big5 Tool User*. This actor has three use cases, i.e. *UC1: predict Big5 Traits, UC2: view Big5 Charts,* and *UC3: view Big5 Map.*

Moreover, the *UC1: predict Big5 Traits* can be implemented by having a sequence of back-end use cases. In other words, *calculate Big5 Indicator* uses data cubes from Big5DW, which is designed and developed by *specify Big5DW* one. However big mobile log data provided by Orange has been extracted, transformed and loaded into the Big5DW by mean of *etl Big5Data* use case.

4.2 Big5 Tool Implementing Results

The Big5 tool is built by Angular framework, Nodejs and Mongodb and PostgreSQL databases. Figure 8 shows on the top the results of the UC2 and UC3, which are using output of UC1. In this context, degree levels of Big5 traits of an user can be predicted by using UC1, and can be displayed in chart format by using UC2: *view Big5 Chart*. Furthermore, using calculated indicators by users (*user_ids*) and arrondissements (*arr_ids*), a Big5 prediction map of Senegal has been specified by UC1 and generated in the context of UC3: *view Big5 Map*.

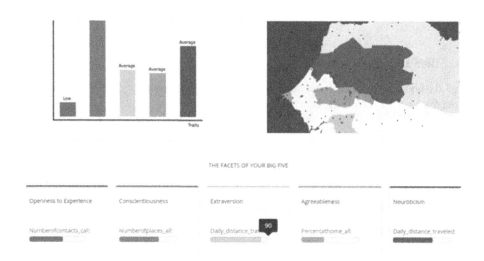

Fig. 8. Big5 tool

5   Conclusion

This paper introduced the concepts of Big5 and its indicators, which can be considered as underline background for predicting a mobile phone user's personality. First, Orange Senegal mobile phone logs data are preprocessed and loaded in our data warehouse in term of *callbyday* and *smsbyday* fact tables. Afterwards, multi dimensional data cubes can be specified as aggregations of the two fact tables. Based on the multi dimensional data model, a set of indicators has been calculated and used for predicting Big5 based on Naive Bayes classification method. In this context, the Big5 tool has been designed in UML and developed to proof of our concepts.

Future work of our approach could then be able to support Big5 predicting in related application domains, e.g. data from other mobile phone providers, data from other sources. Furthermore, we will focus on the implementation of Big5 tool with new features to make use of our concepts.

Acknowledgement. Thanks to Orange Sonatel Senegal and the D4D team for providing the mobile phone data. Support from the Duy Tan University, Vietnam is acknowledged.

References

1. Alam, F., Stepanov, E.A., Riccardi, G.: Personality traits recognition on social network-facebook. In: Proceedings of Workshop on Computational Personality Recognition, pp. 6–9. AAAI Press, Melon Park (2013)
2. CNN: Your phone company is selling your personal data. http://money.cnn.com/2011/11/01/technology/verizonattsprinttmobileprivacy/index.htm
3. de Montjoye, Y.A., Quoidbach, J., Robic, F., Pentland, A.: Predicting personality using novel mobile phone-based metrics. In: Greenberg, A.M., Kennedy, W.G., Bos, N.D. (eds.) Social Computing, Behavioral-Cultural Modeling and Prediction, SBP 2013. Lecture Notes in Computer Science, vol. 7812, pp. 48–55. Springer, Heidelberg (2013). https://doi.org/10.1007/978-3-642-37210-0_6
4. de Montjoye, Y.-A., Smoreda, Z., Trinquart, R., Ziemlicki, C., Blondel, V.: D4D-Senegal: The Second Mobile Phone Data for Development Challenge (2014)
5. de Oliveira, R., et al.: Towards a psychographic user model from mobile phone usage. In: Proceedings of the 2011 Annual Conference Extended Abstracts on Human Factors in Computing Systems. ACM (2011)
6. Chittaranjan, G., Blom, J., Gatica-Perez, D.: Who's who with big-five: analyzing and classifying personality traits with smartphones. In: Proceedings of the 2011 15th Annual International Symposium on Wearable Computers (ISWC 2011). IEEE Computer Society, Washington, pp. 29–36 (2011). https://doi.org/10.1109/ISWC.2011.29
7. Harari, G.M., Lane, N.D., Wang, R., Crosier, B.S., Campbell, A.T., Gosling, S.D.: Using smartphones to collect behavioral data in psychological science: opportunities, practical considerations, and challenges. Perspect. Psychol. Sci.: J. Assoc. Psychol. Sci. **11**(6), 838–854 (2016). https://doi.org/10.1177/1745691616650285
8. Hoang, D.T.A., Ngo, N.S., Nguyen, B.T.: Collective cubing platform towards definition and analysis of warehouse cubes. In: Nguyen, N.-T., Hoang, K., Jędrzejowicz, P. (eds.) ICCCI 2012. LNCS (LNAI), vol. 7654, pp. 11–20. Springer, Heidelberg (2012). https://doi.org/10.1007/978-3-642-34707-8_2
9. Hoang, A.D.T., Nguyen, T.B.: An integrated use of CWM and ontological modeling approaches towards ETL Processes. In: ICEBE 2008, pp. 715–720 (2008)
10. Hoang, A.D.T., Nguyen, T.B.: State of the art and emerging rule-driven perspectives towards service-based business process interoperability. In: RIVF 2009, pp. 1–4 (2009)
11. Oberlander, J., Nowson, S.: Whose thumb is it anyway? Classifying author personality from weblog text. In: Proceedings of the COLING/ACL on Main Conference Poster Sessions (COLING-ACL 2006), pp. 627–634. Association for Computational Linguistics, Stroudsburg (2006)
12. McCrae, R.R., John, O.P.: An introduction to the five-factor model and its applications. J. Pers. **60**(2), 175–215 (1992)

13. Mount, M., Ilies, R., Johnson, E.: Relationship of personality traits and counterproductive work behaviors: the mediating effects of job satisfaction. Pers. Psychol. **59**, 591–622 (2006). https://doi.org/10.1111/j.1744-6570.2006.00048.x

14. Nguyen, T.B., Dang, N.D., Nguyen, T.T.H., Ha, T.T., Phan, T.H.L., Truong, D.H.: Tracking Big5 traits based on mobile user log data. In: The 7th International Conference on Frontiers of Intelligent Computing: Theory And Application (FICTA 2018). Advances in Intelligent Systems and Computing (2018)

15. Nguyen, T.B., Ngo, N.S.: Semantic cubing platform enabling interoperability analysis among cloud-based linked data cubes. In: Proceedings of the 8th International Conference on Research and Practical Issues of Enterprise Information Systems, CONFENIS 2014. ACM International Conference Proceedings Series (2014)

16. Nguyen, T.B., Tjoa, A.M., Wagner, R.: Conceptual multidimensional data model based on metacube. In: Yakhno, T. (ed.) ADVIS 2000. LNCS, vol. 1909, pp. 24–33. Springer, Heidelberg (2000). https://doi.org/10.1007/3-540-40888-6_3

17. Nguyen, T.B., Wagner, F.: Collective intelligent toolbox based on linked model framework. J. Intell. Fuzzy Syst. **27**(2), 601–609 (2014)

18. Nguyen, T.B., Wagner, F., Schoepp, W.: Federated data warehousing application framework and platform-as-a-services to model virtual data marts in the clouds. Int. J. Intell. Inf. Database Syst. **8**(3), 280 (2014). https://doi.org/10.1504/ijiids.2014.066635. ISSN 1751-5858, 1751-5866

19. Nguyen, T.B., Wagner, F., Schoepp, W.: EC4MACS – an integrated assessment toolbox of well-established modeling tools to explore the synergies and interactions between climate change, air quality and other policy objectives. In: Auweter, A., Kranzlmüller, D., Tahamtan, A., Tjoa, A.M. (eds.) ICT-GLOW 2012. LNCS, vol. 7453, pp. 94–108. Springer, Heidelberg (2012). https://doi.org/10.1007/978-3-642-32606-6_8

20. Nguyen, T.B., Wagner, F., Schoepp, W.: GAINS-BI: business intelligent approach for greenhouse gas and air pollution interactions and synergies information system. In: Proceedings of the International Organization for Information Integration and Web-Based Application and Services IIWAS 2008, Linz (2008)

21. Peng, K.-H., Liou, L.-H., Chang, C.-S., Lee, D.-S.: Predicting personality traits of Chinese users based on Facebook wall posts, pp. 9–14 (2015). https://doi.org/10.1109/wocc.2015.7346106

22. Tomlinson, M.T., Hinote, D., Bracewell, D.B.: Predicting conscientiousness through semantic analysis of facebook posts. In: Proceedings of Workshop on Computational Personality Recognition. AAAI Press, Melon Park (2013)

23. Zhang, W., Gao, F.: An improvement to Naive Bayes for text classification. Proc. Eng. **15**, 2160–2164 (2011). https://doi.org/10.1016/j.proeng.2011.08.404. ISSN 1877-7058

24. https://en.wikipedia.org/wiki/Bayes%27_theorem

25. https://en.wikipedia.org/wiki/Big_Five_personality_traits

26. https://en.wikipedia.org/wiki/Naive_Bayes_classifier

27. https://vaciniti.com/mobile-phone-users-worldwide/

Comparative Analysis of Usability of Data Entry Design Patterns for Mobile Applications

Jakub Myka[1], Agnieszka Indyka-Piasecka[1] (ID), Zbigniew Telec[1] (ID),
Bogdan Trawiński[1(✉)] (ID), and Hien Cao Dac[2] (ID)

[1] Faculty of Computer Science and Management,
Wrocław University of Science and Technology, Wrocław, Poland
{agnieszka.indyka-piasecka, zbigniew.telec,
bogdan.trawinski}@pwr.edu.pl
[2] Nguyen Tat Thanh University, Ho Chi Minh City, Vietnam
cdhien@ntt.edu.vn

Abstract. A comparative analysis of usability of data entry design patterns for mobile applications was presented in the paper. For this purpose three versions of a mobile application for a computer store were developed. Each version included the same functions which were designed and programmed using different data entry patterns. The applications were then utilized to conduct usability tests with three groups of users. The research was conducted based on the ISO model of usability. The usability metrics such as task completion time, number of actions, binary task completion rate and percentage of tasks accomplished without any of errors were considered. User satisfaction was measured using the Single Ease Question (SEQ) questionnaire. The results obtained were thoroughly analysed and recommendations for each data entry design patterns for mobile applications were formulated.

Keywords: Usability testing · Data entry · Design patterns ·
Mobile applications

1 Introduction

Nowadays, you can observe the intensive development of the mobile market. People often reach for smartphones, tablets and other mobile devices to search for a variety of information on a daily basis. Virtually every website is created with access to all kinds of mobile devices. Fast insight into the news from the world is just one of the advantages of these modern inventions. Thanks to them, more and more people are starting to use permanent access to the bank as well as contactless payments through NFC (Near Field Communications) technology. Speaking of the applications of mobile solutions, one cannot forget about their use for entertainment purposes, such as: listening to music, watching videos, games or voice, text or video communication.

Due to the rapid spread of mobile technologies in various areas of our lives, the important and necessary action is to adapt mobile applications to the needs of all users. Tablets, smartphones and other devices are used in various places, including those poorly lit or loud. Users are often forced to use them with one hand or by voice, which makes customizing the mobile applications particularly important.

© Springer Nature Switzerland AG 2019
N. T. Nguyen et al. (Eds.): ACIIDS 2019, LNAI 11431, pp. 737–750, 2019.
https://doi.org/10.1007/978-3-030-14799-0_63

According to the standard ISO 9241-11 usability is: "the extent to which a software can be used by specified users to achieve specified goals with effectiveness, efficiency and satisfaction in a specified context of use". The various aspects of the usability of traditional desktop systems have been studied and discussed for a long time, but new challenges emerge in the case of mobile devices such as smartphones and tablets. The use of mobile applications is becoming very common, being the most popular access to information and communication with other users.

The main goal of the paper is to report the results of comparative analysis of usability of data entry design patterns devoted to mobile applications. The research was conducted based on the ISO model of usability. The usability metrics such as task completion time, number of actions, binary task completion rate and percentage of tasks accomplished without any of errors were considered. User satisfaction was also measured using the Single Ease Question (SEQ) questionnaire.

2   Related Works

2.1   Survey of Models, Methods and Metrics

Numerous works presenting usability evaluation of mobile applications with various usability measurements and metrics is evidenced by the large number of studies in which it has been employed. Seffah et al. [26] introduced hierarchical QUIM (**Q**uality in **U**se **I**ntegrated **M**easurement) usability model. Alshehri and Freeman [2] examined various methods for usability evaluations of mobile devices.

Harrison et al. [11] argued that mobile devices require specific usability models, thus they developed the PACMAD usability model by extending well-known Nielsen's or the ISO usability models to the context of mobile applications. The PACMAD model introduces seven characteristics, accomplished with authors definitions, measures, and associations for each of them. The component amounts to effectiveness, efficiency, satisfaction, learnability, memorability, errors and cognitive load. The cognitive load is claimed by the authors to be the main novelty of PACMAD model. However PACMAD guidelines and metrics related to chosen dimension are seen by the community as too general, and also its accuracy for mobile applications should be deeply evaluated.

Saleh et al. [24] extended PACMAD usability model by 21 low level metrics for usability attributes. The main value of GQM model are: task list and user satisfaction questionnaire for collecting objective and subjective data of usability evaluation.

Nayebi et al. [21] examined wide range of usability evaluation and measurement, which are presented in the literature for mobile application usability testing. They emphasized the lack of scientific research concerning requirements for mobile user interfaces, especially shortage of evaluation studies related to iOS mobile application usability measurement.

Similar conclusions are derived at numerous work by Hussain et al. [1, 13, 30]. The existing usability models do not adequately capture the complexities of interacting with applications on a mobile platform and diverse categories of mobile applications require even customizable usability models. In response to those disadvantages the subsequent work of Hussain et al. [12] introduced the mGQM model based on ISO 9241-11 standard usability measurement such as Effectiveness, Efficiency and Satisfaction. As designed especially for evaluating the usability of mobile applications the model is quite comprehensive, however insufficiency of proper descriptions on how to determine adequate usability measurements for a mobile application occurs as inconvenient disadvantage.

The attributes of the most popular usability models, namely ISO, Nielsen, PACMAD, and QUIM ones, are shown in Table 1.

Table 1. Attributes of basic models of usability

ISO	Nielsen	PACMAD	QUIM
Effectiveness	Efficiency	Effectiveness	Effectiveness
Efficiency	Satisfaction	Efficiency	Efficiency
Satisfaction	Learnability	Satisfaction	Satisfaction
	Memorability	Learnability	Learnability
	Errors	Memorability	Productivity
		Errors	Safety
		Cognitive load	Trustfulness
			Accessibility
			Universality
			Usefulness

Shitkova et al. [28] and Coursaris et al. [8] collected the catalogue of 39 usability guidelines for mobile applications and websites as a response to the lack of structured and evaluated usability guidelines for mobile websites and applications in the literature.

As mobile devices has introduced new usability challenges that are difficult to cover by traditional usability models Zhang et al. [31] focused on a number of new issues such as: Mobile Context, Connectivity, Small Screen Size, Different Display Resolution, Limited Processing Capability and Power, Data Entry Methods, which were not consider earlier.

The Standardized User Experience Percentile Rank Questionnaire (SUPR-Q) developed by Sauro and Zarolia [25] is a measure to capture both broad user experience and more specific components of desktop websites by evaluating websites usability, trust, appearance, and loyalty. Authors suggest to use the SUPR-Qm to benchmark the user experience of various mobile applications.

2.2 Heuristic Evaluation of Mobile Applications

Quinones and Rusu [23] presented 73 different studies which provide an overview of several approaches to the creation of usability heuristics and describe the methodologies that were used. They investigated the most current approaches to developing heuristics: adaptation of existing Nielsen's heuristics [32] and using a methodology to create usability heuristics. However, there is no formal specification of the stages or related activities. Moreover, there is currently no clear protocol for heuristics validation.

Promising new heuristics to evaluate usability for mobile applications have been proposed by Gomez et al. [10] for non-experts. The methodology presents clearly defined stages, includes a standard template for specifying heuristics and clear validation methods, and can be applied iteratively. The Authors reuse existing and well-known heuristic guidelines to adapt them to the new mobile-specific usability guidelines.

Joyce et al. [14] empirically investigate that traditional method - Heuristic Evaluation - can be modified and consequently and effectively used at usability issues typical for mobile applications. Comparing three sets of usability heuristics for mobile applications: Nielsen's [32], Bertini's [6] and the last one - defined by the Authors, the Authors set of usability heuristics surfaces more usability issues, in a mobile application, than other sets of heuristics, however improvements to the set are expected still.

Joyce et al. [15] in their latest research proposed SMART (Smartphone Mobile Application heuRisTics) as heuristics for mobile applications. Based on existing heuristics, finally 13 heuristics were developed and positively evaluated by HCI specialists. These heuristics were more suitable to evaluate the usability of mobile applications, but support to apply them was not considered.

Also Othman et al. [22] compared two sets of heuristics, traditional and domain specific for mobile application for cultural heritage sites. Authors contribution provides evidence that domain specific measurement is more comprehensive for a more definite identification of usability issues.

2.3   Usability Testing with Users

Recently, numerous works concerning usability testing of mobile and responsive applications with the users as well as experts have been published [5, 7, 18].

Shirogane et al. [27] evaluated applications and identified problems automatically by analyzing operation histories of 20 users of tablet computer application. From the recorded and analyzed users operation histories, usability problems were identified by the criteria corresponding to the 10 standard usability heuristics by Nielsen [32]. The method automatically analyzed the data obtained at usability evaluations and identified many usability problems. Thus, usability evaluations were performed easily and effectively even by non-usability specialists.

Usability evaluations by questionnaires, observations and video analyses have been combined by Arain et al. [3] or Moumane et al. [20]. Performing usability evaluation of some mobile applications, as an evaluation criteria usability characteristics of ISO/IEC 9126 standards[1], ISO/IEC 25062 standards[2] and ISO 9241 standards[3] were used. While the satisfaction was evaluated with QUIS 7.0 (Questionnaire for User Satisfaction Interaction) [12].

---

[1] ISO/IEC 9126-1:2001, "Software engineering – Product quality –Part 1: Quality model," 2001.

[2] ISO/IEC 25062:2006, "Software engineering – Software product Quality Requirements and Evaluation (SQuaRE) – Common Industry Format (CIF) for usability test reports," 2006.

[3] ISO 9241-11:1998, "Ergonomic requirements for office work with visual display terminals (VDTs) – Part 11: Guidance on usability," 1998.

Silvennoinen et al. [29] examined user experiences and preferences in relation to the visual elements of color and perceived dimensionality analyzing two different mobile application contexts. They focused on the role of visual usability and visual aesthetics in an experimental research where usability data from the users were collected using two online questionnaires. Also the work of Mansar et al. [19] can be mentioned, where usability testing experiment was conducted to adapt the mobile application for weight loss support to the local culture rules and traditions of the Middle East society.

The research by Kortum and Sorber [17] concerning usability measuring of mobile application delivered for two different platform (smartphones vs. tablets) and two different operating systems (iOS vs. Android) adapts System Usability Scale (SUS) and a set of baseline usability measures for mobile application. Earlier also Gatsou et al. [9] aimed their research in the area of usability testing for mobile tablet application.

Performing the usability testing of mobile applications can be done with the users in either the laboratory or the field [4, 16, 31]. Results indicate that conducting a time-consuming field test may not be undertaken when studying user interface or navigation issues to improve user interaction. The field testing is more important when combining usability tests with contextual study where behavior of the user is investigated in an original context.

3   Mobile Application for Usability Testing

Three versions of a mobile application devised for a hardware store were developed to test experimentally the usability of several data entry design patterns. The main functions of the application were: user registration, logging into the account, editing user profile, adding reviews of the product, adding products to the cart, removing products from the cart, changing quantity of products in the cart, completing the order, and browsing order history reports. Each application version included the same functions which were designed and programmed using different data entry patterns:

Email (EM) – three patterns enabled users to enter an email address using one text field (EM1), two separate text fields for username and domain (EM2), and three separate text fields for username, domain, and extension (EM3);

Date (DT) – three patterns facilitated selecting a date with a calendar picker (DT1), scrollers (DT2), and a numeric keyboard (DT3). They are illustrated in Fig. 1;

Time (TM) – three patterns allowed for selecting time by means of a clock-style time picker (TM1), scrollers (TM2), and a numeric keyboard (TM3). They are illustrated in Fig. 2;

Small number (NB) – three patterns enabled users to input a small number with a numeric keyboard (NB1), scroller (NB2), and slider (NB3);

Single-choice list (SL) – three patterns allowed users to select one item from a list using a spinner (SL1), toggle buttons (SL2), and radio buttons (SL3);

Multiple-choice list (ML) – three patterns made possible to select items using checkboxes (ML1), switches (ML2), and by highlighting options (ML3);

Longer text (TX) – three patterns facilitated inputting longer texts through gesture typing (TX1), voice typing (TX2), and predictive text tool (TX3). They are illustrated in Fig. 3.

Fig. 1. Illustration of three versions of date entry patterns including DT1, DT2, and DT3

Fig. 2. Illustration of three versions of time entry patterns including TM1, TM2, and TM3

Fig. 3. Illustration of three ways of entering longer texts including TX1, TX2, and TX3

4 Setup of Usability Tests

The goal of the research was to test usability of three versions of an application for a hardware store in which a number of data entry patterns were implemented. The experiments were carried out with three groups of ten users who completed the same tasks scenarios using the different versions of the same application on smartphones. The basic characteristics of all 30 users are given in Fig. 4.

The study was based on the ISO model of usability which comprises three main attributes: effectiveness, efficiency, and satisfaction. In total four metrics were collected during completion of individual tasks and the Single Ease Question (SEQ) questionnaire was administered after each task was accomplished. The metrics collected during the study are listed in Table 2. In turn, the main steps of the study are shown in Table 3 and the list of task performed by the participants together with the design patterns tested is presented in Table 4.

Table 2. Metrics collected during usability tests

Attribute	Metrics	Description	Unit
Effectiveness	Binary rate	Percentage of participants who completed the task successfully	[%]
	No errors	Percentage of participants able to complete the task with no errors	[%]
Efficiency	Time	Task completion time by a user	[s]
	Actions	Number of clicks, scrolls, taps, swipes, etc. to complete individual tasks	[n]
Satisfaction	SEQ	Single Ease Question score of a questionnaire administered to individual tasks	1–7

Table 3. Main steps of the study

1.	Preparation of the materials for users including a personal form, task list, and questionnaires
2.	Preparation of the environment for conducting research
3.	Welcome the participant, informing him/her about the goals, scope, and course of research
4.	Filling a personal form by the participant
5.	Conducting usability tests with a smartphone
6.	Completing the SEQ questionnaire by the user after having accomplished each task
7.	Completing a satisfaction questionnaire by the user after having accomplished all the tasks
8.	Finishing the usability testing and saying thanks and goodbye

Table 4. List of tasks completed by the participants

Tasks	Patterns tested
T1. Register and edit the user profile	
T2. Edit the email address in the user profile	EM1, EM2, EM3
T3. Edit the birth date in the user profile	DT1, DT2, DT3
T4. Edit of the list of interests in the user profile	ML1, ML2, ML3
T5. Add a review of the product	TX1, TX2, TX3
T6. Add products to the cart and change their quantity	NB1, NB2, NB3
T7. Complete the order	
T8. Change the time of courier arrival in the order summary	TM1, TM2, TM3
T9. Change the shipping method in the order summary	SL1, SL2, SL3

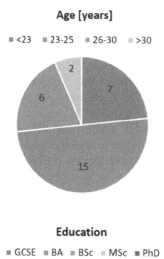

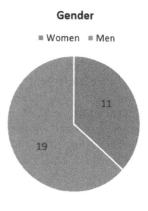

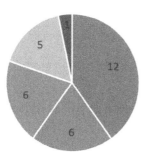

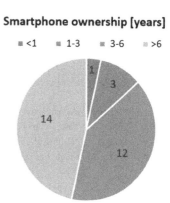

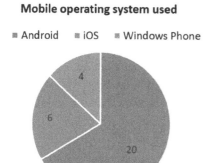

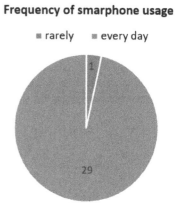

Fig. 4. Descriptive characteristics of usability testing participants

5 Results of Experiment

The results of usability testing of three variants of seven data entry design patterns are depicted in Figs. 5, 6, 7, 8, 9, 10 and 11. The results referring to the effectiveness are placed in the left column. Each bar chart represents the percentage of participants who successfully completed the task and percentage of participants who completed the task committing no errors, respectively. The former is denoted as *Binary rate* and the latter as *No errors*. In turn, the results of efficiency measurement are shown in the right column. In this case bar charts represent time of accomplishing the task expressed in seconds and the number of actions, i.e. the number of clicks, scrolls, taps, swipes, etc. to complete the task. The former is denoted as *Time [s]* and the latter as *Actions [n]*.

The satisfaction was measured using the Single Ease Question (SEQ) questionnaire which was filled by the users after they completed each task. SEQ contained only one item based on the seven-point Likert scale ranging from 1 - *very difficult* to 7 - *very easy*. The average SEQ score provided by all users is placed in Table 5.

Table 5. Results of satisfaction measured with the SEQ questionnaire

Version	EM	DT	TM	NB	SL	ML	TX
1	7.0	7.0	7.0	6.9	7.0	7.0	6.2
2	6.9	7.0	6.9	6.6	7.0	7.0	5.7
3	6.7	7.0	7.0	6.5	7.0	6.9	5.8

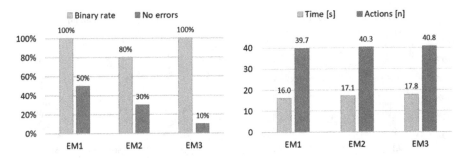

Fig. 5. Results of effectiveness and efficiency for the EM data entry patterns

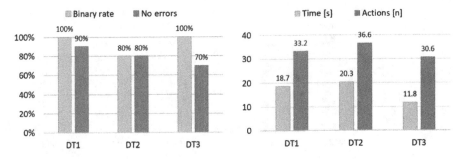

Fig. 6. Results of effectiveness and efficiency for the DT data entry patterns

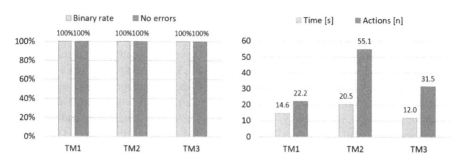

Fig. 7. Results of effectiveness and efficiency for the TM data entry patterns

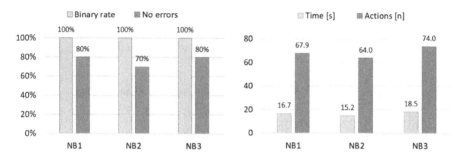

Fig. 8. Results of effectiveness and efficiency for the NB data entry patterns

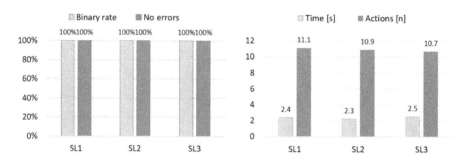

Fig. 9. Results of effectiveness and efficiency for the SL data entry patterns

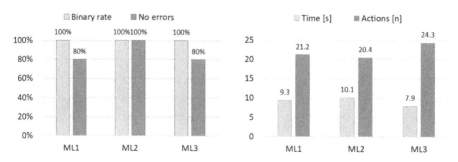

Fig. 10. Results of effectiveness and efficiency for the ML data entry patterns

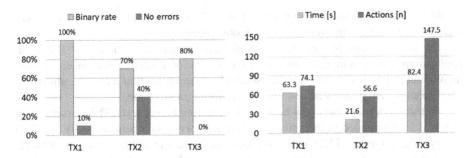

Fig. 11. Results of effectiveness and efficiency for the TX data entry patterns

6 Conclusions

A comparative analysis of design patterns for inputting data in mobile applications was reported in the paper. Three versions of an experimental application were developed using various data entry patterns. A series of usability tests was conducted with 30 users including 10 users per one version. Task scenarios made possible to collect two metrics of efficiency and two metrics of effectiveness. Moreover, the single ease question (SEQ) with a 7-point rating scale was administered to assess how difficult users found individual tasks. Based on the results obtained differences among implemented patterns could be observed. In consequence, some recommendations on how to apply the data entry design patterns in mobile applications were formulated.

Only the first design pattern for inputting email addresses, i.e. EM1, which contains one text field, can be recommended. It produced the best values of metrics of effectiveness, efficiency, and satisfaction. Additional text fields placed in the email address entry pattern decreased its usability.

A calendar picker (DT1) can be recommended for applications where entering a date is important and should be done without any error. The scroller (DT2) could be used in applications where the users are often distracted. However, it should be amended to facilitate more precise and sensitive date selection. Introducing the date with a numeric keyboard (DT3) turned out to be the fastest way but it caused the largest number of errors.

A clock-style time picker (TM1) can be used in the applications where the number of user's actions should be minimized, i.e. in the car navigation systems. The scroller (TM2) cannot be recommended for time entry because it showed the worst performance. The numeric keyboard to input time (TM3) might be applied where the operation speed is the most important factor.

The scroller (NB2) can be recommended to input smaller numbers, e.g. in the range from 1 to 15 whereas the numeric keyboard (NB1) is more suitable for the wider ranges, e.g. from 1 to 50. In turn, the slider (NB3) produced the longest time of inserting the numbers and did not satisfied the users.

The spinner (SL1) is the most suitable to select one option from a list especially when the list contains a larger number of items. The toggle buttons (SL2) ensured the best efficiency but these controls could be ineffective when the number of options would be larger than three. The radio buttons (SL3) need more space on the screen, therefore they are not recommended for mobile devices.

The switches can be applied for implementing multiple-choice lists (ML2) in applications where minimum number of errors is of crucial importance. In turn, highlighting options (ML3) could be useful when the application would be utilized in uncomfortable environmental conditions, e.g. in bright sunshine.

All three design patterns may be useful for entering longer texts. The gesture typing (TX1) is recommended in the situation where the user cannot behave loudly, e.g. in a library. Voice typing (TX2) is the simplest and fastest technique and generates the smallest number of errors. However, if the user needs to input a formal text, which should be grammatically correct, then the predictive text mechanism (TX3) might turn out to be the most advantageous.

Due to the rapid development of mobile technology and emerging new platforms and tools for building mobile applications it is necessary to plan and conduct further study into the usability of data entry design patterns.

Acknowledgments. This paper was partially supported by the statutory funds of the Wrocław University of Science and Technology, Poland.

References

1. Al-Saadi, T.A., Aljarrah, T.M., Alhashemi, A.M., Hussain, A.: A systematic review of usability challenges and testing in mobile health. Int. J. Account. Financ. Report. 5(2) (2015). https://doi.org/10.5296/ijafr.v5i2.8004
2. Alshehri, F., Freeman, M.: Methods for usability evaluations of mobile devices. In: Lamp, J. W. (ed.) 23rd Australian Conference on Information Systems, pp. 1–10. Deakin University, Geelong (2012)
3. Arain, A.A., Hussain, Z., Rizvi, W.H., Vighio, M.S.: Evaluating usability of M-learning application in the context of higher education institute. In: Zaphiris, P., Ioannou, A. (eds.) LCT 2016. LNCS, vol. 9753, pp. 259–268. Springer, Cham (2016). https://doi.org/10.1007/978-3-319-39483-1_24
4. Beck, E., Christiansen, M., Kjeldskov, J., Kolbe, N., Stage, J.: Experimental evaluation of techniques for usability testing of mobile systems in a laboratory setting. In: Proceedings of OzCHI 2003, Brisbane, Australia CHISIG (2003)
5. Bernacki, J., Błażejczyk, I., Indyka-Piasecka, A., Kopel, M., Kukla, E., Trawiński, B.: Responsive web design: testing usability of mobile web applications. In: Nguyen, N.T., Trawiński, B., Fujita, H., Hong, T.-P. (eds.) ACIIDS 2016. LNCS (LNAI), vol. 9621, pp. 257–269. Springer, Heidelberg (2016). https://doi.org/10.1007/978-3-662-49381-6_25
6. Bertini, E., et al.: Appropriating heuristic evaluation for mobile computing. Int. J. Mob. Hum. Comput. Interact. 1(1), 20–41 (2009). https://doi.org/10.4018/jmhci.2009010102
7. Błażejczyk, I., Trawiński, B., Indyka-Piasecka, A., Kopel, M., Kukla, E., Bernacki, J.: Usability testing of a mobile friendly web conference service. In: Nguyen, N.-T., Manolopoulos, Y., Iliadis, L., Trawiński, B. (eds.) ICCCI 2016. LNCS (LNAI), vol. 9875, pp. 565–579. Springer, Cham (2016). https://doi.org/10.1007/978-3-319-45243-2_52
8. Coursaris, C.K., Kim, D.J.: A meta-analytical review of empirical mobile usability studies. J. Usability Stud. 6(3), 117–171 (2011)

9. Gatsou, C., Politis, A., Zevgolis, D: Exploring inexperienced user performance of a mobile tablet application through usability testing. In: Proceedings of the 2013 Federated Conference on Federated Conference on Computer Science and Information Systems FedCSIS 2013, pp. 557–564. IEEE (2013)

10. Gómez, R.Y., Caballero, D.C., Sevillano, J.: Heuristic evaluation on mobile interfaces: a new checklist. Sci. World J. Article ID 434326 (2014). https://doi.org/10.1155/2014/434326

11. Harrison, R., Flood, D., Duce, D.: Usability of mobile applications: literature review and rationale for a new usability model. J. Interact. Sci. 1, 1 (2013). https://doi.org/10.1186/2194-0827-1-1

12. Hussain, A., Hashim, N.L., Nordin, N.: mGQM: evaluation metric for mobile and human interaction. In: Stephanidis, C. (ed.) HCI 2014. CCIS, vol. 434, pp. 42–47. Springer, Cham (2014). https://doi.org/10.1007/978-3-319-07857-1_8

13. Hussain, A., Mkpojiogu, E.O.C.: Usability evaluation techniques in mobile commerce applications: a systematic review. In: AIP Conference Proceedings, vol. 1761, p. 020049 (2016). https://doi.org/10.1063/1.4960889

14. Joyce, G., Lilley, M., Barker, T., Jefferies, A.: Mobile application usability: heuristic evaluation and evaluation of heuristics. In: Amaba, B. (ed.) Advances in Human Factors, Software, and Systems Engineering. AISC, vol. 492, pp. 77–86. Springer, Cham (2016). https://doi.org/10.1007/978-3-319-41935-0_8

15. Joyce, G., Lilley, M., Barker, T., Jefferies, A.: Heuristic evaluation for mobile applications: extending a map of the literature. In: Ahram, T., Falcão, C. (eds.) AHFE 2018. AISC, vol. 794, pp. 15–26. Springer, Cham (2019). https://doi.org/10.1007/978-3-319-94947-5_2

16. Kaikkonen, A., Kekäläinen, A., Cankar, M., Kallio, T., Kankainen, A.: Usability testing of mobile applications: a comparison between laboratory and field testing. J. Usability Stud. 1 (1), 4–16 (2005)

17. Kortum, P., Sorber, M.: Measuring the usability of mobile applications for phones and tablets. Int. J. Hum.-Comput. Interact. 31(8), 518–529 (2015). https://doi.org/10.1080/10447318.2015.1064658

18. Krzewińska, J., Indyka-Piasecka, A., Kopel, M., Kukla, E., Telec, Z., Trawiński, B.: Usability testing of a responsive web system for a school for disabled children. In: Nguyen, N.T., Hoang, D.H., Hong, T.-P., Pham, H., Trawiński, B. (eds.) ACIIDS 2018. LNCS (LNAI), vol. 10751, pp. 705–716. Springer, Cham (2018). https://doi.org/10.1007/978-3-319-75417-8_66

19. Mansar, S.L., Jariwala, S., Shahzad, M., Anggraini, A., Behih, N., AlZeyara, A.: A usability testing experiment for a localized weight loss mobile application. Procedia Technology 5, 839–848 (2012). https://doi.org/10.1016/j.protcy.2012.09.093

20. Moumane, K., Idri, A., Abran, A.: Usability evaluation of mobile applications using ISO 9241 and ISO 25062 standards. SpringerPlus 5, 548 (2016). https://doi.org/10.1186/s40064-016-2171-z

21. Nayebi, F., Desharnais, J.M., Abran, A.: The state of the art of mobile application usability evaluation. In: 25th IEEE Canadian Conference on Electrical and Computer Engineering (CCECE), pp. 1–6 (2012). https://doi.org/10.1109/ccece.2012.6334930

22. Othman, M.K., Sulaiman, M.N.S., Aman, S.: Heuristic evaluation: comparing generic and specific usability heuristics for identification of usability problems in a living museum mobile guide app. Adv. Hum.-Comput. Interact. 2018, 13, Article ID 1518682 (2018). https://doi.org/10.1155/2018/1518682

23. Quinones, D., Rusu, C.: How to develop usability heuristics: a systematic literature review. Comput. Stand. Interfaces 53, 89–122 (2017). https://doi.org/10.1016/j.csi.2017.03.009

24. Saleh, A., Isamil, R.B., Fabil, N.B.: Extension of PACMAD model for usability evaluation metrics using Goal Question Metrics (GQM) Approach. J. Theoret. Appl. Inf. Technol. **79** (1), 90–100 (2015)
25. Sauro, J., Zarolia, P.: SUPR-Qm: a questionnaire to measure the mobile app user experience. J. Usability Stud. **13**(1), 17–37 (2017)
26. Seffah, A., Donyaee, M., Kline, R.B., Padda, H.K.: Usability measurement and metrics: a consolidated model. Softw. Qual. J. **14**, 159–178 (2006). https://doi.org/10.1007/s11219-006-7600-8
27. Shirogane, J., Matsuzawa, M., Iwata, H., Fukazawa, Y.: Usability evaluation method of applications for mobile computers using operation histories. IEICE Trans. Inf. Syst. **E101-D101-D**(7), 1790–1800 (2018). https://doi.org/10.1587/transinf.2017kbp0022
28. Shitkova, M., Holler, J., Heide, T., Clever, N., Becker, J.: Towards usability guidelines for mobile websites and applications. In: Wirtschafsinformatik Proceedings 2015, Paper 107, Association for Information Systems, AIS Electronic Library (2015)
29. Silvennoinen, J., Vogel, M., Kujala, S.: Experiencing visual usability and aesthetics in two mobile application contexts. J. Usability Stud. **10**(1), 46–62 (2014)
30. Zahra, F., Hussain, A., Mohd, H.: Usability evaluation of mobile applications; where do we stand? In: AIP Conference Proceedings, vol. 1891, p. 020056 (2017). https://doi.org/10.1063/1.5005389
31. Zhang, D., Adipat, B.: Challenges, methodologies, and issues in the usability testing of mobile applications. Int. J. Hum.-Comput. Interact. **18**(3), 293–308 (2005). https://doi.org/10.1207/s15327590ijhc1803_3
32. Nielsen, J., Molich, R.: Heuristic evaluation of user interfaces. In: Proceedings of the SIGCHI Conference on Human Factors in Computing Systems, pp. 249–256 (1990)

Author Index